# Lecture Notes in Networks and Systems

## Volume 95

**Series Editor**

Janusz Kacprzyk, Systems Research Institute, Polish Academy of Sciences, Warsaw, Poland

**Advisory Editors**

Fernando Gomide, Department of Computer Engineering and Automation—DCA, School of Electrical and Computer Engineering—FEEC, University of Campinas—UNICAMP, São Paulo, Brazil
Okyay Kaynak, Department of Electrical and Electronic Engineering, Bogazici University, Istanbul, Turkey
Derong Liu, Department of Electrical and Computer Engineering, University of Illinois at Chicago, Chicago, USA; Institute of Automation, Chinese Academy of Sciences, Beijing, China
Witold Pedrycz, Department of Electrical and Computer Engineering, University of Alberta, Alberta, Canada; Systems Research Institute, Polish Academy of Sciences, Warsaw, Poland
Marios M. Polycarpou, Department of Electrical and Computer Engineering, KIOS Research Center for Intelligent Systems and Networks, University of Cyprus, Nicosia, Cyprus
Imre J. Rudas, Óbuda University, Budapest, Hungary
Jun Wang, Department of Computer Science, City University of Hong Kong, Kowloon, Hong Kong

The series "Lecture Notes in Networks and Systems" publishes the latest developments in Networks and Systems—quickly, informally and with high quality. Original research reported in proceedings and post-proceedings represents the core of LNNS.

Volumes published in LNNS embrace all aspects and subfields of, as well as new challenges in, Networks and Systems.

The series contains proceedings and edited volumes in systems and networks, spanning the areas of Cyber-Physical Systems, Autonomous Systems, Sensor Networks, Control Systems, Energy Systems, Automotive Systems, Biological Systems, Vehicular Networking and Connected Vehicles, Aerospace Systems, Automation, Manufacturing, Smart Grids, Nonlinear Systems, Power Systems, Robotics, Social Systems, Economic Systems and other. Of particular value to both the contributors and the readership are the short publication timeframe and the world-wide distribution and exposure which enable both a wide and rapid dissemination of research output.

The series covers the theory, applications, and perspectives on the state of the art and future developments relevant to systems and networks, decision making, control, complex processes and related areas, as embedded in the fields of interdisciplinary and applied sciences, engineering, computer science, physics, economics, social, and life sciences, as well as the paradigms and methodologies behind them.

**\*\* Indexing: The books of this series are submitted to ISI Proceedings, SCOPUS, Google Scholar and Springerlink \*\***

More information about this series at http://www.springer.com/series/15179

Dmitry G. Arseniev · Ludger Overmeyer ·
Heikki Kälviäinen · Branko Katalinić
Editors

# Cyber-Physical Systems and Control

Springer

*Editors*
Dmitry G. Arseniev
Peter the Great St. Petersburg
Polytechnic University
Saint Petersburg, Russia

Heikki Kälviäinen
Machine Vision and Pattern
Recognition Laboratory
Lappeenranta University of Technology
Lappeenranta, Finland

Ludger Overmeyer
Institute of Transport
and Automation Technology
Leibniz Universität Hannover
Garbsen, Niedersachsen, Germany

Branko Katalinić
Institute of Production Engineering
and Photonic Technologies
Vienna University of Technology
Vienna, Austria

ISSN 2367-3370          ISSN 2367-3389 (electronic)
Lecture Notes in Networks and Systems
ISBN 978-3-030-34982-0          ISBN 978-3-030-34983-7 (eBook)
https://doi.org/10.1007/978-3-030-34983-7

This Springer imprint is published by the registered company Springer Nature Switzerland AG
The registered company address is: Gewerbestrasse 11, 6330 Cham, Switzerland

# Preface

The International Conference on Cyber-Physical Systems and Control (CPS&C'2019) was held in Peter the Great St. Petersburg Polytechnic University, which in 2019 celebrates its 120th anniversary.

The CPS&C'2019 was dedicated to the 35th anniversary of the partnership relations between Peter the Great St. Petersburg Polytechnic University and Leibniz University of Hannover.

This conference draws upon the experience of previous major events that focused on information technologies, system analysis, engineering, and control and were hosted by Peter the Great St. Petersburg Polytechnic University in partnership with leading European and Russian academic institutions. The most significant in the series of these events were such annual events as the International Conference on System Analysis in Engineering and Control (since 1998), the Distributed Intelligent Systems and Technologies Workshop (since 2008), the International Scientific Symposium on Automated Systems and Technologies (since 2014), and the International Conference Network Cooperation in Science, Industry and Education (in 2016), each attended by hundreds of participants.

The cyber-physical systems (CPSs) are a new generation of control systems and techniques which help promote prospective interdisciplinary research. A wide range of theories and methodologies are being investigated and developed in this area to tackle various complex and challenging problems. Therefore, CPSs can be considered as a scientific and engineering discipline that is set to make an impact on future systems of industrial and social scale characterised by deep integration of real-time processing, sensing, and actuation into logical and physical heterogeneous domains.

The CPS&C'2019 aimed to bring together researchers and practitioners from all over the world and to reveal cross-cutting fundamental scientific and engineering principles that underline the integration of cyber and physical elements across all application fields.

Participants of the conference represented research institutions and universities from Austria, Belgium, Bulgaria, China, Finland, Germany, the Netherlands, Russia, Syria, Ukraine, the USA, and Vietnam.

The book of proceedings includes 75 papers, arranged into five chapters, namely: Keynote Papers, Fundamentals, Applications, Technologies, and Education and Social Aspects.

Michael Krommer
Dmitry G. Arseniev
Ludger Overmeyer
Conference Co-chairs

# Organisation

## Committees

### Scientific Committee

| | |
|---|---|
| Alexander K. Belyaev | Institute of the Problems of Mechanical Engineering of the Russian Academy of Sciences, Russia |
| Alexey A. Bobtsov | St. Petersburg National Research University of Information Technologies, Mechanics and Optics, Russia |
| David Castells-Rufas | Autonomous University of Barcelona, Spain |
| Joachim Denil | University of Antwerp, Belgium |
| Tatiana A. Gavrilova | St. Petersburg State University, Russia |
| Galina V. Gorelova | Southern Federal University, Russia |
| Stein E. Johansen | Institute for Basic Research, FL, USA |
| Arto Kaarna | Lappeenranta University of Technology, Finland |
| Heikki Kälviäinen | Lappeenranta University of Technology, Finland |
| Nicos Karcanias | City, University of London, Great Britain |
| Branko Katalinić | Vienna University of Technology, Austria |
| Sergey V. Kuleshov | St. Petersburg Institute for Informatics and Automation of the Russian Academy of Sciences, Russia |
| Michael S. Kupriyanov | St. Petersburg Electrotechnical University "LETI", Russia |
| Nikolay V. Kuznetsov | St. Petersburg State University, Russia |
| Dmitriy A. Novikov | Institute of Control Sciences of the Russian Academy of Sciences, Russia |
| Leonid I. Perlovsky | Harvard University, MA, USA |
| Rudolf Pichler | Graz University of Technology, Austria |
| Vladimir A. Polyansky | Institute of the Problems of Mechanical Engineering of the Russian Academy of Sciences, Russia |

| | |
|---|---|
| Oliver Riedel | University of Stuttgart, Germany |
| Andrey L. Ronzhin | St. Petersburg Institute for Informatics and Automation of the Russian Academy of Sciences, Russia |
| Vyacheslav P. Shkodyrev | Peter the Great St. Petersburg Polytechnic University, Russia |
| Hans-Rolf Trankler | Bundeswehr University Munich, Germany |
| Hans Vangheluwe | University of Antwerp, Belgium |
| Violetta N. Volkova | Peter the Great St. Polytechnic University, Russia |
| Valeriy V. Vyatkin | Lulea University of Technology, Sweden |

## Organising Committee

| | |
|---|---|
| Vyacheslav P. Shkodyrev (Chairman) | Peter the Great St. Polytechnic University, Russia |
| Aleksandra V. Loginova | Peter the Great St. Polytechnic University, Russia |
| Vyacheslav V. Potekhin | Peter the Great St. Polytechnic University, Russia |
| Konstantin K. Semenov | Peter the Great St. Polytechnic University, Russia |

# Contents

# Seamless Data Integration in a CPPS with Highly Heterogeneous Facilities - Architectures and Use Cases Executed in a Learning Factory

Rudolf Pichler[1]([envelope]), Lukas Gerhold[2], and Michael Pichler[1]

[1] Graz University of Technology, Graz, Austria
{rudolf.pichler,michael.pichler}@tugraz.at
[2] Siemens AG, Vienna, Austria
lukas.gerhold@siemens.com

**Abstract.** Facing the principal challenges of a Cyberphysical System (CPS) in a manufacturing environment by establishing an appropriate universal and scalable architecture the paper shows two explicit use cases of successfully established communication lines (horizontal and vertical) that integrate facilities derived from highly different domains, this all done at the Learning Factory at Graz University of Technology. In present time effective Cyberphysical Production Systems (CPPSs) live on the pervasive and seamless data integration of its data generators and receivers mainly facilitated by the Linkage Part of a CPPS. The connectivity, its semantic interoperability and the scalability need well-designed concepts and architectures because of the existence of too many standards and protocols. The challenge increases significantly if the network should be set up with facilities from many different suppliers and their proprietary standards. At the Learning Factory of Graz University of Technology the integration of most heterogeneous products at the office floor and at the shop floor is a major part of its research. The paper presents two solutions in form of "Use Cases," representing an innovative concept for both the vertical and the horizontal integration. Usage of an Enterprise Service Bus at the office floor and the installation of the "KEPServerEX"- middleware at the shop floor are selected core approaches for creating a representative CPPS.

**Keywords:** CPS in manufacturing · Cyberphysical production systems · Learning factory · Heterogeneous IoT · Data capturing · Robot control · OPC UA · MindSphere · KEPServerEX · PdM WebConnector

## 1 CPPS – The CPS in Manufacturing Environments

Cyberphysical Systems (CPSs) are engineered systems that are built from and depend upon the integration of computational algorithms and physical components [19]. CPSs enable capability, adaptability, scalability, resiliency, safety, security and usability that will far exceed the simple embedded systems of today [13]. Typical and well-known applications of CPSs are the arenas of Smart Mobility, Smart Health, Smart Grid, Smart Cities and Smart Factory. All of them take their technical and economic advantage out

© Springer Nature Switzerland AG 2020
D. G. Arseniev et al. (Eds.): CPS&C 2019, LNNS 95, pp. 1–10, 2020.
https://doi.org/10.1007/978-3-030-34983-7_1

of the seamless integration and interoperability of their real-world processes and their computed virtual processes. Such configurations are supposed to lead to higher transparency, enable a better understanding and end up in a well-grounded and faster decision taking, this all meant for improving the regarded system.

In concerns of manufacturing, a CPS turns into a Cyberphysical Production System (CPPS), which more or less carries the same principles and characteristics as a CPS but concentrates on the specific tasks of fulfilling future-oriented, competitive production processes. Such CPPS – or Smart Factories – have to regard also the duties of the whole supply chain, the quality management and many other service-oriented processes. A meaningful CPPS need its horizontal integration (shop floor integration) as well as the link to the commercial and supervising levels of a company with the need of a vertical integration (office floor integration).

## 1.1  Typical Setup and Components of a CPPS

The setup of a CPPS always is specific depending on its industrial sector; nevertheless there always can be detected the same three categories of elements: the Physical Part, the Cyber Part and the Linkage Part. Undoubtedly, the latter belongs to the most challenging of them.

Regarding the physical world a CPPS is typically made by tool machines, welding units, robots, shuttles, presses, tools, sensors, and actuators, as well as means of metrology and logistics. It is mainly represented and visible at the shop floor but contains also the IT-infrastructure hardware with computers, servers, monitors, gateways and all kinds of connecting devices.

The cyber part of the CPPS mainly consists of software and assisting tools for planning, modelling, analyzing, simulating and forecasting the relevant processes. These are engineering tools CAD, CAM, and CAE, followed by business administration tools like PLM, ERP, MES and ending up with high-level software applications that provide services for data capturing, cloud computing, safety and security utilities. The "digital twin" – representing either the full process or only an important section out of it – should be mentioned here as one of the most popular cyber elements of a CPPS.

The last and highly essential part for a successful CPPS, the Linkage Part, is the set and the match of communication standards and protocols, the usage of appropriate field busses, middleware and a suitable control system. Only with these enablers based on an appropriate architecture the whole system can turn into a living and effective CPPS. "Interoperability goes beyond technology [28]."

## 1.2  Actual Status and Challenges of a Sound CPPS

Designing the architecture of a CPPS is not bound to the restricted communication lines of the traditional automation pyramid any longer, decentralization and completely new control loops can take place ongoing [17]. Nevertheless, the challenges for the establishment of a well working CPPS did not really decrease. The reasons for this are manifold, the expectations towards it quite often are over exaggerated [16].

The brownfield situation, where facilities of older generations cannot even offer any interface for communication, is not going to be discussed in this paper. However, even the focus on green field applications, where up-to-date technology promises advanced communication features and interoperability between systems, the realization of a fully integrated CPPS likely becomes hard work.

The big variety and complexity of a CPPS already derives from the necessity of so many different components, devices and facilities and consequently the high variety of suppliers. So it is no surprise that for all these future elements of the CPPS, the same standardized programming tools and interfaces are not automatically given. At this stage, the repeatedly mentioned lack of standards has to be commented on. The reality shows that it is quite the opposite: there is a much too high number of upcoming standards that are mainly all incompatible (see e.g.: IEC 61158 with 19(!) different field busses [18]).

So it is even hard for suppliers of such "things" to decide which communication standards there should be offered, not to speak about the customers that are confronted with an uncontrolled growth of possibilities, especially when being at the start of establishing a CPPS.

In the meantime, selected architectures, communication standards and protocols including valid and semantic descriptions of the data [6] come out on top (e.g. OPC-UA, MT-Connect, MQTT). Service-Oriented Architectures (SOAs) have become powerful tools for creating open and scalable CPPSs in order to integrate so many foreign worlds. Nevertheless, there is still a lot of work to do in alignments for achieving the desired full horizontal and vertical integration [29]. Especially the add-ons of a modern structured cognitive production system – this accompanied with the demands of increasingly time-critical, safety- and security-oriented processes – turn out to be more than a challenge [23, 24].

## 2 Research in the CPPS of a Learning Factory

### 2.1 Introduction of the Research Field Smartfactory@tugraz

The smartfactory@tugraz (brand name for the Learning Factory at Graz University of Technology) provides various technological and CPPS-related topics for its researchers and visitors. There is established a full range production line for producing wave gears in diverse variants as objects demonstrator. The production of these variants follows a lot-size-1 sequence in order to show the capabilities of the OT and IT for acting agile. The fundamental facilities of the smartfactory@tugraz are three tool machines, a tool measurement device, a coordinate measurement machine, an assembly line with six robots, one screwing and two pressing units and an AGV for intralogistics services (Fig. 1).

**Fig. 1.** Insight of the Learning Factory smartfactory@tugraz in Graz, Austria

## 2.2 Working with Heterogeneous Facilities and Data Formats

Whilst industrial companies deeply avoid working with heterogeneous vendors of facilities and IT standards because of high integration and education efforts, the smartfactory@tugraz intentionally builds up a highly heterogeneous environment for manufacturing and assembly in order to face exactly these scientific challenges when going into research and development.

As for the shop floor the three tool machines are equipped with PLCs from Mitsubishi once and Sinumerik 840D sl twice. Its data protocols are MT-Connect and OPC-UA. Furthermore, there is a multisource fleet of robots coming from Stäubli, Fanuc, Kuka and Universal Robots, most of them communicate with "OPC-UA," but Fanuc, e.g., carries the drivers for the "GE Ethernet."

At the office floor, the learning factory is working with three major software packages also coming from three different suppliers, the PLM from Siemens, the ERP from proALPHA and the MES from Solidat. Also, in this case there is a confrontation with the problem: how can a seamless data flow be realized starting from the PLM, going down to ERP and MES to the shop floor and back? How can data capturing from such different machinery work? How can these data be transferred to diverse cloud applications for analysis purposes?

## 3  IT-Network Architecture of the Learning Factory

In general, the IT network architecture of the smartfactory@tugraz [29] consists of two major parts: the "Office Floor" and the "Shop Floor". As a principal for creating a scalable and expandable solution, the Service-Oriented Architecture (SOA) was set as the basic approach. As long as both major parts have their specific requirements (e.g. real time execution and safety options preferably at the shop floor, security and outbound connectivity feature more at the office floor), the implementations (see Fig. 2) will be regarded separately.

At the Office Floor an Enterprise Service Bus (ESB) with the name "PdM WebConnector (PWC)" integrates and conducts the data exchange between the software domains of PLM, ERP and MES. This is highly relevant because one of the commands of the Learning Factory is that any master data is allowed to be put into the system only once. Going on this master data must be available for all participating clients in the CPPS in a consistent manner. The PWC enables this required connectivity in executing data mediation, data mapping and data transformation.

At the Shop Floor at a final stage – from the Connectivity Platform upwards – OPC UA is used as a communication standard. The strengths of OPC UA standard definitely lie in the modelling of informations for vertical and horizontal communication by providing semantic interoperability and the advanced data security [4, 18, 22]. Also, at the machine level there are mainly devices with OPC UA protocol but not only, because the Learning Factory intentionally wants also to show the way of integration with heterogeneous participants. This all is done via mighty middleware located at the server of the Connectivity Platform (see Sect. 4.2 for more details).

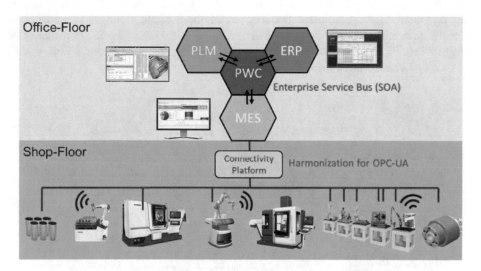

**Fig. 2.** General architecture of the Learning Factory

# 4 Use Cases of Mastering Heterogeneous Environments

For a Learning Factory the research in finding solutions for a consistent data flow and the interoperability of all its CPPS "Things" must end up in proven and robust implementations that could be demonstrated for students and interested industrials. That is why the smartfactory@tugraz has been setting up a couple of so-called "Use Cases" that all follow a certain choreography for best didactic transformations.

In this paper there should be introduced two Use Cases that correlate with the topic "Integration of heterogeneous facilities". One will be a representative for the horizontal integration and a second one for the vertical integration. The titles of these Use Cases are:

- Foreign Domain-Guided Control of Robots;
- Cloud-Oriented Data Capturing at High Diversity.

### 4.1   Foreign Domain-Guided Control of Robots

When buying a robot companies must pay attention to the programming, operating and maintenance skills of the inhouse workers in order to keep education, training and operating efforts low. This consequently leads to a monoculture of infrastructure instead of decisions for an even better and more appropriate type of machinery.

This Use Case is a specialty for working with robots from different suppliers though not being educated or experienced in a broad band of knowledge in all these products. The only requirement for this is to be acquainted with the TIA-Portal (by Siemens) or the programming of Sinumerik 840D sl. With these skills alone, it is possible to actually program and run robots of Kuka, Stäubli and Denso types (Yaskawa is in preparation). Figure 3 shows the set up for achieving the required interoperability with an example connecting a Stäubli robot.

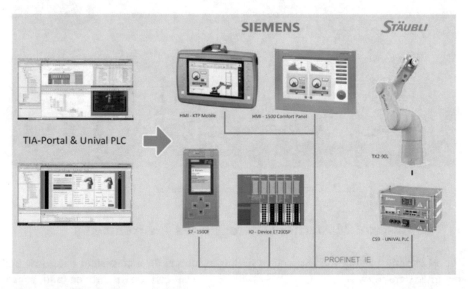

**Fig. 3.** Data flow from user interface at tool machine to the robot

The interoperability between the Stäubli robot and a standard industrial PLC – in this case Simatic or Sinumerik – is based on the Stäubli product "uniVAL PLC". (Kuka, Denso and Yaskawa provide comparable products). It connects the additional necessary components like the PLC (S7-1500F), the I/O device (ET 200SP) and both variants of HMI (small, mobile, with emergency switch OR big and multifunctional) via Profinet. The programming of the working routines is done via either HMI or the TIA Portal. In case of an off-line programming, the program is directly transferred to the PLC. One PLC can control up to five different robots.

In case of running the Stäubli robot from the HMI of a tool machine, the principles and connections are the same. The HMI of the tool machine then is conformed to the "HMI 1500 comfort panel" of the diagram in Fig. 3. This shown case receives realistic importance especially when tool machines are going to be equipped with robots for loading and unloading parts and tools.

## 4.2 Cloud-Oriented Data Capturing from Facilities with a High Diversity in Protocols and Drivers

The goal of this Use Case is successful verification of data capturing though meeting highly heterogeneous data generators at the shop floor and transferring these data into specified clouds with the purpose for ongoing data analytics. Final output of this conception of data capturing is the enhancement of production efficiency and maintenance processes.

Aside the general challenge of allocation and distribution of correct data in a CPPS, another central problem is the standardization of data collection [31]. This task particularly requires new approaches if the addressed data generators are highly heterogeneous like it is done at smartfactory @tugraz with intention. Actually the infrastructure at the shop floor (see Fig. 4) is made up by machinery and robotics with not only the OPC-UA standard but also an MT-Connect protocol and the Fanuc robots, which do not run on Profinet but on the GE Ethernet field bus and its proprietary drivers.

With the target to unify all protocols of the shop floor for the OPC-UA standard (preferred because of its most flexible, powerful and secure features) before finally uploading it to diverse cloud applications there could be found a powerful middleware with the name "KEPServerEX" (see Ref. [11]). It fulfills the desired alignment of data formats by structuring, renaming and converting the raw data from their former exchange format. It provides access to more than 150 protocols (proprietary and IT-protocols) and enables communication with devices and systems from all major automation vendors. This additionally ensures the important scalability of such an architecture. With this tool the add-on of any facility of any origin can be done quite easily without irritation of the existing architecture.

Within the architecture of the smartfactory@tugraz with all this data, there will be addressed two clouds. First, there is the open IoT platform "MindSphere" that collects and saves all defined data from the shop floor in a big data table. Before entering the "MindSphere" cloud all data coming from the "KEPServerEX" Server have to pass a gate called "Mind Connect Box" (see Fig. 4). Access to the data stored in MindSphere is operated via certain Apps which do not necessarily need to be of Siemens origin but can be programmed by any company or user. A second and parallel cloud application with an Apache Hadoop infrastructure is set up by the partner T-Systems in order to provide support in terms of Big Data Analytics.

Figure 4 (right side) shows that even data coming from facilities already equipped with OPC UA protocol run via the KEPServerEX Server though its data would not need any data transformation anyway. The reason for this preferred routing is, firstly, the possible use of the additional KEPServerEX functions in managing the defaults and parameters for the data capturing of the whole shop floor and, secondly, the possibility to have all (!) devices interconnected for data transfer.

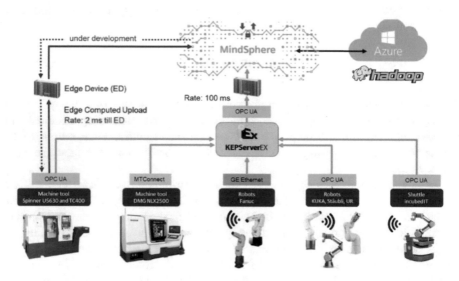

**Fig. 4.** Middleware "KEPServerEx" for harmonizing and translating diverse protocols

The data flow at the outer left side shows a direct connection to the clouds via short-cutting the KEPServerEX Server. This application is bound to additional preconditions. First, the integrated facility must work on Sinumerik 840D sl and, second, there is the need of an additional Siemens Edge Device, mostly applied directly at the machine. This Edge Device includes functionalities of the otherwise usual "Mind Connect Box" as a pre-processor before entering MindSphere. The only reason and advantage of such a solution is the high possible data transfer rate of only 2 ms instead of normally 100 ms.

## 5   Conclusions

Effective CPPSs live on the pervasive and seamless data integration of its data generators and receivers (Things) mainly facilitated by the Linkage Part of a CPPS. The connectivity, its semantic interoperability and the scalability need well designed concepts and architectures not because of lacking standards and protocols but – just the opposite – the existence of too many of them. The challenge increases significantly if the network should be set up with facilities from many different suppliers and their proprietary standards.

At the Learning Factory of Graz University of Technology, the integration of most heterogeneous products at the office floor and at the shop floor is a major part of its research. The paper presents two solutions in the form of "Use Cases", one representing an innovative concept for the vertical and another for the horizontal integration. Usage an Enterprise Service Bus at the office floor and the installation of the middleware "KEPServerEX" in the shop floor are the selected core approaches for creating a representative CPPS.

**Acknowledgements.** This research has been supported by the Know-How of the consortium members of the "IT-Summit" of the smartfactory@tugraz project, by the financial contributions of the Austrian Ministry for Transport, Innovation and Technology and 19 industrial consortium members of the project (see also www.smartfactory.tugraz.at). The research, the planning and the execution of the demonstrators in the Learning Factory has widely been done by the team of the Institute of Production Engineering at Graz University of Technology.

# References

1. Ahmadi, A., Cherifi, C., Cheutet, V., et al.: A review of CPS 5 components architecture for manufacturing based on standards. In: SKIMA, International Conference on Software, Knowledge, Intelligent Management and Applications, Colombo, Sri Lanka. <Hal-01679977>, December 2017
2. Fallah, S.M., Trautner, T., Pauker, F.: Integrated tool lifecycle. In: 12th CIRP Conference on Intelligent Computation in Manufacturing Engineering (2019). Procedia CIRP **79**, 257–262
3. Farahzadi, A., Shams, P., Rezazadeh, J., et al.: Middleware technologies for cloud of things: a survey. Digit. Commun. Netw. **4**(3), 176–188 (2018)
4. Garcia, M.V., Irisarri, E., Perez, F., et al.: An open CPPS automation architecture based on IEC-61499 over OPC-UA for flexible manufacturing in Oil&Gas industry. IFAC PapersOnLine **50–1**, 1231–1238 (2017)
5. Geisberger, E., Broy, M. (eds.): Integrierte forschungsagenda cyber-physical systems. In: Achatech Studies (2012)
6. Gorecky, D., Hennecke, A., Schmitt, M., et al.: Wandelbare modulare Automatisierungssysteme (2017)
7. Reinhart, G. (ed.): Handbuch Industrie 4.0. Carl Hanser Verlag, 567 p. (2017)
8. Hennig, M., Reisinger, G., Trautner, T., et al.: TU Wien pilot factory Industry 4.0. In: 9th Conference on Learning Factories, Braunschweig (2019)
9. Hoppe, S.: Standardisierte horizontale und vertikale Kommunikation: Status und Ausblick. In: Bauernhansl, T., Hompel, M., Vogel-Heuser, B. (Hrsg.) Industrie 4.0 in Produktion, Automatisierung und Logistik, 325 p. (2014)
10. Huber, W.: Industrie 4.0 in der Automobilproduktion, ein Praxisbuch, Standards, pp. 95–116 (2016)
11. KEPServerEX V6 Manual (2018). https://www.kepware.com/getattachment/5759d980-7641-42e8-b4fb-7293c835a2f9/kepserverex-manual.pdf. Accessed 12th Apr 2019
12. Kolberg, D., Berger, C., Pirvu, B.-C., et al.: CyProF – insights from a framework for designing cyber-physical systems in production environments. In: 49th CIRP Conference on Manufacturing Systems (2016). Procedia CIRP **57**, 32–37
13. Lee, J., Bagheri, B., Jin, C.: Introduction to cyber manufacturing. Manuf. Lett. **8**, 11–15 (2016)

14. Lee, J., Bagheri, B., Kao, H.: A cyber-physical system architecture for Industry 4.0-based manufacturing systems. Manuf. Lett. **3**, 18–23 (2015)
15. Moghaddam, M., Cadavid, M.N., Kenley, C.R., et al.: Reference architectures for smart manufacturing: a critical review. J. Manuf. Syst. **49**, 215–225 (2018)
16. Monostori, L.: Cyber-physical production systems: roots, expectations and R&D challenges. Procedia CIRP **17**, 9–13 (2014)
17. Monostori, L., Kadar, B., Bauernhansl, T., et al.: Cyber-physical systems in manufacturing. CIRP Ann. Manuf. Technol. **65**, 621–641 (2016)
18. Munz, H.; Stöger, G.: Deterministische Machine-to-Machine Kommunikation im Industrie 4.0 Umfeld. In: Schulz, T. (ed.) Industrie 4.0 (2017)
19. National Science Foundation. Cyber-Physical Systems (CPS) (2017). https://wwww.nsf.gov/pubs/2017/nsf17529.htm
20. N.N. How do you control a Stäubli robot with a SIMATIC controller? Industry Online Support International (2019). https://support.industry.siemens.com/cs/document//how-do-you-control-a-st%C3%A4ubli-robot-with-a-simatic-controller-?dti=0&lc=en-WW
21. N.N. Driving Stäubli robots with industrial PLCs (2019). https://www.staubli.com/en-at/robotics/product-range/robot-software/val3-robot-programming/unival-solutions/unival-plc/
22. OPC Foundation: OPC Unified Architecture – Interoperabilität für Industrie 4.0 und das Internet der Dinge, white paper (2016)
23. Panetto, H., Iung, B., Ivanov, D., et al.: Challenges for the cyber-physical manufacturing enterprises of the future. Annu. Rev. Control (2019). https://doi.org/10.1016/j.arcontrol.2019.02.002
24. Perez, F., Irisarri, E., Orive, D., Marcos, M.: A CPPS Architecture approach for Industry 4.0 (2015)
25. Schleipen, M., Gilani, S.-S., Bischoff, T., et al.: OPC UA & Industrie 4.0 – enabling technology with high diversity and variability. In: 49th Conference on Manufacturing Systems (2016). Procedia CIRP **57**, 315–320
26. Stojmenovic, I.: Machine-to-Machine communications with in-network data aggregation. Processing and actuation for large scale cyber-physical systems. IEEE Internet Things J. **PP** (99), 1 (2014)
27. Thiede, S., Juraschek, M., Herrmann, C.: Implementing cyber-physical production systems in learning factories. Procedia CIRP **54**, 7–12 (2016). https://doi.org/10.1016/j.procir.2016.04.098
28. Törngren, M., Asplund, F., Bensalem, S., et al.: Characterisation, analysis and recommendations for exploiting the opportunities of cyberphysical systems. In: Song, H., et al. (eds.) Cyber-Physical Systems - Foundations, Principles and Applications (2017)
29. Trabesinger, S., Pichler, R., Schall, D., et al.: Connectivity as a prior challenge in establishing CPPS on basis of heterogeneous IT-software environments. In: 9th Conference on Learning Factories, Braunschweig (2018)
30. Wang, L., Törngren, M., Onori, M.: Current status and advancement of cyber-physical systems in manufacturing. J. Manuf. Syst. **37**, 517–527 (2015)
31. Yli-Ojanperä, M., Sierla, S., Papakonstaninou, N., et al.: Adapting an agile manufacturing concept to the reference architecture model industry 4.0: a survey and case study. J. Ind. Inf. Integr. (2018). https://doi.org/10.1016/j.jii.2018.12.002

# Physics of Mind – A Cognitive Approach to Intelligent Control Theory

Leonid I. Perlovsky[1] and Vyacheslav P. Shkodyrev[2(✉)]

[1] Harvard University, Cambridge, USA
leonid@seas.harvard.edu
[2] Peter the Great St. Petersburg Polytechnic University, Saint Petersburg, Russia
shkodyrev@mail.ru

**Abstract.** Control of structurally-complex industrial and technological objects belongs to the class of problems of intelligent control, which demands making decisions in states of uncertainty. Further development of this industry will be associated with technologies of intelligent control based on knowledge. Such technologies use methods, models, and algorithms extracting and accumulating knowledge needed to find optimal decisions. Intelligent control theory is based on learning surrounding world and adapting to changes in the process of reaching the defined goal. In this paper we consider a cognitive approach to learning developed following the human cognitive ability and a scientific method of physics. The cognitive approach opens new wide directions towards control of industrial objects and situations that are not well structured and difficult to formalize, especially in real-life circumstances with significant uncertainty. A class of cognitive model control agents based on the principles of learning is described in the paper. Cognitive agents are such kind of agents that are learning from their surrounding and modifying their actions to achieve the goals; this type of agents enables solving problems in a wide area of control in the presence of uncertainty.

**Keywords:** Artificial Intelligence · Theory of control · Cognitive models · Cognitive agents · Hierarchy of industrial or technical systems · Cyber-physical system

## 1 Introduction

Artificial Intelligence (AI) and Intelligent System (IS) are central notions in current theory of control system [1]. Intelligent system is capable to function autonomously, by learning its surrounding, adapting to changes, and reaching defined goals [7]. Other researchers consider as key to intelligence the ability to accumulate knowledge, define aims, and plan actions [1]. At present, widely used is the notion of cognitive agent, i.e., such a kind of agent that is learning the surrounding and modifying its actions to achieve the goals [7, 15]. Cognitive agents capable of reaching goals in varying situations are the most perspective class of mathematical models of intelligent control [1, 11–13].

D. G. Arseniev et al. (Eds.): CPS&C 2019, LNNS 95, pp. 11–18, 2020.
https://doi.org/10.1007/978-3-030-34983-7_2

A key principle of intelligent control is control based on knowledge [9]. Existence of knowledge of how to make the best control decisions in the presence of uncertainty is the foundation of intelligence. Thus learning, or accumulation of knowledge is the foundation of intelligent control [8, 14].

There are two aspects of knowledge that agents use for making good decisions. First, the agent must be in a possession of rules for making good decisions. Second, surrounding circumstances are changing, therefore agents should be able to adapt to these changes. Future intelligent systems will combine learning from data by estimating probability densities with learning from a language text. These ideas were previously discussed in the works of the authors [3–8, 10].

It is assumed that the intelligent control agent receives certain information about the current state of the surrounding, defined as situation Si, as well as actively uses the data to interact with the surrounding. Knowledge of regularities, determining the cause-effect relationship between events in a specific situation and enabling to predict various situations or controlled objects development, is the base that a control agent uses to elaborate efficient strategies for making the best – optimal-control decisions. This information exactly refers to the knowledge or representations of control agent cognitive capacities.

Cognitive control agents key characteristics are the autonomous character and purposefulness of actions. This means autonomous commands execution based on a targeted, problem-oriented reasoning. As the main characteristics of a cognitive intelligent agent are also considered autonomous, in which the intelligence is associated with autonomous perception and reasoning, with making decisions and actions in the states of uncertainty of the surrounding. In this case, critical for a cognitive agent becomes its ability to acquire knowledge through learning: that is, the ability to learn. Such ability requires the possibility to extract, accumulate and apply knowledge used for control. Such cognitive agents are able to learn and to be aware of their surrounding and adapt to it, and change it on the account of knowledge accumulated in the functioning process and acquired skills. A cognitive process is a process by which an autonomous artificial system perceives the surrounding, gains experience through learning, predicts the result of the events, acts and adapts to changes in the surrounding.

## 2   Formulation of the Problem

We consider a cyber-physical system that controls the hierarchy of industrial or technical systems [5]. Possible system states are estimated by clustering available data $(x_1, \ldots, x_n)$, where $x_n$ are characteristics of agents and states of technical systems.

A powerful clustering method is dynamic logics (DL), using the Gaussian mixture model [6]. In this model, every cluster m is characterized by a Gaussian likelihood:

$$l(n|m) = \left(\frac{1}{2\pi}\right)^{0.5d} |C_m|^{-0.5} \exp\left[-\frac{1}{2}(x_n - M_m)^T C_m^{-1}(x_n - M_m)\right].$$

Here $M$ and $C$ are the mean and covariance parameters of the Gaussian likelihood. In addition, every cluster is characterized by its rate:

$$r_m = N_m/N,$$

where $N_m$ is the number of data points belonging to the cluster and $N$ is the total number of data points.

DL algorithm for estimating likelihood parameters starts with arbitrary values of unknown parameters $r$, $M$, $C$.

The next step is to compute association variables:

$$f(m|n) = \frac{r_m l(n|m)}{\sum_{m'} r_{m'} l(n|m')}.$$

Using these association variables data points in the cluster and rates are computed:

$$N_m = \sum_n f(m|n), \quad r_m = N_m/N.$$

Next, the mean value is computed:

$$M_m = (1/N_m) \sum_n f(m|n) x_n,$$

as well as the covariance:

$$C_m = (1/N_m) \sum_n f(m|n) (x_n - M_m)^T (x_n - M_m).$$

Having parameters of clusters, it is possible to evaluate the total likelihood of all defined clusters. The total number of clusters will be defined by maximizing the total likelihood.

The clusters make up the system states; they are denoted by:

$$S = <s_1, \ldots, s_m>.$$

These estimated states represent one aspect of knowledge. Another aspect of knowledge consists of selecting control actions $u(t)$ at every moment $t$.

A control action $u(t)$ transforms the state $s(t)$ into $s(t+1)$. Beginning with the initial state $s(1)$ at the moment $t = 1$, the system goes to the state $u(1)s(1) = s(2)$.

The results of actions $u(t)$ at every state si are considered to be known; they are derived from the system model. We also know the system gain $g(i, j)$ derived from transforming any state $s_i$ into $s_j$. The system goes through the following states

$$s(1), u(1)s(1) = s(2), u(2)s(2) = s(3), \ldots, u(T)s(T) = s(final).$$

The optimal control, therefore, consists of maximizing the total gain over the time $T$:

$$G(T) = g(t = 1, t = 2) + g(t = 2, t = 3) + \ldots + g(t = T, t = final).$$

This gain is maximized by selecting actions $u(1), u(2), \ldots, u(T)$ at every moment $t$.

## 3 Cognitive Control Agent

Cognition, considered in the context of the agents' ability to make conclusions about things and events in the world around them, as well as the ability to learn its surrounding, are the most important characteristics of the intelligent control concept [3]. We suppose that cognitive agents that are capable of automatic accumulation and use of knowledge for making better control decisions, represent the next step in the development of distributed control systems. Such agents have adaptive capabilities that provide efficient activity of devices and systems in a dynamically changing surrounding.

An agent's knowledge represents its awareness of the surrounding and itself. We consider the knowledge of the $i$-th agent as its ability to display the current situation $S_t$, defining agent's interaction with the surrounding or a controlled object, as some action $A_t$:

$$\psi_i : S_i \rightarrow A_i,$$

which, in its turn, is directed to the agent's (or system of agents) reaching the defined goal, i.e., target state:

$$S_G = f(S_t, A_i). \tag{1}$$

The current situation $S_i$ is perceived by an agent through its receptors, i.e., sensors, as a certain set of measured during the time interval $t_i, (i = 1, 2, \ldots, m)$ values, i.e., parameters/signs $z_k(t_l), k = 1, \ldots, K$ which are the base for making a certain evaluation $Q_t$ of the current state or the surrounding of the controlled object:

$$\tilde{Q}_t \cong S_i,$$

where estimation of current state $\tilde{Q}_t$ may be evaluated by feature vectors $Z_{ti}$ as:

$$\tilde{Q}_t \equiv Z_{ti} = [z_1(t), \ldots, z_k(t)]^T, \ldots$$

In the essence, transition to a new state reflects agent's achievement of certain equation $\varphi \equiv \{R_1\}$, which is defining the $f$-transformation operator of the current situation $S_i$, represented in a specific way of a characteristics of the "agent-surrounding" state or the "controlled object" $Q = \{q_k\}$, to the control command $U$ triggering the controlling rules. In this context, we consider the knowledge of a $\psi$ agent $A_i$ as a multitude of rules $\{R\}$ defining the displays of "signs of situation" (controlled object states) in action (set of control decisions):

$$\{R_i\} \equiv \{z_1^i \& \ldots \& z_k^i \Rightarrow r_i\}, z_1^i, \ldots, z_k^i, r_i \in L,$$

where $z_1, \ldots, z_K$ are the situation characteristic values, acting as incentives (predicates) triggering the control agent's response to the perception of a current situation; $r_i$ is the response of the agent (predicate), representing (meaning "initiation") a specific control act.

Groups of control agents, interacting with each other and the surrounding in order to achieve a certain target state, from the functional point of view should be considered as a whole, and, therefore, the dividing line between the surrounding and the agent is rather conditional. In this sense, each agent $i \in N$, interacting with the surrounding or the controlled object, at every time $t = 1, 2, \ldots, T$ is assigned a time-varying state:

$$S_{t+1}^i = f(s_t^i, u_t^i),$$

which from a formal point of view is determined (may be considered) as a hypergraph (of classes).

Each control agent of the network from the mathematical point of view can be defined as a cognitive functional module (CFM), which implements the cognitive mapping process of the perceived information in the knowledge of the control strategy:

$$F : X \times U \rightarrow Q,$$

$$q = f(X, U),$$

or, more compactly:

$$\psi_i : S_i \rightarrow U_i,$$

which, in its turn, is directed to the achievement by the agent (or a system of agents) of some preassigned/target state:

$$S_G = f(S_t, U_i).$$

Groups of agents' knowledge accumulation and connection to a network is the realization of cognitive models' abilities through the startup of cognitive mechanisms for transformation of the received data about the current state $S_i$ into the knowledge of the object control strategy, i.e., a sequence to actions the change of controlled object current state $S_0$, and control objective specifications and strategies to achieve it.

Combining of individual network agents' knowledge triggers a cognitive process of forming a qualitatively new properties, i.e., the network emergence. The group of agents' combined knowledge-making is considered as a cognitive process, explains the new quality emergence, i.e., the network emergence. Control agents' group knowledge $\psi_i$, is combined ing by the unitary target control function Eq. (1) into cumulative knowledge $\Psi_\Sigma$.

## 4   Hyper-network Model of Agent's Hierarchy

Control knowledge making cognitive models for practical implementation effective apparatus is a paradigm of self-learning and self-organizing neural network models that illustrate the mechanisms of machine learning.

Primary network adaptive quantization principle we shall apply to create a neural network model of control strategy knowledge-making. A neural network model implementing a cognitive function module (CFM) displaying the controlled object's current status into the control strategy knowledge can be represented as a hypernetwork $HN$ [10]:

$$HN = (V, E, R, P, F, W),$$

where $V = (v_1, v_2, \ldots, v_n)$ is the plurality of nodes (vertexes-states) of the primary network PNet graph; $E = (e_1, e_2, \ldots, e)$ is the plurality of edges corresponding branches (transitions states) of primary network PNet; $R = (r_1, r_2, \ldots, r_n)$ is the plurality of edges corresponding branches of secondary network WNet; $P : E \rightarrow 2^x$ is the mapping which associates each element of the primary network edge $e \in E$ with a set of vertices $P(e) \subseteq V; F : R \rightarrow 2^e_{PNet}$ is the mapping associating each element edge $r \in R$ with corresponding branches of the secondary network; $W: V = (V, R)$ is a plurality of traces $F(r)$ forming a simple route in the graph $PNet = (V, E; P)$.

Each control meta-level hypernetwork agent may be regarded as a cognitive function module that implements the cognitive process of displaying information about the perceived information on the situation $S_i$ into the knowledge of control strategies:

$$\psi_i : S_i \rightarrow U_i,$$

or, considering the knowledge of a control strategy as a certain control operator $F_i$,

$$F : X \times U \rightarrow Q, q = f(X, U),$$

which, in its turn, is directed to the achievement by an agent (or system of agents) of some preassigned/target state:

$$S_G = f(S_t, U_i).$$

The Holonic principle of organization in the coalition interaction says: every agent displays the objective function as a strategy for the group of lower-level agents. The strategy of the accumulation of knowledge by learning in this case is constructed as an iterative process of synaptic connections' $\Delta W$ parameters correction:

$$w_{k+1} = w_k + \Delta w_k,$$

and forms antigradient search direction $-\nabla J(w_k, \pi)$ function (functional) error:

$$J(w_k, \pi) = J(\|S_G - S_k(w_k, \pi)\|),$$

$$\Delta W = -\varepsilon \nabla J(w, \pi),$$

which determines the current discrepancy between the target (desired) $SG$ and the current state of the control object $S_t$. This suggests that at each iteration there is a correction of synaptic connections $W$ matrix, representing the knowledge parameterization. Wherein the antigradient $-\nabla J(w_k, \pi)$ value largely depends on the reference method (parameterization) of the title $S_G$ and the current $S_k(w_k, \pi)$ state.

## 5 Conclusions

Control of the hierarchies of the structurally-complex industrial and technological objects belongs to the class of problems of intelligent control, which demands making decisions in states of uncertainty. The future in this area belongs to the technology of intelligent control, a technology based on the knowledge. This technology uses methods, models, and algorithms extracting and accumulating knowledge needed to find optimal control decisions.

Capabilities of the Control Systems to extract, accumulate and use knowledge for system control requires cognitive abilities of control agents [2]. Cognitive control agents learn their surrounding, adapt to its changes, and modify it, by using accumulated knowledge.

Cognitive approach opens new wide directions towards control of industrial objects and situations that are poorly structured and difficult to formalize, especially in real-life circumstances with significant uncertainties.

Here the authors considered a class of cognitive models control agents based on the principles of learning. These types of control agents enable solving problems in a wide area of control in the presence of uncertainty.

## References

1. Mayorga, R., Perlovsky, L.I. (eds.): Sapient Systems. Springer, London (2008)
2. Perlovsky, L.I., Kozma, R.: Neurodynamics of Higher-Level Cognition and Consciousness. Springer, Heidelberg (2007)
3. Perlovsky, L.I.: A cognitive model of language and conscious processes. In: Pereira Jr., A., Lehmann, D. (eds.) The Unity of Mind, Brain and World, pp. 265–268. Cambridge University Press, New York (2013)
4. Perlovsky, L.I.: Language and cognition – joint acquisition, dual hierarchy, and emotional prosody. Front. Behav. Neurosci. 7, 123 (2013). https://doi.org/10.3389/fnbeh.2013.00123
5. Perlovsky, L.I.: Physics of the mind. Front. Syst. Neurosci. (2016). https://doi.org/10.3389/fnsys.2016.00084
6. Perlovsky, L.I., Deming, R., Ilin, R.: Emotional Cognitive Neural Algorithms with Engineering Applications; Dynamic Logic: From Vague to Crisp. Springer, Berlin (2011)

7. Russell, S., Norvig, P.: Artificial Intelligence. A Modern Approach. Prentice Hall Series, Upper Saddle River (2003)
8. Schoeller, F., Perlovsky, L.I., Arseniev, D.: Physics of the mind: experimental confirmations of theoretical predictions. Phys. Life Rev. **25**, 45–68 (2018)
9. Shkodyrev, V.P.: Technical systems control: from mechatronics to cyber-physical systems. In: Smart Electromechanical Systems, Ser. Studies in Systems, Decision and Control, vol. 49 (2016)
10. Zhang, C., Ren, M., Urtasun, R.: Graph hypernetworks for neural architecture search. In: Proceedings of ICLR (2019). https://arxiv.org/pdf/1810.05749.pdf
11. Runck, B.C., Manson, S., Shook, E., Gini, M., Jordan, N.: Using word embeddings to generate data-driven human agent decision-making from natural language. GeoInformatica **23**(2), 221–242 (2019)
12. Cross, E.S., Hortensius, R., Wykowska, A.: From social brains to social robots: applying neurocognitive insights to human-robot interaction. Philos. Trans. R. Soc. B Biol. Sci. **374** (1771) (2019). https://doi.org/10.1098/rstb.2018.0024
13. Chemchem, A., Alin, F., Krajecki, M.: Improving the cognitive agent intelligence by deep knowledge classification. Int. J. Comput. Intell. Appl. **18**(1) (2019). https://doi.org/10.1142/s1469026819500056
14. Jones, A.T., Romero, D., Wuest, T.: Modeling agents as joint cognitive systems in smart manufacturing systems. Manuf. Lett. **17**, 6–8 (2018). https://doi.org/10.1016/j.mfglet.2018.06.002
15. Milis, G.M., Eliades, D.G., Panayiotou, C.G., Polycarpou, M.M.: A cognitive agent architecture for feedback control scheme design. In: 2016 IEEE Symposium Series on Computational Intelligence, SSCI 2016, Athens, Greece, 6–8 December 2016 (2017). https://doi.org/10.1109/ssci.2016.7850187

# An Overview of Practical Ontology Implementation in Decision Support Systems

Dmitry Kudryavtsev and Tatiana Gavrilova(✉)

Graduate School of Management, St. Petersburg State University,
199004 Saint-Petersburg Volkhovsky per., 3, Russia
{d.v.kudryavtsev, gavrilova}@gsom.spbu.ru

**Abstract.** Ontology has already a rather long history in the computer science. It helps to represent knowledge in the domain of interest and make it available both for humans and machines. For many years, ontologies were mostly considered as an object for academic research, but nowadays they are getting used in a growing number of applications. The suggested paper provides a brief overview of practical ontology implementation in decision support systems. Typical knowledge-intensive tasks were used to organize the overview: one example of a system was provided for each task.

**Keywords:** Ontology · Ontological approach · Decision support system · Management · Knowledge-Intensive tasks

## 1 Introduction

A decision support system (DSS) is an information system, which supports decision-makingin an organization or a business. A DSS lets users sift through and analyze massive reams of data or knowledge and compile information that can help to make better decisions. There are different types of mathematical basis that serves as a formal foundation for the decision support process. Ontological approach to DSS design and development is one of the latest and most promising ones [11].

Ontology is the only knowledge representation model, which is equally suitable for "human–human", "human–computer" and "computer–computer" communication. As a result, many ontologies serve as universal interfaces for intelligent systems.

Ontology-based knowledge representation helps to keep the feed-back and get an answer for "why?" questions that is especially critical and almost absent in the systems which use machine learning.

Ontology and its successor knowledge graph are moving away from laboratories into practical applications and industry.

Ontology is a formal, explicit specification of a shared conceptualization [14]. Understanding the term "ontology" depends on the context and the purposes of its use. In general, an ontology, or a conceptual model of a domain consists of a hierarchy of concepts of the subject domain, links between them and laws that operate within the framework of this model. Ontology is constructed as a network consisting of concepts and connections between them. Connections can be of various types (for example, "is a," "consists of," "is executor of," etc.). The main task of ontology is to serve as a

© Springer Nature Switzerland AG 2020
D. G. Arseniev et al. (Eds.): CPS&C 2019, LNNS 95, pp. 19–26, 2020.
https://doi.org/10.1007/978-3-030-34983-7_3

bridge or a basis for understanding and communication among all participants (people and programs) in the modelled processes of production or other kinds of processes of the subject domain. To fulfill the role of such a common language, ontology includes a dictionary describing the subject domain and a set of explicit intensional definitions defining meaning of the dictionary elements in the logical theory language. The links between the dictionary (signs) and semantics limit the set of possible interpretations of signs. Ontologies have been widely used in knowledge management systems and education during the last 10 years [1, 2, 15].

The term "ontology" intersects with the "knowledge graph," since if we populate ontology with instances, it becomes a knowledge graph (or knowledge base). A knowledge graph is a set of typed entities (with attributes) which relate one to another by typed relationships. The types of entities and relationships are defined in schemas that may be called ontologies [16]. But the main focus of ontology is on classes, properties and axioms (TBox), while knowledge graphs pay more attention to instances (ABox).

## 2 Application of Ontologies for Creating and Running Decision Support Systems

Knowledge base is the main and the most important component of knowledge based systems (KBS). Integrity and consistency of knowledge contained therein determine the KBS's capacity and the quality of provided decisions. Ontology may frame the knowledge base, form the basis for describing main concepts of the field of knowledge and support the integration of databases containing actual knowledge needed for full-fledged functioning of the KBS. Besides, ontology terms may describe expert rules, which enhances their description level and improves "comprehensibility" for expert users. The Decision Support System (DSS) is an interactive automated information and analytical system aiding a decision-maker in using data and models to resolve their underformalized professional tasks. Underformalized tasks pose the necessity to have a detailed and consistent description of the area of concern, where a decision-maker resolves their tasks. Ontology is the essential tool to create such a description.

The majority of DSSs use large-scale data and knowledge arrays. Ontology provides for clear descriptions of data and knowledge semantics; thus, it establishes the basis for their integration and joint use when resolving tasks.

Ontology-based DSS may be designed using ontologies previously developed for another system within the same field of knowledge. It becomes possible to use knowledge tried in practice, which ensures high quality of developed systems, as well as their potential ability to integrate with existing systems.

Ontologies may be used during design and development of DSSs (for ontology based system development [3]), as well as during system functioning as its full-fledged component (for the ontology-based system [3]).

Ontology may be used during DSS design to:

- Shape and record general domain knowledge shared by all experts;
- Directly conceptualise the subject domain, that provides for data semantics description;
- Ensure the opportunity to reuse knowledge;
- Describe functions of a DSS (types of tasks resolved).

Ontology may be used during DSS operation to provide for the following:

- Joint use of diversified data and knowledge within one system;
- Process of resolving tasks being part of system's functions;
- Better understanding of the field of knowledge for system users.

Existing ontology-based applications can be classified using classical tasks for intelligent systems (see Fig. 1). Ontologies can be used both for analysis and synthesis tasks.

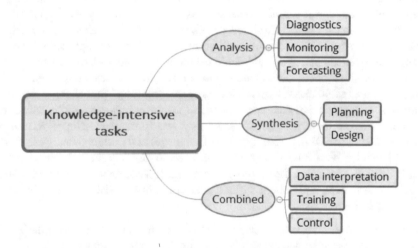

**Fig. 1.** Tasks supported by knowledge-based systems

## 3 Overview of Decision Support Systems Based on Knowledge

This section gives examples of ontology-based expert systems (ES) classified depending on a task resolved: diagnostics, decision support, monitoring, data interpretation, forecasting, training, planning, managing, and design.

**Diagnostics.** Siemens PLMS (Plant Lifecycle Management Systems) COMOS is the industrial facility (power plant, factory…) lifecycle management system supporting its design, development and operation. Siemens specialists drew a paper [6], where they described the PLMS technology and the COMOS system module, which provided for technology implementation and was intended for automated diagnostics and monitoring of developed and operated industrial facilities. Diagnostics mechanism is demonstrated using the example of the developed Simodrive 611 system, a drive control system applied in various automation solutions requiring flexible synchronisation of sophisticated devices. Simodrive 611 consists of a controller and a power source. Diagnostics module identifies two kinds of failures: configuration ones and vibrations of the complete system identified by means of Simodrive 611 components: the controller has a transformation function, and the power source has 2 LED indicators of the status and loading. The system was developed based on ontologies. Symptoms are identified on the basis of failure attributes ("Transformation function," "Power Source Status" and "Power Source Loading"). Each possible fault definition is a sub-class of the "Diagnosis" class. Links between the classes are established based on the relations (for example, "isDiagnosisFor," "hasDiagnosis"). If the controller's transformation function is faulty, and the power source status indicator is off, the diagnosis is the "Faulty Configuration." If both indicators are off, the diagnosis is the "Vibrations."

**Monitoring.** iOSC3 is a desicion support system for monitoring and treatment of patients with acute cardiovascular diseases. System functionality is provided by the specifically developed Critical Cardiac Care Ontology [8]. The fields of knowledge are structured in the system to form a conceptual model: key concepts (like heart rate, infusion pump, medicine, etc.) and their relations "a kind of" (for instance, nitroglycerin > vasodilator > medicine) are organised as a hierarchical taxonomy. The very ontology consists of 40 classes, one object property "atPump" that changes the dose at the infusion pump, and five datatype properties. Decision-making logic is the set of if-then rules linked to the ontology. The ontology was formalised using OWL DL (expanded version of the OWL, Ontology Web Language) and designed using the Protege, SWRL (Semantic Web Rule Language) was used to develop the inference rules. An example of such rules using the common language: IF an infusion pump with amine is connected to a patient, and patient's blood pressure drops, THEN the pump's infusion speed needs to be increased.

**Data Interpretation.** Personal Assistant (PA) is an example of an interpretation system, which is an integrated system interpreting user's requests written in the natural language, capable of performing tasks using a clear interface [10].

Ontologies are used for semantic interpretation of received knowledge and tasks presentation. A statement is interpreted in two stages: (a) grammatical and syntactic analysis and (b) semantic interpretation (the ontology task). The semantic interpretation approach is based on the concept that a statement meaning may be understood by defining its notions and their attributes. Let us review the phrase "Find all articles about agents". It is assumed that the Personal Assistant is integrated in some database. The system uses the ontology and the dictionary to preliminaryly define if the request refers to a specific topic. Then the grammatical and syntactic analysis module brings the matrix containing the list of markers/words and their syntactic classifications. Personal Assistant will determine that the "search" is the item of the "Actions" class (a list of synonyms is also used, for example, "search for," "display" or "find"), the "article" is the class, and agents are its property, or the topic of the article. Thus, the request will be interpreted.

**Forecasting.** Altman Z-score is the company bankruptcy forecasting system based on ontologies. Ontologies contained in the paper [9] were described using the OWL and split into several levels. Upper-level ontologies describe general concepts like time and space, which depend on a specific domain; field of knowledge ontologies describe the vocabulary related to the domain; task ontologies describe the lexis related to the task; application ontologies describe specific notions, i.e., the ontology is a taxonomy: for example, Assets class has sub-classes "Current," "Long-Term" and "Other"; the "Current" class contains sub-classes "Cash" and "Short-Term Investment." A financial report is delivered to the system at the input. When the ontology is drawn, data cluster analysis is carried out, and Edward Altman's formula is used to calculate the Z-score, which is the company's bankruptcy index. Altman calculates the Z-score, the company's bankruptcy index. For instance, if $Z < 2.675$, then it is 95% likely that a company will run bankrupt within one year.

**Training.** SQL-Guide is a kind of the adaptive Web-based educational system (AWBES) for training skills in SQL requests to databases [12]. Tasks for students are based on a range of databases and multiple desired results of request performance. A student needs to make corresponding requests to achieve these results. The system evaluates an answer considering the result correctness and time for request performance and ensures a convenient feedback. SQL ontology in the system describes classes or commands and reserved SQL words (WhereClause is the Clause sub-class, SelectStatement is the Statement sub-class, etc.), as well as their relations (SelectStatement uses WhereClause, WhereClause is UsedIn SelectStatement) (available at http://www.sis.pitt.edu/~paws/ont/sql.owl). Ontology assists task indexing for students and supports development of the adaptive educational content and integration of tutorial systems (for example, SQL Guide and SQL Tutor) [12, 13].

**Planning.** PASER (Planner for the Automatic Synthesis of Educational Resources) is the educational programme automatic synthesis system [4] capable of dynamic development of educational programmes using educational resources and considering individual features, preferences, needs and capabilities of a student. The system consists of (a) a metadata repository storing Learning Object Metadata, student profiles and ontology of the taught field of knowledge; (b) a mechanism of request development and logical display; and (c) the actual planner. The PASER ontology was designed using the RDF (Resource Description Framework) providing for the mechanism of competence and property description. It describes artificial intelligence area and involves 310 competences structured as the isPartOf hierarchy. For instance, the root element has sub-competences like Machine Learning, Planning, Knowledge Representation, etc. They also have their sub-competences, etc. Competences also describe student's goals and achievements. Reasoning engine determines links between objects of study and student's characteristics. Tuition plan is based on the analysis of an ontological description of the student's accumulated knowledge and goals.

**Control.** Telehealth Smart Home is the system of physical and software components ensuring safety of senior patients with cognitive autonomy loss [5]. The system is intended for prevention of possible hazardous situations. A nurse, patient's relatives, a family doctor, a system manager, etc. are included in the control loop. Telehealth Smart Home uses a range of ontologies for the design and configuration of software applications, as well as for human-machine interface. The Habitat ontology describes the accomodation structure (classes "room," "doors," "windows," etc.). The PersonAndMedicalHistory ontology describes the person who needs care, his/her medical record, and an assistant (classes "name," "age," "disease," etc.). The Equipment ontology determines the smart house equipment participating in the system (classes "furniture," "appliances" and "technical equipment"). System software modules are described using the ontology Software Applications. The Task ontology describes system tasks in detail. The Behaviour ontology describes important peculiarities of remote monitoring, like habits and critical physiological parameters. And, finally, the Decision ontology describes the system behaviour in case of a critical situation (for example, a call for emergency aid). Main data is collected using sensors (elements of the TechnicalEquipment class, which is the sub-class of Equipment). Having been processed, these data become part of the Behaviour ontology and two classes of PersonAndMedicalHistory ontology, namely, MedicalHistory and Patient. Equipment classes are linked to Habitat classes through their relations. This provides for the determination of equipment arrangement in a house. Bayesian networks study patient's behaviour (their cognitive disorder) by processing data contained in the ontologies. Then, the Decision on possible activities to ensure patient's safety is made based on the obtained results.

**Design.** Smart Architect is an intelligent system capable of automatic generation of a piece of ancient Chinese architecture [7]. The system applies ontology-based approach to the analysis of various architectural styles and converts primitive geometries into architectural semantic components. The system's ontology consists of three components/levels related to 3 main participants of the design process. They are a designer, a programmer, and a machine. An ontology conceptual model always has a hierarchical structure. For example, the Architecture (Ar) class consists of two sub-classes: Ancient Ar and Modern Ar. The Ancient Ar is the parent class for Chinese Ancient Ar, Indian Ancient Ar, European Ancient Ar and other classes. A user needs to choose 3 parameters, including city maps, semantic components and descriptions of construction styles and structures (these input parameters may be changed in order to identify individual modelling goals) to receive the draft task. These parameters are preliminaryly defined in the designer level ontology, and relations link them to structure-modelling interactive tools. Then the programmer level ontology provides for making models, integration of a structure into the terrain plan and style check. Machine level ontology contains the information on structure visualisation.

## 4　Conclusions

Now the amount of real-life applications of ontologies for developing intelligent systems grows. Ontologies enable to formalize domain knowledge and create a knowledge base for DSS. They help to integrate data in DSS, reuse existing knowledge for creating new systems and support communication with domain experts and users. The paper provided a brief overview of practical ontology implementation in DSS. Typical knowledge-intensive tasks were provided: Diagnostics, Monitoring, Forecasting, Planning, Design, Data interpretation, Training, Control. Then examples of ontology-based systems were provided for these tasks. The examples demonstrated both the value of such systems and the role of ontologies in them.

**Acknowledgement.** The work was supported by the Russian Foundation for Basic Research (#17-07-00228).

## References

1. Abdulla, D., Prado de O Martins, L., Schultz, T.: Decolonising Design Education: Ontologies, Strategies, Urgencies (2018). http://ualresearchonline.arts.ac.uk/id/eprint/12504
2. Gavrilova, T.: Orchestrating ontologies for courseware design. In: Tzanavari, A., Tsapatsoulis, N. (eds.) Affective, Interactive and Cognitive Methods for E-Learning Design: Creating an Optimal Education Experience, IGI Global, USA, pp. 155–172 (2010)
3. Guarino, N.: Formal ontology in information systems. In: Proceedings of the First International Formal Ontology in Information Systems conference FOIS 1998, June 6–8. Trento, Italy: IOS press, pp. 3–15 (1998)
4. Kontopoulos, E., Vrakas, D., Kokkoras, F., Bassiliades, N., Vlahavas, I.: An ontology-based planning system for e course generation. Expert Syst. Appl. **35**(1), 398–406 (2008)
5. Latfi, F., Lefebvre, B., Descheneaux, C.: Ontology-based management of the telehealth smart home, dedicated to elderly in loss of cognitive autonomy. In: Proceeding of the OWLED (2007)
6. Legat, C., Neidig, J., Roshchin, M.: Model-based Knowledge extraction for automated monitoring and control. In: Proceedings of the 18th IFAC World Congress, pp. 5225–5230. Milan (2011)
7. Liu, Y., Xu, C., Zhang, Q., Pan, Y.: The smart architect: scalable ontology-based modeling of ancient Chinese architectures. IEEE Intell. Syst. **1**, 49–56 (2008)
8. Martinez-Romero, M., Vázquez-Naya, J., Pereira, J., Pereira, M., Pazos, A., Baños, G.: The iOSC3 system: using ontologies and SWRL rules for intelligent supervision and care of patients with acute cardiac disorders. Comput. Math. Meth. Med. **2013**, 1–13 (2013)
9. Martin, A., Manjula, M., Venkatesan, V.P.: A business intelligence model to predict bankruptcy using financial domain ontology with association rule mining algorithm. IJCSI Int. J. Comput. Sci. Issues **8**(3), 211–218 (2011)
10. Paraiso, E., Barthès, J.: An Ontology-Based Utterance Interpretation in the Context of Intelligent Assistance, V workshop em Tecnologia da Informacao e da Linguagem Humana, XXVII Congresso da SBC, pp. 1745–1748 (2007)
11. Rospocher, M., Serafini, L.: An ontological framework for decision support. In: Takeda, H., Qu, Y., Mizoguchi, R., Kitamura, Y. (eds.) JIST 2012. LNCS, vol. 7774, pp. 239–254. Springer, Heidelberg (2013). https://doi.org/10.1007/978-3-642-37996-3_16

12. Sosnovsky, S., Mitrovic, A., Lee, D., Brusilovsky, P., Yudelson, M.: Ontology-based integration of adaptive educational systems. In: 16th International Conference on Computers in Education (ICCE 2008 Taipei, Taiwan, 2008), pp. 11–18 (2008)
13. Brusilovsky, P., Sosnovsky, S., Lee, D.H., Yudelson, M., Zadorozhny, V., Zhou, X.: An open integrated exploratorium for database courses. ACM SIGCSE Bull. **40**(3), 22–26 (2008)
14. Studer, R., Benjamins, R., Fensel, D.: Knowledge engineering: principles and methods. Data and Knowl. Eng. **25**(1–2), 161–197 (1998)
15. Sureephong, P., et al.: An Ontology-Based Knowledge Management System for Industry Clusters. Global Design to Gain a Competitive Edge, pp. 333–342. Springer, London (2008). https://doi.org/10.1007/978-1-84800-239-5_33
16. Villazon-Terrazas, B., et al.: Construction of enterprise knowledge graphs (I). Exploiting Linked Data and Knowledge Graphs in Large Organisations, pp. 87–116. Springer, Cham (2017). https://doi.org/10.1007/978-3-319-45654-6_4

# Layout Optimization for Cyber-Physical Material Flow Systems Using a Genetic Algorithm

Nikita Shchekutin[1]([⊠]), Ludger Overmeyer[1],
and Vyacheslav P. Shkodyrev[2]

[1] Leibniz Universität Hannover, Hanover, Germany
Nikita.Shchekutin@icloud.com,
Ludger.Overmeyer@ita.uni-hannover.de
[2] Peter the Great St. Petersburg Polytechnic University, Saint Petersburg, Russia
shkodyrev@mail.ru

**Abstract.** Cyber-physical production systems are a key solution in the Industry 4.0 age. Still, the advantages of using them are not always easily presented with numbers. In this paper, the task of arranging a cyber-physical material flow system is addressed as a multi-objective optimization problem and a genetic algorithm is used to search for a Pareto front of optimal layouts. As an example of such a material flow system, a decentralized modular conveyor, which was developed at the Institute of Transport and Automation Technology at Leibniz University Hannover, is used.

**Keywords:** Logistics · Cyber-physical systems · Genetic algorithm · Layout optimization · Facility layout problem

## 1 Cyber-Physical Production Systems: New Functional Possibilities Lead to New Planning Challenges

The steadily increasing individualization of customer needs demands an increase in flexibility at all stages of production. This has inspired the adoption of Industry 4.0 trends and the application of cyber-physical systems (CPSs) as a key solution [3]. Such an approach implies digitalization and coupling the information with any physical process – whether it is a production step or a logistic operation. This provides the required flexibility in production and allows advanced individual needs of a customer to be addressed, since changes in production layout are now easier and cheaper to implement. Figure 1 presents a sample cyber-physical material flow system, developed at the Fraunhofer Institute for Material Flow. Multishuttle is a decentralized multi-agent material flow system, which is able to switch between continuous and discontinuous modes and autonomously build groups in order to solve complex logistical tasks [8].

Industry 4.0 solutions are becoming an essential part of today's production and the digital transformation affects all stages of production, including supply chain and ERP, but a lack of universal integration concepts remains one of the main problems [2].

© Springer Nature Switzerland AG 2020
D. G. Arseniev et al. (Eds.): CPS&C 2019, LNNS 95, pp. 27–39, 2020.
https://doi.org/10.1007/978-3-030-34983-7_4

This problem is especially important for small and medium enterprises, where the workforce is too small to spend time on integrating such novel solutions; moreover, each wrong solution could bring high costs [7].

**Fig. 1.** Multishuttle - an implementation of a cyber-physical material flow system

The use of CPSs affects planning inside a factory and sets new requirements. Firstly, increased functional flexibility makes it more complex to describe such a system mathematically and extends the potential solution space when searching for an optimal arrangement. Secondly, shortened production cycles add new computational requirements – in the optimal case, such planning must occur in real time [3]. Finally, the role of man in the "man-machine frameworks" must be reduced in all spheres of production and there consequently is a need for self-optimizing solutions. Modern data acquisition approaches allow lots of information to be collected and this information can't be processed manually any longer. Still, the role of experts in the integration of customized Industry 4.0 solutions is essential.

The problem of integrating and optimizing a cyber-physical material flow system has already been described in references [18, 19]; this paper focuses on the implementation of the optimization algorithm. To validate the results of the algorithm in this paper, a decentralized modular conveyor is used as a cyber-physical material flow system. This has been developed at the Institute of Transport and Automation Technology at Leibniz University Hannover. References [10] and [20] present the hardware and software of the conveyor. In the next chapter, the main features are briefly explained.

## 2   Use Case: Arranging a Decentralized Modular Conveyor

The material flow system consists of novel conveyor matrices and conventional belt conveyors. A conveyor matrix has several conveyor modules, which can be combined in arrangements of different shapes and sizes according to a plug-and-play concept. As decentralized control and communication concepts are employed, no extra rearrangement costs are incurred. The decentralized control, communication and modularization concepts are extended to the conventional belt conveyors, so the whole system can easily be rearranged as plug-and-play units. Figures 2 and 3 present the hardware and the corresponding 2D model representations used in this work.

The conveyor matrix has extended functionality compared to conventional belt conveyors: due to the small scale of the conveyor modules, the matrix can change direction, buffer, rotate, sequence, palletize, and combine packets. But as the conveyor matrices are more expensive, they should only be used when the extended functionality is required. The main function of each of the conveyor elements is to transport packets from a source to a goal. Apart from this, there are three optional functions: Input (packet enters the system here); Process (packet can be processed here) and Output (packet leaves the system here).

A similar state-of-the-art problem is the facility layout problem, which involves searching for the optimal arrangement of machines within a single factory [6]. The research on current solution approaches and the selection of an appropriate task formulation approach (quadratic assignment problem) were presented in a previous paper [19]. Figure 4, left, presents the initial state for layout calculation: the positions of the Inputs (green), Processes (blue) and Outputs (red) are shown. A list of available conveyors and their characteristics is also provided. The task is to find an optimal arrangement of conveyor matrices (orange) and belt conveyors (grey) which provide the required routes, see Fig. 4, right, considering the objective values.

Within the scope of this work, the solution matrix Conveyor_Coordinates consist of $xb$ and $yb$ (coordinates of conveyor's bottom-left corner), Orientation (0 for horizontal and 1 for vertical) and the identification number Conveyor_ID, which provides (per request) the remaining information such as length, width, speed, price, and type of conveyor:

$$\text{Conveyor\_Coordinates} = (xb \cdot yb \cdot \text{Orientation Conveyor\_ID}) \qquad (1)$$

To evaluate the quality of a layout, an objective mathematical assessment is required, and the next section addresses this issue.

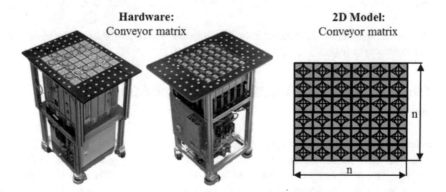

**Hardware:**
Conveyor matrix

**2D Model:**
Conveyor matrix

**Fig. 2.** Conveyor matrix and its 2D-model

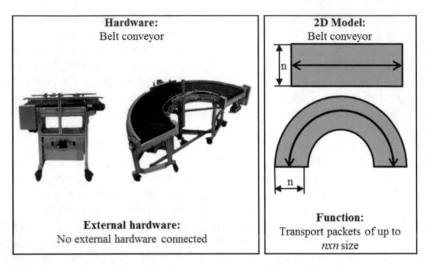

**Fig. 3.** Belt conveyor and its 2D model

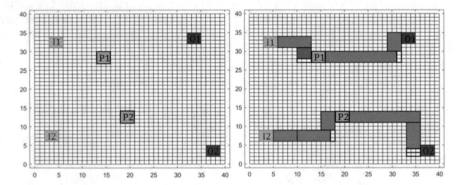

**Fig. 4.** An example of the initial state for the layout calculation (left) and a sample layout for the material flow system (right)

## 3   Objective Function: Challenge of Quantifying Qualitative Advantages

An essential step in addressing any optimization problem is defining objective values. In industrial environment, there always are several conflicting values to optimize – improving one of them often leads to degradating others. Therefore, instead of searching for a single optimal solution, a so-called Pareto front of optimized solutions is sought [14] after.

To optimize a material system, three objective values are selected in this paper. The first two objective values are quantitative: these are transportation costs and transportation time. Calculating these values is not difficult. It can be performed using established equations and is beyond the scope of this paper. But calculating qualitative values such as elasticity and flexibility is more complex, and these are the main advantages of cyber-physical systems and a reason of higher investment costs.

According to reference [21], elasticity describes the ability of a layout to handle changing circumstances (packet size or throughput changes) without being reconfigured. Higher elasticity eliminates the need for reconfiguration and reduces out-of-service losses. In contrast, the flexibility of a layout describes its ability to react to a changing situation by making the cheapest adjustment (for example, by adding extra conveyors).

Such objective values are hard to measure and often depend on the individual needs of an enterprise, so there are no universal equations for them. In this work, they are assessed based on three criteria: type and size of conveyors, number of provided routes and available area for layout extension.

The first summand (type and size of conveyors) rewards a layout for conveyor matrices, because these increase the elasticity of a layout compared to belt conveyors: the packets can successfully pass each other in different directions without any waiting time. Moreover, larger conveyor matrices offer more connection possibilities for additional belt conveyors (see Fig. 5).

The second summand (number of provided routes) rewards a layout for each of the routes provided. Figure 6 demonstrates this principle. The left diagram in this figure demonstrates an example production line with five existing routes; green denotes the conveyors (without showing conveyor types). There are still four routes which could be needed, but are not provided by the current arrangement of the system. Therefore, the second summand for such a layout would be 5/9 (55.6%) of the maximum possible value.

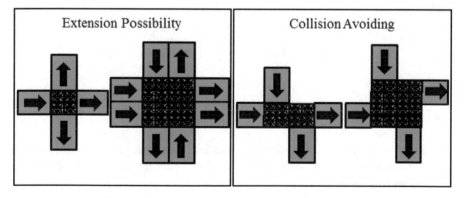

**Fig. 5.** Advantages of larger matrices: more belt conveyors can be connected to these matrices and potential logistical conflicts can be resolved

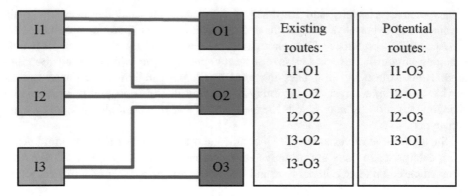

**Fig. 6.** Each available route increases the elasticity of the system

As mentioned above, the third summand rewards the layout for the freely available area, which is calculated as the area of all conveyors in the layout subtracted from the factory area. Summing up, the following three summands determine the elasticity and flexibility of a material flow system layout:

- a special reward $p_i$ is assigned for each $b_i$ type of all m conveyors according to the conveyor's type and size;
- a reward $pbr_j$ is assigned for each route $r_j$ of all $p$ provided routes;
- a reward $p_r$ is assigned for the available area $A_f$, which is calculated as the area $l_i \cdot br_i$ of each conveyor $b_i$ of all $m$ conveyors subtracted from the factory area $l_h \cdot b_h$

Thus, to evaluate the elasticity and flexibility of a layout $F_{EF}$, the following equation is used:

$$\max F_{EF} = \sum_{i=1}^{m} p_i \cdot b_i + \sum_{j=1}^{p} pbr_j \cdot r_j + A_f \cdot p_r; \qquad (2)$$

$$A_f = (l_h \cdot b_h) - \sum_{i=1}^{m} b_i \cdot (l_i \cdot br_i). \qquad (3)$$

In addition to this, the transportation time FTT and transportation costs FTC are used to evaluate the quality of a layout. In the following section, an optimization algorithm will be presented.

## 4   Optimization with a Genetic Algorithm

As a variant of the facility layout problem, the arrangement of material flow systems has the highest computational complexity and belongs to the so-called NP-Hard problems [5]. This means that exact optimization algorithms (convex optimization) are not suitable for solving this problem and metaheuristic algorithms must be applied.

Since a Pareto front is sought after, a population-based algorithm must be used, where several layouts (the population) are optimized at the same time.

The genetic algorithm is the most famous representation of such an optimization approach. An alternative to such an optimization method would be simplification into a single-objective problem, where several objective values are packed into an objective function using weighting coefficients. This is not always possible, because there is normally little a priori information about the problem and weightings are not easy to define. Table 1 presents the most popular algorithms applied to the facility layout problem.

**Table 1.** Metaheuristic algorithms for solving the facility layout problem in current research

| Algorithm | Number of articles | Latest publications |
|---|---|---|
| Genetic algorithms | 50 | [15] |
| Simulated annealing | 37 | [1] |
| Tabu search | 15 | [4] |
| Ant colony optimization | 11 | [13] |
| Particle swarm optimization | 12 | [11] |
| Variable neighborhood search | 6 | [16] |
| Greedy algorithm | 3 | [12] |
| Scatter search | 2 | [9] |

The genetic algorithm for this task consists of three main steps: initialization, solution evaluation and solution adjustment. These steps will be presented further. The algorithm was implemented using a Matlab script and a graphical interface was developed for visualizing the layouts.

## 5  Initialization

At the beginning, a starting population of multiple layouts is created: it is normally called the "0 Generation". There are several important parameters to consider here, such as the size of the initial population, which directly influences the computational duration and the quality of the algorithm's results. The main goal of the initialization is to cover the solution space consistently, which helps to avoid sticking to a local optimum. Figure 7 presents four sample layouts as initial populations for the production scenario presented in Fig. 4. Green blocks denote inputs, blue blocks denote processing machines, red blocks denote outputs; conveyor belts are colored gray and conveyor matrices are colored orange.

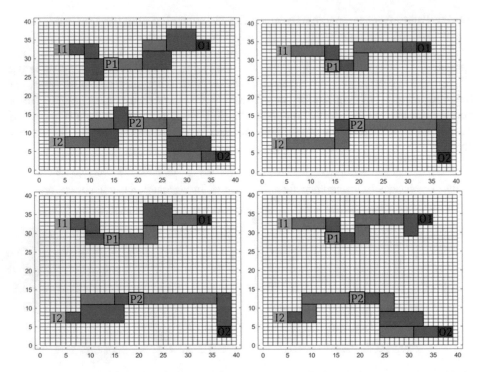

**Fig. 7.** Example of an initial population. All the layouts satisfy the requirements and provide for the same routes, but demonstrate different objective values

## 6   Evaluation of the Solution

In the genetic algorithm, the role of the objective values is to form the basis for selecting the best individuals from a population for further crossover and mutation. Thus, a new generation is always considered current, and a reverse step to a previous generation is not possible. The number of generations and the population size are important algorithm-specific parameters, which are to be set depending on the scenario's complexity.

Three conflicting objective values from the previous section are used for the evaluation of each layout. For a population, a ranking based on Pareto dominance according to references [14, 17] is created: the ranking $F(I)$ of an individual $I$ is calculated by using the number of individuals $n_{dom}(I)$ that dominate this individual in all three objective values:

$$F(I) = 1 - n_{dom}(I). \tag{4}$$

Thus, each layout in the population gets its own fitness value.

# 7   Variate Solution: Selection, Crossover and Mutation

The variation of a solution is achieved by following state-of-the-art steps: selection, crossover, and mutation. Figure 8 demonstrates this iterative process; for each population, several variation cycles are performed and Table 2 explains each of the steps and its outcomes. These methods were chosen and implemented in reference [17], where a comprehensive research into genetic algorithms was performed.

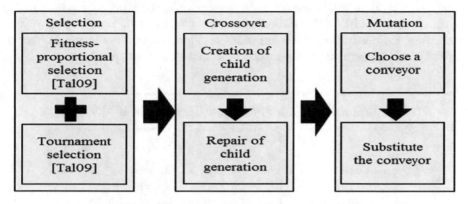

**Fig. 8.** Variation of a solution in the genetic algorithm

This algorithm runs iteratively until a predefined number of generations is reached. The goal is to investigate the solution space, covering as many optimum areas as possible and using as few resources as possible. The danger of sticking in local optima often occurs. That is why, selection routine is tuned in order to sometimes select the solutions with currently worse fitness values; potentially, these solutions could later be a part of the global optimum area. Tuning such algorithms is a special topic of interest and its effectiveness depends on algorithm-specific parameters. In the following section, sample results of the layout optimization are presented.

**Table 2.** Variation steps for the genetic algorithm

| Step | Description | Result |
|------|-------------|--------|
| Selection | Fitness proportional and tournament selection | Mating pool for crossover and mutation |
| Crossover | Creation of child generation out of the best layouts | Partially "broken" layouts |
| | Repairing the child generation | Repaired layouts |
| Mutation | One of three mutation steps are implemented with an equal probability: 1. Changing a conveyor 2. Adjusting the size of conveyor 3. Creating a new route | Layouts of the next generation with improved fitness values |

## 8  Results and Discussion

Figure 9 presents the Pareto front on two axes (transportation time and transportation costs) after 30 generations for the sample production scenario of three routes. In this case, the Pareto front consists of 21 optimized layouts, marked with numbers on the graph and placed according to the corresponding objective values.

For such an example, correlation of several objectives occurs: Layout 13 demonstrates the shortest transportation time, while Layout 2 is the best in terms of transportation costs. Both results are due to the small number of conveyor modules in this layout. Packets tend to move slower on conveyor matrices than on belt conveyors, because of the need to resolve route conflicts. Thus, an increase in the number of matrices leads to increased transportation costs and slower transportation.

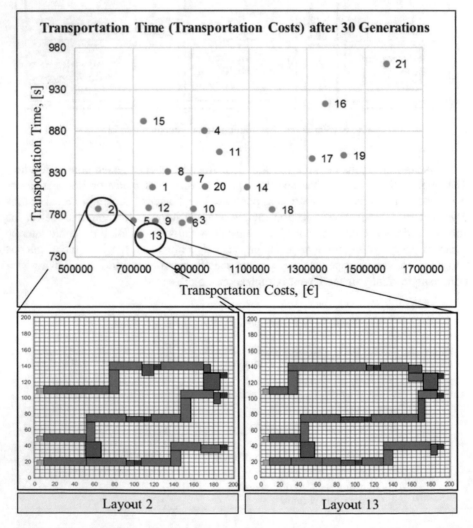

**Fig. 9.** Pareto-front for transportation time against transportation costs after 30 generations

Due to the high cost of creating a conveyor matrix with a new size, the algorithm creates only one if there is no other way to cover the necessary route. Still, creating matrices of different sizes is an essential feature of the genetic algorithm, which is proven during further tests using scenarios that are more complex.

Figure 10 presents the same Pareto front with elasticity and flexibility against the transportation cost. Layout 21 is the best in terms of elasticity and flexibility and is two times better than Layout 5 (the worst in terms of flexibility and elasticity). Still, layout 21 costs almost three times more. Thus, with the help of such a Pareto front, planer can evaluate potential arrangements of available material flow systems and determine the prices for particular qualitative goal values.

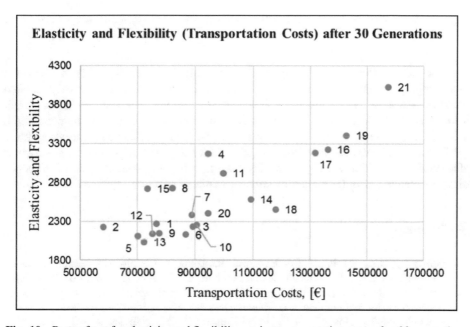

**Fig. 10.** Pareto front for elasticity and flexibility against transportation costs after 30 generations

## 9 Conclusion and Outlook

The genetic algorithm requires more computational time in comparison to trajectory-based optimization techniques (a single layout is optimized throughout the process). Figure 11 presents the comparison of computational effort for the genetic algorithm and simulated annealing approaches. Simulated annealing shows no dependence on the problem size; solution quality gets worse, though, when the problem gets more complex and there are more routes to provide.

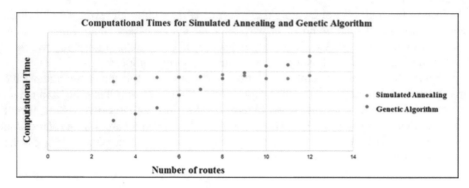

**Fig. 11.** Computational times of a genetic algorithm and simulated annealing, applied to the same problem

The computational time of the genetic algorithm increases quasi-linearly with the increase of problem complexity. The reason is the increased number of generated layouts, which do not satisfy the boundary conditions and need to be repaired. Moreover, the tuning of the algorithm can still be improved: the size of the population, number of generations and selection type are the factors influencing the algorithm's performance. The results prove the feasibility of the approach as applied to a particular decentralized modular conveyor. In future work, a further generalization of the algorithm is planned in order make this concept applicable to other systems.

# References

1. Allahyari, M.Z., Azab, A.: Mathematical modeling and multi-start search simulated annealing for unequal-area facility layout problem. Expert Syst. Appl. **91**, 46–62 (2018)
2. Ardito, L., et al.: Towards Industry 4.0: Mapping digital technologies for supply chain management-marketing integration. Bus. Process Manage. J. 25, 323–346 (2018)
3. Bauernhansl, T., Hompel, M.T., Vogel-Heuser, B. (Hg.): Industrie 4.0, Produktion, Automatisierung und Logistik: Anwendung-Technologien-Migration. Springer, Wiesbaden (2014)
4. Bozorgi, N., Abedzadeh, M., Zeinali, M.: Tabu search heuristic for efficiency of dynamic facility layout problem. Int. J. Advmanuf. Technol. **77**(1), 689–703 (2015)
5. Burkard, R.E., Stratmann, K.H.: Numerical investigations on quadratic assignment problems. Naval Res. Log. Quar. **25**, 129–148 (1978)
6. Drira, A., Pierreval, H., Hajri-Gabouj, S.: Facility layout problems: a survey. Ann. Rev. Control **31**(2), 255–267 (2007). https://doi.org/10.1016/j.arcontrol.2007.04.001,2007
7. Faller, C., Feldmüller, D.: Industry 4.0 learning factory for regional SMEs. Procedia CIRP **32**, 88–91 (2015)
8. Kirks, T., Stenzel, J., Kamagaew, A., Hompel, M.T.: Zellulare Transportfahrzeuge für flexible und wandelbare Intralogistiksyste-me. Logistics J. (2012)
9. Kotothari, R., Ghosh, D.: A scatter search algorithm for the single row facility layout problem. Int. J. Adv. Manuf. Technol. **68**(5–8), 1665–1675 (2013)

10. Kruhn, T.D.: Verteilte Steuerung flächiger Fördersysteme für den innerbetrieblichen Materialfluss. Zugl.: Hannover, Univ., Diss.,. Hg. v. Ludger Overmeyer. Garbsen, Garbsen: PZH-Verl., TEWISS - Technik-und-Wissen-GmbH (Berichte aus dem ITA, 2015, Bd. 1) (2015)

11. Liu, J., Zhang, H., He, K., Jiang, S.: Multi-objective particle swarm optimization algorithm based on objective space division for the un-equal-area facility layout problem. Expert Syst. Appl. **102**, 179–192 (2018)

12. Matai, R., Singh, S.P., Mittal, M.L.: A non-greedy systematic neighbourhood search heuristic for solving facility layout problem. Int. J. Adv. Manuf. Technol. **68**(5–8), 1665–1675 (2013)

13. Ning, X., Qi, J., Wu, C., Wang, W.: A tri-objective ant colony optimization based model for planning safe construction site layout. Autom. Const. **89**, 1–12 (2018)

14. Pardalos, P.M., Migdalas, A., Pitsoulis, L. (eds.): Pareto Optimality, Game Theory and Equilibria (2008)

15. Phanden, R.K., Demir, H.I., Gupta, R.D.: Application of genetic algorithm and variable neighborhood search to solve the facility lay-out planning problem in job shop production system. In: 7th International Conference on Industrial Technology and Management (ICITM), Oxford, pp. 270–274 (2018)

16. Pichka, K., Bajgiran, A.H, Petering, M.E., Jang, J., Yue, X.: The two echelon open location routing problem: Mathematical model and hybrid heuristic. Comput. Ind. Eng. **121**, 97–112 (2018)

17. Poschke, A.: Pareto-Ansatz zur Layoutoptimierung kogntiver Fördersysteme. Master Thesis supervised by, N. Shchekutin, Leibniz Universität Hannover, Institute of Transport and Automation Technology (2018)

18. Shchekutin, N., Zobnin, S., Overmeyer, L., Shkodyrev, V.: Mathematical methods for the configuration of transportation systems with a focus on continuous and modular matrix conveyors. Logistics Journal: Proceedings, Jg. (8) (2015)

19. Shchekutin, N., Sohrt, S., Overmeyer, L.: Multi-objective layout optimization for material flow system with decentralized and scalable control. Logistics Journal: Proceedings, Jg. (10) (2017)

20. Sohrt, S., Heinke, A., Shchekutin, N., Eilert, B., Overmeyer, L., Krühn, T.: Kleinskalige, cyber-physische Fördertechnik, Vogel-Heuser, B., Bauernhansl, T., ten Hompel, M. (Hrsg.): Handbuch Industrie 4.0: Produktion, Automatisierung und Logistik, Springer, Heidelberg (2016)

21. Wascher, G.: Innerbetriebliche Standortplanung bei einfacher und mehrfacher Zielsetzung, Bochumer Beiträge zur Unterneh-mungsführung und Unternehmensforschung. Gabler Verlag, Wiesbaden (1982)

# Cyber-Physical Systems in Complex Technologies and Process Control

Branko Katalinić[1], Dmitry Kostenko[2], Vadim A. Onufriev[2(✉)],
and Vyacheslav V. Potekhin[2]

[1] Vienna University of Technology, Vienna, Austria
branko.katalinic@tuwien.ac.at
[2] Peter the Great St. Petersburg Polytechnic University, Saint Petersburg, Russia
zaba-l@bk.ru, {onufriev_va, slava.potekhin}@spbstu.ru

**Abstract.** In this paper the authors examine two main aspects of complex technological processes control implemented within the paradigm of a cyber-physical management system. These include controlled multi-agent cooperation (such as smart grids) and identification of mathematical models connecting different key performance indicators. The authors of this paper claim that system architecture based on intelligent agents' concept outperforms ad-hoc implementations of distributed systems and architectures. The main advantages of a multi-agent system are the presence of a decision making layer, theory-based fault tolerance and scalability. Advanced distributed coordination and decision-making techniques allow more effective operation and availability. The internal general computational framework presented in the paper can be used for solving business tasks (for example, the task of routine optimization as a part of underlying multi-agent framework). The aforementioned system has been applied to solve a problem of power grid member communication. This multi-agent system was designed to control a power redistribution grid, containing one power plant and multiple zero-energy buildings (ZEBs), connected with a distributed knowledge base.

**Keywords:** Industry 4.0 · Identification · Digital twin · Multicriteria optimisation

## 1 Introduction

Modern trends and challenges in high-tech industries development arise from the development of the infrastructure and the complexity of modern industries, their territorial distribution and a variety of functional purposes, which leads to substantial problems in carrying out the control tasks.

The high structural complexity is defined by rapidly growing production and consumption of increasingly diverse and complex informational structures, following the functioning and development of social systems as a whole and their parts in the global information space.

© Springer Nature Switzerland AG 2020
D. G. Arseniev et al. (Eds.): CPS&C 2019, LNNS 95, pp. 40–54, 2020.
https://doi.org/10.1007/978-3-030-34983-7_5

The worldwide trend is Industry 4.0 [1], the strategy of industrial development based on global digitalisation, data interchange, broad introduction of cyber-physical systems, Internet of Things, cloud computing and artificial intelligence.

With all the aforementioned trends, a theory of network centric control became one of the prospective scientific directions. A network centric control system (NCCS) [2] is a new concept of structurally complex distributed networks, characterised by the principles of openness, self-organisation, weak hierarchy in the decision-making circuit, and the ability to generate objectives inside it.

Network centric control is particularly effective for poorly formalisable or unpredictable situations [3]. The advantage of the NCCS paradigm lies within its ability to achieve synergy opportunities and increase marketability of group management capabilities by integrating the knowledge of individual subsystems and making optimal group-wide decisions [4].

However, implementation of network centric systems requires of us to address the question of effective agents' interaction.

In addition, the management of complex hierarchical processes starts with Key Performance Indicators (KPI) for management [5, 6]. However, this implies that all KPI requirements must be coherently arranged among the agents.

Figure 1 presents a multi-agent system, where Programmable Logic Controllers (PLCs) controlling technological processes of the lower level are located at the bottom. The lower-level agents, associated with the PLCs, are placed directly above them. The upper part of the picture shows the top-level hierarchy members.

Each agent controls its process, associated with a list of KPIs. Each process must be decomposed into sub-processes. The decomposition process gives us three hierarchies: processes, KPI and agents hierarchy.

As a result, complex technologies management comes down to correct interaction of multiple independent agents accounting for the specificity and relationships between all processes and ensuring the achievement of the required values for each KPI. Note that mathematical models for KPIs relationships are unknown, and approaches to the agent interactions are not obvious.

The purpose of this article is to offer integrated approach to the complex processes management based on a cyber-physical system application. The system minimises the difference between the KPIs and desired values at all levels.

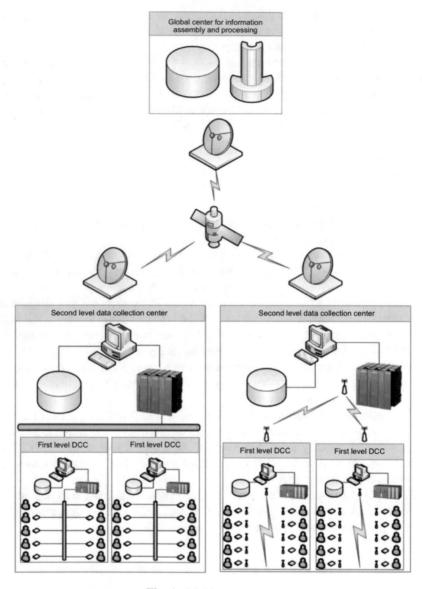

**Fig. 1.** Multi-agent system

# 2   Multi-agent Systems Concept

System architecture based on the intelligent agent concept outperforms ad-hoc implementation of distributed systems and architectures. The main advantages of a multi-agent system include:

- Presence of a decision making layer. Many everyday problems are a form of decision-making. Distributed systems engineering can be simplified by using intelligence rather than ad-hoc distribution and coordination techniques.
- Theory-based fault tolerance and scalability. Advanced distributed coordination and decision-making techniques provide for more effective operation and availability.
- Presence of internal general computational framework that can be used to solve business tasks. For example, same optimization routines being a part of underlying multi-agent framework can be used for solving data mining problems.

The multi-agent paradigm allows different realizations, but we chose a hybrid P2P-backed approach, including both standard server agents and client agents (see Fig. 2). Authors of reference [7] suggest web-based technology for its ease of support and deployment. In order to expand a web-based multi-agent system, the end user needs to open an additional web browser window (a page can also be opened programmatically). Modern standards and APIs [8, 9] allow for implementation of functionalities once considered server-side only. A web-based engine also allows to utilise an existing user's processing power to optimise server load. In fact, all modern systems are web-based.

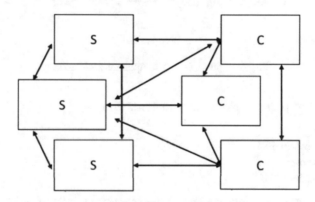

**Fig. 2.** Hybrid multi-agent system topology

Application of different types of agents has its drawbacks. For example, some standard mathematical frameworks are developed to work with only one type of agents. Considering online data exchange, a task of channel source selection arises, which is a problem of both optimal fault-tolerance and load-balancing schemes. The client agent should establish, maintain and support the best subset of possible connections with other client and server agents.

Connection problems can be resolved by lowering the networking activity. The following paragraph describes an example of such solution.

For systems built as a union of separate agents, one may design a layer of integrated self-supervision and control backed by internal intelligence [10]. Partially observable Markov decision processes (POMDPs) model was chosen as a basis of integrated decision-making subsystem. POMDPs are a proven technique of making value-oriented

decisions in presence of observation and action uncertainty. POMDP is defined as a
tuple M = (S, A, O, T, Ω, R), where:

- S is a set of states;
- A is a set of actions;
- O is a set of observations;
- T is a set of conditional transition probabilities;
- Ω is a set of conditional observation probabilities;
- R: A → R is the reward function.

An example of system architecture with a POMDP solver is shown in Fig. 3.

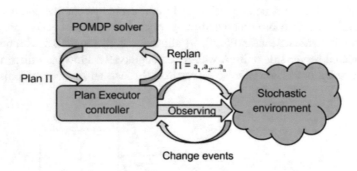

**Fig. 3.** Fault tolerant component architecture

A basic set of possible actions may look like this:

- "Do nothing"
- "Reload component"
- "Turn component on"
- "Turn component off"

Decision-making system/layer reads status messages locally in each component and
performs local supervisory actions. Status messages are inherent observations in the
POMDP model. Internal system state is identified in the process of arriving observations. Considered component states contain the following:

- Down – the component is down and can be reinitialised by means of local
  supervision.
- Crushed – the component cannot be reinitialised without human interaction.
- Incapable – the component is still alive but cannot perform according to
  specifications.
- Working – the component functions according to specifications.

The following paragraphs describe how POMDP can handle component state evolution to enforce system fault tolerance. Application of POMDPs for the recovery process of two components (a) and (b) are shown in Fig. 4.

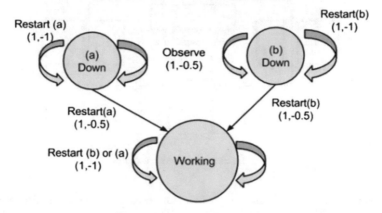

**Fig. 4.** Recovery process scheme

A component can observe its state and take appropriate value-oriented decisions. The POMDP policy is a mapping between state and action for each initial state of the system. The optimal policy, denoted by $p^*$, yields the highest expected reward value for each belief state, compactly represented by the optimal value function $V^*$. This value function is solution to the Bellman optimality Eq. (1):

$$V^*(b) = \max\left[r(b,a) + \gamma \sum \Omega(o|b,a)\, V^*(\tau(b,a,o))\right],\qquad(1)$$

where $\gamma$ is a discount factor, $\tau$ is a belief state transition function, $b$ and $a$ stand for belief state probability. The value function can be found using a variety of algorithms. Partial observability makes the problem hard to solve in a native way, but various other approaches were developed to tackle the complexity of the task: online approaches, approximate offline approaches, hybrid approaches [11–13].

## 3 Applications

The aforementioned system was applied to solve a problem of power grid member communication. This multi-agent system was designed to control a power redistribution grid, containing one power plant and multiple zero-energy buildings (ZEBs), connected with a distributed knowledge base [14]. The distributed knowledge base is essentially a set of agents with the power plant being the server, and ZEBs being the clients (see Fig. 5).

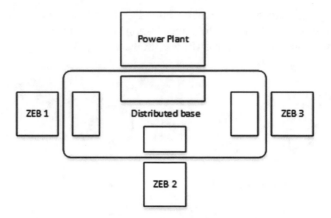

**Fig. 5.** Power network scheme

A power plant and ZEBs have their own low-level KPIs, which regiment smaller-scale parameters. These include temperature, voltage, current, etc., but not the whole power plant or a whole building as part of a smart grid. The low-level KPIs are stored in a register and are initially taken from technical documentation sheets.

Top-level KPIs regiment amount of power generated by a power plant or a ZEB and other similar parameters. In order to count them, one needs to acquire detailed information from the other members of a network (or a network segment).

This information can be provided by a set of special planning units. They compute top-level KPIs and save the data into a register, thus making it available to every member of the network.

Network members also save archive data of their working parameters, so that external computational units could use them to build mathematical models and compute KPIs. Registers are used to send those KPIs back to the network members.

In Fig. 5 the power plant planning unit sends its agents (ZEBs) the following data (see Fig. 6):

- Global set points
- Emergencies
- Global logic

Global set points include alternating current frequency, maximal load for different segments of the network and its support equipment, etc. These are the basic parameters that should not be altered under normal operational circumstances.

Emergencies block covers any type of force majeure circumstances that can change the network layout, e.g., damages of the power lines, failures at power substations, etc. Part of this data can be acquired via member-to-member communication, but having a central hub boosts the system stability.

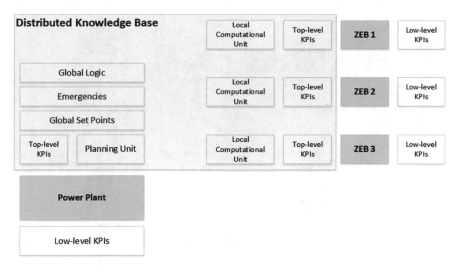

**Fig. 6.** Overall scheme of the distributed knowledge base

Global logic block comprises the most generic information about the grid, such as possible alternate routes, usable lines and the basic communicational parameters. The global logic block is needed to establish the rules of collaboration and give the grid members ability to continue working without a continuous connection to the central hub.

The local computational unit of every grid member is tasked to control both levels of KPIs and collaborate with other members. Data known to every member is written into a corresponding register. Coupled with information from the central hub, it forms a pool of information required to make decisions.

## 4  Obtaining Mathematical Models

The next problem of complex processes management is that the KPI-based control requires of us to know mathematical dependencies between the indicators (multicriteria optimisation problem is not taken into account as being out of scope for this work).

The problem of mathematical models building is closely related to another concept of a digitalised control circuit – the digital twin. The twin preimages the real model, including all basic properties of all produced objects. The digital model and the physical object interact in real time. This information exchange is characterised by the Big Data and can be generated by a multitude of sensors. The digital model is constantly updated. That is, it changes its parameters to better match the current working mode of the physical object [15]. Thus, there is a real opportunity to identify emerging anomalies at the early stages, to predict behaviour of the object and to ensure implementation of algorithms of dynamic optimisation, which ultimately allows to significantly improve the reliability and efficiency of the equipment [16, 17] (Fig. 7).

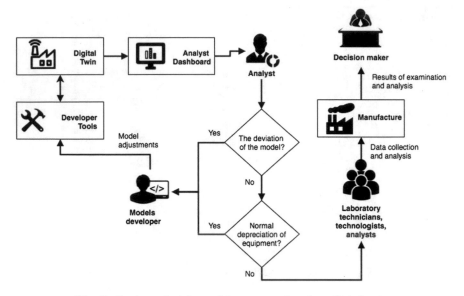

**Fig. 7.** Business decision-making process based on digital twin

In order to get data from the control plant, we can use the distributed control system (DCS) or improved control system (ICS). They are directly connected to the control plant, so they can send control signals to it. We also presume that a protective safety system (PSS) is also placed there, as well as the data base (DB) and a data base management system (DBMS) (see Fig. 8).

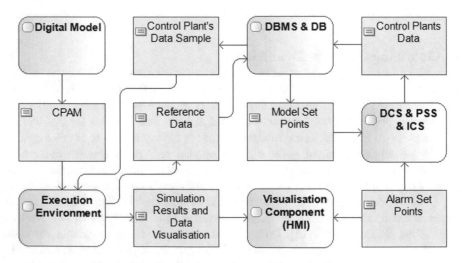

**Fig. 8.** Functional scheme of system dynamic diagnostics

In order to calculate key performance indicators (KPIs) in the situation of the task, with source material or equipment changing, compositional program-analytic models (CPAMs) are used. These are models of a different level of KPIs connections. They are, essentially, expressions, explaining how one KPI depends on another or on the control parameters. There are other types of models, such as functional or system ones. The latter is going to be used in the following paragraphs as an illustration.

The following paragraphs describe an approach that defines which parameters and in what order must be linked. The proposed algorithm assumes that every process is characterised by input and output products, controlling parameters, and restrictions (see Fig. 9).

Input product is the raw material that needs to be transformed during the process. Eventually, the product reaches the desirable condition (according to the process specifications) and becomes the output product. Different refinement states of oil may serve as examples of both input and output products.

Restrictions include natural and artificial conditions required to be met at any moment during the process. Requirements related to the output product are excluded from this list.

Controlling parameters is the only component allowed to vary (within the limits). However, existing restrictions shall not be violated and the output product must meet the requirements set before the beginning of the process.

One must also keep in mind that for every process, one or more target functions can be formulated. This requires of us to take every function into account and to formulate them all before the process starts. Restrictions and controlling parameters must be handled in the same way.

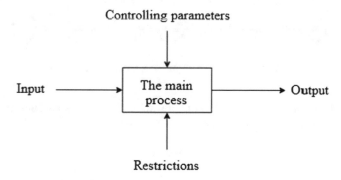

Fig. 9. Identifiable process environment

Taking all the aforementioned guidelines into account, the model creation process can be divided into 7 basic steps:

**Step 1.** Select target functions $P_1$, $P_2$, ...$P_k$ and define uncontrolled parameters $R_1$, $R_2$, ...$R_n$. For example, we will take two target functions – profit $P_1$ and oil export volume $P_2$, and one uncontrollable parameter – resource $R$.

**Step 2.** Mark every top-level process and condition. This step requires of us to list the large processes as "black boxes", inputting and outputting different product states (see Fig. 10). Product states between the processes are marked as $S$.

**Step 3.** Each process must be assigned with a set of controlling parameters. Each target function must be divided into simplified combinations of sub-functions, one combination for every process. Each function must be expressed through its controlling parameter.

For example, resources included into the sub-processes can be viewed as follows:

$$R = R_1 + R_2 + \ldots + R_n = w_1R + w_2R + \ldots + w_nR, \tag{2}$$

where $w_n$ are represented as shares of the common resource $R$.

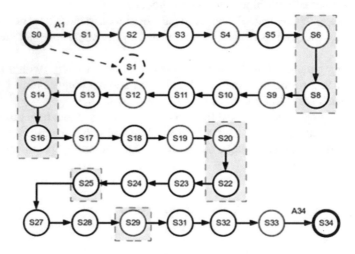

**Fig. 10.** Scheme of a simple interaction sequence

Sub-functions of the target functions are defined via sub-functions from the processes, as seen in following equations:

$$P_1(w) = P_1a(w_1) + P_1b(w_2) + \ldots + P_1m(w_n); \tag{3}$$

$$P_2(w) = P_2a(w_1) + P_2b(w_2) + \ldots + P_2m(w_n). \tag{4}$$

Vector $w$ is composed from the shares of the common resource invested into the process. Letters $a$, $b$ and $m$ represent different stages of the process. For instance, it can be taken as oil prospecting ($a$), oil extraction ($b$), etc.

We must also take into account that every function may depend on more than one parameter. There are two ways to approach it:

1. Function $P_1$ oil_prospecting($w_4, w_5$) is dependent on two parameters and can be decomposed into elementary sub-functions $P_1$ oil_prospecting_1($w_4$) and $P_1$ oil_prospecting_2($w_5$). In case of multiple parameters one may need to execute multiple sequential decompositions.
2. Function $P_1$ oil_prospecting($w_4, w_5$) can be identified. That means defining an analytical dependency between two vector components. This approach is described in Step 5.

**Step 4.** Every condition must be described through its desired attributes, posing as indicators of a specific state. Countable KPIs can function as such indicators.

**Step 5.** Every attribute chosen in step 4 needs to be represented as a function of the controlling parameter $w$.

It means that KPI "Output gasoline volume" must be viewed as a function dependent on $w_1$, where $w_1$ is a controlling parameter. In other words, $w_1$ is a share of $R$, invested into realisation of the whole oil refinement process. It can be formulated as (5):

$$KPI\_output\_gasoline\_volume = f(w_1). \tag{5}$$

To obtain this dependence one needs to use approximation of unknown functions based on known experimental data. This approach requires an experimental data archive. This experimental relationship will be represented by a discrete set of values (see Fig. 11).

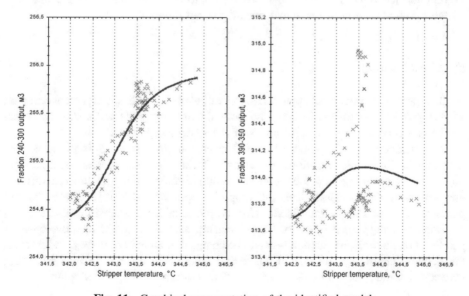

**Fig. 11.** Graphical representation of the identified models

The following identification process can be illustrated with figures from oil recti-fication optimisation project, done with relatively simple backpropagation network. It was trained on statistical data obtained during 24 h of the rectificational column work.

To verify the correlation between the models (lines in Fig. 11) and statistical data (crosses in Fig. 11) correlation coefficients ρx,y have been calculated. The left graph ρx,y = 0,76312 and the right graph ρx,y = 0,90781.

The equation for calculating coefficient is presented below, $\sigma x$ and $\sigma y$ are the mean values of corresponding selections.

$$\rho_{x,y} = \frac{\frac{1}{n}\sum_{t=1}^{n} X_t Y_t - \left(\frac{1}{n}\sum_{t=1}^{n} X_t\right)\left(\frac{1}{n}\sum_{t=1}^{n} Y_t\right)}{\sigma_x \sigma_y}. \tag{6}$$

**Step 6.** Each process present is now described through the target function and control parameters, and can be decomposed into sub-processes. Each process requires steps 2–6 to be repeated. At this level of model complexity, one can take into account such parameters as oil refinement plant workers' salaries.

Thus, the total profit can be formulated as $P_1(w)$, the number of summands in which depends on the degree of decomposition processes. The same can be said about $P_2(w)$.

Algorithms specified in step 5 are able to produce identification based on vector w, without any need for further detalisation or simplification.

**Step 7.** All the sub-functions must be combined together to form an overall main function (or functions), dependent on all controlling parameters. For neural net-work, based algorithm functional linearity is not required.

## 5  Conclusions

Authors present an integrated approach to the complex processes management based on a cyber-physical system application. The method that minimises the difference between the KPIs and desired values at all levels of the system is provided.

An approach for cyber-physical control over complex processes is proposed; the next step is the implementation of a multi-criterial top-level process optimisation. Simultaneous optimisation is often impossible due to contradicting relationships among the processes. At the same time, the proposed approach allows to choose control parameters taking into account hierarchical structure of the managed processes.

Data exchange algorithms for manufacturing scenarios are being developed in parallel. For example, production task can be changed in an unknown way. In this case, the network centric system should generate a solution to reform the load balancing between different pieces of equipment in order to achieve the specified goals.

**Acknowledgements.** The article is published with support of the project Erasmus+ 573545-EPP-1-2016-DE-EPPKA2-CBHE-JP "Applied curricula in space exploration and intelligent robotic systems (APPLE)" and describes the part of the project conducted at Peter the Great St. Petersburg Polytechnic University, St. Petersburg, Russia.

# References

1. Bassi, L.: Industry 4.0: hope, hype or revolution? In: IEEE 3rd International Forum on Research and Technologies for Society and Industry (RTSI), pp. 1–6 (2017)
2. Yefremov, A.Y., Maksimov, D.Y.: Setetsentricheskaya sistema upravleniya – chto vkladyvayetsya v eto ponyatiye? [Network-centric control system - that is embedded in this concept]. In: Trudy 3-y Vserossiyskoy konferentsii s mezhdunarodnym uchastiyem « Tekhnicheskiye i programmnyye sredstva sistem upravleniya, kontrolya i izmereniya » [Proceedings of the 3rd Russian Conference on Technical and Software Control Systems, Control and Measurement] (UKI-2012, Moscow), pp. 158–161. IPU RAN (2012). (in Russian)
3. Skobelev, P.O., Tsarev, A.V.: Setetsentricheskiy podkhod k sozdaniyu bolshikh multiagentnykh sistem dlya adaptivnogo upravleniya resursami v realnom vremeni [The network-centric approach to creation of big multi-agent system for adaptive real-time resource management]. In: Materialy Mezhdunarodnoy nauchno-prakticheskoy multikonferentsii "Upravleniye bolshimi sistemami" [Proceedings of the International Scientific and Practical Multi-conference "Control of Big Systems"], p. 267. IPU RAN Publ. (2011). (in Russian)
4. Skobelev, P.O.: Situatsionnoye upravleniye i multiagentnyye tekhnologii: kollektivnyy poisk soglasovannykh resheniy v dialoge [Situation-driven decision making and multi-agent technology: finding solutions in dialogue] Ontologiya proyektirovaniya [Ontology of design]. Samara: New technology, **8**(2), 26–47 (2013). (in Russian)
5. Gabdrashitova, E.I., Gamilova, D.A.: Otsenka proizvodstvennogo potentsiala nefteservisnykh predpriyatiy [Assessment of the productive capacity of oilfield service companies]. In: Internet-zhurnal « NAUKOVEDENIE » [Online journal "Science studies"], (3) (2014). https://naukovedenie.ru/PDF/30EVN314.pdf. (in Russian)
6. Burenina, I.V., Varakina, V.A.: Sistema yedinykh pokazateley otsenki effektivnosti deyatelnosti vertikalno-integrirovannykh neftyanykh kompaniy [The single indicator system of performance assessment of vertically integrated oil enterprises] Internet-zhurnal « NAUKOVEDENIE » [Online journal "Science studies"], **1** (2014). https://naukovedenie.ru/PDF/12EVN114.pdf. (in Russian)
7. Zobnin, S. S., Potekhin V.V.: P2P architectures in distributed automation systems. In: Proceedings of Symposium on Automated Systems and Technologies, pp. 37–43. PZH Verlag, Hannover (2014)
8. Web Workers (2012). http://www.w3.org/TR/workers/. Accessed 10 Jul 2014
9. WebRTC 1.0. Real-time Communication Between Browsers (2014). http://dev.w3.org/2011/webrtc/editor/webrtc.html. Accessed 10 Jul 2014
10. Fedorov, A.V., Zobnin, S.S., Potekhin, V.V.: Prescriptive analytics in distributed automation systems. In: Proceedings of Symposium on Automated Systems and Technologies, pp. 43–49. PZH Verlag, Hannover (2014)
11. Ross, S., Pineau, J., Paquet, S., Chaib-Draa, B.: Online planning algorithms for POMDPs. J. Artif. Intell. Res. (JAIR) **32**, 663–704 (2008)
12. Pineau, J., Gordon, G., Thrun, S.: Point-based value iteration: an anytime algorithm for POMDPs. IJCAI **3**, 1025–1032 (2003)

13. Spaan, M.T., Vlassis, N.A.: Perseus: randomized point-based value iteration for POMDPs. J. Artif. Intell. Res. (JAIR) **24**, 195–220 (2005)
14. Combin, N.N.: Tehnologii raspredelennogo reestra [Distributed register technologies] In: Nauchnoe soobschestvo studentov XXI stoletija. Tekhnicheskie nauki. Sbornik statei po LI studenchecskoi nauchno-practicheckoi konferencii. [Scientific community of XXI Century. Engineering science. In: Proceedings of the LI student scientific and practical conference], vol. 3, no. 50 (2017)
15. Ponomarev, K., Kudriashov, N., Popelnukha, N., Potekhin, V.: Main principals and issues of digital twin development for complex technological processes. In: Proceedings of the 28th DAAAM International Symposium. DAAAM International, Vienna (2017)
16. Katalinic, B., Kukushkin, I., Pryanichnikov, V., Haskovic, D.: Cloud communication concept for bionic assembly system. Procedia Eng. **69**, 1562–1568 (2014)
17. Gastermann, B., Stopper, M., Kossik, A., Katalinic, B.: Secure implementation of an on-premises cloud storage service for small and medium-sized enterprises. In: Proceedings of DAAAM International Symposium on Intelligent Manufacturing and Automation, DAAAM 2014, Vienna, Austria, vol. 100, no. C, pp. 574–583 (2015)

# Creation of Physical Models for Cyber-Physical Systems

Nataliya D. Pankratova[✉]

Igor Sikorsky Kyiv Polytechnic Institute, Kyiv, Ukraine
natalidmp@gmail.com

**Abstract.** Creation of physical models for cyber-physical systems (CPSs) with consideration of the concept, features and properties of CFS is proposed. The model is based on the general problem of multi-factor risks, the margin of permissible risk, the forecast of the destabilizing dynamics of risk factors, principles, hypotheses, and axioms that are directly related to the analysis of abnormal situations, accidents and disasters. This model is the foundation of the complex technical systems (CTS) functioning. The communication with computational systems and different types of sensors is implemented online in real time. Joint actions of CTS components determine the properties and special features of the mode of functioning of a complex system at any moment of time. A case of the proposed model implementation is given at the example of a real complex technical system.

**Keywords:** Model · Multifactor risks · Margin of permissible risk · Reliability · Forecast · Safety · Computational systems · Abnormal situations

## 1 Introduction

Today's cyber-physical systems have not received an unambiguous and generally accepted definition, since these systems are simultaneously located at the intersection of several fields of activities. Their main common feature is the interaction between physical and computational processes, complexity, uncertainty, and connection with the Internet of things. Thus, we can assume that a cyber-physical system is an elaborate system of computational and physical elements that constantly receives data from the environment.

It is taken into account that a CPS is an elaborate system consisting of various natural objects, artificial subsystems and controllers which allow representing of such alliance as a single whole. A CPS ensures close communication and coordination between computational and physical resources, which demand the creation of two types of models. On the one hand, these are engineering models, and on the other, computer models. This paper focuses on the engineering model in which computational elements interact with sensors, providing for monitoring of the performance and maintenance of the technical system. This model is the foundation of the CTS operation. An attempt was made to improve the quality of the survivability and safety of CTS operation with the account of the concept, features and properties of cyber-physical systems. The CTS information platform includes a model in the form of a set of principles, hypotheses,

© Springer Nature Switzerland AG 2020
D. G. Arseniev et al. (Eds.): CPS&C 2019, LNNS 95, pp. 55–63, 2020.
https://doi.org/10.1007/978-3-030-34983-7_6

axioms, methods and techniques; a system of sensors at critical points of a physical system that is providing for the data in the course of operation, and a computational system that brings all data into a unified format; data analysis software that allows to perform the further control of physical elements.

## 2  Review of the Literature

In cyber-physical systems, computing elements interact with sensors that monitor cyber-physical indicators and with actuators that introduce changes to the cyber-physical environment. The cyber-physical systems carry out computational procedures inside their distributed structure, they include "smart nodes" and make it possible to reconfigure flows in the network depending on the conditions. Thus, cyber-physical systems are distributed systems with the possibility of intelligent processing and reconfiguration of flows at the account of intelligent control [1].

An overview of some history that for more than 40 years (since 197, has been connected with the development of cyber-physics systems, with computers that interact directly with the physical world, was considered in [2]. The recent explosion of the interest in, hype up, and fear of artificial intelligence (AI), data science, machine learning, and robotics have focused a spotlight on software engineers. Business magnate Elon Musk called for regulations and the President of Russia Vladimir Putin declared that the domination in the world will come as a result of AI mastering. Are software engineers responsible for these outcomes? The author of [3] claims that software engineers have less control over their designs than they, most likely, realize. Instead, software technologies are evolving in a Darwinian way or, more precisely, they are co-evolving with the human culture.

One of the biggest challenges in the cyber-physical system (CPS) design is its intrinsic complexity, heterogeneity, and multidisciplinary nature. Emerging distributed CPSs integrate a wide range of heterogeneous aspects, such as physical dynamics, control, machine learning, and error handling. Furthermore, system components are often distributed over multiple physical locations, hardware platforms and communication networks. While model-based design (MBD) has tremendously improved the design process, CPS design remains a difficult task. Models are meant to improve understanding of a system, yet this quality is often lost when models become too complicated. In the paper [4] it was shown how to use aspect-oriented (AO) modeling techniques in MBD, as a systematic way to segregate domains of expertise and cross-cutting concerns within the model.

The role of modeling in the engineering of cyber-physical systems is considered in Reference [5]. It is argued that the role that models play in engineering is different from the role they play in science, and that this difference should invite us to use a different class of models, where simplicity and clarity of the semantics dominate over accuracy and detail. It is argued that determinism in models that are used for engineering is a valuable property and should be preserved whenever possible, regardless of whether the system under modeling is deterministic. There are three classes of fundamental limits on modeling, that is chaotic behavior, the inability of computers to numerically handle a continuum, and the incompleteness of determinism.

The paper [6] is about a better design of cyber-physical systems (CPSs) using better models. Deterministic models have historically proven to be extremely useful and arguably form the basis of the industrial revolution, as well as the digital and IT revolutions. Key deterministic models that have proven to be successful include differential equations, synchronous digital logic and single-threaded imperative programs. Cyber-physical systems, however, combine these models in such a way that determinism is not preserved. Two projects show that deterministic CPS models with exact physical realizations are possible and practical. A new system science that is jointly physical and computational is proposed [7]. In the author's understanding, the embedded computers and networks monitor and control physical processes, usually with feedback loops where physical processes affect computations and vice versa. The integrated simulation tool using a simulator of computer architecture is presented in Reference [8]. In this paper, the simulating computer architecture has many potential use cases as a cyber-physical system, including simulation of side channels and software-in-the-loop modeling and simulation.

The future development of the society is associated with the creation of the Internet of Things, which will allow creating dynamic networks consisting of billions and trillions of such things communicating among themselves. This will ensure a fusion of the digital and physical worlds, for which applications, services, middleware components and end devices are things [9].

The existing diagnosing technologies are oriented at the exposure of failures at early stages, before the appearance of serious malfunctioning in a certain place and class [10–12]. The approach for diagnosing the technical state of a system before a failure, taking into account uncertainties related to the time of the fault, its location and class is considered in [13]. The issues of designing and creating complex anthropogenic systems which satisfy the required level of guaranteed quality (reliability, durability and safety) under conditions of incompleteness of the original information for forecasting technical systems' conditions are investigated in Reference [14].

## 3   Model of Survivability and Safety of CTS Functioning

The proposed model is based in the replacement of a typical principle of the operability detection turning into the inoperability state based on the detection of failures, malfunctioning, and faults of an object by a qualitatively new principle. The essence of the proposed principle is timely identification and elimination of the causes of undesirable events occurrences and prevention of the transition from normal to an abnormal mode. The strategy of this principle is based on the system analysis of multifactor risks of abnormal situations, a credible estimation of the margin of the permissible risk for different modes of operation of a CTS, and a forecast of the main indicators of operability of an object during the assigned operating period [15].

We shall formulate the main problem of the system analysis of multifactor risks in generalized form [16]. The $M_0$ set of risk factors $\rho_q$ is known from the data of testing a complex system of arbitrary nature and other a priori information

$$M_0 = \left\{ \rho_q \mid q = \overline{1, n_0} \right\}.$$

Each risk factor $\rho_q \in M_0$ is characterized by a set $L_q$ of attributes $l_{qj}$:

$$L_q = \left\{ l_{qj} \mid q \in N_0; \quad j = \overline{1, n_q} \right\}, \quad N_0 = [1, \, n_0].$$

Each attribute $l_{qj} \in L_q$ is defined by the information vector

$$I_{qj} = \left\{ x_{qj} \mid x_{qj} = \langle x_{qjp} \mid p = \overline{1, n_{qj}} \rangle; \quad x_{qjp} \in H_{qjp}; \quad q \in N_0; \quad j \in N_q \right\},$$

$$H_{qjp} = \left\{ x_{qjp} \mid x_{qjp}^- \leq x_{qjp} \leq x_{qjp}^+ \right\}; \quad N_q = [1, \, n_q].$$

Based on $I_{qj}$ sets, the information vector is formed for each risk factor $\rho_q$.

$$I_q = \left\{ I_{qj} \mid q \in N_0; \quad j = \overline{1, n_q} \right\},$$

$$I_q = \left\{ x_{qj} \mid x_{qj} = \langle x_{qjp} \mid p = \overline{1, n_{qj}} \rangle; \quad x_{qjp} \in H_{qjp}; \quad q \in N_0; \quad j = \overline{1, n_q} \right\}.$$

The set $M_0$ corresponds to a definite, a priori predicted set $S_0$ of risk situations. In the functioning of a CTS, new risk factors affect it and are revealed, and the properties and indicators of a priori known risk factors $\rho_q \in M_0$ are changed. This results in quantitative and qualitative changes in the set of risk factors that determine the necessity to form a sequence of embedded sets of the form

$$\begin{aligned} M_0 \subset M_1 \subset \ldots \subset M_\tau \subset \ldots, \\ S_0 \subset S_1 \subset \ldots \subset S_\tau \subset \ldots, \end{aligned} \tag{1}$$

where $M_\tau$, $S_\tau$ are sets of risk factors and risk situations respectively at the moment $T_\tau \in T^\pm$, and $T^\pm$ is an assigned or predicted period of functioning of a CTS. Sets $M_\tau$, $S_\tau$ are defined as

$$M_\tau = \left\{ \rho_q^\tau \mid q \in \overline{1, n_\tau} \right\}, \quad S_\tau = \left\{ S_k^\tau \mid k \in \overline{1, K_\tau} \right\}.$$

Each situation $S_k^\tau \in S_\tau$ is characterized by set $M_k^\tau \in M_\tau$ of risk factors $\rho_{qk}^\tau$.

$$M_k^\tau = \left\{ \rho_{qk}^\tau \mid q_k \in \overline{1, n_k^\tau} \right\}.$$

Each factor $\rho_{qk}^\tau \in M_k^\tau$ is characterized by set $L_{qk}^\tau$ of attributes $l_{q_k j_k}^\tau$:

$$L_{qk}^\tau = \left\{ l_{q_k j_k}^\tau \mid q_k \in N_k^\tau; \quad j = \overline{1, n_{qk}^\tau} \right\}, \quad N_k^\tau = [1, \, n_k^\tau].$$

Each attribute $l^\tau_{q_k j_k} \in l^\tau_{q_k j_k}$ is revealed based on the information obtained and processed by a diagnostic system. Information at the moment of measurement $T_\tau$ is characterized by its incompleteness, uncertainty and inaccuracy.

In the process of controlling CTS functioning on a true scale of the set moments of time $T_\tau$ or with a certain time interval $\tilde{T}_\tau \in T^\pm_\tau$, $T^\pm_\tau = \{\tilde{T}_\tau \mid T_\tau < \tilde{T}_\tau < T_{\tau+1}\}$, it is required to carry out a multifactor estimation of risk of any situation $S^\tau_k \in S_\tau$ and, based on the obtained results, to form and implement a decision on preventing and/or minimizing undesired consequences before the critical moment $T_{cr}$ comes.

In the general case, risk factors $\rho_q$ include the following parameters: risk degrees $\eta_i$ as the probability of occurrence of undesirable consequences of the impact of any risk factors at any moment of time $T_i \in T^\pm$ in the process of CTS functioning; risk level $W_i$ as the size of damage caused by the influence of any risk factors at any point in time $T_i \in T^\pm$ and the margin of permissible risk $T_0$ as the duration of complex system functioning period in a certain mode, when the risk degree and risk level will not exceed the a priori assigned permissible values under the possible influence of risk factors.

We point out a number of fundamentally important peculiarities of the formulated problem [17]:

- sets of risk factors and sets of situations are largely unlimited;
- a threshold restriction of time for decision forming is a top priority;
- the problem is not completely formalized;
- indicators of a multifactor risk estimation are not determinate;
- criteria of a multipurpose risk minimization are not determinate;
- the set of risk situations in principle cannot be a complete group of random events.

Indeed, the problem is presented in a generalized statement that gives the decision-maker certain freedom in adapting it to practical needs in a specific subject domain by the concrete definition of the aforementioned indicators and criteria. Based on the decomposition principle, the general problem of an analysis of the multifactor risk is represented as a sequence of the following system of coordinated, informationally interconnected problems [15]:

- System multifactor classification of revealed and predicted risk situations;
- System multifactor recognition of revealed and predicted risk situations;
- System multicriterion ranking of situations;
- Multipurpose risk minimization of a predicted set of abnormal situations;
- Rational multipurpose optimization of the informedness level in recognition of abnormal situations in the process of complex system functioning;
- Rational coordination of the margin of permissible risk of a predicted set of abnormal situations;
- Determination of a level of rational informedness under the threshold time limitation in the process of complex system functioning;
- System estimation of margin of permissible risk under the dynamics of abnormal mode.

# 4 Survivability and Safety of Ambulance Operations

*Substantial Statement of the Problem.* The work of an ambulance which moves in the operational mode, i.e., with a patient on board, is considered. Patient's life is supported by the medical equipment, which is powered from the ambulance's onboard electrical system. The charging current is limited at the level that corresponds to the power extracted from the generator, that is equal to 200 W. The ambulance must travel a distance of 70 km with a particular velocity profile determined by the situation on the road.

It is required to ensure the supply of the electric power for the medical equipment, which is located in the main cabin. Since the motion occurs at night, additional internal and external illumination needs to be provided.

Depending on the speed, the transmission ratio changes, therefore, the frequency of the crankshaft rotation of the main internal combustion engine (ICE1) changes too. In the beginning of the trip, there are 47 L of fuel in the tank. Both engines (ICE1 and ICE 2) are supplied from the same tank. In a normal situation, the car would safely drive the patient for 11,700 s (3 h and 15 min). In this case, the battery voltage does not drop below 11.85 V. At the end of the trip, there are 4.1 L of fuel left in the tank.

The transition into an abnormal mode is caused by the malfunction of the charger, i.e., the voltage sensor RB. It is assumed that the sensor gives out false information that the battery is fully charged. Since no recharging of RB is being made, then with the lapse of time, the battery gets discharged and, consequently, the voltage of the on-board network is also getting decrease during the generator outages (when switching gears, ICE1 idling). Due to the deep discharge, the mode is occurred when the RB output voltage is not enough to maintain the medical equipment operability, and this is an emergency situation.

*Recognition of an Abnormal Situation.* The recognition of an abnormal situation occurs in accordance with prescribed critical values.

- For the voltage in the on-board network: the abnormal voltage amounts to 11.7 V, while the emergency one is 10.5 V.
- For the amount of fuel: the abnormal value is 21, and the emergency value is 11.
- For the voltage in the rechargeable battery: in the abnormal situation, it is 11.5 V. This way, in the case of the decrease of the function value below one of the set values, the operation of the ambulance goes into an abnormal mode of functioning.

*Critical Variables*

- Board voltage (depending on the parameters of the RB, generator's condition and load current). If the board voltage drops below the trip level of medical equipment, this could lead right into an emergency.

- Fuel level. Depends on the power, which is taken from the main engine (in proportion to the rotation speed). Decline below a certain point can lead to an abnormal (when you can call another car equipment from an RB) or emergency mode (when the car had to make a stop for a long time without charging).
- Voltage RB (depending on the generator's condition, the total electricity consumption).

The diagnostics unit, which is the basis for ensuring the survivability and safety of complex technical objects functioning, is developed as an information platform for engineering diagnostics [15, 18]; it contains the following modules:

- acquisition and processing of the initial information during the CTO operation;
- recovery of functional dependences (FDs) from the empirical discrete samples;
- quantization of the discrete numerical values;
- identification of sensor failures;
- timely diagnosis of abnormal situations;
- forecast of non-stationary processes;
- generation of the process of engineering diagnostics.

Some results of an ambulance functioning during the first 7000 s are shown in Fig. 1 as diagrams of voltage distribution in the onboard network; the amount of fuel in the tank; and the rechargeable battery voltage. The transition into an abnormal mode happens due to the failure of the battery voltage sensor. The voltage sensor outputs false information to the RB. So far, as long as the battery recharging is not implemented, the battery is discharged with the lapse of time and, consequently, the voltage in the on-board network within 6500–7400 s. is also decreasing and transits into abnormal mode. When the voltage of the on-board network is lower than 11.7 V, the situation becomes abnormal. After lowering the level below 10.5 V, the equipment of the ambulance is turned off and the situation goes into an accident case. The fuel level, which depends on the capacity of the internal combustion engine, is also reduced. The driver stops the car, incorporates a backup generator and troubleshoots the charger. The situation transfers into a normal mode. The period of emergency situation amounts to 120 s: from the moment when the equipment is switched off to the start of the backup generator. After troubleshooting, the driver restarts the motion, without disconnecting the backup oscillator.

At any time of the program operation, the user has the ability to look at the operator's scoreboard, which displays a series of indicators that reflects the state of the CEO of the ambulance functioning. These indicators include: readings of the sensors of the accumulator battery voltage, amount of fuel in the tank, voltage of the on-board network, the state of the system, the risk of damage, causes of the abnormal or emergency mode, as well as the readings of indicator of the danger level for the system operation and possible failures of sensors.

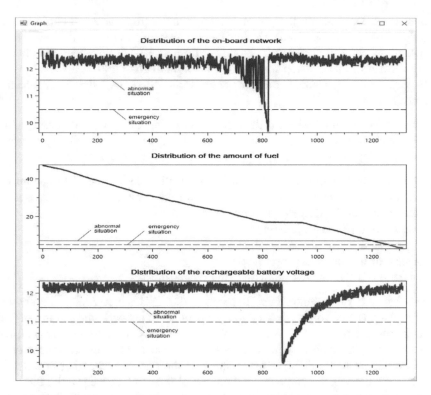

**Fig. 1.** Voltage distribution of the on-board network, the amount of fuel in the tank, the rechargeable battery voltage in accordance of time $t$ sec

## 5  Conclusion

The proposed strategy of systemic coordination of survivability and safety for technical systems' operation is one of the physical models of a cyber-physical system. The proposed strategy for the operation of the CTS ensures survivability and safety of the system thanks to the timely detection of abnormal situations, assessment of their degree and level of risk, and determination of the margin of acceptable risk in the process of forming decisions on operational actions. Combining a number of similar models into a single network will allow to carry out a rational distribution of the required resources among different consumers online. To solve this problem, it is necessary to develop computational processes, take into account the heterogenity of the data obtained from various applications and devices, develop models and methods for collecting, storing and processing large data, analyze the results obtained from the timely made decisions.

# References

1. Tsvetkov, V.Y.: Cyber physical systems. Int. J. Appl. Fundam. Res. **6**(1), 64–65 (2017)
2. Lee, E.A.: Modeling in Engineering and Science. Commun. ACM **62**(1), 35–36 (2019). Viewpoint
3. Lee, E.A.: Is it a smart design or evolution? Commun. ACM **61**(9), 34–36 (2018). Viewpoint
4. Akkaya, I., Derler, P., Emoto, S., Lee, E.A.: Systems engineering for industrial cyber-physical systems using aspects. IEEE Proc. **104**(5), 997–1012 (2016)
5. Lee, E.A.: Fundamental limits of cyber-physical systems modeling. ACM Trans. Cyber Phys. Syst. **1**(1), 3 (2016)
6. Lee, E.A.: The Past, present, and future of cyber-physical systems: a focus on models. Sensors **15**(3), 4837–4869 (2015)
7. Lee, E.A.: Cyber-physical systems – are computing foundations adequate? In: Position Paper for NSF Workshop on Cyber-Physical Systems: Research Motivation, Techniques and Roadmap, Austin, TX (2006)
8. Kim, H., Wasicek, A., Lee, E.A.: An integrated simulation tool for computer architecture and cyber-physical systems. In: Design, Modeling, and Evaluation of Cyber Physical Systems. Lecture Notes in Computer Science. LNCS, vol. 11267 (2019)
9. Chernyak, L.: The Internet of Things: new challenges and new technologies. Open Syst. DBMS **4**, 14–18 (2013)
10. Huang, S.-R., Huang, K.-H., Chao, K.-H., Chiang, W.-T.: Fault analysis and diagnosis system for induction motors. Comput. Electr. Eng. **54**, 195–209 (2016)
11. Pennacchi, P., Vania, A.: Diagnostics of a crack in a load coupling of a gas turbine using the machine model and the analysis of the shaft vibrations. Int. J. Mech. Syst. Sign. Process. **22**(5), 1157–1178 (2008)
12. Chao, K., Chiang, W., Huang, S., Huang, K.: Fault analysis and diagnosis system for induction motors. Comput. Electr. Eng. **54**, 195–209 (2016)
13. Kulik, A.S., Luchenko, O.A., Firsov, S.N.: Algorithmic providing of the diagnostics modules and restore functionality satellite orientation and stabilization system. Int. J. Radio Electron. Inform. Control **1**, 112–122 (2012)
14. Kotelnikov, V.G., Lepesh, G.V., Martyschenko, L.A.: System analysis of quality and reliability of complex anthropogenic complexes. Int. J. Tech. Technol. Probl. Serv. **4**(2), 35–41 (2013)
15. Pankratova, N.D.: System strategy for guaranteed safety of complex engineering systems. Int. J. Cybern. Syst. Anal. **46**(2), 243–251 (2010)
16. Zgurovsky, M.Z., Pankratova, N.D.: System Analysis: Theory and Applications. Springer, Heidelberg (2007)
17. Pankratova, N.D.: The integrated system of safety and survivability complex technical objects operation in conditions of uncertainty and multifactor risks. In: Proceedings of conference IEEE, Kyiv, Ukraine, no. 50, pp. 1135–1140 (2017)
18. Pankratova, N.D., Radjuk, A.N.: Guaranteed safety operation of complex engineering systems. In: Continuous and Distributed System, Theory and Application, pp. 313–326. Springer (2014)

# Metrological Assurance of Environmental Estimation of Soil Pollution by Heavy Metals by a Geoinformation Cyber-Physical System

Marina Meshalkina[1], Valerii Tsvetkov[1(✉)], Nadezhda Kryzhova[1], and Elena Sokolova[2]

[1] Peter the Great St.Petersburg Polytechnic University, Saint Petersburg, Russia
meshalkina_mari@mail.ru, sva_1946@mail.ru
[2] RiverD International B.V., Rotterdam, The Netherlands

**Abstract.** The article is devoted to estimation of the concentration of heavy metals in soils. A Geo-Information Cyber-Physical system was used to visualize the measurement results. The metrological assurance of the reliability of pollution boundaries selection has been evaluated.

**Keywords:** Heavy metals · Geo-information system · Cyber-physical system · Metrological assurance

## 1 Introduction

The purpose of this work is to measure the concentration of heavy metals, such as lead, zinc, copper, nickel, chromium, arsenic, iron and manganese in the soils of the western part of the Polytechnichesky Park, adjacent to Polytechnicheskaya Street. The Geo-Information System (GIS) was used to visualize the measurement results. The article conducted a metrological assessment of the reliability of the measurement results application on the map.

The park of Peter the Great St. Petersburg Polytechnic University is one of the most important environmental water regulating complexes in the northern part of St. Petersburg. It is classified by the City Administration as an especially valuable cultural heritage site of the Russian Federation.

The university park is a place of rest for students, teachers and residents of the surrounding areas. However, it is located next to a busy transport hub, and close to various industrial enterprises. This can have a negative impact on the environmental situation in the park.

Tests of the soil from the part of the park that adjoins Polytechnicheskaya Street have been performed to assess the pollution levels. The condition of the soils gives the most reliable picture of the ecological state of the area studied. Heavy metals are considered as the main pollutants of the soil. The accumulation of heavy metals in soils have been analyzed in a number of studies [1–10].

This article presents the results of a study of heavy metals concentration in the soils of the Polytechnichesky Park. The methodology includes sampling, preparation, measuring of the concentration of metals in the samples using X-ray fluorescence

© Springer Nature Switzerland AG 2020
D. G. Arseniev et al. (Eds.): CPS&C 2019, LNNS 95, pp. 64–73, 2020.
https://doi.org/10.1007/978-3-030-34983-7_7

method, analyzing the reliability of measurements, and presenting results using a geographic information system.

## 2 Organization of the Experiment

Sampling in the park was carried out in compliance with the State Standards (GOST17.4.4.02-2017) [11]. The study area was divided into 17 base sections in accordance with the grid. Distances between the sampled spots were determined. In our case, because of the terrain relief, the grid was uneven with a grid spacing of 19.5 m. The point spot samples were taken using an auger. Samples were taken with the envelope method at five spots from different depths between the surface and 20 cm down at each base segment of the grid. The sample material from each selected spot was mixed to obtain a combined sample. The mass of the combined sample was no less than 400 g. Total 17 combined samples were obtained from different sites. The breakdown by sections is shown in Fig. 1.

The sample preparation procedure included drying the samples to the air-dry state at 105 °C. Then the dried samples were ground so that the maximum particle size would not exceed 1 mm. The mass of a ground sample for analysis was at least 100 g.

The analysis was performed using X-ray fluorescence spectrometry [12]. This method was also recommended by EPA USA [13] and has been used in many soil studies by various researches [14–19].

Measurements of heavy metals concentration in the soil were carried out using the Spectroscan device. It is intended for element analysis of various objects. The tests were performed in the accredited laboratory of D.I. Mendeleyev Institute of Metrology. The uncertainty of the measurements was 25%. The priority in measurements was given to such elements as lead, zinc, copper, nickel, chromium, arsenic, iron and manganese. The measurement results are shown in Table 1.

**Table 1.** The results of measurements of the concentration of metals (mg/kg) in the soil by X-ray F Fluorescence spectrometry method

| • o | • b | • n | • u | • i | • r | • s | • e | • n |
|---|---|---|---|---|---|---|---|---|
| 1 | 152.0 | 200.0 | 0.5 | 22.0 | 72.0 | 28.0 | 2.7 | 686 |
| 2 | 31.0 | 576.0 | 0.5 | 23.0 | 69.0 | 55.0 | 2.7 | 753 |
| 3 | 94.0 | 173.0 | 6.1 | 36.0 | 73.0 | 18.1 | 3.0 | 859 |
| 4 | 102.0 | 264.0 | 28.0 | 71.0 | 95.0 | 19.4 | 4.1 | 1228 |
| 5 | 179.0 | 244.0 | 0.5 | 32.0 | 72.0 | 33.0 | 3.4 | 1153 |
| 6 | 101.0 | 163.0 | 3.1 | 36.0 | 86.0 | 19.0 | 3.1 | 875 |
| 10 | 135.0 | 175.0 | 0.5 | 23.0 | 64.0 | 26.0 | 2.4 | 621 |
| 13 | 48.0 | 180.0 | 0.5 | 40.0 | 75.0 | 10.4 | 3.4 | 932 |
| 14 | 63.7 | 169.6 | 74.3 | 1.4 | 62.0 | 4.1 | 3.0 | 700 |
| 15 | 24.0 | 91.0 | 20.6 | 22.0 | 65.0 | 6.4 | 2.5 | 613 |
| 16 | 36.0 | 144.0 | 26.0 | 44.0 | 72.0 | 7.8 | 3.0 | 931 |
| 17 | 100.0 | 165.0 | 5.0 | 34.0 | 88.0 | 19.0 | 3.0 | 909 |

## 3 Results and Discussion

A map of the university's campus territory with applied GIS was used to visualize the measurements. The sites, samples from which have been taken for analysis, were mapped. The ranges were set manually based on the values presented in Table 1. The red color was assigned to the site with the highest concentration of pollutants. Dark green color was assigned to the site with the lowest concentration. All other areas were marked by intermediate colors. Figure 1 shows a map of chromium content in mg/kg in the soils of the Polytechnichesky Park.

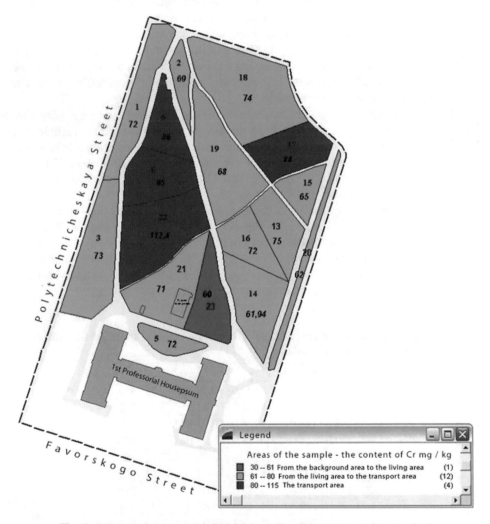

**Fig. 1.** Map of chromium (mg/kg) in the soils of the Polytechnichesky Park

The map of the indicator of total pollution in soils was constructed in accordance with the existing methodology [20]. The map is shown in Fig. 2. The total indicator of pollution $Z_c$ is the sum of the exceedances of concentrations of chemical elements that accumulate in man-made anomalies. It is calculated by the equation:

$$Z_c = \sum_{c=1}^{n} K_c - (n - 1), \tag{1}$$

where $K_c$ is the concentration ratio of a chemical substance. It is determined by the assignment of its actual content in the soils to the background:

$$K_c = \frac{C}{C_f}, \tag{2}$$

$n$ is the number of assumable elements.

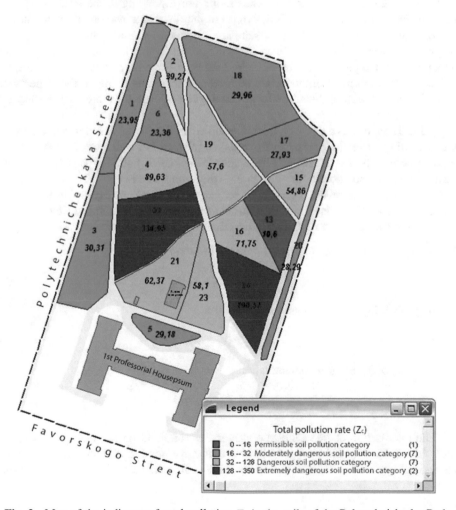

**Fig. 2.** Map of the indicator of total pollution $Z_c$ in the soils of the Polytechnichesky Park

The concentrations found in land plots not exposed to technogenic impact have been used as background concentrations. Usually, these are areas that are about 50 km away from major industrial cities.

The environmental map (EM) should give a reliable display of the regional environmental situation. The characteristic of the degree of contamination at any point of the terrain in the EM must correspond to the actual contamination with known to the user accuracy. Conduction of the metrological support for the creation of aEM determining the boundaries within which the uncertainty of EC information exists with a given probability is needed.

The form of expression of the accuracy and uncertainty of the EM information must be determined. If the EM was compiled in the form of a table in which each point of the terrain was assigned to a value of the degree of contamination (concentration of a harmful element or a generalized pollution index), then it would be logical to indicate each such value with an interval of uncertainty corresponding to the established confidence level. However, the EM is drawn up in the form of a map on which the degree of contamination is ranked by the established boundaries, and zones with the same degree of contamination are outlined by isolines. An isoline is a line connecting on a geographical map places with the same indicators of some magnitude. In this case, the indication for each point of the terrain interval of uncertainty is impossible. Therefore, it is advisable to ensure the accuracy of the display of each local area, each zone and EM as a whole.

A local area means a territory, the pollution characteristic of which is established on the basis of one sample. This is usually a rectangular grid in which one sample is taken at each point. The site is assigned to a certain degree of pollution k = 1, ... n. This means that at all points of this area, the concentration of the i-element is $z_i \in [W_k, W_{k+1}]$, where $W_k$ is the lower concentration limit corresponding to the pollution degree k. The reliability of the local area of the map is estimated by the probability of these and opposite events. Therefore, the indicators will be:

- the probability of correct display area on the map

$$F_{1i} = P\{W_k \leq z_i \leq W_{k+1}\};$$ (3)

- the probability of undetected contamination

$$F_{2i} = P\{z_i \geq W_{k+1}\};$$ (4)

- the probability of false contamination

$$F_{3i} = P\{z_i \leq W_k\}.$$ (5)

The distribution of these indicators over the zone and on the map as a whole is inexpedient, because different probabilities will correspond to different sections. In the absence of systematic errors in the determination of these probabilities, on the isolines bounding the zones will always be equal to 0.5. Therefore, in practice, one can only find their average estimates of the type:

$$F_l = \sum_{i=1}^{N} F_{li}, \qquad (6)$$

where $N$ is the number of local sites in the zone or region as a whole.

The following approach is considered to be more correct. Ensuring the reliability of an EM is determined by the fulfillment of two requirements:

- establishing all significantly monre polluted sites or, on the contrary, significantly cleaner sites on the general background of contamination of adjacent territories. Even small (by arca) spots, significantly different by the degree of contamination from the surrounding areas should not be missed.
- application of the contour lines should be done as accurately as possible.

At the first glance, it seems that these requirements contradict with each other. The first requirement involves a continuous study of the entire area with the smallest possible coordinate grid throughout the study area in the absence of prior information. This means a mapping based on a uniform grid. Fulfillment of the second requirement with limited research suggests the use of a non-uniform grid, where more frequent mapping is applied to the areas containing isolines than to other areas.

However, this contradiction is apparent. First, we are talking about different properties of a EM determining its quality. Second, because EC is, essentially, a measuring GIS. This can be shown by drawing an EC analogy with the measuring instrument. The first requirement is the resolution requirement (sensitivity) of the EM. The minimum division price is a characteristic of the sensitivity of the measuring instrument. The minimum sizes of objects indicated on the map are characteristics of the EM. The second requirement is the accuracy requirement of the EM. With limited funds for the creation of an EM, as well as any measuring instrument, it is necessary to provide for a higher level of one property by the detriment of another, or to find their optimal combination.

This consideration allows us to determine the range of basic metrological characteristics of an EM:

- the resolution characteristic is the minimum linear dimension of the pollution spots indicated on the map (with an uneven grid, in each of its zones);
- accuracy characteristics are characteristics of uncertainty of the isolines.

The accuracy characteristics include the standard uncertainty u and the uncertainty interval $\Delta$. $\sigma$ is an analogue of the mean square error. $\Delta$ is the interval covering the isoline with a known probability, that is, an analogue of the confidence limits of uncertainty. The indication of the contour uncertainty intervals should become mandatory for EM. The boundaries of the territory will be in this case clearly defined. For example, the boundary of a territory with probability $P$ close to 1n, k and $(k + 1)$ degrees of pollution, and the areas between them, the degree of contamination of which with a sufficiently significant probability can be k and $(k + 1)$ will be defined.

Along with the result of measurements in the form of nominal isolines satisfying the equation

$$W(x, y) = W_k, \qquad (7)$$

their confidence limits are determined with the account of all monitoring uncertainties. In this case, the uncertainty interval will equal to

$$\Delta(x, y) = \left[ \left( x_{1-a/2} - x_{a/2} \right)^2 + \left( y_{1-a/2} - y_{a/2} \right)^2 \right]^{0,5}, \qquad (8)$$

where $x$, $y$ are the coordinates of the nominal isoline, $x_{a/2}$, $y_{a/2}$, $x_{1-a/2}$, $y_{1-a/2}$ are the coordinates of its confidence limits. Since any contour separates two zones with different pollution indices, there will be two confidence limits for each side. These boundaries should be united by constructing an envelope according to the criterion of maximum uncertainty. The combination of these envelopes determines the uncertainty bands of all EM isolines and, therefore, characterizes the accuracy of the EM as a whole. EM uncertainty can serve as an integral indicator of EM accuracy. It is defined as the maximum value of the ratio of uncertainty $\Delta(x,y)$ to the corresponding half-width of the zone $l(x,y)$ among all EM zones:

$$u(EM) = \max_{x,y} [\Delta(x, y) \, / \, l(x, y)]. \qquad (9)$$

Areas of uncertainty can be in the middle of the zones that do not have common points with nominal isolines. The probability of correct display area with coordinates $(x,y)$ on the map is

$$F_{li} = P\{W_k \le z(x, y) \le W_{k+1}\} = \int_{W_k^{-z(x,y)}}^{W_{k+1}^{-z(x,y)}} f(\xi) d\xi, \qquad (10)$$

where $z(x,y)$ is the result of determining the concentration of an element in this area, $f$ ($\xi$) is the density of the error distribution of this definition. Under the natural assumption of a normal distribution of uncertainty,

$$F_{1l} = \hat{O}\left\{ \frac{W_{k+1} - z(x, y) - m(z)}{\sigma(z)} \right\} - \hat{O}\left\{ \frac{W_k - z(x, y) - m(z)}{\sigma(z)} \right\}, \qquad (11)$$

where $m(z)$ is the expectation of the position of the isoline, and $\sigma(z)$ is the standard deviation uncertainty of the concentration determination; $\Phi(\xi)$ is the Laplace function.

Equating $F_{1l}$ to the given value of the confidence probability $P$, we obtain an expression to determine the boundaries of the reliable display of the zone:

$$W_k - m(z) - \lambda_P \cdot \sigma(z) \le z(x, y) \le W_k - m(z) + \lambda_P \cdot \sigma(z), \qquad (12)$$

where $\lambda_P$ is the quantile of the normal distribution corresponding to the probability $P$. The set of points $(x, y)$ at which the condition is violated will constitute the internal region of uncertainty of a given zone, and its boundary will pass through points $(x, y)$ that satisfy one of the conditions:

$$z(x, y) = W_k - m(z) - \lambda_P \cdot \sigma(z) \text{ or } z(x, y) = W_k - m(z) + \lambda_P \cdot \sigma(z). \qquad (13)$$

The area bounded by the isoline satisfying the first condition will be the area of risk of undetected contamination. The area, the isoline of which satisfies the second condition, is a zone of risk of false pollution.

Such approach allows not only to obtain information on the contamination of the studied area, but also to formulate probabilistic indicators of the efficiency of EM. The average confidence index of an EM can be determined. It characterizes the probability on average of the fact that the environmental situation at an arbitrary point in the region is correctly estimated in an EM and is equal to:

$$F_1(EM) = 1 - \sum_{i=1}^{n} S_i / S_{EM}, \qquad (14)$$

where $\sum_{i=1}^{n} S_i$ is the total area of the bands and areas of EM uncertainty, SEM is the area of the territory covered by the EM.

A serious limitation of the proposed approach is the possible omission of local point graves. This is quite possible with large areas. When studying the integral background soil contamination, the above approach makes it possible to build isolines of contaminated sites while simultaneously displaying the accuracy data of both the measurements and the boundaries of these sites.

## 4   Conclusions

In conclusion, it is necessary to point out that for the first time, a complex evaluation of the Polytechnichesky Park ground heavy metals contamination has been done by using standard methods and available Spectroscan instrument. The investigations showed that heavy metal ground contaminations in the Polytechnichesky Park exceed the ecological norms. The extended measurement uncertainty did not exceed 25%. Taking into consideration the fact that the measurements have been done not continuously, but in local points, this is possible to say that the most interesting result is in the evaluation of the accuracy of forming isolines. In the result, the index of the ecological control average reliability was determined. The determination of this index will give the possibility to calculate the minimal number of the sites for taking probes on a given territory in the future.

# References

1. He, B., Yun, Z.J., Shi, J.B., Jiang, G.B.: Research progress of heavy metal pollution in China: sources, analytical methods, status, and toxicity. Chin. Sci. Bull. **58**(2), 134–140 (2013)
2. Peña-Fernández, A., Lobo-Bedmar, M.C., González-Muñoz, M.J.: Annual and seasonal variability of metals and metalloids in urban and industrial soils in Alcalá de Henares (Spain). Environ. Res. **136**, 40–46 (2015)
3. Shi, G., Chen, Z., Xu, S., Zhang, J., Wang, L., Bi, C., Teng, J.: Potentially toxic metal contamination of urban soils and roadside dust in Shanghai, China. Environ. Pollut. **156**(2), 251–260 (2008)
4. Zhao, Z., Hazelton, P.: Evaluation of accumulation and concentration of heavy metals in different urban roadside soil types in Miranda Park, Sydney. J. Soils Sediments **16**(11), 548–556 (2016)
5. Curran-Cournane, F., Lear, G., Schwendenmann, L., Khin, J.: Heavy metal soil pollution is influenced by the location of green spaces within urban settings. Soil Res. **53**(3), 306–315 (2015)
6. Gall, J.E., Boyd, R.S., Rajakaruna, N.: Transfer of heavy metals through terrestrial food webs: a review. Environ. Monit. Assess. **187**(4), 201 (2015)
7. Hernández, A.J., Pastor, J.: Relationship between plant biodiversity and heavy metal bioavailability in grasslands overlying an abandoned mine. Environ. Geochem. Health **30**(2), 127–133 (2008)
8. Pehluvan, M., Karlidag, H., Turan, M.: Heavy metal levels of mulberry (Morus alba L.) grown at different distances from the roadsides. J. Anim. Plant Sci. **22**(3), 665–670 (2012)
9. Nabulo, G., Oryem-Origa, H., Diamond, M.: Assessment of lead, cadmium, and zinc contamination of roadside soils, surface films, and vegetables in Kampala City, Uganda. Environ. Res. **101**(1), 42–52 (2006)
10. Alloway, B.J.: Heavy Metals in Soils. Environmental Pollution, pp. 11–50. Springer, Dordrecht (2013). In: Alloway, B.J. (ed.) Ch. Chapter 2
11. Russian State Standard 17.4.4.02-2017 Nature protection. Soils. Methods for sampling and preparation of soil for chemical, bacteriological, helmintological analysis [GOST 17.4.4.02-2017 Ohrana prirodyi. Pochvyi. Metodyi otbora i podgotovki prob dlya himicheskogo, bakteriologicheskogo, gelmintologicheskogo analiza]
12. Russian State Standard 33850-2016 Soils. Determination of chemical composition by X-Ray fluorescence spectrometry [GOST 33850-2016 Pochvyi. Opredelenie himicheskogo sostava metodom rentgenofluorestsentnoy spektrometrii]
13. USEPA Method 6200: field portable x-ray fluorescence spectrometry for the determination of elemental concentrations in soil and sediment. http://www.epa.gov/osw/hazard/testmethods/sw846/pdfs/6200.pdf
14. Chakraborty, S., Man, T., Paulette, L., Deb, S., Li, B., Weindorf, D.C., Frazier, M.: Rapid assessment of smelter/mining soil contamination via portable X-ray fluorescence spectrometry and indicator kriging. Geoderma **306**, 108–119 (2017)
15. Hu, W., Huang, B., Weindorf, D.C., Chen, Y.: Metals analysis of agricultural soils via portable x-ray fluorescence spectrometry. Bull. Environ. Contam. Toxicol. **92**(4), 420–426 (2014)
16. Jang, M.: Application of portable X-ray fluorescence (pXRF) for heavy metal analysis of soils in crop fields near abandoned mine sites. Environ. Geochem. Health **32**(3), 207–216 (2010)

17. Kalnicky, D.J., Singhvi, R.: Field portable XRF analysis of environmental samples. J. Hazard. Mater. **83**(1–2), 93–122 (2001)
18. Peinado, F.M., Ruano, S.M., González, M.G.B., Molina, C.E.: A rapid field procedure for screening trace elements in polluted soil using portable X-ray fluorescence (PXRF). Geoderma **159**(1–2), 76–82 (2010)
19. Paulette, L., Man, T., Weindorf, D.C., Person, T.: Rapid assessment of soil and contaminant variability via portable x-ray fluorescence spectroscopy: Copşa Mică. Romania. Geoderma **243–244**, 130–140 (2015)
20. Russian Methodical instructions 2.1.7.730-99 Hygienic evaluation of soil in residential areas [MU 2.1.7.730-99 Gigienicheskaya otsenka kachestva pochvyi naselennyih mest]

# A Comparison of Tuning Methods for PID-Controllers with Fuzzy and Neural Network Controllers

Clemens Gross$^{(\boxtimes)}$ and Hendrik Voelker

Peter the Great St. Petersburg Polytechnic University, Saint Petersburg, Russia
{gross.yak, voelker.h}@edu.spbstu.ru

**Abstract.** Conventional approaches to control systems still present a reasonable solution for a variety of different tasks in control engineering problems. Controllers based on the PID approach are used in a wide range of applications due to their easy handling, realization and set up, as well as their modest need of computational resources during the runtime. In order to heuristically find near-optimal parameters for the controller design, different approaches to tuning PID controllers have been developed. The Ziegler–Nichols methods are still commonly used despite that they have long been known, though modern methods, such as the T-Sum method, have also emerged. In this work, a comparison of the tuned PID controllers with a Mamdani-Fuzzy-Logic controller and an adaptive neural network controller is offered. A unified step response is used to classify the performance of controllers. It is shown that a PID control can work just as well as a fuzzy logic or neural network control for simple applications with time-invariant parameters or in applications where the parameters only change slightly and no strict constancy of the plant output is necessary.

**Keywords:** Control engineering · PID · Fuzzy control · Neural network control · PID tuning · Fuzzy logic

## 1 Introduction

Although well known since the 1930s, the proportional–integral–derivative (PID) controllers are popularly used in many applications of control engineering even today. Huge efforts are made in order to optimize its control behaviour and to optimally adjust a PID controller's parameters to a specific use case. Therefore, classical PID control is still subject to many ongoing projects of applied research.

To optimally adjust the PID controller's parameters, an accurate model of the plant which should be controlled should be known or developed. Depending on the level of abstractness of a specific process, this step can pose a problem. Thus, different tuning methods for PID controllers exist. With the help of these methods, such as the Ziegler-Nichols (ZN) [1] or T-Sum [2] methods, it is possible to heuristically find close enough parameters to achieve a good performance of the process. Kumar et al. [3] as well as Wang et al. [4] compare in their works the various tuning paradigms with respect to

© Springer Nature Switzerland AG 2020
D. G. Arseniev et al. (Eds.): CPS&C 2019, LNNS 95, pp. 74–86, 2020.
https://doi.org/10.1007/978-3-030-34983-7_8

different plant and controller topographies. The current research offers an online adaption of PID controllers in dependence from time-variant parameters to overcome one of the major downsides in the PID control. Approaches to this have been available since the 1990s, as is shown, for example, by Sung et al. in [5]. Despite that, two modern and more sophisticated approaches for controllers are now used more and more often, to control not only nonlinear, time-variant processes. The Fuzzy Set Theory [6], or Fuzzy-Logic (FL) was first applied to control problems in 1974 by Mamdani [7], and a general overview of the possible industrial applications of Fuzzy Control as compared to PID controllers is given in [8]. By its help the system behaviour and the desired control reaction can be described by linguistic variables and, therefore, is easy to adapt to a plant the exact parameters of which are unknown. The second modern control approach is based on Artificial Neural Networks (ANN). Even though the theory behind simple ANN is also fairly old, first postulated in 1943 by McCulloch and Pitts [9], only recently it was actually applied to different problems in engineering and information technology. Different principles of Neural Network Control (NNC) have evolved since the 1990s as is shown in [10]. Current research in control theory focuses on merging the advantages of classical PID control with FL (Deng et al. [11]) and ANN approaches in order to adapt PID parameters during the process runtime. Application of NNC to robotic manipulators is shown in [12] by Shuzhi et al., whereas Potekhin et al. [13] investigate a Fuzzy Neural Network for controlling autonomous decentralized energy grids.

This work shows a comparison between the PID tuning methods, FL and ANN control. The aim of this work is to indicate which of the methods has the best performance as along with the lowest implementation effort. For the PID tuning methods, the ZN and T-Sum ones, are chosen, while a Mamdani-FL controller and another one based on Feed-Forward-ANN is used for simulation of the controllers and the process.

## 2 Different Control Approaches

Three classes of controllers are compared in this work. PID Controllers and those based on FL and ANN are the main control paradigms used in closed-loop control applications. This section gives an overview of the theory and systematic schemes behind the different types of controllers and their respective tuning methods.

### 2.1 PID-Control

A PID controller is composed of three individual parts, hence is the name of proportional–integral–derivative controller. Linking these three blocks in parallel results in the general transfer function for a PID controller in the Laplacian domain

$$K_{PID}(s) = K_P + \frac{K_I}{s} + K_D s = K_P \left(1 + \frac{1}{T_I s} + T_D s\right), \tag{2.1}$$

where $T_I = K_P/K_I$ is called the reset time and $T_D = K_D/K_P$ is called the lead time [14]. By transforming Eq. (2.1) back to the time domain as shown in Eq. (2.2), the dependence of the controller's output $u(t)$ from the control error $e(t)$ is easily derived:

$$u(t) = K_P e(t) + \frac{K_P}{T_I} \int_0^t e(\tau) d\tau + K_P T_D \frac{de(t)}{dt}. \qquad (2.2)$$

A graphical presentation of a PID controller with the control error and control output in the Laplacian domain can be seen in Fig. 1. If the plant is well known and can be modelled mathematically without further difficulties, numerical methods to set the parameters are preferred. However, it is not always practical to invest a lot of time and resources into developing an adequate mathematical model of the respective plant. Therefore, different tuning methods exist which make it possible to heuristically find near-optimal solutions for the parameters of the PID controller.

### 2.1.1 Ziegler-Nichols Tuning Method

While the Ziegler-Nichols method [1] has been known since 1942, it is still used today, mainly for strongly delayed processes. The ZN method proposes two different approaches; one of the two being presented here. The plant is approximated as a first-order plus time delay model. The time delay of the plant $T_{dead}$, the time constant $T$ and the stationary amplification $K_s$ have to be known or determined experimentally from the system's step response by adding an inflexion tangent to the step response of the system. The time between the zero point of the time scale, when the step was applied to the plant, and the intersection of the inflexion tangent with the time axis can be considered equal to $T_{dead}$, while the time between the intersection of the inflection tangent with the x-axis and with the stable output of the plant is equal to $T$. The PID controllers' parameters are then adjusted in the following way: $K_P = 1.2/K_s \cdot T/T_{dead}$, $T_I = 2 \cdot T_{dead}$ and $T_D = 0.5 \cdot T_{dead}$. Other tuning methods, like Chien-Hrones-Reswick [15], are based on the second method of ZN.

### 2.1.2 T-Sum Tuning Method

The T-Sum method was first introduced in 1995 by Kuhn [2]. Together with other approaches, like the one proposed by Åström and Hägglund [16], it stands for a modern and more sophisticated approach to PID tuning. It can be used for plants which can be characterized by a low-pass behaviour and have the transfer function as:

$$G(s) = K_s \frac{(1 + T_{U,1}s)(1 + T_{U,2}s)\ldots(1 + T_{U,m}s)}{(1 + T_{L,1}s)(1 + T_{L,2}s)\ldots(1 + T_{L,n}s)} e^{-sT_{dead}} \qquad (2.3)$$

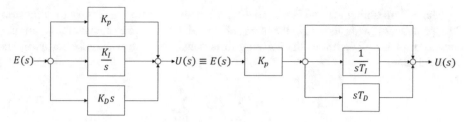

**Fig. 1.** Block diagram of a PID controller

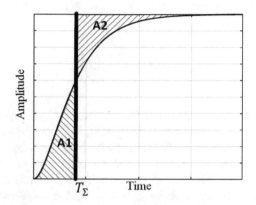

**Fig. 2.** Graphical determination of the T-Sum

**Table 1.** Parameters for the T-Sum method (Original)

| Parameter | Value |
| --- | --- |
| $K_P$ | $K_s$ |
| $T_I$ | $0.66 \cdot T_\Sigma$ |
| $T_D$ | $0.167 \cdot T_\Sigma$ |

**Table 2.** Parameters for the T-Sum method (Fast)

| Parameter | Value |
| --- | --- |
| $K_P$ | $2/K_s$ |
| $T_I$ | $0.8 \cdot T_\Sigma$ |
| $T_D$ | $0.194 \cdot T_\Sigma$ |

The T-Sum tuning method is named after the main operation necessary to adjust the PID parameters with its help, the sum of the delaying time constants $T_L$ in the numerator of Eq. (2.3) minus the deriving time constants in the denominator and plus the dead time:

$$T_\Sigma = T_{dead} + \sum_{i=1}^{n} T_{L,i} - \sum_{j=1}^{m} T_{U,j}. \tag{2.4}$$

If the transfer function of the plant is unknown, the T-Sum shown in Eq. (2.4) can also be obtained from the experimentally found step response of the plant, either arithmetically (2.5)

$$T_\Sigma = \int_{0}^{\infty} \left(1 - \frac{y(t)}{K_s}\right) dt \tag{2.5}$$

or graphically from the step response of the system, which is denoted as $y(t)$ in (2.5). Graphically, the value of $T_\Sigma$ can be derived by introducing a straight line, perpendicular to the time axis. This straight line divides the area between it and the y-axis under the step response and above the x-axis from the area above the step response and below the steady state of it between this straight line and the steady state of the step response. When these two areas are equal ($A_1 = A_2$ in Fig. 2), the value for $T_\Sigma$ can be directly seen from the position of this straight line on the time axis, as shown in Fig. 2. The values for the PID's parameters can then be directly seen from Table 1 the original T-Sum method and from Table 2 the fast T-Sum method, which compromises a higher overshoot for a faster settling time.

## 2.2   Fuzzy Control

Unlike PID controllers, which are parametrized either by mathematical modelling or experimental knowledge, fuzzy control relies more on previous knowledge in terms of heuristic IF <condition> THEN <action> rules. Therefore, designing a fuzzy controller is actually similar to the experimental method for constructing a table of inputs and corresponding output values called a ruleset in Fuzzy Logic. What makes FL interesting, though, is that due to the fuzzification of the inputs, not only the situations actually listed in the respective ruleset result in an action, but also all intermediate values which at least partially correspond to one of the if-conditions. To achieve this, the reference input $e(t)$ of the FL controller is mapped to a value in the range [0, 1].

This certain value is called degree of belief in FL and is achieved by applying the fuzzy membership functions (see Fig. 3) to the crisp input $e(t)$. A typical FL controller is composed of the fuzzification of the reference input, which is then led to the Fuzzy Inference System (FIS). Depending on the respective if-then rules in the rule base, the FIS infers a certain fuzzy control input $u(t)$, which is then defuzzified and applied to the plant of the control loop [17]. This can be seen schematically in Fig. 4.

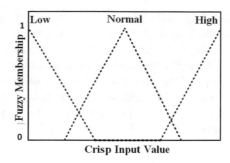

**Fig. 3.** Exemplary membership functions

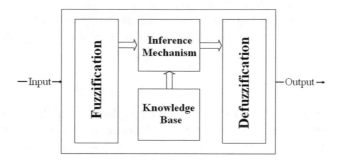

**Fig. 4.** Schema of a fuzzy controller

A big advantage of fuzzy controllers is their ability to work with plants with uncontinuous transfer functions and their robustness when a plant with time-variant behaviour is controlled [17]. The first applications of FL to control problems were proposed by Sugeno in [18].

## 2.3 Neural Network Control

Neural Network Control is another approach to control systems that has emerged lately from the huge interest in ANN. In the general ANN research, ANN are classified by their structure, which are feed-forward, radial-basis functions, convolutional or long-short term memory ANN among others. For NNC, though, Agarwal [10] tried to classify different controllers. The basic principle of ANN relies on small, independent neurons, inspired by the biological processes of the brain of living organisms, which multiply $n$ input signals, by the respective weights $w_1, \ldots, w_n$. The weighted input signals are then summed up with a possible bias of the neuron and fed to a specific activation function, which gives back the output of the neuron. Literature sources, e.g., [19] or [20], present a more detailed introduction to ANN.

One of the main advantages of NNC is the ability of feed-forward ANN with nonlinear, differentiable activation functions to approximate any given function, even if the function is nonlinear [21]. This is used in the first step towards building a NNC, when the plant's transfer function is approximated with the help of ANN, instead of being analytically modelled manually. The second step, the actual NNC, profits from one of the main reasons why the ANN are so popular nowadays: i.e., their ability to learn. Due to this fact, it is possible to build adaptive controllers with the help of ANN [22]. However, the necessity of the structure of the ANN to be determined a-priori, the control task can pose a problem because it can lead to an overdetermined network structure resulting in high computational complexity, or to an underdetermined structure causing poor performance of an actual controller [21].

## 3  Experimental Set-Up

In the course of this work, three different control paradigms are simulated in Matlab/Simulink R2016B. The control reference $e(t)$ is implemented by a unified step at the time $t = 0$. The obtained unified step response of the plant for different controllers is then compared.

### 3.1  Plant

In order to compare the performance of the controllers, a plant for the control loop should be chosen. Due to its wide range of applications, a DC motor is selected; it can be modelled according to the Newton (3.1) and Kirchhoff (3.2) laws

$$J\frac{d^2\theta}{dt^2} = t - b\frac{d\theta}{dt};$$
(3.1)

$$L_a\frac{di}{dt} = -R_a i + V - e$$
(3.2)

as a second order plant. A DC motor can be modelled in more detail as well, the representation in the form of a linear PT2 element is chosen due to the possibility to model a variety of different control problems in this form of a second order plant. Figure 5 represents the model of the plant which is created in Simulink. In order to study the impact that time-variant behaviour of the plant produces on the different controllers, the values of the plant parameters are altered after the controllers have been adjusted to it. This adjustment lies in the range of plus or minus 20% for each parameter and is, therefore, by all means within the range of deviations that the parameters of individual components of the same series may have or may acquire over the time [23]. The changes applied to the plant parameters after the optimization of the controllers can be seen in Table 3.

### 3.2  Controllers

A PID controller is realized by the use of respective gain blocks, integrators and derivative blocks, just as shown in Fig. 1 The different parameters of the PID controller are then derived from the step response of the plant, using the $2^{nd}$ ZN method, as well as the original and fast adjustment values from the T-Sum method. This results in the following parameters for PID controllers that are shown in Table 4.

**Table 3.** Change of plant parameters to simulate change over the time

| | | |
|---|---|---|
| $L_{a,new} = 0.9 \cdot L_a$ | $R_{a,new} = 0.9 \cdot R_a$ | $K_{t,new} = 1.2 \cdot K_t$ |
| $J_{new} = 0.8 \cdot J$ | $b_{new} = 1.1 \cdot b$ | $K_{e,new} = 1.05 \cdot K_e$ |

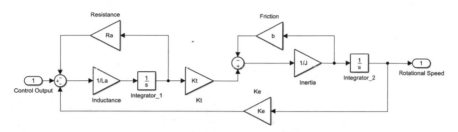

**Fig. 5.** DC motor modelled as a linear second order plant in simulink

The fuzzy controller as well as the neural network controller is chosen from the respective toolbox in Simulink. For the fuzzy controller, a Mamdani approach with two input variables (control reference $e(t)$ and fed-back plant output $y(t)$) and the control output $u(t)$ is applied. Five triangular membership functions (a positive and negative ones for the control reference and a positive, negative and zero ones for the feedback signal) and three output membership functions are implemented with a respective ruleset. The membership functions for the input and output variables can be seen in Fig. 6.

Similarly, a controller from the neural network toolbox in Simulink is chosen to characterize the implementation and the behaviour of such a type of controller. The chosen NN predictive controller [24] has a specific control horizon of discrete steps in which the plant's behaviour is predicted and the necessary control actions are performed (the control horizon). These parameters are set to ten and eight respectively. The controller is trained by random control reference inputs in a given range of values and time. During a training session with the presented plant, a series of 5000 discrete time steps is run over the NN with a batch size of 10. The layers consist of eight neurons each, and the weights are optimized by using a 1-dimensional backtracking algorithm in Matlab. Figure 7 shows the architecture of the NN used in the controller.

**Table 4.** Parameters for the PID controller

| Parameter | 2$^{nd}$ Ziegler-Nichols | T-Sum (original) | T-Sum (fast) |
|---|---|---|---|
| $K_p$ | 4.505 | 1.000 | 2.000 |
| $T_I$ | 1.334 | 1.815 | 2.200 |
| $T_D$ | 0.333 | 0.459 | 0.535 |

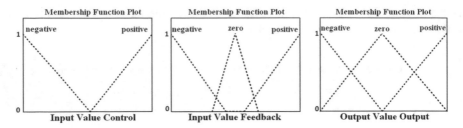

**Fig. 6.** Membership functions of the implemented fuzzy controller

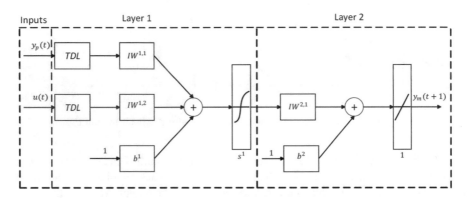

**Fig. 7.** Architecture of the NN predictive controller

## 4   Results

The simulation is first carried out with the plant's parameters for which the controllers were optimised and it can be seen in Fig. 8. In order to assess the performance of different controllers, a unified method to quantify the plant's output is necessary. Important parameters used to evaluate controller's performance with the use of the step response are the overshoot of the plant's output $y(t)$ over the reference signal $e(t)$, the settling time (in dependence from a certain error criteria), and the stationary error. In this work, the settling time is evaluated with respect to a deviation of 0.8% from the stationary signal. In order to include all of these parameters in the estimation, the root mean squared error (RMS) of the discrete signals $y$ and $e$ (4.1) is used:

$$\text{RMS} = \sqrt{\frac{\sum_{i=1}^{n} (e_i - y_i)^2}{n}}. \tag{4.1}$$

The RMS, the overshoot and the settling time are shown in the following Table 5 for different controllers. None of the controllers showed a stationary error. After the plant's parameters were altered, the simulation is carried out again. The results can be seen in Fig. 9 and Table 6 respectively. In order to be able to quantify the influence that the change of the plant's parameters has on the controller's performance, the difference of the RMS from Table 5 $RMS_{old}$ and the RMS with the changed plant's parameters $RMS_{new}$ is computed (4.2):

$$\Delta_{RMS} = \frac{RMS_{old} - RMS_{new}}{RMS_{old}} \cdot 100\% \tag{4.2}$$

It can be seen that the results of the different control paradigms differ. In view of the minimal error criteria, the ZN-tuned PID controller shows the best result, even though it has the highest overshoot exceeding those of the controller with the parameters set by the fast T-Sum method by more than the factor three. The FL controller shows the best behaviour in terms of the overshoot but in terms of the RMS error criteria it is at the bottom together with the NN predictive controller. This would not even change when the plant's parameters are changed to simulate time-variant plant behaviour. If a low overshoot is required, though, the examined FL controller has a feasible behaviour.

**Table 5.** Key performance indicators for the controllers, original plant parameter

| Parameter | PID Control | | | Fuzzy Control | NN pred. Control |
|---|---|---|---|---|---|
| | 2$^{nd}$ ZN | T-Sum (or.) | T-Sum (fast) | | |
| RMS | 0.232 | 0.301 | 0.259 | 0.312 | 0.317 |
| Overshoot [%] | 40.65 | 6.81 | 12.37 | 2.39 | 6.65 |
| Settling t. [s] | 8.12 | 11.4 | 8.8 | 10.24 | 9.24 |

**Table 6.** Key performance indicators for the controllers, changed plant parameter

| Parameter | PID Controller | | | Fuzzy Controller | NN pred. Control |
|---|---|---|---|---|---|
| | 2$^{nd}$ ZN | T-Sum (or.) | T-Sum (fast) | | |
| RMS | 0.212 | 0.285 | 0.244 | 0.285 | 0.293 |
| $\Delta_{RMS}$ [%] | 8.352 | 5.357 | 5.946 | 8.665 | 7.556 |
| Overshoot [%] | 35.52 | 0.46 | 7.43 | 6.79 | 11.47 |
| Settling t. [s] | 7.05 | 10.4 | 7.95 | 8.41 | 10.41 |

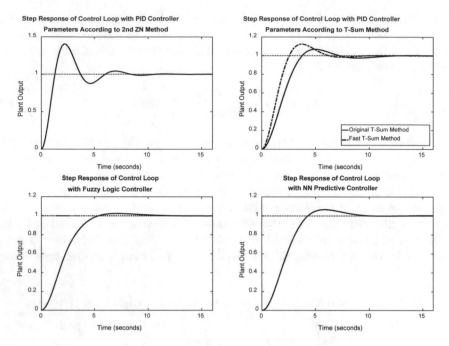

**Fig. 8.** Comparison of the step responses of the control loop with the original parameters

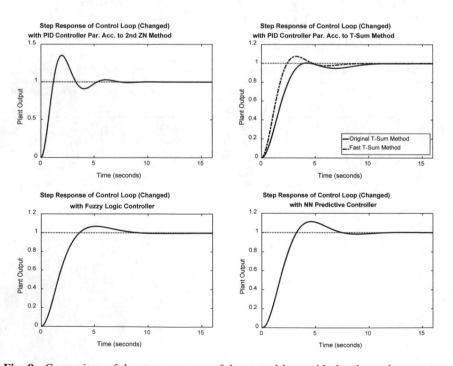

**Fig. 9.** Comparison of the step responses of the control loop with the changed parameters

# 5 Conclusion

Controllers – and different tuning methods – can be compared to each other by the basic criteria that are mandatory for every control system. All of the tested controllers showed stable behaviour, had good accuracy as measured by the resting control deviation, and could be considered robust, as all the controllers were able to satisfactorily control the plant, even when the parameters of the plant altered over time. The coherence of speed and attenuation of control loops is shown in this work again. The controllers which have a lower setting time, as a general rule, have a higher overshoot. Therefore, a trade-off has to be made when choosing the right controller for a control application. For the set-up used in this work, the ZN-tuned PID controller showed the fastest settling time, whereas the FL controller had the lowest overshoot.

It has to be said, though, that the behaviour of the FL controller highly depends on the chosen ruleset and membership functions, and a different characteristic for the step response can, therefore, be obtained with an FL controller. This is also true for the NN predictive controllers, where the chosen architecture of a controller, as well as the training data, influences the future control action. Therefore, these two certain types of controllers stand only prototypical for the respective control approaches, as no generalized methods of bestpractice design of these controllers exists so far.

It should be noted that for small control problems that can be linearized with sufficient accuracy, like the one examined in this work, neither a fuzzy controller nor a neural-network controller are of certain advantageous character. In all performance indices shown in Tables 5 and 6, the PID controller with a specific tuning method can keep up with more sophisticated control approaches. By choosing other tuning methods for a PID controller and manual fine tuning of the parameters, the shown behaviour can be changed if necessary. Especially when taking into account the lesser implementation expenses and the lower computational complexity of a PID controller, fuzzy and NN controllers cannot be considered practicable for such a control application.

# References

1. Ziegler, J.G., Nichols, N.B.: Optimum settings for automatic controllers. Trans. ASME **64**, 759–768 (1942)
2. Kuhn, U.: Eine praxisnahe Einstellregel für PID-Regler: Die T-Summen-Regel. Automatisierungstechnische Praxis **5**, 10–16 (1995)
3. Kumar, R., Singla, S., et al.: Comparison among some well known control schemes with different tuning methods. J. Appl. Res. Technol. **13**, 409–415 (2015)
4. Wang, L., Freeman, C., Roger, E.: Experimental evaluation of automatic tuning of PID controllers for an electro-mechanical system. IFAC PapersOnLine **50**(1), 3063–3068 (2017)
5. Sung, S.W., Lee, I.-B.: On-line process identification and PID controller autotuning. Korean J. Chem. Eng. **16**(1), 45–55 (1999)
6. Zadeh, L.A.: Fuzzy sets. Inf. Control **8**(3), 338–353 (1965)
7. Mamdani, E.H.: Application of fuzzy algorithms for control of simple dynamic plant. Proc. Inst. Electr. Eng. **121**(12), 1585–1588 (1974)
8. Lugli, A.B., et al.: Industrial application control with fuzzy systems. Int. J. Innov. Comput. Inf. Control **12**(2), 665–676 (2016)

9. McCulloch, W., Pitts, W.: A logical calculus of ideas immanent in nervous activity. Bull. Math. Biophys. **5**(4), 115–133 (1943)
10. Agarwal, M.: A systematic classification of neural-network-based control. IEEE Control Syst. Mag. **17**(2), 75–93 (1997)
11. Deng, J., et al.: Self-tuning PID-type fuzzy adaptive control for CRAC in datacenters. In: Springer IFIP Advances in Information and Communication Technology (AICT), vol. 419, no. 1, pp. 215–225 (2014)
12. Shuzhi, S.G., Hang, C., Woon, L.: Adaptive neural network control of robot manipulators in task space. IEEE Trans. Industr. Electron. **44**(6), 746–752 (1997)
13. Potekhin, V.V., Pantyukhov, D.N., Mikheev, D.V.: Intelligent control algorithms in power industry. EAI Endorsed Trans. Energy Web **3**(11), e5 (2017)
14. Lunze, J.: Regelungstechnik 1 – System-theoretische Grundlagen, Analyse und Entwurf einschleifiger Regelungen. Springer, Heidelberg (2010)
15. Chien, K.L., Hrones, J.A., Reswick, J.B.: On the automatic control of generalized passive systems. Trans. ASME **74**, 175–185 (1952)
16. Åström, K.J., Hägglund, T.: Automatic tuning of simple regulators with specifications on phase and amplitude margins. Automatica **20**(5), 645–651 (1984)
17. Passino, K.M., Yurkovich, S.: Fuzzy Control. Addison Wesley Longman Inc., Menlo Park (1998)
18. Sugeno, M.: Industrial Applications of Fuzzy Control. Elsevier Science Inc., New York (1985)
19. Haykin, S.: Neural Networks - A Comprehensive Foundation. Pearson Education (Singapore) Pte. Ltd., Delhi (2005)
20. Patterson, J., Gibson, A.: Deep Learning - A Practicioner's Approach. O'Reilly Media Inc, Sebastopol (2017)
21. Liu, G.: Nonlinear Identification and Control - A Neural Network Approach. Springer, London (2001)
22. Rovithakis, G.A., Christodoulou, M.A.: Adaptive Control with Recurrent High-Order Neural Networks. Springer, London (2000)
23. Halfmann, C., Holzmann, H.: Adaptive Modelle für die Kraftfahrzeugdynamik. Springer, Heidelberg (2003)
24. Soloway, D., Haley, P.: Neural generalized predictive control. In: Proceedings of the 1996 IEEE International Symposium on Intelligent Control, Dearborn, MI, USA (1996)

# Measuring of the Growth Capacity of a Semantic Model of a Defence-Oriented Industrial Enterprise Using a Cyber-Physical System

Tatyana S. Katermina[1], Efim Karlik[2], Ermin Sharich[3,4],
and Daria Iakovleva[4(✉)]

[1] Nizhnevartovsk State University, Nizhnevartovsk, Russia
nggu-lib@mail.ru
[2] St. Petersburg State University of Economics, Saint Petersburg, Russia
fmkarlik@mail.ru
[3] DOO Energoinvest, Sarajevo, Bosnia and Herzegovina
[4] St. Petersburg State University, Saint Petersburg, Russia
{st062696, st062671}@student.spbu.ru

**Abstract.** Given that there are no methods for evaluating growth capacity of a company within its management functions, the article discusses the important problem of measuring the growth capacity of a defense-oriented industrial enterprise. It presents an innovative multi-agent system of measures for the growth capacity of an enterprise based on a semantic model, which takes into account both the financial and production outcomes of the company. It also focuses on innovative capacity and human capital of the company, which is of great importance in contemporary technological environment. Without such a method, it is virtually impossible to diagnose, detect and mitigate the effects of the changing micro and macro environment. By estimating the growth capacity, it is possible to facilitate the process of resource management coordination and thus provide sustainable innovative development of the enterprise and its socio-technological environment.

**Keywords:** Growth capacity · Management · Intellectual capital · Socio-technological systems · Strategy · Innovations

## 1 Introduction

In the modern economy, the role of intellectual capital and innovative capacity in the defense-oriented industry is increasing as a factor of scientific and technological progress; the personnel of enterprises or the human capital asset develops the ways of levelling off the industry's dependence on the import of equipment (especially IT facilities). The main challenge is to identify and diagnose the factors contributing to the improvement of interdependent indicators that characterize the growth capacity of defense-oriented enterprises and the conditions specific for the industry and affecting

© Springer Nature Switzerland AG 2020
D. G. Arseniev et al. (Eds.): CPS&C 2019, LNNS 95, pp. 87–97, 2020.
https://doi.org/10.1007/978-3-030-34983-7_9

production activities and innovation. The resulting effect should be an increase in innovations as this factor is the basis for the functioning of domestic production, investment, and business processes.

The aim of the study is to structure the concept of enterprise capacity by management functions and show the correlation between the growth capacity of an enterprise and its development, innovation and financial results. The research study also looks into the problem of goal-setting in the strategic management system, and introduces the concept of innovative progress and measuring the capacity scale on the basis of a semantic model.

## 2 Characteristics of the Approaches Used

If viewed retrospectively, the key theories of management, the basic socio-technological criteria for estimating the opportunities of an enterprise mutated depending on the conceptual approaches to functioning of an enterprise and assessing its efficiency in achieving strategic goals. Thus, in the production concept (1860–1920–2005) in the characteristics of an enterprise's capabilities the emphasis was put on the ability to convert resources and produce goods and services. The estimated indicators were the volume of production, the fixed and working capital, and the number of employees. Further on, in commodity (1920–1930) and marketing (1930–1950) concepts, the attention was attracted to the ability of an enterprise to produce qualitative goods and to promote them through marketing efforts. The concept of marketing (1960–1980–2005) first of all focuses on the ability of an enterprise to meet consumers' needs. The socio-technological concept of marketing (1980–1995–2005) and the modern concept of management focus on the rational, more precise and optimal use of resources in the production of products and meeting the needs of consumers, and interests of business partners. With such an approach, the goals of social development in practice are substituted with maximizing profits rather than with increasing the quality of life. Nevertheless, at the present stage of the neoclassical theory there is a necessity to assess the capacity not only from the angle of value-based management, but also from the angle of systematization functions of management of an enterprise as prompters of social and technological systems. Therefore, it can be said that the principle of estimation of the enterprise's capacity comprises both maximization of the market value of the enterprise for its owners, and socially-oriented purposes of the development of each person, the workforce, and society as a whole. One of the criteria of technological efficiency of capital use (transformation of the capacity into a result and its estimation) is positive value added (VA), which takes into account the operating profit of the company, tax rates, financial alternative costs of the capital, and the results of the formation of intellectual capital. The development of the concept is characterized with the management aimed at creating (increasing) the market value of an enterprise through qualitative improvement of strategic and operational decisions at all levels of the organization, which depends on the efficiency of coordination among all functions of management (planning, organization, coordination and control). If the efforts of all decision-makers are concentrated on the key levels of management, information and communication technologies (ICTs), implementation of ERP Systems, artificial intelligence (AI), and take into account the industry specifics, it contributes to the intention to create long-term competitive advantages in a constantly changing environment.

In the 1980s, the technologies of information support systems management of complex socio-technological and technical systems were developed by system analysts, cyberneticians Optnerer, Uemov, Pospelov, Denisov, Kukor, Mikhail Ignatiev, etc., who helped to describe in advance and formalize the idea of problems emerging in management in the case of risk and uncertainty, variability and dynamic environment. Logical-linguistic models have been playing a big role in the theory of situational management for many years. They have been constantly evolving and expanding, and are introduced as technological factors that allow to simulate management of complex social and technological systems [3, 8].

Multi-agent systems are used to build management models for complex socio-technological systems as part of the implementation of the theory of situational management [1, 2, 5]. The implementation of multi-agent systems is primarily related to the possibility to integrate the latest developments in the fields of information technology and information systems, which sequentially facilitate the task of solving problems in management in the conditions of risk, uncertainty and dynamism of socio-technological systems.

In the present study, the authors propose to apply the methodical approach of situational management to such a problematic situation as "measuring the growth capacity of a defence-oriented enterprise" on the basis of management theory, system analysis, and mathematical and logical-linguistic modelling.

## 3  Problems of Capacity Measuring

Analysis of the problem suggests approaches to its solution. The most comparable meaning of growth capacity is the innovative capacity, as it is innovation that provides for growth [6, 9]. The innovative capacity is defined as "the ability of an enterprise to achieve its goals while maximizing the use of available resources". The innovative capacity of the enterprise can be estimated by means of enterprise life cycle, through analyzing and forecasting financial results and expected production.

Various characteristics of approaches to determining the innovative capacity of an enterprise were researched by such scientists as E. Voronkov, V. Katkalo, E. Brooking, Yu. Lomakin, J. Fox, S. Shishkin, A. Matvejkin, V. Trifilova, V. Arterchuk, and A. Babkin. Most of the authors define investment capacity as an opportunity to achieve goals using a processing approach during evaluation. For a successful outcome, the enterprise should be able to create competitive advantages, both at the expense of resource provision and by taking into account market conditions, requirements and opportunities. It should also replace its business model n with a more innovative one if necessary.

By highlighting the core management functions, such as planning, organization, coordination and control, growth capacity in the enterprise management system can be determined by the use of the functional approach in CRM (Customer Relationship Management) systems and resource approach in the modern ERP (Enterprise Resource

Planning) system, the conditions of the digital transformation, the use of BIG data and data science in real time. The quality of implementation of the strategic development plan of a defense-oriented industrial enterprise will be considered as the measure of the growth capacity.

The diagnosis of each system's characteristic covers the following stages.

• Establishing the planned indicators, including target regulations, which will be the basis for control. According to the target regulations, the system will be evaluated for a certain period. There are three forms of the targeted standards: numerical, scale, and descriptive.
• Measurement procedure: the emphasis should be placed on the measurement frequency, lag parameters.
• Control mechanism, which has a hierarchical structure with feedback channels.
• Evaluation and analysis of the efficiency (control).

In the example below, in order to measure the capacity of Kozitsky Plant, the standard financial statements of the enterprise are used. They do not take into consideration the innovative progress and interaction between the human and artificial intelligence. That is why it is important to develop such a scale for measuring the growth capacity, which would include innovation as a fundamental factor of enterprise development.

The indicators must be calculated for estimating the development capacity of a defense-oriented industrial enterprise by the method of distances. We use the main operational, investment and financial indicators presented in Table 1.

**Table 1.** The summary table of the standardized indicators of the capacity measurement of Kozitskiy plant

| Indicator | Resource potential | Indicator | Growth potential |
|-----------|-------------------|-----------|------------------|
| Operational indicators | | | |
| X1 | 1,003 | Y1 | −1,14 |
| X2 | 0,85 | Y2 | 0,26 |
| X3 | 1,087 | Y3 | 0,24 |
| Investment indicators | | | |
| X4 | 0,278 | Y4 | 0,411 |
| X5 | 0,26 | Y5 | −0,4 |
| X6 | 0,21 | Y6 | 1 |
| Financial indicators | | | |
| X7 | 0,41 | Y7 | 0,87 |
| X8 | 1,164 | Y8 | 0,967 |
| X9 | 8,63 | Y9 | 0,43 |

Further on, we will use the method of distances in order to find integral coefficients of the enterprise's growth capacity. We will calculate the integral index of resource capacity of the enterprise growth. This index integrates financial, operational and investment indicators of the company's activity. The resource capacity is 7.72. Now we can calculate the growth capacity of the enterprise by using the appropriate indicators given in Table 1. The capacity of growth is 2.62.

Due to the reason that both values of indicators are greater than one, the resource capacity and the growth capacity of the company Kozitskiy Plant are stable.

However, the calculated integrated indicators of different capacitys do not yet give the full confidence that the further work of the enterprise will be successful due to the obvious interconnection: the resource capacity affects the growth capacity and it does not ensure that these two indexes show improvement in the quality of life such as the average wage, the level of education, chosen in this study as the main indicators of growth. The strength of this relation should be determined by the semantic model of the subject area, which means a multi-agent system.

The visual semantic model of the subject area, in which the horizontally depicted agents influence its kinetic equilibrium, which reflects causal relations between elementary objects is presented in Fig. 1. In this figure, the multi-agents are enterprises, society, external, political environment, knowledge, cyberagents, resource complexes, state defense orders, the Ministry of Defense, federal budget (taxes), finance, labor, population, and the target, i.e., quality of life.

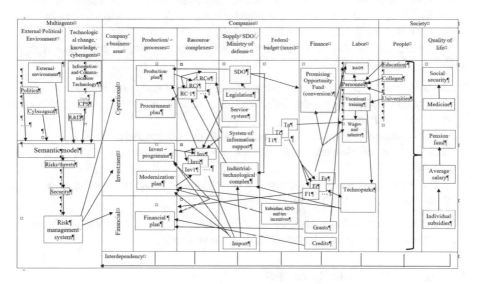

**Fig. 1.** Semantic model of the Domain

The criteria of efficiency in the given subject are the streamlining of innovations, efficient usage of available (limited) resources of all types, qualification of employees, average salary, working conditions and the environment. The purpose of measuring the capacity is to ensure the sustainability of the enterprise's system influenced by external factors and dynamic socio-technological changes.

The problem of managing socio-technological processes at the enterprise and its subunits is very topical for management. Although huge statistical data sets have been accumulated, it is not always possible to understand how this information can be applied in order to assess the consequences of a particular management decision.

Management decisions must be made in accordance with some concepts. Therefore, it is necessary to assess the consequences of taking a decision in advance; to evaluate the mutual influence of different subsystems on each other, taking into account the influence of the external environment. The controlling effect on a subsystem of an enterprise can lead to negative consequences for other subsystems. It is necessary to try to minimize such consequences.

To meet these challenges, it is necessary to build a model that would include the interaction of enterprise subsystems with each other and the external environment, and make it possible to predict socio-technological processes depending on the management decisions.

At the moment, there are many models of socio-technological processes. These can be models based on the application of graph theory, cognitive and neural network models, models constructed by applying logistical analysis, methods of multivariate analysis, system-cognitive analysis, etc. Models differ and, quite often, describe only one part of enterprise functioning, or a social and technological process. Thus, there is no unified model that with a predetermined degree of approximation would show the interaction of a multitude of subsystems selected by experts.

Summarizing the above, we can list the following challenges associated with the modeling of an enterprise.

- A large number of elements of the simulated system and links between them
- The lack of knowledge about individual elements of the system and the processes occurring in them
- A large number of input and output parameters of the system
- Identification of dependencies between system parameters is difficult if at all possible.
- It is not always possible to express the dependencies between elements of a system in the language of mathematical formulas.

Further, a toolkit that allows to create models of enterprises will be offered; its application can reduce the number of management errors.

## 3.1   Lingua-Combinatorial Models

Modeling of weakly formalized systems, including socio-technological ones, is one of the priority areas of science. Development of effective tools for modeling weakly formalized systems will simplify the task of managing such systems.

Any weakly formalized system [10] can be described in natural language words. The problem is to find a way to move from the description in the words of a natural language to a formalized language, the language of mathematical formulas. The method of linguistic-combinatorial modeling [10, 11] offers that opportunity.

The linguistic-combinatorial algebraic model is as follows.

If we have a phrase in a natural language:

$$A_1 + A_2 + A_3 \tag{1}$$

where $A_1$, $A_2$, $A_3$ are individual words, there is a need to introduce the concept of the meaning of words based on the context. Usually, in a natural language, words are designated and meanings are implied. Meanings can be introduced like

$$A_1 E_1 + A_2 E_2 + A_3 E_3 \tag{2}$$

where $E_1$, $E_2$, $E_3$ are the meanings of words based on the context (1).

Equation (2) can be solved by introducing a third group of variables, i.e.,—arbitrary coefficients

$$\begin{aligned} A_1 &= U_1 E_2 + U_2 E_3, \\ A_2 &= -U_1 E_1 + U_3 E_3, \\ A_3 &= -U_2 E_1 - U_3 E_2, \end{aligned} \tag{3}$$

where $U_1$, $U_2$, $U_3$ are arbitrary coefficients. All these operations are performed in the framework of an algebraic ring [10].

Similarly, we can solve Eq. (2) with respect to the meanings; all of this applies to any number of phrases and words in a natural language, a universal sign system, which is important.

Arbitrary coefficients $U_1$, $U_2$, $U_3$ can be used to solve various problems on a manifold (3).

For example, if we want to reach a maximum on the surface $F(A_1, A_2, A_3) = 0$ of variable $A_3$, then we can assign arbitrary coefficients

$$\begin{aligned} U_2 &= -bA_1, \\ U_3 &= -bA_2, \end{aligned} \tag{4}$$

and hence

$$\begin{cases} E_1 = U_1 A_2 - bA_1 A_3, \\ E_2 = -U_1 A_1 - bA_2 A_3, \\ E_3 = b(A_1^2 + A_2^2), \end{cases} \tag{5}$$

If $b > 0$, variable $A_3$ steadily tends to a maximum, and the coefficient remains to manipulate the trajectory $U_1$.

In the structure of the equivalent equations of systems with structured uncertainty, there are arbitrary coefficients that can be used to adapt the system to various changes in order to improve the accuracy and reliability of functioning of systems and their survivability in the flow of change.

The number of arbitrary coefficients is a measure of uncertainty and adaptability. Thus, linguistic-combinatorial modeling can be used to build models and extract meanings in any field of human knowledge [10].

A system with indefinite coefficients has two possibilities to adapt to changes in the environment: by manipulating arbitrary coefficients or by using additional conditions, i.e., a learning system.

Consider the state of the enterprise model built by identifying keywords in this area.

To do this, we will make a list of key words: "enterprise collective", "motivation", "territory", "production", "environment and safety", "finance", "external relations".

In accordance with the above methodology, the equation of the enterprise will be as follows:

$$A_1E_1 + A_2E_2 + \ldots + A_7E_7 = 0, \tag{6}$$

where $A_i$ are concepts from the relevant field, and $E_i$ are the meanings of corresponding words.

$A_1$ are characteristics of employees, which includes the characteristics of health, education, employment; $A_2$ – the characteristic of motivation, collective aspirations, people have the freedom of choice when making decisions, and this choice is important, which is assessed by a sociological analysis; $A_3$ – description of the territory, including land and underground structures; this unit may be a geographic information system; $A_4$ – characteristics of the production, including evaluation of various types of activities, e.g., scientific, industrial, transport, trade, etc.; $A_5$ – characteristic of the environment and safety; $A_6$ – characteristics of finance, financial flows and stocks; $A_7$ – characteristics of the external relations of the enterprise, including the assessment of incoming and outgoing flows of people, energy, materials, information, finance; $E_1, \ldots, E_7$ – changes in these characteristics, respectively.

Equivalent equations will be:

$$\begin{cases} E_1 = U_1A_2 + U_2A_3 + U_3A_4 + U_4A_5 + U_5A_6 + U_6A_7, \\ E_2 = -U_1A_1 + U_7A_3 + U_8A_4 + U_9A_5 + U_{10}A_6 + U_{11}A_7, \\ E_3 = -U_2A_1 - U_7A_2 + U_{12}A_4 + U_{13}A_5 + U_{14}A_6 + U_{15}A_7, \\ E_4 = -U_3A_1 - U_8A_2 - U_{12}A_3 + U_{16}A_5 + U_{17}A_6 + U_{18}A_7, \\ E_5 = -U_4A_1 - U_9A_2 - U_{13}A_3 - U_{16}A_4 + U_{19}A_6 + U_{20}A_7, \\ E_6 = -U_5A_1 - U_{10}A_2 - U_{14}A_3 - U_{17}A_4 - U_{19}A_5 + U_{21}A_7, \\ E_7 = -U_6A_1 - U_{11}A_2 - U_{15}A_3 - U_{18}A_4 - U_{20}A_5 - U_{21}A_6, \end{cases} \tag{7}$$

In the system of differential equations equivalent to the enterprise equation, there also are arbitrary coefficients $U_1$, $U_2$, ... $U_{21}$, which are designed to control and solve various problems on a manifold. The number of blocks in the linguistic-combinatorial model of an enterprise may be different. From the point of view of accuracy of modeling, the more blocks are used the better, but at the same time, the visibility of the model and its perception by decision-makers deteriorate. When modeling an enterprise, it is important to consider the entire hierarchy of the systems of which it consists.

A seven-block model can be used to model various enterprises where people work, and the block structure for each type of enterprise will be different.

Consider the management of the model. The algebraic model constructed by the method of linguistic-combinatorial modeling has its own ways of adapting to the environment and learning. The higher are the adaptation capabilities of the system, the more uncertain coefficients it includes [10]. With the help of undefined coefficients, one can change the trajectory of movement according to given varieties [10].

Any system goes through various stages: upsurge, stagnation, recession. "Finance" in our system reflects variable $A_6$, its change $-E_6$. In order to bring the economy of the system out of a state of stagnation, it is necessary for the "finance" characteristic to start growing, and its change must be strictly positive.

$$U_5 = a * A_1, \ U_{10} = b * A_2, \ U_{14} = c * A_3,$$

$$U_{17} = d * A_4, \ U_{19} = e * A_5, \ U_{21} = f * A_7,$$

where $a, b, c, d, e, f$ are the gain factors, and $a < 0, b < 0, c < 0, d < 0, e < 0, f < 0$. Then $A_6$ will grow steadily. The task is, in a specific situation depending on the real state of finance, i.e., variable $A_6$, to assign coefficients $a, b, c, d, e, f$, so that the optimal growth of finances is accompanied by an acceptable changes of the other variables. For this, you can also use the remaining undefined factors.

For example, it is necessary to ensure sustainable growth of two variables: $A_4$ – to the variable ("production") and $A_6$ to the variable ("finance").

For this, we assume $U_{17} = 0$. We assign coefficients $U_3, U_5, U_8, U_{10}, U_{12}, U_{14}, U_{16}, U_{18}, U_{21}$, so that variables $E_4$ and $E_6$ are strictly positive. Other variables are $U_5 = a \cdot A_1, \ U_{10} = b \cdot A_2, \ U_{14} = c \cdot A_3, \ U_{19} = d \cdot A_5, \ U_{21} = e \cdot A_7, \ U_3 = f \cdot A_1, \ U_8 = g \cdot A_2, U_{12} = h \cdot A_3, U_{16} = i \cdot A_5, U_{18} = j \cdot A_7$.

When $a < 0, b < 0, c < 0, d < 0, e > 0, f < 0, g < 0, h < 0, i > 0, j > 0$.

## 3.2 Analysis of the General Structure of the Equations Obtained by the Method of Linguistic-Combinatorial Modeling

Let's consider the general structure of the systems of equations obtained by the method of linguistic-combinatorial modeling. Using the method of linguistic-combinatorial modeling, one can obtain the equations in the form

$$A_1 \cdot E_1 + A_2 \cdot E_2 + A_3 \cdot E_3 = 0, \tag{8}$$

where $A_i$ is the characteristic of a concept, and $E_i$ is the change in the corresponding characteristic. In addition, there may be more than one equation of this sort; then we get a system of equations whose behavior must be investigated in terms of interference in the computing system.

We define an algorithm for constructing an extended system of equivalent equations from the data.

The construction of any computational process is associated with the selection of output variables among the studied basic relations and with the organization of such a process that these variables would be calculated as a function of the input parameters. The assignment of output variables and input parameters in the basic relations is carried out in the process of problem statement. In this case, the systems considered are described by finite equations or the Pfaff equation, and the number of variables in them is equal to or greater than the number of coupling equations. The Pfaff equations are quasilinear equations for the differentials of variables. Finite equations can be reduced to a quasilinear form either by differentiation or by selecting output variables among the groups of terms in the equations. Thus, to determine the structure of equations with the systems of redundancy, it is necessary to study quasilinear systems with uncertainty, which is a nontrivial task.

Indefinite coefficients in the structure of these equations can be used to satisfy initial and boundary conditions and to determine the values of the unknown function at the points of interest.

It should be noted that in the construction of a general solution of the equations by the method of redundant variables, all linear combinations of particular solutions of these equations were used. The structure of the obtained general solution is equivalent with respect to the minors of the coefficients of the original equations, which allows, in particular, to specify any motions on the original manifolds.

### 3.3  Model Adaptation Capabilities

We considered the systems in which undetermined coefficients play an important role, since their use allows the system to adapt to the manifestations of negative environmental factors, reduce the influence of interference signals, and improve the accuracy of the system. The number of arbitrary coefficients is a measure of the uncertainty and adaptability of the system [10]. As shown in [10], the adaptive capabilities of the system are the higher the more uncertain factors which can serve both to ensure the growth or reduction of a variable, and to change the trajectory of movement for given varieties are in it.

### 3.4  Application of the Cluster Analysis

Huge volumes of accumulated statistical data allow to create an enterprise model, so that later, using the model, to be able to predict socio-technological processes depending on the management decision made, observe the interaction between the elements of the system and the ongoing processes with the help of the model.

The decision on how to select subsystems and elements for modeling should be made by experts. But since the volumes of existing statistics are so large that a person often cannot cover all of them; it is necessary to use tools of such a field of science as data mining to identify the main elements and the links between them. In particular, it is intended to use clear and fuzzy clustering tools.

## 4   Conclusions

The innovative multi-agent system of estimating the growth capacity of an enterprise is based on a semantic model which takes into account both financial and production outcomes of the company. The strength of the bonds between multi-agents should be determined by the semantic model of the subject area, which means by a multi-agent system. Without an innovative multi-agent system, it is nearly impossible to diagnose, detect and mitigate the effects of the changing micro- and macro- environment. Using the multiagent system, it is possible to determine the system's empirical state and measure the growth capacity. The multi-agent system includes the innovative capacity and human capital of the company, which is of great importance in today's technological environment. Growth capacity is determined in order to facilitate the process of resource and management coordination and provide sustainable innovative development of the enterprise and its socio-technological environment.

**Acknowledgements.** Work is performed with the support of the Russian Foundation for Basic Research, grant 19-010-00257 A. Methodology for the analysis of the industrial enterprises and intangible industries in the conditions of information society and digitalization.

## References

1. Volkova, V.N., Denisov, A.A.: Theory of Systems and System Analysis Textbook, 679 p. Jurajt, Moscow (2010)
2. Kukor, B.L., Pytkin, A.N., Klimenkov, G.V.: Fundamentals of strategic management in the regional economy. (Construction of situational management systems on the basis of logic-linguistic modelling) Kukor, B.L. (ed.), 337 p. Perm, Publishing house "Nika" Ltd. (2009)
3. Kukor, B.L., Kuzmin, N.A.: Communicative mechanism of the system of strategic adaptive management of the regional economy. System analysis in design and management, pp. 159–168. Scientific, SA (2016)
4. Ripe, D.A.: Situational management: Theory and practice, 288 p. Science, Moscow (1986)
5. Ujomov, A.I.: Systemic approach and general theory of systems, Moscow, 272 p. (1978)
6. Yakovlev, E.A., Kozlovsky, E.A., Boyko, Yu.: Estimation of innovative potential of the enterprise as growth potential on the basis of cost approach. Issues Innov. Econ. **8**(2) (2018)
7. Federal law of July 21, 2005 No 94-FZ "On placing orders for the supply of goods, performance of works, provision of services for state and municipal needs"
8. Karlik, A.E., Maksimtsev, I.A., Platonov, V.V.: Intellectual management systems for the RF State Program Science and Technology Development for years 2013–2020. In: Proceedings of the 19th International Conference on Soft Computing and Measurements, SCM, pp. 500–502. IEEE (2016)
9. Platonov, V.V., Dukeov, I.I., Ulitin, D.B., Maksimov, D.N.: Navigator of innovative development of oil and gas industry. Oil Ind. (Neftyanoe Khozyaystvo) **10**, 59–63 (2017)
10. Ignatiev, M.B., Katermina, T.S.: System analysis of cyberphysical structures/System analysis in design and management. In: Proceedings of XXI International Scientific and Practical Conference, pp. 15–24 (2017)
11. Ignatiev, M.B., Karlik, A.E., Iakovleva, E.A., Platonov, V.V.: Linguo-combinatorial model for diagnosing the state of human resources in the digital economy. In: Proceedings of 2018 17th Russian Scientific and Practical Conference on Planning and Teaching Engineering Staff for the Industrial and Economic Complex of the Region, PTES, pp. 201–205 (2018)

# Parametric Control of Oscillations

Leonid Chechurin[1,2]([⊠]), Sergej Chechurin[2], and Anton Mandrik[2]

[1] LUT University, Lapeenranta, Finland
Leonid.Chechurin@lut.fi
[2] Peter the Great St. Petersburg Polytechnic University, Saint Petersburg, Russia

**Abstract.** Any oscillating system is described by certain parameters, and very often these parameters can be dynamically changed in a certain way to reach control goals. We overview a number of designs in which periodic variation of parameters in linear time-variant and nonlinear systems is the main control paradigm. We use frequency analysis and one-frequency approximation as the mathematical instrument. The approach that is also known as stationarization uses equivalent transfer functions for each time-variant and nonlinear element and reduces the stability analysis to classical Nyquist plot. The study presents in a unified framework several problems that have been solved in the last decades and new ideas, such as parametric synchronizing of oscillation. As the approach uses si mple mathematics, it can be used by field engineers for inventive oscillation control design for cranes, ships, rotors and many other vibrating systems.

**Keywords:** Parametric control · Oscillations

## 1 Introduction

It can be easily proven that when we talk about automation, automatic control or control in general, we mean mostly the feedback control. Indeed, this paradigm has been most probably introduced first by the patent of Black on the feedback amplifier, followed by the even more famous work of his colleague Nyquist on the stability of feedback system. There are many factors explaining the success of this concept in practice, but the most visible is its problem-solving potential: feedback control is the conceptual idea on how to improve the performance of a machine or device even when to change its design or parameters is very hard or impossible.

The latter restriction can be effective in many practical situations. For instance, in mass production or when the plant is too complex, it could be too expensive to change the design of an object, but adding a feedback controller can help to improve its performance. Or else, in unique autonomous systems the feedback loops can help to reach specific goals or to cope with internal or external uncertainties.

At the same time, the feedback control became very popular in academia. Mathematicians realized that a large piece of the theory of differential equations can be packaged as control theory. The latter is what control theory became associated with: a set of certain mathematical models of plants to be controlled (mostly differential equations) and methods to design feedback controller (mostly described by another set of differential equations) that would guarantee certain properties of closed loop systems.

D. G. Arseniev et al. (Eds.): CPS&C 2019, LNNS 95, pp. 98–107, 2020.
https://doi.org/10.1007/978-3-030-34983-7_10

In all the cases *the control means the controller in the feedback loop* and this layout became a symbol of control in general. But this concept is not the only one to be considered by a control engineer to reach design specification. The paper presents initial speculations on the alternative control paradigm called *parametric control* that can enrich the toolkit of solutions for designer.

## 2   Definitions

We are interested exclusively in periodic motions and periodic parameter variations. Thus we focus on various forms of oscillatory stability and instability as the most popular phenomena and subject of analysis in dynamic systems. Most research focused on the mathematical treatment of the state space time-variant equations by Lyapunov function and optimal control instruments. As it comes to periodic parameter variations and oscillation control, periodic parameters' variation is used for modulation of signals and synchronization of oscillations [1, 3, 4, 6]. Here we need several definitions to conduct the analysis within the control framework.

*(Program) parametric control* is the control by assigning the variation of a parameter a certain program, typically periodic, of certain period/frequency and magnitude. The profile of parameter change $x(t)$ is considered as the exogenous signal. The block representation of the control scheme for a plant with transfer function $W(p)$ is given in Fig. 1.

The idea employs a known from the 1930s physical phenomenon called parametric resonance. The control perspective of the scheme above is first discussed in [2]. It is shown which range of program signal $x_0(t)$ in terms of frequencies and magnitudes may lead to instability of the system. It is also shown how an external additive disturbance to the system (forced oscillation case) influences the excitation conditions. It is a fundamental fact that if we set the parameter variations to the excitation conditions, the system responses by instability in the form of parametric resonance. In other words, the oscillations in the system automatically synchronize with the parameter variation in such a manner that the output is unstable. It is also interesting that although the representation in Fig. 1 looks like a traditional feedback scheme, we are not able to think of a transfer of a function from $x_0$ to $x_n$, since the given $x_0$ is a harmonic signal, while $x_n$ either vanishes to zero (no parametric resonance) or grows ad infimum (parametric resonance).

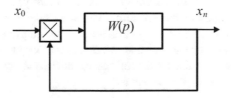

**Fig. 1.** Program parametric control scheme

*Feedback/automatic parametric control* is the control where the variation of a parameter depends on certain system's output coordinates. Typically, the phase of parameter variation is set at certain value with respect to the phase of system's output. Or else, the magnitude of the parameter variation is proportional to the magnitude of the system output. The profile of parameter change is defined by the output variables in a feedback matter, so we can speak about *parametric regulator x(y)* in this case. Thus, the block representation of the control scheme for a plant with transfer function $W$ ($p$) becomes (Fig. 2).

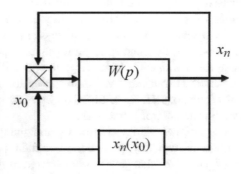

**Fig. 2.** Parametric regulator scheme

In contrast with Fig. 1, this scheme enables to play with phase shift of parametric signal injection that, in addition to the magnitude and frequency of the parameter variation, can influence the system behavior. The patent (Mandrik et al. [5]) presented initial ideas on parametric feedback $x_n(x_0)$ realization based on time delay. Let us consider the following general parametric regulator.

$$x_n(t) = x_0(t) \, k \, \text{sign}\left\{\frac{d}{dt}[x_0(t-\tau)]^2\right\}. \tag{1}$$

The physical meaning of this expression is the following. The square of the output signal doubles the frequency of parameter oscillations $\Omega = 2\omega$. The differentiation eliminates the constant component. The sign-function forms the square feedback signal form and thus removes the influence of the magnitude and frequency of coordinate oscillations on the magnitude of parameter variation. Finally, the variable gain value k controls the magnitude parameter variations and the delay $\tau$ delivers the phase shift between the periodic variation of the parameter and the output coordinate. This phase shift is the critical parameter we will use to control the output oscillations.

We are going to stay in the framework of mono-frequency approximation in our analysis, so let us assume $x_0(t) = A \, \sin \omega t$.

Then (2) takes the form

$$x_n(t) = (A\ \sin \omega t)\ k\ \mathrm{sign}[A^2\omega\ \sin 2\omega(t-\tau)] = (A\ \sin \omega t)\ k\ \mathrm{sign}\ \sin 2\omega(t-\tau) \cong$$
$$\cong \tfrac{4k}{3}(A\ \sin \omega t)\ \sin 2\omega(t-\tau) \cong \tfrac{2kA}{3}\ \sin(\omega t - 2\psi),$$
$$\psi = \omega\tau.$$

Here, the square periodic variations are substituted by their first harmonic components, whereas the third and higher harmonics are neglected. As one can see, the parametric controller is linear in the given approximation, as well as the whole system becomes linear parametrically time-variant. According to Chechurin and Chechurin, (Chechurin, Chechurin 2018), to avoid the first parametric resonance in such a system, the following condition should be held.

$$|W^{-1}(j\omega)| < a/2, \tag{2}$$

where $W(j\omega)$ is the frequency response of the stationary part of the system on the complex plane (magnitude-phase-frequency plot), $a$ is the magnitude of mono-frequency parameter's harmonic variations of the frequency $\Omega$ that is doubled frequency of the object output coordinate oscillations $\omega$.

Therefore, no parametric resonance exists if the controller transfer function gain $2k/3$ is less than the gain stability margin of the object, and the phase shift $\omega\tau$ is less than the phase stability margin.

## 3 Example: Parametric Pendulum Control

If we consider a simple pendulum, the realization of the parametric control would mean the controllable variation of the pendulum length $l=l(t)$ as the function of the sway angle $\alpha$ (Fig. 3). But we can vary the length of the pendulum to damp the oscillation faster than if they were free. In other words, we plan to find "parametric antiresonance" conditions.

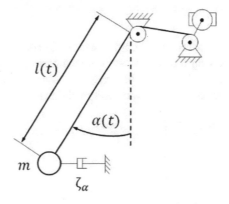

**Fig. 3.** The parametric pendulum control setup

The linearized model of pendulum is

$$ml^2\ddot{\alpha} + \varsigma_\alpha\dot{\alpha} + mgl\alpha = M, \qquad (3)$$

where $M$ is the input torque. We take mass $m = 1$ kg, length $l = 1$ m, damping factor $\varsigma_\alpha = 0.05$N rad/s, gravity constant $g \cong 10$ m/s$^2$ and arrive at the pendulum transfer function

$$W(s) = \frac{\alpha(s)}{M(s)} = \frac{1}{s^2 + 0.05s + 10}.$$

We will compare the oscillation control results by the time constant evaluation. The time constant $T$ for the damped motion is known to be defined from the graphs. For this, the tangent line to the oscillation envelop at $t = 0$ is to be extended until it meets the time axis. The intersection point provides the time constant $T$. In the case of free pendulum oscillations $T$ is about 40 s (Fig. 4). Let us compare this value in coordinate and parametric control cases with initial conditions given by the vector $[\dot{\alpha}(0) \quad \alpha(0)] = [0, \text{rad/s} \quad 0.1, \text{rad}]$.

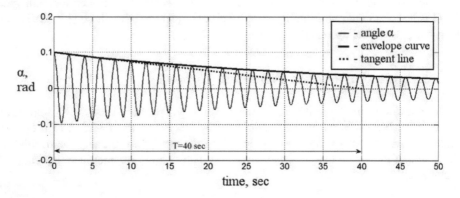

**Fig. 4.** The uncontrolled pendulum oscillations

Now let us turn to the parametric control, where the control algorithm follows (1). The controller gives the length of the pendulum small periodic variations $\Delta l(t)$. While the physical parameters are the same (the mass is chosen to be equal the mass of the rotor from the coordinate control scheme), the mathematical model changes

$$m[l + \Delta l(t)]^2\ddot{\alpha} + \varsigma_\alpha\dot{\alpha} + mg[l + \Delta l(t)]\alpha = 0 \qquad (4)$$

assuming $\Delta l(t) \ll l$ .

Let us divide (4) by $[l + \Delta l(t)]^2$ and expand the resulting functions $l(t)$ in Taylor series near $l_0$. Retaining the linear terms only, we obtain

$$s^2 + \left[ \frac{\varsigma_\alpha}{ml_0^2} + \frac{d}{dl}\left(\frac{\varsigma_\alpha}{ml^2}\right)\Big|_{l=l_0} \Delta l \right] s + \left[ \frac{g}{l_0} + \frac{d}{dl}\left(\frac{g}{l}\right)\Big|_{l=l_0} \Delta l \right] = 0$$

or

$$s^2 + \left[ \frac{\varsigma_\alpha}{ml_0^2} - 2\frac{\varsigma_\alpha}{ml_0^3} \Delta l \right] s + \left[ \frac{g}{l_0} - \frac{g}{l_0^2} \Delta l \right] = 0.$$

Since $\Delta l(t)$ is supposed to be periodic, we use its stationary model $W(j\varphi)$. If $\Delta l(t)$ is sinusoidal, then

$$s^2 + \frac{\varsigma_\alpha}{ml_0^2} s + \frac{g}{l_0} = \left[ 2\frac{\varsigma_\alpha}{ml_0^3} s + \frac{g}{l_0^2} \right] W(j\varphi),$$

where

$$W(j\phi) = \frac{\Delta l}{2} e^{-j\varphi}.$$

Now the bounds for the first parametric resonance excitation can be found from the equation

$$W(s) = \frac{2\frac{\varsigma_\alpha}{ml_0^3} s + \frac{g}{l_0^2}}{s^2 + \frac{\varsigma_\alpha}{ml_0^2} s + \frac{g}{l_0}} = W^{-1}(j\varphi).$$

Thus, the linear approximation of (4) yields the following transfer function from $\Delta l$ to the parametrically controlled pendulum for the given parameters.

$$W(s) = \frac{\alpha(s)}{\Delta l(s)} = \frac{0.1s + 10}{s^2 + 0.05s + 10}. \tag{5}$$

The numerical simulation scheme is given in Fig. 5. The controlled pendulum oscillations vanish faster with $T$ about 8 s (Fig. 6).

The paradigm of automatic parametric control provides one more interesting perspective of analysis as well as an idea for oscillation control. More precisely, it is a tool to support inventive design of oscillation systems. Let us think of the feedback parametric controller as of a physical unit that provides the required delay, frequency doubling, and/or gain modulation. Once such a physical embodiment is found, the "closed loop system" can have the required damping. And if the embodiment is passive (does not use any external sources of energy), the design, most probably, is quite simple.

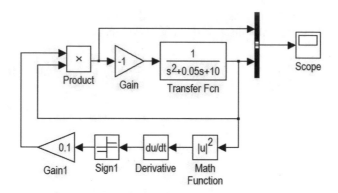

**Fig. 5.** The Simulink model of the parametric stable pendulum control

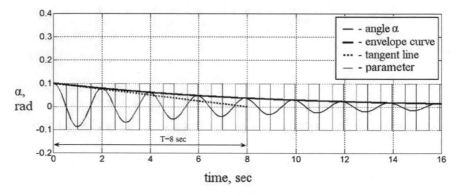

**Fig. 6.** Parametrically controlled oscillations, the time constant $T \cong 8$ s

Let us consider a spring pendulum in gravitation field that also has sway oscillations (Fig. 7). This system can be seen from an interesting perspective now. Indeed, there is "natural" (passive) parametric stabilizing control. The centrifugal force reaches its maximum as the pendulum sway crosses the vertical axis and extends the length of the pendulum, since there is a spring. Thus, the length of the pendulum depends on the angular position like in parametric control. The equation of the motion and its frequency analysis is given in [6]. The signal flowchart in Simulink is given in Fig. 8 with the following parameters of the pendulum: mass $m = 1$ kg, length $l = 1$ m, damping factor $\varsigma_\alpha = 0.1$N sec/rad, stiffness of the spring $c = 40$ N/m, damping factor of radial movement $\varsigma_x = 1$ N s/m. Its output is presented in Fig. 9.

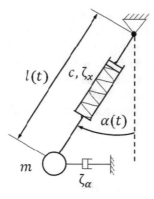

**Fig. 7.** The Simulink model of the spring pendulum

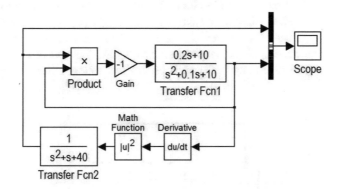

**Fig. 8.** The simulink model of the spring pendulum

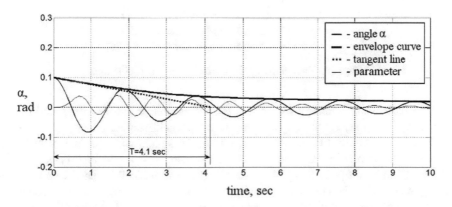

**Fig. 9.** The signal output of uncontrolled spring pendulum, the time constant $T \cong 4.1$ s

Thus, we conclude that parametric control and parametric regulators can be used to implement a possible simple alternative to classical feedback oscillation control scheme. It is important to point out that (1) is not the only possible parametric feedback form. And passive parametric feedback in Fig. 7 is also just an illustration. In general, we need to think of two or more oscillating systems in which oscillation of a coordinate in one system is the periodic parameter variation for another. Deeper analysis of stability conditions in this case requires a separate study and relates to the analysis of nonlinear combinatory oscillations that may have very interesting forms of behavior, for example, the flutter.

## 4  Synchronization of Self-oscillations with Parameter Variations

Next design can be used in a wide class of problems where we need to reach synchronization of a self-oscillating system with a certain periodic system. A typical framework to approach this problem is the one when the external signal enters the oscillator loop additively. We suggest alternative layout depicted in Fig. 3, where the synchronizing signal modulates the oscillations in the system consisting of the linear part $W(p)$ and a nonlinear unit $\varphi(y)$.

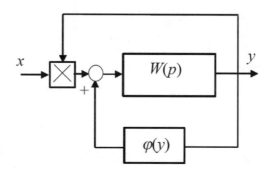

**Fig. 10.** Parametric synchronization of self-oscillations

Chechurin and Chechurin [2] present the evaluation of the frequency and magnitude of $x(t)$ that may result in synchronization with self-oscillations $y(t)$.

Finally, the ability to use parametric control to synchronize (as in Fig. 10) or to generate oscillations in the form of parametric resonance (as in Fig. 1) enables the following important possible application. The problem basically related to power grid management, when an autonomous generator of alternative current is to be connected to a loaded network. Obviously, it means that the frequency and phase of the generator are to be synchronized with those of the grid. The signal in the grid is basically used as the reference to vary the parameters of the generator until they are identical to it. The new idea is to multiplicatively inject the reference signal of double frequency into the generator loop, formed as a parametric resonance circuit. The variation of the parameter will lead to the parametric resonance and growing oscillation of the generator to the required frequency and phase. When they reach the required magnitude, the generator can be connected to the grid.

## 5  Conclusions

We suggest a new framework for control design called parametric control and demonstrate a number of popular engineering problems that can be solved in this framework. Depending on the limitations of a specific design case, the parametric control can be considered as an alternative to classical (additive) feedback regulators or used along with them. Moreover, it can be helpful in inventive design of simple passive damping regulators. It can hopefully enrich the means of control engineers in control design. The approach uses mono-frequency approximation as the main tool to prove the declared benefits, and it seems to be the only limitation of the results. Future work will be done to develop practical recommendations for parametric regulators design and new application problems.

## References

1. Amer, Y.A., Ahmed, E.E.: Vibration control of a nonlinear dynamical system with time-varying stiffness subjected to multi external forces. Int. J. Eng. Appl. Sci. (IJEAS) **5**(4), 50–64 (2014)
2. Chechurin, L., Chechurin, S.: Physical Fundamentals of Oscillations, p. 264. Springer, Cham (2018). https://doi.org/10.1007/978-3-319-75154-2
3. Eissa, M., Kamel, M., El-Sayed, A.T.: Vibration reduction of a nonlinear spring pendulum under multi external and parametric excitations via a longitudinal absorber. Mechanica **46**, 325–340 (2011)
4. Insperger, T., Stépán, G.: Optimization of digital control with delay by periodic variation of the gain parameters. In: Proceedings of IFAC Workshop on Adaptation and Learning in Control and Signal Processing, and IFAC Workshop on Periodic Control Systems, Yokohama, Japan, pp. 145–150 (2004)
5. Mandrik, A.V., Chechurin, L.S., Chechurin, S.L.: Method for Stabilizing of Output Signal of Oscillating System. Patent RU2393520 (2010)
6. Mandrik, A.V., Chechurin, L.S., Chechurin, S.L.: Frequency analysis of parametrically controlled oscillating systems. IFAC-Papers OnLine. In: Proceedings of the 1st IFAC Conference on Modelling, Identification and Control of Nonlinear Systems MICNON. vol. 48, no. 11, pp. 651–655 (2015)

# Pilot-Induced Oscillations
# and Their Prevention

Boris Andrievsky[1,3(✉)], Dmitry G. Arseniev[2,3],
Nikolay V. Kuznetsov[1,2,3,4], and Iuliia S. Zaitceva[5]

[1] Institute for Problems of Mechanical Engineering, Russian Academy of
Sciences, St. Petersburg, Russia
boris.andrievsky@gmail.com, nkuznetsov239@gmail.com
[2] Peter the Great St. Petersburg Polytechnic University, Saint Petersburg, Russia
vicerector.int@spbstu.ru
[3] St. Petersburg State University, Saint Petersburg, Russia
[4] University of Jyväskylä, Jyväskylä, Finland
[5] ITMO University, 197101 Saint Petersburg Kronverkskiy Pr., 49, Russia
julia.zaytsev@gmail.com

**Abstract.** The paper is devoted to such an important problem of piloted aircraft control as an unfavorable aircraft-pilot coupling. Mainly, the so-called Pilot-Induced Oscillations, i.e., long uncontrolled irregular oscillations which may happen due to the controlling surface magnitude and rate limitations when pilot attempts to control the aircraft with the high rate and accuracy. The simulation and in-flight test results of several PIO events are described, and the literature overview of the PIO prevention method is given.

**Keywords:** Pilot-induced oscillations · Piloted aircraft · Oscillation prevention

## 1 Introduction

Pilot-Induced Oscillations (PIO) are long uncontrolled irregular oscillations that occur as a result of the efforts of the pilot to control the aircraft with high rate and accuracy. While the PIO can be easily detected by post-flight data analysis, pilots often do not know that PIO condition has arisen: from their point of view, the aircraft seems to be "having a breakdown", or is faulty [1]. When approaching the stability boundary, the behavior of the linear system deteriorates in a manner predictable for the pilot. When the system exhibits significant non-linearities, instead of a gradual deterioration of the aircraft behavior, a sudden, abrupt change may occur, resulting in the so-called "*flying qualities cliff*" [2]. Loss of control of the aircraft when approaching a disruption may occur after minor warning signs about its behavior. The investigators believe that the main non-linear factor leading to the PIO is the speed limitations on deviation of the aircraft control elements (the aerodynamic control surfaces of the aircraft). This restriction may cause a delay in the response of the aircraft to the pilot's commands. If the plane does not react as expected to signals from the cockpit, the pilot can perform intensive movements of the control handles. Ultimately, an excessive reaction of the aircraft appears, which causes the pilot to change the direction of the control signal,

© Springer Nature Switzerland AG 2020
D. G. Arseniev et al. (Eds.): CPS&C 2019, LNNS 95, pp. 108–123, 2020.
https://doi.org/10.1007/978-3-030-34983-7_11

and, due to the delay, also with excessive intensity. This is how the famous test pilot, Hero of the Soviet Union S.A. Mikoyan describes this phenomenon in his memoirs [3]: "Giving a handle against any unwanted movement of the aircraft along the pitch, for example, raising the nose, the pilot, without immediately feeling the effect due to the delayed reaction of the aircraft, still added steering deflection, which was already superfluous. Then, feeling the excessive movement of the aircraft, the pilot rejected the controls in the opposite direction. Again, without immediately noticing the effect, he added more steering than needed. Thus, acting 'with overshoot' and at random due to the delay, the pilot himself rocked the plane, leaving at the same time a large overload". Further Mikoyan points out: "even knowing the nature of these oscillations and waiting for them to occur, the pilot at first reflexively makes two or three movements to parry them, exacerbating the buildup, and only then stops the control stick. On fighter planes designed for high overload, this usually does not lead to trouble. However, on heavy planes whose overload limit is much smaller, enhancing it with increasing vibration amplitude can lead to structural failure".

The rest of the paper is organized as follows. An essence of the PIO phenomenon is described in Sect. 2. In Sect. 3, a brief literature overview of PIO prevention methods is presented. Concluding remarks and the future work intentions are given in Sect. 4.

## 2   Essence of the PIO Phenomenon

Unfavorable aircraft-pilot coupling (APC) events include a broad set of undesirable and sometimes hazardous phenomena in the relationship between the pilot and the aircraft; these are rare, unexpected and unintended deviations in the spatial position of the aircraft and the flight path caused by anomalous interaction between the pilot and the aircraft [4]. The temporal nature of these *pilot-vehicle system* (PVS) excursions may be oscillatory or divergent (non-oscillatory, a-periodic). The pilot's interactions with the aircraft can form either a closed or open-loop system, depending on how closely the pilot's responses and signals from the aircraft are related. When the dynamics of the aircraft (including the flight control system (FCS) and the dynamics of the pilot) connect, the result is called the *Pilot-Induced* (or *Pilot-Involved*) Oscillations (PIO).

Although it is often difficult to pinpoint the cause of specific PIO events, most of the major APC events are the result of aircraft design flaws (especially with respect to the FCS), which lead to an unfavorable connection of the pilot with the aircraft. In certain circumstances, this unfavorable connection causes unintended fluctuations or discrepancies when the pilot tries to precisely maneuver the aircraft. If the PVS instability takes the form of oscillations, the unfavorable APS event is called PIO. PIO differ from aircraft vibrations caused by deliberate periodic control movements imposed by the pilot, such as rapid oscillatory movements of the control knob in an open-loop circuit. Forced oscillation in an open circuit does not correspond the PIO. If the unstable movements of the closed-loop aircraft system have a divergent rather than an oscillatory nature, they are called APC events or non-oscillating PIO events.

PIO events can occur if the pilot is operating in a behavioral mode that is not suitable for the task being performed, and such events are properly attributed as the pilot's error. However, the Aviation Safety Committee believes [4] that the most serious events of the PIO, due to a pilot's error, are the result of an unfavorable APC, which misleads the pilot who takes actions that aggravate the situation. After that, one can carefully analyze the event and determine the sequence of actions that the pilot could have taken to overcome the shortcomings of the aircraft design. However, as a rule, the pilot cannot determine and perform the necessary actions in real time.

The phenomena of PIO have a wide range. On the one hand, these are short-term oscillations, easily corrected bursts of low amplitude, which pilots often encounter getting accustomed to new configurations during training. This type of oscillations can occur on any aircraft at one time or another. On the other end of the scale, there are quite rare events of high-amplitude oscillations, which jeopardize the safety of the aircraft, crew and passengers. After the notable events involving the PIO with the participation of military aircrafts, as well as incidents with civilian aircrafts, a number of politicians, test pilots, technical managers and engineers drew attention to this problem. It was found that almost all new aircrafts with the technology of *fly-by-wire* (FBW) FCSs, demonstrated PIO events for some time during the development, which suggests that some side effects have not been fully studied or foreseen.

PIO events usually occur when a pilot is involved in a very complex control task in a closed control loop. For example, many of the reported PIO events occurred during airborne refueling and landing operations, especially if the pilot is concerned about the low fuel level, adverse weather conditions, emergency situation or other circumstances. Under these conditions, pilot's participation in a closed control loop is intense, and a quick response and accurate execution of aircraft commands are necessary. Despite this, these operations usually occur without PIO problems. PIO events do not occur unless there is a transient triggering event that interrupts the already demanded operations of the aircraft or requires an even higher level of accuracy. Typical triggers include shifts in the dynamics of an efficient aircraft caused by an increase in the amplitude of pilot commands, changes in prop, minor mechanical problems, or severe atmospheric disturbances. Other triggers may occur due to a mismatch between pilot's expectations and reality.

PIOs were part of the history of aviation from the beginning of manned flights, and serious PIO persist, despite significant efforts to eliminate them. When a certain type of PIO occurs, usually unexpectedly, it causes corrective actions. This experience is useful in the sense that when designing a new aircraft, conditions that are believed to underlie this type of PIO are usually excluded. Since other PIO occur under different circumstances, the cycle repeats. Over the time, understanding improves, and some causes are eliminated, but the occurrence of oscillations in a closed loop remains, and only details change as the aircraft and the FCS technology changes.

From the pilot's point of view, there are three types of oscillations, ranging from mild during training and ending with serious and potentially dangerous ones. Slow wiggles are overcome by exiting the pilot from the closed control loop. Conversely, in many serious PIO, the pilot becomes tied up to the behavior that maintains hesitation, even though the pilot often feels completely disconnected from the system. If the deficiencies in the effective dynamics of the aircraft are substantially linear in nature, such as excessive time delay in response to the pilot's control signal, then PIO Category I may occur. If the effective dynamics of the aircraft changes depending on the amplitude of the pilot's command or the change of the FCS modes, thereby creating a non-linear sudden change (failure) in the effective dynamics of the aircraft, the resulting PIO is assigned either the Category II (when the dominant nonlinearities are related to the speed limit or the position of the control surfaces) or Category III (when non-linear changes are more complex) [4]. The PIO Categories II and III are especially insidious, because the effective dynamics of the aircraft and the flight qualities associated with it can stay decent right up to the start of the PIO. Identifying the potential for these PIO, which almost always occur in unusual conditions when the air vessel system is operating near the limits of sustainability, is a serious problem for pilots and test engineers. Intensive experiments and research are needed to identify and not lose sight of the tendency of the Categories II and II. The non-oscillating APC events are not as well defined and understood as PIO. Even if the pilot is extremely active in control, the aircraft does not necessarily respond in an oscillatory manner. Instead, the accumulation of delays in the response of aircraft controls to pilot's commands may ultimately lead to deviations from the intended movement of the aircraft. As in the case of serious PIOs, pilots in these cases report a feeling of detachment from the aircraft's behavior in terms of both awareness of what is happening and the point of view of the temporal connections between the pilot's control and aircraft response.

The development of new technologies in aviation, on the one hand, simplifies and automates the work of pilots, and on the other hand, it introduces new errors and unforeseen failures in the operation of flight control systems. In aviation, the problems of anomalous interaction were often associated with the introduction of new technologies, functionalities, or complications. This partly explains why the problems of anomalous interaction are more common in military aviation, which traditionally introduces advanced technologies, and less common in civil aviation, which, as a rule, accepts new technologies only after they have been approved in military aviation. In addition, the nature of military operations often includes maneuvers that require higher pilot response rates than are commonly used on civil aircraft.

In [5], a general approach is proposed for the analysis of nonlinear PIO phenomena. This approach involves the identification of such a significant nonlinear effect as the *Hopf bifurcation* [6, 7], which leads to significant changes in the structural stability of the aircraft-pilot system. As an example, the occurrence of PIO caused by the speed and limitation of the executive drive of an experimental rocket plane X-15 is considered. The research results showed the presence of a significant jump in the amplitude of the limiting cycle, indicating a significant change in the structural stability of the system when approaching the situation of PIO. The global behavior of the system is investigated in the work on the basis of the so-called "*methodology of the theory of bifurcations and chaos*". The study includes finding equilibrium states and stable limit cycles, global geometry of equilibrium surfaces and bifurcations, catastrophes associated with nonlinear geometry, structural stability, and feedback control. Since the aircraft system is subject to external influences, for applying this methodology, the system model is complemented by a nonlinear oscillator with an asymptotically stable periodic solution representing (specifying) the command effect, which makes it possible to consider an autonomous system in further analysis. Since there is no feedback from the airplane system to the command signal adjuster, the global properties of the original and modified (expanded) systems are the same. The following system is used as one of the oscillators in [5] (*Bautin equation*):

$$\dot{x}_1 = x_1 + \beta x_2 - x_1\left(x_1^2 + x_2^2\right); \tag{1}$$

$$\dot{x}_2 = -\beta x_1 + x_2 - x_2\left(x_1^2 + x_2^2\right),$$

where $\beta$ is the frequency of the command signal, $x_1$, $x_2$ are the oscillator state variables.

Finding limit cycles and analysis of bifurcations are performed as follows. The extended system of an airplane-pilot (with a generator (1)) is represented by the equations

$$\begin{bmatrix} \dot{x}_1 \\ \dot{x}_2 \end{bmatrix} = \begin{bmatrix} f_1(x_1; \beta, A) \\ f_1(x_1, x_2; K_P, \tau) \end{bmatrix} \tag{2}$$

where $x_1$, $x_2$ are the state vectors of the nonlinear oscillator (1) and the airplane systems $\beta$, $A$ are the frequency and amplitude of the command signal (it is assumed that it can be represented by a finite set sine wave); $K_P$ – "transfer gain" of the pilot; $\tau$ – effective lag time. In this model, $\beta$, $A$, $K_P$, $\tau$ are the parameters of the bifurcation analysis problem, and the behavior of the pilot is described by a pure lag link with the transfer function $K_P e^{-s\tau}$. The paper proposes to monitor the equilibrium surface depending on the parameters of the model of the pilot. When a Hopf bifurcation is detected, limit cycles and the associated oscillation frequency of the PIO are determined. For the numerical study, in [5] the following transfer function of the aircraft X-15 is used (the transfer function from the elevator deviation $\delta_{at}$ to the pitch angle $\theta$):

$$W_\delta^\theta(s) = \frac{86.9(s + 0.0292)(s + 0.883)}{(s^2 + 0.38s + 0.01)(s^2 + 1.68s + 5.29)(s + 25)}. \tag{3}$$

It is accepted that the rudder deflection speed is limited to 15 deg/s, and the steering dynamics in the linear region is described by an aperiodic link with the time constant 0.02 s. The paper presents the dependences of the amplitude of the limit cycle on the coefficient $K_P$ for different $\tau \in [0\ldots0.09]$ sec, from which a jump-like increase in the oscillation amplitude is seen as $K_P \approx 2.3$ (for $\tau = 0.09$ s) and for $K_P \approx 2.7$ (for $\tau = 0$). The authors of [5] also analyze the possibility of indicating the occurrence of PIO in real time through the procedure of identifying a closed system, considering it as linear with varying parameters, which, when a case of PIO occurs, "goes" to the oscillatory stability limit. For identification, the recursive version of the least squares method is used.

## 2.1  Examples of PIO Events

### 2.1.1  Simulated PIO in Research Airplane *X-15* Pitch Control

Consider the process of controlling the pitch angle of a research aircraft X-15, whose transfer function from the angle of deflection of the elevator to the pitch angle $\vartheta$ is [5, 8, 9]

$$W_\delta^\theta(s) = \left\{\frac{\theta}{\delta_e}\right\} = \frac{86.9(s+0.883)(s+0.0292)}{(s+25)(s+0.3516)(s+0.02845)(s^2+1.68s+5.29)}, \quad (4)$$

where $\delta_e$ is the elevator deviation from the balancing value, $\vartheta(t)$ is the pitch angle (the numerical values of all variables are given in the SI system).

The drive is modeled by a first-order lag unit with a restriction on speed deviation of the rudder:

$$\dot{\delta}_e(t) = \text{sat}_{\bar{\omega}}\big(T^{-1}(u(t) - \delta_e(t))\big), \quad (5)$$

where $\text{sat}_{\bar{\omega}}(\cdot)$ is the saturation function at the level of $\bar{\omega}$ (for simplicity, it is assumed that the drive has a unit static gain).

The pilot is usually modeled as a sequentially included element in the control loop [10–13]. The models of the pilot are often taken in the form of a static (inertialess) element. In the papers [5, 8, 9, 14, 15], the pilot model with a transfer coefficient of $K_P$ acting on the mismatch signal on the pitch is used, so that

$$u(t) = K_p(\vartheta^*(t) - \vartheta(t)), \quad (6)$$

or, in a more complex form, as an integrating-differentiating unit with the delay, cf. [10, 16–19]:

$$W_P(s) = \left\{\frac{u}{\Delta\vartheta}\right\} = K_p \frac{T_L s + 1}{T_I s + 1} e^{-\tau_e s}, \quad (7)$$

where $\Delta\vartheta$ is a pitch tracking error; $u(t)$ is the pilot's control action on the steering servo regulator; $K_P$ is static pilot transfer coefficient; $T_L$ is differentiation time constant; $T_I$ is time delay constant; $\tau_e$ is an effective time delay, including transport delay and neuromuscular delays in the high-frequency region.

As stated in [10–13, 19–22], the pilot tries to adjust the time constants of differentiation and deceleration so that the system's sensitivity to them in the lower frequency segment has small changes, leaving an effective time delay as its primary value for monitoring the stability of a closed loop and the dominant modes.

Following [23], let us numerically study the behavior of the closed system (4), (5), (6), (7) assuming that the action of the pilot $u(t)$ is obtained as a feedback on the mismatch in a pitch. Following [5, 8], let us assume that there is a rudder speed limit $\bar{\omega} = 15/57.3$ deg/s. The setting time (5) is assumed to be $T = 0.02$ s, and the pilot model is taken as a static transfer gain (6). Thus, let's consider the system (4), (5), (6).

Free Motion Dynamics. Let the value $K_P = 2.8$ be taken from (6). The linearization of (4), (5), (6) in the neighborhood of the equilibrium state shows that a closed system is asymptotically stable in a certain neighborhood of zero. The eigenvalues $\lambda_i$ of the linearized system model have the values $\lambda_i = \{-50; -26; -0.36 \pm 3.7i; -0.72; -0.03\}$. However, applying the method described in Sect. 2, we obtain the existence of a hidden attractor in a closed-loop system, as is illustrated by the graphs in Figs. 1 and 2. The figure shows phase trajectories in the space $(\vartheta, \omega_z, \delta_e)$ and transient processes in the (4), (5), (6) systems for different initial values of $\delta_e(0)$ (initial values of the remaining state variables are taken to be zero). The value $\delta_e(0) = 12°$ (with zero initial conditions for other variables) can be considered as a certain boundary of the hidden attractor, i.e., trajectories starting at lower values of $\delta_e(0)$ tend to the state of stable equilibrium. Corresponding transients in the (4), (5), (6) system at $K_P = 2.8$ for $\delta_e(0) \in \{8; 12; 14\}$ deg are shown in Fig. 2.

The Nyquist curves of the linear part of the system shown in Fig. 2 show that, according to the harmonic balance method, the value of $K_P = 2.09$ is a certain limit, below which there are no hidden oscillations and the trajectories of free movement of the system tend to the origin. The Nyquist curve in Fig. 2 allows one to conclude that for $K_P \geq 2.09$ there can be two limit cycles in the system: a stable and an unstable ones.

Dynamics of Forced Movements. The behavior of a non-autonomous system is of a substantially more complex nature, since it depends not only on the initial conditions, but also on external influences that may be of the most diverse nature. But, nevertheless, in this case a hidden attractor may appear.

Figure 3 shows the response of the (4), (5), (6) system to a piecewise constant driving force in pitch $\vartheta^*(t)$ with zero initial conditions and $K_P = 2.80$, $K_P = 2.09$. It can be seen from the graphs that if $\vartheta^*$ is big enough, then for $K_P = 2.80$, oscillations arise in the system, which are not there when $K_P = 2.09$.

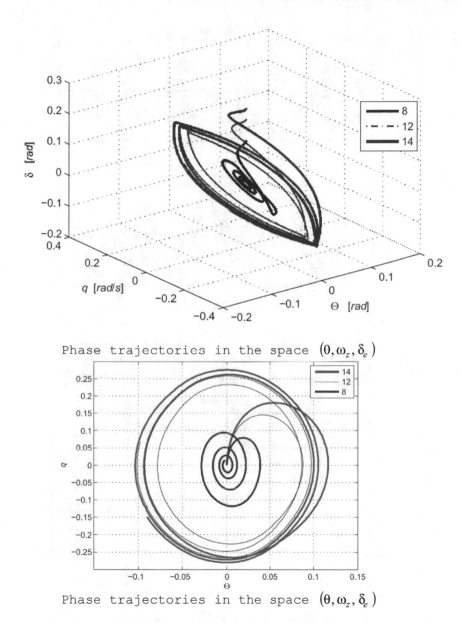

Phase trajectories in the space $(\theta, \omega_z, \delta_e)$

Phase trajectories in the space $(\theta, \omega_z, \delta_e)$

**Fig. 1.** Projections of the phase paths of the system (4), (5), (6) with $\delta_e(0) \in \{8; 12; 14\}$ deg. $K_P = 2.8$ to the subspaces $(\theta,\ \omega_z,\ \delta_e)$ and $(\theta,\ \omega_z)$. The presence of a hidden attractor is visible.

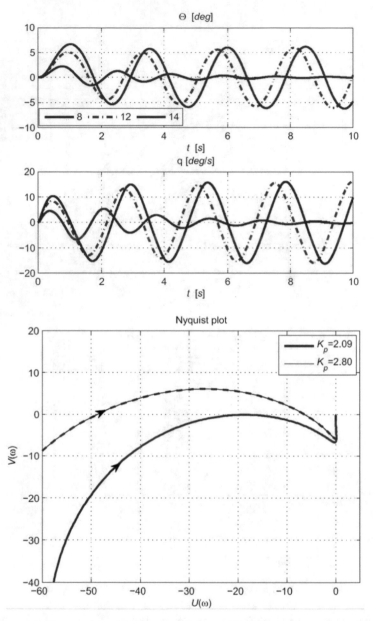

**Fig. 2.** Transients on the angle and angular velocity of pitch in the system (4), (5), (6) at $\delta_e(0) \in \{8; 12\}$ hail, $K_P = 2.8$. Nyquist curves of the linear part of the system with $K_P \in \{2.09; 2.80\}$

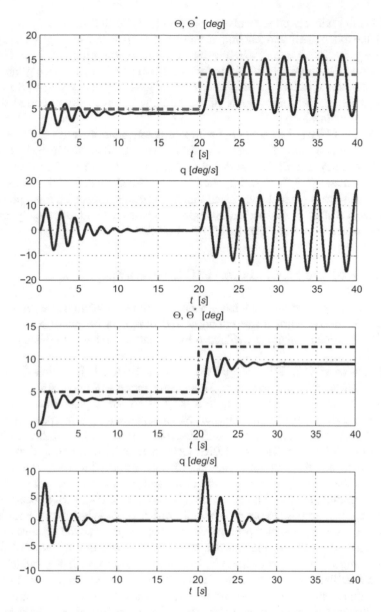

**Fig. 3.** Transients in the (4), (5), (6) system with a piecewise constant driving force $\vartheta^*(t)$

### 2.1.2    Training Jet Northrop T-38 Talon

An example of recorded during a real flight situation PIO event, which was preceded by a low-amplitude, high-frequency oscillation involving only the pitch axis of the aircraft and the simple adaptive control (SAC), is described in [24], and the time histories are presented. The force the pilot applied to the stick was zero during the pre-PIO phase. In other words, the initial oscillation was an instability of the SAS-aircraft in combination with no pilot involvement. To eliminate the oscillation, the pilot disengaged the pitch SAS and entered the control loop in an attempt to counter the resulting upset. These events and the intervention of the pilot led to a very rapid development of oscillations with frequency of 1.2 Hz (7.4 rad/s). During the cycle or so, the oscillations reached an amplitude of 5 g, gradually increasing to 8 g, dangerously approaching the design limits of the aircraft. The recovery occurred when the pilot left the control loop.

In [25], the PIO studies carried out with a remotely controlled sub-scale aircraft, are reported. Various existing PIO analysis criteria, such as Bandwidth/Pitch Rate Overshoot, Neal-Smith and Smith-Geddes for Category I and Open Loop Onset Point criteria for Category II PIO were applied to the testbed system. The results of theoretical analysis are compared with actual flight test results. PIO and human pilot control characterization study performed using flight data collected with a remotely controlled unmanned research aircraft "Phastball" are presented in [26].

## 3 Survey of the Methods for PIO Prevention

The oscillations caused by PIO have been a serious problem since the advent of aviation, and the efforts of many scientists and designers for many years have been directed at eliminating them, as evidenced by the large number of publications on this subject.

The authors of [14, 27] note that the JAS 39 Gripen fighter aircraft type flight control system usually provides a phase stability margin of about 45°, so the speed limit control surfaces, which can cause a large phase shift, pose a serious problem. When the speed limiter used in autopilot to prevent saturation in speed of hydraulic drives becomes saturated, the phase shift drastically reduces the stability limits of the closed loop, and increases the risk of PIO occurrence. In [14, 27], to limit the speed, it is proposed to use the traditional method of anti-induct correction, which provides feedback on the error signal between the input and output of the saturation link. But, since the introduction of an integrator for this structure leads to the appearance of an offset, the authors propose to use an aperiodic link instead. When the rate of change of the input signal $u$ exceeds the specified limit $r$, the feedback signal becomes negative, which reduces the value of the speed limiter input. When the direction of the input signal is changed, the output changes its direction almost immediately, which leads to small-phase shifts. If $|\dot{u}| < r$, then the feedback signal quickly fades out. In [15, 23], frequency (harmonic linearization) and time (computer simulation) areas are analyzed, showing the advantage of the proposed speed limiting method compared to the traditional one. The paper also discusses how to eliminate the drawback of this scheme in case of high-frequency disturbances and provides information about the positive results of flight tests on JAS 39 Gripen.

In [28], a nonlinear phase compensator for eliminating the second-type PIO (associated with limiting the speed of movement of control surfaces) is proposed. Phase compensation is achieved by calculating the mismatch between the desired and true rudder position, which is passed through the phase-ahead corrective link $G_p(s)$, the output of which is added to the measured rudder position. The received signal is fed to the speed limiter to ensure the required speed saturation. The results of numerical analysis for the longitudinal motion X-15, described by the transfer function (3), are given. As in [5], the rudder deflection rate is limited to 15 deg/s. The behavior of the pilot is modeled by coefficient $K_p$. The harmonic linearization method is used to analyze the occurrence of self-oscillations in the original system (without phase correction), showing the appearance of the PIO at $K_p \geq 2.6$. The introduction of a compensator with $G_p(s) = \frac{(s+2.4)^2}{(s+4.5)^2}$ as shown by calculations and modeling, makes it possible to eliminate the limiting cycle without deteriorating the flight quality of the system.

In the civil aviation industry, in order to avoid trends to the appearance of PIO, a number of agreements have been adopted concerning the shape of the fuselage, the size, shape and location of the propulsion unit, and so on. As noted in [1], these agreements lead to very stable aircraft in flight, which leads to a decrease in energy efficiency and maneuverability, therefore, in the concepts of developing future aircraft, there might be a departure from many currently accepted agreements. Studies show that reducing static stability is a key method for reducing fuel consumption and emissions in flight. One of the concepts of designing a *blended wing body aircraft* is a further step towards the development of statically unstable aircraft equipped with a closed control loop to ensure stability, just as it is done for fighter jets. Methods should be developed to reduce the effect of the PIO effect and allow the aircraft to implement the existing potential in energy saving and environmental friendliness without well-established constraining design principles.

The authors of [1] have outlined the following ways to avoid and eliminate PIO:

- sensors are required to measure values related to the PIO;
- assessment methods are needed, based on the processing the data from sensors, which allow detecting or predicting the onset of unfavorable aircraft behavior (for example, approaching a break in flight performance);
- high speed control organs are required to ensure sufficient control energy at the time of low weight, creating a small braking force and not requiring a source of energy of considerable power;
- a pilot interface should be developed, including means of displaying visual and auditory information, cockpit controls to inform the pilot of the current situation and develop recommendations for him on the appropriate course of action;
- control laws are needed to find suitable pilot's actions or of the autopilot, to prevent the onset of undesirable effects;
- on-board computers must be fast enough to avoid a significant lag in computing.

The authors of [29] provide a detailed and easy-to-follow method for developing flight control systems. It allows the use of a linear inversion of the dynamics for a low-order aircraft model, which allows to simplify the finding of controller parameters depending on the flight mode. Design the control law is carried out on the basis of the *Quantitative feedback theory*, QFT) [30]. The program for analyzing the "pilot-aircraft" system allows one to meet the requirements for the pilot qualities of the system and the limitations associated with the PIO for both linear and non-linear modes (associated with restrictions on the magnitude and speed of movement of the rudders).

In [31], an anti-windup correction method with speed limits in the actuator and an algorithm for adjusting the correction parameters are proposed; those allow a compromise between the control quality and the size of the estimated assessment of the segment of attraction is reached. The application of the method is demonstrated on a realistic example of flight control for a nonlinear model of the longitudinal and transverse motion of an experimental aircraft ATTAS (*Advanced Technology Testing Aircraft*) [32] used by German aerospace center (*Deutsche Zentrum fur Luft und Raumfahrt*, DLR). The possibility of applying anti-winding correction to reduce the sensitivity of the aircraft to the phenomenon of PIO is shown. The results of the work received further confirmation in a number of test flights. The synthesis of SU, analysis of its characteristics and flight test data for the experimental aircraft ATTAS, as well as a comparative analysis of low-order dynamic anti-winch compensators for determining the significance of various design parameters are performed in [33, 34].

The authors of [35, 36] propose the so-called *Control Allocation* (to recover from) *Pilot-Induced Oscillations*, CAPIO. The main idea of the CAPIO method is to minimize the phase lag introduced into the system by limiting the rudder speed by minimizing the derivatives of the mismatch between the desired and existing rudder deviations, as well as minimizing the mismatch in them using optimization techniques with constraints. The control distributor redistributes the general control signal of a $p$-dimensional vector control signal $v$ produced by the controller between the components of the excess $m$-dimensional vector of deviation of control surfaces ($m > p$). The target function is the weighted sum of the norms of deviations: $J = \|Bu - v\| + \varepsilon\|u - u_p\|$, where, $u$ is the $p$-dimensional vector of deviation of control surfaces ($m > p$), $B$ is ($p \times m$)-matrix, $\varepsilon > 0$ is a weight coefficient, $u_p$ is the "preferred" rudder position. The advantages of the CAPIO method are demonstrated at the example of a multiply connected system for controlling the angular motion of an aircraft according to the ADMIRE model from [37].

The method of control allocation, which helps the pilot to eliminate the PIO, is also presented in [38]. The technique uses real-time optimization to distribute controls to reduce the phase lag caused by speed constraints on the steering.

The problem of distribution of controls is also considered in [39], which focuses on maintaining system stability and eliminating the PIO in the event of a steering failure. An approach for building fault detection and diagnostics system based on artificial neural networks (NN), automatic training method for such systems and investigating various aspects of this method is presented in [40].

For preventing unfavorable APC events caused by major control actions of the pilot, a nonlinear pre-filter of the lever deflection signal is often used [41]. It is a low-pass filter of the first order with the output signal rate limitation. Its level is adjustable dependently on the flight modes using gain $K(q)$, where $q$ denotes the velocity head. The magnitude of the speed limit of the output signal of the pre-filter is selected in accordance with the maximum allowable rate of deflection of the lever for each flight mode.

Sequential nonlinear correction for PIO prevention is also used in [15, 23], where the so-called *pseudo-linear* corrective devices (PLCD) [42–46] are employed. Namely, the following *Nonlinear Phase Predicting Filter* (NPPF) is used:

$$y = k\,|u|\,\text{sign}(x), \tag{8}$$

$$A(p)x = B(p), u, \tag{9}$$

where $p = d/\,dt$ is the time differentiation operator, $A(p)$, $B(p)$ are operator polymials, such that $W(s) = B(s)/A(s)$ is the transfer function of a properly chosen linear predicting filter. In [15], application of $W(s)$ in the form of the first-order lead-lag unit is considered. Preliminary study gives that for the system (4)–(6) or (4), (5), (7), the phase prediction produced by this filter is not sufficiently large, and the following second-order lead-lag unit is taken:

$$W(s) = \frac{T_1^2}{T_2^2} \cdot \frac{(T_2 s + 1)^2}{(T_1 s + 1)^2}, \tag{10}$$

where $0 < T < T_2$ are chosen time constants (the design parameters). The phase shift, introduced by filter (10), is as

$$\varphi(\omega) = 2\arctan \omega T_2 - 2\arctan \omega T_1 > 0 \quad \forall \omega > 0. \tag{11}$$

The results obtained show that this nonlinear correction allows to increase the admissible "pilot gain" in several times and, therefore, to make possible for a pilot to act in a more aggressive manner, from the one hand, and to prevent PIO, ensuring the flight safety, on the other.

## 4   Conclusions

In the paper, the PIO phenomenon is described and its main reasons are analyzed. The simulation and in-flight test results of several PIO events are described. The literature overview of the PIO prevention method is given. In the future work it is planned to extend application of nonlinear correction method to more realistic pilot-aircraft models and to analyze possible benefits of the adaptive approach to the PIO prevention problem.

**Acknowledgements.** This work was performed in the SPbSU and supported by the Russian Science Foundation (project 19-41-02002).

## References

1. Acosta, D.M., Yildiz, Y., Klyde, D.H.: In: Samad, T., Annaswamy, A. (eds.) The Impact of Control Technology – 2nd Ed., (IEEE CSS 2014). http://ieeecss.org/sites/ieeecss.org/files/CSSIoCT2Update/IoCT2-RC-Acosta-1.pdf
2. Military standard – Flying qualities of piloted vehicles, MIL-STD-1797A (1990)
3. Mikoyan, S.A.: Vospominaniya voennogo letchika-ispytatelya. Tehnika – molodezhi, Moscow (2002) (in Russian)

4. McRuer, D.T., Warner, J.D. (eds.): Aviation Safety and Pilot Control: Understanding and Preventing Unfavorable Pilot-Vehicle Interactions. Committee on the Effects of Aircraft-Pilot Coupling on Flight Safety Aeronautics and Space Engineering Board Commission on Engineer-ing and Technical Systems National Research Council. National Academy Press, Washington, DC (1997). http://www.nap.edu/catalog/5469.html

5. Mehra, R., Prasanth, R.: In: Proceedings International Conference Control Applications (CCA 1998), vol. 2, pp. 1404–1408 (1998). https://doi.org/10.1109/cca.1998.721691

6. Shilnikov, L.P., Shilnikov, A.L., Turaev, D.V., et al.: Metody kachestvennoj teorii v neline-jnoj dinamike. Part 2. Izhevsk, NIC « Regulyarnaya i xaoticheskaya dinamika » , Institut kompyuternyx issledovanij, (2009). http://www.ni.gsu.edu/∼ashilnikov/pubs/volume2s.pdf (in Russian)

7. Leonov, G.A., Kuznetsov, N.V.: Int. J. Bifurcat. Chaos. 23(1), 1330002 (2013). https://doi.org/10.1142/s0218127413300024

8. Alcalá, I., Gordillo, E., Aracil, J.: In: Proceedings American Control Conference (ACC 2004). AACC, Boston, Massachusetts, USA, pp. 4686–4691 (2004)

9. Andrievsky, B., Kuznetsov, N., Kuznetsova, O., Leonov, G., Mokaev, T.: SPIIRAS Proc. 49 (6), 5 (2016). https://doi.org/10.15622/sp.49.1

10. McRuer, D.T., Jex, H.R.: IEEE Trans. Hum. Factors Electron. HFE 8(3), 231 (1967)

11. Byushgens, G.S., Studnev, R.V.: Aerodinamika samoleta: Dinamika prodolnogo i bokovogo dvizheniya. Mashinostroenie, Moscow (1979). (in Russian)

12. Byushgens, G. (ed.): Aerodinamika, ustojchivost i upravlyaemost sverxzvukovyx samoletov. Nauka. Fizmatlit, Moscow (1998). (in Russian)

13. Efremov, A.V., Ogloblin, A.V., Predtechenskij, A.N., Rodchenko, V.V.: Letchik kak dinamich-eskaya sistema. Mashinostroenie, Moscow (1992). (in Russian)

14. Rundqwist, L., Ståhl-Gunnarsson, K.: In: Proceedings International Conference Control Applications (CCA 1996). IEEE Press, Piscataway, NJ, Dearborn, MI, USA, pp. 19–24 (1996)

15. Andrievsky, B., Kuznetsov, N., Kuznetsova, O., Leonov, G., Seledzhi, S.: In: Proceedings 9th IEEE European Modelling Symposium on Mathematical Modelling and Computer Simulation (EMS 2015), Madrid, Spain (2015). http://uksim.info/ems2015/start.pdf

16. Barbu, C., Reginatto, R., Teel, A.R., Zaccarian, L.: In: Proceedings American Control Conference (ACC 1999), AACC, vol. 5, pp. 3186–3190 (1999)

17. Efremov, A.V., Ogloblin, A.V.: In: Grant, I. (ed.) Proceedings 25th International Congress of Aeronautical Sciences, ICAS 2006, ICAS, Hamburg, Germany, ICAS 2006–6.9.1 (2006). http://www.icas.org/ICAS\_ARCHIVE/ICAS2006/PAPERS/175.PDF

18. Lone, M., Cooke, A.: Aerosp. Sci. Technol. 34, 55 (2014)

19. Efremov, A., Koshelenko, A., Tyaglik, M., Tyumencev, Y., Tyan, V.: Izvestiya vysshix uchebnyx zavedenij. Aviacionnaya texnika. 34(2) (2015) (in Russian)

20. Zhabko, A., Chizhova, O., Zaranik, U.: Cybern. Phys. 5(2), 67 (2016)

21. McRuer, D.T., Krendel, E.S.: J. Franklin Inst. 267, 381 (1959)

22. McRuer, D., Graham, D., Krendel, E., Reisener, W., J. Human pilot dynamics in compensatory systems-theory, models, and experiments with controlled element and forcing function variations. Tech. Rep. AFFDL-TR-65-15, Franklin Inst. (1965)

23. Andrievsky, B., Kravchuk, K., Kuznetsov, N., Kuznetsova, O., Leonov, G.: IFAC-PapersOnLine, vol. 49, p. 30 (2016). https://doi.org/10.1016/j.ifacol.2016.07.970, https://www.sciencedirect.com/science/article/pii/S2405896316312587

24. McRuer, D.T., Warner, J.D. (eds.): Aviation Safety and Pilot Control: Understanding and Preventing Unfavorable Pilot-Vehicle Interactions. Committee on the Effects of Aircraft-Pilot Coupling on Flight Safety Aeronautics and Space Engineering Board Commission on Engineering and Technical Systems National Research Council National Academy Press, Washington, DC, (1997). http://www.nap.edu/catalog/5469.html

25. Mandal, T., Gu, Y., Chao, H., Rhudy, M.B.: In: Proceedings AIAA Guidance Navigation and Control Conference (GNC 2013), Boston, MA (AIAA, 2013), 1–15, (2013). https://doi.org/10.2514/6.2013-5010

26. Mandal, T., Gu, Y.: Aerospace 3(42), 1 (2016). https://doi.org/10.3390/aerospace3040042. https://www.mdpi.com/2226-4310/3/4/42

27. Rundqwist, L., Stahl-Gunnarsson, K., Enhagen, J.: In: 1997 European Control Conference (ECC), pp. 3944–3949 (1997)

28. Alcala, I., Gordillo, F., Aracil, J.: In: Proceedings American Control Conference (ACC 2004), vol. 5, pp. 4687–4691 (2004)

29. Siwakosit, W., Snell, S., Hess, R.A.: IEEE Trans. Contr. Syst. Technol. 8(3), 483 (2000). https://doi.org/10.1109/87.845879

30. Horowitz, I.: Int. J. Control 53(2), 255 (1991)

31. Sofrony, J., Turner, M.C., Postlethwaite, I., Brieger, O., Leissling, D.: In: Proceedings 45th IEEE Conference Decision & Control (CDC 2006), IEEE Press, Piscataway, NJ, San Diego, CA, USA, pp. 5412–5417 (2006)

32. Bauschat, J.M., Duus, G., Hahn, K.U., Heine, W., Willemsen, D. In: RTO MP-051, Proceedings RTO AVT Symposium. Active Control Technical Enhanced Performance Operational Capabilities of Military Aircraft, Land Vehicles and Sea Vehicles. RTO AVT, Braunschweig, Germany (2000)

33. Brieger, O., Kerr, M., Postlethwaite, J., Sofrony, J., Turner, M.C.: In: American Control Conference (ACC 2008), pp. 1776–1781 (2008)

34. Brieger, O., Kerr, M., Leißling, D., Postlethwaite, I., Sofrony, J., Turner, M.: Aerosp. Sci. Technol. 13(2–3), 92 (2009)

35. Yildiz, Y., Kolmanovsky, I. In: Proceedings American Control Conference (ACC 2010). Baltimore, MD, USA, pp. 516–523 (2010)

36. Yildiz, Y., Kolmanovsky, I., Acosta, D.: In: Proceedings American Control Conference (ACC 2011), pp. 444–449 (2011). https://doi.org/10.1109/acc.2011.5991270

37. Harkegard, S.O., Glad, T.: Automatica 41, 137 (2005)

38. Acosta, D.M., Yildiz, Y., Craun, R.W., Beard, S.D., Leonard, M.W., Hardy, G.H., Weinstein, M.: J. Aircraft 52(1), 130 (2015)

39. de Lamberterie, P., Perez, T., Donaire, A.: In: Australian Control Conference (AUCC), pp. 284–289 (2011)

40. Arseniev, D., Lyubimov, B., Shkodyrev, V.: In: Proceedings IEEE International Conference on Control Applications (CCA 2009), St. Petersburg, Russia, IEEE Press, pp. 1815–1819 (2009). https://doi.org/10.1109/cca.2009.5281003

41. Efremov, A., Zaxarchenko, V., Ovcharenko, V., Suhkanov, V.: Dinamika poleta: uchebnik dlya studentov vysshix uchebnyx zavedenij. Mashinostroenie, Moscow (2011). (in Russian)

42. Popov, E.P. (ed.): Nonlinear Corrective Devices in Automatic Control Systems (Nelinejnye Korrektirujushhie Ustrojstva v Sistemah Avtomaticheskogo Upravlenija). Mashinostroenie, Moscow (1971). (in Russian)

43. Zel'chenko, V., Sharov, V.: Nonlinear Correction of Automatic Control Systems (Nelinejnaja korrekcija avtomaticheskih sistem). Sudostroenie, Leningrad (1981) (in Russian)

44. Taylor, J.H.: In: Proceedings American Control Conference (ACC 1983), San Francisco, CA, USA, pp. 141–145 (1983)

45. Taylor, J.H., O'Donnell, J.R.: In: Proceedings American Control Conference (ACC 1990), San Diego, USA, pp. 2217–2222 (1990)

46. Nassirharand, A., Firdeh, S.R.M.: Int. J. Control Autom. Syst. 6(3), 394 (2008)

# Quality Assessment in Cyber-Physical Systems

Sergey G. Redko$^{(\boxtimes)}$ and Alexander D. Shadrin

Peter the Great St. Petersburg Polytechnic University, Saint Petersburg, Russia
redko_sg@spbstu.ru

**Abstract.** The article considers the subjective side of quality as the degree to which a set of characteristics of objects and subjects of a cyber-physical system fulfils the requirements of all parties concerned. The authors suggest an interpretation of the term of a "cyber-physical system" with consideration of its performance quality assessment. The study proves that an adequate assessment model requires development of special algorithms and significant computational resources. Its practical implementation is possible via digital models or digital twins.

**Keywords:** Quality · Cyber-physical system · Quality assessment model

## 1 Introduction

A cyber-physical approach is invariant to the objects concerned. It deals with the structures and functions of control systems, and its logic, reasoning, argumentation and methods are detached from any specific area and provide *optimum orderliness* (or *objective function optimization*) of a system based on coordination between people, computational resources and physical processes. We believe that a system can be considered a cyber-physical system (CPS) only if it is optimally object-oriented.

CPS involved in goods and services production (hereafter referred to as companies) are comprised of two subsystems: process control systems (PCS) and administration systems which are now more commonly known as management systems (MS).

PCS deal with daily equipment control, goods or services production and the formation of product or service features, their quality in particular. MS operate using the entirety of objectives and methods of human resources administration [1] – including hierarchy, allocation of responsibilities, short-term and long-term planning and relationships with the natural and the social environment – based on the performance outcome of PCS.

These two subsystems aim at the realization of the objectives of a company which often contradict one another. Such objectives might be the reliability of a product and its production cost, or the amount of the company's tax payments and its contributions to the development fund and payroll.

The combination of all contradictory objectives of the two subsystems of a company is reflected in the notion of quality.

D. G. Arseniev et al. (Eds.): CPS&C 2019, LNNS 95, pp. 124–130, 2020.
https://doi.org/10.1007/978-3-030-34483-7_12

# 2  Analysis of Approaches

Researchers have been preoccupied with the quality notion complexity for almost a hundred years. Walter Shewhart identified two aspects of quality. One aspect is related to the perception of a product quality as an objective reality. The other considers consumers' feelings and attitudes to a product quality. In production, quality standards need to be established in a quantitative manner. However, subjective quality assessments do not lose their importance. Quite the opposite, they are commercially interesting [2].

The first aspect determined by W. Shewhart – product or service quality as an objective reality – is *directly* ensured through PCS. The process optimization methods are realized on the basis of such cybernetic principles as feedback, main factor exposure and mathematical methods and models [3].

The second aspect – subjective quality assessments – is provided by management systems. They also take into account objective features of a certain product or service. However, these subjective assessments are ultimately influenced by many individual characteristics of an interested party, as well as by economic, political, regional, and other factors.

The arrangement and optimization methods in MS are realized through such cybernetic principles as feedback, the division of a whole system into subsystems, main factor exposure, the law of requisite variety, statistical probability, management hierarchy and automatic performance, and emergence [3, 4].

The approach of mathematical methods and models is used in MS less frequently. As Norbert Wiener stated, sociology and anthropology are primarily sciences of communication, therefore they are included in cybernetics. Economics, being a particular branch of sociology, is distinguished from other branches by a more precise usage of numbers for values considered [5].

An important factor of the impact of cybernetics on management is that the implementation of isomorphism has resulted in the development of a range of international and national management standards [6]. The content of these standards is fully based on the practical realization of cybernetic principles. Particularly, Subclause 2.4.1.1 General of ISO 9000:2015 highlights that 'organizations share many characteristics with humans as a living and learning social organism' [7].

The named standard efficiently combines the objective and the subjective aspects of *quality* in a precise definition, stating that quality is '*the degree to which a set of inherent characteristics of an object fulfils requirements*'.

However, a proper understanding of the definition requires explanation of two more terms given in [7]:

*an object* is an '*entity, item, anything perceivable or conceivable*', for example, a product, service, process, person, organization, system, resource;

*requirement* is a '*need or expectation that is stated, generally implied or obligatory*'; it should be noted that requirements '*can be generated by different interested parties*'.

The standard specifies *all* interested parties, such as owners, people in an organization, customers, partners (providers or bankers), and society.

The definition of quality demonstrates that the term implies not only objective features of a product or service stated in contracts or standards but also subjective characteristics of interested parties which assess its quality.

Hence the term quality in its current understanding describes the outcome of interaction between an object's features and all interested subjects. It could, therefore, be stated that providing adequate quality, or such quality that sufficiently fulfils the requirements of all interested parties, should be the only objective of a company.

Thus, profit that is often mentioned as the main objective of any company is, in fact, not its objective but a means for meeting the needs of all five interested parties. Moreover, profitability is included in a set of inherent characteristics of a company's performance; in other words, it is used to measure the quality of a company's performance.

The same statement can be made regarding a company's production quantity. If a company produces two pieces of bread, it meets the requirements of interested parties to a degree different from when it produces one piece of bread. Therefore, with respect to the given definition, different quantities of output have different qualities.

The definition of the term of quality taken from ISO 9000 can explain mixed assessments of quality. The key word in it is *degree*.

It is commonly understood that quality is not a characteristic of an object that can be either present or absent. A degree to which a set of characteristics of a product or service fulfils requirements is different for any subject and varies from one extreme that is presence to the highest degree to the other that is a lack of presence even to the lowest degree. These variations may occur not only among different subjects but also in one certain subject over time, as individual characteristics of a person or a company tend to change constantly, hence their assessments change as well.

However, as for the Russian theory and practice, to the date of publication, most authors continue to consider the terms of quality and product quality synonymous, although it is an obvious and gross mistake, as it has been proven above.

*The views of the authors of the article regarding the feasible practical quality assessment methodology in a cyber-physical system based on ISO 9000 are summarized hereunder.*

Due to peculiarities of subjects performing assessments or objects under assessments, various outcomes are possible. It should be noted that objects under assessments are divided into two fundamentally distinct types - bulk and single products or services.

When assessing bulk products, statistical methods are applied. Next various cases of assessing a single product will be considered (a car, a pork chop, an article, a play, a piece of equipment, etc.).

Case A. A single product and one interested party (one person). The degree to which a set of product's characteristics meets the requirements can be presented as a point on the axis from a very negative to a positive assessment.

Case B. A single product and a few interested parties. In this case, the graphic form of quality assessment is showcased as a frequency polygon.

Case C. A single product and multiple interested parties (hundreds of them or more; these include situations of food products assessment by customers, government quality assessment by citizens, etc.).

A sufficient graphic form of the given assessments in this case may be presented as a probability distribution plot of a continuous random variable. The random variable here is quality assessment.

Figure 1 shows the quality assessment of a single product in the form of a normal probability plot, although in practice a distribution graph may have various forms and shapes.

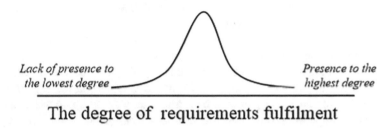

*Lack of presence to*          *Presence to the*
*the lowest degree*          *highest degree*

## The degree of requirements fulfilment

**Fig. 1.** A quality assessment model of a single product by multiple interested parties

Therefore, when assessing quality in a cyber-physical system, three aspects need to be considered.

**First Aspect.** Quality is intangible. In other words, quality of a product or service is information about its features. The quality perception process resembles the perception process of any kind of information. Complete objective information about an object is perceived through physical, semantic and pragmatic filters any subject possesses [8].

It is the possession of these three information filters that causes the quality problem and continuous disputes about the quality assessment of a product or service.

**Second Aspect.** When assessing quality, it is crucial to consider Maslow's five-tier hierarchy of needs.

In his works, Maslow highlighted that all five types of needs are present within a human being concurrently and permanently. He gave the following example to support this statement. According to his findings, only 85% of physiological needs, 70% of safety needs, 50% of social belonging, 40% of self-esteem and 10% of self-actualization of an average person are met [9]. It is apparent that here Maslow implies the degree to which the needs are fulfilled. However, these figures are not universal and differ for each person, yet they can serve as centers, or statistical expectations of respective distributions.

**Third Aspect.** When assessing quality, it should be noted that objective quality of a product or service consists of many different features. A vehicle is characterized by such features as its size, weight, engine power, color, load capacity, etc. A hotel service has such features as the area of a room or rooms, number of towels, personnel qualification, availability of an air conditioner, Wi-Fi, etc.

In qualimetry, *a figure of merit (FOM)* is a quantitative characteristic of one or several features of a product or service related to its quality that is assessed with respect to the specific conditions of production or consumption of a product or service.

If only one feature of a product or service is concerned, *a single figure of merit (SFOM)* is applied. *An integrated figure of merit (IFOM)* is used to calculate the ration of the combined performance and consumption efficiency of a product or service to its combined production and performance costs [10].

The same product or service with the same set of features can fulfill various needs of the same person to various degrees.

At the same time, each feature of an object characterized by a SFOM can fulfill various needs of various people to various degrees.

The practical implementation of the proposed quality assessment model requires collecting and processing vast quantities of information regarding the degree to which the requirements of interested partied are fulfilled, as well as appropriate interfaces, algorithms and computing powers aimed at ensuring that the data is received relatively quickly and it is full, sound and available to interested parties.

This approach corresponds to the current cyber-physical systems development concept that suggests the integration of computing resources into physical objects of any kind [11]. In this case, computing resources are distributed throughout the whole physical environment. In such system all components are connected through the whole value chain. They interact with one another using standard networking protocols for prediction, self-adjustment and adaptation to changes.

Figure 2 shows a flowchart that is also an algorithm of quality assessment in a cyber-physical system with $M$ subjects.

Implementing digital models, or digital twins, helps to eliminate error probabilities and increase accuracy when estimating figures.

A digital twin is a digital replica of a physical product, a group of products or a process, whose purpose is to collect and reuse digital information. A digital twin performance is not limited to collecting data at the stage of product development and production only. It can aggregate data for its whole operational lifetime. The data may be related to a product's state, sensors' readings, operation logs, factory default and service configurations, software version, etc. A digital twin stores the whole history of working data, which adds more opportunities for assessing and improving quality of a product and shows the entire timeline of its quality development [12].

For a long time, the level of technology hindered the implementation of such processes, as it was impossible to transfer information between systems and use it on different platforms. However, today these technological obstacles are disappearing. The Internet of Things (IoT) provides great amounts of data received from sensors that control the use, performance and quality of a product. All data can be added to a digital twin to increase its accuracy.

A digital twin can be regarded as an electronic datasheet including information about raw materials, materials, performed operations, checkouts and laboratory tests, as well as single and integrated figures of merit. Based on these assessments, a digital twin can monitor quality and guarantee it as a requirements fulfilment degree.

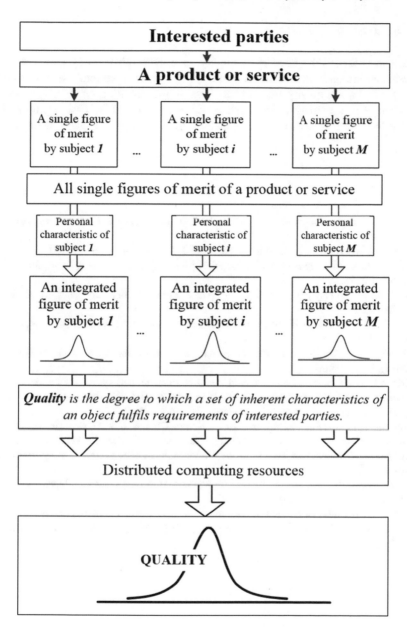

**Fig. 2.** Quality assessment flowchart in a cyber-physical system

## 3  Conclusion

The article analyzed the concept of quality in a cyber-physical system. The authors suggested an interpretation of the term of a "cyber-physical system" with consideration of its performance quality assessment. The implementation of quality assessment methodology in a cyber-physical system according to ISO 9000 was proposed; that included the use of its technological capabilities, particularly, digital models or digital twins.

## References

1. Glushkov, V.M.: The Basis of Paperless Informatic. Science Publication, Moscow (1987). (in Russian)
2. Hoyer, R.W., Brooke, B.Y.: What is Quality? J. Stand. Qual. **3**, 97 (1997)
3. Dumler, S.A.: Management and Cybernetics. Mechanical Engineering Publication, Moscow (1969). (in Russian)
4. Beer, S.: Mozgfirmy [Brain of the Firm]. Radio and Communication Publ, Moscow (1993). (in Russian)
5. Wiener, N.: I Am a Mathematician. Cited after Soviet Radio Publication, Moscow (1964). (in Russian)
6. Shadrin, A.D.: On the concept of management based on standards. J. Stand. Qual. **10**, 74 (2013). (in Russian)
7. ISO 9000:2015 International Standard. Quality management systems – Fundamentals and vocabulary; the Russian version: GOST R ISO 9000:2015 The Russian National Standard. Quality management systems – Fundamentals and vocabulary
8. Shadrin, A.D.: Quality and information. J. Stand. Qual. **4,5**, 30 (1996). (in Russian)
9. Maslow, A.H.: Motivation and Personality. Eurasia Publication, St. Petersburg, (1999), (in Russian). = Maslow, Abraham H. Motivation and Personality (2nd ed.), Harper&Row, New York (1970)
10. Oglezneva, L.A.: Qualimetry: Tutorial. Publishing House of Tomsk Polytechnic University, Tomsk (2012). (in Russian)
11. Alur, R.: Principles of Cyber-Physical Systems. MIT Press, Cambridge (2015)
12. Borovkov, A.I., Maruseva, V.M., Ryabov, Yu.A.: Digital Production: Methods Ecosystems Technologies, 24–43 (2018)

# Data Shuffling Minimizing Approach
# for Apache Spark Programs

Maksim Popov$^{(\boxtimes)}$ and Pavel D. Drobintsev

Peter the Great St. Petersburg Polytechnic University, Saint Petersburg, Russia
popovmk97@gmail.com

**Abstract.** This article discusses a way to optimize the Apache Spark program by reducing the number of transformations with wide dependencies and, as a result, the number of data shuffles. This is achieved by combining sequential data processing algorithms in chains based on common key fields, as well as grouping the data which is stored in resilient distributed structures i.e., Spark SQL Datasets, according to the keys by which the processing takes place.

**Keywords:** Apache spark · Big data · Hadoop · Optimization · Scala

## 1 Introduction

The main features of the Apache Spark framework [1] are associated with powerful transformations that Spark programmers use for data grouping, convolution, sorting and filtering. The transformations are performed on RDD-like structures (Resilient Distributed Datasets), which are stored and processed on slave nodes of clusters. There are two types of these transformations: transformations with narrow dependencies and transformations with wide dependencies [2, 3].

Transformations with narrow dependencies are transformations where each partition of the parent RDD is used by no more than one partition of the child RDD. So, dependencies of the child sets are simple, finite, and can be determined at the design time [4]. These transformations can be performed in arbitrary sections when there is no information about the other parts of RDD. Thus, transformations that have narrow dependencies are fast [5, 6].

Transformations with wide dependencies are transformations where each partition of the parent RDD may be used by multiple child partitions. These transformations require specific data partitioning, for example, according to their key. Because of this, transformations with wide dependencies cannot be performed in arbitrary sections. If Spark does not know how data is partitioned until the execution, in most cases, wide transformations lead to data shuffling, which consists of moving data from one cluster node to another to group this data in a specified way [7]. The cost of such shuffles is almost always extremely high, and it increases as the volume of data increases [8, 9] and as the proportion of data being transferred from its total volume increases.

Thus, reducing the number of transformations with wide dependencies, as well as reducing their complexity, is one of the most effective ways to optimize programs that are written using the Apache Spark framework.

© Springer Nature Switzerland AG 2020
D. G. Arseniev et al. (Eds.): CPS&C 2019, LNNS 95, pp. 131–139, 2020.
https://doi.org/10.1007/978-3-030-34983-7_13

## 2  Related Works

The Apache Spark framework is a very powerful and modern tool. And the topic of reducing the number of transformations with wide dependencies is extremely relevant nowadays.

There are many researches with attempt to do different logical optimizations and to optimize the analysis, physical planning, and code generation [10]. The most useful of these works is Spark SQL's Catalyst Optimizer [11–13]. But this optimizer can only combine transformations with wide dependencies slightly reducing their number.

There are also researches about general patterns of code planning [14, 15]. In these articles researchers describe how to optimize the memory usage, how to tune data structures and garbage collection and describe typical mistakes of code planning.

But the Spark SQL's Catalyst Optimizer, as well as the articles about general patterns of code planning, cannot do strong optimization of Spark programs if in the source code there are repeatable complicated algorithms that use the same data. So, this work will discuss the ways to optimize an Apache Spark program by reducing the number of transformations with wide dependencies in such situations.

## 3  Background and Goals

This section describes an existing program that needs to be optimized by developing a method of program code planning and using this method to refactor the code of the program.

### 3.1  Specification

There is a system designed to match records with a complex structure - a key and a large number of attributes, on the basis of which the matching process takes place and which are updated during the operation of the program.

This system is implemented using the Spark framework (Spark SQL module) in Scala language [16]. The matching module works according to the following algorithm.

- Reading records from an Apache Hive database [17]. Selection of records to be matched.
- Matching records by various algorithms and creating corresponding pairs.
- Calculation of attributes on the basis of the received pairs and records related to these pairs.
- Updating the records with the received attributes.
- Writing updated records and pairs to tables in Apache Hive and Apache Phoenix databases [18].

These pairs can be formed based on the matches (or mismatches) of different fields of records. Also, each record can act as an agent or as a contractor. So, there are many matching algorithms that use a common pattern: each algorithm selects from all records those ones that can participate in this algorithm as an agent or as a contractor. After that, the algorithm selects a corresponding key for each record. The key is a set of fields that are required for the current algorithm. Next is an attempt to make a pair of agent and contractor records for the received keys. The algorithms can use the same keys, but they will use and update different attributes.

The following example contains four records. For an algorithm that requires matching "invoice_date" and "invoice_no" fields, records with rowkeys = 1201, 3212 will create a pair (Fig. 1).

| ROWKEY | INVOICE_KPR_EFF | INVOICE_DATE | INVOICE_NO | JOURNAL_PART_ID | AMOUNTS_VAT | INVOICE_SELLER_NUM | MATCH_ID | (OTHER FIELDS) |
|--------|-----------------|--------------|------------|-----------------|-------------|--------------------|----------|----------------|
| 1201 | 156821 | 01.01.2018 | AUT-MATCH-NM | 1 | 50.00 | 12345671 | 2 | ... |
| 4918 | 51221 | 09.07.2018 | AUT-MATCH-Replaced2 | 4 | 10.00 | 56181892 | 3 | ... |
| 3212 | 156122 | 01.01.2018 | AUT-MATCH-NM | 3 | 100.00 | 16267421 | 2 | ... |
| 2002 | 445122 | 02.01.2018 | AUT-MATCH-Replaced1 | 4 | 60.00 | 20100002 | 0 | ... |

**Fig. 1.** Example of records (with one pair)

The matching module should work for an acceptable time for daily launch with industrial data volumes (less than 8 h). This limit for industrial data volumes is displayed in the module operating time limit on test data volumes: less than 20 min (switching from an industrial system to a test one preserves the ratio of the number of allocated cores and memory, so the time display is almost linear). Due to the fact that the matching module works for more than 70 min on the test data (accordingly, it will work more than a day on the real data), this module requires a strong optimization.

Thus, the purpose of this work is to develop a method for optimizing the program code, due to which the data will be used more optimally, and the operation time of the matching module will be reduced.

To achieve this goal, it is necessary to reduce the number of transformations with wide dependencies, which entail extremely costly data shuffling. So, it is necessary to make the maximum possible number of transformations and calculations based on the data of each record that has appeared in RAM. For this purpose, it is necessary to avoid situations when records stored on one node are moved to another node (for example, the situation when the pair to some record is on another node).

Each algorithm matches all records, so transformations with wide dependencies will occur after each algorithm (and may also occur during the execution of some of them).

Thus, we can formulate a list of tasks for this work.

## 3.2 Task List

- Create a table of all algorithms for finding common fields in the keys of these algorithms.
- Make chains of the algorithms on the basis of common key fields of these algorithms.
- Group the records for each chain so the records with common fields would be on the same node of the cluster, and most of the operations will happen without data shuffling.
- Execute algorithms for each chain of grouped records.

# 4  Implementation

## 4.1  Splitting of Algorithms

In order to avoid data shuffling, which would occur after each matching algorithm, all algorithms were analyzed to find the common key fields of the records. Algorithms that have similar fields in the keys and follow each other were combined into chains. The following table shows the common fields of agent and contractor records. Chains of algorithms were created with the usage of these fields (Fig. 2).

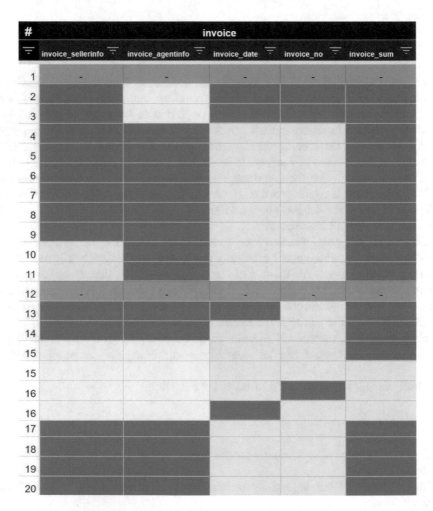

**Fig. 2.** Common fields in the algorithm keys for the records

Algorithms are marked with corresponding numbers. As we can see from the table, we can create several chains that combine the majority of algorithms that use common fields in their keys.

We can create the following chains:

- Algorithms 4–9; during the execution of this chain of algorithms, the records will be grouped by the invoice_date and invoice_no fields.
- Algorithms 10–11, grouped by the invoice_seller_info, invoice_date and invoice_no fields.
- Algorithms 13–14, grouped by the invoice_no field.
- Two rules in the 15th algorithm, grouped by the invoice_date and invoice_no fields.
- Algorithms 17–20, also grouped by the invoice_date and invoice_no fields. After the algorithms are chained together, it is necessary to group the records according to the corresponding fields that form the common key for the algorithms: that will help to avoid data shuffling after each algorithm inside the chains (Fig. 3).

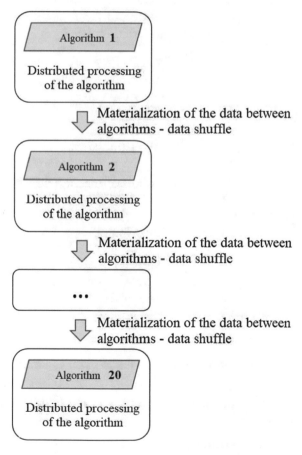

**Fig. 3.** Scheme of the matching module before optimization

In the Spark framework, you have to use RDD-like structures for data storage (such as Dataset, DataFrame and RDD itself), and in this work we used Dataset. When you use transformations with wide dependencies, as well as the materialization of data, the data will be mixed in most cases. So, it is necessary to group the data in Datasets by the corresponding keys and then process the records using the usual Scala transformations of dynamic arrays. In this case, we used the "List" (Fig. 4).

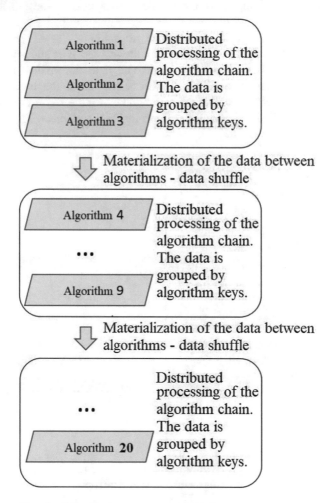

**Fig. 4.** Scheme of the matching modules after optimization

## 4.2    An Example of the Implementation of the Chain in the Matching Module's Code

The following code shows an example of combining algorithms into a chain, as well as data grouping and processing by these algorithms.

```
def selectAndMatchChain(invLeft: Dataset[Invoice],
invRight: Dataset[Invoice]): Unit = {
import spark.implicits._
//1
val grouped = (invLeft union invRight).groupByKey(inv =>
(inv.invoiceDate, inv.invoiceNo.getOrElse("")))
//2
.mapGroups((_, invIterator) => {
var selected = (invIterator.toList, List[In-voice](),
List[Invoice](), List[MatchPair]())
var updatedInvoices = (List[Invoice](), List[In-voice](),
List[Invoice]())
//3
(algos.map(algo => {
selected = algo.selectAndMatch(selected._1 union updatedInvoic-
es._1)
updatedInvoices = algo.resultCalcula-tor.calcMatchAttr(algo.alg,
selected._2, selected._3, selected._4)
(updatedInvoices._2, updatedInvoices._3, se-lected._4)}), select-
ed._1 union updatedInvoices._1)
})
val leftUpdated = grouped.flatMap(_._1.flatMap(_._1))
val rightUpdated = grouped.flatMap(_._1.flatMap(_._2))
val notSelected = grouped.flatMap(_._2)
val newPairs = grouped.flatMap(_._1.flatMap(_._3))
//4
updateInvoices(leftUpdated, rightUpdated, notSelected)
updatePairs(newPairs)
}
```

These blocks of code that implement one of the chains of the algorithms contain:

- Grouping data by fields invoice_date and invoice_no;
- Using lambda functions on each group. The data is converted to List type inside the lambda function;
- This block applies each algorithm to the resulting lists;
- After processing all groups, the records are updated and pairs are added to the result.

## 5   Results and Conclusions

The following image shows the operation time of the matching module before optimization (Fig. 5):

| Result: | ⊘ Tests passed: 222 |
| Time: | 16 Dec 18 04:50 - 06:02 (1h:12m) |
| Branch: | refactor/match |

**Fig. 5.** The operation time of the matching module before optimization

After optimization, the matching module works less time on the same test data (Fig. 6):

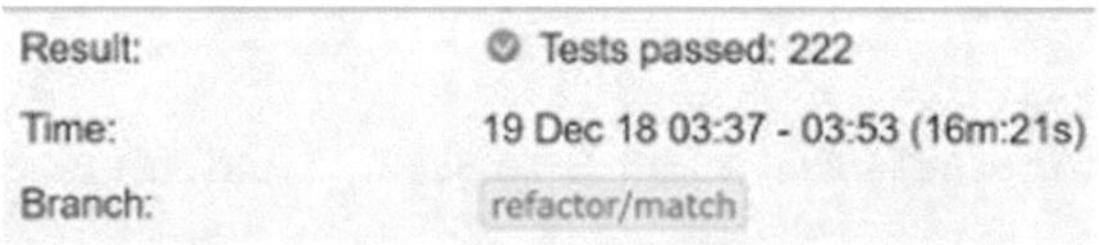

| Result: | ⊘ Tests passed: 222 |
| Time: | 19 Dec 18 03:37 - 03:53 (16m:21s) |
| Branch: | refactor/match |

**Fig. 6.** The operation time of the matching module after optimization

As a result, the goal of this work was achieved: we have developed a general method of program code planning, thanks to which it is possible to significantly reduce the execution time of a program running on scalable systems, so we have reduced the operation time of the matching module by optimizing the usage of data.

To do this, it is necessary to make chains of algorithms based on their common fields. Next, you need to group the data for each chain, so records with common fields will be in the same node of the cluster, and most of the operations will be performed without data shuffling. Thus, the algorithms of each chain will be executed on grouped records, and it will reduce the program execution time.

# References

1. Apache Spark. http://spark.apache.org/
2. Karau, H., Warren, R.: High Performance Spark (2017)
3. Banerjee, S.: Apache Spark and Amazon s3 gotchas and best practices (2016)
4. Chambers, B.: Spark: The Definitive Guide (2017)
5. Wide vs Narrow Dependencies (2017). https://github.com/roh-gar/scala-spark-4/wiki/Wide-vs-Narrow-Dependencies
6. Zaharia, M., Wendell, P., Konwinski, A., Karau, H.: Learning Spark (2015)
7. Cloudera: Spark Execution Model. https://www.cloudera.com/docu-mentation/enterprise/5-9-x/top-ics/cdh_ig_spark_apps.html#spark_exec_model
8. Rizay, S., Leserson, U., Owen, S., Wills, J.: Spark for Professionals: Modern Patterns of Big Data Processing (2018)
9. Khazaei, T.: Spark Performance Tuning: A Checklist (2017)
10. Hu, R., Wang, Z., Fan, W., Agarwal, S.: Cost Based Optimizer in Apache Spark 2.2 (2017)

11. Armbrust, M., Huai, Y., Liang, C., Xin, R., Zaharia, M.: Deep Dive into Spark. SQL's Catalyst Optimizer (2015)
12. Armbrust, M., Xin, R.S., Lian, C., Huai, Y.: Spark SQL: Relational Data Processing in Spark (2015)
13. Catalyst Optimizer. https://databricks.com/glossary/catalyst-optimizer
14. Moisan, Y.: Spark performance tuning from the trenches (2018)
15. Apache. «Tuning Spark». https://spark.apache.org/docs/latest/tuning.html
16. The Scala Programming Language. https://www.scala-lang.org/
17. Apache Hive. https://hive.apache.org/
18. Apache Phoenix. https://phoenix.apache.org

# Operators of Bounded Locally Optimal Controls for Dynamic Systems

Vladimir N. Kozlov and Artem A. Efremov$^{(\boxtimes)}$

Peter the Great St. Petersburg Polytechnic University, Saint Petersburg, Russia
eartm@mail.ru

**Abstract.** The problems of locally linear and quadratic optimal stabilization in finite-dimensional and functional spaces based on the projection method were studied in a number of papers [1, 2, 4, 5], as well as in a series of other studies. In this paper, the problem of quadratic locally optimal program stabilization in functional space is formulated, from which follows the problem of quadratic locally optimal stabilization of the equilibrium state of a dynamical system.

**Keywords:** Locally optimal controls · Finite-dimensional optimization projectors

## 1 Formulation of the Problem

Let the dynamics of a controlled object in the space of vector functions $\mathbf{R}^n(L^2[0, T])$ be represented by a linear differential operator

$$x'(t) = Ax(t) + Bu(t), \quad x(t_0) = x_0, \tag{1}$$

where the vector functions class is given by

$$x(t) \in \mathbf{R}^n\big(L^2[0, T]\big), \quad u(t) \in \mathbf{R}^m\big(L^2[0, T]\big).$$

The numeric matrices of the control object (1) belong to the spaces

$$A \in \mathbf{R}^{n \times n}, \quad B \in \mathbf{R}^{n \times m}.$$

For example, let's suppose that for the matrices the rank control criterion according to R. Kalman is satisfied, guaranteeing the existence of a nonempty set of controls, stabilizing the studied control object. In other words, these matrices form an asymptotically stabilized pair.

It is required to synthesize the vector of optimal controls in a given class of functions for optimal stabilization of the equilibrium position of this system in terms of minimum of a given local functional

$$\varphi = \|z(t)\|^2_{L^2[0, T]} \in \mathbf{R}, \tag{2}$$

D. G. Arseniev et al. (Eds.): CPS&C 2019, LNNS 95, pp. 140–145, 2020.
https://doi.org/10.1007/978-3-030-34983-7_14

which is given on a generalized vector

$$z(t) = [x(t)|u(t)]^T \in \mathbf{R}^{n+m}(L^2[0, T]),$$

which includes the coordinate vector of states and controls of the model of a dynamic object of type (1).

## 2  General Formulation of Problems for the Synthesis of Locally Optimal Controls

For the synthesis of stabilizing controls, the inverse operator for the operator (1) will be used in the form of the Cauchy integral operator, which determines the predictions of the optimized state coordinates and controls based on the solution of the problem for operator (1) in the form

$$x(t) = e^{At}x^0 + \int\limits_0^t e^{A(t-\tau)}Bu(\tau)d\tau. \tag{3}$$

The problem of synthesis is formulated based on the minimization of the functional

$$\varphi = \|z(t) - C_z(t)\|^2_{\mathbf{R}^n(L^2[0, T])} = \int\limits_0^T [z(t) - C_z(t)]^T[z(t) - C_z(t)]dt \tag{4}$$

with constraints like equalities and inequalities

$$Az(t) = b(t), \quad \|z(t)\|^2_{\mathbf{R}^n(L^2[0, T])} \le r^2, \tag{5}$$

which in $\mathbf{R}^n(L^2[0, T])$ are determined by the intersection of the linear manifold and a ball. Moreover, in the space of generalized vectors

$$z(t) = [x(t)|u(t)]^T \in \mathbf{R}^{n+m}(L^2[0, T])$$

linear manifold is defined by the following relation

$$Az(t) = A\left[x(t)| - \int\limits_0^t e^{A(t-\tau)}Bu(\tau)d\tau\right] = e^{At}x^0 = b(t) \tag{6a}$$

of the finite-dimensional space of vector-functions.

The ball of constraints-inequalities of this space is determined by the quadratic inequality

$$\|z(t)\|^2_{\mathbf{R}^n(L^2[0,\,T])} = \int\limits_0^T z^T(\tau)z(\tau)d\tau \leq r^2 < \infty. \tag{6b}$$

Then the quadratic quality functional defined on the generalized vectors for the problem of locally optimal program stabilization will take form

$$\varphi = \|z(t) - C_z(t)\|^2_{\mathbf{R}^{n+m}(L^2[0,\,T])} = \left\| [x(t)|u(t)]^T - C_{xu}(t) \right\|^2_{\mathbf{R}^{n+m}(L^2[0,\,T])}, \tag{7a}$$

where $C_z(t) = C_{xu}(t) \in \mathbf{R}^n(L^2[0,\,T])$ is the target stabilized generalized vector for the problem of locally optimal program stabilization. If the vector-function of program action satisfies the condition

$$C_z(t) = C_{xu}(t) = 0_{n+m} \in \mathbf{R}^{n+m}(L^2[0,\,T]), \tag{7b}$$

then the task (3) transforms from the "quadratic grogram stabilization problem" to the "problem of optimal stabilization of the zero equilibrium position" of the dynamic object (1) for a system with feedback.

Thus, on the basis of relations (3)–(7a, 7b), we can formulate two main tasks for calculating controls:

- The task of synthesis of controls for a locally optimal system of program stabi- lization in discrete-continuous time, which can be constructed as finite-dimensional (approximating) optimization problems, and for this version, this problem is for- mulated based on relations of type (3)–(7a, 7b) and is solved further in Sect. 3.
- The tasks of calculating the controls for local or interval optimal stabilization of the zero position or a given program vector $C_z(t)$, which are also constructed on the basis of relations (3)–(7a, 7b) in an infinite-dimensional space to stabilize the zero equilibrium position, analyzed in Sect. 4.

Based on the objectives of the problem of locally optimal stabilization, a synthesis problem is formulated: to calculate the optimal generalized vector of the type $z(t)$, determined by the equality (2), defined on the object's trajectories (3) with the con- straints (4) which have the form (6a, 6b) in $\mathbf{R}^{n+m}(L^2[0,\,T])$.

## 3   Calculation of Optimal Controls Based on Finite-Dimensional Optimization Projectors

Optimal controls are calculated based on the object's operator, which for piecewise constant controls is

$$u(kh) = \sum_{s=0}^{kh} u(t_s), \tag{8}$$

and taking into account the additive property (3), takes form

$$x(t_k) = x(kh) = e^{Akh}x^0 + \int_0^{kh} e^{A(kh-\tau)}d\tau\, B\sum_{s=0}^{kh} u(t_s)$$

$$= e^{At}x^0 + A^{-1}\left(e^{Akh} - E_n\right)B\sum_{s=0}^{kh} u(t_s). \tag{9}$$

As a result, the synthesis problem as a problem of finite-dimensional conditional minimization has the form: to calculate the countable set of controls in the form of piecewise constant vector functions $u_*(ph)$, which provide for a minimum of the functional

$$u_*(ph) = T_u \arg\min\left\{ \begin{array}{l} \|z(ph) - C_z(ph)\|^2_{\mathbf{R}^n(L^2[0,\,T])}\Big|Az(ph) = b(z(ph)) \\ x(ph) - A^{-1}\left(e^{Aph} - E_n\right)B\sum_{s=0}^{ph} u(t_s) = e^{Aph}x^0, \quad \|z(ph)\| < r^2 \end{array} \right\}, \tag{10}$$

in a discrete prediction interval $[0,\,p] \in N$, where the "filtering" matrix $T_u \in \mathbf{R}^{m\times(n+m)}$ "selects" from the generalized vector $z(ph)$ of controls $u_*(ph)$.

The finite-dimensional conditional minimization problem (10) for stabilizing the vector of program actions can be solved on the basis of reduction to the extremum problem and projection operators of finite-dimensional minimization of type (12a) and (12b), considered in [3, 4].

As noted above, quasi-analytic optimization operators, delivering solutions in the form of finite relations, will be used to calculate the optimal controls. The minimization operator for solving a finite-dimensional non-classical extremum problem has the form:

$$x_* = T_u \arg\min\left\{ \varphi(x) = \|x - C\|^2_2 \Big|Ax = b,\; A \in \mathbf{R}^{m\times n},\; \mathrm{rang}\,A = m,\right.$$
$$\left. x^T x \le r^2\right\} \in \mathbf{R}^n, \tag{11}$$

where the vector of program control actions is separated from zero, i.e. satisfies the inequality

$$0 < \delta_c^2 \le \|C\|^2_2 \le \bar{\delta}_c^2 \le r^2 - r_c^2,$$

separating systems resources for program and stabilizing components of the control. In problem (10), the quality functional $\|x - C\|^2_2 = (x - C)^T(x - C) \in \mathbf{R}$ is given by the square of the Euclidean norm, and the admissible set is determined by a non-empty intersection of the linear manifold and the ball approximating a parallelepiped (ball) [3, 4].

The solution of the optimization problem (11) due to the "principle of boundary Lagrangian extremes" and "narrowing of the admissible region" is represented by a convex linear combination of patterns for regularized orthogonal projectors [1–5]

$$x_* = (1 - \theta_*)x_3(\eta_+) + \theta_* x_3(\eta_-) = P^+ b(t) + (1 - 2\theta_*)P^0 C(t)\eta. \qquad (12a)$$

Lagrange vectors $x_3(\eta_+)$ and $x_3(\eta_-)$ in (12a), belonging to the intersection of the linear manifold (subspace) and the sphere as the boundary of the ball in (11), are defined by orthogonal projectors

$$x_{3\pm} = x_3(\pm\eta) = P^+ b(t) \pm P^0 C(t) \, \rho^{-1}\sqrt{\alpha/\rho}, \qquad (12b)$$

where operators have the form

$$P^+ = A^T (AA^T)^{-1}; \; P^0 = E_n - P^+ A; \; \eta = \sqrt{\alpha/\rho};$$

$$\rho = \left\| P^0 C \right\|_2, \alpha = r^2 - \left\| P^+ b \right\|_2^2; \; \rho = \left\| P^0 C \right\|_2^2.$$

Then, as shown in [4, 5, 6], for the optimal solutions (12a) and (12b), it is necessary and sufficient that the optimal parameter in (12a, 12b) is given by the equality

$$\theta_* = P(\theta_*) = 0.5(|\theta_0| - |\theta_0 - 1| + 1) \in [0, \, 1], \qquad (12c)$$

where

$$\theta_0 = 0.5(1 - \eta^{-1}); \; \eta^{-1} = \sigma = \sqrt{\rho/\alpha}; \; \alpha = r^2 - b^T (AA^T)^{-1}b; \; \rho = C^T P^0 C.$$

The operators formulated in the statement are further used for the projection-operator representation of the problems of optimal stabilization of the equilibrium position of program assignment studied in this paper.

## 4    On the Calculation of Controls for the Problem of Calculating Locally Optimal Stabilizing Equilibrium Controls

In this case, based on relations (3)–(7a, 7b), it is necessary to use the Lagrange function, which for program optimal stabilization has the form [5, 6]

$$L = \|z(t)\|_2^2 + \lambda_0 \left( x(t)| - \int_0^t e^{A(t-\tau)} Bu(\tau)d\tau - e^{At}x^0 \right) + \lambda \left( \|z(t)\|_2^2 - r^2 \right). \qquad (13)$$

For the problem of locally optimal stabilization of the equilibrium position, the quality functional follows from (11) for $C(t) = 0$. As shown in [4, 5], the optimal control vector is calculated by the operator following from (12a), (12b), having the form

$$x_* = P^+ b(t), \tag{14}$$

since the second term in Eq. (12b) is zero due to $C(t) = 0$.

## 5  Conclusion

Thus, the principles of calculating optimal controls in the "state-control" spaces are formulated for objects with models like (1) or (3) for problems of optimal stabilization of the equilibrium position and problems of optimal stabilization of program actions. In contrast to the classical optimization methods, the optimal controls are calculated on the basis of solving mathematical programming problems, which are represented in quasi-analytical projection projectors that allow performing a qualitative analysis of the dynamics of systems with feedback.

The research results are used to optimize the frequency and active power control systems of energy associations, to control the transfer of hydrocarbons along the lines of main pipeline networks, as well as in studying the processes of multilayer thermal conductivity in solid multilayer objects.

## References

1. Kozlov, V.N.: Method of non-linear operators in automated design of dynamic systems, 166 p. Leningrad state university named after A.A. Zhdanov, Leningrad (1986). (in Russian)
2. Kozlov, V.N., Kupriyanov, V.E., Zaborovskiy, V.S.: Computational methods for the synthesis of automatic control systems, 220 p. Leningrad state university named after A.A. Zhdanov, Leningrad (1989). (in Russian)
3. Kozlov, V.N.: The method of minimization of linear functionals based on compakt sets. In: Proceedings of the 12th International Workshop on Computer Science and Information Technologies (CSIT 2010), Moscow – St. Petersburg, vol. 2, pp. 157–159 (2010). (in Russian)
4. Kozlov, V.N.: Smooth systems, operators of optimization and stability of energy systems, 177 p. St. Petersburg Polytechnic University, St. Petersburg (2012). (in Russian)
5. Kozlov, V.N.: A projection method for optimizing optimal limited controls of dynamic systems, 190 p. Publishing and Printing Association of Higher Education Institutions St. Petersburg (2018). (in Russian)
6. Kozlov, V.N., Efremov, A.A.: Introduction to functional analysis, 79 p. Publishing and Printing Association of Higher Education Institutions, St. Petersburg (2018). (in Russian)

# The Concept of an Open Cyber-Physical System

Yury S. Vasiljev, Violetta N. Volkova$^{(\boxtimes)}$, and Vladimir N. Kozlov

Peter the Great St.Petersburg Polytechnic University, Saint Petersburg, Russia
saiu@ftk.spbstu.ru, violetta_volkova@list.ru

**Abstract.** The purpose of this article is the substantiation of the concept of cyber-physical system (CPS) as of an open system with active elements. The development takes place due to the exchange with material objects, energy, information (system openness), the environment and, due to the presence of active elements, which (1) initiate the implementation and interaction of industrial (advanced industrial) and modern information (digital) technologies resulting in a new quality; there arises a law of emergence which is the basement of negentrophic tendencies opposing the law of increasing entropy in closed systems (i.e., the second physical law of thermodynamics), which ensures the development in the open systems theory of L. von Bertalanffy, and (2) maintaining the stability and ability to resist the undesirable results that may arise from simultaneous implementation of heterogeneous innovative technologies. The possibility of using the proposed concept of the CPS as a means for sustainable development of an enterprise is ensured by the fact that CPS is developed not as a separate introduction of technologies, but as a system in which digital technologies have to be included in the enterprise life cycle (in the technological process and organizational management system), and the task is to preserve control functions for the active elements that are part of decision-makers on the selection and implementation of innovations. Developing models for analysis and control the choice of innovative technologies for creating cyber-physical systems for specific enterprises and organizations, we are proposing to use systems theory and systems analysis methods.

**Keywords:** Innovations · Innovative technologies · Cyber-physical system · Methods for complicated expertise organizing · Model · Industrial enterprise · Industrial revolutions · Systems theory · Technologies · Emergence

## 1 Introduction

Nowadays, the intense progress in technologies has initiated the appearance of new concepts and terms that help understanding systems' = development. The term of "Cyber-Physical System" is becoming broadly widespread. It means an information technological concept that implies the integration of computing resources into physical processes.

The term "open system", which was proposed by the Austrian biologist L. von Bertalanffy as the basic of the organismic approach concept, became the basis of a general theory of systems, which allowed to obtain a number of useful results for studying processes in systems of various classes—technical, biological, and socio-economical.

© Springer Nature Switzerland AG 2020
D. G. Arseniev et al. (Eds.): CPS&C 2019, LNNS 95, pp. 146–158, 2020.
https://doi.org/10.1007/978-3-030-34983-7_15

However, the "open system" term is also used in other scientific areas, e.g., thermodynamics, mechanics, chemistry, quantum mechanics, where it reflects the exchange of explored objects with the environment of substances and energy. Recently, this term has been proposed for using in computer sciences; therefore, the concept of an open information system (OIS) originated.

We'll try to show that the concept of an open can promote the formation of the theory of cyber-physical systems.

## 2 Initial CPS Concept

The term of a "cyber-physical system" was proposed in the USA. In a number of papers (for example, [1, 2]) there is information that the term was authored in 2006 by Helen Gill while she was the director for embedded and hybrid systems at the US National Science Foundation. The term was proposed as a title for a seminar created at the initiative of the Department of Advanced Research Projects of the U.S, Department of Defense (DARPA: Defense Advanced Research Projects Agency). Itstask is to manage the development of new technologies for the military forces in order to preserve the technological superiority of the U.S.; prospective studies, bridging the gap between basic research and their use in the military sphere.

It is necessary to implement such grandiose goals in order to create integrated multifunctional complexes consisting of various natural objects, artificial systems and controllers that are combined into a unified whole and include: automated controls, embedded real-time systems, distributed computing systems, systems for technical processes and objects, wireless sensor networks.

Such complexes essentially are automated systems, larger and more intricate than the existing ones, where computers are integrated or built-in into certain physical devices or systems. The study of such complexes requires harmonious coexistence of two types of models: traditional engineering models (mechanical, construction, electrical, biological, chemical, etc.) and computer models, which provide automated management of these processes.

It is for the unification of such heterogeneous models that the theory of cyber-physical systems appeared as an interdisciplinary scientific theory, attempting to combine two independently developing schools:

- The computational (Computer Science), based on mathematical linguistics and theory of algorithms, and
- Of automatic control using integro-differential equations to simulate dynamic processes.

Thus, the essential task was to create a new interdisciplinary concept, which is reflected in the common definition:

"Cyber-physical system (cyber-physical system) is an information technology concept that implies the integration of computing resources into physical entities of any kind, including biological and man-made objects. In cyber-physical systems, the computing component is distributed over the entire physical system, which is its carrier, and is synergistically linked with its constituent elements".

The English-language Wikipedia states that CPSs use interdisciplinary approaches, combining cybernetics, mechatronics, theory and practice of control of specific processes. CPSs differ from the usual embedded systems by better developed connections between computational and physical elements; therefore, an architecture similar to the Internet of Things (IoT) is used.

Consequently, the concept of a CPS should be considered as a new interdisciplinary direction, which combines models of computational and physical processes. In this case, any processes, i.e., manufacturing, servicing, transportation, organizational management, etc. are understood as physical processes.

At the same time, in many interpretations (for example, [3–7] etc.), CPSs are defined as "systems consisting of various natural objects, artificial subsystems and controllers that allow to represent such entity as a whole …", i.e., not as a concept, but as some products and real-life systems.

This is explained by the dialectics of the subjective and objective in such a generic scientific concept as a "system". This term may mean both a philosophical and methodological manifestation of the desired and also the real object. See, for example, V.G. Afanasyev: "… objectively existing systems—and the concept of a system; the concept of a system used as a tool for cognizing a system, and—again a real system, the knowledge of which is enriched by our systemic concepts: such is the dialectics of the objective and the subjective in the system" [8].

This is also explained by the fact that the term of cyber-physical system was used in the German government's project on the computerization of industry, i.e., Industry 4.0, and is associated with the Fourth Industrial Revolution. Exactly after this, it began to be used widely.

However, with such an interpretation, there is a danger of a simplified understanding of what the cyber-physical system is. Therefore, for a deeper oneness of the term cyber-physical system, it is necessary to apply methods and models of the general systems theory.

## 3    Formalized Definition and CPS Modeling Methods

Any system can be represented by a generalized set-theoretic model [21] that can be interpreted for a CPS:

$$S \underset{def}{\equiv} <Z, SiF, TECH, SR, COND, N> \tag{1}$$

where $Z = \{z\}$ is a set of goals or a structure of goals and functions of a system;

$SiF$ is the content of the process of the system and forms of its implementation. This component can be represented as a CPS structure;

$TECH = \{meth, means, alg, …\}$ is a technology set (methods $meth$, $means$, algorithms $alg$ и т. п.) that create a CPS;

$SR$ is the environment with which the system interacts;

$COND = \{\varphi_{ex}, \varphi_{in}\}$ are the conditions in which the development management of an system is exercised; that is, factors affecting the process of managing the operation and development of the system $\varphi_{ex}$ are the external and $\varphi_{in}$ the internal);

$N$ are the observers (according to W. R. Ashby) or stakeholders (according to R. Akkof), that is, those who carry out the structuring of goals, the choice of methods and means of modeling, the development of algorithms and programs, offer new technologies, organize decision-making processes for generating control actions, CPS operators, etc.

As a means of implementing a CPS, the most widely used are

PR – industrial robots; and coordinating their interaction devices;

CV – computer vision systems (image processing, machine vision, visualization, etc.),

3D – 3D printing for prototyping and prototyping; 3D modeling;

AR (augmented reality) – augmented reality technology to create visual "instructions-tips" in the workplace;

VR (virtual reality) – virtual reality technologies for creating physical models, for advertising, promoting product sales;

ABD – big data analysis technologies (Big Data) for on-line decision support;

IIoT (Industrial Internet of Things) – Industrial Internet of Things;

CRM – an automated customer relationship management system ("supplier-customer") and its integration into a single loop of managing end-to-end business processes and data exchange;

M2 M – a set of technologies that allow machines to exchange information with each other, or transmit it unilaterally; these can be wired and wireless monitoring systems for sensors or any device parameters (temperature, inventory, location, etc.), and other technologies.

Technological trends underlying cyber-physical systems are already used individually in different areas, but being integrated into a single whole, they change the existing relationships between manufacturers, suppliers and buyers, as well as between the man and machine.

Therefore, it is natural to assume, that the development of CPSs should be based on the theory of systems, the use of methods and models of system analysis.

Representation of a model in a formalized form helps to preserve the integral representation of the organization of the growth process and the sustainable development of CPSs, not to miss all the components of this process.

The most common methods for modeling CPSs are currently developed in the theory of object-oriented modeling of complex dynamic systems based on the formation of hybrid or integrated dynamic systems which are hierarchical, event-driven systems of variable structure.

With the help of a graph, processes are described by a sequence of local states and behaviors, the change of which takes place under the influence of events.

Formally, a hybrid automaton is described by a set of components:

$$H = \{t, G, V, C, P, A, F\}, \tag{2}$$

where $\{G = S, s_0, E\}$ is a directed graph with vertices mapped to elements of the set of discrete states of the automaton $\{S = s_i | i = 1 \ldots m\}$, and the arcs are possible transitions of the automaton from one state to another $E: S \rightarrow S$. One of the states – $s_o$ is initial; $t \in T \subset R$ – is an independent variable that determines the value of continuous time;

V is a set of variables, including:
C is a set of continuous mappings;
P is the set of logical predicates;
A is a set of instant actions;

F is a mapping that associates a set of continuous mappings to a set of states (graph vertices); $Fp: P \rightarrow E$ is a mapping that associates a set of predicates to a set of transitions (arcs of a graph); $Fa: A \rightarrow E$ is a mapping that associates a set of instant actions with a set of transitions (arcs of a graph).

Thus, a hybrid automaton is a directed graph, with the vertices of which some quality states of the system are associated and continuous actions that are performed while this state is current are assigned.

A hybrid automaton can be viewed as a description of a sequential composition of continuous components that interact through initial conditions.

The most promising is the use of the open system concept of L. von Bertalanffy.

# 4   Justification of the Concept of a Cyber-Physical System as an Open System with Active Elements

The concept of an open system was proposed by von Bertalanffy [9], because he found out from the study of living organisms that the mechanical concept underlying classical science, based on theoretical physics and the laws of thermodynamics, cannot explain the extraordinary order, organization, regulation, continuous changes observed in living organisms, and a new understanding of the problems of controllability and sustainable development of systems is needed. In open systems, in the contrast to closed ones (isolated from the environment), thermodynamic laws manifest, which contradicts with the second law of thermodynamics. In accordance with this origin, the general course of physical events in closed systems occurs in the direction of increasing entropy and reaching the state of maximum disorder. At the same time, in open systems in which the transfer and transformation of substance takes place, in accordance with the concept of L. von Bertalanffy: "...e the introduction negentropy is quite possible that is, a decrease in entropy; ... such systems can maintain their high level of order and even develop in the direction of increasing the order of complexity" [10, p. 42].

L. von Bertalanffy has, actually, discovered a new law for open systems, i.e., "the ability to resist entropic (destructive) tendencies and to show negentropical tendencies", which contradicts with the second law of thermodynamics.

An open system, in the contrast to a closed (isolated) one under particular conditions, according to Bertalanffy, reaches a state of mobile equilibrium, in which its structure remains constant. But in contrast to the usual equilibrium, this constancy is maintained in the process of continuous exchange and movement of matter [10, p. 42].

In parallel, Bauer investigated one of the fundamentally important for understanding of the development of living systems pattern of fundamental non-equilibrium, that is, the urge to maintain a stable imbalance and use the energy to maintain a non-equilibrium state. E. Bauer explains this by the fact that all structures of living cells at the molecular level are pre-charged with "extra" energy compared to the same-type non-living molecule, and the organism receives external energy not for work, but for maintaining its non-equilibrium structure [11].

E. Bauer formulated the principle of sustainable imbalance of living systems as "... living systems are never in equilibrium and at the account of their own free energy constantly perform work against the equilibrium compulsory by the laws of physics and chemistry under the existing external conditions" [12].

This principle serves to make a fundamental difference between a working living system and a working mechanical system or machine, which is expressed in inequality of potentials and in the electric gradient created, whereas in an inanimate closed system, any gradients are distributed evenly according to the entropy. This "extra" energy that exists in living cells at any level Bauer calls "structural energy" and means deformation and imbalance in the structure of a living molecule.

The meaning of the principle of sustainable disequilibrium lies in the biophysical aspects of the area of energy movement in living systems. E. Bauer argues that the work done by this living cell structure is performed only by disequilibrium, and not by the energy coming from outside, while in a machine the work is done directly from an external energy source. The body uses the energy coming from outside not for working but only for maintaining the "excess energy" in living cells. "Consequently, in order to preserve them, that is, the conditions of the system, it is necessary to constantly renew them, that is, to constantly expand work. Thus, the chemical energy of food is consumed in the body to create a free energy structure, to upgrade, preserve this structure, and not to directly turn it into work" [12].

Due to the Bertalanffy law and E. Bauer principle, the system exhibits the ability to resist entropic (destructive) tendenciesand show adaptability, i.e., the ability to adapt to changing environmental conditions and interference, both external and internal; the ability to develop behaviors and change their structure (if necessary) while maintaining the integrity and basic properties; the ability to pursuit goal setting.

The cited features have a variety of manifestations, which can sometimes be identified as independent characteristics. The features are contradictory. In most cases, they are dual in the nature, are both positive and negative, desirable and undesirable for a socioeconomic system. On the one hand, there are among them the properties that are useful for the existence of the system, e.g., its adaptability to changing environmental conditions, but at the same time, these features cause uncertainty, non-stationarity of parameters, instability of the system functioning, unpredictability of it behavior.

Studies have shown that the features of open systems and the laws that explain them are cause by the presence of active elements that stimulate the exchange of material, energy and information products with the environment and appear to be their own "initiatives", i.e., an active principle. Due to this, the regularity of entropy increase in such systems is infringed and negentropic tendencies can be observed, i.e., self-organization and development

Thus, the development of an open system occurs (1) thanks to the exchange of information, energy, material components (i.e. due to openness of the system) with the environment and (2) due to active elements that initiate their own innovations and ensure the interaction of innovations.

Understanding of the features of open systems allows us to understand that the term of "design" does not in principle apply to living objects and systems that we refer to as open systems. A living system, the study of which became the basis of Bertalanffy's organismic approach, cannot be "assembled" from parts. Thus, a system can only "grow", develop, adjust, and we may influence the process of moving towards the desired state by controlling this process.

Therefore, the role of the cybernetic (managerial) component adopted in a number of works on CPSs is important, which should ensure the new quality of production management, the transition to a qualitatively new automatics of machine-to-machine exchange among everyone in the framework of CPS projects, which is provided by new technologies, i.e., intelligent sensors, wireless communication, etc.

Taking into account the peculiarities of cyber-physical systems, they can be considered as an chance to get the effect of emergence, that is, the emergence of new quality of processes in a system, and at the same time, resist undesirable results when introducing emergent technologies, since a CPS does not develop as an isolated technology, and let us see how the systems, in which the management of a complex of technologies is supposed are. It can be expected that the active element, e.g., a person will retain the control functions and decision- making role concerning the feasibility of their implementation.

The most promising form ensuring the development of an enterprise is engineering in the original understanding of this term, which arose in Europe in the 16th century (engineering, from the Latin ingenium – inventiveness, invention, knowledge), i.e. not only computer engineering (software for engineering analysis and design), but first of all the use of scientific and technical knowledge to create systems, devices, materials, organization of production processes and management of production processes and enterprises in general.

In this understanding, engineering can be considered as the living cells likeness, which is containing some "excess energy", according to E. Bauer, and in socioeconomic systems – information that initiates innovations for the development of an enterprise "organism", i.e., is implemented on the basis of L. von Bertalanffy organismic approach.

The principle of sustainable disequilibrium of E. Bauer, with regard to enterprises, can be interpreted as the need to constantly maintain the existence of energy or, rather, information in certain structures, constantly taking relevant actions on this, which is the task of engineering.

## 5   Using Open Systems Theory Laws in the Research on the Problem of Sustainable Development of CPSs of an Enterprise

The development of intellectual technologies will lead to the fact that more and more functions that previously only people could perform would be transferred to artificial intelligence systems. At the same time, as a result, unpredictable consequences may arise, among which may be such that would affect the development of the enterprise both positively and negatively, or even be dangerous. Schwab [13] predicts that technologies open up new opportunities, but at the same time he also predicts the opposite effect of the introduction of technologies, especially in the context of their combination and the occurrence of an emergent effect.

According to K. Schwab, in its scale, dimensions and complexity, "The fourth industrial revolution has no analogues in the previous experience of mankind. New technologies unite the physical, informational and biological worlds; they are capable of creating, on the one hand, enormous opportunities, and on the other hand be a potential threat". [13]. At the same time, K. Schwab predicts that initially innovative technologies might be used separately, but "the turning point would soon come when they would begin to develop, stratifying and strengthening each other, representing the interweaving of technologies from the world of physics, biology and digital realities" [13]. "Technologies will help find solutions to many problems that we are facing today, but they will exacerbate some of these problems" [14, p. 273].

Research conducted in the theory of innovation showed that any innovation disrupts the normal functioning of enterprises and organizations, create a situation of "creative destruction" (by Schumpeter [15] and Sombart [16]); Christensen [17] even introduces the term of "disruptive innovations". Therefore, it is necessary to develop models for managing the sustainable development of enterprises and organizations in the conditions of introducing and using innovations, especially fundamentally new emergent technologies (for example, [18–20]).

Studies of entropic-negentropic processes in open systems show that those manifest themselves ambiguously. On the one hand, the negentropic tendencies realized in the form of innovations are the basis of development, but at the same time, they destabilize the system, introduce irregularity and disorder ("creative destruction" according Schumpeter [15] and Zombart [16]). Entropic tendencies, which were considered as manifestations of disorder, on the contrary, stabilize the state of the system, since the minimal energy state to which entropic processes lead is the most stable.

The laws of the theory of systems help to understand these contradictions, and allow to evaluate the degree of manifestation of entropic and negentropic tendencies in a system. The main ones are as follows:

1. The law of integrity or emergence which can be formulated in the following way: the properties of a system $Q_s$ is not a simple sum of the properties of its elements (parts) $q_i$.
2. The law of the additivity or the summability, which characterizes the decay of a system into parts.

3. The law of hierarchy or hierarchical ordering leads to an intensification of the process of the emergence of new properties of any system, including unpredictable and uncontrolled properties. This law was among the first laws of the theory of systems which founder von L. Bertalenfy believed that "... living systems can be defined as hierarchically organized open systems that preserve themselves or develop in the direction of achieving a state of mobile equilibrium" [10, p. 41].

To control the sustainable development of a system it is necessary to monitor the state of preserving its integrity and sustainability. In this paper it is proposed to develop the models for estimation the integrity of a system using the information approach of Denisov [18]. In the Denisov's information approach comparative quantitative appraisals of hierarchical structures are introduced from the point of view of the degree of integrity ($\alpha$) and the coefficient of utilization of the elements of a system as a whole ($\beta$) (see Eqs. 5 and 6).

In accordance with the information approach the information complexity of a system ($C$) can be estimated as follows:

$$C = J \cap H, \tag{3}$$

where $J$ is the information of perception;

$H$ is the information essence (potential), in other words, it is the importance of the measured information for functioning and development of a system.

Information of perception $J$ and potential $H$ can be measured probabilistically and deterministically [18, 21].

The different types of the information complexity of a system in the approach of A. A. Denisov are disclosed in the following equation:

$$C_s = C_i + C_m, \tag{4}$$

where $C_s$ is an estimate of the system complexity;

$C_i$ is an estimate of the intrinsic complexity of a system;

$C_m$ is an estimate of the mutual complexity of a system.

The degree of integrity of a system ($\alpha$) and the coefficient of utilization of the elements of a system as a whole ($\beta$) can be defined as follows:

$$\alpha = -C_m / C_i, \tag{5}$$

$$\beta = C_s / C_i. \tag{6}$$

Researches of A. A. Denisov showed that any developing system is in an intermediate state between the state of absolute integrity ($\alpha = 1$) and the state of absolute freedom of elements ($\beta = 1$):

$$\alpha + \beta = 1. \tag{7}$$

Integrity provides stability, and it can be assumed that the increasing integrity of a system should increase the efficiency of existence and development of this system.

However, studies of the processes of interaction between systems as a whole and their elements showed that the efficiency of a system increases with the increasing of regulation (the degree of integrity) only to a certain limit, and when a system becomes over-regulated, its efficiency begins to decline. The reason it occurs is because the initiatives that promote the development of a system are suppressed, and it negatively affects the efficiency of a system, and, in addition, the suppression of active elements causes their resistance, which reduces the safety of a system, and in the future may lead a system to destabilization and even to death.

To study the problems of managing the development of CPSs, it is useful to apply other laws of the theory of systems (e.g., the potential effect of B. Fleischman, the law of historicity).

It is predicted that the new technologies for CPSs, which are becoming increasingly fantastic, the automatic machine-to-machine exchange (M2M) systems for everyone with everyone will create a unified information space of an enterprise with practically simultaneous information support of all components of the system, which can lead to continuous transformation of structures and to unpredictable consequences, which can both positively and negatively affect the development of the enterprise. Therefore, when working on modernization projects for production automation systems, it is important to make tests on interoperability of automated equipment and software of new technologies; it is necessary to develop models for managing the sustainable development of the enterprise.

In view of this, a problem of comparative analysis and selection of innovative technologies arises, taking into account their features, capabilities, usefulness and consequences of the latter and their implementation. This begins to be realized and considered in developing models for the selection and management of innovative technologies introduction (for example, [22–24]), control of sustainable development [25, 26], training of the personnel for working in the new information environment [27]. In developing these models, methods and models of systems theory are used (for example, [21, 28], and others).

## 6 Conclusions

The article aims to substantiate the concept of the cyber-physical system as an open system with active elements, ensuring the sustainable development of an enterprise.

The proposed concept of a cyber-physical system is based on the idea that CPSs, on the one hand, can be viewed as an opportunity to get the effect of emergence, that is, the emergence of a new quality of production processes, from the combination of industrial and information technologies and, on the other hand, to avoid undesirable results when combining technologies, since CPSs, combining digital technologies and physical components, are developed not as separate technologies but as a system in which technologies are incorporated into a technological process or even a full life cycle (from the order to sales) of an enterprise; we can hope that management functions will be retained by decision-makers regarding the choice and implementation of technologies, ensuring sustainable development.

Despite the predicted development of intelligent technologies, there are management functions, such as purpose-setting, planning, developing a system of values and criteria ensuring the sustainable development of an enterprise, that currently cannot be automated, and some researchers, who initially hoped to automate such intellectual functions, begin to doubt about the possibility of automation of those in the future. This allows us to hope that human beings will keep control over the development of CPSs, meaning individuals who develop control actions for the development of the enterprise's CPS.

Since the development takes place not only on the basis of the exchange with the environment but also through active elements, the further research involves the study of the forms and methods of using the idea of engineering, ensuring the development of the production process, and the enterprise as a whole through the use of active elements of technical knowledge for creating products and organizing production and management processes, analyzing enterprise's development goals and developing criteria for assessing the emergence effect of integration and information and advanced industrial technologies, both in terms of increasing the productivity of labor and the quality of products, which significantly depends on the specific types and conditions of production processes.

When developing models for analyzing and managing the development of enterprises' CPSs, it is proposed to use the laws of the theory of open systems and methods of system analysis.

# References

1. Cyber-physical system. https://ru.wikipedia.org
2. Chernyak, L.: The Internet of Things: New Challenges and New Technologies. Open Systems DBMS (4), 14–18 (2013). http://www.osp.ru/os/2013/04/13035551
3. Lee, E.A., Seshia, S.A.: Introduction to Embedded Systems. A Cyber-Physical Systems Approach. (2011). LeeSeshia.org
4. Colombo, A.W., et al. (eds.): Industrial Cloud-Based Cyber-Physical Systems. Springer, Cham (2014). https://doi.org/10.1007/978-3-319-05624-1
5. Kupriyanovskiy, V.P., Namiot, D.E., Sinyagov, S.A.: Kiberfizicheskiye sistemy kak osnova tsifrovoy ekonomiki [Cyber-physical systems as the basis of the digital economy]. Int. J. Open Inf. Technol. 4(2), 18–25 (2016)
6. Dobrynin, A.P., et al.: Tsifrovaya ekonomika - razlichnyye puti k effektivnomu primeneniyu tekhnologiy [Digital economy - various ways to efficiently applied technology] (BIM, PLM, CAD, IOT, Smart City, BIG DATA ets). Int. J. Open Inf. Technol. 4(1), 4–11 (2016)
7. Lee, E.: Cyber Physical Systems: Design Challenges. University of California, Berkeley Technical Report No. UCB/EECS-2008-8. Retrieved 2008-06-07 (2008)
8. Voprosy filosofii [Questions of philosophy], 6, 62–78 (1980)
9. von Bertalanffy, L.: General system theory. Gen. Syst. 1, 1–10 (1956)
10. von Bertalanffy, L.: Obshchaya teoriya sistem: kriticheskiy obzor [General Systems Theory: A Critical Review], Issledovaniya po obshchey teorii sistem. Moscow, Progress, pp. 23–82 (1969). (in Russian)

11. Bauer, E.S.: Fizicheskiye osnovy v biologii [Physical fundamentals in biology], Izd. Mosoblzdravotdel, Moscow, 103 p. (1930)
12. Bauer, E.S.: Teoreticheskaya biologiya [Theoretical Biology], VIEM, Moscow-Leningrad, 206 p. (1935)
13. Schwab, K.: The Fourth Industrial Revolution. Portfolio/Penguin, 184 p. (2017)
14. Schwab, K., Devis, T.: Tekhnologii chetvertoy prpomyshlennoy revolyutsii [Shaping the Fourth Indusyrial Revolution], 320 p. "E" Publication, Moscow (2018). (in Russian)
15. Shumpeter, I.: Teoriya ekonomicheskogo razvitiya [The Theory of Economic Development]. Progress Publication, Moscow (1982)
16. Zombart, V.: Sobraniye sochineniy. Tom 3. Roskosh' i kapitalizm. Voyna i kapitalizm [Collected Works. Volume 3. Luxury and capitalism. War and capitalism]. Izd-vo Vladimir Dal', Moscow (2008)
17. Kristensen, K.: Dilemma innovatora: Kak iz-za novykh tekhnologiy pogibayut sil'nyye kompanii [Innovator's Dilemma: How Strong Companies Perish Due to New Technologies]. Al'pina Business Books, Moscow (2004)
18. Denisov, A.A.: Modern Problems of System Analysis: A Textbook. SPb.: Publishing house Polytechnic. University, 304 p. (2008). (in Russian)
19. Volkova, V.N., Loginova, A.V., Leonova, A.Y., Chernyy, Y.Y.: Podkhod k sravnitel'nomu analizu i vyboru tekhnologicheskikh innovatsiy tret'yey i chetvertoy promyshlennykh revolyutsiy [An approach to comparative analysis and selection of technological innovations of the third and fourth industrial revolutions]. XXI Mezhdunarodnaya konferentsiya po myagkim vychisleniyam i izmereniyam (SCM-2018), pp. 373–376. Sb. doklady, St. Petersburg (2018)
20. Volkova, V.N., Kudryavtseva, A.S.: Modeli dlya upravleniya innovatsionnoy dey-atel'nost'yu promyshlennogo predpriyatiya [Models for managing the innovation activities of industrial enterprises]. Otkrytoye obrazovaniye **22** (4), 64–73 (2018). https://doi.org/10.21686/1818-4243-2018-4-64-73, (in Russian)
21. Volkova, V.N., Denisov, A.A.: Teoriya sistem i sistemnyy analiz: uchebnik [Systems Theory and Systems Analysis], 616 p. Izd-vo Yurayt, Moscow (2014). (in Russian)
22. Volkova, V.N., Gorelova, G.V., Efremov, A.A.: Modelirovanie sistem i protsessov. Prak-tikum [Modeling Systems and processes], 295 p. Izdatel'stvo Yurait Publ., Moscow (2016). (in Russian)
23. Volkova, V.N., Vasiliev, A.Y., Efremov, A.A., Loginova, A.V.: Information technologies to support decision-making in the engineering and control. In: Proceedings of 2017 20th IEEE International Conference on Soft Computing and Measurements, SCM 2017, vol. 20, pp. 727–730 (2017)
24. Volkova, V.N., Efremova, A.A. (eds.): Informatsionnyye tekhnologii v sisteme upravleniya [Information technology in control systems], 408 p. Polytechnic University Publishing House, St. Petersburg (2017). (in Russian)
25. Volkova, V.N., Loginova, A.V., Chernenkaja, L.V., Romanova, E.V., Chernyy, Y.Y., Lankin, V.E.: Problems of sustainable development of socio-economic systems in the implementation of innovations. In: Proceedings of the 3rd International Conference on Human Factors in Complex Technical Systems and Environments, Ergo 2018, vol. 3, pp. 53–56 (2018)

26. Volkova, V.N., Lankin, V.Y.: Problema ustoychivosti sotsial'no-ekonomicheskoy sistemy v usloviyakh vnedreniya innovatsiy chetvertoy promyshlennoy revolyutsii [The problem of the sustainability of the socioeconomic system in the context of innovation of the fourth industrial revolution]. Ekonomika i upravleniye: problemy i resheniya **77**(6), 25–29 (2018)
27. Volkova, V.N., Kozlov, V.N., Karlik, A.E., Iakovleva, E.A.: The impact of NBIC-technology development on engineering and management personnel training. In: Strategic Partnership of Universities and Enterprises of Hi-Tech Branches (Science Education Innovation), vol. 6, pp. 51–54 (2018)
28. Volkova, V.N., Kozlov, V.N. (eds.): Modelirovanie sistem i protsessov [Modeling Systems and Processes], 450 p. Yurait Publ., Moscow (2016). (in Russian)

# Distributed Knowledge Base of Intellectual Control in Smart Grids

Dmitry Kostenko, Dmitry G. Arseniev, and Vadim A. Onufriev$^{(\boxtimes)}$

Peter the Great St. Petersburg Polytechnic University, Saint Petersburg, Russia
onufriev_va@spbstu.ru

**Abstract.** The article proposes a solution for low-efficiency and inadequate power distribution in zero-energy building grids. Distributed knowledge base application enables power grid segments with self-sustainability and amplification of zero-energy building efficiency. Proposed methodology of smart grids development is headed towards the optimal power generation and consumption control, adjusted power spike handling and improvements in operational reliability. Key performance indicators of various levels and hardware restrictions are taken into account. The system is able to function based on already existing equipment with minor adjustments. According to the proposed algorithm, a computer model is written. The model creates randomly generated power grid segments consisting of zero-energy buildings, non-producing buildings and supportive equipment. Each individual building receives its own power consumption and production levels. Those parameters are measured and analyzed within two series of experiments. The conducted tests showed efficiency of the proposed logic and measure potential power savings within the worst-case and the best-case scenarios and use cases.

**Keywords:** Smart grid · Distributed knowledge base · Intellectual control · Zero energy building

## 1 Introduction

Digitalization became a world-wide trend of industrial development. It includes such concepts as Industry 4.0, Industrial Internet of Things [1], Smart Grid [2], Digital Twin and Multiagent Systems [3].

However, fields like finance and paperwork management are beginning to popularize another direction called Distributed Knowledge Bases, also referred to as Distributed Registers or Blockchains [4]. The latter can be used to decentralize complex control systems to make them faster and more reliable. Removal of a central hub from a control system forces its distributed parts to communicate directly to each other, thus increasing the speed of cooperation.

Still, the technology of Distributed Registers fits the aforementioned conception of Industry 4.0, as it directly includes control system decentralization and self-sustainability of its agents. That's why the Blockchain technology can be used as a tool of practical implementation of the distributed control systems [5].

D. G. Arseniev et al. (Eds.): CPS&C 2019, LNNS 95, pp. 159–168, 2020.
https://doi.org/10.1007/978-3-030-34983-7_16

The smart grid concept is seen as a high-priority advancement for the electricity industry. It suggests that every part of an electricity supply system, including electricity power producers, consumers and supportive intermediate units, are able to operate without any need for a central control hub. All the agents communicate directly and make effective collaborative decisions. Effectiveness of the aforementioned decisions can be evaluated using the Key Performance Indicators (KPI), such as voltage, frequency, etc.

The task of the distributed control system is to keep all the KPIs within the scope given. In order to grant it, every agent of the system must acquire a complete set of information about its role. A reliable information flow between the agents is vitally important to support an effective collaboration towards the global goal.

Figure 1 shows an informational model. It is a functional scheme of a control system digital twin [6].

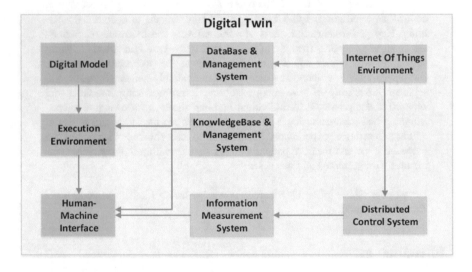

**Fig. 1.** A functional scheme of a digital twin of a control system

The most important parts are the knowledge base control system and the database control system.

Digital twin's knowledge base is a place to store the set points. The data base is used to store both actual and archived production data, which can be analysed to produce a digital model.

The aim of this work is to propose a working algorithm for the distributed control system of an energy delivery grid and to test a model of it to prove its effectiveness. Effectiveness is measured according to (1), where *ppc* stands for power plant consumption (amount of electricity bought from the power plant), c stands for power consumption of buildings, p for power production of buildings, and S is a set of vectors, representing connections between the buildings to allow for building-to-building power redistribution.

$$ppc = f(c, p, S) \rightarrow \min \tag{1}$$

The *ppc* value is set to be minimized and the effectiveness shows the difference between an ordinary power consumption (100%) and optimized power consumption.

## 2    Distributed Logic

However, Fig. 1 gives no information about the way distributed components function. To arrange an informational flow between electricity producers and consumers, distributed registers are used [7].

Figure 2 shows a configuration of an electricity distribution system, containing one power plant and multiple zero-energy buildings (ZEB).

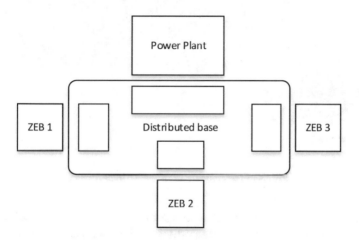

**Fig. 2.** Power network scheme

Power plant and ZEBs have their own low-level KPIs, which regiment smaller-scale parameters. These include temperature, voltage, current, etc., but not the whole power plant or a whole building as a part of a smart grid. Low-level KPIs are stored in a register and are initially taken from technical documentation sheets.

Top-level KPIs regiment amounts to power generated by a power plant or a ZEB and other similar parameters. In order to count them, one needs to acquire detailed information from the other members of a network (or a network segment).

This information can be provided by a set of special planning units. They compute top-level KPIs and save the data into a register, thus making it available to every member of the network.

Network members also save archive data of their working parameters, so that external computational units could use it to build mathematical models and compute KPIs. Registers are used to send those KPIs back to the network members.

On Fig. 2 power plant's planning unit sends the following data to its agents (ZEBs) (see Fig. 3):

- Global set points
- Emergencies
- Global logic

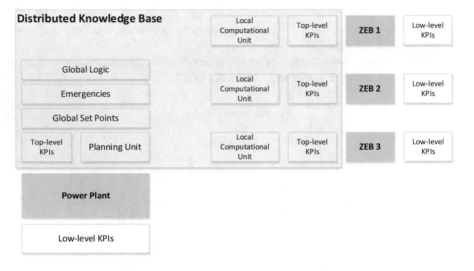

**Fig. 3.** Overall scheme of a distributed base.

Global set points include alternating current frequency, maximal load for different segments of the network and its supportive equipment, etc. These are the basic parameters that should not be altered under normal operational circumstances.

Emergencies block covers any type of force majeure circumstances that can change the network layout: power lines being damaged, power substations failures, etc. A part of this data can be acquired via member-to-member communication, but having a central hub boosts the system's stability.

Global logic comprises the most common information about the grid, such as possible alternate routes, usable lines and the basic communicational parameters. The global logic block is needed to establish the rules of collaboration and give the grid members ability to continue working without a constant connection to the central hub.

The local computational unit of every grid member is tasked to control both levels of KPIs and collaborate with other members. Data known to every member is written into a corresponding register. Coupled with information from the central hub, it forms a pool of information required to make decisions.

## 3 Network Control Algorithm

The goal of every ZEB control system is to minimize the difference between the amount of power taken and returned back to the power grid, and to nullify it if possible. The proposed network control algorithm consists of the following steps.

1. Every agent that has been newly connected to the grid gets a set of global set points. Until that the agent is forbidden to give electricity into the grid for safety reasons.
2. After that the agents compute their own KPIs.
3. An agent subscribes to track changes in both Emergencies and Global Logic pools. Received data is sent to the agent's local computational unit. A power plant's planning unit receives a confirmation.
4. Power plant's Global logic set is rebuilt with the following steps:

   - Building a graph of agents. ZEBs and substations (and other supportive equipment) form the graph's nodes. Power lines form the graph's edges
   - A newly-built graph is divided into groups of preferable collaboration. These groups are formed based on geographical position of their members and the grid's supportive equipment capabilities (see Fig. 4). Energy exchange inside these groups is the most effective in minimizing the power transportation losses

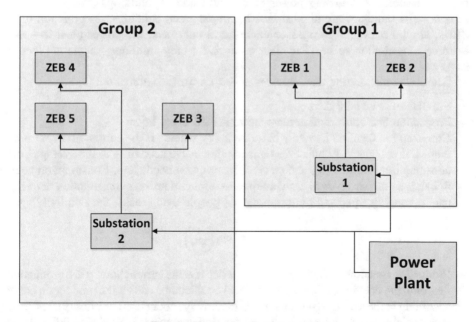

**Fig. 4.** Groups localization scheme

   - Given enough agents, they can be organized into multiple levels of hierarchy
   - Members of the newly-formed groups exchange their top-level KPIs. Local computational units store this information and subscribe for updates

- Collaboration between members of different groups is possible via simplified "request energy" – "propose energy" tickets exchange, but in this case the KPIs are not shared
- After all the groups have been formed, agents can change their affiliation based on better KPI compatibility or known statistical data
- Graph edges are complemented with information about accessible supportive equipment to account for transmission capacitance, power cable condition, safety restrictions, etc.
- After the groups are formed, the Global Logic becomes available for every agent of the grid
- In case of emergency, local computational units combine information about the power grid damages with Global Logic statements.

# 4  Computer Simulation

In order to conduct a computer simulation, a customizable power grid model has been written in C# (see Fig. 4). The model contains three types of buildings: a power plant that produces electricity; two substations that stand between the power plant and the power consumers conducting power conversion; and 20 buildings (the power consumers). The buildings are divided into 2 groups: green buildings, also referred to as ZEBs, are the ones able to produce energy and only need the power plant to cover extreme consumption spikes. The other group is for grey buildings, unable to produce energy at all.

The following starting parameters were set up for the simulation:

- Simulation covers 90 days
- Production and consumption rates are refreshed every hour
- Consumption function for every building is (2), where $t$ is the current hour, $3.7$ is an empirical coefficient to offset the consumption curve according to the averaged out consumption statistics [8], and $cc$ is a consumption coefficient. Consumption coefficient is a number given to each building to differentiate their consumption rates and can be roughly equalized to the amount of people living inside the said building:

$$c = cc \cdot \sin\left(\frac{t}{3.7}\right) \qquad (2)$$

- Production function for every ZEB is (3), where $t$ is the current hour, $tf$ is a production randomizing coefficient, 3.7 is an empirical coefficient to offset the production curve according to the averaged out solar panel electricity production statistics [9], and $pc$ is an electricity production coefficient. Production coefficient is needed to differentiate power generating equipment possibilities between different ZEBs.

$$p = -pc \cdot \cos\left(\frac{t+tf}{3.7}\right). \qquad (3)$$

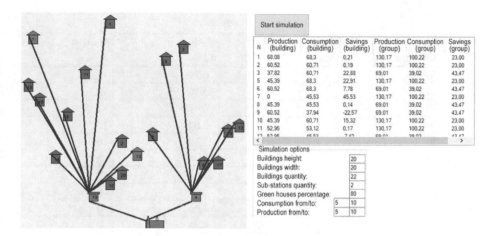

**Fig. 5.** Power grid simulation

As seen in Fig. 5, each building counts its production, consumption and savings. Formula (4) shows how the power saving value ($s$) is calculated based on production per hour ($ph$) and consumption per hour ($ch$) values. Here $m$ is the amount of hours, and for a 90 day long simulation it equals to 2160. Group production, consumption and savings are the sums of individual building parameters.

$$\sum s = \sum (ph_n + ph_{n+1} + \ldots + ph_m) - \sum (ch_n + ch_{n+1} + \ldots + ch_m) \qquad (4)$$

According to the aforementioned limitations, two series of experiments have been conducted. The first series included 25 pseudo-random power grid configurations, consisting of 1 power plant, 2 substations and 20 consuming buildings. Consuming buildings were differentiated into "green" (ZEBs) and "grey" (non-producers). Consumption and production coefficients were adjusted for average consumption to exceed average production by 20%.

**Table 1.** Lowered production simulation

| No | Total buildings | Non-producing (grey) buildings | Savings in group 1, % | Savings in group 2, % |
|----|-----------------|-------------------------------|----------------------|----------------------|
| 1 | 20 | 6 | 5.36 | 0 |
| 2 | 20 | 3 | 0 | 7.15 |
| 3 | 20 | 3 | 8.48 | 12.53 |
| 4 | 20 | 2 | 5.92 | 0 |
| 5 | 20 | 3 | 3.78 | 51.90 |
| 6 | 20 | 6 | 0 | 37.70 |
| 7 | 20 | 4 | 21.47 | 4.65 |

(continued)

**Table 1.** (*continued*)

| No | Total buildings | Non-producing (grey) buildings | Savings in group 1, % | Savings in group 2, % |
|----|-----------------|--------------------------------|-----------------------|-----------------------|
| 8  | 20 | 4 | 6.76  | 7.67  |
| 9  | 20 | 4 | 15.10 | 9.36  |
| 10 | 20 | 2 | 0     | 17.19 |
| 11 | 20 | 1 | 3.84  | 15.37 |
| 12 | 20 | 4 | 2.16  | 6.61  |
| 13 | 20 | 5 | 13.75 | 5.75  |
| 14 | 20 | 3 | 0     | 0     |
| 15 | 20 | 4 | 0     | 6.86  |
| 16 | 20 | 6 | 4.61  | 0     |
| 17 | 20 | 4 | 4.35  | 0     |
| 18 | 20 | 2 | 3.72  | 7.35  |
| 19 | 20 | 4 | 0     | 2.32  |
| 20 | 20 | 2 | 17.70 | 5.87  |
| 21 | 20 | 5 | 9.87  | 0     |
| 22 | 20 | 4 | 4.42  | 0     |
| 23 | 20 | 4 | 0     | 2.95  |
| 24 | 20 | 3 | 54.71 | 3.32  |
| 25 | 20 | 3 | 14.72 | 0     |
|    |    | Average: | 8.03% | 8.18% |

The "Average" cell numbers were computed according to expression:

$$a_{avg} = \frac{1}{25} \sum_{i=1}^{25} a_i. \tag{5}$$

The first series of experiments (see Table 1) shows that within the given conditions we were able to lower the amount of electricity bought from the power plant or *ppc* value (see Eq. 1) by 8%.

The second series of experiments includes equal coefficients of power production and consumption. It means that at average ZEBs are able to produce enough energy to cover their needs.

**Table 2.** Equal production simulation

| No | Total buildings | Non-producing (grey) buildings | Savings in group 1, % | Savings in group 2, % |
|----|-----------------|--------------------------------|-----------------------|-----------------------|
| 1  | 20 | 2 | 38.78 | 29,70 |
| 2  | 20 | 4 | 23.91 | 5,69  |

(*continued*)

**Table 2.** (*continued*)

| No | Total buildings | Non-producing (grey) buildings | Savings in group 1, % | Savings in group 2, % |
|----|-----------------|-------------------------------|----------------------|----------------------|
| 3  | 20 | 4 | 26.99 | 7,94 |
| 4  | 20 | 3 | 38.30 | 0 |
| 5  | 20 | 5 | 20.13 | 0 |
| 6  | 20 | 2 | 57.95 | 0 |
| 7  | 20 | 6 | 31.49 | 16,38 |
| 8  | 20 | 5 | 3.04 | 40,29 |
| 9  | 20 | 4 | 40.96 | 19,86 |
| 10 | 20 | 4 | 24.61 | 27,10 |
| 11 | 20 | 4 | 14.00 | 11,41 |
| 12 | 20 | 3 | 20.89 | 31,51 |
| 13 | 20 | 5 | 10.34 | 33,31 |
| 14 | 20 | 3 | 23.79 | 7,90 |
| 15 | 20 | 2 | 35.94 | 46,09 |
| 16 | 20 | 5 | 13.80 | 0 |
| 17 | 20 | 4 | 37.26 | 20,26 |
| 18 | 20 | 5 | 24.93 | 17,18 |
| 19 | 20 | 5 | 7.24 | 8,84 |
| 20 | 20 | 5 | 4.10 | 24,60 |
| 21 | 20 | 6 | 7.06 | 34,63 |
| 22 | 20 | 4 | 19.44 | 16,54 |
| 23 | 20 | 5 | 27.23 | 34,68 |
| 24 | 20 | 4 | 16.11 | 26,65 |
| 25 | 20 | 9 | 4.06 | 1,79 |
|    |    | Average: | 22,89% | 18.49% |

According to Table 2, *ppc* savings are ranging from 18.49% to 22.89%. Compared to the first series, we can also tell that the amount of non-generating buildings is less impactful to the overall power savings.

It must be noted that the aforementioned statistics are not accounting for the inevitable loss of power during the transmission from one ZEB to another, converting voltages, etc.

## 5  Conclusion

During the work we have analyzed existing methods of creating digital twins and looked into schemes of creating a distributed knowledge base for a power grid, consisting of a power plant and ZEBs. A computer model was created based on the proposed logic and its efficiency was measured during a series of experiments. Experiment results have been evaluated under the list of aforementioned assumptions.

The model can be extended by taking into account specific power grid parameters, such as alternate routes (for more effective power transmission), transmission power losses, supportive equipment requirements, reservation and failures. More sophisticated model can help improving the overall logic.

# References

1. Weyrich, M., Ebert, C.: Reference architectures for the internet of things. IEEE Softw. **33**(1), 112–116 (2016)
2. Anku, N., Abayatcye, J., Oguah, S.: Smart grid: an assessment of opportunities and challenges in its deployment in the Ghana power system. In: 2013 IEEE PES Innovative Smart Grid Technologies Conference (ISGT), Washington, DC, pp. 1–5 (2013)
3. Bulgakov, S.V.: Primenenie multiagentnykh sistem v informacionnykh sistemakh [Application of multi-agent systems in information systems]. Perspektivy nauki i obrazovaniya [The prospects of science and education] **5**(17), 136–140 (2015). (in Russian)
4. Fu, Y., Zhu, J.: Big production enterprise supply chain endogenous risk management based on blockchain. IEEE Access **7**, 15310–15319 (2019)
5. Miller, D.: Blockchain and the internet of things in the industrial sector. IT Prof. **20**(3), 15–18 (2018)
6. Kostenko, D., Kudryashov, N., Maystrishin, M., Onufriev, V., Potekhin, V., Vasiliev, A.: Digital twin applications: diagnostics, optimisation and prediction. In: Proceedings of the 29th DAAAM International Symposium. DAAAM International, pp. 574–581 (2018)
7. Combin, N.N.: Tehnologii raspredelennogo reestra [Distributed register technologies] Nauchnoe soobschestvo studentov XXI stoletija. Tekhnicheskie nauki. Sbornik statei po LI studenchecskoi nauchno-practicheckoi konferencii. [Scientific community of XXI Century. Engineering Science. In; Proceedings of the LI student Scientific and Practical Conference], vol. 3, no. 50 (2018). (in Russian)
8. Analiz sutochnogo grafika potrebleniya elektricheskoi energii naseleniem [Analysis of the daily electricity consumption of the population]. http://www.energosbit.net/day.html. Accessed 15 Mar 2019. (in Russian)
9. Chernenko, A.N., Kryukov, P.V.: Energosberezhenie i malaya solnechnaya energetika dlya mnogokvartirnogo doma v usloviyakh RF [Energy saving and small solar power for multifamily homes in Russia]. Sovremennye problemy nauki i obrazovaniya [Modern problems of science and education] **1**(1) (2015). (in Russian)

# Methodology of Complex Objects Structural Dynamics Proactive Management and Control Theory and Its Application

Boris Sokolov[1,2]($\boxtimes$), Vyacheslav Zelentsov[1], Nikolay Mustafin[1], and Vadim Burakov[1]

[1] St. Petersburg Institute for Informatics and Automation of the Russian Academy of Sciences, Saint Petersburg, Russia
sokolov_boris@inbox.ru, v.a.zelentsov@gmail.com, nikolay.mustafin@gmail.com, v.v.burakov@gmail.com
[2] St. Petersburg State University of Aerospace Instrumentation, Saint Petersburg, Russia

**Abstract.** Methodological and methodical fundamentals of the complex objects (CO) proactive management and control theory based on the fundamental results obtained in the interdisciplinary field of system knowledge are proposed. The paper provides information on the developed innovative multiple-model complexes, combined methods, algorithms and techniques for solving various classes of problems of operational, structural and functional synthesis and management of the development of the regarded classes of CO. The tasks of controlling the structural dynamics of CO belong to the structural and functional synthesis class of problems and the formation of appropriate programs for managing and control of their development. The main difficulty and a special feature of the solution of the regarded problems is as follows. Determination of optimal control programs for the basic elements and subsystems of CO can be performed only after all functions and algorithms of information processing and control that should be implemented in these elements and subsystems are known. In its turn, the distribution of functions and algorithms by the elements and subsystems of CO depends on the structure and parameters of the control laws of these elements and subsystems. The difficulty of resolving this controversial situation is ex-acerbated by the fact that under the influence of various reasons, the composition and structure of the CO at different stages of their lifecycle changes over time. The given examples of solving practical problems for such subject areas as spacecrafts, logistics, and industrial production.

**Keywords:** Complex objects · Proactive management and control theory · Multiple-model complexes · Combine methods · Algorithms · Structure dynamic control

D. G. Arseniev et al. (Eds.): CPS&C 2019, LNNS 95, pp. 169–177, 2020.
https://doi.org/10.1007/978-3-030-34983-7_17

# 1 Introduction

The main subject of our research is complex objects (CO). By complex objects we mean such objects that should be studied using polytypic models and combined methods. In some instances, investigations of complex objects require multiple methodological approaches, many theories and disciplines, and interdisciplinary researches. Different aspects of complexity can be considered for distinguishing between a complex system and a simple one, for example: structure complexity, operational complexity, complexity of the choice of behavior, complexity of development [1–3, 10, 16, 18].

Classic examples of CO are: control objects for various classes of moving objects, such as surface and air transport, ships, space and launch vehicles, etc.; geographically distributed heterogeneous networks, flexible computerized manufacturing [2–6, 13, 18].

One of the main features of modern CO is the changeability of their parameters and structures caused by objective and subjective reasons at different stages of the CO life cycle. In other words, in practice we always come across the CO structure dynamics [10].

Under the existing conditions, the CO potentialities for increment (stabilization) or degradation (reducing) makes it necessary to perform the CO structure management and control (including the management and control of reconfiguration of structures). There are many possible variants of CO structure dynamics management and control. For example, they can be [5–7, 11–18]: *alteration of CO functioning means and objectives; alteration of the order of observation tasks and control tasks solving; redistribution of functions, problems, and control algorithms between CO levels; reserve resources control; control of motion of CO elements and subsystems; reconfiguration of CO structures.*

According to the contents of the structure-dynamics management and control problems, they belong under the class of the CO structure, i.e., functional synthesis problems and problems of program construction providing for the CO development.

As applied to CO, we distinguish the following main types of structures: the structure of CO goals, functions and tasks; organization structure; technical structure; topological structure; structure of special software and mathematical tools; technology structure (the structure of CO control technology).

By structure dynamics management and control we mean a process of control inputs producing and implementing for the CO transition from the current macro-state to a given one [10]. So, in our paper, we propose a new applied theory of CO structure dynamics management and control (SDMC).

# 2 Results

The main aim of our research is to prove the need of integrated modeling for parallel structural-functional synthesis of CO under dynamic conditions. Moreover, the main idea of our approach is to use fundamental results of structural-dynamic management and control theory [10] for multiple-model description of CO functioning and research.

## 2.1   Methodology

During our research we describe the main classes of CO integrated modeling tasks. For these aims, we use SDMC theory. Methodological basics of this theory include: the methodologies of generalized system analysis and the modern optimal control theory for CO with reconfigurable structures [2–6, 13, 18]. Moreover, these basics are related with the concepts of proactive control of their structure dynamics, the concept of complex preemptive modeling of the specified objects and their functioning processes, concepts of integration of knowledge, information and data, as well as the concept of model (rather than algorithm) priority when constructing relevant proactive monitoring and controlling systems. Our research has shown that these concepts have received implementation in a number of principles. Those are: the principle of non-terminal decisions, diversity absorption, hierarchical compensation, and complementarity; the principle of self-recursive description and modeling of the research objects; the homeostatic balance of interaction, overcoming the separation principle; the principles of multiple-model and multi-criteria approaches;, the principles taken as a basis for ontology creation, the principles of decomposition and aggregation; the principle of a rational multi-criteria compromise providing unavoidable threshold information and time limitations [8, 9, 11–16].

## 2.2   Methodology

As provided by the concept of CO multiple-model description, the proposed general model includes particular dynamic models: the dynamic model of CO motion control (Mg model); dynamic model of CO channel control (Mk model); dynamic model of CO operations control (Mo model); dynamic model of CO flows control (Mn model); dynamic model of CO resource control (Mp model); dynamic model of CO operation parameters control (Me model); dynamic model of CO structure dynamic control (Mc model); dynamic model of CO auxiliary operation control (Mv model) [6, 7, 10]. Figure 1 illustrates a possible interconnection of the models.

CO structure-dynamic control problem has some specific features in comparison with classic optimal control problems. *The first feature* is that the right parts of the differential equations undergo discontinuity at the beginning of interaction zones. The considered problems can be regarded as control problems with intermediate conditions. *The second feature* is the multi-criteria nature of the problems. *The third feature* is concerned with the influence of uncertainty factors. *The fourth feature* is the form of time-spatial, technical, and technological non-linear conditions that are mainly considered in control constraints and boundary conditions. On the whole, the constructed model is a non-linear non-stationary finite-dimensional differential system with a reconfigurable structure. Different variants of model aggregation were proposed. These variants produce a task of model quality selection that is the task of model complexity reduction. Decision-makers can select an appropriate level of model thoroughness in the interactive mode. The level of thoroughness depends on the input data, external conditions, and required level of solution validity.

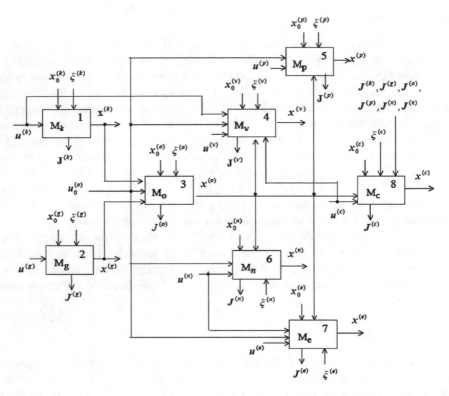

**Fig. 1.** Structural-dynamics management and control multiple-model description of complex objects

The proposed interpretation of CO structure dynamics control processes provides the advantages of modern optimal control theory for CO analysis and synthesis. Procedures of structure-dynamics problem-solving depend on the variants of transition and output functions (operators) implementation. Various approaches, methods, algorithms and procedures of coordinated choice through complexes of heterogeneous models have been developed by now.

As results of our investigations, the main phases and steps of a program-construction procedure for optimal structure-dynamics control in CO were proposed.

*At the first phase,* forming (generation) of allowable multi-structural macro-states is performed. In other words, a structure-functional synthesis of a new CO make-up should be fulfilled in accordance with an actual or forecast situation. Here, the *first-phase* problems come to CO structure-functional synthesis.

*At the second phase,* a single multi-structural macro-state is selected, and adaptive plans (programs) of CO transition to the selected macro-state are constructed. These plans should specify transition programs, as well as programs of stable CO operation in intermediate multi-structural macro-states. *The second phase* of program construction is aimed at solving multi-level multi-stage optimization problems.

One of the main opportunities of the proposed method of CO SDMC program construction is that besides the vector of program control, we receive a preferable multi-structural macro-state of CO at the end point. This is the state of CO reliable operation in the current (forecast) situation.

A method of multi-optional prediction of multi-structural macro-states of GSB MGDO based on the construction and approximation of attainability domains of the logic and dynamic models describing the structural dynamics of these objects was developed. In addition, it was shown that the orthogonal projection of the goal set (a set of required values of quality indicators of proactive control over CO) on the specified attainability set evidently allows to receive a set of non-dominated alternatives (Pareto set) in the virtual space of system and technical parameters characterizing multi-structural macro-state of CO. This result is based on the theorem proved by Professor L.A. Petrosyan in1982. In the course of rescarch, it was also found that this set can be considered as a set of non-terminal decisions, the capacity of which allows to judge about the potentials of the CO control system (in other words, structural controllability of this category of objects).

## 2.3   First Prototype

The multiple-model description of CO structure-dynamics management and control proccsscs is the base of integrated analytical-simulation technologies and simulation systems. Figure 2 illustrates the general structure of a simulation system, which was used for CO structure-dynamics control simulation. We assume the simulation system to be a specially organized complex. This complex consists of the following elements: simulation models (the hierarchy of models); analytical models (the hierarchy of models) for a simplified (aggregated) description of objects studied; informational subsystem that is a system of data bases (known as ledge bases); control-and-coordination system for interrelation and joint use of prcvious clements and interaction with the user (decision-maker).

The components of the simulation system were thc main parts of the developed program prototypes in the course of our investigation. The processes of CO structure-dynamics control are hierarchical, multi-stage, and multi-task ones. The structure of simulation system (SIS) models conforms the features of control processes. There are three groups of models in SIS: models of CO CS and OS functioning (subsystem I of SIS); models of evaluation (observation) and analysis of structural states and CO CS structure-dynamics (subsystem II of SIS); decision-making models for control processes in CO CS (subsystem III of SIS). The subsystem of models for CO CS and OS functioning includes: models of CO functioning, models of CO classes functioning, and models of CO system functioning (subsystems 1, 2, 3 of SIS); models of CO interacting station (IS) functioning (subsystem 4 of SIS), models of functioning for control center (CC), central control station (CCS), and control station (CST) (subsystems 5, 6 of SIS); models of CO CS subsystems interaction and models of interaction between CO CS and OS (subsystem 7 of SIS); models of objects–in-service (OS) functioning (subsystem 8 of SIS); models of environmental impacts on CO CS (subsystem 9 of SIS); simulation models of CO CS goal directed applications under conditions of environmental impact (subsystem 10 of SIS).

In general, CO functioning includes informational, material, and energy interaction with OS, with other CO, and with the environment. Along with the interaction, the facility functioning, resource consumption (replenishment), and CO motion are to be considered via functioning models.

The subsystem of CO CS structure-dynamics evaluation (observation) models and analysis models includes: models and algorithms of evaluation (observation) and analysis of states of CO motion; facilities, interactions and resources (subsystem 11 of SIS); models and algorithms of evaluation (observation) and analysis of SO states (subsystem 12 of SIS); models and algorithms of situation evaluation and analysis (subsystem 13 of SIS).

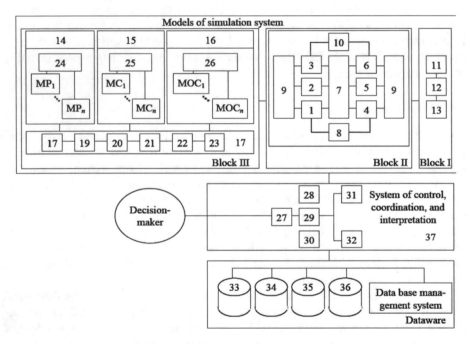

**Fig. 2.** Structure of simulation system

The subsystem of decision-making models includes: models and algorithms of CO and CO CS long-range and operational planning (sub-system 14 of SIS); models and algorithms of control for CO CS topologic, organizational, technology, and technical structures; models and algorithms of control for CO CS structures of software and dataware tools structures (subsystems 16, 17, 18, 19, 20, 21 of SIS); models and algorithms of correction for CO CS long-range and operational plans (subsystem 15 of SIS); models and algorithms of coordination for functioning of CO CS subsystems at stages of planning, correction, and operational control (subsystem 15 of SIS) (subsystems 24, 25, 26 of SIS); models and algorithms of operational control in CO CS (subsystem 16 of SIS).

In Fig. 2, the following notations where used: MP1,..., MPn; MC1,..., MCn; MOC1,..., MOCn are the models of planning, correction, and operational control for CO of (1,...,n) types correspondingly. Figure 2 also shows the system of control, coordination and interpretation containing user interface and general control subsystem (27 subsystem of SIS), local systems of control and coordination (28 subsystem of SIS), subsystem of data processing, analysis, and interpretation for planning, control and modeling (30 subsystem of SIS), subsystem of modeling scenarios formalization (31 subsystem of SIS), subsystem of software para-metric and structural adaptation (32 subsystem of SIS), subsystem of recommendations producing for decision-making and modeling (29 sub-system of SIS).

The data-ware includes data bases for CO states (33 subsystem), for CO CS states and general situation (35 subsystem), for SO states (34 subsystem), and data bases for analytical and simulation models of decision-making and of CO CS functioning (36 subsystem).

The main feature of integrated modeling is the coordination of different models constructed via formal or non-formal decomposition of tasks. Various approaches, methods, algorithms and procedures of a coordinated choice through complexes of heterogeneous models have been by now developed [7, 8].

# 3 Conclusion

Methodological and methodical basis of CO structure-dynamics management and control theory have been developed by now. This theory can be widely used in practice. It has an interdisciplinary basis provided by the classic control theory, operations research, artificial intelligence, systems theory and systems analysis.

The presented multiple-model complex, as compared with known analogues, have several advantages. It simplifies decision-making in CO structure-dynamics management and control, because it allows seeking for alternatives in finite dimensional spaces rather than in discrete ones. The complex permits to reduce dimensionality of CO structure-functional synthesis problems to be solved in a real-time operation mode. Moreover, the proposed approach to the problem of CO structural dynamics management control enables:

- common goals of CO functioning to be directly linked with those implemented in CO control process;
- a reasonable decision and selection (choice) of adequate consequences of problems solved and operations fulfilled are related to structural dynamics to be created (in other words, to synthesize and develop a CO control method);
- a compromise distribution (trade-off) of a restricted resources appropriated for a structural dynamics management and control to be found voluntarily.

A more detailed information about CO structure dynamics management and control theory implementation in different applied areas is placed on the web site http://litsam.ru.

**Acknowledgments.** The research described in this paper was partially supported by the Russian Foundation for Basic Research (grants 16-29-09482-ofi-m, 17-08-00797, 17-06-00108, 17-01-00139, 17-20-01214, 17-29-07073-ofi-i, 18-07-01272, 18-08-01505, 19–08–00989), state order of the Ministry of Education and Science of the Russian Federation №2.3135.2017/4.6, state research 0073–2019–0004, and International project ERASMUS+ , Capacity building in higher education, # 73751-EPP-1-2016-1-DE-EPPKA2-CBHE-JP.

# References

1. Baldonado, M., Chang, C.-C.K., Gravano, L., Paepcke, A.: The stanford digital library metadata architecture. Int. J. Digit. Libr. **1**, 108–121 (1997)
2. Bruce, K.B., Cardelli, L., Pierce, B.C.: Comparing object encodings. In: Abadi, M., Ito, T. (eds.) Theoretical Aspects of Computer Software. Lecture Notes in Computer Science, vol. 1281. Springer, Heidelberg, pp. 415–438 (1997)
3. Van Leeuwen, J. (ed.): Computer Science Today. Recent Trends and Developments. Lecture Notes in Computer Science, vol. 1000. Springer, Heidelberg (1995)
4. Michalewicz, Z.: Genetic Algorithms + Data Structures = Evolution Programs, 3rd edn
5. Becerra, G., Amozurrutia, J.A.: Rolando García's "Complex Systems Theory" and its relevance to sociocybernetics. J. Sociocybern. **13**(15), 8–30 (2015)
6. Bir, T.: Cybernetics and Production Control. Fizmatlit, Moscow (1963). (in Russian)
7. von Foerster, H.: Cybernetics. In: Shapiro, S.C. (ed.) Encyclopedia of Artificial Intelligence. Wiley, New York (1987)
8. Hyötyniemi, H.: Neocybernetics in Biological Systems. Helsinki University of Technology, Control Engineering Laboratory, Report 151, 273 p., August 2006
9. Ignatyev, M.B.: Semantics and selforganization in nanoscale physics. Int. J. Comput. Anticip. Syst. **22**, 17–23 (2008)
10. Ivanov, D., Sokolov, B., Pavlov, A.: Optimal distribution (re)planning in a centralized multi-stage network under conditions of ripple effect and structure dynamics. Eur. J. Oper. Res. **237**(2), 758–770 (2014)
11. Kalinin, V.N., Sokolov, B.V.: Multi-model description of control processes for space facilities. J. Comput. Syst. Sci. Int. **1**, 149–156 (1995)
12. Mancilla, R.G.: Introduction to Sociocybernetics (Part 1) Third-order cybernetics and a basic framework for society. J. Sociocybern. **9**(1–2), 35–56 (2011)
13. Mancilla, R.G.: Introduction to sociocybernetics (Part 2) Power culture and institutions. J. Sociocybern. **10**(1–2), 45–71 (2012)
14. Okhtilev, M.Y., Sokolov, B.V., Yusupov, R.M.: Intellectual Technologies of Monitoring and Controlling the Dynamics of Complex Technical Objects. Nauka, Moscow (2006). (in Russian)
15. Maruyama, M.: The Second cybernetics. Deviation amplifying mutual causal process. Am. Sci. **5**(2), 164–179 (1963)
16. Sokolov, B.V., Zelentsov, V.A., Yusupov, R.M., Merkuryev, Y.A.: Multiple models of information fusion process quality definition and estimation. J. Comput. Sci. **13**(15), 18–30 (2014)
17. Skurikhin, V.I., Zabrodsky, V.A., Kopeychenko, Yu.V.: Adaptive control objects in machine-building industry. Mashinostroenie, Moscow (1989). (in Russian)

18. Verzilin, D.N., Maximova, T.G.: Models of social actors' reaction on external impacts. St. Petersburg State Polytech. Univ. J. Comput. Sci. Telecommun. Control Syst. **120**(2), 140–145 (2011)
19. Verzilin, D.N., Maximova, T.G.: Time attributes reconstruction for processes of states changing in social medium. St. Petersburg State Polytech. Univ. J. Comput. Sci. Telecommun. Control Syst. **126**(3), 97–105 (2011). (in Russsian)
20. Wang, S., Wang, D., Su, L., Kaplan, L., Abdelzaher, T.F.: Towards cyber-physical systems in social spaces. The data reliability challenge. In: Real-Time Systems Symposium (RTSS), 2–5 December 2014, pp. 74–85. IEEE (2014)
21. Wiener, N.: The Human Use of Human Beings Cybernetics and Society. Da Capo Press, Boston (1950)
22. Zhuge, H.: Semantic linking through spaces for cyber-physical-socio intelligence. A methodology. Artif. Intell. **175**, 988–1019 (2011)

# "Moment" Representation of "Fast Decreasing" Generalized Functions and Their Application in Stochastic Problems

Andrey N. Firsov[✉]

Peter the Great St. Petersburg Polytechnic University, Saint Petersburg, Russia
anfirs@yandex.ru

**Abstract.** This paper describes the process of building a special space of generalized functions, its properties and applications. Presented applications are: constructive solution of Kolmogorov-Feller type equation with polynomial drift coefficient; proof of the exponential nature of equilibrium establishment in rarefied gas, described by Boltzmann equation of kinetic theory of gases.

**Keywords:** Generalized functions · Moment representation · Kolmogorov-Feller equation · Boltzmann equation

## 1 Introduction

The article briefly describes the construction, properties and applications of a special space of generalized functions (see the terminology in [1]), proposed by Firsov [2, 3] and developed together with Koval [4]. As examples of applications, the following ones are presented: a constructive solution of an equation of Kolmogorov-Feller type with a polynomial drift coefficient, as well as proof of an exponential in time nature of establishing equilibrium in rarefied gas, described by the Boltzmann kinetic equation.

Equations of the Kolmogorov-Feller type can be found, in particular, in control theory, communication theory, and stellar dynamics. In the literature devoted to analytical methods for solving such equations, the case of a linear dependence of the drift coefficient on the coordinate is usually considered. In this case, it is possible to use the Fourier transform. In this paper, we consider solution of the Kolmogorov-Feller equation with quadratic and cubic drift coefficients, i.e., the cases in which the application of the Fourier transform is ineffective.

We also study the asymptotic (for large values of time) properties of solutions of the linearized Boltzmann equation.

The Boltzmann kinetic equation underlies the substantiation and analysis of mathematical models of transfer processes (mass, energy, momentum) in gases, since the derivation of this equation is based on a more natural understanding of gases as a set of interacting molecules, and not as a continuous medium. Such an approach is especially preferable when studying the motion of bodies in the upper layers of the atmosphere, where the idea of gas as a continuous medium is inadequate. The main problem that arises here is the complexity of the Boltzmann equation, which comprises in the structure of the integral collision an operator associated with the microstructure

© Springer Nature Switzerland AG 2020
D. G. Arseniev et al. (Eds.): CPS&C 2019, LNNS 95, pp. 178–186, 2020.
https://doi.org/10.1007/978-3-030-34983-7_18

of the medium and the nature of the interaction of substance molecules. This is one of the reasons for the relatively small number of studies devoted to an analytically rigorous formulation of solvability conditions and the analysis of solutions of the Boltzmann equation. The latter concerns not only the complete non-linear version of the Boltzmann equation but also its linear approximation.

## 2 Theory of "Rapidly Decreasing" Generalized Functions [2, 3]

*Definition 1.* Let $s > 0$. Let $E_s$ denote the space of (complex-valued) functions $\varphi \in C^\infty(R^\nu)$, such that for any $p > 0$

$$|D^q \varphi(x)| \leq C(s+p)^{|q|} e^{(s+p)|x|}, \quad x \in R^\nu.$$

Here $C$ is a constant depending, generally speaking, on $\varphi$, $s$ and $p$, but not dependent of $q \in Z_0^\nu$.

Note that polynomials $P(x_1, x_2, \ldots, x_\nu)$ and functions $e^{s_1 x_1 + \ldots + s_\nu x_\nu}$, $e^{i(s_1 x_1 + \ldots + s_\nu x_\nu)}$ belong to $E_s$, where $s > \max(|s_1|, \ldots, |s_\nu|)$.

Let us introduce a countable system of norms into $E_s$

$$\|\varphi\|_s^{(p)} = \sup_{q,x} \left[ \frac{|D^q \varphi(x)|}{(s+p)^{|q|}} e^{-(s+p)|x|} \right], \quad p = 1, 2, \ldots \tag{1}$$

*Theorem 1.* The space $E_s$ endowed with the system of norms (1) is a complete countable-normed space.

As far as $E_{s+1} \supset E_s$ and from the convergence of a sequence $\{\varphi_n\}$ in $E_s$ follows its convergence in $E_{s+1}$, you can enter a countable union $E = \bigcup_{s=1}^{\infty} E_s$; the convergence in $E$ is determined in the usual way (see [3]). The space $E$ is obviously complete in the sense of corresponding convergence. The following property of spaces $E_s$ is the main one for the sequel.

*Theorem 2.* Let $\varphi \in E_s$, $a = (a_1, a_2, \ldots, a_\nu) \in R^\nu$. Then:

- Taylor series for $\varphi$

$$\varphi(x) = \sum_{l=0}^{\infty} \sum_{|q|=l} \frac{\varphi^{(q)}(a)}{q!} (x - a)^q$$

converges for all $x \in R^\nu$;

- partial sums

$$S_m(x) = \sum_{l=0}^{m} \sum_{|q|=l} \frac{\varphi^{(q)}(a)}{q!}(x-a)^q$$

converge to $\varphi$ in terms of convergence in $E_s$.

The following lemmas point to a number of other properties of the space $E$ (see details in [2, 3]).

*Lemma 1.* For $\varphi, \psi \in E$, we have $\varphi \cdot \psi \in E$.

*Lemma 2.* If $\varphi_n \to \varphi$, $\psi_n \to \psi$ in $E$, then $\varphi_n \cdot \psi_n \to \varphi \cdot \psi$ in $E$.

*Lemma 3.* If $\psi \in E$, then $D^q \varphi \in E_s$.

*Lemma 4.* If $\varphi_n \to \varphi$ in $E_s$, then $D^q \varphi_n \to D^q \varphi$ in $E_s$.

The space $E'$ is introduced in the standard way as a conjugate to $E$. It defines linear operations, the operation of multiplication by functions from $E$ and differentiation in the usual way. These operations, as follows from the previous results, are continuous in the sense of convergence in $E'$ (i.e., in the sense of weak convergence). By the theorem on the completeness of a space conjugated to a complete countably normalized space, this space will be complete (relatively weak convergence) – see [1].

We immediately note that the stock of regular functionals in $E'$ is big enough. So, function $f(x) \in L_1(R^v)$, satisfying

$$f(x) = O\left(e^{-\alpha|x|^{1+\varepsilon}}\right); \ |x| \to \infty; \ \alpha > 0, \ \varepsilon > 0,$$

generates a functional $\tilde{f} \in E'$ by the formula

$$(\tilde{f}, \varphi) = \int_{R_v} f(x)\varphi(x)dx, \ \varphi \in E.$$

*Lemma 5.* If $f \in L_1(R^v)$, and for all $\varphi \in E$ $\int_{R^v} f(x)\varphi(x)dx = 0$, then $f(x) = 0$ almost everywhere.

Note also that if an "ordinary" function $f(x)$ is differentiable in the usual sense, and both f and $f^{(q)}$ generate regular functionals $\tilde{f}$ and $\tilde{f}^{(q)}$, then $f^{(q)} = \tilde{f}^{(q)}$, where on the right side we have the derivative of the functional $\tilde{f}$ in the sense of differentiation in space $E'$. Finally, the delta function $\delta_a = \delta(x - a)$, defined in the usual way, i.e., $(\delta_a, \varphi) = \varphi(a)$, $\varphi \in E$ also belongs to $E'$ and is a singular functional.

We now formulate the basic properties of generalized functions from $E'$ (see details in [2, 3]):

### Theorem 3

1. Any generalized function $f \in E'$ at an arbitrary point $a \in \mathbf{R}^n$ is representable as

$$f = \sum_{k=0}^{\infty} \sum_{|q|=k} C_a^{(q)} \delta_a^{(q)},$$

where the coefficients $C_a^{(q)} = (-1)^q \frac{(f,(x-a)^q)}{q!}$.

2. Generalized function $f \in E'$ if and only if when the row

$$\sum_{k=0}^{\infty} s^k \sum_{|q|=k} |C_a^{(q)}| \qquad (2)$$

converges for any $s > 0$.

3. Let $f, g \in E'$ and $C_a^{(q)}$, $d_a^{(q)}$ are coefficients of "moment" decomposition. Then the following statements are true:

- $\alpha f + \beta g = \sum_{k=0}^{\infty} \sum_{|q|=k} (\alpha C_a^{(q)} + \beta d_a^{(q)}) \delta_a^{(q)}$, where a, $\beta \in R$;

- $D^k f = \sum_{k=0}^{\infty} \sum_{|q|=k} C_a^{(q)} \delta_a^{(q)}$;

  - if $\psi \in E$, then $\psi f = \sum_{k=0}^{\infty} \sum_{|q|=k} h_a^{(q)} \delta_a^{(q)}$, where

$$h_a^{(q)} = \sum_{r=0}^{\infty} (-1)^r \sum_{\substack{|n| = r + |q| \\ n \geq q}} \binom{n}{n-q} C_a^{(q)} \psi^{(q)}(a).$$

In the space $E'$ you can use the usual procedure to determine the convolution of two generalized functions. Such a convolution always exists (unlike other spaces of generalized functions), has the usual properties, and the following is true for it.

*Theorem 4.* Let $f, g \in E'$ and $C_a^{(q)}$, $d_a^{(q)}$ are the corresponding coefficients of the "moment" decomposition. Then

$$f * g = \sum_{k=0}^{\infty} \sum_{|q|=k} h_a^{(q)} \delta_a^{(q)},$$

where

$$h_a^{(q)} = \sum_{i+j<q} (-1)^{|q-i-j|} \frac{a^{q-i-j}}{(q-i-j)!} C_a^{(i)} d_a^{(j)}.$$

In particular, when $a = 0$, the coefficients of "moment decomposition" are

$$h_0^{(q)} = \sum_{i+j=q} C_0^{(i)} d_0^{(j)}.$$

## 3   Solving the Kolmogorov-Feller Equation with a Quadratic Drift Coefficient [4]

The one-dimensional stationary Kolmogorov-Feller equation with quadratic drift coefficient has the form:

$$\frac{d}{dx}[\alpha(x)W(x)] + v \int_{-\infty}^{+\infty} p(A)W(x-A)dA - vW(x) \tag{3}$$

Let the desired function $W(x)$ and given function $p(A)$ be generalized functions from $E'$. Then, according to the first paragraph of Theorem 3, the function $W(x)$ can be represented as $W(x) = \sum_{q=0}^{+\infty} C_0^{(q)} \delta_0^{(q)}$. Similarly, $p(A) = \sum_{q=0}^{+\infty} d_0^{(q)} \delta_0^{(q)}$, where $d_a^{(q)} = (-1)^q \frac{(p, A^q)}{q!}$. Since the meaning of the function $p(A)$ is probability density, it should be considered that $(p, 1) = 1$.

Let's consider separately each of the items on the left side of Eq. (3). Using obvious transformations and the assertions of the third item of Theorem 3, we obtain

$$\frac{d}{dx}[(\alpha x + \beta x^2)W(x)] = \sum_{q=0}^{+\infty} k_0^{(q)} \delta_0^{(q)} + \sum_{q=0}^{+\infty} l_0^{(q)} \delta_0^{(q)},$$

where $k_0^{(q)} = \alpha C_0^{(q)} - 2\beta(q+1)C_0^{(q)}$, $l_0^{(q)} = \beta(q+1)(q+2)C_0^{(q)} - \alpha(q+1)C_0^{(q)}$.

According to Theorem 2, we transform the convolution in Eq. (3) to

$$p * W = \sum_{q=0}^{+\infty} \left( \sum_{i=0}^{q} d_0^{(q-i)} C_0^{(i)} \right) \delta_0^{(q)}.$$

Thus, Eq. (3) is reduced to the form

$$\sum_{q=0}^{+\infty} k_0^{(q)} \delta_0^{(q)} + \sum_{q=0}^{+\infty} l_0^{(q)} \delta_0^{(q)} + v \sum_{q=0}^{+\infty} \left( \sum_{i=0}^{q} d_0^{(q-i)} C_0^{(i)} \right) \delta_0^{(q)} - v \sum_{q=0}^{+\infty} C_0^{(q)} \delta_0^{(q)} = 0,$$

which is equivalent to

$$\sum_{q=0}^{+\infty} \left[ k_0^{(q)} + l_0^{(q)} + v \sum_{i=0}^{q} d_0^{(q-i)} C_0^{(i)} - v C_0^{(q)} \right] \delta_0^{(q)} = 0.$$

Using item 1 of Theorem 3, we obtain that

$$k_0^{(q)} + l_0^{(q)} + v \sum_{i=0}^{q} d_0^{(q-i)} C_0^{(i)} - v C_0^{(q)} = 0 \quad \forall q \in \mathbf{Z}_0.$$

As a result, after appropriate transformations, given the property $(p, 1) = 1$, we obtain a recurrent relation from which all coefficients $C_0^{(q)}$ can be successively determined:

$$C_0^{(q+1)} = \frac{\alpha}{\beta(q+1)} C_0^{(q)} - \frac{v}{\beta q(q+1)} \sum_{i=0}^{q-1} d_0^{(q-i)} C_0^{(i)}. \tag{4}$$

To obtain a unique solution to the chain of Eq. (4), it is necessary to set the coefficients $C_0^{(0)}$ и $C_0^{(1)}$. Thus, to solve the problem of the existence of a solution of Eq. (3) in space $E'$, one should verify the fulfillment of Theorem 3, Sect. 2.

*Theorem 5.* To carry out part 2 of Theorem 3, it is sufficient that the sequence $\left\{ C_0^{(q)} \right\}_{q=0}^{+\infty}$ from (4) is limited.

The proof of Theorem 5 is based on the following lemma.

*Lemma 6.* Let there be such $Q > 0$, that for any $0 \le q \le Q$ the inequality $\left| C_0^{(q)} \right| \le \left| C_0^{(Q)} \right|$ is satisfied. Let us, moreover, consider to be fulfilled $\left| \frac{\alpha}{\beta} \right| + \left| \frac{v}{\beta} \right| D \le Q + 1$, where $D = \max_{q \ge 0} \left| d_0^{(q)} \right|$. In this case $\left| C_0^{(q)} \right| \le \left| C_0^{(Q)} \right|$ for any $q^3 Q$.

To prove the lemma, the principle of complete mathematical induction is used.

We give an example of solving the Eq. (3) using (4). Let $p(A)$ be a regular functional, which is a normal distribution with expectation 0 and variance $\sigma$. In this case, the coefficients of the "moment" decomposition are

$$d_a^{(q)} = (-1)^q \frac{(p, A^q)}{q!} = \begin{cases} 0, & q = 2i+1, \quad i \in \mathbf{Z}_0, \\ \frac{\sigma^q}{i!!}, & q = 2i, \quad i \in \mathbf{Z}_0. \end{cases}$$

Then the moments of the desired function $W(x)$ can be found using the ratio

$$C_0^{(q+1)} = \frac{\alpha}{\beta(q+1)} C_0^{(q)} - \frac{v}{\beta q(q+1)} \sum_{i=0}^{q-1} \frac{\sigma^q}{2 \cdot 4 \cdot \ldots \cdot q} C_0^{(i)}, \quad q = 2j, \ j \in \mathbf{Z}_0.$$

## 4   Solving the Kolmogorov-Feller Equation with a Cubic Drift Coefficient

The one-dimensional stationary Kolmogorov-Feller equation with a cubic drift coefficient is

$$\frac{d}{dx}\left[(\alpha x + \beta x^2 + \gamma x^3)W(x)\right] + v \int_{-\infty}^{+\infty} p(A)W(x-A)dA - vW(x). \tag{5}$$

The chain of Eqs. (4) will replace the relations:

$$C_0^{(q+2)} = -\frac{\alpha}{\gamma(q+1)(q+2)}C_0^{(q)} + \frac{\beta}{\gamma(q+2)}C_0^{(q+1)} + \frac{v}{\gamma q(q+1)(q+2)}\sum_{i=0}^{q-1}d_0^{(q-i)}C_0^{(i)} \tag{6}$$

There is a theorem similar to Theorem 5:

*Theorem 6.* Let the sequence of coefficients (6) $\left\{C_0^{(q)}\right\}_{q=0}^{+\infty}$ be bounded; then condition 2 of Theorem 3 is satisfied.

The proof of this theorem is based on a lemma similar to Lemma 6.

## 5   Analysis of the Asymptotic of Solutions of the Linearized Boltzmann Equation for Large Values of Time [3]

Consider the Cauchy problem for the linearized Boltzmann equation of the kinetic theory of gases [5]:

$$\begin{array}{c}\frac{\partial f}{\partial t} + \bar{u}\frac{\partial f}{\partial \bar{x}} = L[f], \quad L[f] = K[f] - vf, \\ f = f(\bar{x}, \bar{u}, t), \quad \bar{x} \in \mathbf{R}^3, \quad \bar{u} \in \mathbf{R}^3, \quad t \geq 0, \\ f|_{t=0} = f_0(\bar{x}, \bar{u}). \end{array} \tag{7}$$

Here $f(\bar{x}, \bar{u}, t)$ – linearized distribution of molecules by coordinates $\bar{x}$ and speeds $\bar{u}$ at the moment of time $t$. $K[f]$ – linear bounded operator acting in $f$ as a function of $\bar{u}$; $v = v(u) = O(u^\beta)$ when $u \to \infty$, $0 < \beta \leq 1$, $u = |\bar{u}|$. Properties of the function $v$ (u) end on the specific model of intermolecular interaction, taken in the derivation of kinetic equations. See details in [5].

It is known [6] that the solution of problem (7) in the case of "hard" potentials of intermolecular interaction $U = C_k r^{-k}$, $k > 5$ has at $t \to \infty$, in general, the power asymptotics of the form $O\left(\frac{1}{1+t^\mu}\right)$, $\mu > 0$. This result is obtained under the assumption that $f(\bar{x}, \bar{u}, t)$ at $x = |\bar{x}| \to \infty$ behaves like a function from $L_p(\mathbf{R}_x^3)$, $p > 1$.

It turns out that if you impose more stringent conditions on the behavior of $f(\bar{x}, \bar{u}, t)$ when $x \to \infty$, for example, require that $f(\bar{x}, \bar{u}, t)$ satisfies (uniformly for $\bar{u}$, $t$):

$$f(\bar{x}, \bar{u}, t) = O\left(\exp\left(-\alpha|\bar{x}|^{1+\varepsilon}\right)\right), \ |\bar{x}| \to \infty, \ \alpha > 0, \ \varepsilon > 0.$$

then equilibrium in time occurs exponentially fastly.

The idea of the proof is as follows. We'll seek $f(\bar{x}, \bar{u}, t)$ in a class of functions such where for almost all $\bar{u} \in \mathbf{R}^3$ and all $t > 0$ $f(\bar{x}, \bar{u}, t) \in E'_x$, that is, the function $f$ can be represented as

$$f(\bar{x}, \bar{u}, t) = \sum_{l=0}^{+\infty} \sum_{|q|=1} C_a^{(q)}(\bar{u}, t) \delta_a^{(q)}(\bar{x}). \tag{8}$$

Substituting this expression into (7) and taking into account the results of paragraph 2 of this paper, we obtain infinite "hooking" system of equations for the coefficients $C_a^{(q)}(\bar{u}, t)$:

$$\frac{\partial C_a^{(0)}}{\partial t} = L\left[C_a^{(0)}\right]$$
$$\vdots \tag{9}$$
$$\frac{\partial C_a^{(q)}}{\partial t} = L\left[C_a^{(q)}\right] - \left[u_1 C_a^{(q-I_1)} + u_2 C_a^{(q-I_2)} + u_3 C_a^{(q-I_3)}\right], \ |q| \neq 0,$$

where $I_1$, $I_2$, $I_3$ denotes multi-indices $(1, 0, 0)$, $(0, 1, 0)$ and $(0, 0, 1)$ respectively.

Equations (9) are inhomogeneous equations of the form

$$\frac{\partial C_a^{(q)}}{\partial t} = L\left[C_a^{(q)}\right] - g_q(\bar{u}, t), \ |q| \neq 0,$$

where $g_q(\bar{u}, t)$ is a known function (its own at each step). Thus, the properties of functions $C_a^{(q)}(\bar{u}, t)$ depend on the properties of the operator $L$. The latter has been thoroughly studied. (see [3, 5–9]). In particular, the operator $L$ on a subspace of functions $w(\bar{u}, t)$, that are orthogonal (in the sense of $L_2(\mathbf{R}_u^3)$) to the subspace of additive invariants (which is essentially equivalent to fulfilling the classical conservation laws for gas), generates a semigroup $T(t)$, $t > 0$ of bounded operators that give a solution to the abstract Cauchy problem for Eq. (7); and the inequality holds: $\|T(t)\| \leq \mathrm{const} \cdot e^{-\mu t}$, $\mu > 0$. Method of mathematical induction similar to that used in [6, 7], gives for solutions of Eq. (9) an estimate of the form (the norm is understood in the sense of $L_2(\mathbf{R}_u^3)$): $\|C_a^{(q)}(t)\| \leq \mathrm{const} \cdot e^{-\gamma t}$, $\gamma > 0$, where the const depends on the initial distribution function $f_0(\bar{x}, \bar{u})$ and parameters of the operator $L$. The latter estimate, taking into account (8), allows us to make a conclusion about the establishment of equilibrium exponentially fast (in time) in the system described by task (7).

# 6    Results

The article describes the construction, properties and applications of a special space of generalized functions proposed by the author [2, 3]. The appendices include: a constructive solution of a Kolmogorov-Feller type equation with quadratic and cubic drift coefficients, as well as proof of the exponential in time character of equilibrium in rarefied gas described by the Boltzmann kinetic equation.

The proposed method of "momentary" representation of generalized functions can be quite effectively used for solving differential equations with polynomial coefficients and convolution equations.

# References

1. Gel'fand, I.M., Schilov, G.E.: Spaces of basic and generalized functions. In: Generalized Functions. Academic Press, New York (1968). (in Russian)
2. Firsov, A.N.: The method of moments in the theory of generalized functions and its applications in problems of system analysis and control. Theory Fundamentals. St. Petersburg State Polytech. Univ. J. Comput. Sci. Telecommun. Control Syst. (6), 74–81 (2010). (in Russian)
3. Firsov, A.N.: Generalized mathematical models and methods for analyzing dynamic processes in distributed systems. Polytechnical University Publishers, St. Petersburg (2013). (in Russian)
4. Firsov, A.N., Koval', A.B.: Solution of the Kolmogorov-Feller equation in the space of "rapidly decreasing" generalized functions. System analysis in design and control. In: Proceedings of the XVIII International Scientific and Practical Conference, vol. 1, pp. 128–132. Polytechnical University Publishers, St. Petersburg (2014). (in Russian)
5. Chercignany, C.: Theory and applications of the Boltzmann Equation. Scottish Academic Press, Edinburgh-London (1975)
6. Maslova, N.B., Firsov, A.N.: Solution of the Cauchy problem for the Boltzmann equation. I, II. Leningrad Univ. News (19), 83–88 (1975). (1), 97–103 (1976). (in Russian)
7. FIrsov, A.N.: On a Cauchy problem for the nonlinear Boltzmann equation. – Aerodynamics of rarefied gases. No. 8. Leningrad University Publishers, Leningrad, pp. 22–37 (1976). (in Russian)
8. Maslova, N.: Nonlinear Evolution Equations. Kinetic Approach. Series on Advances in Mathematics for Applied Sciences, vol. 10. World Scientific, Singapore (1993)
9. Arsen'ev, A.A.: The Cauchy problem for the linearized Boltzmann equation. Zh. Vychisl. Mat. Mat. Fiz. **5**(5), 864–882 (1965). U.S.S.R. Comput. Math. Math. Phys. **5**(5), 110–136 (1965). (in Russian)

# Knowledge Processing Method with Calculated Functors

Vasily Meltsov[1]($\boxtimes$), Alexey Kuvaev[1], and Natalya Zhukova[2]

[1] Vyatka State University, Kirov, Russia
`meltsov69@gmail.com`, `aleksey-kuvaev@mail.ru`
[2] St. Petersburg Institute for Informatics and Automation of the Russian
Academy of Sciences, Saint Petersburg, Russia
`nazhukova@mail.com`

**Abstract.** Intellectual inference methods are among of the convenient tools for solving certain classes of knowledge processing tasks. One of the current areas in which the application of logical inference methods and engines can lead to new results is the field of cyber-physical systems that has been actively developing in recent years, including the control of unmanned vehicles and aircrafts, intelligent mechatronics and robotics. But this requires the operations of processing numerical information to enter into the logical conclusion procedure. The high-performance method of parallel output based on the disjunct (clauses) division is selected as the basic method of logical inference. To implement arithmetic operations, this method is proposed to be supplemented with a special mechanism of calculated functors. The developed modified inference method differs from the known methods by a number of important advantages. Firstly, it will significantly expand the use of artificial intelligence methods in cyber-physical systems. Secondly, inferences and arithmetic operations can be performed in parallel. And thirdly, it is an opportunity to use for the arithmetic calculations the available special processors of logical inference on the FPGA for autonomous intelligent systems for various purposes.

**Keywords:** Intelligent systems · Knowledge processing · Logical inference · Calculated functors

## 1 Introduction

Currently, there is a steady trend towards the intellectualization of computers and their software [15]. The tasks of intelligent data and knowledge processing include: logical prediction [20], enterprise management, transport logistics [13], business analytics [2, 12], medical and technical diagnostics [1, 10, 21], computer security [8, 14, 22, 24], etc. One of the current areas in which the application of inference methods and engines can lead to new results is the field of cyber-physical systems that has been actively developing in recent years, including the management of unmanned cars and aerial vehicles, intelligent mechatronics and robotics [8], and various real-time systems [3].

Developments in the field of AI is advancing by leaps and bounds. This is explained by the fact that during the transition to a new generation of systems, it is necessary to

D. G. Arseniev et al. (Eds.): CPS&C 2019, LNNS 95, pp. 187–194, 2020.
https://doi.org/10.1007/978-3-030-34983-7_19

focus not so much on the element base as on the transition from a data-oriented architecture to a knowledge-oriented architecture [11, 15, 17, 18]. Hypothetically, a knowledge processing system should consist of three parts: an intelligent interface that allows communication with the user; a data and knowledge storage subsystem; a data and knowledge processing subsystem, which includes an inference engine (Fig. 1).

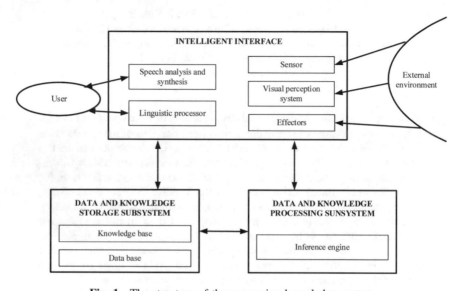

**Fig. 1.** The structure of the processing knowledge system

Robots that must be autonomous and sufficiently independent in their actions, be able to perform the necessary set of operations in dynamically changing production conditions, i.e., they cannot work only in accordance with a set of hard-conduct programs embedded into them, are used in modern robotized industries. The intellectual level of such robots must be at a sufficiently high level; therefore, their control system must include a special scheduler unit, whose task is to draw up a sequence of actions of the robot in those environmental conditions that are currently fixed by the robot's receptor system.

## 2 Formal System of the First-Order Logic

One of the most interesting and promising areas of science, studying both methods and algorithms for modeling reasoning, is mathematical logic [17]. The use of logical inference as a mathematical basis for creating intelligent systems is becoming increasingly relevant [5–7]. Here, probably, the method of resolutions developed by J. Robinson should be considered as the starting method [4, 23]. The idea of procedural interpretation of logical inference, which gave rise to the actual logical programming as a language, technology and paradigm that was developed by R. Kowalski, was for the resolution method [4].

*Formal system (FS)* is a collection of purely abstract objects, which presents the rules for operating with a multitude of characters in a purely syntactic interpretation, without taking into account the semantics [19]. FS is determined if:

- The finite *alphabet* is set (finite set of characters) – *T*.
- The procedure for constructing *formulas* is defined – *H*.
- The variety of formulas is set called *axioms* – *A*.
- A finite set of relations between formulas is given - a set of *inference rules* – *R*.

Inference rules *R* are also called inference conclusion rules.

Different variety of sets *T*, *H*, *A*, *R* define various formal systems [16, 19]. The formal system in which the IMDD is implemented is described in the papers [11, 19].

In the proposed FS, the problem is formulated as follows: for a given set of sentences (assumptions) $A_1, \ldots, A_n$ determine whether the sentence (conclusion) B is derived from these assumptions.

## 3    Accelerated Knowledge Processing Method

As mentioned above, the first order predicate calculus is the most popular formal knowledge representation system. Due to its simplicity, it forms the basis of many other formal systems. One of the fastest methods in the logic of first-order predicates today is a parallel inference method of disjunct division (IMDD) (inference method of disjunct division) [11, 19].

The accelerated inference method can be divided into 4 different procedures:

- inference procedure (V-procedure);
- procedure for the finite remainders formations (N-procedure);
- procedure for filling and analyzing the matrix of partial derivatives (M-procedure);
- unification procedure (U-procedure).

The mechanism of inference can be represented as the repeated use of some V-procedure (inference step):

$$V = \langle M_p, M_f, r, Q, r^* \rangle, \tag{1}$$

where $M_p$ – set of initial rules $R_i$; $M_f$ – set of facts $F_j$; $r$ – inferencing (provable) rule; $Q$ – sign of inference ending; $r^*$ – a new inferencing rule.

Thus, it can be said that the inference process is the iterative use of the V-procedure, and the results of the *i*-th call of the V-procedure act as input data for the $(i + 1)$ call. Consider an inference procedure for IMDD on the next well-known example about a robot (in some sources, about a monkey) and bananas [4, 11].

*Example.* It is assumed that there is some room, on the ceiling of which a bunch of bananas is suspended. Also, in the room there are a robot (monkey) and a box. The robot cannot get bananas without the help of a box. With a capital letter, we will write variables, and with a lowercase letter - constants.

Initial state: $S(a, b, c, \text{robot})$ – robot at point $a$, box at point $c$, bananas at point $b$, the robot is ready for operation ($S$ is a state of the system).

Initial *premises-rules*

1. The robot, being at any point $X1$, can go to point $c$.

$$S(c, Y1, c, \text{go}(X1, c, Z1)) : -S(X1, Y1, c, Z1).$$

2. The robot, being in the same point $X2$ as the box, can move it to point $Y2$.

$$S(V2, Y2, V2, \text{move}(X2, V2, Z2)) : -S(X2, Y2, X2, Z2).$$

3. The robot, being in the same point $X3$ as the box, can stand on it.

$$S(X3, Y3, X3, \text{climb}(Z3)) : -S(X3, Y3, X3, Z3).$$

4. If the robot, the box and the bananas are at one point $b$, and the robot can stand on the box, then it can take bananas.

$$A(\text{climb}(Z4)) : -S(b, b, b, \text{climb}(Z4)).$$

*Purpose*: if the initial state of all objects is known and the robot is activated, then it is necessary to find the value of the variable ACT satisfying predicate A(ACT).

We write the rule system in the IMDD notation.

$$1 \mapsto S(c, Y1, c, \text{go}(X1, c, Z1)) \vee \overline{S}(X1, Y1, c, Z1);$$
$$1 \mapsto S(V2, Y2, V2, \text{move}(X2, V2, Z2)) \vee \overline{S}(X2, Y2, X2, Z2);$$
$$1 \mapsto S(X3, Y3, X3, \text{climb}(Z3)) \vee \overline{S}(X3, Y3, X3, Z3);$$
$$\underline{1 \mapsto A(\text{climb}(Z4)) \vee \overline{S}(b, b, b, \text{climb}(Z4));}$$
$$1 \mapsto A(ACT) \vee \overline{S}(a, b, c, \text{Robot}).$$

*Step 1.*
Unifying substitutions: $X1/a$; $Y1/b$; $Z1/\text{Robot}$; $ACT/\text{Stand}(Z4)$.
New inference rule received:

$$r^* = S(b, b, b, \text{stand}(Z4)) \vee S(c, b, c, \text{go}(a, c, \text{Robot})).$$

*Step 2.*
Unifying substitutions: $Z4/\text{move}(c, Y2, Z2)$; $Y2/b$; $Z2/\text{move}(a, c, \text{Robot})$.
Inference ended successfully: $ACT = \text{stand} \ (\text{move} \ (c, b, \text{go}(a, c, \text{Robot})))$.
Thus, the solution will be an expression of the form:

$$\text{stand}(\text{move}(c, b, \text{go}(a, c, \ \text{Robot}))),$$

and the robot must go from point $a$ to point $c$; move the box from point $c$ to point $b$; climb on the box, and take the bananas.

Solving this problem using the SLD-resolution method embedded in the Prolog programming language will require 5 steps [4].

Despite the significant performance of the proposed method, it still has one considerable drawback that prevents the introduction of this method into intelligent cybernetic systems. This drawback is the narrowing of classes of tasks due to the lack of the possibility of describing and performing arithmetic operations. The following is an approach to eliminate this disadvantage for IMDD.

## 4  Introduction of Calculated Functors

To implement the possibility of performing arithmetic and logical operations in the process of inference, it is necessary to expand the formal IMDD system and introduce additional mechanisms and procedures into the algorithm of the logical inference engine.

The following changes are made to this formal system:

- If $f_0(t_1, \ldots, t_n)$ is functor, then $[f_0(t_1, \ldots, t_n)]$ is calculated functor.
- If $t$ is term, then $[t]$ is term. Designation $[t]$ implies a value-result of calculations over $t$. The operator [] itself will be called the calculator;
- If $[t]$ is the value-result of calculations over the term $t$, and there is a matching pair of terms $[t]$ and $z$, where $z$ is a constant, then $t^*$ is a constant substituted for the result of matching $[t]$ and $z$ and iteratively determined when performing the procedure step of inference.

It is also necessary to supplement the axiom system with the basic axiom of the calculated functor: $P(\ldots, [t_i], \ldots) \rightarrow P[\ldots, t_i^*, \ldots]$, where $t_i^*$ is a constant substituted for the functor.

The formal formulization of the problem for the modified inference remains the same, but two additional conditions are introduced:

- formulas of the conclusions should not contain calculated functors;
- all calculated functors used to perform the inference step must be replaced by subject constants.

The replacement of a calculated functor with a concretized constant occurs on a specialized device. At this stage of the research, only the simplest arithmetic operations (addition, subtraction, multiplication, division) over integer values are implemented, therefore the device is located inside the inference engine executed independently of the computer on the Altera Cyclone V GTX FPGA [9]. For more complex calculations and floating point numbers, you may need the help of a central processor.

Consider the advantages of introducing the mechanism of calculated functors in the previous example. Previously, in the solution of this logical problem, it was indicated that the robot must move from point $A$ to point $B$ without a route of movement. How to fix this?

Let's form a room map: $P_0(0, 0, \text{robot})$, $P_0(10, 10, \text{bananas})$, $P_0(20, 30, \text{box})$ (Fig. 2).

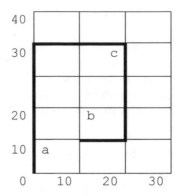

**Fig. 2.** Map of the room with the coordinates of objects

Let's "teach" the robot to the simplest moves, allowing it to step forward, right, back, and left

$$1 \mapsto \overline{P}_0(x1, y1, \text{step}(\text{forward}, N)) \text{ v } P(x1, [\text{add}(y1, 10)], N);$$
$$1 \mapsto \overline{P}_0(x2, y2, \text{step}(\text{right}, N)) \text{ v } P([\text{add}(x1, 10)], y1, N);$$
$$1 \mapsto \overline{P}_0(x2, y2, \text{step}(\text{back}, N)) \text{ v } P([x1, [\text{sub}(y1, 10)], N);$$
$$1 \mapsto \overline{P}_0(x2, y2, \text{step}(\text{left}, N)) \text{ v } P([\text{sub}([x1, 10)], y1, N);$$

Now, at the first step of inference, we will not only determine that the robot should go from the starting point $P_0(0, 0, \text{robot})$ to the box $P_0(30, 20, \text{box})$, but also get a program for the robot to move inside of the room. Of course, there will be several ways to reach the end point (box) for the parallel method, but by introducing the priorities for each box and the strategy for traversing the decision tree, the first minimum route can be got. For example, such as:

$$N = \text{step}(\text{right}, (\text{step}(\text{right}, (\text{step}(\text{forward}, (\text{step}(\text{forward}, (\text{step}(\text{forward}, \text{robot}))))))))) $$

"Read" a similar entry is necessary from the very first (internal) unifying substitution (robot). Of course, you can teach the robot to turn, bypass obstacles, and so on. It is also planned to implement a mechanism for performing the basic logical operations - more, less, equal, etc.

## 5   Conclusion

The method of accelerated inference has a number of distinctive features:

- It uses the strategy "first in breadth", which contributes to an increase in the number of simultaneously processed processes. Also, this method, unlike its counterparts, which use the "first in breadth" strategy, initiates a two-way tree traversal (bidirectional), which makes it possible to get a concrete answer in one pass.

- It allows to widely parallelize the process of computing at various levels of processing.
- The formal description of the problem for this method if free from syntactic structures and non-logical control operators that change the inference sequence.

The method of logical inference, supplemented by the possibility of calculations, not only will allow to construct formulas in one logic program, but also to calculate their values in parallel.

A separate subsystem of arithmetic (and logical) calculations allows you to use the most convenient representations of the source data; reduce the volume of logic programs; increase the speed of solving many applied logical problems containing arithmetic operations; more fully use the computational capabilities of traditional computers in the software implementation of the logical inference system on them.

Comparative analysis showed that the introduction of the possibility of describing arbitrary arithmetic calculations allows you to maintain the full declarativeness of the method, in contrast to the infix operators used in Prolog [4].

# References

1. Arseniev, D.G., Lyubimov, B.E., Shkodyrev, V.P.: Intelligent fault detection and diagnostics system on rule-based neural network approach. In: Proceedings of the IEEE International Conference on Control Applications, CCA 2009, no. 5281003 (2009)
2. Arseniev, D.G., Shkodyrev, V.P., Yarotsky, V.A., Yagafarov, K.I.: The model of intelligent autonomous hybrid renewable energy system based on Bayesian network. In: Proceedings of the IEEE 8th International Conference on Intelligent Systems, pp. 758–763 (2016)
3. Bonci, A., Carbonari, A., Cucchiarelli, A., Pirani, M., Vaccarini, M.: A cyber-physical system approach for building efficiency monitoring. Autom. Constr. **102**, 68–85 (2019)
4. Bratko, I.: Prolog Programming for Artificial Intelligence. Addison-Wesley Longman Ltd., Boston (2001)
5. Dolzhenkova, M.L., Meltsov, V.Yu., Strabykin, D.A.: Method of consequences inference from new facts in case of an incomplete knowledge base. Indian J. Sci. Technol. **9**(39), 100413 (2016)
6. Dyachenko, O., Zagorulko, Y.: A collaborative development of ontology-based knowledge bases. Commun. Comput. Inf. Sci. **468**, 219–228 (2014)
7. Gavrilova, T., Onufriev, V.: Conceptual modelling: common students' mistakes in visual representation. In: 20th International Conference on Interactive Collaborative Learning, ICL 2017. Advances in Intelligent Systems and Computing, vol. 716, pp. 199–209 (2018)
8. Hammoudeh, M., Parizi, R., Dehghantanha, A., Xu, Z., Choo, K.-K.R. (ed.): Conference review. In: International Conference on Cyber Security Intelligence and Analytics, CSIA 2019. Advances in Intelligent Systems and Computing, vol. 928 (2019)
9. Levin, I., Dordopulo, A., Fedorov, A., Kalyaev, I.: Reconfigurable computer systems: from the first FPGAs towards liquid cooling systems. Supercomput. Front. Innov. **3**–1, 22–40 (2016)
10. Mamoutova, O.V., Shirokova, S.V., Uspenskij, M.B., Loginova, A.V.: The ontology-based approach to data storage systems technical diagnostics. In: E3S Web of Conferences. Topical Problems of Architecture, Civil Engineering and Environmental Economics, TPACEE 2018, vol. 91, no. 080182018 (2019)

11. Meltsov, V.: High-Performance Systems of Deductive Inference. Science Book Publishing House, Yelm (2014)
12. Meltsov, V., Lesnikov, V., Dolzhenkova, M.: Intelligent system of knowledge control with the natural language user interface. In: Proceedings of the 2017 International Conference IT and QM and IS 2017, St. Petersburg, pp. 671–675 (2017)
13. Mikhailov, S., Kashevnik, A.: An ontology for service semantic interoperability in the smartphone-based tourist trip planning system. In: 23rd Conference of Open Innovation Association, FRUCT 2018, pp. 239–245 (2018)
14. Noor, U., Anwar, Z., Amjad, T., Choo, K.: A machine learning-based FinTech cyber threat attribution framework using high-level indicators of compromise. Future Gener. Comput. Syst. **96**, 227–242 (2019)
15. Norvig, P., Russell, S.: Artificial Intelligence: A Modern Approach. Pearson Education Limited, Edinburgh (2011)
16. Osipov, G.S., Panov, A.I.: Relationships and operations in a sign-based world model of the actor. Sci. Techn. Inf. Process. **45**(5), 317–330 (2018)
17. Pospelov, D.: Modeling of deeds in artificial intelligence systems. Appl. Artif. Intell. **7**, 15–27 (1993)
18. Rahman, S.A., Haron, H., Nordin, S., Bakar, A.A., Rahmad, F., Amin, Z.M., Seman, M.R.: The decision processes of deductive inference. Adv. Sci. Lett. **23**(1), 532–536 (2017)
19. Strabykin, D. Inference in knowledge processing systems, St. Petersburg (1998). (in Russian)
20. Strabykin, D.: Logical method for predicting situation development based on abductive inference. J. Comput. Syst. Sci. Int. **52**(5), 759–763 (2013)
21. Strabykin, D., Meltsov, V., Dolzhenkova, M., Chistyakov, G., Kuvaev, A.: Formal verification and accelerated inference. In: 5th Computer Science On-line Conference, CSOC 2016. Advances in Intelligent Systems and Computing, vol. 464, pp. 203–211 (2016)
22. Sychugov, A.A., Meltsov, V.Yu., Kuvaev, A.S., Grishin, V.M.: Network intrusions detection and prevention method using a team of intelligent agents. J. Mech. Eng. Res. Dev. **42**(2), 14–17 (2019)
23. Vagin, V., Derevyanko, A., Kutepov, V.: Parallel-inference algorithms and research of their efficiency on computer systems. Sci. Tech. Inf. Process. **45**(5), 368–373 (2018)
24. Vagin, V., Antipov, S., Fomina, M., Morosin, O.: Application of intelligent data analysis methods for information security problems. In: 2nd International Conference on Intelligent Information Technologies for Industry, IITI 2017. Advances in Intelligent Systems and Computing, vol. 679, pp. 16–25 (2018)

# Graph Model Approach to Hierarchy Control Network

Dmitry G. Arseniev, Dmitry Baskakov,
and Vyacheslav P. Shkodyrev[✉]

Peter the Great St. Petersburg Polytechnic University, St. Petersburg, Russia
{vicerector.int, baskakov.de, shkodyrev}@spbstu.ru

**Abstract.** In this paper, we propose to consider the possibility of using graph models for control and monitoring systems. The central element of a control system is a neural network, which will be built using the Directed Acyclic Graph. The novelty of the approach lies in the fact that the proposed software implementation of this approach is using deep learning packages, for example, *TensorFlow* or *Keras*. In this case, it is useful to use the mathematical apparatus of the Directed Acyclic Graph to construct such complex multi-level hierarchical models of deep learning.

**Keywords:** Directed Acyclic Graph · Distributed ledger · Cyber-physical systems · Robotics · Multi-agent systems · Smart contracts · Virtual Power Plant

## 1 Introduction

Modern approaches to control based on artificial intelligence are undergoing a pro found transformation, due to significant changes in the technological plan above all. With increasing computing power, it became possible to use deep-learning models to build very deep neural networks. At the same time, researchers often had to use typically standard methods and learning models. In this article, the authors propose to use a completely new concept based on the so-called API on the ground of Ten-sorFlow, for example. As a result, we are able to build neural networks and learning algorithms exclusively at the individual level, which is very productive and in the future will affect the result of activities.

New society is undergoing a profound transformation of all processes and spheres of human activities. The main trends are focused, in our opinion, in the following major areas.

- Digital production.
- The widespread exclusion of man from the decision-making contours.
- Increase of data processing speed while increasing security.
- Deep transformation of business models.

The authors suppose that Distributed Ledger Technology has every chance of becoming a new paradigm that will affect the entire landscape of human life. From the

D. G. Arseniev et al. (Eds.): CPS&C 2019, LNNS 95, pp. 195–211, 2020.
https://doi.org/10.1007/978-3-030-34983-7_20

point of view of the theory, the Distributed Ledger Technology has all signs of a paradigm, namely [1]:

- A paradigm rarely is a copy.
- The use of this technology allows you to achieve success better than the use of competing ways to solve same problems.
- The use of this technology allows you to solve potential problems, even if it is not fully known in advance what these problems will be like.

It is important to note that the Distributed Ledger paradigm has already been used one way or another.

Robots that do not interact with humans, that is, autonomous robots, are becoming more common. Autonomous robots perform their tasks without human control. At the same time, the fact that such robots can interact with each other becomes increasingly important and even critical. In this case, mathematical algorithms that allow to build models of interaction between such robots go in the first place. At the same time, according to the authors, the development of this concept goes in the framework of the algorithms of System 1 and System 2 [2]. More information about these systems of thinking and decision-making can be found in [3].

It should be noted that the task of teaching robots or other cyber-physical agents is very nontrivial. In addition to teaching on already tagged data, it is important to use robots' self-learning algorithms, as well as the ability to transfer this knowledge to other robots or agents. In this case, we get some kind of a closed-knowledge system, where system agents extract and formalize knowledge from the interaction with environment, transfer knowledge to other system participants via secure communication channels, and also control such transfer using the central arbiter. In fact, we get intelligent machines that are able to communicate with each other and transmit knowledge, that is, learn from each other [4]. And here we come to the next stage in the development of artificial intelligence systems, which we call Machine Reasoning (MR) [5].

## 2 Hierarchical Control

### 2.1 Objective Function

Control systems for complex objects are characterized by a large number of hierarchies and control loops, which makes it necessary to rethink modern concepts of managing complex systems and data processing [6]. The interaction between management levels occurs through a variety of objective functions for each hierarchy. The objective function is a real or integer function of several variables, that is, to be optimized (minimized or maximized) in order to solve an optimization problem. In addition to the objective function in the optimization problem for variables, constraints in the form of a system of equalities and inequalities can be specified. Thus, the objective function is the concept of achieving one of the particular goals included in the vector of goals in the process of managing. It is important to understand that if the signals about the state of objects are continuous functions of time, then such control systems are called continuous control systems. In the case of the formation of signals in discrete time, we get control systems also discrete. Consider the control function of the hierarchical system in the general case (Fig. 1):

$$\begin{cases} F_1(x_1, x_2, x_3, \ldots, x_m) = 0, \\ F_2(x_1, x_2, x_3, \ldots, x_m) = 0, \\ \ldots\ldots\ldots\ldots\ldots\ldots\ldots\ldots\ldots \\ F_N(x_1, x_2, x_3, \ldots, x_m) = 0. \end{cases} \tag{1}$$

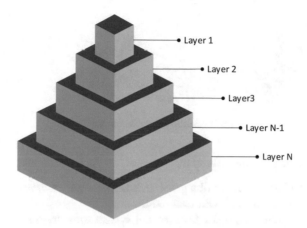

**Fig. 1.** Hierarchical system

In this case, the task of managing the entire system in the simplest case will be reduced to minimizing the objective function:

$$S = \sum_{j=1}^{N} F_j^2(x_1, x_2, x_3, \ldots, x_m). \tag{2}$$

The task of control over such a hierarchical system can be reduced to solving a multicriteria optimization problem, namely:

$$\min\{F_1(\mathbf{x}), F_2(\mathbf{x}), \ldots, F_N(\mathbf{x})\}, \quad \mathbf{x} \in X. \tag{3}$$

In this case, a valid solution $\tilde{\mathbf{x}} \in X$, called effective Pareto or Pareto optimal, if there is no other solution in $\mathbf{x} \in X$ is such, that $F_k(\mathbf{x}) \le F_k(\tilde{\mathbf{x}})$ for all $k = 1, \ldots, p$ and $F_i(\mathbf{x}) \le F_i(\tilde{\mathbf{x}})$ for at least one $i = 1, \ldots, p$.

Modern control systems are hierarchical structures with many possible vertices, as it is shown in the Fig. 2 below [7]:

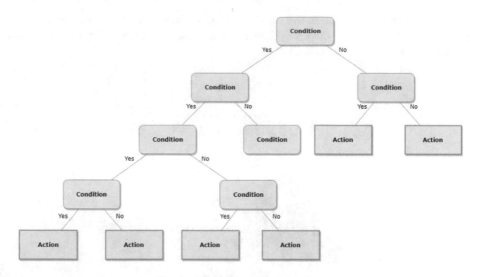

**Fig. 2.** Multi-hierarchical structure

## 2.2   Hypergraph Systems Modeling

A hypergraph is a generalization of a graph in which not only any two vertices, but also any subsets can be connected with each edge.

In this formulation, a complex hierarchical system turns in the mathematical sense into a hypergraph $H_n$, which consists of many vertices $V_n(H_n)$ and a hyper scraper $E_n(H_n)$ [8]. An example of a hypergraph is shown in the Fig. 3 below:

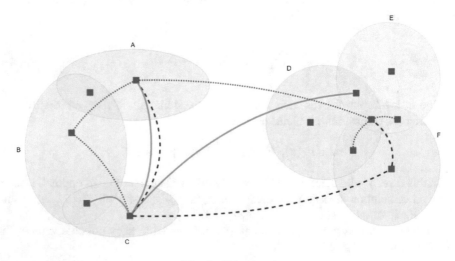

**Fig. 3.** Hypergraph

Directed hypergraph $H$ or Dihypergraph is a hypergraph with oriented hyperedges (hyperarcs). The extremities of a hyperedge have a very specific sense. A directed hyperedge is defined as a couple $E_j = (A, B)$, where $A$ and $B$ are two disjoint sets of vertices (Fig. 4).

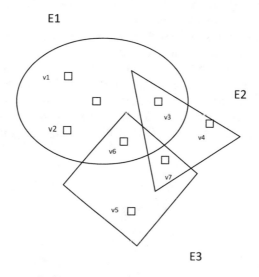

**Fig. 4.** Hypergraph composed of three hyperedges $E_1, E_2, E_3$.

Hypergraphs of various types are becoming increasingly common in the systems for managing complex hierarchical structures, which makes it possible to use this mathematical apparatus applied as in the Distributed Ledger Technologies. [9]. The set of possible solutions consists of overlapping sets of optimization functions at each level of the management hierarchy presented above (Fig. 5):

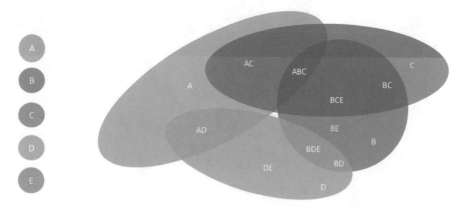

**Fig. 5.** Sets of possible solutions

A key factor in decision-making in such complex control systems is time, which significantly restricts the use of traditional blockchain paradigms [10].

## 2.3   Directed Acyclic Graph (DAG)

The authors of this paper believe that the use of a Directed Acyclic Graph (DAG) seems to be one of the promising areas for using the Distributed Ledger Technologies. In this type of graphs, there are no directed cycles, but there may be parallel paths that can lead to the end node but by different ways. It should be noted that directed acyclic graphs are widely used, including that in the tasks of artificial intelligence, together with, for example, Neural Networks (NN), in statistics and machine learning. The key factor in the use of direct acyclic graphs is precisely the speed of transactions, which is extremely important in real-time control systems with the use of a Distributed Ledger [11]. An example of a directed acyclic graph is shown below (Fig. 6):

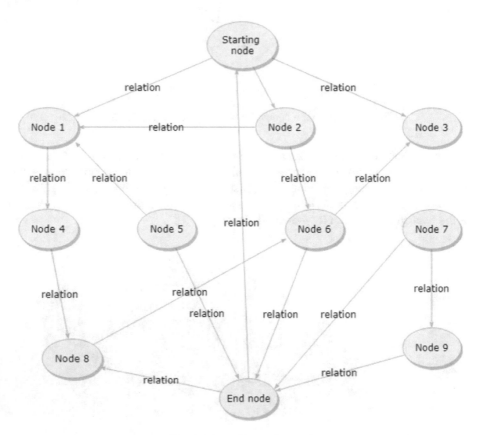

**Fig. 6.** Directed acyclic graph

In our opinion, use of a Directed Acyclic Graph as a mathematical tool is justified,, especially in the problems of multi-critical optimization. It should be also remembered that the DAG is a co-generalization of trees, that is, a forest. Every acyclic digraph has an acyclic ordering of vertices [12].

## 3   Optimization

The control problem using the mathematical apparatus of graph theory is a complex multi-level optimization problem, which consists of several subtasks, namely:

- Construction of the graph and its nodes
- Writing for each of node its own objective function, the solution of which will determine the further direction
- Solution of the optimization problem of finding the shortest path on a graph [13].

To achieve a sufficient level of generality, the utility of a path is defined by a global utility function built as a normalized weighted sum of the following partial utility functions [13]:

- A topology-oriented partial utility function, for which the utility of a path is equal to the opposite of the sum of the cost of all nodes belonging to the path.
- A cost-related partial utility function, for which the utility of a path is the opposite of the sum of the cost all nodes belonging to path.
- A quality-oriented partial utility function, for which the utility of a path is equal to the minimum quality of a node in the path.

Setting the control task is as follows. The matrix of this form is:

$$
D_j = \begin{bmatrix} F_{1,1} & \cdots & F_{1,N} \\ \vdots & \ddots & \vdots \\ F_{M,1} & \cdots & F_{M,N} \end{bmatrix}, \tag{4}
$$

where $F_{1,1}, \ldots, F_{1,N}$ — control objectives functions at the 1–level of the hierarchy and $F_{M,1}, \ldots, F_{M,N}$ — control objectives functions at the $M$–level of the hierarchy.

We can reduce the solution of this problem to the problem of optimizing Neural Networks or to the problem of optimizing deep models. In fact, in management and control tasks, there can often be no single solution, and thus we arrive to a possible set of input data $D_1, \ldots, D_J$. This set of control input matrices is mapped into a certain set of output data, e.g., using Neural Networks [14].

That is, we can consider the task of finding the optimal solution using both DAG and Neural Networks. At the same time, we believe that sharing of these two concepts can bring a significantly useful result [15].

## 4    TensorFlow. API

Many neural architectures developed recently require nonlinear networking when the network has a structure-oriented acyclic graph. The inception family of networks (developed by Christian Szegedy and colleagues forGoogle), for example, relies on inception modules in which input data is processed by several parallel convolution branches, the outputs of which are then combined into a single tensor.

Directed Neural Network give us a language for describing structured probabilistic models. Another popular language is that of undirected models, otherwise known as Markov Random Fields (MRFs) or Markov networks. As their name implies, undirected models use graphs with undirected edges.

Directed models are most naturally applicable to situations where there is a clear reason to draw each arrow in one particular direction. Often these are situations where we know the causality and it flows in only one direction. One of such situations is the relay race example. Earlier runners affect the finishing times of later runners; later runners do not affect the finishing times of earlier runners [16] (Figs. 7 and 8).

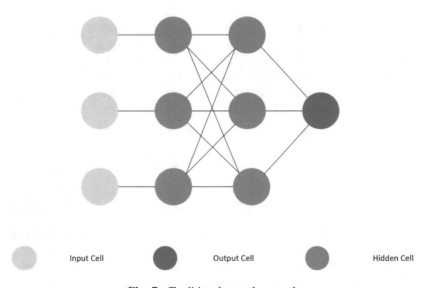

Input Cell        Output Cell        Hidden Cell

**Fig. 7.** Traditional neural network

To solve problems of such complexity by building individual networks, we recommend using the API TensorFlow.

For example, it is well known that models with several inputs, models with several outputs, and graph-like models cannot be implemented using only the class of Sequential models from Keras. In our opinion, there is another, a much more universal and flexible way to use Keras, i.e., to use the functional API. The TensorFlow or Keras functional API allows you to directly manipulate tensors and use levels as control functions that accept and return tensors. In this case, using the functional API, you can not only create models with multiple inputs and outputs, but also construct networks

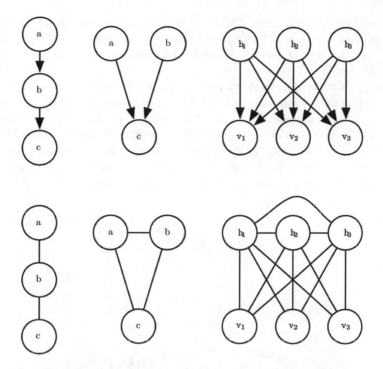

**Fig. 8.** Examples of converting directed models (top row) to undirected models (bottom rows) [16]

with a complex internal topology. Keras allows you to create neural networks organized as arbitrary oriented acyclic graphs of layers. It is known that such acyclic graphs have no closed routes. Tensor 1 cannot serve as an entrance to one of the layers that generate 1. The only allowable processing cycles (that is, recurrent constraints) are the cycles inside the recurrent layers.

There are several typical components of neural networks, implemented as graphs. The most famous of these are the *Inception* and residual links modules.

It is important to note that the functional API allows you to use models as layers - in fact, models can be thought of as "large layers". This is true for both classes: the Sequential and Model. In other words, you can call a model, pass it the input tensor, and get the output tensor:

$$y = \text{model}(x).$$

If the model accepts multiple input tensors and returns several output tensors, it should be called with lists of tensors:

$$y1, y2 = \text{model}([x1, x2]).$$

By invoking an instance of a model, you reuse its weights, just like when you invoke an instance of a layer. Calling any instance – a layer or a model – always entails reusing an existing received instance representation, which is completely understandable.

A simple example of the practical use of a reusable instance of a model is the model of view, which uses a dual camera as an input: two parallel cameras spaced one inch apart. Such a model can perceive depth, which can be useful in many applications. You do not need to create two independent models to extract visual signs from the images transmitted by the left and right cameras before combining them in two streams. Low-level processing of two input streams can be performed together, that is, using layers that share the same weights and, accordingly, representations.

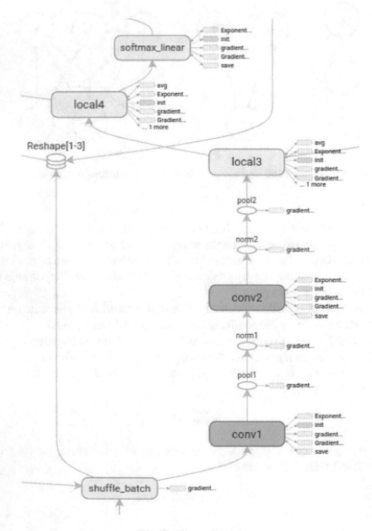

**Fig. 9.** TensorBoard.

For fruitful research or development of good models, it is necessary to have versatile, frequently updated information about what is happening inside the model during experiments. The main goal of the experiments is to get as much information as possible about how well the model works. Moving forward is iterative or cyclical in nature: you start with an idea and develop an experiment plan that confirms or disproves it. You run an experiment and process the information received. This gives an impetus to the birth of a new idea. The more iterations in this cycle you complete, the more perfect and powerful your ideas will become. Keras will help you move from idea to experiment as soon as possible, and a quick GPU will help you get the results of the experiment as quickly as possible. But what about processing of the results? In this case we recommend to useTensorBoard (Figs. 9 and 10).

**Fig. 10.** TensorBoard. Visualization of the learning process

# 5  Smart Systems with Neural Network. Future

The authors see the use of Smart Systems with neural network where human participation is minimized; they are seen as an extremely promising direction for the development of the Distributed Ledger Technologies. The use of this technology in the activities of energy companies in the following areas looks promising.

## 5.1  Energy

Using the Distributed Ledger technology in the energy sphere seems to be extremely promising for a number of reasons. Below are the main ones.

Integrated Process Automation [17]. The Distributed Ledger technology allows you to transform the landscape of traditional business with well-established markets and relationships. We also assume a significant reduction in the risks of certain transactions within the framework of traditional business processes.

Intellectual billing. Using Distributed Ledger technology along with Smart Contracts allows you to automate calculations and measurements in Distributed Energy Systems [18].

Sales and marketing. The key success factor here is the use of Distributed Ledger technology in conjunction with artificial intelligence systems (Machine Learning and Deep Learning). This symbiosis allows to ensure the construction of an individual consumer profile with the transfer of all data via secure communication channels.

Communication with intelligent devices via secure channels significantly changes data transfer algorithms, as well as processes in energy companies.

Security. The use of cryptography significantly increases the security of transactions and provides an unsurpassable level of trust between the participants of the system. Also, the use of Distributed Ledger technology allows a better control of data privacy and identity management.

Transparency achieved using a Distributed Ledger technology.

Competition. The use of a DL technology makes it possible to qualitatively simplify the task of switching between energy suppliers, thereby increasing the flexibility of the choice of suppliers for the consumer. There are few works on this subject, but from the point of view of game theory, in any case, the consumer has flexibility, which increases competition [19].

Algorithms and business models. Automation of processes in the energy sector using the Distributed Ledger technology allows you to qualitatively change the traditional management paradigms. In fact, with the advent of the Distributed Ledger

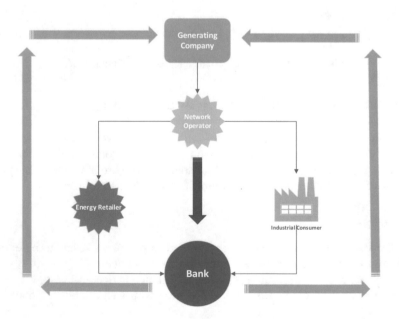

**Fig. 11.** Traditional market

technology, we will observe a breakdown of established development trends and business models. Traditional companies will become Decentralized Autonomous Corporations (DACs), which operate under completely different business laws [20]. This paradigm is quite new; the number of publications on this subject is extremely small [21] (Figs. 11 and 12).

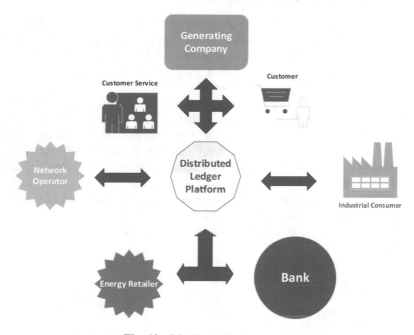

**Fig. 12.** Distributed ledger market

## 5.2   Robotics

The use of a Distributed Ledger technology in cyber-physical systems, such as robotics, is a very promising direction, especially in the field of interaction and intellectual training of elements of such systems. Let's briefly consider the main directions and prospects for the possible use of Distributed Ledger technologies in such systems:

- Intellectual interaction between robots. In this context, it refers to the ability to share information, participate in training and decision-making without human intervention [22].
- Distributed solution of a problem. In this case, we use a Distributed Ledger technology to collect distributed control robots and control the overall goal.
- Distributed Ledger technology for voting [23].
- Task optimization. In this case, a Distributed Ledger technology is used to optimize the problem being solved from the viewpoint of controlling the interaction among all other intelligent agents and robots.
- Control of integrity, access, confidentiality and integrity.

The main problem in describing these advantages now is that all solutions are rather prototypical in nature, without any introduction into real activities. But all this can be justified by the innovativeness of the Distributed Ledger technology, the development of which is fraught with certain risks and nuances.

### 5.3     Virtual Power Plants (VPP). Definition

Recently, the concept of a Smart Grid (SG) of power supply has become very popular. A much more impressive continuation of this concept, according to the authors, is the concept of Virtual Power Plants. By introducing the Artificial Intelligence (AI) function into this concept, we would be able to generate and sell energy in more flexible realities, taking into account possible supply and demand, which should allow the theory to maximize profits of energy suppliers not through the high prices, but rather by reducing generation of losses. At the same time, from the viewpoint of the development of this direction, authors do not yet have any well-established terminology. Different authors define this promising direction in their own way. In fact, VPP is defined as a combination of several players, namely, generating companies, consumers of different levels, other players who both produce and generate energy on demand or demand in accordance with requests or demand with minimal human participation [24]. There are alternative definitions, which, under VPP, understand a certain system that is capable to generate and accumulate energy demand and supply in one way or another [25] (Fig. 13).

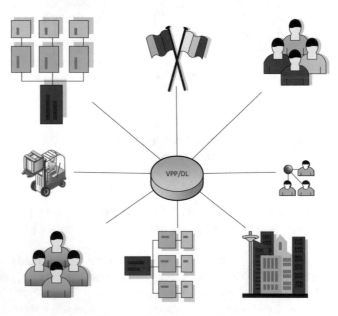

**Fig. 13.** Virtual power plants

# 6  Algorithmic, Management and Technological Limitations

In the process of developing a Distributed Ledger technology, a number of obstacles and limitations must be overcome. The key factor here is the human factor, since this paradigm substantially rebuilds the landscape of traditional thinking and current business processes, e.g., [26]. Many authors completely do not take into account the human factor, which can be a significant constraint on the way of implementing these solutions. The next factor that can have a serious impact on the implementation of technologies in the near future may be legislative and legal restrictions that would not allow the Distributed Ledger technology to develop relatively rapidly. In this case, various industry standards and regulations would not allow to see rapid and explosive growth in the near future. The authors look positively at this paradigm. Of course, we are aware that the development of this concept will require serious efforts and time. An important, in our opinion, restriction may be the algorithmic and mathematical apparatus of this concept. Classical algorithms are not always applicable to real industrial automation solutions [27]. Many algorithms and approaches require a more detailed study in the process of implementation in real smart devices, which hinders the development of automation processes based on a Distributed Ledger technology. This requires a serious engineering adaptation of solutions in terms of process automation [28]. At the same time, the authors purposefully do not consider open and closed platforms in this case, since the use of the latter will require serious improvements, taking into account the features of each specific implementation. Well, an extremely important factor from the point of view of automation of management processes in real cyber-physical systems is the need to build target functions and their multi-criteria optimization [29]. All this will require extremely serious efforts, both from the point of view of mathematics and of the final engineering implementation [30].

Nevertheless, the authors believe that the development of Distributed Ledger technologies will continue, which will lead to a profound transformation of all existing and future business processes.

# 7  Conclusion

In our opinion, the Distributed Ledger technology is a unique technology of the future. Its peculiarity lies in the complex influence on the processes taking place in society. The breadth of future solutions is truly impressive. At the same time, most of the works focus on the technological nuances of the technology, going into the field of classical Computer Science, which, in our opinion, is not entirely correct [31]. In fact, it seems to us that the problem lies not so much in the technology as in the management and changing paradigms of doing business [32].

# References

1. Kuhn, T.: The Structure of Scientific revolutions. The University of Chicago Press (2012)
2. Peysakhovich, A.: Reinforcement learning and inverse reinforcement learning with system 1 and system 2 (2019)
3. Kahneman, D.: Thinking. Fast and Slow, Farrar, Straus and Giroux, New York (2013)
4. Lazaridou, A., Peysakhovich, A., Baroni, M.: Multi-agent cooperation. In: ICLR 2017 (2017)
5. Bottou, L. (2011). https://arxiv.org/ftp/arxiv/papers/1102/1102.1808.pdf
6. De Domenico, M., et al.: Mathematical formulation of multilayer networks. Phys. Rev. X **3** (4), 041022 (2013)
7. Boccaletti, S., et al.: The structure and dynamics of multilayer networks. Phys. Rep. **544**(1), 1–22 (2014)
8. Fong, B., Spivak, D.I.: Hypergraph categories. J. Pure. Appl. Algebra **223**(11), 4746–4777 (2019)
9. Di Francesco, D., Maesa, A.M.: Laura Ricci. Detecting artificial behaviours in the Bitcoin users graph, Online Social Networks and Media (2017)
10. Kotilevets, I.D., Ivanova, I.A., Romanov, I.O., Magomedov, S.G., Nikonov, V.V., Pavelev, S.A.: Implementation of directed acyclic graph in blockhain network to improve security and speed of transactions. In: IFAC 2018 (2018)
11. Quiterio, T.M., Lorena, A.C.: Using complexity measures to determine the structure. Appl. Soft Comput. **65**, 428–442 (2018)
12. Bang-Jensen, J., Gutin, G.: Classes of Directed Graphs. Springer, Heidelberg (2018). https://doi.org/10.1007/978-3-319-71840-8
13. Comuzzi, M.: Optimal directed hypergraph traversal with ant-colony optimisation. Inf. Sci. **471**, 132–148 (2018)
14. Zhanga, Z., Chen, D., Wang, J., Bai, L., Hancock, E.R.: Quantum-based subgraph convolutional neural networks. Pattern Recogn. **88**, 38–49 (2019)
15. Narayan, A., O'N Roe, P.H.: Learning graph dinamics using deep neural networks. In: IFAC 2018 (2018)
16. Goodfellow, I., Bengio, Y., Courville, A.: Deep Learning. The MIT Press, Cambridge (2016)
17. Blockchain in energy and utilities use cases, vendor activity. Indigo Advisory Group (2019). https://www.indigoadvisorygroup.com/blockchain
18. Why the energy sector must embrace blockchain now. Ernst & Young Global Limited (2019). https://www.ey.com/en_gl/digital/blockchain-s-potential-win-for-the-energy-sector. Accessed 09 Apr 2019
19. Mengelkamp, E., Gärttner, J., Rock, K., Kessler, S., Orsini, L., Weinhardt, C.: Designing microgrid energy markets. a case study: the Brooklyn microgrid. Appl. Energy **210**, 870–880 (2018)
20. Hsieh, Y.-Y. The rise of decentralized autonomous organizations: coordination and growth within cryptocurrencies. https://ir.lib.uwo.ca/cgi/viewcontent.cgi?article=7386&context=etd. Accessed 10 Apr 2019
21. Kypriotaki, K.N., Zamani, E.D., Giaglis, G.M.: From bitcoin to decentralized autonomous corporations. In: Proceedings of the 17th International Conference on Enterprise Information Systems (ICEIS-2015) (2015)
22. Afanasyev, I., Kolotov, A., Rezin, R., Danilov, K., Kashevnik, A.: Blockchain solutions for multi-agent robotic systems: related work and open questions (2019). https://arxiv.org/pdf/1903.11041.pdf. Accessed 10 Apr 2019

23. Pawlak, M., Poniszewska-Maranda, A., Kryvinska, N.: Towards the intelligent agents for blockchain e-voting system. In: The 9th International Conference on Emerging Ubiquitous Systems and Pervasive Networks (EUSPN 2018), Leuven, Belgium (2018)
24. Ramos, S.: Demand response programs definition supported by clustering and classification techniques. In: 16th International Conference on Intelligent System Applications to Power Systems, Hersonissos, Greece (2011)
25. Pereira, F., Faria, P., Vale, Z.: The influence of the consumer modelling approach in demand response programs implementation. In: 2015 IEEE Eindhoven PowerTech, Eindhoven, Netherlands (2015)
26. Turk, Ž., Klinc, R.: Potentials of Blockchain technology for construction management. In: Creative Construction Conference 2017, CCC 2017, Primosten, Croatia (2017)
27. Rubio, M., Alba, A., Mendez, M., Arce-Santana, E., Margarita. A.: Consensus algorithm for approximate string matching. In: 2013 Iberoamerican Conference on Electronics Engineering and Computer Science, San Luis Potosí, S.L.P., México (2013)
28. Mathias, S.B., Rosset, V., Nascimento, M.C.: Community detection by consensus genetic-based algorithm for directed networks. In: 20th International Conference on Knowledge Based and Intelligent Information and Engineering Systems (2016)
29. Liua, S., Papageorgiou, L.G.: Multi-objective optimisation for biopharmaceutical manufacturing under uncertainty. Comput. Chem. Eng. **119**, 383–393 (2018)
30. Xu, C.: A big-data oriented recommendation method based on multi-objective. Knowl. Based Syst. **177**, 11–21 (2019)
31. Viriyasitavat, W., Hoonsopon, D.: Blockchain characteristics and consensus in modern business. J. Ind. Inf. Integr. **13**, 32–39 (2018)
32. Angelis, J., da Silvac, E.: Blockchain adoption: a value driver perspective. Bus. Horiz. **62**(3), 307–314 (2018)

# Application of Methods for Optimizing Parallel Algorithms for Solving Problems of Distributed Computing Systems

Yulia Shichkina$^{(\boxtimes)}$, Mikhail Kupriyanov,
and Al-Mardi Mohammed Haidar Awadh

St. Petersburg Electrotechnical University "LETI", St. Petersburg, Russia
strange.y@mail.ru

**Abstract.** Today, various researchers have developed a set of methods for optimizing parallel algorithms for systems with distributed memory. These methods are optimized for various parameters and taking into account various properties of the algorithm. A distributed computing system has its own characteristics, such as heterogeneity of computing nodes, network bandwidth and others. The studies conducted by the authors of this article show that these characteristics do not interfere with the application of these methods to solving problems in a distributed computing environment. The article shows that there is no need to modify and adapt optimization methods for the use in distributed computing systems. However, it is necessary to take into account the properties of such systems contributed to the emergence of iteration in the application of optimization methods and the increase of the complexity of the process of analysis and optimization of the initial parallel algorithm. The article also describes ways to solve the problem of reducing the time complexity of the iterative application of optimization methods to the initial parallel algorithm. The results of the authors' research is a method for constructing a special type of graph for a parallel algorithm that takes into account properties of a given computing system and approaches to constructing the schedule of the algorithm.

**Keywords:** Schedule · Optimization · Algorithm · Information graph · Network bandwidth · Execution time · Operation · Process · Node interconnection graph

## 1 Introduction

There are many important tasks whose solution requires the use of large computational resources, which are often inaccessible to modern computing systems. One of the indicators of the quality of a parallel program is the load density of computational nodes. Even the small time delays in transferring data via communication channels from one processor to another are summed up during the program operation time, and as a result, there is obtained a long delay and this greatly increases the overall algorithm operation time [1, 2].

The purpose of the construction and equivalent transformation of the information graph is to minimize the total program execution time due to the best distribution of

D. G. Arseniev et al. (Eds.): CPS&C 2019, LNNS 95, pp. 212–224, 2020.
https://doi.org/10.1007/978-3-030-34983-7_21

tasks among processors and organization of the sequence of tasks on each processor [3, 4]. The construction of the task solution schedule is done in the way preserving prioritization among all tasks performed in a distributed environment [5].

The beginning of the analysis of the scalability of parallel programs and computer systems was laid in 1967, when an employee of IBM Gene Amdahl published an article [6], which later became classic. Further evaluation of the increase in the speed of execution of computational algorithms, the use of resources and the performance of computing systems was obtained in many other studies [7–10]. An analysis of the effectiveness of parallel computing, the effect of performance variability on the accuracy and efficiency of the optimization algorithm, and a strategy for minimizing the effect of this variability is given in [11]. The problem of load balancing processors is considered in [12, 13].

The problem of constructing a schedule of a parallel algorithm with optimal execution time has been presented in various studies starting of the 1980s. Since then, various scheduling algorithms have been created, based on the methods for finding the maximum flow in the transport network, based on branches and boundaries, dynamic programming, and various iterative algorithms (for example, genetic algorithms) [14–17]. Most of these algorithms, in particular, based on finding the maximum flow in the transport network, have pseudopolynomial complexity [18, 19].

Problems of parallelization of individual fragments of algorithms, usually the most difficult to parallelize, are considered in [20, 21].

One of the drawbacks of the existing methods for constructing parallel algorithms is that they do not take into account the transfer of data between computational nodes. Initially, it was assumed that computing modules – processors or computers – could be interconnected so that the solution of problems on a new computing system would speed up as many times as the number of computing modules are used in the system. However, it quickly became clear that for complex tasks, such acceleration is usually impossible to achieve due to two reasons:

- any task is parallelized only partially - there always are parts that cannot be parallelized;
- the communication environment, which links individual parts of a parallel system, is much slower than the processors', so that the transfer of information significantly delays the calculations.

This article proposes several solutions to the problem of obtaining a schedule for executing an algorithm that is efficient in a number of parameters, such as node power, network bandwidth, data transfer time, etc., in a distributed computing environment using optimization methods developed for distributed memory computing systems [22].

## 2  Terminology

The information graph of the algorithm is an oriented acyclic multigraph, in which the vertices correspond to the operations of the algorithm, and the edges are the data transfer between operations.

Suppose that all operations of an algorithm are divided into groups, and the set of groups is ordered so that each operation of any group depends either on the initial data or on the results of operations in the previous groups. The presentation of the algorithm in this form is called the parallel form of the algorithm. And the information graph corresponding to the parallel form of the algorithm is called the parallel form of the information graph.

The tier of an information graph is a group of vertices in the parallel form of an information graph corresponding to the operations of the algorithm with the same start execution time.

The height of the information graph is the number of tiers, i.e., graph operation time.

The width of the information graph is the maximum number of operations in tiers.

The algorithm execution schedule is an indication of which processes and at what time the algorithm operations should be performed.

An operation is a set of elementary actions, which is a whole in the framework of a given algorithm. An elementary operation is the simplest action in a machine language performed by a computer, that is, an action that cannot be represented by a set of simpler actions.

The process, according to ISO 9000: 2000, is a set of interrelated and interacting actions that convert input data into output data. Each computational process can be implemented as an operating system process or computational processes can be a set of execution threads within one process of the operating system.

Interprocess communication is a set of ways to transfer data between processes. Processes can be run on one or more computers connected by a network [23].

## 3   Formulation of the Problem

Suppose there is a computer system $S$ consisting of a limited number of computing nodes $n$: $S(n)$.

Let the algorithm correspond to the information graph $G(k, t, v)$ uniquely defining the number of operations $k$, the execution time of operations $t$ and the connection between operations $v$ by data.

Then the parallel form of the graph is an information graph with additional information on the start time of each operation $t_0$ (or the number of the tier corresponding to each operation in the information graph) and information about the computing system: $G' = G'(G(k, t, v), t_0, P)$. Obviously, each algorithm will correspond to a set of equivalent graphs in a parallel form $G'(G(k, t, v), t_0, P)$.

Suppose a set $X_{S(n),G}$ is a set of all possible schedules $x$ for the execution of a given algorithm with a given information graph $G(k, t, v)$ on a given computing system $S(n)$. The function $f(x)$ is a function of the execution time of the algorithm using the schedule $x \hat{I} X_{S(n),G}$.

Then, as an algorithm optimization problem, we will understand the problem of finding among the $x$ elements forming the set $X$ of such an element $x^*$, for which the function $f(x)$ has the minimum value, i.e.

$$f(x) \longrightarrow \min_{x \in X_{S(n),G}}$$

with the following limitation:

$$m \leq n.$$

Since $x = t_0 = (t_{01}, t_{02}, \ldots, t_{0k})$, then the optimization problem can be formulated as follows:

find $G'' \in \{G'(G(k, t, v), t_0, P)|f(t_{01}, t_{02}, \ldots, t_{0k}) = \min f(x), x \in X_{S(n),G}, m \leq n\}$.

Since the tass of finding all the graphs $G'(G(k, t, v), t_0, P)$ for an algorithm, even on a limited number of nodes, is difficult to solve, then the suboptimal optimization problem is most often to be solved. The task of suboptimal optimization is the task of finding the optimal schedule among the set of possible schedules limited by the rules of search, i.e.:

$$G'' \in \{G'(G(k, t, v), t_0, P)|f(t_{01}, t_{02}, \ldots, t_{0k}) = \min f(x), x \in X_{S(n),G}, m \leq n, R\},$$

where $R$ is a set of rules according to which the schedules are selected.

In this article, we consider approaches to solving this problem. We also describe the method of supplementing the information graph with new parameters of the algorithm and the computing system.

## 4   The Parameters of the Algorithm and Computer Systems

Most often, when there is a creation schedule for the execution of a parallel algorithm, it takes into account such parameters of algorithms as:

- time of execution of operations;
- connections between the operations according to data.

But, in addition to these characteristics, there is a number of other parameters that can significantly affect the execution time of the algorithm, for example, these are:

- heterogeneity of computational nodes;
- the volume of transmitted data;
- the length of communication between two nodes;
- network bandwidth, etc.

All these parameters can be divided into two classes:

- parameters of the algorithm;
- parameters of the computing environment.

**Parameters of the Algorithm**
All parameters of the algorithm can be assembled in one total parameter. Most often, this total parameter is "*time*". So, if during interprocess communication, the volume of data transferred varies in a very wide range, then when transferring a large volume of

data, the time spent on transferring this data is added to the execution time of the operation that delivers data or operations that receive data.

In this case, a large volume of data means a volume that requires time $t > t_{cr}$ to be transferred from a process to a process. In this case, the $t_{cr}$ parameter can be determined based on the specific conditions for solving the problem. One of the approaches to the determination of $t_{cr}$ can be a statistical approach, when $t_{cr}$ can be defined as the standard deviation from the expectation $t$ for all operations ($t_{cr} = \bar{t} + \sigma = \frac{1}{n}\sum_{i=1}^{n} t_i +$

$\sqrt{\frac{1}{n-1}\sum_{i=1}^{n}(t_i - \bar{t})^2}$, where $n$ is the number of operations of the algorithm) or standard deviation from the minimum value of $t$ for all operations ($t_{cr} = t_{min} + \sigma = \min_{i=1,...n} t_i +$

$\sqrt{\frac{1}{n-1}\sum_{i=1}^{n}(t_i - \bar{t})^2}$).

It should be noted that the time obtained as a result of the optimization method described in Sect. 6 will be equal to or greater than the actual execution time of the algorithm. This is due to the fact that when adding the data transfer time to the execution time of the operation, it is not fixed anywhere for which operation this addition was performed. And if, during the execution of the algorithm optimization method, the operation following the current operation will be set in the schedule of the same process as the operation that transfers data to it and has an added time, then the next operation can be performed faster, because, in fact, the data will not be transmitted.

**Parameters of the Computer System**
The main distinguishing feature of the methods developed for parallel computing is the condition that all computational nodes perform computations with the same power and, therefore, the execution time of the same operation on different nodes is assumed to be the same. In distributed computing, this is often not the case. Therefore, for algorithms in distributed computations, a number of additional parameters are immediately added to the properties of the information graph of the algorithm [24]. These parameters may be different depending on the network architecture. For example, they may include:

- The node power. The methods by which the power is calculated can be very different. Their essence does not affect the optimization of algorithms. It is only important that for all nodes the power was calculated using the same methodology and, therefore, the ranking of nodes by this parameter was correct.
- The network bandwidth. Computers exchange information using networks of different physical nature: cable, fiber optic, etc. Network bandwidth is the maximum speed of information transmission over the network per unit of time. The speed of information transfer is the amount of information transmitted per unit of time.
- The transmission time. It will depend on the physical parameters of the network and the length of the transmission path.

It is not important how many and what parameters these will be. The essence of the problem is that there will be more than one of them and they will be numeric.

## 5  Projection of the Information Graph on the Graph of Node Relationships

To create a schedule for the execution of the algorithm in a distributed environment is not enough to create an information graph of the algorithm. Computing nodes of the system, as well as the operations of the algorithm, are interconnected, and the number and quality of these connections can be very different. Therefore, to create an algorithm execution schedule in a distributed environment, there is a need to create two graphs.

Graph of interconnections of nodes of a distributed computing system. This graph is a weighted undirected one. At the same time, it will not just be weighted, but will have a set of weights $(P_1, P_2, \ldots, P_m)$ (where $P_i = \{p_{ij}\}$, $i = 1..m$, j $= 1..n$, $m$ is the number of weights, $n$ is the number of vertices of the graph). Each of the weights will affect the schedule of execution of the algorithm. An example of such a graph is shown in Fig. 1, where $Y_i$ – nodes, $P_i = \{p_i, l_i, d_i\}$, $p_i$ – node performance, $l_i$ – network bandwidth, $d_i$ – path length.

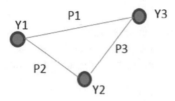

**Fig. 1.** Example of a weighted graph of a distributed computing system consisting of three nodes.

Information graph of the algorithm in a parallel form on which it is possible to uniquely reproduce the schedule for the execution of the algorithm.

The whole complexity of this approach lies in the fact that in order to create a second graph (information graph of the algorithm in parallel form), it is necessary to produce a multiple projection of the initial information graph of the algorithm on the graph of the interconnections of nodes.

Example. Suppose there is a graph presented in Fig. 2a. After applying the algorithm for the distribution of vertices on tiers [22] to this graph, it is possible to obtain the following groups of vertices on tiers: $M_1(1, 2, 3)$, $M_2(5, 4, 6)$, $M_3(8, 7)$, $M_4(9)$. In accordance with these tiers, an information graph will be created in a parallel form in the first approximation, i.e., without taking into account the execution time of operations and features of a distributed system. This graph is shown in Fig. 2b.

Let there be operations corresponding to the vertices of the graph, executed according to the time $T = \{t_i\} = \{3, 2, 4, 5, 2, 3, 4, 3, 3\}$, where $i$ is the number of the vertex of the graph, $i = 1..9$.

For the operations, the maximum amount of data obtained at the output of the operation is also known: $V = \{v_i\} = \{13, 4, 6, 1, 3, 6, 9, 5, 8\}$, where $i$ is the number of the vertex of the graph, $i = 1..9$.

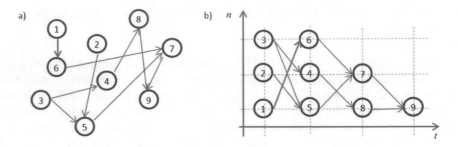

**Fig. 2.** An example of an information graph of the algorithm, (a) the initial graph, (b) a graph reduced to parallel form

The following information is available by nodes:

- the performance of the $j$-th node $P = \{p_j\} = \{7, 5, 3\}$;
- the length of the path between the $j$-th and $(j + 1)$-th nodes $D = \{d_j\} = \{5, 7, 15\}$;
- the network bandwidth between the $j$-th and $(j + 1)$-th nodes $L = \{l_j\} = \{10, 8, 9\}$,

where $j$ is the number of node, $j = 1..3$.

Then, at the initial moment of time, there is a possible next projection (Fig. 3) of the information graph from Fig. 2b onto the graph of interconnections of nodes with Fig. 1.

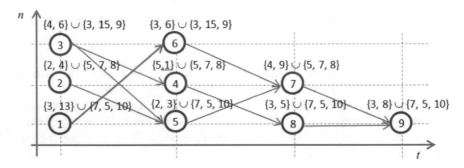

**Fig. 3.** An example of the projection of the information graph of the algorithm on the graph of the interconnections of nodes

It is possible to make simple calculations. Suppose that on the $j$-th tier there are $nj$ vertices, the graph of interrelations of nodes consists of $ny$ nodes.

Option 1. Let $nj < ny$. Then on each tier there are $A_{ny}^{nj} = \frac{ny!}{(ny-nj)!}$ of all possible combinations of nodes and operations. Then the total number of projections of the information graph on the graph of interrelations of nodes will be

$$G_p = \prod_{j=1}^{m} A_{ny}^{nj},$$

where $m$ is the number of tiers.

Option 2. Let $nj = ny$. Then on each tier there can be $P_{ny} = 1$ all possible combinations of nodes and operations. Then the total number of projections of the information graph on the graph of interrelations of nodes will be

$$G_p = \prod_{j=1}^{m} ny! = (ny!)^m,$$

where $m$ is the number of tiers.

Option 3. Let $nj > ny$. Then it is necessary to apply the method of optimization of the parallel algorithm taking into account the constraints on computing resources and reduce the parallel form of the algorithm to the form for the case $nj = ny$.

Example. For the information graph from Fig. 2b and the graph of interrelations of nodes from Fig. 1 the number of projections will be

$$G_p = (3!)^4 = 1296.$$

For a graph as small as in Fig. 2, the number of projections is very large. It is impossible to analyze all these projections in order to find the most effective projection in terms of execution time and the number of involved resources using simple manual search. Obviously, with the increasing number of vertices of the information graph and the number of nodes with different power, location, and other parameters, the number of projections will quickly increase. And for an information graph with more than 100 vertices, and a computing system with more than 10 nodes, the number of projections will be about 1 million. It will take a lot of time to analyze such a large number of projections even with the use of computer technology.

## 6  Approaches to Reducing the Number of Iterations

There are several solutions to this problem.

*Solution A.* Reducing the number of projections by reducing for each vertex the list of possible nodes. For example, the result of an operation corresponding to vertex 7 is a data set of volume $v_7 = 9$. If this operation is performed on 1 node, then on the next tier it is possible to exclude node number 2 from the destination of our data. The capacity of the network segment between the 1-st and 2-nd nodes is only 8. The disadvantages of this approach are:

- The complexity of the calculations of the number of projections that remain in the search for the optimal schedule.
- Complicating the method of finding the optimal schedule by adding $Gp$ conditions to check whether the node is subject to consideration or not.
- A small decrease in the number of projections. For example, for graphs with Figs. 1 and 2, $Gp$ will decrease by no more than 10.

- Impracticality. The number of conditions added to check the compliance of the node for each operation will be much higher than the number of projections that are removed from consideration. The complexity of this way will increase.

*Solution B.* Pre-processing of the information graph without taking into account the time of execution of operations but taking into account interprocess transfers. This will reduce the number of transfers and remove from consideration all the projections on which data is not transmitted from node to node.

As a preprocessing of an information graph, for example, the algorithm optimization method by volume of interprocessor data transfer is suitable. The method is as follows:

1. To base groups of vertices corresponding to tiers obtained in any way. It is better if these groups are obtained in a formalized way, for example, by optimizing the information graph in width using a matrix or adjacency list;
2. The process of moving vertices begins with the last group. Suppose, there are m groups altogether, then the number of the next group $k = m$;

   First of all, it is necessary to put in the schedule (taking into account the groups) vertices for which a binary connection is used, i.e., they only receive data from one previous operation, and only then to arrange the vertices with multiple connections, because in this case the existence of a downtime is inevitable.

3. In the $k$-th group, select the first vertex. Consider the position number of this vertex in the group equal to 1: $i = 1$;
4. Compare the vertex $M_{ki}$ (where $i$ is the position number of the vertex in the group) with the vertices of the previous $(k\text{-}1)$-th group. If in the $(k\text{-}1)$-th group there is a vertex $(M_{k-1,j}$, where $j$ is the position number of the vertex in the group, $j^3 i)$ directly connected in the information graph by an edge with a given vertex; then the vertex $M_{k-1,j}$ must be moved in its group to the $i$-th position. If there is no vertex in the $(k\text{-}1)$-th group associated with the vertex $M_{ki}$, then step 6;
5. If $k > 2$, then $k = k\text{-}1$ and step 4;
6. If not all vertices are considered in the $m$-th group, then $k = m$, $i = i+1$ and step 4;
7. If in the last group all vertices are considered and $m > 2$, then $m = m\text{-}1$ and step 6;
8. If $m = 1$, then the end of the method.

*Example.* After applying the optimization method to the information graph from Fig. 2b, we obtain the graph from Fig. 4.

As a result, data transfer does not require operations corresponding to vertices 1, 2, 4, 6, 7, 9. Let's call the set of these vertices $M_0 = \{1, 2, 4, 6, 7, 9\}$.

Table 1 shows the values of combinations of operations and nodes for each vertex of the graph.

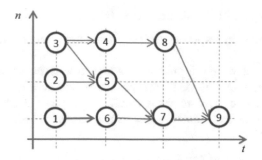

**Fig. 4.** Information graph optimized by the number of interprocess data transfers

Consider the operation on each tier. The first tier – operations 1, 2, 3. They do not depend on anything and, therefore, each can be executed on any node. Therefore, in the table line the *"combination"*, they correspond to number 3 (the operation can be performed on any of the 3 nodes).

**Table 1.** The number of combinations of operations and nodes for different vertices of the information graph

| Vertices | 1 | 2 | 3 | 4 | 5 | 6 | 7 | 8 | 9 |
|---|---|---|---|---|---|---|---|---|---|
| Combinations | 3 | 3 | 3 | 1 | 1 | 1 | 1 | 1 | 1 |

Operation 6 depends on operation 1, which is the initial vertex in the binary connection and, therefore, belongs to the group M0. Since operation 6 depends on operation 1 from the M0 group, then operation 6 must be performed on the same node as operation 1. Therefore, in Table 1 in the line *"combination"* it corresponds to the number 1 (can be performed only on one specific node without any choice). The situation is similar for operation 8.

Operation 5 depends on operation 2, which belongs to the group M0. The second operation with number 3, on which operation 5 depends, is not included in the group M0. Therefore, operation 5 does not have the right to select a node for execution. Consequently, in Table 1 in the line *"combination"* it corresponds to number 1 (can be performed only on one specific node without a choice). The situation is similar for operations 7, 9.

For operation 4, there is simply no choice, since two nodes are already taken from the three available nodes.

Accordingly, the number of combinations of vertices of the graph, and, consequently, the number of projections, will be equal to:

$$G_p = 3 \cdot 3 = 9.$$

Thus, to select the optimal schedule, it is necessary to evaluate only 9 graphs. This is 162 times less than the total number of projections.

The disadvantages of this solution are as follows:

- Not every information graph can achieve such a significant reduction in the number of projections.
- Often, when transferring information to a higher-performance node, the execution time of the operation may be less, even taking into account the cost of the data transfer process compared to the execution time on the same node, but low-powered. Therefore, this solution gives a local optimal schedule, which may not be equal to the global.

*Solution C.* Development of a method for optimizing the schedule "by tier": selection of nodes on the first two tiers and then the choice of nodes taking into account those that have already been fixed for the vertices of the previous tiers.

This method will be more effective in terms of complexity compared with the method of enumerating all projections, but its disadvantage is also the lack of a guarantee that the schedule found will be the global optimum.

# 7  Conclusions

This article discusses the features of a distributed computing system. In the article there is a description of a graph of interconnections of nodes of the computer system and the information graph of the algorithm. It is shown that the complexity of designing the schedule of the algorithm in a distributed environment lies in the fact that to create an information graph of the algorithm in a parallel form, it is necessary to make a multiple projection of the initial information graph of the algorithm on the graph of interconnections of nodes. As a result, a very large number of graphs will need to be processed, and this will immediately affect the schedule construction time.

One of the solutions to this problem is to pre-process the information graph without taking into account the execution time of operations but only the interprocess data transfers. This will reduce the number of transfers and remove from consideration all the projections on which data is not transmitted from node to node.

Another approach is to use the method of optimizing the "by-tier" schedule. This method has already been developed by the authors and is currently under testing. This method is not included in this article due to the limitations on the amount of material in one paper. The method is used to select nodes on the next tier, taking into account those that have already been fixed for the vertices of the previous tiers.

For clarity, all the methods presented in this article are illustrated with examples. It should be noted that the solutions described in the article were all tested with the help of a special program on more than 100 test cases.

The future plans of the authors include an analysis of the possibilities of combining the method "by tier" of scheduling optimization with other methods of optimizing parallel algorithms and the possibilities of applying these methods to queries in distributed databases of various types.

**Acknowledgments.** The paper was prepared within the scope of the state project "Initiative scientific project" of the main part of the state plan of the Ministry of Education and Science of the Russian Federation (task № 2.6553.2017/8.9 BCH Basic Part) and was funded by RFBR according to the research project № 19-07-00784.

# References

1. Abramov, O.V., Katueva, Y.: Multivariant analysis and stochastic optimization using parallel processing techniques. Manage. Probl. **4**, 11–15 (2003)
2. Jordan, H.F., Alaghband, F.: Fundamentals of Parallel Processing, p. 578. Pearson Education Inc., Upper Saddle River (2003)
3. Voevodin, V.V., Voevodin, V.V.: Parallel Computing, p. 608. BHV-Petersburg, St. Petersburg (2002)
4. Drake, D.E., Hougardy, S.: A linear-time approximation algorithm for weighted matchings in graphs. ACM Trans. Algorithms **1**, 107–122 (2005)
5. Hu, C.: MPIPP: an automatic profileguided parallel process placement toolset for SMP clusters and multiclusters. In: Proceedings of the 20th Annual International Conference on Super-Computing. New York, NY, USA, pp. 353–360 (2006)
6. Amdahl, G.M., Reston, V.A.: Validity of the single processor approach to achieving large-scale computing capabilities. In: Proceedings AFIPS Spring Joint Computer Conference, pp. 483–485 (1967)
7. Grama, A., Gupta, A., Karypis, G., Kumar, V.: Introduction to Parallel Computing, 2nd edn. Addison Wesley, USA (2003)
8. Gergel, V.P., Strongin, R.G.: Parallel Computing for Multiprocessor Computers. NGU Publishing, Nizhnij Novgorod (2003). (in Russian)
9. Quinn, M.J.: Parallel Programming in C with MPI and OpenMP, 1st edn. McGraw-Hill Education, New York (2003)
10. Wittwer, T.: An Introduction to Parallel Programming, VSSD uitgeverij (2006)
11. Tiwari, A., Tabatabaee, V., Hollingsworth, J.K.: Tuning parallel applications in parallel. Parallel Comput. **35**(8–9), 475–492 (2009)
12. Mubarak, M., Seol, S., Lu, Q., Shephard, M.S.: A parallel ghosting algorithm for the flexible distributed mesh database. Sci. Program. **21**(1–2), 17–42 (2013)
13. Kruatrachue, B., Lewis, T.: Grain size determination for parallel processing. IEEE Softw. **5** (1), 23–32 (1988)
14. Rauber, N., Runger, G.: Parallel Programming: For Multicore and Cluster Systems, p. 450. Springer, Heidelberg (2010). https://doi.org/10.1007/978-3-642-37801-0
15. Gergel, V.P., Fursov, V.A. Lectures of Parallel Programming: Proc. Benefit, p. 163. Samara State Aerospace University Publishing House (2009)
16. Liu, C.L., Layland, J.W.: Scheduling algorithms for multiprogramming in hard real-time environment. J. ACM **20**(1), 46–61 (1973)
17. Marte, B.: Preemptive scheduling with release times, deadlines and due times. J. ACM **29**(3), 812–829 (1982)
18. Burns, A.: Scheduling hard real-time systems: a review. Softw. Eng. J. **6**(3), 116–128 (1991)
19. Stankovic, J.A.: Implications of Classical Scheduling Results for Real-Time Systems. IEEE Computer Society Press, Los Alamitos (1995)
20. Tzen, T.H., Ni, L.M.: Trapezoid self-scheduling: a practical scheduling scheme for parallel compilers. IEEE Trans. Parallel Distrib. Syst. **4**, 87–98 (1993)
21. Sinnen, O., Sousa, L.A.: Communication contention in task scheduling. IEEE Trans. Parallel Distrib. Syst. **16**(6), 503–515 (2005)

22. Shichkina, Y., Kupriyanov, M.: Creating a schedule for parallel execution of tasks based on the adjacency lists. In: Galinina, O., Andreev, S., Balandin, S., Koucheryavy, Y. (eds.) NEW2AN/ruSMART -2018. LNCS, vol. 11118, pp. 102–115. Springer, Cham (2018). https://doi.org/10.1007/978-3-030-01168-0_10
23. Liedtke, J.: On Micro-Kernel Construction. In: Proceedings of the 15th ACM Symposium on Operating System Principles. ACM, December (1995)
24. Tanenbaum, A., Woodhull, A.: Operating Systems Design and Implementation, 3rd edn, pp. 197–495. Prentice Hall, Eaglewood Cliffs (2006)

# Complex System and Value-Based Management and Control: Multi-attribute Utility Approach

Yuri P. Pavlov and Rumen D. Andreev[(✉)]

Institute of Information and Communication Technologies,
Bulgarian Academy of Sciences, Sofia, Bulgaria
{yupavlov15, rumen}@isdip.bas.bg

**Abstract.** Inclusion of personal human knowledge and intuition or even the social needs and understanding in the system modeling is a challenge and puts a new level of challenges. The utility theory is mathematical approach to measurement and utilization of qualitative, conceptual information; it permits the inclusion of the decision-maker (or the technologist, manager) in a complex model like "Technologist dynamical model" in mathematical terms. The subject of this paper is the design of a methodology and algorithms for evaluation of expert utility (value) function: and that permit development of value-driven control in systems where the individual human choice is decisive for the solution. It presents a methodology and machine-learning algorithms for evaluation of expert utility function used in the development of value-driven control where the human preferences are the basic or important part for the solution. This approach is illustrated by modeling exhaustible forest resource management and timber production based on the optimal control of exhaustible forest resource. The principles of rationality and market efficiency need economically effective resource management. A multi-attribute utility function is included in the dynamical model and by this, the optimal control solution becomes synchronous with the consumers' preferences.

**Keywords:** Human preferences · Utility · Learning algorithm · Eco systems · Optimization · Control

## 1 Introduction

Contemporary management or control of complex systems including human beings as an element of them requires an analytical description of human knowledge and intuition. Following the system analysis, a complex system includes active or decisive human participation in the description and choice of the final decision [10, 16]. In the context of system analysis, it is a "human-process" system. The main approaches for reducing a complex system to a simpler one are:

- to accumulate sufficient information in order to achieve structural and parametrical clarity [14];
- to seek interpretation and expression of the different aspects of a complex system through expert analysis and evaluation of the experts' preferences [5].

© Springer Nature Switzerland AG 2020
D. G. Arseniev et al. (Eds.): CPS&C 2019, LNNS 95, pp. 225–234, 2020.
https://doi.org/10.1007/978-3-030-34983-7_22

Mathematical modeling of "human–process" systems needs analytical representation of quantitative information at the levels of human preferences and concerns value-driven modeling [18]. This approach is based on measurement of human's preferences on the basis of utility and measurement theory, and uses an iterative learning process as an essential part of its implementation [1, 18, 20, 21]. It is the first step in carrying out a human-centered value-driven design process and decision-making that avoids the contradictions in human decisions and permits mathematical calculations. Such modeling is a contemporary trend in scientific research [4, 16]. The preferences are usually of cardinal type and contain uncertainty, which expresses points to the stochastic approximation theory.

The objective of the paper is to present a mathematical approach for estimation and modeling of human preferences as machine-learning in the process of construction of models of complex systems with human participation. This approach is illustrated by modeling of exhaustible forest resource management and timber production based on the optimal control of exhaustible forest resource.

## 2  Mathematical Preliminary

The representation of a human's preferences as value or utility function will mathematically allow the inclusion of the decision-maker (DM) in the complex system modeling [13]. Common understanding is that the *Value-based design* is a systems engineering strategy which enables multidisciplinary design optimization [8]. From practical point of view, the basic empirical system of human preferences is a system of the relations "prefer" ($\succ$) and "indifference" ($\approx$) (indifferent or equivalent). The "indifference" relation ($\approx$) is based on ($\succ$) and is defined by $((x \approx y) \Leftrightarrow \neg((x \succ y) \vee (x \prec y)))$. Let $\mathbf{X}$ be the set of alternatives ($\mathbf{X} \subseteq \mathbf{R}^m$). A Value function is a function ($u^* : \mathbf{X} \to \mathbf{R}$) for which is fulfilled [9, 11]:

$$((x, y) \in \mathbf{X}^2, x \succ y) \Leftrightarrow (u^*(x) > u^*(y)).$$

The assumption of existence of a value function $u(.)$ leads to the "negatively transitive" and "asymmetric" relation ($\succ$): a "weak order". A "strong order" is a "weak order" for which $(\neg(x \approx y) \Rightarrow ((x \succ y) \vee (x \prec y)))$ is fulfilled. The existence of a "weak order" ($\succ$) over $\mathbf{X}$ leads to the existence of a "strong order" over $\mathbf{X}/\approx$. The assumption of existence of a value function $u(.)$ leads to the existence of asymmetry $(((x \succ y) \Rightarrow (\neg(x \prec y))))$, transitivity, $(((x \succ y) \wedge (y \succ z) \Rightarrow (x \succ z))$ and transitivity of the "indifference" relation ($\approx$). The measurement scale is defined with accuracy to monotone functions [15]. A transformation with an arbitrary monotonous function leads to another equivalent ordinal scale and it is impossible to tell the distance between the different alternatives.

According to utility theory, let **X** be the set of alternatives and **P** be a set of probability distributions over **X**, and **X** $\subseteq$ **P**. A Utility function $u(.)$ will be any function for which the following is fulfilled [9]:

$$(p \succ q, (p,q) \in \mathbf{P}^2) \Leftrightarrow (\int u(.)dp > \int u(.)dq).$$

To every decision choice and action corresponds a probability distribution of appearance of final alternatives (results). The notation ($\succ$) expresses the preferences of DM over **P** including those over **X** (**X** $\subseteq$ **P**). The interpretation is that the integral of the utility function $u(.)$ is a measure concerning the comparison of the probability distributions $p$ and $q$ defined over **X**.

There are different systems of mathematical axioms that give satisfactory conditions of a utility function existence. The most famous of them is the system of Von Neumann and Morgenstern's axioms [9]:

(A.1) The preferences relations ($\succ$) and ($\approx$) are transitive, i.e., the binary preference relation ($\succ$) is a weak order;

(A.2) Archimedean Axiom: for all $p$, $q$, $r \in$ **P** such that $(p \succ q \succ r)$, there is an $\alpha$, $\beta \in$ (0, 1) such that $((\alpha p + (1-\alpha)r) \succ q)$ and $((q \succ (\beta p + (1-\beta)r)))$;

(A.3) Independence Axiom: for all $p$, $q$, $r \in$ **P** and any $\alpha \in (0, 1]$, then $(p \succ q)$ if and only if $(((\alpha p + (1-\alpha)r) \succ (\alpha q + (1-\alpha)r)))$.

Axioms (A1) and (A3) cannot give a solution. Axioms (A1), (A2) and (A3) give a solution in the interval scale (precision up to an affine transformation):

$$((p \succ q) \Leftrightarrow (\int v(x)dp \succ \int v(x)dq)) \Leftrightarrow (v(x) = au(x) + b, \text{ where } a, b \in \mathbf{R}, a > 0, x \in \mathbf{X}).$$

Following from this proposition, the measurement of the preferences is in the interval scale (temperature scale) [19]. The gambling approach is used to construct the utility function in the sense of von Neumann. The reason is that to be in the interval scale, the set of the discrete probability distributions P has to be convex. The same holds true in respect of the set **X**. The utility function is evaluated by the "Gambling approach" [13]. This approach consists of the comparisons between lotteries. A "lottery" is called every discrete probability distribution over **X**. We denote as $<x, y, a>$ the simplest lottery: a is the probability of the appearance of the alternative $x$ and $(1-a)$ - the probability of the alternative $y$.

The weak points of the gambling approach are the violations of the transitivity of preferences and the-so called "certainty effect" and "probability distortion" [7, 12]. The violations of the transitivity of the relation equivalence ($\approx$) also lead to declinations in the utility assessment. All these difficulties explain the DM behavior observed in the Allais Paradox [3]. This uncertainty of the preference expressions determined our orientation to preference evaluations with the use of stochastic approximation theory [1, 2].

## 3  Utility Function Evaluation

We know that the utility function $u(.)$ over $\mathbf{X}$ is determined with the accuracy of an affine transformation [9]. This property is essential for the application of utility theory, since it allows a decomposition of the multiattribute utility into simple functions. Starting from the gambling approach, we use the following sets [18]:

$$\mathbf{A}_{u^*} = \{(x, y, z, \alpha)/(\alpha u^*(x) + (1-\alpha)u^*(y)) > u^*(z)\},$$

$$\mathbf{B}_{u^*} = \{(x, y, z, \alpha)/(\alpha u^*(x) + (1-\alpha)u^*(y)) < u^*(z)\}.$$

The notation $u^*(.)$ is the DM's empirical utility assessment. Through stochastic recurrent algorithms we approximate functions recognizing the two sets above. The process is machine-learning, based on the DM's preferences. This is a probabilistic pattern recognition because $(\mathbf{A}_{u^*} \cap \mathbf{B}_{u^*} \neq \emptyset)$ and the utility evaluation is a stochastic approximation with noise (uncertainty) elimination.

The following presents the evaluation procedure. The DM compares the "lottery" $<x, y, \alpha>$ with the simple alternative $z, z \in \mathbf{Z}$ ("better $-\}$, $f(x, y, z, \alpha) = 1$", "worse $-\{$, $f(x, y, z, \alpha) = (-1)$" or "can't answer or equivalent $- \sim$, $f(x, y, z, \alpha) = 0$", $f(.)$ denotes the qualitative DM's answer). This determines a learning point $((x, y, z, \alpha), f(x, y, z, \alpha))$. The following recurrent stochastic algorithm constructs the utility polynomial approximation:

$$u(x) = \sum_i c_i \mathbf{\Phi}_i(x),$$

$$c_i^{n+1} = c_i^n + \gamma_n \left[ f\left(t^{n+1}\right) - \overline{(c^n, \mathbf{\Psi}(t^{n+1}))} \right] \mathbf{\Psi}_i\left(t^{n+1}\right),$$

$$\sum_n \gamma_n = +\infty, \quad \sum_n \gamma_n^2 < +\infty, \; \forall n, \; \gamma_n > 0.$$

In the formula the following notations are used (based on $A_u$): $t = (x, y, z, \alpha)$, $\mathbf{\Psi}_i(t) = \mathbf{\Psi}_i(x, y, z, \alpha) = \alpha \mathbf{\Phi}_i(x) + (1 - \alpha)\mathbf{\Phi}_i(y) - \mathbf{\Phi}_i(z)$, where $(\mathbf{\Phi}_i(x))$ is a family of orthogonal polynomials. The line above the scalar product $\bar{v} = \overline{(c^n, \mathbf{\Psi}(t^{n+1}))}$ means: $(\bar{v} = 1)$, if $(v > 1)$, $(\bar{v} = -1)$ if $(v < -1)$ and $(\bar{v} = v)$ if $(-1 < v < 1)$. The coefficients $c_i^n$ take part in the following polynomial presentation as a scalar product:

$$g^n(x) = \sum_{i=1}^{n} c_i^n \mathbf{\Phi}_i(x),$$

$$(c^n, \mathbf{\Psi}(t)) = \alpha g^n(x) + (1 - \alpha)g^n(y) - g^n(z) = G^n(x, y, z, \alpha).$$

The mathematical procedure describes the following assessment process. The expert relates intuitively on the "learning point" $(x, y, z, a))$ of the set $\mathbf{A}_{u^*}$ with probability $D_1(x, y, z, \alpha)$ or of the set $\mathbf{B}_{u^*}$ with probability $D_2(x, y, z, \alpha)$.

The probabilities $D_1(x, y, z, \alpha)$ and $D_2(x, y, z, \alpha)$ are mathematical expectations of $f(.)$ over $\mathbf{A}_{u^*}$ and $\mathbf{B}_{u^*}$ respectively, $(D_1(x, y, z, \alpha) = M(f/x, y, z, \alpha))$ if $(M(f/x, y, z, \alpha) > 0)$, $(D_2(x, y, z, \alpha) = (-)M(f/x, y, z, \alpha))$ if $(M(f/x, y, z, \alpha) < 0)$. Let $D'(x, y, z, \alpha)$ be a random value: $D'(x, y, z, \alpha) = D_1(x, y, z, \alpha)$ if $(M(f/x, y, z, \alpha) > 0)$; $D'(x, y, z, \alpha) = (-D_2(x, y, z, \alpha))$ if $(M(f/x, y, z, \alpha) < 0)$; $D'(x, y, z, \alpha) = 0$ if $(M(f/x, y, z, \alpha) = 0)$. We approximate $D'(x, y, z, \alpha)$ by a function of the type $G(x, y, z, \alpha) = (\alpha g(x) + (1-\alpha)g(y) - g(z))$, where $g(x) = \sum_i c_i \Phi_i(x)$. The coefficients $c_i^n$ take part in the approximation of function $G(x, y, z, \alpha)$:

$$G^n(x, y, z, \alpha) = \left(c^n, \Psi\left(t^{n+1}\right)\right) = \alpha g^n(x) + (1 - \alpha)g^n(y) - g^n(z),$$

$$g^n(x) = \sum_{i=1}^{n} c_i^n \Phi_i(x).$$

Function $G^n(x, y, z, \alpha)$ is positive over $\mathbf{A}_{u^*}$ and negative over $\mathbf{B}_{u^*}$ depending on the degree of approximation of $D'(x, y, z, \alpha)$. Function $g^n(x)$ is the approximation of the utility function $u(.)$.

The multiattribute utility function could be decomposed to a combination of single attributed, utility functions normed between 0 and 1 [13]. These facts permit determination of the coefficients in the multiattribute decomposition with an easy procedure. The coefficients in the utility formula are determined by comparisons of lotteries of the following type [18]:

$$
\left\{
\begin{array}{c}
U(x_1) \times \alpha \\
+ \\
U(x_2) \times (1 - \alpha)
\end{array}
\right\}
\begin{array}{c}
\succ \\
\approx \\
\prec
\end{array}
U(x_3).
$$

Here, $x_1$, $x_2$ and $x_3$ are fixed and $(x_1 \succ x_3 \succ x_2)$ is fulfilled. The questions to the decision-maker are like lotteries in which we vary only the probability $\alpha \in [0,1]$. The alternatives $x_1$ and $x_2$ are with fixed values (utilities) and the aim is to determine the value $U(x_3)$. Let $\alpha$ be a uniformly distributed random value in $[0, 1]$. We define the following random vector $\chi = (\eta_1, \eta_2, \eta_3)$, where:

If $(\succ) \Rightarrow \chi = (\eta_1 = 1,\ \eta_2 = 0,\ \eta_3 = 0)$;

If $(\prec) \Rightarrow \chi = (\eta_1 = 0,\ \eta_2 = 0,\ \eta_3 = 1)$;

If indiscernibility $(\approx) \Rightarrow \chi = (\eta_1 = 0,\ \eta_2 = 1,\ \eta_3 = 0)$.

Let $\chi^n$ be a learning sequence of independent random values with equal to $\chi$ distribution. The stochastic recurrent procedure is the following:

$$\left(\lambda_1^{n+1}, \lambda_2^{n+1}, \lambda_3^{n+1}\right) = \Pr_P\left[\left(\lambda_1^n, \lambda_2^n, \lambda_3^n\right) + \gamma_n\left(\left(\eta_1^n, \eta_2^n, \eta_3^n\right) - \left(\lambda_1^n, \lambda_2^n, \lambda_3^n\right)\right)\right]$$

$$\sum_1^{\infty} \gamma_n = \infty, \quad \sum_1^{\infty} \gamma_n^2 < \infty, \gamma_n \geq 0,\ \forall n \in N,$$

for example

$$\gamma_n = \frac{1}{(n+1)}$$

with initial point is 0.

The notation $\Pr_P$ has the meaning of projection over the set:

$$P = \{(\lambda_1, \lambda_2, \lambda_3)/\lambda_1 \geq 0, \ \lambda_2 \geq 0, \ \lambda_3 \geq 0, \ \lambda_1 + \lambda_2 + \lambda_3 = 1\}.$$

The searched value is determined in the end as the following:

$$U(x_3) = U(x_2) + (\lambda_1 + \lambda_2/2)(U(x_1) - U(x_2)).$$

The procedure and its modifications are machine-learning. The learning points $((x, y, z, \alpha), f(x, y, z, \alpha))$ are set with a Sobol's pseudo-random sequence [22].

## 4  Exhaustible Forest Resource Management and Production Control

The case of optimal use of exhaustible natural resources is investigated by Clarke [6]. The investigation provides an understanding of the optimal control strategy and allows for numerical approaches to be employed in this control.

Still there remains the question of how to deal with complex logging processes accounting for the social and environmental factors. A multiattribute utility function has been built, taking into account three important factors pertaining to a forest system [17]:

- $X_2$ – factor (supporting services – environmental effect) [1–200 number of species per hectare];
- $X_3$ – factor (percentage of employed locals in the forestry sector – social effect) [1-30%];
- Independence of the utility was found between the following factors in the process of investigation: $X_2$ from $X_1$; $X_2$ from $X_3$; $X_3$ from $X_1$; $X_3$ from $X_2$.

Utility independence of $X_2$ from the other two factors means the following. The DM's preferences between the lotteries for $x_2$, $x_2 \in X_2 \in$ at different values of $x_1$, $x_1 \in X_1$ and $x_3$, $x_3 \in X_3$ do not change, suggesting utility independence of $x_2$ from the changes of other two factors. Using the theory for decomposition of a multi-attribute utility to simpler functions the following multi-attribute utility structure is determined:

$$U(x_1; x_2; x_3) = k_1 U\left(x_1; x_2^0; x_3^0\right) + f_2(x_1) \times \left[U\left(x_1^0; x_2; x_3^0\right)\right] + f_3(x_1) \times \left[U\left(x_1^0; x_2^0; x_3\right)\right] + \\ f_{23}(x_1) \times \left[U\left(x_1^0; x_2; x_3^0\right)\right] \times \left[U\left(x_1^0; x_2^0; x_3\right)\right],$$

where $U\left(x_1^0; x_2^0; x_3^0\right) = 0$ and $U\left(x_1^*; x_2^*; x_3^*\right) = 1$.

In the formula above $X^0 = \left(x_1^0; x_2^0; x_3^0\right) = (10, 1, 1)$ and $X^* = \left(x_1^*; x_2^*; x_3^*\right) = (300, 200, 30)$. Functions $f_2$, $f_3$ and $f_{23}$ have the forms:

$$f_2(x_1) = U\left(x_1; x_2^*; x_3^0\right) - k_1 U\left(x_1; x_2^0; x_3^0\right),$$

$$f_3(x_1) = U\left(x_1; x_2^0; x_3^*\right) - k_1 U\left(x_1; x_2^0; x_3^0\right),$$

$$f_{23}(x_1) = U\left(x_1; x_2^*; x_3^*\right) - f_2(x_1) - U\left(x_1; x_2^0; x_3^*\right).$$

Each of these functions is evaluated based on the DM's preferences. In Fig. 1 utility functions $U(x_1; 132; x_3)$ and $U(x_1; 20; x_3)$ are shown.

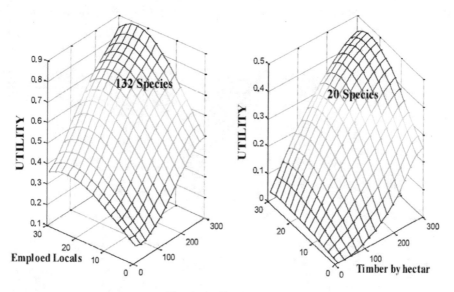

**Fig. 1.** Utility functions

The described value modeling is part of the Smart Forest Ecological Management System ForEco which is based on Open Data and FIWARE. This utility model can be used for value-based modeling and value-based optimal control, taking into account the environmental, social and financial impact on the local population. It uses a logging model of the following type [6]:

$$\begin{aligned}
&\dot{y} = -\min(z, v) r(y), \\
&\dot{z} = \gamma w, \\
&y(0) > 0, \ z(0) > 0, \\
&0 \le v, \ 0 \le w, \ (v + w) \le L(t).
\end{aligned}$$

In the formula, $y(t)$ is marks the "extracted woods per hectare" – $\left[m^3 ha^{-1}\right]$; $z$ is the capital invested in the equipment and $v$ is the labor as a workforce (one man – one shovel).

Both dimensions are in standard units (for example, money, utils, etc.). With ($w$) is denoted the labor used in the production of the equipment. As a criterion for optimal control, is chosen the integral:

$$\int_0^T \exp(-\delta t)(\pi\min(z, v)r(y) - \alpha v - \beta w)dt.$$

This integral has the meaning of net income for the user in a competitive market. $T$ is indicating the period of work, $\pi$ is the price of the timber, $\delta$ is the devaluation factor over time, $\alpha$ and $\beta$ are the value of the labor force and capital investments respectively. The determination of the optimal solution requires a special technique, namely, non-smooth optimization [6]. The function $r(y)$ is the utility function $u(X_1; 20; 5)$, $y \in X_1$, shown on Fig. 2:

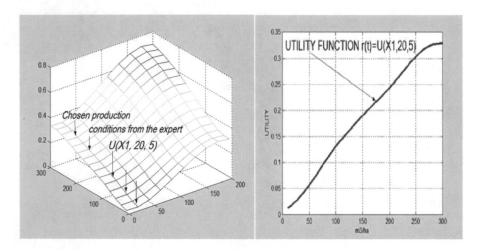

**Fig. 2.** Utility function and $r(y)$

The modeling is performed with parameters: $y(0) = 150$ $(m^3 ha^{-1})$, $T = 10$ (months), $L = 80$ (thousand-BGNs), $\alpha = 30$ (BGNs), $\beta = 20$ (BGNs) and devaluation factor $\delta = 0.3$. For the initial value of the investments it is assumed $z(0) = 25$ (thousand-BGNs), and for the coefficient of technology efficiency $\gamma$ is accepted as $\gamma = 1.7$. The yield in cubic meters per hectare, the capital for equipment, and the labor employed in the production of the equipment-parameter w are shown in Fig. 3.

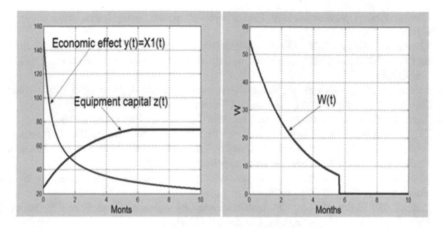

**Fig. 3.** Optimal control

The utility function $u(X_1; 20; 5)$, is used for determination of the derivative $r(t)$ in the dynamical model.

## 5   Conclusions

The calculations demonstrate the possibility for mathematical representation of a complex production process taking into account social and economic effects. Value-based models permit effective management of timber resources in accordance with DM's preferences. That is, the DM's preferences are analytically reported in the model and indirectly in the optimal control.

The presented methodology and mathematical procedures allow for the design of individually-oriented decision support systems. Such systems may be autonomous or parts of larger intelligent information or decision support systems and can permit reasonable optimal solutions and prognoses.

## References

1. Aizerman, M., Braverman, E., Rozonoer, L.: Theoretical foundations of the potential functions method in pattern recognition learning. Autom. Remote Control **25**, 821–837 (1964). (in Russian)
2. Aizerman, M.A., Braverman, E., Rozonoer, L.: Potential Function Method in the Theory of Machine Learning. Nauka, Moscow (1970). (in Russian)
3. Allais, M.: Le comportement de l'hommerationneldevant le risque: Critique des postulats et axiomes de l'école américaine. Econometrica **21**, 503–546 (1953)
4. Bandura, A.: *Social foundations of thought and action: A social cognitive theory*. Prentice-Hall, NJ (1986)
5. Baron, J.: Thinking and Deciding, 4th edn. Cambridge University Press, Cambridge (2008)

6. Clarke, F.H.: Optimization and Non-Smooth Analysis. Wiley-Interscience, New York (1983)
7. Cohen, M., Jaffray, J.-Y.: Certainty effect versus probability distortion: an experimental analysis of decision making under risk. J. Exp. Psychol. Hum. Percept. Perform. **14**(4), 554–560 (1988)
8. Collopy, P., Hollingsworth, P.: Value-driven design. In: AIAA, paper 2009-7099. American Institute of Aeronautics and Astronautics, Reston, VA (2009)
9. Fishburn, P.: Utility Theory for Decision-Making. Wiley, New York (1970)
10. Gig, J.: Applied General Systems Theory, vol. 1–2. Moscow, Mir (1981). (in Russian)
11. Herbrich, R. et al.: Supervised learning of preferences relations. In: Proceedings Fachgruppentreffen Maschinelles Lernen, pp. 43–47 (1998)
12. Kahneman, D., Tversky, A.: Prospect theory: an analysis of decisions under risk. Econometrica **47**, 263–291 (1979)
13. Keeney, R., Raiffa, H.: Decision with Multiple Objectives: Preferences and Value Trade-Offs. Cambridge University Press, Cambridge (1999)
14. Keeney, R.L.: Value-driven expert systems for decision support. Decis. Support Syst. **4**, 405–412 (1988)
15. Krantz, D.H., Luce, R.D., Suppes, P., Tversky, A.: Foundations of Measurement. vol. I–III. Academic Press. New York (1971, 1989, 1990)
16. Larichev, O., Olson, D.: Multiple Criteria Analysis in Strategic Sitting Problems, 2nd edn. Springer, Boston (2010). https://doi.org/10.1007/978-1-4757-3245-0
17. Lyubenova, M., Pavlov, Y., Chikalanov, A., Spassov, K., Novakova, G., Nikolov, R.: Smart forest ecological management system. J. Balkan Ecol. **18**(4), 363–375 (2015)
18. Pavlov, Y., Andreev, R.: Decision Control, Management, and Support in Adaptive and Complex Systems: Quantitative Models. IGI Global, Hershey (2013)
19. Pfanzagl, J.: Theory of Measurement. Physical-Verlag, Wurzburg-Wien (1971)
20. Raiffa, H.: Decision Analysis. Addison-Wesley Reading Mass, New York (1968)
21. Mandel, I.: Causality modeling and statistical generative mechanisms. In: Rozonoer, L., Mirkin, B., Muchnik, I. (eds.) Braverman Readings in Machine Learning. Key Ideas from Inception to Current State. LNCS (LNAI), vol. 11100, pp. 148–186. Springer, Cham (2018). https://doi.org/10.1007/978-3-319-99492-5_7
22. Sobol, I.: Monte-Carlo Numerical Methods. Nauka, Moscow (1973). (in Russian)

# Estimating the Accuracy Increase During the Measuring Two Quantities with Linear Dependence

Vladimir Garanin and Konstantin Semenov$^{(\boxtimes)}$

Peter the Great St. Petersburg Polytechnic University, St. Petersburg, Russia
semenov.k.k@gmail.com

**Abstract.** Increasing of measurement accuracy is always a relevant goal. Applied to cyberphysical systems, it helps to improve their operation and quality of decision-making. One of the methods to achieve this is to use relations between quantities to be measured – if such relationships exist and are known at least approximately. At present, there are not so many published articles that describe metrological applications which use this kind of information about measured quantities to get a better accuracy. It seems that the small amount of practical realizations is due to the necessity to use rather sophisticated mathematical approaches based on the probability theory and mathematical statistics along with numerous simulations to make a conclusion about the potential increase of accuracy. This paper presents a simplified approach producing a set of clear enough equations and indicators, which are helpful for engineers during preliminary estimation of the potential increase of accuracy from the known interconnections between the measured quantities. A case of linear dependency between the measured quantities is analyzed to show how the approach works.

**Keywords:** Measurement results · Inaccurate data · Accuracy increase · Relationships between measurands

## 1 Introduction

The modern state of technologies tends to the complexity and miniaturization of measuring units and their wider usage. Cyberphysical systems include intelligent sensors and take into account different other sources of objective information on object under control to make more optimal decisions. All possible information is collected, because it helps to achieve better efficiency in cyberphysical systems operation. The accuracy of the used data is an important factor to be considered because the relevancy and quality of decisions made depend on it. The main goal of this paper is to show on simple example how available information can be used to increase accuracy of measurement results.

The information about the relationships between the measured quantities can be successfully applied to improve the accuracy of the measurements in many practical situations [1–18]. However, if even this relationship is known for a specific application accurately, a corresponding deriving for a better estimate of the measurement result based on this information requires an obligatory analytical study that is often too labor-intensive for typical engineering tasks. As an example, a set of methods based on the

© Springer Nature Switzerland AG 2020
D. G. Arseniev et al. (Eds.): CPS&C 2019, LNNS 95, pp. 235–246, 2020.
https://doi.org/10.1007/978-3-030-34983-7_23

maximum likelihood estimation (MLE) can be mentioned [1, 13]: these methods can provide for the increasing accuracy of measurements results, and their use requires additional analytic work – even for simple equations describing interconnections between the measurands. The lack of non-complicated equations that could be used directly to estimate the accuracy increase is an obstacle for a wide use of information about dependencies between the measured quantities.

This paper presents a simple theoretical construction, which allows to quickly estimate the benefits from using the information about linear relations between two measurable variables to improve the accuracy of measurements results.

The way to find the compromise between math's complexity using for accuracy increase and achieved improvements of measurements results is described. The foundation to construct simplified approach is that in many cases even simplified mathematical processing of measurement data provides a significant increase in accuracy. Based on this idea, the simple approximation was obtained from MLE approach for the case of the linear dependence of the two measured parameters.

Several simple enough expressions and equations are presented for typical measurement characteristics, which help making a decision about using or not using information on measurands interconnections without significant calculations or premodeling.

## 2   The MLE-Based Approach

For the possibility of generalization, we implicitly assume that the error in the measurement results has a normal distribution. Multivariate normal distribution $P(\mathbf{X})$ of measured values $\mathbf{X}$ will be:

$$P(\mathbf{X}) = \frac{1}{\sqrt{(2\pi)^N |\mathbf{V}|}} e^{-\frac{1}{2}(\mathrm{X}-\mathrm{M})^{\mathrm{T}}\mathbf{V}^{-1}(\mathrm{X}-\mathrm{M})}, \tag{1}$$

where $N$ is a number of measured values, $\mathbf{X}$ is a vector of arguments of the multivariate normal distribution and potential values of measurements results, $\mathbf{M}$ is a vector of mathematical expectations of measurement results (the true values of measurands in the case if there are no systematic errors in measurement results), $\mathbf{V}$ is a covariance matrix for $\mathbf{X}$.

Let $\mathbf{X}_i$ be the concrete vector of measurement results obtained during multiple measurements. Let us denote $n$ as a total number of repeated measurements: $i = 1$, 2, ... $n$.

Vectors $\mathbf{X}_i$ and $\mathbf{M}$ and matrix $\mathbf{V}$ have the following form:

$$\mathbf{X}_i = \begin{pmatrix} x_{1i} \\ x_{2i} \\ \cdots \\ x_{Ni} \end{pmatrix}, \quad \mathbf{M} = \begin{pmatrix} \mu_1 \\ \mu_2 \\ \cdots \\ \mu_N \end{pmatrix}, \quad \mathbf{V} = \begin{pmatrix} \sigma_1^2 & \cdots & \mathrm{cov}_{1N} \\ \vdots & \ddots & \vdots \\ \mathrm{cov}_{N1} & \cdots & \sigma_N^2 \end{pmatrix},$$

where $x_{ji}$ is the $i$-th result of the repeated measurement of the $j$-th interconnected value $x_j$; $\mu_j$ is the mathematical expectation of the measurement result of the quantity $x_j$; $\sigma_j^2$ is the variance of the measurement result of this quantity; $\mathrm{cov}_{lm}$ is the covariance of the measured values of the related quantities $x_l$ and $x_m$ (where $l$ and $m$ vary from 1 to $N$).

Thus, according to MLE, we obtain the likelihood function for the values of the vector $\mathbf{X}_i$:

$$L = \sum_{i=1}^{n} \ln(P(\mathbf{X}_i)) = -\frac{Nn}{2}\ln(2\pi) - \frac{n}{2}\ln(|\mathbf{V}|) - \frac{1}{2}\sum_{i=1}^{n}(\mathbf{X}_i - \mathbf{M})^{\mathrm{T}}\mathbf{V}^{-1}(\mathbf{X}_i - \mathbf{M}).$$

(2)

It is known that the estimation of the values $\mu_1, \dots \mu_N$ suggests finding of such their values for which the likelihood function reaches its maximum [11].

$$\mathbf{M}^* = \arg\max_{\mathbf{M}} L.$$

Selecting the part of the expression that determines the position of the maximum of the function, we get that the vector $\mathbf{M}^*$ of the estimates of $\mu_1, \dots \mu_N$ is equal to

$$\mathbf{M}^* = \arg\min_{\mathbf{M}} \sum_{i=1}^{n}(\mathbf{X}_i - \mathbf{M})^{\mathrm{T}}\mathbf{V}^{-1}(\mathbf{X}_i - \mathbf{M}).$$

If the covariance between the measured quantities is zero (a common situation during repeated measurements without systematically changing errors), then

$$\mathbf{M}^* = \arg\min_{\mathbf{M}} \left( \sum_{i=1}^{n}\frac{(x_{1i} - \mu_1)^2}{\sigma_1^2} + \sum_{i=1}^{n}\frac{(x_{2i} - \mu_2)^2}{\sigma_2^2} + \dots + \sum_{i=1}^{n}\frac{(x_{Ni} - \mu_N)^2}{\sigma_N^2} \right).$$

Let's assume that only two parameters are analyzed: $x_1$ and $x_2$. Moreover, there is an equation $x_2 = f(x_1)$ of a known form that describes the interconnection between measured quantities. Then estimate of the first parameter should be

$$\mu_1^* = \arg\min_{\mu_1} \left( \sum_{i=1}^{n}\frac{(x_{1i} - \mu_1)^2}{\sigma_1^2} + \sum_{i=1}^{n}\frac{(x_{2i} - f(\mu_1))^2}{\sigma_2^2} \right).$$

Let us denote the functional to be minimized as $\tilde{L}$. The extremumsincluding the desired minimum of this quadratic form can be found if we set the partial derivative of $\tilde{L}$ with respect to the estimated variable to zero.

$$\frac{\partial \tilde{L}}{\partial \mu_1} = -\frac{2}{\sigma_1^2}\sum_{i=1}^{n}(x_{1i} - \mu_1) - \frac{2f'(\mu_1)}{\sigma_2^2}\sum_{i=1}^{n}(x_{2i} - f(\mu_1)) = 0;$$

$$\frac{1}{\sigma_1^2}(\mu_1 - \bar{x}_1) + \frac{f'(\mu_1)}{\sigma_2^2}\sum_{i=1}^{n}(f(\mu_1) - \bar{x}_2) = 0,$$

where $\bar{x}_1$ and $\bar{x}_2$ are mean values calculated using the repeated measurement results for quantities $x_1$ and $x_2$ correspondingly.

So, the estimation $\mu_1^*$ of the quantity $x_1$ is

$$\mu_1^* = \bar{x}_1 + \frac{\sigma_1^2}{\sigma_2^2}f'(\mu_1)(\bar{x}_2 - f(\mu_1)). \tag{3}$$

The Eq. (3) is common for any interconnection equations and can be used for different types of dependencies between two measured quantities.

A linear relationship is one of the most common types of dependencies. It is typical for most areas from electricity to social sciences. For that reason, it seems rational to consider an example with a linear dependence between two measured quantities in the form

$$x_2 = a\,x_1,$$

where $a$ is a coefficient of proportionality.

In that case, after simple derivations, we can obtain the estimation of the quantity $x_1$ in the form

$$\mu_1^* = \frac{\sigma_2^2}{\sigma_2^2 + \sigma_1^2 a^2}\bar{x}_1 + \frac{\sigma_1^2 a}{\sigma_2^2 + \sigma_1^2 a^2}\bar{x}_2. \tag{4}$$

The variance of the random error of this estimate is

$$\sigma_{\mu_1^*}^2 = \frac{\sigma_{\bar{x}_1}^2 + \sigma_{\bar{x}_2}^2 a^2}{\left(1 + \frac{\sigma_1^2}{\sigma_2^2}a^2\right)^2}. \tag{5}$$

where $\sigma_{\bar{x}_1}^2$ and are $\sigma_{\bar{x}_2}^2$ the variances of the values $\bar{x}_1$ and $\bar{x}_2$ correspondingly.

We can see that this variance for some situations will be smaller than the variance $\sigma_{\bar{x}_1}^2$ of the simple mean value; this indicates the benefits of using the information about interconnected variables. The same derivations can be produced for the second variable $- x_2$.

During the statistical modeling, Eq. (4) was used for multiple estimation of the $x_1$ value and expression (5) was applied for analytical evaluation of the variance.

# 3   Useful Parameters for Analyzing Possible Accuracy Increase

As it was mentioned, a set of easily computable parameters for a certain class of tasks can be useful for solving engineering problems. The main purpose of this section is to describe the characteristics that will help to make decisions on the fly about possible benefits from using information on relationships between measured quantities.

Let us analyze the quantity $x_1$. For the quantity $x_2$ the situation will be similar. Because of the leading role of $x_1$, let us call $x_2$ the auxiliary variable during this consideration.

To compare the statistical variation of the results of the MLE-based estimate with the variation of the estimate obtained by simple averaging of repeated measurements, the dimensionless parameter K was introduced:

$$K = \sigma_{\bar{x}_2} / \sigma_{\mu_1^*}, \tag{6}$$

where $\sigma_{\bar{x}_1}$ is a standard deviation for estimation of $x_1$ values by arithmetic meaning; $\sigma_{\mu_1^*}$ is a standard deviation for estimation by MLE approach (i.e., using equation describing interconnections between measurands).

The parameter K is designed to evaluate the benefits of using the interconnections equation for the increase of accuracy. If K is less or equal to 1, there would be no benefits in using Eqs. (3) or (4).

The next step is to describe measurement factors that have significant influence on the estimation accuracy in dimensionless form. It is obvious that there are two most important parameters: the measurement results accuracy of the auxiliary quantity ($x_2$) and the accuracy of the relationship information.

To estimate the influence of the first parameter, the dimensionless characteristic $\gamma$ was introduced. It is defined as follows: if we have two related quantities $x_1$ and $x_2$ with equation $x_2 = f(x_1)$ describing their interconnection their ratio would be

$$k = \frac{x_2}{x_1} = \frac{f(x_1)}{x_1},$$

The traditional way to represent statistical variation as a relative part of the corresponding variable value is to use the deviation coefficient, which is equal to the ratio between the variable and its standard deviation. So, to measure relative contribution of measurements' random errors in values of interconnected quantities $x_1$ and $x_2$, the coefficient $\gamma$ can be used:

$$\gamma = \frac{\sigma_{\mu_2^*}}{k\,\sigma_{\mu_1^*}},$$

where $\sigma_{\mu_1^*}$ is the standard deviation of the estimated quantity, $\sigma_{\mu_2^*}$ is the standard deviation of the auxiliary quantity.

For a linear interrelation equation the ratio parameter $\gamma$ can be simplified to

$$\gamma = \frac{\sigma_{\mu_2^*}}{a\,\sigma_{\mu_1^*}}.$$

To assess the influence of the possible inaccuracy of the relation model (interconnection equation), a dimensionless characteristic $\eta_{mod}$ is introduced. It connects the error of the model with measurement errors in the following manner. Let $\varepsilon_{mod}$ be the absolute error of the proportionality coefficient $a$ and, as a consequence, the interconnection equation in reality takes the form

$$x_2 = f(x_1) = a\,x_1 = (a_{true} + \varepsilon_{mod})x_1,$$

where $a_{true}$ is the true value of the linear coefficient $a$.

Then $\eta_{mod}$ expresses the ratio between $\varepsilon_{mod}$ and a standard deviation $\sigma_a$ of proportionality parameter $a$ if we decide to calculate it from the measured values $x_1$ and $x_2$ in the situation without interconnection equation error:

$$\eta_{mod} = \varepsilon_{mod}/\sigma_a.$$

Since $a$ is supposed to be calculated as the ratio $x_2/x_1$, to estimate $\sigma_a$, let us consider the equation:

$$a_{true} + \varepsilon_a = \frac{\mu_2 + \varepsilon_{x_2}}{\mu_1 + \varepsilon_{x_1}},$$

where $\varepsilon_{x_1}$ and $\varepsilon_{x_2}$ are random errors of measurement results of $\mu_1$ and $\mu_2$.

If errors $\varepsilon_{x_1}$ and $\varepsilon_{x_2}$ are small enough, then it can be derived that

$$\sigma_a = a_{true}\sqrt{\frac{\sigma_{x_2}^2}{\mu_2^2} + \frac{\sigma_{x_1}^2}{\mu_1^2}}, \tag{7}$$

where $\sigma_{x_1}^2 = n\,\sigma_{\bar{x}_1}^2$ is the variance estimate of each of repeated measurements of the quantity $x_1$, $\sigma_{x_2}^2 = n\,\sigma_{\bar{x}_2}^2$ is the variance estimate of each of repeated measurements of the quantity $x_2$; $n$ is the number of repeated measurements.

The value of $\sigma_a$ represents the characteristic of totalized error of performed measurements of $x_1$ and $x_2$ through the prism of their relationship.

So, $\eta_{mod}$ can be described as

$$\eta_{mod} = \frac{\varepsilon_{mod}}{a_{true}\sqrt{\frac{\sigma_{x_2}^2}{\mu_2^2} + \frac{\sigma_{x_1}^2}{\mu_1^2}}}.$$

The value $\eta_{mod}$ changes from 0 to 1 and determines the measure of the ratio between $\varepsilon_{mod}$ and $\sigma_a$.

There is another parameter that influences the accuracy of the results of estimation by MLE-based approach. It helps to answer the question: what if the a priori information about the width of the uncertainty interval for the auxiliary quantity $(x_2)$ is not completely true?

As we can see from Eq. (4), the standard deviation $\sigma_{x_2}$ of the quantity $x_2$ is used for the MLE-based estimation. To evaluate what would happen if the real $\sigma_{x_2}$ is bigger than expected, the unaccounted error $\varepsilon_{x_2\text{extra}}$ of the auxiliary quantity was introduced. It is expressed as a part of the standard deviation $\sigma_{x_1x_2}$ that corresponds to the aggregated estimate of measurand $x_2$ by using measurement values for $x_2$ along with those for $x_1$. In fact, we know that $x_2 = a\,x_1$. So, if we want to take into account all obtained information on the quantity $x_2$, we can estimate it from its direct measurement results and, indirectly, from the measurement results for $x_1$ if we multiply them by $a$. Let us use the average between direct and indirect measurement results as the final estimate for $x_2$. Then its standard deviation $\sigma_{x_1x_2}$ would be equal to

$$\sigma_{x_1x_2} = \sqrt{\sigma_{x_2}^2 + a^2\sigma_{x_1}^2}/2.$$

The unaccounted error $\varepsilon_{x_2\text{extra}}$ is described as

$$\varepsilon_{x_2\text{extra}} = \eta_{x_2\text{extra}}\sqrt{\sigma_{x_2}^2 + a^2\sigma_{x_1}^2}/2,$$

where $\eta_{x_2\text{extra}}$ changes from 0 to 1 and determines the measure of the ratio between the unaccounted measurement error $\varepsilon_{x_2\text{extra}}$ of the auxiliary parameter $x_2$ and standard deviation $\sigma_{x_1x_2}$ of its aggregated measurement estimate.

All the characteristics described above are obtained for random errors, but in case of a systematic error they will have a similar form. The influence of both types of errors was evaluated during the statistical modeling (standard deviations were calculated relatively to the true values of the measured values, so the systematic errors were taken into account).

## 4   Simulation Results

To evaluate the influence of different types of errors on the accuracy of the proposed approach, a multiple modeling by Monte Carlo method with $N = 1000$ iterations was performed. The number of repeated measurements was $n = 100$. All samples for the quantity $x_1$ were generated with the same standard deviation $\sigma_{x1} = 0.001 \cdot x_1$ (hereinafter all random errors have normal distribution), while the standard deviation (SD) of the auxiliary quantity $x_2$ was varied over a range for each simulation with the specified measurement parameters $\varepsilon_{\text{mod}}$ and $\varepsilon_{x_2\text{extra}}$.

Figure 1 shows the dependency between the degree of accuracy increase and the factor that represents normalized statistical precision of performed measurements of the quantities $x_1$ and $x_2$.

As it can be seen from the simulation results, in the absence of additional unaccounted errors, the accuracy of proposed approach results is rapidly increasing when

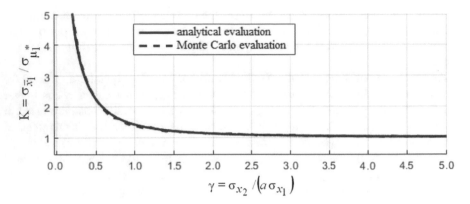

**Fig. 1.** Dependency between the accuracy of the estimate obtained with the proposed approach and SD of auxiliary quantity value

the uncertainty of the auxiliary quantity decreases. This observation is valid for any type of interconnection equations, but forms of the presented dependency may differ.

It is also obvious that there is no sense in using information about the relationship if the SD of the measurement results of the auxiliary quantity $x_2$ is two (or more) times bigger than the SD of the estimated quantity $x_1$ multiplied by the proportionality coefficient $a$. In that case, an estimation result would be equal to simple averaging of the measured values of $x_1$.

### 4.1    The Case of the Imprecise Model

In Fig. 2a and b the families of curves for various systematic (a) and random (b) errors of the linear relationship parameter $a$ are shown.

These results confirm that the available quantity and accuracy of a priori information about the relationship between measured values directly affects the uncertainty of the results of their refinement.

The systematic and random errors have a similar effect on the evaluation results. For both types of errors, there is a model error threshold. When the values of $\varepsilon_{mod}$ are more than 10% of the totalized measurement error $\sigma_a$, the proposed approach provides lesser accuracy than a simple averaging of the estimated quantity sample.

### 4.2    The Case of an Unaccounted Error of the Auxiliary Quantity

As it can be seen from Eq. (4), the standard deviation of the auxiliary quantity is one of the estimation equation's parameters. Given the simplicity of the interconnection equation, we can expect that its inaccuracy has similar influence on the estimation results as the model error.

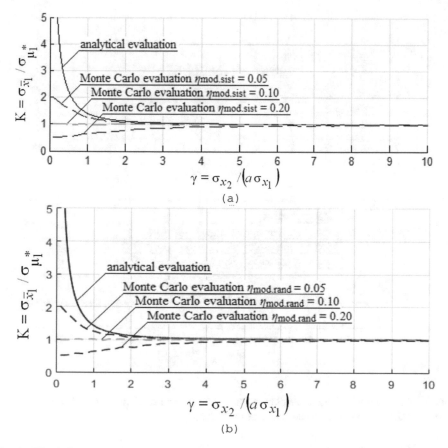

**Fig. 2.** The influence of the systematic (a) and random (b) errors of the model on the standard deviation of estimate $\mu_1^*$

Statistical modeling for the three values of the unaccounted error $\varepsilon_{x_2\text{extra}}$ was produced. Figures 3a and b show how the dependency between relative SD of the auxiliary quantity and the accuracy of the proposed approach changes when the value of the auxiliary parameter's standard deviation is more than expected.

Comparing the experimental curves in Figs. 2 and 3, we can conclude that the additional unaccounted error of the auxiliary parameter has half lesser effect on the estimation accuracy than the error of the interrelation model. However, the type of accuracy reduction is the same: the efficiency of the proposed approach decreases in the low auxiliary quantity values' area. The value about 20% of the total measurement error is the threshold at which the increase of accuracy becomes impossible.

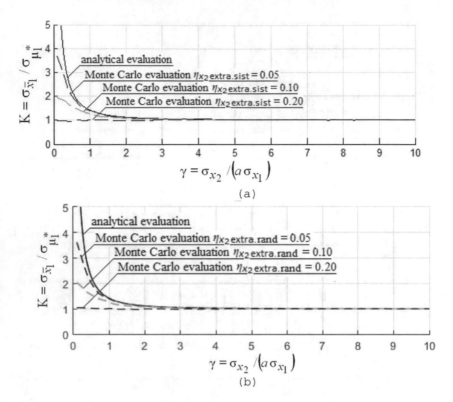

**Fig. 3.** The influence of the unaccounted systematic (a) and random (b) errors of the auxiliary quantity value on the standard deviation of estimate $\mu_1^*$

## 5  Conclusions

Simple equations are presented to estimate the accuracy increase during the measuring of two quantities with linear dependence. Using a statistical experiment as the proposed approach for analyzing the effectiveness of using the relationship between the measured quantities was examined. The common-case Eq. (4) for the estimation by the MLE approach for any interconnection equation with two quantities was proposed. To calculate a variance of this estimate for linear dependency, Eq. (5) was derived.

It was observed that at least for linear relations, the comparison between the estimates based on using the equation of the relationship with the estimate by averaging the measured sample proved its usefulness. The convenience is in the fact that with the growth of standard deviation of the auxiliary parameter, the accuracy of the relationship-using approach decreases down to the quality of estimation by averaging, while the coefficient K from Eq. (6) tends to 1.

To help evaluate the possible increase in accuracy without pre-modeling, statistical experiments were performed with different types of errors. The curves obtained allow us to draw the following conclusions. Firstly, for the case of two linearly dependent quantities, an increase in accuracy can be achieved if the relationship equation's error does not exceed $\pm 10\%$ of the total measurement error (see Eq. (8)). Secondly, the unaccounted error of the auxiliary quantity should be smaller than $\pm 20\%$ of the total measurement error due to the same reason. And thirdly, for a significant increase in accuracy (at least 10%), the uncertainty of the auxiliary quantity should be no more than doubled uncertainty of the quantity to be estimated.

# References

1. Reznik, L.K., Solopchenko, G.N.: Use of priori information on functional relations between measured quantities for improving accuracy of measurement. Measurement **3**, 98–106 (1985)
2. Franceschini, F., Galetto, M.: A taxonomy of model-based redundancy methods for CMM online performance verification. Int. J. Technol. Manage. **37**, 104–124 (2006)
3. Reznik, L., Kluever, K.A.: Improving measurement accuracy in sensor networks by an object model generation and application. In: Proceedings of the 6th Annual IEEE Conference on Sensors, pp. 371–374 (2007)
4. Sturza, M.A.: Navigation system integrity monitoring using redundant measurements. Navigation **35**, 483–501 (1988)
5. Marquet, F., Company, O., Krut, S., Pierrot, F.: Enhancing parallel robots accuracy with redundant sensors. In: Proceedings of the IEEE International Conference on Robotics and Automation, vol. 4, pp. 4114–4119 (2002)
6. Frank, P.M.: Fault diagnosis in dynamic systems using analytical and knowledge-based redundancy: a survey and some new results. Automatica **26**, 459–474 (1990)
7. Guerrier, S.: Improving accuracy with multiple sensors: study of redundant MEMS-IMU/GPS configurations. In: Proceedings of the 22nd International Technical Meeting of the Satellite Division of the Institute of Navigation, pp. 3114–3121 (2009)
8. Zachár, G., Simon, G.: Radio interferometric tracking using redundant phase measurements. In: Proceedings of 2015 International IEEE Conference on Instrumentation and Measurement Technology, pp. 2003–2008 (2015)
9. Borovskikh, L.P.: A priori information usage for improving measurement accuracy for the parameters of the objects with multi-element equivalent circuit. Sensors and systems **12**, 22–25 (2006). [Ispol'zovanie apriornoy informacii dlya povysheniya tochnosti izmereniya parametrov ob'ectov s mnogoelementnoy skhemoy zamescheniya] (in Russian)
10. Jafari, M.: Optimal redundant sensor configuration for accuracy increasing in space inertial navigation system. Aerosp. Sci. Technol. **47**, 467–472 (2015)
11. Wahrburg, A., Robertsson, A., Matthias, B., Dai, F., Ding, H.: Improving contact force estimation accuracy by optimal redundancy resolution. In: IEEE/RSJ International Conference on Intelligent Robots and Systems, pp. 3735–3741 (2016)
12. Kretsovalis, A., Mah, R.S.: Effect of redundancy on estimation accuracy in process data reconciliation. Chem. Eng. Sci. **42**, 2115–2121 (1987)
13. de Oliveira, E.C., Frota, M.N., de Oliveira Barreto, G.: Use of data reconciliation: a strategy for improving the accuracy in gas flow measurements. J. Nat. Gas Sci. Eng. **22**, 313–320 (2015)
14. Guo, S., Liu, P., Li, Z.: Inequality constrained nonlinear data reconciliation of a steam turbine power plant for enhanced parameter estimation. Energy **103**, 215–230 (2016)
15. Guo, S., Liu, P., Li, Z.: Data reconciliation for the overall thermal system of a steam turbine power plant. Appl. Energy **165**, 1037–1051 (2016)

16. Alhaj-Dibo, M., Maquin, D., Ragot, J.: Data reconciliation: a robust approach using a contaminated distribution. Control Eng. Pract. **16**, 159–170 (2008)
17. Dubois, D., Fargier, H., Ababou, M., Guyonnet, D.: A fuzzy constraint-based approach to data reconciliation in material flow analysis. Int. J. Gen Syst. **43**, 787–809 (2014)
18. Menendez, A., Biscarri, F., Gomez, A.: Balance equations estimation with bad measurements detection in a water supply net. Flow Meas. Instrum. **9**, 193–198 (1998)

# Situation Awareness in Modeling Industrial-Natural Complexes

Alexander Ya. Fridman[1]([⊠]) and Boris A. Kulik[2]

[1] Institute for Informatics and Mathematical Modelling,
Kola Science Centre of RAS, Apatity, Russia
fridman@iimm.ru
[2] Institute of Problems in Mechanical Engineering of RAS,
St. Petersburg, Russia
ba-kulik@yandex.ru

**Abstract.** The concept of situation awareness (SA) is adapted and concretized for the previously developed situational conceptual model of an industrial natural complex. The features of the presented approach consist in quantitative assessing of the three main levels of achieving SA and taking into account the possibility of transferring the modelling object from the normal operation mode to an abnormal or emergency situation. Analysis of SA is performed in a discretized state space of this object with an expertly synthesized metric. The research allows to objectify measuring SA level for decision makers involved in managing complex's components.

**Keywords:** Situation awareness · Conceptual model · Industrial-natural complex · Situation awareness level · Measuring · Decision making

## 1 Introduction

At present, the concept of situation (situational) awareness (for example, [1–3]) describes the most general principles of preparing and processing information for implementing a situational approach in dynamic subject areas. This concept has not become widespread in Russia yet; there are only few ideas for applying this approach to solve specific problems [4–6].

Situation awareness includes consciousness of what is happening in the environment in order to understand how information, events and personal actions will affect goals and objectives in the current moment and in near future. Insufficient or incorrect SA is considered to be one of the main factors associated with accidents caused by the "human factor" [1]. Thus, SA is especially important in professional activities, where the flow of information can be quite high, and bad decisions can lead to serious consequences. This is especially evident in high-dynamic subject areas (for example, piloting an aircraft, military actions, handling seriously ill or wounded patients, etc.). Nevertheless, to our minds, it also needs to be taken into account in the considered below problems of modelling industrial natural complexes (INCs) [7], where the decision making time is long enough, while other aspects of SA are quite significant.

© Springer Nature Switzerland AG 2020
D. G. Arseniev et al. (Eds.): CPS&C 2019, LNNS 95, pp. 247–256, 2020.
https://doi.org/10.1007/978-3-030-34983-7_24

Some researchers criticize this approach for being too generic (see, in particular, [3]). It seems that the general principles of SA really become realistic only in relation to a specific model of decision-making in a particular subject area. Therefore, they are further interpreted for the situational conceptual model (SCM) [7–9], developed for studying spatial dynamic systems, INCs in particular. In our opinion, the main advantage of extending the SA paradigm to the area of situational modelling is in objectifying SA measures by their deriving from the structure and the current state of a situational model. For the SCM, such measures can be obtained by analyzing structural and quantitative deviations of model elements' states from their nominal values (positions in the state space).

For better perception, main statements of SA are rendered below in *italics*, the usual font describes their features in the SCM. Due to the size limits of this paper, we avoid any proofs and equations; those can be found in our referenced publications along with the SCM terminology.

## 2   Levels of SA Achievement

*To achieve SA, it is necessary to ensure correct processing of information at three levels: perception of environmental elements, comprehension of the current situation, and projection of future status* [2]. In the SCM, perception is modelled by setting an initial situation, comprehension is provided by formation of the corresponding full situation and determining the organizational level for solving the problem, and projection is implemented by choosing a desired class of situations and simulating the behaviour of an INS in the selected class. To combat the "information explosion" at the levels of perception and comprehension, effective means are needed to determine essential factors from the set of available measurements and observations. Within the SCM, this task is naturally solved in the course of generalizing situations by searching for alternatives (sufficient situations), which can maximally satisfy the decision-maker's task of transferring the INC from its current class of situations into a given new class.

In the works of Endsley, the founder and leader of the SA paradigm, it is repeatedly noted (for example, [10]) that *a high level of SA is most often achieved by experts in solving a particular task, and not by beginners* (although she always notes that miracles do happen). In the SCM, the acuteness of this problem is removed by the fact that although decision-makers (DMs) are not always experts in all aspects of functioning of their subordinate part of an INC, the model is, ideally, created by experts; therefore, a sufficient level of SA looks quite achievable.

*Correct setting of priorities during selection and subsequent analysis of available information is one of the main problems in achieving an acceptable level of SA* [10]. To this end, SCM comprises a number of situation processing methods [7]: tools for detailed control of the correctness of SCM at all stages of its life cycle; classification and generalization algorithms for situations, including those developed in the framework of the cognitive approach [11]; means for evaluating effectiveness of available alternatives for implementation of an INC, etc. The cognitive method of classifying situations based on the proposed semantic hierarchical metric of proximity of situations

[11] makes it possible to take into account expert assessments of importance for different levels of the hierarchical model of an INC.

# 3   SA and Decision-Making

*The input to the SA attainment procedures is the state of the environment; these procedures are followed by decision-making and performing certain actions* [12].

Since in the SCM the model of hazardous situations and emergencies is an extension of the model of normal functioning [13], each process/object must have at least one executor of the normal functioning model and at least one executor to model its functioning in dangerous situation(s). The latter must be linked and executed when some specified variables come out of their safe ranges (SRs) [13, 14]. Technologically, this is done by analyzing certain expertly formed conditions and the degree of danger of getting out of an SR with connection of the corresponding executor of the model element that is affected by SR violations. The degree of danger of the current situation must be taken into account when assessing SA. This approach also allows to search for critical processes and objects, problems in which greatly reduce SA.

In contrast to "classical" areas of SA application, in SCM it is possible not to only detect the moment of transition of some object from the mode of normal operation to an abnormal mode, but also to calculate the degree of SA loss during such transitions, if the simulation is done in a unified discretized space [15] that allows for the presence of both numeric and string variables in the state vector of an object. Thus, the "usual" conceptual space (for example, [16–19]) expands and becomes applicable also for hazardous operating modes of an INC.

*Sets of critical cues (parameters) allow for using a mental scheme (model) to indicate, instantly classify and understand prototypical situations* [10]. In the SCM, this is a significant and fast raising of the index of generalized expenses above unity for an element of the model, which indicates inefficiency of this element's functioning (assuming correct values of its input resources) [20], as well as getting values of certain essential resources [13] beyond their safe ranges: that displays a possibility of occurring of some events initiating emergencies and accidents [14].

*Situation awareness is an internal mental model of the state of the environment comprehended by an operator (in our case, a DM). It is built by applying system knowledge, knowledge about interfaces of some software means and the world around it. SA, decision-making and effective performance are different steps that interact within a continuous cycle, which can be broken by other factors* [10, 12].

Applying SA for the SCM results in choosing an initial situation (this is system knowledge of the most important parameters characterizing the state and behaviour of the INC under study) and, after analyzing the corresponding complete situation, in choosing a new (or conserving the former) class of situations for the further functioning of the INC or any part of it, as well as in selecting among proposed alternatives (sufficient situations) to implement the chosen class of situations.

## Dynamic Aspects of SA

*When analyzing the temporal characteristics of SA, M. Endsley points out importance of considering the speed of change in the surrounding world, which should correspond to the speed of decision-making by the operator* [10].

For this purpose, the SCM allows to use values of gradients (increments) [20] of the quality criteria for the object at which a DM is located, and the means of analyzing sensitivity of the supposed decisions to changes in the parameters of these criteria. The values of the gradients of the criteria naturally show the possible time interval for predicting (projecting) the behaviour of the modelling object: the larger these gradients are, the shorter is the interval of a reliable forecast due to the inevitable uncertainties in assessing characteristics of the INC and the environment.

## SA Measurement

In original papers on the SA measurement, the general idea is based on test polls of operators working on simulators either in real time or with resuming simulation to fill in polling lists. The first technique is implemented by SPAM (Situation Present Assessment Method) [21]; its main drawback is seen in the restricted number of test questions. To apply the second approach, the software system SAGAT (SA Global Assessment Technique) [22, 23] was created.

According to the SA authors [10, 22–24], SA measures and metrics should ensure that:

- Measures are focused on the construct that is needed, and not a reflection of other processes
- SA depends on states, not on processes
- For SA, relative values are more important than absolute values
- The ideal SA is the exact knowledge of all relevant aspects of the situation.

Most of these requirements are automatically satisfied in the SCM due to its independence of human factors; the rest ones can be achieved to the extent defined by adequacy and detalization of the model. Sure, this model (just as any other model) cannot reveal situations, which were not provided in it for, but it gives opportunities to investigate SA for different modes of INC operation.

Besides the above-introduced similarities, we have found other analogies between various aspects of SA in the "classical" interpretation and characteristics of the SCM, but it has already been stated enough to conclude that application of the SA paradigm for INCs situational modeling is promising. However, to facilitate understanding of quantitative measures of SA in the SCM, first of all, it is necessary to briefly outline principles for handling situations in this modeling system.

## Basics of Handling Situations in the SCM

The SCM describes three types of elements (entities) of the real world, namely, objects, processes and data (or resources).

*Objects* (constituent parts of the INC under study) form a hierarchy that reflects the organizational and spatial structure of the researched object, each of them can be associated with a set of *processes* that describe procedures for converting a subset of resources input to the process under consideration into another subset called output resources.

Resources characterize the state of the INC. They are used in execution of processes and model results of their implementation. Execution of any process changes the data and yields a transition of the system from one state to another [4]. In general, each SCM element is assigned an *executor* who ensures its implementation during simulation. The type of an executor determines, for example, the programming language in which the corresponding process is implemented, and the type of this executor in an algorithmic language.

Characteristics and main types of situations specified in the SCM are summarized in Table 1.

**Table 1.** Properties of situations in the SCM

| Name | Format | Purpose in the SCM | Basic specifications |
|------|--------|--------------------|----------------------|
| Fact | Sublist of values for a resource | Elements of a situation | Number of values, sign of their combining |
| Situations: – initial | List of facts | Task setting | Number of facts, list of resources |
| Situations: – initial | List of facts | Task setting | Number of facts, list of resources |
| – complete | List of facts and a corresponding SCM | Preparation for simulation, generation of DBs | Number of facts, list and values of resources |
| – sufficient | fragment | Determining the level of a task | Root object, class of the situation |
| – controlling | | Structural change of the SCM | Dominant alternative |
| Scenario | An ordered list of sufficient situations | Preparation for simulation | Time interval |
| Fragment | A connected subgraph of the object tree | Implementation of a complete situation | Root object, list of leaves |

The performance quality criterion (PQC) for each object is calculated as:

$$\Phi ::= \left( \frac{1}{m} \sum_{i=1}^{m} \left( \frac{a_i - a_{i0}}{\Delta a_i} \right)^2 \right)^{1/2} ::= \left( \frac{1}{m} \sum_{i=1}^{m} \delta a_i^2 \right)^{1/2}, \tag{1}$$

where: $a_i$ are signals from the list of output parameters of this object, their total number is $m$;

$a_{i0}$ and $\Delta a_i > 0$ are adjusting parameters that reflect requirements of the parent object to the nominal value $a_i$ and its allowable deviation $\Delta a_i$ from this value, respectively;

$$\delta a_i ::= \frac{a_i - a_{i0}}{\Delta a_i}$$

is the relative deviation of the actual value of the signal $a_i$ from its nominal value $a_{i0}$.

If we assume that $a_i$ are scalar PQCs of a model element whose nominal values are $a_{i0}$, then (1) is a generalized criterion (see, for example, [14]) with importance coefficients inversely proportional to the allowable deviations of scalar criteria, which corresponds to common sense: the more important this criterion is for the decision-maker, the less acceptable are its deviations from the nominal value.

The value of the criterion will be equal to unity if values of all its arguments are on the verge of tolerances:

$$\Phi = 1, \text{ if } |a_i - a_{i0}| = \Delta a_i, \ i = \overline{1, m}, \tag{2}$$

This value does not exceed unity if all arguments are within tolerances.

The specific value of a change in the criterion (1) when one of its arguments changes:

$$\delta \Phi ::= \frac{\partial \Phi / \partial a_i}{\Delta a_i} = m \ \Phi \delta a_i, \tag{3}$$

characterizes the relative sensitivity of the quality criterion (1) to a change in this argument. Assuming equal importance of all resources to achieve the goal of functioning for an SCM element, the specific value of the generalized expenses for each argument of the criterion (1) is estimated by

$$\eta_i ::= \frac{1}{m} \delta \Phi. \tag{4}$$

Equations (1)–(3) yield that when the argument $a_i$ is within acceptable limits, the value of $\eta_i$ does not exceed unity. This value is proposed to be used as an indicator of own expenses of an SCM element for generating a particular resource in comparative analysis of various structures for implementation of a particular complete situation. If this element consumes any (material) resources from other elements of the model, it is necessary to consider expenses for obtaining input resources and add them to its own expenses. Then (4) takes the form:

$$\eta_i ::= \Phi \delta a_i + \frac{1}{m} \sum_{j=1}^{n} \eta_j, \tag{5}$$

where: $n$ is the number (length of the list) of input resources of a given model element;

$\eta_j$ are the specific expenses for obtaining input resources calculated similarly to (5) for a decision-making object (DMO) in which the decision-maker is located. Either the root object of the fragment constructed for the studied complete situation, or any superobject of this object, up to the global element of the SCM, can serve as a DMO.

The principle of situations classification in the SCM is given by the following definitions.

*Definition 1.* Two sufficient situations from the same fragment of the SCM with the same DMO belong to the same class of situations, if both of them have the minimum value of the specific expenses (5) for the same output resource $a_i$ of this DMO (let us call this criterion dominant compared to other criteria). For two sufficient situations within one class of situations, the more preferable one is that with the lower value of expenses (5).

*Definition 2.* The optimal sufficient situation from a given class is the one with the minimal value of the specific expenses (5).

Since sufficient situations contain no redundancy by definition, any variant of the above method for calculating absolute expenses provides an unambiguous calculating of the own and absolute expenses for obtaining all resources of an SCM fragment and classifying sufficient situations based on the dominance of one of scalar criteria in the expenses at the DMO output. Moreover, conditions (2) significantly simplify searching for the reason why the model's performance parameters have overshot their tolerances: it is enough to determine the most underlying object, at the output of which the indicator values (4) (or (5) if it is not a leaf object) significantly exceed unity; its malfunctioning is the source of the unacceptable increase in expenses.

Generalized expenses were also proposed to be used when solving the tasks of coordinating controls in the SCM (for example, [15]).

## SA Measures in the SCM

In an INC, parts of which are controlled by several DMs, it becomes possible to introduce and calculate the achieved degree of SA for each of the DMs.

Since Endsley and her colleagues have repeatedly shown (for example, [10]) that not absolute, but relative values of SA are important, we assume that values of the total degree of SA (SAD) and each of its three components (perception of environmental elements, comprehension of the situation and forecast of future status) are characterized by a non-negative number with the maximal value of 1. For a quantitative assessment of the degree of SA, currently achieved by each decision-maker that has a given area of responsibility (DA), we propose the following equation:

$$SAD_i = PD_i * CSD_i * FD_i, \tag{6}$$

where: $PD_i$ is the degree of perception of the environment that depends on the ratio between the set power of input resources for the area of responsibility of a DM, which can change without their participation, the number of output resources controlled by this DM, and their share in the total number of resources essential to functioning of the $DA_i$ [25]. Thus, we propose that

$$PD_i = \frac{n}{n+m} \tag{7}$$

where $n$ is the number of the input resources for the $DA_i$, m is the number of resources generated by the $DA_i$;

$CSD_i$ is the degree of comprehension of the situation; it is a synthesis of the elements of the $PD_i$ (measure of proximity of the current state of the system to the ideal state) [10]. In the SCM, it depends on the vector of residuals of the own PQC for the

$DA_i$ $\delta a_i^{co6} = \sqrt{\sum_{j=1}^{n} \delta a_j^2}$ and the vector of residuals for the input resources of the

$DA_i$ $\delta a_i^{ex} = \sqrt{\sum_{k=1}^{m} \delta a_k^2}$ ;

FD$_i$ is determined by the change rate for the current situation in the $DA_i$, i.e., by increments of the quality criterion (1) for the DMO of this zone. The greater is the magnitude of these increments in time, the shorter becomes the interval of reliable forecast due to inevitable uncertainties in assessment of characteristics of the INC and the environment.

When developing formulas for estimating components of SA, it is necessary to take into account their desirable asymptotic properties.

For CSD$_i$ they are as follows:

when $\delta a_i^{ex} \to 0$ and $\delta a_i^{co6} \to 0$, CSD$_i \to 1$;

when $\delta a_i^{co6} \gg 1$, CSD$_i \to 0$;

when $\delta a_i^{ex} \gg 1$, CSD$_i \to 0$;

Hence, a permissible formula for evaluating $CSD$ is:

$$CSD_i = \frac{2 - \delta a_i^{co6} - \delta a_i^{ex}}{2 - (\delta a_i^{co6})^2 - (\delta a_i^{ex})^2} \tag{8}$$

Asymptotic properties of FD$_i$ are:

with $\Phi_i \to 0$ and $\Delta\Phi_i \to 0$ FD$_i \to 1$, $T_i \to \infty$, where $T_i$ is the interval of a reliable forecast;

at $|\Delta\Phi_i| \to \infty$ FD$_i \to 0$, $T_i \to 0$;

at $\Phi_i \gg 1$ FD$_i \to 0$, $T_i \to 0$;

So, a valid formula for $FD$ is:

$$FD_i = 1 - e^{-\tilde{T}_i}, \tag{9}$$

where $\tilde{T}_i$ is the smoothed value of $T_i$,

$$T_i = \frac{\alpha}{\Phi_i |\Delta\Phi_i|}, \tag{10}$$

and $\alpha > 0$ sets the time scale (dynamics) of operation of the $DA_i$.

If the used in (4) and below presupposition regarding equal importance of all involved resources for the performance of SCM elements is not acceptable, this importance can be made different by weighing shares of resources with some expert-estimated coefficients similar to moving totals techniques. Additional options to make formulas more flexible and adjustable appear within the discretized state space [15], where an expert can consider different degree of danger for each value of every resource by assigning weights to distances between those values.

# 4   Conclusion

The proposed correlations make it possible to objectively evaluate importance of decisions of every DM and take this importance into account when searching for a balance of interests of all DMs that affect functioning of an INC, in order to coordinate their actions and eliminate conflicts.

Besides investigation of INCs, the introduced approach looks prospective for estimating SA in SCM-based supply chain management (for instance, [26]) and development networks of intelligent situational centres [25].

**Acknowledgments.** This work was partially supported by grants from the Russian Foundation for Basic Researches (projects No. 16-29-12901, 18-29-03022, 18-07-00132, 18-01-00076, and 19-08-0079).

# References

1. Lundberg, J.: Situation awareness systems, states and processes: a holistic framework. Theor. Issues Ergon. Sci. **16**, 447–473 (2015)
2. Endsley, M.R.: Toward a theory of situation awareness in dynamic systems. Hum. Factors **37**(1), 32–64 (1995)
3. Banbury, S., Tremblay, S.: A Cognitive Approach to Situation Awareness: Theory and Application, pp. 317–341. Ashgate Publishing, Aldershot (2004)
4. Popovich, V.V., Prokaev, A.N., Sorokin, P.P., Smirnova, O.V.: On recognizing the situation based on the technology of artificial intelligence. In: SPIIRAS Proceedings, Issue 7, Science, St. Petersburg, Russia, pp. 93–104 (2008). (in Russian)
5. Afanasyev, A.P., Baturin, Yu.M., Yeremchenko, E.N., Kirillov, I.A., Klimenko, S.V.: Information-analytical system for decision-making on the basis of a network of distributed situational centres. Inf. Technol. Comput. Syst. **2**, 3–14 (2010). (in Russian)
6. Yampolsky, S.M., Kostenko, A.N.: Situational approach to management of organizational and technical systems in operations planning. Sci. Intensive Technol. Space Res. Earth **8**(2), 62–69 (2016). (in Russian)
7. Fridman, A.Ya.: Situational Control of the Structure of Industrial Natural Systems. Methods and Models. LAP, Saarbrucken (2015). (in Russian)
8. Fridman, A.Ya., Kurbanov, V.G.: Formal conceptual model of industrial natural complexes as a means of controlling computational experiments. SPIIRAS Proc. **6**(37), 424–453 (2014). (in Russian)
9. Artemieva, I.L., Fridman, A.Ya.: Ontologies in the automation problem for situational modelling. In: Proceedings of the 2018 3rd Russian-Pacific Conference on Computer Technology and Applications (RPC), IEEE, pp. 48–53 (2018)
10. Endsley, M.R.: Theoretical underpinnings of situation awareness: a critical review. In: Endsley, M.R., Garland, D.J. (eds.) Situation Awareness Analysis and Measurement, LEA, Mahwah, NJ, pp. 3–32 (2000)
11. Fridman, A.: Cognitive categorization in hierarchical systems under situational control. In: Advances in Intelligent Systems Research, Atlantis Press, vol. 158, pp. 43–50 (2018)
12. Endsley, M.R.: Final reflections: situation awareness models and measures. J. Cognitive Eng. Decis. Making **9**(1), 101–111 (2015)

13. Fridman, A.Ya., Kurbanov, V.G.: Situational modelling of reliability and safety in industrial-natural systems. Inf. Manage. Syst. **4**(71), 1–10 (2014). (in Russian)
14. Yakovlev, S.Yu., Isakevich, N.V., Ryzhenko, A.A., Fridman, A.Ya.: Risk assessment and control: implementation of information technologies for safety of enterprises in the murmansk region. In: Barents Newsletter on Occupational Health and Safety, Helsinki, vol. 11, no. 3, pp. 84–86 (2008)
15. Fridman, A.Ya.: Expert space for situational modelling of industrial natural systems, herald of the Moscow University Named after S.Y. Witte, **1**(4), 233–245 (2014). (in Russian)
16. Sowa, J.F.: Conceptual Structures – Information Processing in Mind and Machines. Addison-Wesley Publishing Company (1984)
17. Gärdenfors, P.: Conceptual Spaces: The Geometry of Thought. A Bradford Book. MIT Press, Cambridge (2000)
18. Zenker, F., Gärdenfors, P.: Applications of Conceptual Spaces. The Case for Geometric Knowledge Representation, Synthese Library, Springer, vol. 359 (2015)
19. Decock, L., Douven, I.: What is graded membership? Noûs **48**, 653–682 (2014)
20. Fridman, A., Fridman, O.: Gradient coordination technique for controlling hierarchical and network systems. Syst. Res. Forum **4**(2), 121–136 (2010)
21. Loft, S., Morrell, D.B.: Using the situation present assessment method to measure situation awareness in simulated submarine track management. Int. J. Hum. Factors Ergon. **2**(1), 33–48 (2013)
22. Endsley, M.R.: Situation awareness measurement in test and evaluation. In:, O'Brien, T.G., Charlton, S.G. (eds.) Handbook of Human Factors Testing & Evaluation, Lawrence Erlbaum, Mahwah, NJ, pp. 159–180 (1996)
23. Endsley, M.R.: Direct measurement of situation awareness in simulations of dynamic systems: validity and use of SAGAT. In: Garland, D.J., Endsley, M.R. (eds.) Experimental Analysis and Measurement of Situation Awareness, Embry-Riddle University, Daytona Beach, FL, pp. 107–113 (2000)
24. Endsley, M.R.: Situation awareness misconceptions and misunderstandings. J. Cognitive Eng. Decis. Making **9**(1), 4–32 (2015)
25. Oleynik, A., Fridman, A., Masloboev, A.: Informational and analytical support of a network of intelligent situational centers in the russian arctic, IT&MathAZ 2018 information technologies and mathematical modeling for efficient development of arctic zone. In: Proceedings of the International Research Workshop on Information Technologies and Mathematical Modeling for Efficient Development of Arctic Zone, Yekaterinburg, Russia, 19–21 April 2018, pp. 57–64 (2018)
26. Sokolov, B., Ivanov, D., Fridman A.: Situational modelling for structural dynamics control of industry – business processes and supply chains. In: Intelligent Systems: From Theory to Practice. Studies in Computational Intelligence, pp. 279–308. Springer, Heidelberg (2010)

# The Development of Soft Defined Distributed Infocommunication Systems Architecture Based on the Active Data Technology

Sergey V. Kuleshov, Alexandra A. Zaytseva[✉],
and Andrey L. Ronzhin

St. Petersburg Institute for Informatics and Automation of the Russian Academy
of Sciences, St. Petersburg, Russia
{kuleshov, cher, ronzhin}@iias.spb.su

**Abstract.** The active data technology allows developing the architecture of soft defined systems based on the new principles. The most interesting implementation is to build self-organized networks that consist of UAV and robotic complexes. This paper is dedicated to the new architecture of mobile communication networks proposed by the authors. The analytical review of existing architectures of wireless self-organizing networks is given; research of available solutions' weaknesses allowed to put forward ideas for the development of a new concept. The proposed concept of node reconfiguration was developed; it helps to provide the required network structure with appropriate characteristics. The authors introduced the technology that was developed to organize a transmission channel for networks with mobile nodes based on the active data concept. The suggested architecture allows to arrange data transmission channels in the areas where it is difficult to deploy a network of ground communications nodes.

**Keywords:** Network · Architecture · Mobile distributed system · Mobile communications · Wireless self-organizing networks · Optimization · Active data concept

## 1 Introduction

In fact, the task of communication and the associated problem of data transmission is a key to eliminating situational ignorance in any mobile distributed system. This is especially true in the environment, when the user is in a remote place, not covered by the modern infrastructure of communication networks. In such cases, it becomes relevant to use a wireless data network, e.g., to provide a data transfer channel with mobile users; quickly deploy a network in the area not equipped with other telecommunication channels; provide data transfer in the area that is not suitable for the fixed infrastructure (water surface, highlands), etc.

Thus, the use of autonomous mobile communication centers (elements of a robotic technical complex) as repeaters or intermediate nodes in a data transmission network sets a problem of organizing their interaction in the context of continuously changing environmental factors. It is known that works in this sphere have been undertaken by

© Springer Nature Switzerland AG 2020
D. G. Arseniev et al. (Eds.): CPS&C 2019, LNNS 95, pp. 257–265, 2020.
https://doi.org/10.1007/978-3-030-34983-7_25

large companies, for example, by Boeing (see reference [1]). Over the past decade, unmanned aerial vehicle (UAV) technology has significantly advanced in such areas as autonomous onboard processor's computing power, flight control, communications capabilities. Initially, due to the technical complexity of the implementation, UAVs were mainly used in the military sphere. However, with the progress of technology, the scope of applications has gradually expanded to other areas of activity: from agriculture, observation of large industrial facilities and rescue actions in emergency situations to the delivery of parcels. This makes it possible to use UAVs as small mobile platforms [2].

## 2  Existing Wireless Self-organizing Network Architectures

Let's consider the existing architecture and technology of building wireless self-organizing networks and their features.

By the definition given in the reference [3], mesh networks are networks with a mesh topology, consisting of wireless fixed (georeferenced) routers that create a data transmission channel and a service area (coverage area) of subscribers with access to one of the routers. The "STAR" topology with a random connection of reference nodes is used.

Ad hoc networks that implement decentralized control of random stationary subscribers in the absence of base stations or reference nodes are employed. A fixed network with random connection of nodes is used as the topology.

**MANET**
MANET (Mobile Ad Hoc NETworks) are the networks implementing fully decentralized management of random mobile subscribers in the absence of base stations or reference nodes.

Compared to traditional MANET networks, FANET networks (networks built on the basis of flying nodes) have a number of unique functions, such as high mobility and frequent topology changes that create problems for users in connecting to the network [4]. According to the reference [5], UAVs have a speed of 30–460 KPH. This raises the problem that the traditional routing protocols developed for MANET (for example, AODV, DSR, OLSR) cannot adapt well enough if they are applied directly to the FANET. As a result, it is extremely important to use a specialized reliable routing protocol specific to FANET [6].

**FANET**
FANET (Flying Ad Hoc Networks) self-organizing networks are ad hoc networks. The main difference of such networks from the wired ones [7] is that they are decentralized and have no routers, i.e., packets of information are sent on the fly, and Ad Hoc networks are routed dynamically. In reference [8] it is indicated that there are two main types of communication between UAVs in FANET networks.

FANET networks are networks for organizing the interaction of aircrafts =. Unlike VANET and MANET, they are distinguished by high mobility and rapid changes in the network topology. In such networks, there is an access point that provides traffic to all UAVs of the network. This is the advantage of FANET networks over other networks, since using points and channels can be reserved and, accordingly, the necessary distribution and traffic control.

The first FANET systems were implemented on the basis of the existing TCP and UDP network transport protocols [9]. However, as practice has shown, such protocols are poorly suited for data transmission.

## JAUS

JAUS (Joint Architecture for Unmanned Systems) is an emerging standard for messaging between unmanned systems. The JAUS architecture has proven to be effective due to its active use in a large number of projects; for that reason, military, civil and commercial organizations plan to implement it in robotic complexes of various scales, including FANET networks [7, 9].

One of the emerging types of networks for controlling UAVs from the ground when interacting with FANET networks is sensor networks of WSN (Wireless Sensor Networks).

## WSN

WSN (Wireless Sensor networks) are the distributed, self-organizing networks of multiple sensors and actuators, interconnected via a radio channel. The coverage area of such a network can range from several meters to several kilometers due to the ability to relay messages from one node to another. The combination of a sensor network and a FANET network assumes the presence of two segments: i.e., a ground and flying ones, which interact with each other. The ground segment, as a rule, is a distributed network of self-organizing sensory nodes [7, 10]. The flying segment represents one or more UAVs. WSN is a self-building system with many capabilities. For example, it can be transformed into a network capable of adapting to a changing external environment. Such adaptation is carried out by means of self-organized changes in the topology of network connections. The established intra-network rules (protocols) will ensure the reciprocal (shuttle) distribution of information in a heterogeneous network using the multi-hop method.

The dynamically adaptable WSN communications architecture is able to provide solutions for such tasks as attaching new nodes, expanding the spatial area occupied by the network, self-restoration (continuing the previous operation of the network in case of failure of individual nodes), etc. WSN can provide information gathering and data transferring over large areas for a long period of time.

In fact, all considered types of networks are an infrastructure for communication between nodes for solving the tasks of this group (in particular, FANET for organizing UAV interaction). As an implementation of a scenario for solving the problem of data transmission through a grouping of nodes as through a distributed infocommunication environment, we consider three options (strategies) for using UAV as a repeater:

(1) UAVs are advancing to predetermined positions in a controlled or unmanned mode and holding for the required time in hover mode [11];
(2) UAV are moving according to a predefined flight plane [12, 13];
(3) about group of UAVs in the mode of automatic search for locations most suitable for maintaining uninterrupted transmission is deploey with continuously changing environmental conditions taken into account [14–17].

Strategies 1 and 2 have been widely used for a long time in both military and civilian spheres; they have obvious limitations of applicability and, therefore, are of no interest for further consideration. Strategy 3 is a challenging task of ensuring interaction of multiple UAVs, which multiplies the workload of UAV operators in maintaining situational awareness, reducing their effectiveness, which, in its turn, leads to the loss of efficiency and possible incidents [13, 15]. Recently, several approaches have been proposed to ensure autonomous functioning of UAV groups, for example, the principles of swarm behavior [4].

In references [16, 17], methods for optimizing the mutual positions of several UAV repeaters with the aim of obtaining the required communication characteristics are considered. However, managing UAV groups is a challenging task for operators due to amplified workload and increasing situational awareness requirements. The operator has to assess the situation, make a decision and give the right control commands, while analyzing information flows coming from several UAVs at the same time. Information overload of the operator can lead to a decrease in the efficiency of work, and to potential emergencies.

In reference [18], the main problems of the organization of dynamic self-organizing networks based on mobile devices and methods for their solution are considered. It is indicated that the core issue is the definition of a data transfer route. Since the network topology and, moreover, the location of its nodes are irregular, determining the direction in which the receiving node is located becomes very difficult. To solve this problem, it is possible to propose several approaches to the construction of the architecture, which are distinguished by the complexity of implementation, on the one hand, and data delivery efficiency, on the other. The following approaches are applicable to transferring data to a static and mobile nodes:

- Static node is using devices without GPS module
- Static node is using devices with a GPS module
- Static node is based on statistical data on the relative location of network nodes
- Mobile node is using devices without a GPS module: (a) using an additional channel to transmit the coordinates of a node; (b) using statistical information
- Mobile node is using devices with a GPS module: using an additional channel to transfer the coordinates of a finite node; using statistical information about the relative location of network nodes
- Broadcasting.

In addition, many researchers focus on communication protocols and the organization of the network nodes interaction without focusing on the control of a single UAV or group of UAVs, while the specific location of the repeaters or the pattern of their movement often depends on the nature of the transmitted content and may change during the infocommunication session.

In general, all the considered methods of autonomous control of a UAV group are aimed at solving only one specific task, while the introduction of a software-defined technology in combination with the active data approach [19] to control of a UAV group can significantly expand the range of tasks and adaptability to environmental conditions due to the possibility of reprogramming network nodes on the fly.

# 3   The Concept of Active Data

The analysis of approaches and architectures has shown that most of the approaches are aimed at creating routing protocols and solving problems of traffic in mobile data networks optimization.

Developers offer new communication protocols and communication modules based on them, or use standard modules to build a mobile network, paying more attention to the behavior of network nodes.

Unlike analogous approaches, the authors propose to use an architecture that uses the active data approach [20, 21] (application of executable codes transmitted via data transfer channels and executed on communication nodes), which allows dynamic updating of a program component for the behavior of a group of UAVs (see Fig. 1).

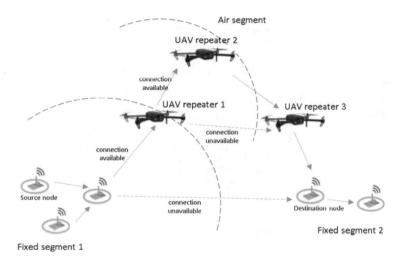

**Fig. 1.** Architecture of an infocommunication system based on a network with mobile nodes using the concept of active data (AD)

Active data (AD), according to [22], configure the software-defined equipment required for their propagation through the communication environment.

It is possible to distinguish the main groups of actions implemented by the means of the active data:

- Change of the operating frequencies, types of modulation or manipulation, topology of the network of radio devices through the re-initialization of communication modules
- Change of the data transfer formats, protocols, coding types by dynamically replacing the software component in communication modules
- Generating commands for changing the position of the UAV to control the spatial configuration of the repeater groups
- Reconfiguration of the flight controller by replacing the software component.

The listed action scenarios used in the tasks of information communication greatly increase the flexibility of data transmission channels based on mobile repeater networks.

## 4    Place of the Active Data in the UAV Control System

The proposed architecture partially changes the hardware structure of the UAV on-board equipment [23], as well as the software stack (see Fig. 2).

Let the flight plan be defined by the program, which determines not only the flight order for the given waypoints, but also the actions that need to be performed at these points or the allowed reactions to certain conditions that arise during the flight.

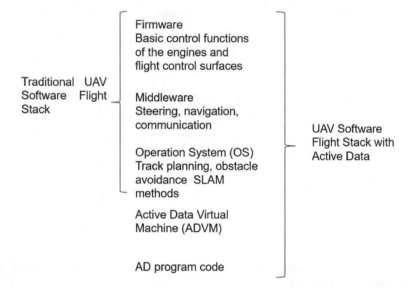

**Fig. 2.**  UAV control system software stack

The use of an active data architecture makes it possible to control the behavior of a repeater UAV using the data flow which itself is allowing for a much greater functionality and scalability of the system.

## 5    Conclusion

An analytical review of existing architectures of wireless self-organizing networks, including the concepts and principles of MESH, MANET, and other distributed networks, showed that all considered types of networks are realizations of the communication infrastructure needed to solve the tasks of an autonomous mobile UAV network, and the researchers mainly concentrate on communication protocols and organizing interaction between network nodes without focusing on the reconfiguration of the network and its individual nodes.

To eliminate these shortcomings, a concept of node reconfiguration was developed to provide the required network structure with the required characteristics. To achieve this goal, a technology was developed to organize a transmission channel for the networks with mobile nodes based on the active data concept, which extends the range of possible operations due to active adaptation to changing conditions.

The proposed architecture provides for the organization of data transmission channels in the areas where it is difficult to deploy a network of ground communication nodes, as well as in emergency situations.

In contrast to the well-known works [1, 4, 6, 14, 24–27], it is proposed to use UAV controlled by AD as a mobile node where the software component of AD, being launched on each mobile node, analyzes the communication environment to make a decision on data delivery to the destination node.

**Acknowledgments.** The research was granted by the State Contract from Ministry of Science and Education of the Russian Federation N 0073-2019-0005.

# References

1. Quick, D.: Boeing demonstrates swarm technology. https://newatlas.com/uav-swarm-technology/19581/. Accessed 15 Apr 2019
2. Chmaj, G., Selvaraj, H.: UAV cooperative data processing using distributed computing platform. In: Advances in Intelligent Systems and Computing, vol. 1089, pp. 455–461. Springer, Cham (2015)
3. Pavlov, A.A., Daticv, I.O.: Routing protocols in wireless networks [Protokoly marshruti-zatsii v besprovodnykh sctyakh]. In: Proceedings of the Kola Science Center of the Russian Academy of Sciences [Trudy Kol'skogo nauchnogo tsentra RAN], vol. 5, no. 24 (2014). (in Russian)
4. Bekmezci, İ., Sahingoz, O.K., Temel, Ş.: Flying ad-hoc networks (FANETs): a survey. Ad Hoc Netw. **11**(3), 1254–1270 (2013)
5. Kuiper, E., Simin, N.-T.: Mobility models for UAV group reconnaissance applications. In: International Conference on Wireless and Mobile Communications (ICWMC 2006), vol. 33 (2006)
6. Li, X., Yan, J.: LEPR: link stability estimation-based preemptive routing protocol for flying ad hoc networks. In: Proceedings – IEEE Symposium on Computers and Communications (ISCC), Heraklion, pp. 1079–1083 (2017)
7. Kucheryavy, A.E., Vladyko, A.G., Kirichyok, R.V.: Flying sensor networks - a new Internet of Things App. Actual problems of information telecommunications in science and education. In: IV International Scientific-Technical and Scientific-Methodical Conference: A Collection of Scientific Articles in 2 Volumes, vol. 1, pp. 17–22 (2015). (in Russian)
8. Camp, T., Boleng, J., Davies, V.: A survey of mobility models for ad hoc network research. Wireless Commun. Mob. Comput. (WCMC) **2**(5), 483–502 (2002). Special issue on Mobile Ad Hoc Networking: Research, Trends and Applications

9. Leonov, A.V., Chaplyshkin, V.A.: FANET networks. Omskiy nauchnyy vestnik **3**(143), 297–301 (2015). (in Russian)
10. Rusakov, A.M.: Model of flow control of measuring information of sensors in wireless sensor networks. Promyshlennyye ASU i Kontrollery **4**, 37–40 (2010). (in Russian)
11. Jawhar, I., Mohamed, N., Al-Jaroodi, J., et al.: Communication and networking of UAV-based systems: classification and associated architectures. J. Netw. Comput. Appl. **84**, 93–108 (2017)
12. Floreano, D., Wood, R.J.: Science, technology and the future of small autonomous drones. Nature **521**(7553), 460–466 (2015). https://doi.org/10.1038/nature14542
13. Samad, T., Bay, J.S., Godbole, D.: Network-centric systems for military operations in urban terrain: the role of UAVs. J. Proc. IEEE **95**(1), 4118473, 92–107 (2007). https://doi.org/10.1109/jproc.2006.887327
14. Lysenko, O.I., Valuiskyi, S.V., Tachinina, O.M., Danylyuk, S.L.: A method of control by telecommunication Airsystems for wireless AD HOC networks optimization. In: IEEE 3rd International Conference Actual Problems of Unmanned Aerial Vehicles Developments (APUAVD) Proceedings, Kyiv, Ukraine, 13–15 October 2015, pp. 182–185 (2015)
15. Kirichek, R., Kulik, V.: Long-range data transmission on flying ubiquitous sensor networks (FUSN) by using LPWAN protocols. In: Vishnevskiy, V., Samouylov, K., Kozyrev, D. (eds.) Distributed Computer and Communication Networks (DCCN 2016). Communications in Computer and Information Science, vol. 678, pp. 442–453. Springer, Cham (2016)
16. Ono, F., Ochiai, H., Miura, R.: A wireless relay network based on unmanned aircraft system with rate optimization. IEEE Trans. Wireless Commun. **99**, 7562472 (2016). https://doi.org/10.1109/TWC.2016.2606388
17. Aksyonov, K., Antonova, A., Goncharova, N.: Choice of the scheduling technique taking into account the subcontracting optimization. In: Thampi, S.M., Krishnan, S., Corchado Rodriguez, J.M., Das, S., Wozniak, M., Al-Jumeily, D. (eds.) Advances in Signal Processing and Intelligent Recognition Systems, SIRS, Advances in Intelligent Systems and Computing, vol. 678, pp. 297–304. Springer, Cham (2018)
18. Shishaev, M.G., Potaman, S.A.: Modern technologies of ad-hoc type networks and possible approaches to the organization of peer-to-peer telecommunication networks based on mobile devices with a short range of action. Trudy Kol'skogo nauchnogo tsentra RAN **3**, 70–74 (2010). (in Russian)
19. Alexandrov, V.V., Kuleshov, S.V., Zaytseva, A.A.: Active data in digital software defined systems based on SEMS structures. In: Gorodetskiy, A. (ed.) Smart Electromechanical Systems. Studies in Systems, Decision and Control, vol. 49, pp. 61–69. Springer, Cham (2016)
20. Kuleshov, S.V., Tsvetkov, O.V.: Active data in digital software-defined systems. J. Informatsionno-izmeritel'nyye i upravlyayushchiye sistemy **6**, 12–19 (2014). (in Russian)
21. Alexandrov, V.V., Kuleshov, S.V.: Esterification and terminal programs. J. Informatsion-noizmeritel'nyye i upravlyayushchiye sistemy **10**(6), 50–53 (2008). (in Russian)
22. Tang, Y.-R., Li, Y.: The software architecture of a reconfigurable real-time onboard control system for a small UAV helicopter. In: 8th International Conference on Ubiquitous Robots and Ambient Intelligence (URAI), pp. 228–233. IEEE, Incheon (2011)
23. Kuleshov, S.V., Zaytseva, A.A., Aksenov, A.Y.: The conceptual view of unmanned aerial vehicle implementation as a mobile communication node of active data transmission network. Int. J. Intell. Unmanned Syst. **6**(4), 174–183 (2018). https://doi.org/10.1108/IJIUS-04-2018-0010

24. Quick, D.: Aerial swarming robots to create communications networks for disaster relief. https://newatlas.com/smavnet-robot-swarm/16499/. Accessed 25 Apr 2019
25. Owano, N.: Airborne robot swarms are making complex moves (w/video). https://phys.org/news/2012-02-airborne-robot-swarms-complex-video.html. Accessed 25 Apr 2019
26. Created a "flying repeater" for the ESU TK: Army Journal "Army Gazette" (2012). https://armynews.ru/2012/10/sozdan-letayushhij-retranslyator-dlya-esu-tz/
27. The complex ESU TZ: the desired and real: Army Journal "Army Gazette" (2010). https://armynews.ru/2010/11/kompleks-esu-tz/

# Cloud System for Distributing Multimedia Content in Cyber-Physical Systems

Dmitriy Levonevskiy$^{(\boxtimes)}$, Anton Saveliev, Ilya Duboyskiy, and Pavel Drugov

St. Petersburg Institute for Informatics and Automation of the Russian Academy of Sciences, St. Petersburg, Russia
DLewonewski.8781@gmail.com

**Abstract.** This paper considers building a cloud system for distributing multimedia content in cyber-physical environments. The paper describes overall system architecture, gives detailed description of the users. involved in content distribution. Their typical actions are defined in use case diagrams. UML-diagrams are given, describing modules for content management and transmission are presented. Distributed data storage and management are also thoroughly described here. Various output channels and display modes are compared in terms of client-server interactions to be considered when displaying multimedia content. An essential part of the cyber-physical system described here is mobile device management and data coordination among such devices. The presented systems can be easily implemented with the Unity engine (C#) and GraphQL query language. This paper ultimately gives a high-level perspective of content delivery networks in a modern enterprise.

**Keywords:** Multimedia · Human-machine interactions · Distributed systems · Content delivery systems · Cyber-physical systems

## 1 Introduction

Content delivery belongs to the most important tasks of information systems. Using interactive information and advertising kiosks, chatbots, mobile platforms, and mobile apps allows to enhance user's experience during human-machine interaction, to make it more convenient and efficient. Usage of distributed systems also simplifies content administration and allows users with administrative privileges to manage a large-scale content delivery system via a unified user interface.

However, for efficient content delivery, certain criteria should be met; in particular:

- Reliability of the delivery system and possibility of information transfer via telecommunication channels
- Guarantees of comprehensive content reproduction on output devices with sufficient output quality
- Consistency between the content, delivery system requirements and user expectations.

© Springer Nature Switzerland AG 2020
D. G. Arseniev et al. (Eds.): CPS&C 2019, LNNS 95, pp. 266–274, 2020.
https://doi.org/10.1007/978-3-030-34983-7_26

## 2  Related Work

Content delivery in cyber-physical systems (CPSs) is performed using various devices and modalities. Such devices include, in particular, stationary and mobile information kiosks, mobile robotic systems, and mobile devices [2]. Multimedia content can also be streamed into digital signages, TV networks, e.g., enterprise TV networks [1], and displayed on web sites with different information and advertising blocks, in desktop and mobile apps. Content consumer can manage the streaming video via mobile apps and multimodal interfaces [4]. Integration of different information systems and communication interfaces enables establishment of cyber-physical interactive augmented reality [14–16, 20]. Architecture of software-defined smart content management systems based on the Internet of Things (IoT) paradigm [17] is proposed in the reference [18].

Scientific sources pay much attention to content delivery in distributed systems. Several studies propose content delivery protocols, where improved load balancing is under research, ensuring offloading of critical system links in different settings, e.g., in the Internet of Vehicles [5, 8]. Various issues of network topology organization are considered to ensure optimal data transfer modes [6] and fault-tolerance of systems [10]. Furthermore, there are some studied systems of consolidated content delivery from multiple sources and approaches to routing in such systems [7, 12].

Nevertheless, content delivery processes in interactive human-machine interactions using multimodal interfaces [13] are less studied. At the same time, content delivery and streaming is required by development of smart help systems and chatbots. One such enterprise-grade information and navigation system is described in [11]. The authors of that work consider some approaches to human-machine interactions but omit content delivery aspects. Possible user interaction scenarios in the context of this system are described in [19]. This work, on the contrary, considers the same kind of systems from another point of view focusing on content delivery and takes into account the properties of the content and devices (multiple modalities, different capabilities of broadcasting channels).

## 3  Content Distribution Model

Multimedia content is distributed among devices, registered in a cloud environment as access points. Providing content in a cloud environment, it is reasonable to enable centralized management of this content to maintain it up-to-date, as well as to control the streaming process according to defined user roles. The content distribution process is defined by necessary tradeoffs between content provider requirements, end device features, and user satisfaction [3].

The following types of actors are involved in the content distribution process:

Users that are entitled, depending on their roles, to provide and consume the content, register and maintain access points and manage content delivery.

Cloud environment ensuring the content storage and distribution across access points.

Access points: CPS elements included into content distribution platforms and delivering the content to the end users.

Terminal access devices visualizing the content are employed for interacting with users. Examples of such devices are smartphones, tablets, desktop PCs, digital signages. To obtain the content, end devices leverage the access infrastructure (sites, signage software), which connects to access interfaces (cloud system API). Content workflow is shown in Fig. 1.

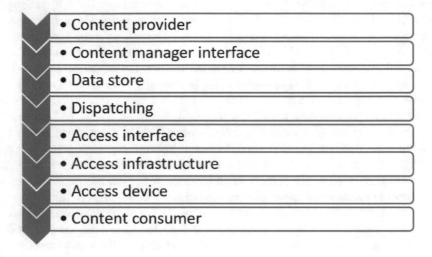

**Fig. 1.**  Content delivery sequence

Let's consider the high-level interaction pattern, involving users, cloud environment and access points, depicted in Fig. 2. The diagram clearly shows that the users themselves manage content distribution, and they can be divided into 3 main categories: content providers, dispatchers and platform owners. Let's examine closer every role and the processes within the cloud environment attributed to this role.

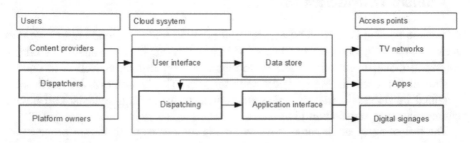

**Fig. 2.**  Interaction between users, cloud environment and access points

Providers upload content to the cloud environment. That content is characterized by such features as a set of modalities (text, sound, image), size (timing, information volume), a set of categories (keywords). Content providers create repositories (content item sets), where the content is stored. Every content item is an entity, comprising the content itself and data on its properties and location. Depending on the type, the content item has a set of properties, defining its features and options of its delivery to access points.

Use case diagram for content manager is presented in Fig. 3.

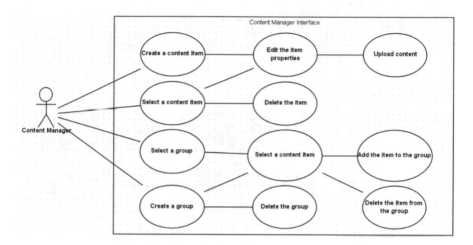

**Fig. 3.** Use case diagram for content manager

In Fig. 4, a class diagram is presented, featuring types and properties of program entities describing the content. Common properties, pertinent to any content items, are described by the ContentItem class. Content properties, concerning various features and modalities, are described by abstract classes, i.e., Visual Content Item, Audio Content Item, etc. These classes allow determining requirements to access points, necessary for content streaming, e.g., bitrate parameters for audio content, resolution, color depth for visual content, as well frame rate for video. Typical content objects, containing audio, video and text, are inherited from the abstract classes.

All content types have the location property, i.e., whether a file is stored on the server or it is a torrent file. The content is also characterized by a set of metadata (categories, links, topics) to facilitate its distribution across client devices. These properties are defined by the group attribution of the content (ContentGroup class), which could be explicitly defined by users with administrator privileges for further ease of administration.

Dispatching is, essentially, the process of content distribution across client devices. Dispatching could be performed automatically, respecting how the client devices meet requirements to display certain types of content, and also manually. The respective use case diagram is shown in Fig. 5.

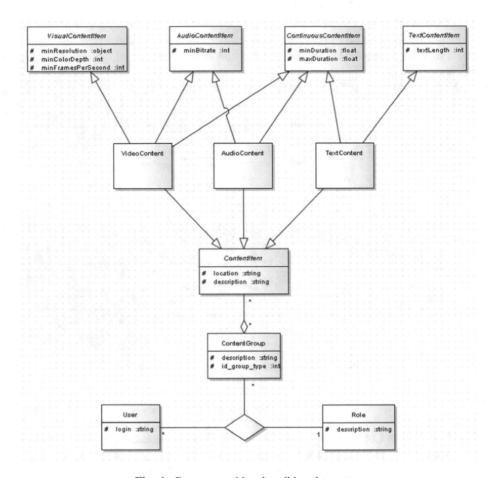

**Fig. 4.** Program entities describing the content

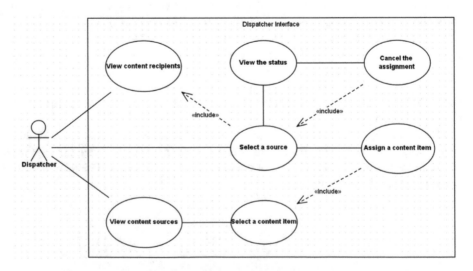

**Fig. 5.** Dispatcher use case diagram

Streaming control is achieved via creation and management of Broadcast entities (see Fig. 6).

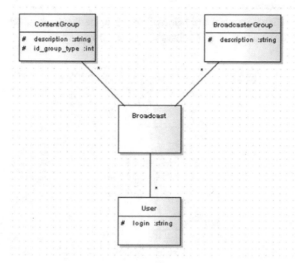

**Fig. 6.** Programmatic dispatch entities

When distributing the content among end devices, the compatibility of the content with the end device should be considered; the device should support any modalities of the content and provide the quality of data transfer at least as high, as is specified in the content metadata. The streaming should be well targeted, what is ensured by handling content metadata during the dispatch. The end device also should have sufficient resources (data transfer channels).

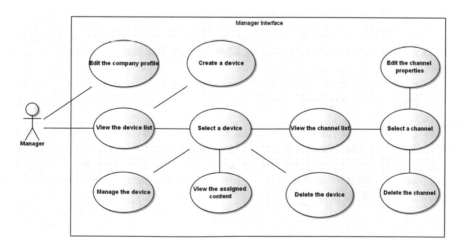

**Fig. 7.** Access point management

Content is streamed by CPS devices, registered in the cloud environment as access points. Examples of such devices particularly are: digital signages, TV networks, including enterprise ones, sites (featuring information and advertising blocks), desktop and mobile applications.

Devices and their properties are controlled by a manager. A use case diagram for such activity is presented in Fig. 7, class diagram in Fig. 8.

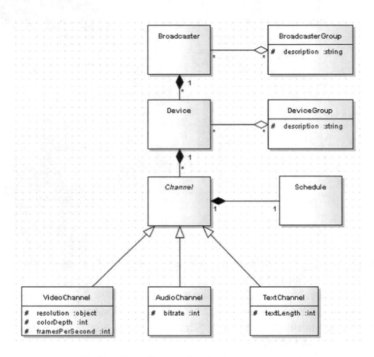

**Fig. 8.** Class diagram for device management

For example, let's consider a front-end module downloading and streaming multimedia entities in the interactive mode. This module has been developed using the Unity engine in C# language.

At first, the algorithm of client-server interaction builds a GraphQL query to request content and its streaming mode. Let's consider media display case on mobile devices (see Fig. 9).

Upon touching the screen, coordinates of the touch location are determined. The coordinates of the last detected gesture are also fetched. If two values of X-coordinate differ, it means that the user has swept some slides. If the coordinate of the latter location is less than the coordinate of the former, then the next slide will be displayed, elsewise, the previous one.

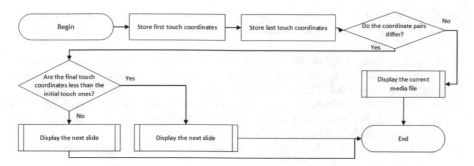

**Fig. 9.** Algorithm of media display on mobile devices

## 4 Conclusions

The presented models give a high-level perspective of content delivery system in CPS. The impact of the proposed models consists in the application of content delivery processes and data structures to the conditions of multi-modal human-computer interaction. The dispatch mechanism allows to control the quality of service and properly manage different content types and devices, ensuring mutual compatibility between them.

Different technological stacks could be used in the proposed architecture. For example, distributed data storage systems, particularly, peer-to-peer networks [9], ensure some advantages in improving system reliability and resource saving. The following advantages are particularly important for:

Server disk space saving: the server should store only the content registry, not the content itself (for example, as a set of torrent files).

Reducing the network load, because the files are downloaded from different sources.

Improved fault-tolerance of the system, as applications could still download content if the server is temporarily unavailable.

## References

1. Levonevskiy, D., et al.: Processing models for conflicting user requests in ubiquitous corporate smart spaces. In: MATEC Web Conference 13th International Scientific-Technical Conference on Electromechanics and Robotics "Zavalishin Readings", vol. 161, article no. 03006. EDP Sciences (2018). https://doi.org/10.1051/matecconf/201816103006
2. Prishchepa, M.V., Ronzhin, A.L.: Modeli interaktivnogo vzaimodejstviya s podvizhnym informacionno-navigacionnym kompleksom [Interactive models with a mobile information and navigation system]. Doklady TUSUR 2(28) (2013). (in Russian)
3. Sarwar, G., et al.: QoS and QoE aware N-screen multicast service. J. Sens. **2016**, 11 (2016). article id 8040138
4. Jucker, A.H., et al.: Doing space in face-to-face interaction and on interactive multimodal platforms. J. Pragmatics **134**, 85–101 (2018). https://doi.org/10.1016/j.pragma.2018.07.001. ISSN:0378-2166

5. Meneguette, R.I., et al.: A novel self-adaptive content delivery protocol for vehicular networks. Ad Hoc Netw. **73**, 1–13 (2018). https://doi.org/10.1016/j.adhoc.2018.02.005. ISSN:1570-8705
6. Zheng, Z., Zheng, Z.: Towards an improved heuristic genetic algorithm for static content delivery in cloud storage. J. Comput. Electr. Eng. B, 422–434 (2018). https://doi.org/10.1016/j.compeleceng.2017.06.011. ISSN: 0045-7906
7. Hashemi, S.N.S., Bohlooli, A.: Analytical modeling of multi-source content delivery in information-centric networks. J. Comput. Netw. **140**, 152–162 (2018). https://doi.org/10.1016/j.comnet.2018.05.007. ISSN:1389-1286
8. Silva, C.M., et al.: Designing mobile content delivery networks for the internet of vehicles. J. Veh. Commun. **8**, 45–55 (2017). https://doi.org/10.1016/j.vehcom.2016.11.003. ISSN: 2214-2096
9. Anjum, N., et al.: Survey on peer-assisted content delivery networks. J. Comput. Netw. **116**, 79–95 (2017). https://doi.org/10.1016/j.comnet.2017.02.008. ISSN:1389-1286
10. Natalino, C., et al.: Infrastructure upgrade framework for content delivery networks robust to targeted attacks. J. Opt. Switching Netw. **31**, 202–210 (2019). https://doi.org/10.1016/j.osn.2018.10.006. ISSN:1573-4277
11. Levonevskiy, D.K., et al.: MINOS multimodal information and navigation cloud system for the corporate cyber-physical smart space [Mnogomodal'naya informacionno-navigacionnaya oblachnaya sistema minos dlya korporativnogo kiberfizicheskogo intellektual'nogo prostranstva]. J. Programmnaya inzheneriya **3**, 120–128 (2017). https://doi.org/10.17587/prin.8.120-128. (in Russian)
12. Liu, Y., Yu, S.Z.: Network coding-based multisource content delivery in content centric networking. J. Netw. Comput. Appl. **64**, 167–175 (2016). https://doi.org/10.1016/j.jnca.2016.02.007. ISSN:1084-8045
13. Budkov, V.Y., et al.: Multimodal human-robot interaction. In: Proceedings of 2010 International Congress on Ultra Modern Telecommunications and Control Systems and Workshops (ICUMT), pp. 485–488. IEEE (2010). https://doi.org/10.1109/ICUMT.2010.5676593
14. Davies, N., et al.: Pervasive displays: understanding the future of digital signage. Synth. Lect. Mobile Pervasive Comput. **8**(1), 1–128 (2014). https://doi.org/10.2200/S00558ED1V01Y201312MPC011
15. She, J., et al.: Smart signage: a draggable cyber-physical broadcast/multicast media system. IEEE Trans. Emerg. Top. Comput. **1**(2), 232–243 (2013). https://doi.org/10.1109/TETC.2013.2282618
16. Fu, H.: Smart signage: a cyber-physical interactive display system for effective advertising, Dissertation, Hong Kong University of Science and Technology (2013)
17. Perera, C., et al.: The emerging internet of things marketplace from an industrial perspective: a survey. IEEE Trans. Emerg. Top. Comput. **3**(4), 585–598 (2015). https://doi.org/10.1109/TETC.2015.2390034
18. Kim, E., et al.: Efficient contents sharing between digital signage system and mobile terminals. In: Proceedings of the 15th International Conference Advanced Communication Technology (ICACT), pp. 1002–1005. IEEE (2013)
19. Vatamaniuk, I., et al.: Scenarios of multimodal information navigation services for users in cyberphysical environment. In: Proceedings of the 18 International Conference Speech and Computer (SPECOM 2016), pp. 588–595, 23–27 August 2016. Springer, Budapest (2016). https://doi.org/10.1007/978-3-319-43958-7_71
20. Schwab, K.: The Fourth Industrial Revolution, World Economic Forum (2017)

# Static Force Analysis of a Finger Mechanism for a Versatile Gripper

Ivan I. Borisov[1,2(✉)], Sergey A. Kolyubin[1], and Alexey A. Bobtsov[1]

[1] Faculty of Control Systems and Robotics, ITMO University,
49 Kronverksky Pr., St. Petersburg 197101, Russia
borisovii@itmo.ru
[2] Center for Technologies in Robotics and Mechatronics Components,
Innopolis University, Innopolis, Russia

**Abstract.** In the development of a mechanism, it is important to know the magnitudes, directions, and locations of the constraint forces between the connected links of the kinematic chain in order to design a mechanism with desired characteristics. This paper presents an approach of a static force analysis of the novel complex mechanism consisting of 8 links, which belongs to the VI class of the Assur group. It means that it can be only separated into an input link and a system of 6 links which cannot be divided into smaller groups; thus, traditional methods for graphical analyses cannot be used. The mechanism is used to implement a finger of a versatile bio-inspired industrial gripper, which can change the degree of freedom (DOF) in order to change the mode of grasping. It is possible to change DOF via breaking/reconnecting the kinematic chain of the finger. When the mechanism is intact, it has only 1 DOF and it represents a fully kinematically defined structure that allows performing a precision grasp. When the kinematic chain is broken, the finger gets underactuated, thus it has 2 DOF, and an underactuated power grasp can be performed. The finger represents different types of a mechanism in precision and power grasps. Force analyses of the finger in both modes were carried out in order to get information about the relationship between the torque applied to a driving link and forces applied to surfaces of phalanges. The paper is concerned with the force analysis and a design of a prototype of the gripper.

**Keywords:** Grasping · Grippers · Mechanisms · Underactuation · Robotics

## 1 Introduction

Articulated robots became a universal automation tool for manufacturing processes. They are intended to increase the quality of products, decrease production costs by reducing scrap, and increase quantities of products since they can work non-stop for long periods of time. Robots are broadly used in order to reduce routine low-skilled manual labor and in cases of frequent changes in objects to produce.

However, a robot alone is not able to perform any task without an end effector. Therefore, a design of the performance device at the end of a robotic arm, designed to interact with the environment, is very important. There are two basic categories of end effectors: a gripper to grasp and hold objects for pick-n-place operations or a tool to

D. G. Arseniev et al. (Eds.): CPS&C 2019, LNNS 95, pp. 275–289, 2020.
https://doi.org/10.1007/978-3-030-34983-7_27

perform various manufacturing processes. This paper is concerned with the first category of end effectors. An analysis of the finger mechanism and a design of the versatile gripper are presented.

A lot of grippers are available on the market. There are electric, pneumatic, magnetic grippers and suction cups which can be used to grasp and hold objects [1]. But when it comes to universal devices, electric grippers seem to be the best option, since they are capable to

- Control the position of the gripper finger using encoded motors. Thus, position control can be performed
- Detect the finger position during a contact with an object using encoders
- Control the grip force and speed through detecting current supply
- Use the electric gripper without additional pneumatic hardware.

The challenge for researches is to create a universal gripper which can perform all or almost all manufacturing operations. The idea is to equip all robots with the same device or a narrow range of similar devices.

Since human beings have been a kind of textbook for the development of robots because they perform the tasks that humans routinely do [2], a real human hand is proposed as an example of a "true universal gripper". The level of dexterous manipulation by robots is currently far from that of human beings. It is possible to improve the abilities of robots by transferring human functions to robotic manipulation. The structure of a human finger and hand plays an important role for dexterous manipulation. Let us suppose that a gripper which acts like a human hand will be the best option. There are several grasping styles, which are presented here [3], but the most popular and fundamental ones are the precision and power grasps [2]. For the precision grasp, only fingertips are used for grasping, in-hand movements are available. For the power grasp, internal areas of all parts of the hand are utilized to envelope an object, the hand and the object can be considered as one rigid body. Since a real human hand can perform a great number of grasps [4], the robotic hand also must be able to perform both the precise pinch and power encompassing grasps.

This paper is devoted to an approach of a static force analysis of a novel complex mechanism consisting of 8 links, which can be separated only into an input link and a system of 6 links which cannot be divided into smaller parts. The mechanism is applied to implement a finger for an industrial anthropomorphic gripper, which can perform both precision grasps and power grasps. The concept and the first prototype of the industrial gripper is described in [5]. The paper is organized as follows. Section 2 describes commonly used strategies to develop a finger mechanism for a bio-inspired gripper or artificial hand. Section 3 proposes the finger mechanism and closing sequences in both modes. Section 4 is devoted to the force analysis of the finger mechanism. Section 5 proposes a mechanical design of the gripper and describes the operation process. The conclusion is the final part of the paper.

# 2  Current Status and Challenges in the Development of Versatile Grippers

A lot of the electrical grippers have been created over the past two decades. But despite everything, these devices are still far away from human-like movements, functionality, and dexterity of a real human hand. Since a human hand is composed of a fixed palm and actuated digits, it is enough to create a mechanism for fingers. There are several commonly employed strategies to implement a mechanism for a digit of an anthropomorphic gripper or artificial hand, which are described underneath.

## 2.1  Coupling Linkage Mechanism

Coupling linkage mechanism is the one that is fully kinematically defined and has only one degree of freedom. A finger represents a closed kinematic chain with one link fixed. This type of mechanism is used to create an artificial hand to achieve human-like movements [6]. Precision grasp can be performed since the motions of all points on the linkage can be measured with respect to the fixed link. However, no adaptive grasps can be performed.

## 2.2  Multi DOF Mechanism

Multi DOF mechanism is the one that has more than one degree of freedom. The mechanism represents an open kinematic chain. Each link is actuated by its own motor. Thus, both grasp and adaptive grasp and precision can be performed. However, the payload capacity is limited by motors. There are commercial robotic hands, such as Barrett [7] or Schunck SDH [8]. They both are equipped with three 2-phalange fingers and each of their joints is actuated by a servo DC motor; as a result, these grippers are very precise, but payload capacity is low.

## 2.3  Underactuated Mechanism

One of the most famous and suitable variants of finger mechanisms is the underactuated one. An underactuated mechanism is the one that has fewer actuators than degrees of freedom [9]. A mechanical finger based on an underactuated mechanism makes the gripper self-adaptive. Such fingers envelop the objects to grasp them and automatically adapt to the shape of the objects having only one motor. According to [10], underactuation is able to help to overcome mechanical complexity that is often caused by the need to independently actuate and control each DOF individually. The authors state that underactuated mechanisms are an interesting approach because they offer increased functionality ensuring good adaptability. According to [11], the concept of underactuation allows to reduce a number of actuators without affecting DOFs by using passive elements, such as springs and mechanical limits. Such approach is able to ensure the capability of self-adaptation without any help from control algorithms when the object is grasped. But this type of the mechanism has its own disadvantages too. It is reported that a precision grasp is difficult to achieve by underactuated fingers [9]. It is stated, that grasping of small objects by underactuated hands is impossible, unless specific design modifications are applied [12].

## 2.4  Hybrid Mechanism

There are devices which can produce both modes of operation. The Multi-Modal (M2) Gripper [13] can produce both underactuated and fully actuated behaviors. The hand can adaptively grasp objects of varying geometries, pinch-grasp smaller items, and perform some degree of in-hand manipulation. An underactuated soft gripper utilizing fluid fingertip based on the hybrid soft and hard structure of human fingertips was developed to implement versatile grasping [14]. The gripper can perform pinching grasp and enveloping grasp.

## 2.5  Proposed Design

The main feature of the proposed mechanism for a gripper finger is the ability to lock or run free a special link of the finger mechanism that allows the finger to either be fully defined kinematically, or underactuated (the gripper design is the subject for PTC patent application PCT/RU2019/000167 19.03.2019). The closing/opening of the kinematic chain is implemented with a BK special link (see Figs. 2, 3 and 4), which can fix its length or make it variable. The fixing/unfixing can be implemented in plenty of ways, such as via electromagnet, pneumatic pin-to-hole setup or with an elastic element with variable stiffness. This approach can be used to create grippers for different tasks both, for industrial grippers and prostheses. This feature allows to choose between a power-adaptive encompassing grasp or a pinch precision grasp. The ability to fix the number of degrees of freedom can help to prevent unpredictable behavior of underactuated modes, resulting in a more stable and accurate pinch grasp.

## 3  Finger Implementation

Let us consider the mechanism of the finger which is composed of 8 links (see Figs. 2, 3 and 4). The finger represents a six-class Assur mechanism which consists of a driving link AB and an Assur group of class VI. The Assur group is a closed kinematic chain with zero degrees of mobility; it cannot be divided into smaller groups.

$$W = 3n - 2P_5 = 0,$$

where $W$ is the DOF, $n$ is the number of movable links, $P_5$ is the number of kinematic pairs with 5 constraints. If $W$ is zero, then $n$ is even. In this case, there are only a few combinations of $n$ and $P_5$ (see Table 1).

**Table 1.**  The Assur groups

| n | 1 | 2 | 4 | 6 |
|---|---|---|---|---|
| P5 | 1 | 3 | 6 | 9 |
| Class of Assur group | I | II | IV | VI |

The input link with $W = 1$ is an exception (see Fig. 1a). All other Assur groups are even. The typical Assur group class VI is shown in Fig. 1d. The class of the whole mechanism equals the class of the highest group of the mechanism.

The finger mechanism is shown in Fig. 4. Points $O$, $A$ and $Q$ are frames and represent the eighth fixed link. The link $AB$ is a binary driving link, which is connected with the frame in the point $A$ and with ternary link $BC$ in the point $B$. The link $BC$ is ternary. In Fig. 2, there are two joints in the point $B$ since it connects three links. The ternary link $BC$ is connected to the binary link $BK$ also in the joint $B$. The finger consists of two phalanges. The distal phalanx $CDH$ is a binary link which connects with the link $BC$ in the point $C$ and with the proximal phalanx $DEO$ in the point $D$. The proximal phalanx $DEO$ is connected with the frame in the joint $O$ and with the binary link $EF$ in the joint $E$. The ternary link $KFQ$ connects with the frame in the joint $Q$, with the binary link $EF$ in the joint $F$, and with the binary link $BK$ in the joint $K$.

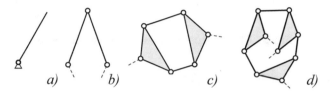

**Fig. 1.** Examples of Assur groups: (a) the I class (an input link), (b) the II class, (c) the IV, (d) the VI class of Assur group

A desirable closing sequence is as follows: the distal phalanx $CDH$ has to keep perpendicular orientation with respect to the palm surface during the whole closing sequence. A closing sequence of the finger in the pinch mode is shown in Fig. 2. Figure 2a presents the finger in the initial position, Fig. 2b presents the finger is the transitional position, and Fig. 2c presents the finger in the final position. As one can see, the orientation of the distal phalanx to the palm surface is constant during the closing sequence of the finger, which allows performing a pinch grasp.

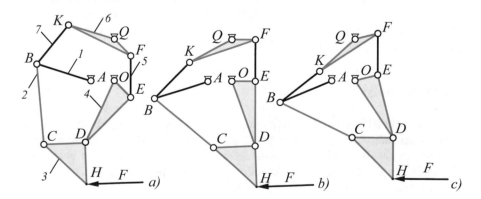

**Fig. 2.** Initial, transitional and final positions of the finger in the pinch mode

A grip force is applied only to the distal phalanx during a pinch grasp. The links are supposed to be rigid. It means that there can be no relative motion between two randomly chosen points in the same link.

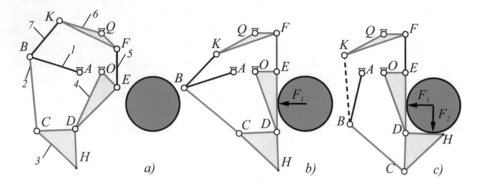

**Fig. 3.** Closing sequence in an underactuated mode

When a kinematic chain of a digit is closed (see Fig. 2), it has only one *DOF*:

$$W = 3n - 2P_5 = 3 \cdot 7 - 2 \cdot 10 = 1.$$

Thus, a pinch grasp can be performed. But if the kinematic chain of a digit is opened by breaking the kinematic chain, the digit will have two DOFs:

$$W = 3n - 2P_5 = 3 \cdot 6 - 2 \cdot 8 = 2.$$

That allows accomplishing power and adaptive underactuated grasps of unknown random objects. Closing/opening of the kinematic chain is implemented with a special link with an electromagnet that can fix its length or make it variable. Thus, a digit can choose between different operation modes of the closing sequence. Figure 3 presents the closing sequence in an underactuated mode. Figure 3a presents the finger in the initial position, Fig. 3b shows the first contact of the proximal phalanx with the object to grasp, and Fig. 3c shows the finger in the final position. As it can be seen, the magnitude of the special link *BK* has changed.

## 4  Force Analysis

It is necessary to understand the relationships between the geometry and motions of the parts of a mechanism and forces that produce these motions [15]. The finger represents a complex mechanism. It has to be established, what is the relationship between the motion of the input link and the motion of output links or phalanges, what are the loads on the surfaces. The synthesis of the proposed mechanism was presented in [16]. The topic of the paper is to present the approach that allows examining the proposed design.

## 4.1    Force Analysis of the Finger in Pinch Mode

In the pinch mode, the mechanism represents a coupling linkage mechanism with only one degree of freedom. The links are supposed to be rigid bodies. The finger mechanism is shown in Fig. 4a. Several forces are applied to the mechanism: a force of the contact with the object $F$ at the point $H$ and gravitational forces of each link. Since the motion of the finger is slow and the mechanism has got a fixed structure during the grasp, it is possible to use only the static equations of equilibrium, without taking into account D'Alembert's principle. It means, there is no need to calculate inertial forces and inertial torques, which makes the calculations much easier.

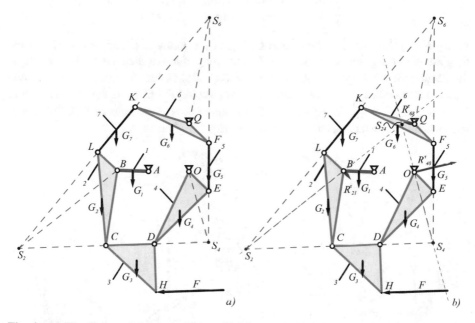

**Fig. 4.** (a) The finger mechanism with applied forces, (b) The finger mechanism with applied forces and reaction forces in fixed points $A$, $O$ and $Q$, and the position of point $S_{24}$

For the initial position of the finger, we found all the constraint forces and their reactions necessary for this to be a position of equilibrium. The force analysis was carried out using Assur group theory [17–20]. However, since the finger is a class VI Assur mechanism, it can only be divided into three parts: the frame, the input link, and the VI class group. Since it is impossible to separate the class VI group into smaller parts, the whole group should be considered.

Let us consider the equilibrium of the class VI group. It is impossible to calculate full reaction forces before the preparation. The task of the first step is to calculate tangential constraint forces of each link of the group. In order to do this, each link must be considered in the following order 3, 5, 7, 4, 6, 2. Then the equilibrium of whole group should be considered. The moment at point $S_{24}$, which is an interaction of lines $BS_2$ and $OS_4$ (see Fig. 4b), should be found in order to find the full reaction force in joint $Q$. The final step is to consider each link once again and get full reactions in each joint using the method of vector diagrams.

The graphic solution for the mechanism is presented:

1. Let's begin drawings of the free-body diagram of the link 3 or the distal phalanx, which is shown in Fig. 5a. There are 4 forces: the applied force **F**, the gravitation force **G**$_3$ and two tangential reaction forces $R_{34}^\tau$ and $R_{32}^\tau$. Therefore, taking counterclockwise moments as positive, we obtain the static equations of equilibrium for link 3:

$$\sum M_C = -G_3 l_{C_3} + R_{34}^\tau l_{CD} - F l_{DH} = 0,$$
$$\sum M_D = G_3 l_{D_3} + R_{32}^\tau l_{CD} - F l_{DH} = 0,$$
(1)

where $M_C$ is the moment about joint $C$, $M_D$ is the moment about joint $D$, $G_3$ is the magnitude of the gravitation force **G**$_3$, $F$ is the magnitude of the applied force **F**, $R_{34}^\tau$ is the magnitude of tangential reaction force $R_{34}^\tau$, $R_{32}^\tau$ is the magnitude of tangential reaction force $R_{32}^\tau$, $l_{C_3}$ is the moment arm of **G**$_3$ about joint $C$, $l_{D_3}$ is the moment arm of **G**$_3$ about joint $D$, $l_{DH}$ is the length of $DH$, and $l_{CD}$ is the length of $CD$. The counterclockwise moments taken as positive, the Eq. (1) give:

$$R_{34}^\tau = (F l_{DH} + G_3 l_{C_3})/l_{CD},$$
$$R_{32}^\tau = (F l_{DH} - G_3 l_{D_3})/l_{CD}.$$
(2)

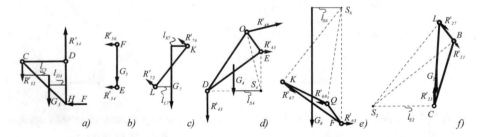

**Fig. 5.** Free-body diagrams of the links 3, 5, 7, 4, 6 and 2 with applied forces

2. Next we proceed to a free-body diagram of the link 5, which is shown in Fig. 5b. There are 3 forces: the gravitation force **G**$_5$ and two tangential reaction forces $R_{56}^\tau$ and $R_{54}^\tau$. The counterclockwise moments taken as positive, the static equations of equilibrium give:

$$\sum M_E = -G_5 l_{E_5} + R_{56}^\tau l_{EF} = 0,$$
$$\sum M_F = G_5 l_{F_5} - R_{54}^\tau l_{EF} = 0,$$
(3)

which yields to

$$R_{56}^\tau = G_5 l_{E_5}/l_{EF},$$
$$R_{54}^\tau = G_5 l_{F_5}/l_{EF}, \tag{4}$$

where $M_E$ is the moment about joint $E$, $M_F$ is the moment about joint $F$, $l_{E_5}$ is the moment arm of $G_5$ about joint $E$, $l_{F_5}$ is the moment arm of $G_5$ about joint $F$, $l_{EF}$ is the length of $EF$.

3. Also, 3 forces are applied to link 7: the gravitation force $G_7$ and two tangential reaction forces $R_{76}^\tau$ and $R_{72}^\tau$ (see Fig. 5c). The counterclockwise moments taken as positive, the static equations of equilibrium give:

$$\sum M_L = -G_7 l_{L_7} + R_{76}^\tau l_{KL} = 0,$$
$$\sum M_K = G_7 l_{K_7} - F_{72}^\tau l_{KL} = 0, \tag{5}$$

which yields to

$$R_{76}^\tau = G_7 l_{L_7}/l_{KL},$$
$$R_{72}^\tau = G_7 l_{K_7}/l_{KL}, \tag{6}$$

where $M_L$ is the moment about joint $L$, $M_K$ is the moment about joint $K$, $l_{L_7}$ is the moment arm of $G_7$ about joint $L$, $l_{K_7}$ is the moment arm of $G_7$ about joint $K$, $l_{LK}$ is the length of $LK$.

4. Let us consider the equilibrium of the ternary link 4, which is shown in Fig. 4d. The gravitation force $G_4$ and three tangential reaction forces $R_{48}^\tau$, $R_{43}^\tau$ and $R_{45}^\tau$ are applied to link 4. The method for continuation is to measure the moment arms about point $S_4$, which can be found as an interaction of lines $CD$ and $FE$ (see Fig. 4a).

$$\sum M_{S_4} = -R_{45}^\tau l_{ES_4} + R_{43}^\tau l_{OS_4} + G_4 l_{S_4} - R_{48}^\tau l_{OS_4} = 0, \tag{7}$$

which yields to

$$R_{48}^\tau = (-R_{45}^\tau l_{ES_4} + R_{43}^\tau l_{OS_4} + G_4 l_{S_4})/l_{OS_4} = 0. \tag{8}$$

Where $l_{ES_4}$ is the moment arm of $R_{45}^\tau$ about point $S_4$, $l_{OS_4}$ is the moment arm of $R_{43}^\tau$ about point $S_4$, $l_{S_4}$ is the moment arm of $G_4^\tau$ about point $S_4$, and $l_{OS_4}$ is the moment arm of $R_{48}^\tau$ about point $S_4$.

5. We proceed next to link 6. Applied forces $G_6$, $R_{67}^\tau$, $R_{68}^\tau$ and $R_{65}^\tau$ are shown in Fig. 5e. The magnitude of the reaction force $R_{68}^\tau$ can be calculated like this:

$$\sum M_{S_6} = R_{67}^\tau l_{KS_6} + R_{65}^\tau l_{FS_6} - R_{68}^\tau l_{QS_6} + G_6 l_{S_6} = 0, \tag{9}$$

which yields to

$$R_{68}^\tau = (R_{67}^\tau l_{KS_6} + R_{65}^\tau l_{FS_6} + G_6 l_{S_6})/l_{QS_6} = 0, \tag{10}$$

here $l_{KS_6}$ is the moment arm of $R_{67}^\tau$ about point $S_6$, $l_{FS_6}$ is the moment arm of $R_{65}^\tau$ about point $S_6$, $l_{QS_6}$ is the moment arm of $R_{68}^\tau$ about point $S_6$ and $l_{S_6}$ is the moment arm of $\mathbf{G}_6$ about point $S_6$. Point $S_6$ can be found as an interaction of lines $LK$ and $FE$ (see Fig. 4a).

6. Figure 5f shows the equilibrium of the last link 2. Forces $\mathbf{G}_2$, $R_{23}^\tau$, $R_{27}^\tau$ and $R_{21}^\tau$ are applied to the ternary link 2. The magnitude of the reaction force $R_{21}^\tau$ is as follows:

$$\sum M_{S_2} = -R_{21}^\tau l_{BS_2} - R_{27}^\tau l_{LS_2} + R_{23}^\tau l_{CS_6} - G_2 l_{S_2} = 0, \tag{11}$$

$$R_{21}^\tau = (-R_{27}^\tau l_{LS_2} + R_{23}^\tau l_{CS_2} - G_2 l_{S_2})/l_{BS_2} = 0, \tag{12}$$

where $l_{BS_2}$ is the moment arm of $R_{21}^\tau$ about point $S_2$, $l_{LS_2}$ is the moment arm of $R_{27}^\tau$ about point $S_2$, $l_{CS_2}$ is the moment arm of $R_{23}^\tau$ about point $S_2$ and $l_{S_2}$ is the moment arm of $\mathbf{G}_2$ about point $S_2$. Point $S_2$ is an interaction of lines $LK$ and $CD$ (see Fig. 4a).

That is the way how the magnitudes of tangential reaction forces can be calculated.

7. At this stage, the equilibrium of the chain of links $BCDOEFQKL$ should be considered. Since the action and reaction forces have equal magnitudes and opposite directions, the sum of the action and reaction forces equals zero. When the moment arms of gravitation forces, the grip force and reaction forces $R_{21}^\tau$, $R_{48}^\tau$, $R_{68}^\tau$ and $R_{68}^n$ are found, Eq. (13) should be measured. Therefore, taking counterclockwise moments as positive, we obtain:

$$\sum M_{S_{24}} = G_2 l_{S_{24}G_2} + G_3 l_{S_{24}G_3} + G_4 l_{S_{24}G_4}$$
$$- G_5 l_{S_{24}G_5} + G_6 l_{S_{24}G_6} + G_7 l_{S_{24}G_7} - F l_{S_{24}F} \tag{13}$$
$$+ R_{21}^\tau l_{S_{24}B} + R_{48}^\tau l_{S_{24}O} - R_{68}^\tau l_{S_{24}Q} - R_{68}^n l_{S_{24}Q} = 0,$$

which yields

$$R_{68}^n = -(G_2 l_{S_{24}G_2} + G_3 l_{S_{24}G_3} + G_4 l_{S_{24}G_4}$$
$$- G_5 l_{S_{24}G_5} + G_6 l_{S_{24}G_6} + G_7 l_{S_{24}G_7} - F l_{S_{24}F} \tag{14}$$
$$+ R_{21}^\tau l_{S_{24}B} + R_{48}^\tau l_{S_{24}O} - R_{68}^\tau l_{S_{24}Q})/l_{S_{24}Q},$$

where $R_{68}^n$ is the normal reaction force of frame $Q$, $l_{S_{24}G_i}$ is the moment arm of $G_i$ about point $S_{24}$, $l_{S_{24}F}$ is the moment arm of $F$, $l_{S_{24}B}$ is the moment arm of $R_{21}^\tau$, $l_{S_{24}O}$ is the moment arm of $R_{48}^\tau$, $l_{S_{24}Q}$ is the moment arm of $R_{68}^\tau$ about point $S_{24}$.

All others normal reaction forces of each link can be calculated by vector diagrams. At this stage, we proceed to vector diagrams showing the graphic solution for each link. Let's begin drawings of the vector diagram for each link.

The equation of the static equilibrium for links 6, 7, 5, 4, 3, and 2, respectively gives:

$$\overset{n\checkmark}{R_{67}} + \overset{\tau\checkmark}{R_{67}} + \overset{n\checkmark}{R_{68}} + \overset{\tau\checkmark}{R_{68}} + \overset{\circ\checkmark}{G_6} + \overset{n\checkmark}{R_{65}} + \overset{n\checkmark}{R_{65}} = 0, \tag{15}$$

where the first symbol before $\checkmark$ above each vector indicates whether the magnitude is known or unknown respectively; the second symbol before $\checkmark$ or above each vector indicates whether the direction is known or unknown, respectively. $\overset{n\checkmark}{\overset{\circ\checkmark}{R_{67}}}$ means that the direction of the vector is known, but the magnitude is unknown. A two-dimensional vector equation can be easily solved for 2 unknown variables (2 magnitudes, 2 directions, or 1 magnitude and 1 direction) [12].

The similar is true for all other links:

$$\overset{\checkmark\checkmark}{R_{76}} + \overset{\checkmark\checkmark}{G_7} + \overset{\tau\checkmark}{R_{72}} + \overset{n\checkmark}{R_{72}} = 0, \tag{16}$$

$$\overset{\checkmark\checkmark}{R_{56}} + \overset{\checkmark\checkmark}{G_5} + \overset{\tau\checkmark}{R_{54}} + \overset{n\checkmark}{R_{54}} = 0, \tag{17}$$

$$\overset{n\checkmark}{R_{43}} + \overset{\tau\checkmark}{R_{43}} + \overset{\checkmark\checkmark}{R_{45}} + \overset{\checkmark\checkmark}{G_4} + \overset{\tau\checkmark}{R_{48}} + \overset{n\checkmark}{R_{48}} = 0, \tag{18}$$

$$\overset{\checkmark\checkmark}{R_{34}} + \overset{\checkmark\checkmark}{G_3} + \overset{\checkmark\checkmark}{F} + \overset{\tau\checkmark}{R_{32}} + \overset{n\checkmark}{R_{32}} = 0, \tag{19}$$

$$\overset{\checkmark\checkmark}{R_{23}} + \overset{\checkmark\checkmark}{R_{27}} + \overset{\checkmark\checkmark}{G_2} + \overset{\tau\checkmark}{R_{21}} + \overset{n\checkmark}{R_{21}} = 0. \tag{20}$$

The vector diagrams for each link is shown in Fig. 6. Thus, it is possible to calculate full constraint forces in all joints and to figure out the torque, which has to be applied to the input link to get the equilibrium position of the finger mechanism during a pinch mode.

## 4.2 An Example of Calculations

To grasp an object with the weight $G_0$, the magnitude of the grip force $F$ is:

$$F = G_0 2\mu = 1\,kg \cdot 9.8\,m/s^2 \cdot 2 \cdot 0.6 = 8.2\,N, \tag{21}$$

where $\mu$ is the coefficient of friction between surfaces of the distal phalanx and an object, the mass of the object to grasp is 1 kg. We proceed next to a torque applied to the input link. The moment of $R_{21}$ about point $A$:

$$M_1 = R_{21}h_{21} = 15.854\,N \cdot 0.014\,m = 0.22\,Nm, \tag{22}$$

where $h_{21}$ is the moment arm of the $R_{21}$. Thus, the information about the relationships between the torque applied to a driving link and forces applied to surfaces of phalanges are obtained.

### 4.3 Note for the Static Force Analysis of the Finger Mechanism in an Underactuated Mode

The finger mechanism in an underactuated mode has two degrees of freedom. The first one is responsible for the input link *AB*, which drives the proximal phalanx, the second is responsible for the distal phalanx. After the first contact with the object (see Fig. 3b), the degree of freedom is decreasing to 1. Thus, the finger mechanism after the first contact represents a typical four-bar mechanism. A static force analysis for such kind of mechanisms is presented in [12].

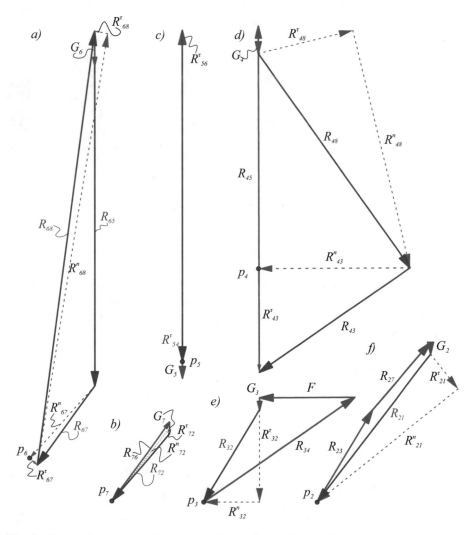

**Fig. 6.** Vector diagrams showing the graphical solution for each link: (a) Link 6, (b) Link 7, (c) Link 5, (d) Link 4, (e) Link 3 and (f) Link 2

## 5   Design of the Gripper

The grasping tool consists of two fingers and a palm. Both fingers are actuated by one worm gear DC motor with encoder and both fingers are equipped with clutches to break/reconnect the kinematic chains of the fingers. The torque of the motor is 0.2 Nm. The main feature of the gripper is the ability to change the mode of the finger mechanism in order to use power or precision grasps. The gripper was designed to perform variable types of grasps to grip objects of variable shapes and sizes using the suggested hybrid structure. Figure 7 shows the gripper performing both kinds of grasps. The sizes of the gripper and the working area are shown in Fig. 8. The gripper opening varies from 0 to 120 mm. The gripper weight is 2 kg. Object diameter for encompassing is 10–100 mm. Max payload in the adaptive grip is 4 kg and 1 kg in the pinch mode. The prototype of the gripper was manufactured via 3D printer from ABS plastic. Bars, which are located inside joints, are made from steel. The gripper has a flange to be connected as an end effector to an industrial robot.

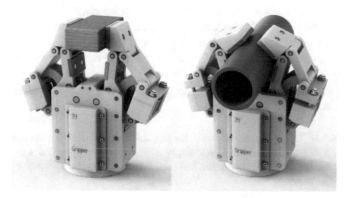

**Fig. 7.** The design of the gripper. The pinch mode is on the left, the underactuated mode is on the right

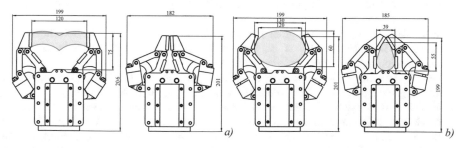

**Fig. 8.** The sizes of the gripper and the working area of the fingers in pinch mode (a) and underactuated mode (b)

## 6    Conclusion

In this paper, a graphical approach of static force analysis of a finger mechanism of versatile bio-inspired industrial gripper is presented. The proposed mechanism represents a complex structure consisting of 8 links. In performing the static analysis, the proposed mechanism can only be separated into an input link and a system of 6 links which cannot be divided into smaller parts. The proposed approach allows examining a static force analysis for such complex mechanisms.

The force analysis is needed to calculate the magnitudes and directions of the constraint forces between the connected links of the mechanism and to understand the relationship between the motion of the input link and the motion of phalanges. According to the obtained magnitudes, adjustments can be made to the geometry of the mechanism to optimize the magnitudes of constraint forces. In order to design a finger mechanism with desired characteristics, it is necessary to know the relationship between the geometry and motions of the parts of a mechanism and the forces that produce these motions. The force analysis was carried out using Assur group theory. The finger, which is class VI Assur mechanism, can be only divided into three parts: the frame, the input link and the class VI group. Since it is impossible to separate the class VI group into smaller parts, the equilibrium of the whole class VI group has been discussed. Such kind of mechanisms is needed to enable the finger to change a DOF in order to switch the mode of grasping through breaking/reconnecting the kinematic chain of the finger. The prototype of the gripper was created in order to apply the proposed mechanism to a real device.

**Acknowledgment.** This work is supported by the Russian Science Foundation grant (project №17-79-20341). The authors would like to express their deepest appreciation to TRA Robotics Ltd. Company for the technical assistance and support of this study.

## References

1. Guillaume, R.: How to choose the right end effector for your application, Robotiq, September 2013. https://blog.robotiq.com/bid/67331/New-Ebook-How-to-Choosethe-Right-End-Effector-for-your-Application
2. Watanabe, T.: Chapter 1 - background: dexterity in robotic manipulation by imitating human beings. In: Watanabe, T., Harada, K., Tada, M. (Eds.) Human Inspired Dexterity in Robotic Manipulation, pp. 1–7. Academic Press (2018)
3. Feix, T., Romero, J., Schmiedmayer, H., Dollar, A.M., Kragic, D.: The grasp taxonomy of human grasp types. Proc. IEEE Trans. Hum.-Mach. Syst. **46**(1), 66–77 (2016)
4. Gonzalez, F., Gosselin, F., Bachta, W.: Analysis of hand contact areas and interaction capabilities during manipulation and exploration. IEEE Trans. Haptics **7**(4), 415–429 (2014)
5. Borisov, I.I., et al.: Versatile gripper as key part for smart factory. In: Proceedings of 2018 IEEE Industrial Cyber-Physical Systems (ICPS), St. Petersburg, Russia, pp. 476–481, May 2018
6. Wang, X.-Q., et al.: Design and control of a coupling mechanism-based prosthetic hand. J. Shanghai Jiaotong Univ. (Sci.) **15**(5), 571–577 (2010)

7. Townsend, W.: The barreth and grasper – programmable flexible part handling and assembly. Ind. Robot: Int. J. Robot. Res. Appl. **27**(3), 181–188 (2000)
8. Sdh servo-electric 3-finger gripping hand (2018). http://www.schunk-modular-robotics.com
9. Lalibert'e, T., Birglen, L., Gosselin, C.: Underactuation in robotic grasping hands. Mach. Intell. Robot. Control **4**(3), 1–11 (2002)
10. Suarez-Escobar, M., Gallego-Sanchez, J.A., Rendon-Velez, E.: Mechanisms for linkage-driven underactuated hand exoskeletons: conceptual design including anatomical and mechanical specifications. Int. J. Interact. Des. Manufact. (IJIDeM) **11**(1), 55–75 (2017)
11. Yoon, D., Choi, Y.: Underactuated finger mechanism using contractible slider-cranks and stackable four-bar linkages. IEEE/ASME Trans. Mechatron. **22**(5), 2046–2057 (2017)
12. Kragten, G.A., Baril, M., Gosselin, C., Herder, J.L.: Stable precision grasps by underactuated grippers. IEEE Trans. Robot. **27**(6), 1056–1066 (2011)
13. Ma, R.R., Spiers, A., Dollar, A.M.: M2 gripper: extending the dexterity of a simple, undcractuated gripper. In: Advances in Reconfigurable Mechanisms and Robots II, pp. 795–805. Springer, Cham (2016)
14. Watanabe, T.: Chapter 7 - hand design — hybrid soft and hard structures based on human fingertips for dexterity. In: Watanabe, T., Harada, K., Tada, M. (Eds.) Human Inspired Dexterity in Robotic Manipulation, pp. 115–147. Academic Press (2018)
15. Uicker, J., Pennock, G., Shigley, J.: Theory of Machines and Mechanisms. Oxford University Press, Oxford (2011)
16. Borisov, I.I., et al.: Novel optimization approach to development of digit mechanism for bio-inspired prosthetic hand. In: Proceedings of the 2018 7th IEEE International Conference on Biomedical Robotics and Biomechatronics (Biorob), pp. 726–731 (2018)
17. Shai, O.: Topological synthesis of all 2D mechanisms through Assur graphs. In: Proceedings of the 2010 International Design Engineering Technical Conferences and Computers and Information in Engineering Conference, American Society of Mechanical Engineers, pp. 1727–1738 (2010)
18. Crossley, F.R.E.: Mechanisms in modern engineering design – a handbook for engineers designer and inventors. In: Artobolevsky, I.I. (Ed.), vols. 1, 2 (Part 1) and 2 (Part 2). (Trans: Russian into English by Weinstein, N.) MIR, Moscow (1976). Mechanism and Machine Theory, vol. 14, Issue 2 (1979)
19. Artobolevsky, I.I.: Mechanisms in Modern Engineering Design, 2814 p. Mir Publ., Moscow (1979)
20. Chu, J., Cao, W.: Systemics of Assur groups with multiple joints. Mech. Mach. Theory **33**(8), 1127–1133 (1998)

# Open Source File System Selection for Remote Sensing Data Operational Storage and Processing

Andrei N. Vinogradov[1]([⊠]), Evgeny P. Kurshev[2], and Sergey Belov[3]

[1] Department of Information Technologies,
Peoples' Friendship University of Russia (RUDN University),
Miklukho-Maklaya str. 6, Moscow 117198, Russia
vinogradov-an@rudn.ru
[2] Ailamazyan Program Systems Institute of RAS (PSI RAS),
Petra-I st. 4a, Veskovo, Pereslavl District, Yaroslavl Region 152021, Russia
epk@epk.botik.ru
[3] Creation and Transfer of Technologies JSC,
Dmitrovskoe Hw. 60A, Moscow 127474, Russia
s.belov@cttgroup.ru

**Abstract.** Significant increase in the number of Earth remote sensing devices requires improvement of ground-based means of automatic stream data processing, which would allow the implementation of a mass remote sensing service in real time. The experience of choosing a freely distributed parallel file system working on the Linux operating system for solving problems of operational storage and processing of remote sensing data (RSD) is presented. RSD processing is organized according to the technology which excludes multiple copying of data between the processing steps and the delivery of the program code to the data. The processed data array is a collection of both target data files with the length of up to tens of gigabytes, and a set of short files associated with them with service data of tens of kilobytes in length. Technology Hadoop is used to implement the complex. To improve data processing performance, it was decided to replace the standard HDFS file system with a more efficient one. As a result of analysis and testing, the parallel OrangeFS file system was chosen. Instead of the previously used AVRO format, HDF5 is used as the internal technological format of data exchange. The issues of optimizing the settings of the data access mechanisms stack are considered, including the mode with dynamic selection of the I/O scheduler. The use of a more advanced parallel file system made it possible to increase the processing speed up to 40% relative to the standard Hadoop file system.

**Keywords:** Earth remote sensing · ERS · Computer cluster · Distributed file system · Parallel file system · Remote direct memory access · RDM · Asynchronous input-output · AIO · I/O scheduler

© Springer Nature Switzerland AG 2020
D. G. Arseniev et al. (Eds.): CPS&C 2019, LNNS 95, pp. 290–304, 2020.
https://doi.org/10.1007/978-3-030-34983-7_28

# 1   Introduction

Earth remote sensing technologies, which primarily included the survey of the Earth's areas with optical-electronic devices (cameras) of high resolution, are an indispensable tool for solving many pressing national economic and defense problems. The main areas of application of remote sensing data are cartography, security, agriculture, monitoring of emergency situations, natural resource management and others.

In recent years, small spacecraft have been used throughout the world to solve remote sensing problems; examples of Russian devices of this class are the Canopus and Aist series [1]. The low cost of development, production and launch of such devices will allow to create groups from tens to hundreds of spacecraft in a short time. In Russia, within the Federal Space Program, by 2025 it is planned to bring the constellation of a remote sensing satellites up to 23 pcs. [2]. Meanwhile, within the developed "Sphere" Federal Target Program it is planned to launch more than 600 spacecrafts, some of which will be used for remote sensing tasks [3]. The presence of such satellite constellations requires the introduction of modern methods of automatic stream processing of remote sensing data that will provide for the possibility of operational remote sensing services to thousands of consumers in real time.

Processing of remote sensing data is a resource-intensive task, requiring the use of high-speed computing, storage and processing of huge amounts of data. The target data stream from the high resolution remote sensing satellite is about 300 GB/day, while the remote sensing data banks contain tens to hundreds of Pb data.

The task of remote sensing data processing is one of the types of BigData processing, since a typical task is to process volumes of data that do not fit in the RAM of the processing computer (server), and the processing itself requires large amounts of computational resources.

# 2   Problem Formulation

For the organization of remote sensing data processing, high-performance computing systems are used, in particular, of cluster type, providing distributed data processing. One of the important tasks of creating such complexes is to optimize the operational storage of and access to data. This task is one of the key tasks to ensure processing performance, data storage reliability, and scalability of the computing complex. As an operating system of high-performance computing systems (nodes), Linux is currently used in many cases. However, the choice of a file system that would provide distributed data storage in the computing complex is not a trivial task. Currently, dozens of distributed file systems that can be used on Linux based computing systems [4] have been developed. Such a high number of developments indirectly confirms the fact that it is impossible to choose the optimal solution "for all occasions". The optimal solution depends on the type and features of the problem.

This paper presents the experience of choosing a freely distributed parallel file system for solving problems of high-speed remote sensing data processing.

**A Typical Task of Remote Sensing Data Operative Storage and Processing**

Typical data constraints in remote sensing tasks are: the size of the array of the source of processed target information ("survey route") is up to 10–12 GB, unpacked up to 100 GB, stored as one or several files. The size of the file with the data of one "scene", i.e., the "frame" of shooting is about 100 MB. The target information is accompanied by service information, including various survey parameters necessary for the correct processing of target data. The service data files are about 32 KB in size, and about 20 service data files fall on one survey route file. Thus, the remote sensing data includes a collection of a large number of both extra-large files (up to 12 GB) and small files (32 KB).

Processing of remote sensing data includes a large number of operations - preliminary, standard, thematic processing. Standard processing includes radiometric calibration, geo-referencing, geometric correction, orthorectification. A number of operations are performed using "reference" information located in the service databases [5].

## 3   Technological Solution

The traditional approach to the remote sensing data processing (see Fig. 1) is to split the processing algorithm into a number of stages, with the processed data array divided into parts before each of those; when completed, it is stitched and transferred to the next stage.

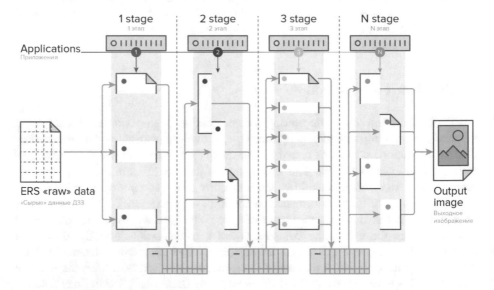

**Fig. 1.** Traditional approach to the remote sensing data processing

In this approach, as practice shows, up to 40% of the processing time is spent on procedures of segmentation, stitching and copying data. To reduce these overheads, we have implemented a technology for processing remote sensing data, in which the processed data is segmented before the first stage of processing, distributed to compute nodes and remain at the same places during all processing steps, while processing software modules and service data files that are significantly smaller than the data being processed. The result is stitched at the last stage before the issuance to the consumer. If necessary, the consumer can receive a piece of data from an intermediate or final stage of processing, without waiting for the formation of the complete array (see Fig. 2).

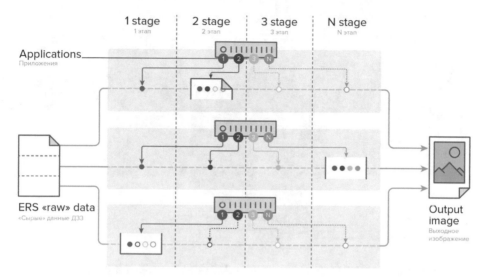

**Fig. 2.** Receiving data from an intermediate stage of processing, without waiting for the complete array

### 3.1   Implementation Experience of Using Hadoop

To implement the described technology in one of the projects of our company [6], the freely distributed technology platform Hadoop and the distributed file system HDFS [7] were used.

This technology provides the implementation of processing using horizontally scalable standard low-cost processing units (Fig. 3). Files containing processed remote sensing data are segmented and placed on various computing nodes of the cluster with a replication ratio (duplication) of at least 3–4 x. Duplication, on the one hand, ensures data integrity in case of failures of computation nodes, and, on the other hand, it facilitates dispatching of the computational processes in order to ensure distributed processing and uniform loading of computational resources.

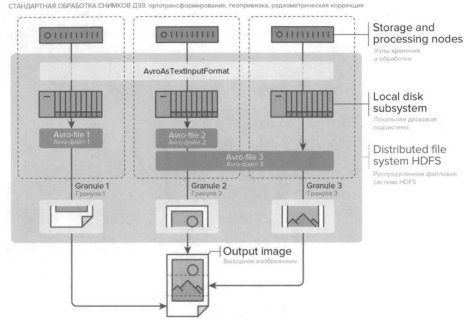

**Fig. 3.** Hadoop cluster using horizontally scalable standard low-cost processing units

For internal technological data exchange, the avro format was used: this is a text-based data exchange format and a library for working with it. It allows you to economically convert the transmitted data into a sequence of bits (serialization), along with a description of their structure in JSON format. This format is more concise than XML.

We applied these technologies in 2014–2015, during the implementation of the ACT processing of remote sensing data of meteorological satellites [8]. In the course of implementing the aforementioned STR, it was possible to ensure an increase in processing productivity by 35–40% as compared to the traditional approach.

The use of a set of software tools Hadoop has provided high manufacturability, convenience, clarity when creating open source software, but the "side effect" of choosing this platform is a certain reduction in data processing performance, which is a "payback" for the universality. In addition, it was confirmed that the inability to edit individual fragments of files, which is a feature of Hadoop technology, significantly affects the decrease in overall performance for the tasks we implement.

To further increase the speed of processing remote sensing data, it was decided to replace the standard HDFS file system with a more productive, compatible with Hadoop [9] one.

## 3.2   Choosing a Parallel File System

To improve the overall processing performance of remote sensing data, it is advisable to use a parallel file system that provides independent parallel data access for various applications, including fast data access for stream processing tasks and efficient access to a large number (tens of thousands) of both large and short files to update them.

File systems that do not provide for the ability to control the placement of file fragments on computing nodes were excluded from consideration, since this would not allow implementing the technology of "delivering" processing programs to data. For example, the POSIX standard does not provide for this capability, but a number of file systems make available special functions for managing data placement.

Based on the analysis of the features of the application area and tasks to be solved, the following criteria were considered when comparing parallel file systems:

- Open source
- Compatibility with Hadoop products, including the ability to control the placement of data on computing nodes
- Processing speed of the aggregate of both very large and small files
- The ability to edit file fragments without overwriting the entire file
- Support for asynchronous I/O (AIO)
- Support for dedicated server metadata.

When choosing a parallel file system, the following most characteristic representatives of this class of file systems were considered: CephFS [10], GlusterFS [11], Luster [12], MoosFS [13], BeegFS [14], and OrangeFS [15].

CephFS, GlusterFS and MooseFS are general-purpose file systems whose main features are high scalability and the ease of administration.

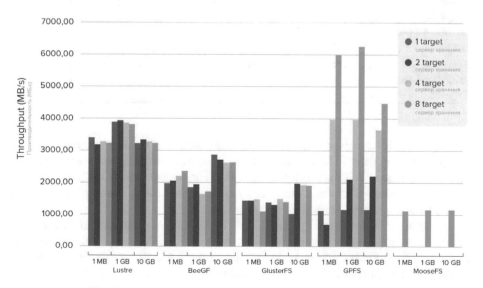

**Fig. 4.** Performance of pre-selected file systems testing results

Luster, OrangeFS, BeegFS are parallel file systems for the use in high-performance computing systems and supercomputers. Their main advantage is the parallel high-speed access in the recording mode and support for high-speed data exchange technologies (Infiniband, etc.) [16].

The results of testing the performance of pre-selected file systems are shown in Fig. 4.

Also, the commercial parallel file system GPFS [17], used in high-performance complexes, was added to the comparative table, although it was not suitable for the application in question because of its high cost and code closeness.

For comparison, the original distributed HDFS file system, originally focused on BigData processing is also shown.

The results of the file system comparison are shown in Table 1.

**Table 1.** Comparative characteristics of distributed file systems

| Feature/file system | HDFS | GPFS | CephFS | GlusterFS | Lustre | MoosFS | BeeGFS | OrangeFS |
|---|---|---|---|---|---|---|---|---|
| Open-source code | Yes | No | Yes | Yes | Yes | Yes | Yes | Yes |
| Hadoop compatibility | Yes | Yes | Yes | Yes | Yes | No | Yes | Yes |
| Large file processing speed | High | High | Avg. | Avg. | High | Avg. | High | High |
| Small file processing speed | Low | High | Low | Avg. | High | Avg. | High | High |
| Ability to change file fragments | No | Yes | Yes | Yes | Yes | Yes | Yes | Yes |
| RDMA support | No | Yes | No | Yes/No* | Yes | No | Yes | Yes |
| AIO support | Avg. | Good | Avg. | Avg. | Good | Low | Average | Good |
| Dedicated Metadata Server Support | Yes | No | Yes | No | Yes | Yes | Yes | Yes |

CephFS, a file system developed by Red Hat, Inc., uses an object storage network that provides both the file and block access interfaces and supports work with very large storages containing petabytes of data. It is compatible with POSIX standards.

The disadvantage is the slower data access speed (both reading and writing) compared to the other file systems we reviewed.

GlusterFS is a distributed, parallel, linearly scalable file system with failure protection. However, this file system does not support the RDMA mode with the required quality; using it, the computing node can access the RAM of another node, which is necessary for high-speed processing of remote sensing data. In addition, in this file system, the metadata server is not supported and in the result of that there are large delays in the processing files introduced by the notification and replication mechanisms. Because of this, in particular, there is a big delay when it is necessary to start processing a file just created by a node on one of other nodes.

Luster is a distributed file system for mass-parallel computing (MPP), typically used for large-scale cluster computing systems. It can support tens of thousands of client systems, tens of petabytes (PBs) of memory for data storage, and I/O throughput of hundreds of gigabytes per second (GB/s). It provides concurrent reading and writing access, and a unified namespace for all files and data in the file system using standard POSIX semantics. Luster uses a dedicated matadata server and efficiently uses the technology of remote direct memory access (RDMA) to compute nodes. High availability is provided by a robust fault tolerance and recovery mechanism that provides for a transparent server reboot during failures.

However, Luster creates high load on the processors, "taking away" the computational resources from the tasks of data processing itself, and in our applications a large amount of computation is required.

MooseFS: Resilient, Distributed Network File System

This file system:

- Supports files of very large size
- Preserves POSIX attributes, i.e., permissions, access time and modifications
- Has a hierarchical structure, i.e., a directory tree
- Supports special files: block, sign systems, sockets and pipes

However, the effectiveness of MooseFS manifests itself in the presence of a large number of data storage servers and the need to ensure quick access to them, while in our case, processing occurs mainly on the same nodes where the data is located.

BeeGFS is a good "working" option that satisfies all the necessary requirements for our task. Testing has shown that in this parallel file system parallel data access is well organized both by means of RW, in which data is loaded into cache memory, and in MMAP technology, in which several processes simultaneously work with the memory where data is cached from disk. It implements asynchronous input/output (AIO) mode, providing for parallel access to data without blocking operations. To work with metadata, a dedicated server is used.

OrangeFS is even more suitable for the reviewed and tested options for solving our problems. The parallel OrangeFS file system has all the functionality of the previous version, with the asynchronous input-output (AIO) mechanism implemented in it more efficiently. In addition, several metadata servers are supported, each of which serves as part of the computing or storage servers, which increases the speed of data processing and expands the scalability.

Thus, the OrangeFS file system was chosen as the most promising parallel file system for processing remote sensing data.

The specified open source file system has the following features:

- The ability to distribute data between file servers
- Metadata processing is provided by several collectively interacting servers
- Support for distributed directories, file search tools for metadata

- Support simultaneous access of multiple clients
- Use of existing local file systems and access methods for storing data and metadata
- Implementation of the file system in user space with the removal of the basic functions in the kernel module
- Support for the Hadoop technology platform (can be used as a direct replacement for HDFS)
- Support MPI technology
- Support for caching the most requested data on the client side
- Existence of a flexible access control mechanism based on access control lists (ACL) and authenticated accounts
- Adjustable level of data storage redundancy for different files, allowing you to find a balance between fault tolerance, performance and hardware costs.

### 3.3 Choosing an Internal Data Exchange Format

The experience of using avro format for internal data exchange has shown that standard "embedded" processing procedures lead to excessive delays. They have to be abandoned and rewritten. Subsequently, for the development of an advanced STR for the processing of remote sensing data for the purposes of internal data exchange, the HDF5 format was chosen [18]; it is designed to store a large amount of digital information. Libraries for working with the format and related utilities are available for use under a free license. The contents of HDF5 files are organized like a hierarchical file system, and paths similar to POSIX syntax are used to access data. Metadata is stored as a set of named attributes of objects.

The format is convenient for the presentation of remote sensing data and ensures their rapid formation and conversion during the exchange between various processing programs.

The format allows organizing parallel access to data for both reading and writing. For our task, the effectiveness of an increase of the number of used parallel data streams to four was experimentally verified.

### 3.4 Configure Data Access

The data access architecture in a distributed computing system includes the following elements:

- Application
- System calls
- Virtual file system
- Parallel file system
- Local file system
- Volume manager (logical)
- Block I/O interface
- I/O driver
- I/O controller (including cache)
- Hard drives.

The data access architecture is shown in Fig. 5.

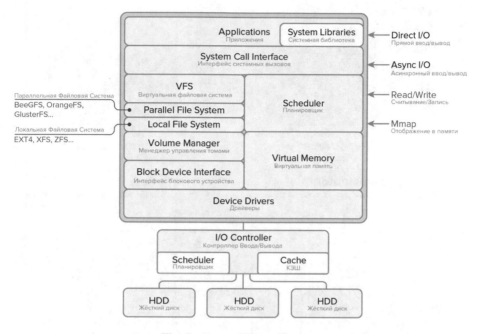

**Fig. 5.** Data access architecture

The correct choice and tuning of the above components significantly affect the efficiency of the entire system, and the loss of productivity in case of inconsistent use of mechanisms at various levels can reach hundreds of percent.

An example of the influence of the choice of scheduler on the speed of execution of I/O operations is shown in Fig. 6, which displays graphs of the time that the data blocks have been in the queue when using various Linux schedulers (Noop, Deadline, Anticipatory, CFQ [19]) with "default" settings.

An example of the results of experiments showing the influence of the HDD connection technology on the speed of performing I/O operations is shown in Tables 2, 3 and 4.

In particular, the analysis of the data shows that when using the Noop and Deadline schedulers, applications are quickly processed by the operating system, but they stay in the queue for the disk for a long time. In this case, it is advisable to use a disk with the least "intelligent" controller, for example, SATA. With a more "intelligent" controller, it makes sense to use a more complex scheduler, e.g., CFQ.

With a single data stream, the simplest schedulers (Noop) are more efficient; with several streams, it is advisable to use the CFQ scheduler, while in case of a very high load Deadline would be preferable.

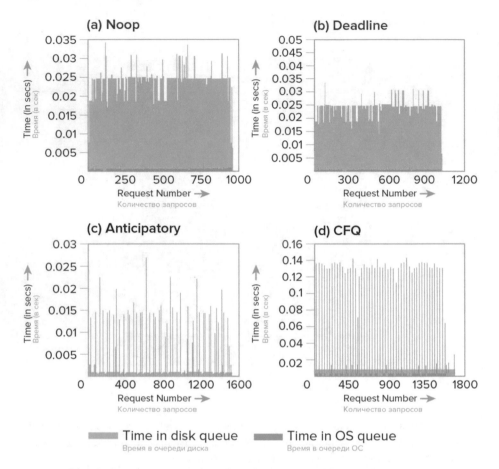

**Fig. 6.** Speed of execution of I/O operations with different scheduler type

**Table 2.** Process Running Time (1 process)

| Disk type | Process | Scheduler type | | | |
|---|---|---|---|---|---|
| | | Anticipatory | cfq | Deadline | Noop |
| SATA | Sequent. read | 4,96 | 4,75 | 4,57 | 4,56 |
| | Sequent. write | 3,38 | 2,55 | 1,65 | 1,52 |
| SCSI | Sequent. read | 6,86 | 5,31 | 7,02 | 7,03 |
| | Sequent. write | 4,02 | 5,33 | 2,70 | 2,53 |

Considering that the used scheduler can be changed during the operation of the computing system, we applied the "dynamic" mode for selecting the type of the scheduler; when the intensity of disk usage is less than 60%, the CFQ scheduler is used, and with a higher disk load, the Deadline one.

**Table 3.** Process Running Time/Sequential read (4 concurrent proc)

| Disk type | Process | Scheduler type | | | |
|---|---|---|---|---|---|
| | | Anticipatory | cfq | Deadline | Noop |
| SATA | Sequent. read 1 | 8,18 | 8,36 | 16,98 | 17,32 |
| | Sequent. read 2 | 8,85 | 8,60 | 17,02 | 17,35 |
| | Sequent. read 3 | 7,32 | 8,37 | 16,98 | 17,25 |
| | Sequent. read 4 | 8,78 | 8,40 | 17,00 | 17,34 |
| SCSI | Sequent. read 1 | 32,15 | 33,50 | 41,25 | 42,44 |
| | Sequent. read 2 | 35,66 | 33,42 | 42,42 | 42,08 |
| | Sequent. read 3 | 35,09 | 33,49 | 41,95 | 41,51 |
| | Sequent. read 4 | 34,88 | 33,85 | 41,65 | 42,65 |

**Table 4.** Process Running Time/Sequential write (4 concurrent proc)

| Disk type | Process | Scheduler type | | | |
|---|---|---|---|---|---|
| | | Anticipatory | cfq | Deadline | Noop |
| SATA | Sequent. write 1 | 9,82 | 9,91 | 9,42 | 9,48 |
| | Sequent. write 2 | 9,73 | 9,78 | 9,72 | 10,22 |
| | Sequent. write 3 | 7,89 | 9,42 | 9,38 | 8,59 |
| | Sequent. write 4 | 9,59 | 9,30 | 7,48 | 8,62 |
| SCSI | Sequent. write 1 | 18,08 | 26,98 | 30,23 | 29,26 |
| | Sequent. write 2 | 27,32 | 31,84 | 30,96 | 30,60 |
| | Sequent. write 3 | 31,06 | 31,81 | 30,22 | 34,27 |
| | Sequent. write 4 | 30,32 | 32,98 | 30,23 | 23,66 |

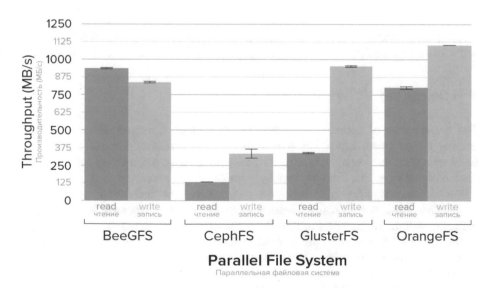

**Fig. 7.** Orthorectification functions performance using different parallel file systems.

The obtained values of the speed of reading and writing a data segment in the course of performing orthorectification functions when using different parallel file systems (CephFS, GlusterFS, BeeGFS, OrangeFS) are shown in Fig. 7. The best results are obtained when using BeeGFS and OrangeFS with a slight advantage of the latter.

## 4 Results

Measurements of the performance in the processing of remote sensing data were carried out on test examples of a computing cluster similar to the one on which the initial development was carried out, including 5 computing nodes with the following characteristics: 2x CPU Intel Xeon E5-2660 v2 (10 cores, 25 MB cache, 2.2 GHz), 128 GB, 10x HDD 900 GB/10K, InfiniBand HCA.

The results of measuring the achieved remote sensing data processing performance are shown in Table 5.

**Table 5.** Remote sensing data processing performance after data access system upgrade

| Processed dataset of Earth remote sensing data | Data size, Gb | Initial version (HDFS) | Upgraded version (OrangeFS) |
|---|---|---|---|
| Data files from type 1 equipment, covering a raster of 230 thousand lines with the width of 1572 pixels in three channels | 3,8 | 160 s | 100 s |
| Data file from type 2 equipment covering a raster in 80–100 thousand lines with the width of 7926 pixels in three channels | 2,0 | 285 s | 175 s |

In general, using the parallel file systems BeeGFS/OrangeFS makes it possible to improve the processing performance of remote sensing data relative to HDFS by 35–40%.

## 5 Conclusion

The organization of optimal data access when creating a remote sensing data processing cluster under Linux OS is considered. The processed remote sensing data includes a great number of both large files containing target information and small files containing service data. Data processing is implemented according to the principle of minimizing segmentation-stitching-copying operations and "delivering" program code to the data. Data storage and processing technology is implemented on the Hadoop platform with the replacement of the HDFS file system with the OrangeFS file system in parallel. HDF5 is used as the internal technological format of data exchange. All system tools used refer to the open source software.

The use of a more advanced parallel file system made it possible to increase the processing speed up to 40% as compared to the standard Hadoop file system.

In general, the use of these technologies and system tools made it possible to increase the speed of processing remote sensing data by more than 2 times comparative to the traditional approach with copying and data segmentation at each stage of processing.

**Acknowledgments.** Authors are grateful to D.N. Golubev for the help in preparing the material. The publication was prepared with the support of the state program AAAA-A19-119020690042-2 "Research and development of data mining methods".

# References

1. Anikeyeva, I.A.: Comparative analysis of space imagery materials obtained from the Kanopus-V and AIST-2D spacecraft. In: Proceedings of the 10th International Scientific and Practical Conference "Geodesy, Mine Surveying, Aerial Photography" [Materialy X mezhdunarodnoj nauchno-prakticheskoj konferencii "Geodezija, Markshejderija, Ajeros'emka"], 14–15 February 2019, Moscow, Russia (2019). (in Russian)
2. Federal Space Program 2016–2025 key points, Osnovnye polozhenija Federal'noj kosmicheskoj programmy 2016–2025. https://www.roscosmos.ru/22347/. (in Russian)
3. The new satellite system "Sphere" will launch in 2022, Kommersant newspaper, 19 July 2018. http://rusletter.com/articles/the_new_satellite_system_sphere_will_launch_in_2022. (in Russian)
4. List of file systems. https://en.wikipedia.org/wiki/List_of_file_systems#Distributed_file_systems
5. Chandra, A., Ghosh, S.: Remote Sensing and Geographical Information System. Narosa Publishing House, New Delhi (2006)
6. Malyshevsky, A.A.: Using the Hadoop ecosystem in the earth remote sensing data processing. In: Proceedings of the National Supercomputer Forum (NSCF 2015) [Sbornik dokladov Nacional'nogo superkomp'yuternogo foruma (NSKF 2015)], 24–27 November 2015, Pereslavl-Zalessky, Russia (2015). (in Russian)
7. Apache Hadoop. Apache software foundation. http://hadoop.apache.org/
8. Moroz, V., Belov, S., Nikonov, O., Ermakov, V.: Automated complex of the meteor-M-2 satellite payload data quality analysis. In: Proceedings of the 18th International Scientific and Technical Conference "FROM IMAGERY TO DIGITAL REALITY: ERS & Photogrammetry", 24–27 September 2018, Crete, Greece (2018). http://conf.racurs.ru/conf2018/eng/program/Conference_Proceedings.pdf
9. Belov, S.A., Smirnov, V.N., Kleyev, A.V., Mikhailyukova, P.G.: Distributed computing environment for solving problems of the earth remote sensing (RSD), Collection of presentations of reports and articles NSKF 2018. [Sbornik prezentacij i statej dokladov NSKF 2018], 27–30 November 2018, Pereslavl-Zalessky, Russia (2018). http://2018.nscf.ru/prezentacii/. (in Russian)
10. Weil, S.A., Brandt, S.A., Miller, E.L., Long, D.D.E., Maltzahn, C.: Ceph: a scalable, high-performance distributed file system. In: Proceedings of the 7th Symposium on Operating Systems Design and Implementation (OSDI), pp. 307–320 (2006)
11. Davies, A., Orsaria, A.: Scale out with GlusterFS. Linux J. **235**, 1 (2013)
12. Sun Microsystems, Inc., Santa Clara, CA, USA, Lustre file system – High performance storage architecture and scalable cluster file system (2007)
13. Konopelko, P.: MooseFS 3.0 User's Manual, 7 January 2017. https://moosefs.com/Content/Downloads/moosefs-3-0-users-manual.pdf

14. BeeGFS Technology Overveiw Brochure. ThinkParQ. Trippstadter Str. 110, Kaiserslautern, Germany. https://www.beegfs.io/docs/BeeGFS_Flyer.pdf
15. OrangeFS Overview. Omnibond Systems LLC (2018). http://orangefs.com/#services
16. Blomer, J.: Experience on File Systems. Which is the best file system for you? CHEP, Okinawa, Japan (2015)
17. General Parallel File System (GPFS) product documentation, IBM Corporation 1990–2017. https://www.ibm.com/support/knowledgecenter/SSFKCN/gpfs_content.html
18. Introduction to HDF5, The HDF Group, 1–15 May 2019. https://portal.hdfgroup.org/display/HDF5/Introduction+to+HDF5+-+PDF
19. Seelam, S., Romero, R., Teller, P.: Enhancements to Linux I/O scheduling. In: Proceedings of the Linux Symposium, Ottawa Linux Symposium, vol. 2, pp. 175–192, July 2005

# Digital Twins of Open Systems

Boris F. Fomin[1](✉), T. L. Kachanova[2], and Oleg B. Fomin[2]

[1] Peter the Great St. Petersburg Polytechnic University, Saint Petersburg, Russia
bfomin@mail.ru
[2] St. Petersburg Electrotechnical University "LETI", Saint Petersburg, Russia

**Abstract.** This report is devoted to a new area of physics of open systems. In this area, methods and technologies were created to provide cognition, scientific understanding and rational explanation of states and properties inherent in open systems. In doing so, these systems can be represented by hundreds, thousands, and tens of thousands of variables. On this basis, a multidimensional knowledge-centric analytics of open systems considered at their natural scale and real complexity has appeared. Presently, the formation of a new cyber-physical paradigm of systems research and development goes on. The report includes a review of possibilities for applying this paradigm to automatic generation of digital twins of open systems in complex subject matter areas.

**Keywords:** Open systems · Physics of open systems · Systems' eigen qualities · Systems states · Reconstructions of systems' states · Digital twins

## 1 Introduction

Physics of open systems (POS) has arisen on the basis of statistical physics and a synergistic paradigm [1, 2]. In POS, open systems are mathematical dynamic models. POS understands openness of systems as a principle, and considers their complexity as a complexity of movement. Theory of complexity is the subject-matter of POS. The main purpose of POS is to scientifically understand the relationship between systems' complexity and laws of nature.

In the middle of the 1990s, a new area of POS appeared, in which formation of a cyber-physical paradigm of open systems' cognition is going on. It must be said that these systems are defined by their empirical descriptions [3, 4]. This report is devoted to this area of POS.

Open systems (natural, anthropogenic, social, and technical ones) are objects of POS. In the process of exploration, POS takes them at their natural scales and real complexity. POS develops scientifically proven knowledge about the system from huge multidimensional sets of semi-structured polymodal heterogeneous data. POS provides scientific understanding and rational explanation of the obtained knowledge, explores its value (correctness, fullness, completeness), and (on the basis of knowledge), investigates the properties, states, and evolutions of open systems [5].

The solving of the general problem of systems' reconstructive analysis becomes the foundation for POS [6–8]. That offeres great opportunities for obtaining scientifically proven knowledge about systems ontology. For POS development, a creation of

© Springer Nature Switzerland AG 2020
D. G. Arseniev et al. (Eds.): CPS&C 2019, LNNS 95, pp. 305–314, 2020.
https://doi.org/10.1007/978-3-030-34983-7_29

systems language was of fundamental importance. This language made it possible to understand the inner systems code manifested in ontological knowledge [9, 10]. And reconstructive analysis of systems obtained (through this language) the status of a scientific theory. The design of systems states has become yet another key consequence of POS, that provides rational explanation of ontological knowledge, and a scientific answer to the question: "… how is ontological knowledge related to key concepts in the real world of systems, such as: variables, states, properties of variables, properties of states, variables' variability, variability of states, and variability of properties?" [9, 11, 12].

These three fundamental milestones of POS (i.e., reconstructive analysis, systems language, and design of systems' states) provided an opportunity to build the digital twins of systems (DTS).

## 2   Unified Technology

A unified technology (UT) of POS is developed. It provides for creating DTS and includes the following parts:

- a system of informational and associated cognitive technologies of POS;
- software for automatic producing and exploiting scientifically proven knowledge about open systems on the basis of POS technologies;
- software for multidimensional knowledge-centric system analytics of POS;
- competences, techniques, and best practices that relate to the usage of scientific methods, technologies, and software tools of POS in research, as well as in analytical and project activities.

## 3   Technological Platform

Creation and testing of a UT of POS were carried out in parallel with developing a technological platform (TP) of POS. R&D works were carried out at different stages of this project for promising applications. In the course of these works, numerous tests and examinations have been performed. At the beginning of the current year, the TP possesses the following opportunities:

- It automatically discovers scientifically proven knowledge about ontology of open systems from a huge amount of multidimensional polymodal heterogeneous empirical data.
- It automatically provides scientific understanding and rational explanation of obtained ontological knowledge about properties, states, and variability mechanisms of open systems that include hundreds and thousands of variables.
- It automatically explores the value of obtained knowledge (its correctness, fullness, completeness, and significance).

- It automatically generates resources of system knowledge (informational, intellectual, cognitive, and technological) that are needed in analytic, research, and projective activity.
- It automatically creates normative documented reports about results of cognition, scientific understanding, and rational explanation of ontological knowledge.
- It automatically presents the obtained knowledge.
- It automatically carries out a subject examination for system knowledge resources, and on this basis, it automatically generates resources for solving specific subject system problems.
- It supports the development of programs that are solvers for general system problems, and also it supports the creation of technological R&D clusters for target areas of research and development.

TP operates according to a unified scenario and covers the full cycle of automatic generating and verifying scientific knowledge about systems. Further, on the basis of obtained knowledge, it forms appropriate DTS, and uses them to solve complex system problems, both scientific and applied ones.

## 4  Generation of System Knowledge

**System in Data.** Empirically observed reality manifests in every system investigated. Having been involved in the process of cognition, it generates ontological knowledge on the basis of general methodological tools of POS. It must be noted that the existing empirical description of a system determines forms and volume of this knowledge [6]. A system in its natural scale, real complexity and real environment is initially represented by "Object – Property" table as a system in data. Every row of the table (object) is an actual single state of this system. Each column of the table (property) is a variable with a unique name, the variable that describes a single property of a system or property of its environment. The number of rows (tens, hundreds, and thousands) defines a representative set sampled from the system's actual states. The number of columns (hundreds, thousands, and tens of thousands) presents a full set of properties describing a system's state, taking into account system environment. A system is characterized by its empirical and statistical portraits. The empirical portrait consists of general external evaluations for the system's initial representation, while the statistical portrait characterizes (integrally) the variability of all variables of the system and its environment.

**System in Relations.** The system as a whole is presented by a connection graph (CG) as a system in relations [6]. All graph vertices are variables of the system, and all edges are paired relationships between the variables. Each relationship is defined independently and is a semantic part of the system being considered as a whole. Each relationship has attributes, such as: sign, force, significance, and complexity. The system given in relations is characterized by its structural portrait. It contains integral evaluations through which the quality of expression forms, and the manifestation fullness of the system's essence can be assessed. This assessment is based on sets of attributes determined for paired relationships, all star graphs and special two-layer sub-graphs of CG, and also for CG as a whole.

**System in Qualities.** The system considered in all its complexity and unity of the whole is presented as system in qualities [6–8]. It can be defined by a full recon-structive set of system models (SMs) and by a full set of models of intrasystem interactions (interaction models IM). Each SM is a model of one certain eigen quality of the system. The system in its every eigen quality is a part of the whole and, simultaneously, is a whole in the context of the part. A variety of SMs presents a full range of the system's eigen qualities. Intrasystem relationships defined for a set of system qualities are represented by models of the IM-family. They describe mecha-nisms forming unity and integrity of the system. The system taken in its qualities expresses scientifically proven knowledge about the system's general ontology and reveals complexity of the system through an ensemble of its unique eigen qualities. Knowledge about the ontology of the system is contained in its system portrait. It includes integral evaluations of the revealed complexity of the system.

**System in Standards.** Scientific understanding of knowledge about the system's ontology is provided by the language of systems [9]. Scientifically understood knowledge contains a vision of the system through standards. In this vision of the system, images of all system's eigen qualities that can be implemented in the actual states are specified [10, 11]. A system's ability to transfer its own senses and actualize them is expressed through absolute evaluations of whether the' system eigen qualities are being shaped and whether they are homogeneous. Such evaluations can be obtained in comparing each SM with the base sample which integrates the best properties inherent in all models of the SM family. Utmost (normative) semantic forms of each eigen quality are given through the standards of system's states. The "Knowledge Essence" and "Fullness of Knowledge" reports include results assessing the extent to which the quality of knowledge about the system as a whole, as well as the volume of knowledge and main aspects of knowledge expressed in standards, have been scien-tifically understood. The "Fullness of knowledge" report represents the opportunity to generate complete knowledge about the system's ontology (evaluations of heteroge-neous essence of the system) and its ability to discover all semantic forms of the system and implement them correctly into empirical facts (evaluations of the multi-qualitative essence of the system). The "Knowledge Essence" report characterizes the ability to convey the understood system's senses outwardly (evaluations of the understood types of qualities), specifies organizational completeness of the system's semantic forms (evaluations of standard states), and establishes the finality of the system's standards verification (evaluations of the semantic carriers).

**System in Models of Standards Implementation Forms.** Semantic figures of facts express a transition from the sense to a fact, and how a sense is actualized in an empirical fact. Each eigen quality of the system is manifested in either its actual state and is represented there by means of some certain standard of the state of this quality. Standards implementations can be blocked by characteristic forms of IM types. Each standard corresponds to a cluster of actual system states, each of which has internalized the sense of this standard. Variables of the cluster objects are assessed through their ability to manifest the sense of a given standard. Each standard is manifested in a fact by a variety of forms of its implementation. The system's sense internalized by the fact is expressed through system's presentation in the models of standards' implementation forms [11, 12]. This presentation of the system can be described by a set of characteristics (standards' clusters; variability areas of implemented standards; multi-differences of system's semantic forms; the extent to which semantic forms have been factually implemented and the potential feasibility of these semantic forms; and the consistency of system and specificsubject senses of the system's variables) that have integral evaluations included in the report dedicated to such presentation of the system.

**System in States.** Any single actual system state is represented by an assemblage of standards. Each assemblage is a reconstruction of a certain state; it is its formal model, a carrier of knowledge about the state and also emergent properties of the system in this state. A reconstruction explains the regularity of joint, coordinated variability of variables of the system's state [5, 11, 12]. Variables in states reconstructions are determined by a certain system mechanisms forming the values' levels of variables and potentials of their variability. Through the states reconstructions, the system as a whole is represented as a system in states. This representation of the system includes a full range of knowledge about the system as a whole, about the system's standards and its states. The system in the states is a completed (shaped) image of the system's essence where:

- Synthesis of both sense and fact is finished and their commensuration is achieved.
- Values of all variables in all states are determined and main mechanisms responsible for forming states and for their mobility are manifested.
- All actual states initially specified by the system in data are described through formal reconstructions of these states.
- Integrally, obtained reconstructions are characterized by the reconstructions number, average of the standards for one state, and by the percentage of variables that are involved in reconstructions; by the quality of modeling real values of variables on the basis of levels of their values, and by classes of mechanisms forming the levels of values of variables and the variability of states, and, finally, by the distribution of reconstructions over specific groups of states and over the order parameters.
- The scientific understanding and rational explanation of actual states are represented by overall evaluations that are final.

# 5   Analysis of the Value of System Knowledge

**Process of Value Estimation.** The value of knowledge about the system's ontology is the subject of study. The process activities of estimating the value create evaluative judgments about all forms of system's representations obtained when system knowledge is being produced. Evaluative judgments provide a comprehensive analysis of the finality of representation forms; evaluate the fullness of transferring senses from one representation form to another one, and investigate truth, determinacy, theoretical and empirical validity of ontological knowledge [9–12]. On their basis, the changes improving forms of system's representation are created and performed, as well as the recommendations for improving the full system context are developed.

**Qualimetry of System Knowledge.** Knowledge value is expressed in evaluative statements [12]. Each statement can be characterized by the following: the object (element of knowledge), type of evaluation (absolute/comparative), basis for evaluation (substantive aspect of object value), and the subject of estimating (method of POS). Absolute evaluation applies to one object, and uses the concept of an ideal (norm). Comparative evaluation applies to at least two objects or two states of a single object. Value assessments have qualitative and quantitative nature. A special scale (i.e., linguistic scale, ordinal (rank), and quantitative (metric) scale) is applied to each aspect of estimation and for each evaluation. All objects estimated on the basis of one and the same aspect are comparable. Evaluations characterize both completeness of the process of estimating value and quality of obtained system knowledge. Own apparatus of preprocessing, scaling, aggregation, and visualization is used by POS to calculate them. Value assessments link the POS's scientific method and its practical application. Emphasis is placed on practice.

**Resources of System Knowledge.** The degree of trust in the results of cognitive processes that produce ontological knowledge is formed by the process of value estimating. This process endows all knowledge elements by attributes of correctness, completeness, fullness, quality, significance, and applicability. Results of the analysis of the knowledge value are presented in reports dedicated to informational, intellectual, and technological resources of system knowledge [12]. Each report contains the evaluative judgments represented by estimation criteria and seems of systems' language. These criteria arise in the result of scaling evaluations related to concepts' qualities, i.e., those concepts that correspond to categories of "Techno-Cubes" of cognitive processes which aim at scientific understanding and rational explanation of obtained ontological knowledge [6, 11]. In the formation of knowledge resources, such characteristics as variability, multiplicity, system predestination of variables, expressiveness of relations structures, structural invariants, and the sense carriage; variables' properties, implementation forms of standards, states, and intrasystem interactions are used; all of them must be estimated and these evaluations should be scaled. Levels of senses' manifestations and their translation together with variables' ranks and SM ranks are assessed through words and semes of the systems' language.

# 6   Digital Twin of System

TP of POS supports collaboration between system analysts and subject experts who are working on forming the multidimensional sets of empirical data needed to create the initial empirical context of the system, the context that corresponds to the requirements of POS. On this basis, TP of POS automatically generates the DTS without resorting to expert knowledge, subjective analysis, and interpretations.

A DTS concept means a complete computer model of the system, a model that is based on scientific knowledge. This model consists from the following components: empirical reality of the system (DTS-reality), a digital twin of eigen qualities of the system (DTS-quale), digital twin of system states (DTS-state), a digital twin of intrasystem interactions (DTS-interactions), and a digital twin of the system taken as a whole (DTS-whole).

**DTS-Reality.** The system in data, system in relations, and also information resource of system knowledge form the basis of the DTS-reality component. This component defines, shapes, and provides scientifically proven knowledge about variables that characterize the system phenomenon as empirical reality. This knowledge is expressed through:

- The sample sets of observable (measured) values of the system's variables and its environment; descriptive statistics of the variables; assessments of representativeness that together with assessments of representativeness of "the typical" and "the special" characterize the ability of variables to manifest the system's complexity in empirical facts.
- Elements of state vector; evaluations of states objectification and evaluations of detecting the states, i.e., the evaluations that specify fullness, representativeness, uniqueness and reproducibility of the system in data.
- Variables participating in attributed binary relations; evaluations of correlativity and evaluations of expression of intrasystem interactions; these evaluations define attributed structures of relations as sources of information about interactions in the system.

**DTS-Quale.** The system in qualities, system in standards, and the intellectual resource of system knowledge form the basis of a DTS-quale component. This component defines, shapes, and provides scientifically proven knowledge about eigen system's qualities considered as independent entities. This knowledge is represented by:

- System roles of variables; semantic activities of variables; evaluation of system conditionality, evaluation of multiplicity of variability forms, and evaluations of system distinguishability, preferences, and ladenness of variables taken as carriers of system senses.
- Full spectrum of eigen system qualities; evaluation of expressiveness of system senses, evaluations of coordinated separability and organizational completeness – all these evaluations characterize correctness, fullness, and completeness of senses' figuration in structural invariants of the system.
- Qualimetric formedness evaluations and homogeneity ones of each SM.
- Invariants of structures and states (behavior stereotypes, standards of states); evaluations of system and subject definiteness, and evaluations for areas of variability and ladenness specify the quality and potential of invariants' verification.

**DTS-State.** The system in models of standards implementation forms, system in states, and also the technological resource of system knowledge form the basis of a DTS-state component. This component defines, shapes, and provides scientifically proven knowledge about reconstructions of actual system states. This knowledge is expressed in:

- Clusters of carriers in which the sense of eigen system qualities is implemented (representativity of clusters, system and subject roles of variables); full volume of implemented sense, evaluation of proximity of sense's carriers to standards, evaluation to what degree the verification's potential is implemented; and evaluation to what extent standards cores of states are homogeneous.
- Reconstructions of actual system states; attributes of variables and also state's standards of eigen system qualities; evaluation of fullness of states description, evaluation of reconstructions volume and explainability evaluation for levels of variables values.

**DTS-Interaction.** The IM-family (interaction types based on singlets, doublets, and triplets) forms the basis of a DTS-interaction component. This component defines, shapes, and provides scientifically proven knowledge about intrasystem mechanisms. This knowledge is expressed through:

- Types of interaction models (singlets, doublets, and triplets) of each SM where this SM participates; evaluations of SM participation in different types of interactions; the environment where the interaction is manifested, together with the environment where the interaction is disseminated.
- Actual forms of interaction types, cleanliness assessments of manifesting forms of interaction, and ways to actualize system mechanisms of interaction for each form (similarity, switching, feedback, blocking, restriction, ability to oscillate, influence of parameters of local and global action, etc.).

**DTS-Whole.** DTS-whole component describes the system as an organized whole. It defines, shapes, and provides scientifically proven knowledge about the system as a whole (when the system is understood as a real-world phenomenon) and evaluates whether this knowledge is full. This knowledge is represented through:

- "Assemblages" of mechanisms of intrasystem interaction that determine variability of values' levels for each order center of the whole system.
- "Assemblages" of mechanisms' models of intrasystem interaction that explain all actual states of the whole system, taking into account its environment.

The cyber-physical paradigm of POS has provided real opportunities to produce DTS automatically, directly from empirical descriptions of open natural, social, anthropogenic, and technical systems with hundreds, thousands, and tens of thousands of variables. TP of POS automatically converts huge multidimensional arrays of empirical data into "smart" models (virtual analogues) of systems, i.e., the models that give complete scientific knowledge about the revealed, understood, and explained complexity of the system to researchers and analysts [5].

# 7 Conclusion

Researchers from different scientific areas, who work with large data sets, can directly apply DTS in order to cognize, understand, and explain (exclusively on a rigorous scientific basis) the organization, states, intrasystem variability mechanisms, and emergent properties of systems being studied [13–17].

Due to POS, the researchers and analysts using advanced methods of data science (machine learning, predictive models, statistical learning, deep learning, neural and probabilistic networks) [18–24], have obtained new multidimensional system analytics which has no analogues. It arose within TP of POS as a result of implementing automatic subject examination of knowledge obtained from DTS, automatic explication of this knowledge onto the levels of subject ontologies and ontologies of subject problems, and also automatic production of informational, intellectual, and technological resources for solving system problems on the basis of DTS and specific-subject knowledge about the system and system problems [5]. At the present time, multidimensional POS-analytics has methods that are the leading in natural system classification, in defining types of system effects of multifactor influences, in identifying events, states, situations, and evolution forecast of open systems, and also in system comparativistics [5, 25–29].

# References

1. Klimantovich, Y.L.: Introduction to Physics of Open Systems. Janus-K, Moscow (2002)
2. Klimantovich, Y.L.: Statistical Theory of Open Systems, Volume 1: A Unified Approach to Kinetic Description of Processes in Active Systems. Kluwer Academic Publishers, Dordrecht (1995)
3. Kachanova, T., Fomin, B.: Foundations of the Systemology of Phenomena. ETU ("LETI") Publishing Center, St. Petersburg (1999)
4. Kachanova, T., Fomin, B.: Physics of systems as a postcybernetic paradigm of systemology. In: International Symposium "Science 2.0 and Expansion of Science: S2ES" in the context of the 14th World-Multi-Conference "Systemics, Cybernetics and Informatics" (WMSCI 2010), Orlando, FL, USA, pp. 244–249 (2010)
5. Kachanova, T., Fomin, B., Fomin, O.: Generating scientifically proven knowledge about ontology of open systems. multidimensional knowledge-centric system analytics. In: Ciza, T. (ed.) Ontology in Information Science, InTech, Rijeka, Croatia, pp. 169–204 (2018)
6. Kachanova, T., Fomin, B.: Technology of System Reconstructions. Politechnika, St. Petersburg (2003)
7. Kachanova, T., Fomin, B.: Physics of open systems: generation of system knowledge. J. Syst. Cybern. Inform. 11(2), 73–82 (2013)
8. Kachanova, T., Fomin, B.: Cognition of ontology of open systems. Procedia Comput. Sci. J. 103, 339–346 (2017)
9. Kachanova, T., Fomin, B.: Introduction to the Language of Systems. Nauka, St. Petersburg (2009)

10. Fomin, B.F., Kachanova, T.L., Turalchuk, K.A., Fomin, O.B.: Scientific understanding of ontological knowledge about open systems that is automatically mined from big data. In: 2018 IEEE Conference on Data Science: Challenges of Digital Transformation (2018 IEEE DSDT), IEEE Russia North West Section, 15 June 2018, Peter the Great St. Petersburg Polytechnic University, St. Petersburg, Russia (2018)

11. Kachanova, T., Fomin, B.: The Methods and Technologies for Generating a Systemic Knowledge: A Manual for Masters and Postgraduate Students. ETU ("LETI") Publishing Center, St. Petersburg (2012)

12. Kachanova, T., Fomin, B.: Qualitology of System Knowledge: A Manual for Masters and Postgraduate Students. ETU ("LETI") Publishing Center, St. Petersburg (2014)

13. Fomin, B., Kachanova, T., Khodachenko, M., et al.: Global system reconstructions of the models of solar activity and related geospheric and biospheric effects. In: Favata, F., Sanz-Forcada, J., Gimenez, A. (eds.) Proceedings of 39th ESLAB Symposium "Trends in Space Science and Cosmic Vision 2020", pp. 73–82, Noordwijk, the Netherlands (2006)

14. Ageev, V., Fomin, B., Fomin, O., Kachanova, T., Chen, C., Spassova, M., Kopylev, L.: Physics of open systems: a new approach to use genomics data in risk assessment. In: Tyshenko, M. (ed.) The Continuum of Health Risk Assessments, InTech, pp. 135–160 (2012)

15. Ageev, V., Fomin, B., Fomin, O., et al.: Physics of open systems: effects of the impact of chemical stressors on differential gene expression. J. Cybern. Syst. Anal. **50**(2), 218–227 (2014)

16. Ageev, V., Araslanov, A., Kachanova, T., Turalchuk, K., Fomin, B., Fomin, O.: Generation of system knowledge on the problems of social tension in Russia's regions. Sci. Tech. Sheets SPbSPU **2–1**(147), 300–308 (2012)

17. Ageev, V., Kachanova, T., Fomin, B., Fomin, O.: Analytical preparation for reengineering of manufacturing the metal products on the basis of system knowledge. Sci. Tech. Sheets SPbSPU **4**(159), 141–155 (2012)

18. Murphy, K.P.: Machine Learning: A Probabilistic Perspective. The MIT Press, Cambridge (2012)

19. Forrester, A., Sobester, A., Keane, A.: Engineering Design via Surrogate Modeling. A Practical Guide. Wiley, New York (2008)

20. Bernstein, A.V., Kuleshov, A.P.: Mathematical methods of metamodeling. In: Works of 3rd International Conference "System Analysis and Information Technologies", Zvenigorod, Russia, pp. 756–768 (2009)

21. Vapnik, V.: The Nature of Statistical Learning Theory. Springer, New-York (2000)

22. Goodfellow, I., Bengio, Y., Courville, A.: Deep Learning. MIT Press, Cambridge (2016)

23. Jensen, F.V., Nielsen, T.D.: Bayesian Networks and Decision Graphs. Springer, New-York (2001)

24. Pearl, J.: Causality: Models, Reasoning, and Inference. Cambridge University Press, New York (2000)

25. Kachanova, T., Turalchuk, K., Fomin, B.: Class reconstruction in the space of natural system classification. Procedia Comput. Sci. **150**, 140–146 (2019)

26. Kachanova, T., Fomin, B.: System ontology of classes. Izvestiya SPbGETU ("LETI") **7**, 25–36 (2015)

27. Kachanova, T., Fomin, B., Turalchuk, K., Ageev, V.: Natural classification of acute poisoning with organophosphorus substances. Izvestiya SPbGETU ("LETI") **8**, 8–17 (2015)

28. Kachanova, T., Fomin, B.: System effects of multifactorial influences in open system. Izvestiya SPbGETU ("LETI") **1**, 28–38 (2017)

29. Kachanova, T., Fomin, B.: Applying the method of determining typology of system effects of multifactorial influences. Izvestiya SPbGETU ("LETI") **2**, 19–29 (2017)

# Synthesis of the Coordinated Control Algorithms for a Biaxial Manipulator

Valeriy Lyubich[✉] and Aron Kurmashev

Peter the Great St. Petersburg Polytechnic University, Saint Petersburg, Russia
valeriy.lyubich@gmail.com

**Abstract.** Objectives of this work were to synthesize a coordinated control algorithm and a reduced coordinated control algorithm for the biaxial manipulator and for a typical (straight line and circular arc) and non-typical (parabola) trajectories. The authors synthesized the coordinated control algorithm (CCA) based on A.D. Kurmashev algorithm and proposed the reduced coordinated control algorithm (RCCA) for a biaxial manipulator model. Two synthesized algorithms and an uncoupled system on different trajectories and different contour speeds were compared using mathematical modeling. Utilizing the coordinated control algorithms leads to an increase in the minimal quality factor and may lead to a decrease in the contour speed error in comparison to the uncoupled system. The synthesized algorithms for the manipulator moving along the typical and non-typical trajectories eliminate contour and contour speed errors better than the uncoupled system. The RCCA allows to control manipulator by using information only from position and speed sensors. Therefore, it is possible to implement the RCCA on existing equipment without any significant modification of it like adding additional sensors, loops, etc.

**Keywords:** Coordinated control algorithm · Biaxial manipulator · Mathematical modelling · Synthesized algorithm

## 1 Introduction

The contour control of a manipulator has a relatively low-quality factor, i.e., the ratio of the contour speed to contour error (see Ref. [1]) because it has non-linear parameters, dynamic moments of resistance, and cross-linkages between the axes. Furthermore, the contour control implements an uncoupled structure [1, 2] (see Fig. 1) with the following drawbacks [1–4]: omission of the contour error – the system minimizes contour error indirectly through minimization of the coordinate errors; omission of a nonlinearity of the plant; it is impossible to choose the ratio of the contour error and contour speed error, hence it is impossible to set control priorities – in many cases, a low contour error is preferable than a low contour speed error.

Systems with the coupled structure (see Fig. 2) do not have these drawbacks. Due to the active coordination of axes [2, 4], they allow to account contour error and nonlinearity of the plant. Furthermore, using the contour regulator allows us to choose control priorities [1, 2].

© Springer Nature Switzerland AG 2020
D. G. Arseniev et al. (Eds.): CPS&C 2019, LNNS 95, pp. 315–325, 2020.
https://doi.org/10.1007/978-3-030-34983-7_30

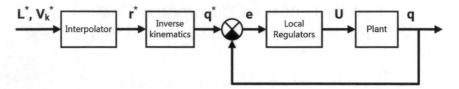

**Fig. 1.** The uncoupled system structure: $L^*$ – the target trajectory; $V_k^*$ – the target velocity; $r^*$ – the target position vector in Cartesian coordinates; $q^*$ – the target position vector in generalized coordinates; $e$ – the error vector; $U$ – the control vector; $q$ – The generalized coordinates vector

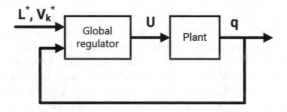

**Fig. 2.** The coupled system structure

Examples of systems with the coupled structure: the Path Precompensation Method [5–7], the Cross-Coupled System [8, 9], methods based on passive velocity field control [1, 2, 10–12], and [1, 13–16]. In this paper, we applied the coordinated control algorithm by A.D. Kurmashev [17], which is hereinafter referred to as the coordinated control algorithm (CCA). We also proposed the reduced coordinated control algorithm (RCCA), which is the CCA with reduced order of the differential equation.

The aim of this study is to synthesize the coordinated control algorithm and the reduced coordinated control algorithm for the biaxial manipulator and for a typical (straight line and circular arc) and non-typical (parabola) trajectories.

## 2   Synthesis of the Control Algorithms for a Biaxial Manipulator

In this chapter, the procedure of synthesizing control algorithms for a biaxial manipulator model (see Fig. 3) is shown.

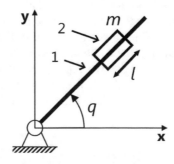

**Fig. 3.** Scheme of the biaxial manipulator model: 1 – the crank, 2 – the slider

The mathematical model of the manipulator looks like:

$$\begin{cases} m*l^2*\ddot{q} - m*g*l*\cos(q) = \tau_1 \\ m*\ddot{l} - m*g*\sin(q) = \tau_2 \end{cases} \tag{1}$$

where $m$ is the mass of the end effector; $q$, $l$ are generalized coordinates; $\tau_1$, $\tau_2$ are axes torques.

Figure 4 shows the manipulator structure scheme. The electro-mechanic part of the robot is represented as a moment source with saturations and gains ($K_{eq}$, $K_{el}$).

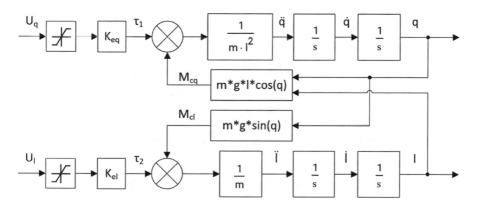

**Fig. 4.** Structural scheme of the manipulator

First step is to transform the mathematic model (1) into the matrix form:

$$\dot{X} = A*X + B*U + F; \tag{2}$$

$$X = \begin{bmatrix} q & l & \dot{q} & \dot{l} \end{bmatrix}^T; \; \dot{X} = \begin{bmatrix} \dot{q} & \dot{l} & \ddot{q} & \ddot{l} \end{bmatrix}^T; \; U = \begin{bmatrix} u_q & u_l \end{bmatrix}^T. \tag{3}$$

Then it is required to choose a set of residuals to form the control algorithms:

1. The first residual is the contour error or the equivalent contour error ($\varepsilon_k$) [18]. This residual stabilizes movements along the target trajectory. The first residual should be zero at the trajectory and have no special points (minimums, maximums, and saddle points) outside the trajectory. Examples of contour errors are:

Contour error for a straight line ($y = kx + b$):

$$\varepsilon_k = \frac{y - k \cdot x - b}{\sqrt{1 + k^2}}. \tag{4}$$

Contour error for a circular arc $((x - x_c)^2 + (y - y_c)^2 = R^2)$:

$$\varepsilon_k = (x - x_c)^2 + (y - y_c)^2 - R^2. \tag{5}$$

Equivalent contour error for a parabola $(y = a \cdot (x - x_c)^2 + b)$:

$$\varepsilon_k^* = a \cdot (x - x_c)^2 + b - y. \tag{6}$$

2. The second residual is the contour speed error (7). This residual stabilizes the contour speed.

$$\varepsilon_{Vk} = V_k^{*2} - V^2. \tag{7}$$

Let's set the dynamics of the residuals:

$$\begin{cases} \ddot{\varepsilon}_k + \alpha_1 \cdot \dot{\varepsilon}_k + \alpha_2 \cdot \varepsilon_k = 0 \\ \dot{\varepsilon}_{Vk} + \beta_1 \cdot \varepsilon_{Vk} = 0 \end{cases} \tag{8}$$

And transform the dynamics equation of the contour error (8) into a matrix form:

$$\dot{\varepsilon} = \Lambda * \varepsilon; \dot{\varepsilon} = \begin{bmatrix} \dot{\varepsilon}_k \\ \ddot{\varepsilon}_k \end{bmatrix}; \Lambda = \begin{bmatrix} 0 & 1 \\ -a_2 & -a_1 \end{bmatrix}. \tag{9}$$

We can find the contour error derivatives as:

$$\dot{\varepsilon} = E * \dot{X}. \tag{10}$$

and the contour speed error derivatives:

$$\dot{\varepsilon}_{Vk} = B_1 * \dot{X}. \tag{11}$$

We synthesize the first control vector $(\overline{U}_k)$ by using Eq. (14) from Reference [17], which minimizes the contour error;

$$B^+ = (B^T * B)^{-1} * B^T; \tag{12}$$

$$E^+ = E^T * (E * E^T)^{-1}; \tag{13}$$

$$U_k = B^+ * [E^+ * \Lambda * \varepsilon - (A * X + F)]; \tag{14}$$

where $B^+$ and $E^+$ are pseudo-inverse matrixes.

We synthesize the second control vector $(\overline{U}_V)$ by using Eq. (16) from Reference [17], which minimizes the contour speed error.

$$B_1^+ = B_1^T * (B_1 * B_1^T)^{-1}; \tag{15}$$

$$U_V = -B^+ * [A * X + F + B_1^+ * \beta_1 * \varepsilon_{Vk}]; \tag{16}$$

where $B^+$ is a pseudo-inverse matrix.

We can compose the CCA:

$$U = U_k + U_V \tag{17}$$

The chosen coordinated control method – as opposed to other coordinated control methods [14, 15] – allows reducing the differential equation order by equating the highest order differentials to zero. This significantly reduces the cost of the algorithm execution and allows doing that on the existing equipment without any significant modifications if it.

We can find the control vector for the proposed algorithm – the RCCA – Eq. (19) by equating the highest order differentials to zero:

$$\begin{cases} \ddot{q} = 0 \\ \ddot{i} = 0 \end{cases} \tag{18}$$

$$U^* = U_k^* - U_V^*. \tag{19}$$

## 3   Experiment

The synthesized algorithms were compared with the reference uncoupled system [19] (US) (see Fig. 5).

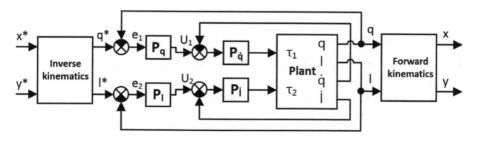

**Fig. 5.** The reference uncoupled system structure

Quantization time, in reality, is usually restricted. To consider that, we added quantization time constant ($T_k = 0.0005$ s) for the US and the coordinated control regulators.

## 3.1   Main Parameters

The plant parameters are as follows:

$$m = 0.5 \, \text{kg}; \, K_{eq} = 1 \, \text{C}; \, K_{el} = 1 \, \text{C}. \tag{20}$$

The saturations limits are:

$$S_{q\_lim} = \pm 100 \, \text{V}; \, S_{l\_lim} = \pm 100 \, \text{V}. \tag{21}$$

The CCA and the RCCA dynamics constants are:

$$T_{ek} = 0.0009 \, \text{s}; \, \xi_{ek} = 2.5; \, T_{evk} = 0.0008 \, \text{s}; \tag{22}$$

$$\alpha_2 = \frac{1}{T_{ek}^2}; \, \alpha_1 = 2 * \frac{\xi_{ek}}{T_{ek}}; \tag{23}$$

$$\beta = \frac{1}{T_{evk}}; \tag{24}$$

where $T_{ek}$ and $T_{evk}$ are the time constants of the residuals transient response (for the contour error and contour speed error respectively); $\xi_{ek}$ is the oscillation constant for the contour error transient response.

## 3.2   Start Conditions

$$\begin{aligned} |V_0| &= 0 \, \text{m/s}, \\ |\xi_{VK}| &\neq 0 \, \text{m/s}, \\ |\xi_K| &= 0 \, \text{m}. \end{aligned} \tag{25}$$

## 3.3   Trajectory Parameters

The coordinates of the start position of the end effector are $(x_0, y_0)$.
$T_{\text{sim}}$ is the simulation time.
The straight line is described by the following parameter values:

$$k = -1; \, b = 2 \, \text{m}; \, x_0 = -4 \, \text{m}; \, y_0 = -6 \, \text{m}; \, T_{\text{sim}} = 8 \, \text{s}. \tag{26}$$

The circular arc is described by the following parameter values:

$$R = 5 \, \text{m}; \, x_c = 2 \, \text{m}; \, y_c = 1 \, \text{m}; \, x_0 = 0 \, \text{m}; \, y_0 = 5.58 \, \text{m}; \, T_{\text{sim}} = 11 \, \text{s}. \tag{27}$$

The parabola is described by the following parameter values:

$$\alpha = 0.07; \; x_c = 1\,\text{m}; \; b = 2\,\text{m}; \; x_0 = -4\,\text{m}; \; y_0 = -3.52\,\text{m}; \; T_{\text{sim}} = 8\,\text{s}. \qquad (28)$$

## 3.4    The Quality Factor Estimation for the Parabola

As we used the equivalent contour error to synthetize the CCA for the parabola, we also needed to estimate the real contour error which we made by using Eq. (29). The estimation was made with the following assumption: the end effector is always within a small neighborhood of the trajectory.

$$\varepsilon_k = \frac{L^*(x_2, y_2)}{\sqrt{\left(\frac{\partial L^*}{\partial x}\right)^2 + \left(\frac{\partial L^*}{\partial y}\right)^2}} = \frac{a * (x - x_c)^2 + b - y}{\sqrt{4 * a^2 * (x - x_c)^2 + 1}}. \qquad (29)$$

where $L^*$ is the trajectory equation in a parametric form; $(x_2, y_2)$ are the coordinates of the end effector position.

## 3.5    The US Regulators Adjusting

Parameters of the US (see Table 1) were adjusted on the contour error local minimum by using the gradient descent algorithm (for contour speed 0.5 m/s).

Table 1.  The US parameters

| Path | P$\dot{q}$ | P$\dot{i}$ | Pq | Pl |
|---|---|---|---|---|
| Line | 19.1 | 10.4 | 1.0 | 1.1 |
| Circle | 24.4 | 23.0 | 1.0 | 1.0 |
| Parabola | 19.6 | 17.2 | 1.0 | 1.1 |

## 3.6    Synthesis of the Coordinated Control Algorithms

Using the Eqs. (20–29) we can synthesize the full-order coordinated control algorithm (17) and the reduced-order coordinated control algorithm (19).

Since optimization of the coordinated control algorithms is not in the priority of the paper, parameters of the algorithms were tuned once at the local minimum (for the line and the contour speed 0.5 m/s) and never changed.

## 3.7    The Assessment Criteria

1. The minimal quality factor on a trajectory is described by the equation:

$$V_{\min} = \frac{V_k^*}{e_{k\_max}}; \qquad (30)$$

where $e_{k\_max}$ is the maximum contour error on a trajectory; $V_k^*$ is the target velocity.
The average quality factor on a trajectory is described by the equation:

$$\langle V \rangle = \frac{V_k}{\langle e_k \rangle};$$  (31)

where $\langle e_k \rangle$ is the root mean square contour error on a trajectory.
The root mean square contour speed error is described by the equation:

$$\langle e_{Vk} \rangle = \sqrt{\frac{\sum_i^N e_{Vk}^2(i)}{N}}.$$  (32)

## 3.8   Experimental Results

The results of the experiment are presented in Table 2.

**Table 2.**  The experimental data

| Path | Vk [m/s] | The US | | | The CCA | | | The RCCA | | |
|---|---|---|---|---|---|---|---|---|---|---|
| | | vmin [1/s] | \<v\> [1/s] | \<evk\> [m/s] | vmin [1/s] | \<v\> [1/s] | \<evk\> [m/s] | vmin [1/s] | \<v\> [1/s] | \<evk\> [m/s] |
| The straight line | 0.3 | 5.60E+00 | 4.89E+05 | 0.07 | 4.55E+03 | 3.36E+06 | 0.02 | 4.63E+03 | 3.36E+06 | 0.02 |
| | 0.5 | 3.07E+01 | 8.15E+05 | 0.12 | 7.59E+03 | 1.06E+07 | 0.02 | 7.63E+03 | 8.21E+06 | 0.02 |
| | 0.7 | 1.41E+01 | 1.14E+06 | 0.17 | 9.37E+03 | 7.92E+06 | 0.03 | 1.07E+04 | 1.16E+07 | 0.03 |
| The circular arc | 0.3 | 6.78E+00 | 1.61E+06 | 0.06 | 6.38E+01 | 3.25E+06 | 0.11 | 6.35E+01 | 3.25E+06 | 0.11 |
| | 0.5 | 1.70E+01 | 2.69E+06 | 0.11 | 5.23E+01 | 7.94E+06 | 0.12 | 5.09E+01 | 7.94E+06 | 0.11 |
| | 0.7 | 2.51E−01 | 3.77E+06 | 0.22 | 4.78E+01 | 1.11E+07 | 0.11 | 4.64E+01 | 1.11E+07 | 0.11 |
| The parabola | 0.3 | 2.85E+01 | 9.18E+05 | 0.07 | 1.54E+03 | 1.15E+07 | 0.01 | 1.54E+03 | 6.27E+06 | 0.01 |
| | 0.5 | 1.06E+01 | 1.53E+06 | 0.12 | 1.68E+03 | 1.15E+07 | 0.02 | 1.68E+03 | 7.18E+06 | 0.02 |
| | 0.7 | 4.40E+00 | 2.14E+06 | 0.17 | 1.21E+03 | 1.67E+07 | 0.02 | 1.21E+03 | 1.15E+07 | 0.02 |

Using the experimental data (see Table 2) we plotted relations of the average
quality factor on the trajectories to contour speed (see Fig. 6).

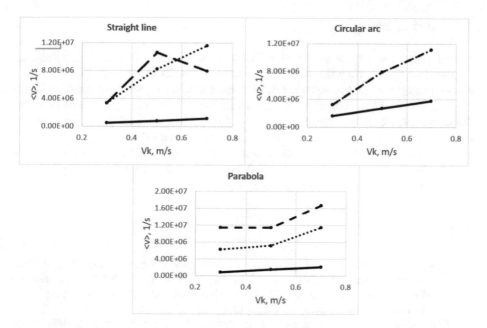

**Fig. 6.** The relations of the average quality factor on the trajectories to the counter speed for the straight line, the circular arc, and the parabola. The solid lines (-) – the US; the dashed lines (—) – the CCA; The dotted lines (…) – the RCCA

### 3.9 The Vector Fields Examples

The parabola vector fields for the control vector components are presented in Fig. 7.

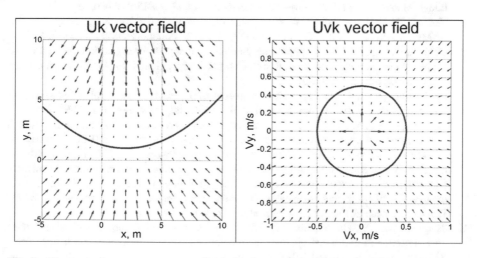

**Fig. 7.** The control components vector fields for the parabola: $U_k$ (on the left) and $U_{vk}$ (on the right)

# 4  Conclusion

In this paper, the authors proposed a RCCA algorithm and described in details the synthesis of the coordinated control algorithms – the CCA and the RCCA – for a biaxial manipulator and three trajectories: a straight line, a circular arc, and a parabola.

The results of the experiment are shown in Table 2 and Fig. 6; it can be seen that the coordinated control algorithms improve the quality factor of the trajectories and reduce the contour speed error as compared to the uncoupled system.

The RCCA algorithm has similar criteria to the CCA algorithm. However, the RCCA algorithm is much cheaper to implement in practice, as it can be implemented with the existing equipment without its modification.

# References

1. Beliaev, A.N., Kurmashev, A.D., Sokolov, O.A.: Zamknutye sistemy CHPU robotami [CNC coupled systems for robots], Voronezhskii Politech. Institut, Voronezh (1989). (in Russian)
2. Shapovalov, A.A., Kurmashev, A.D.: Contouring system of coordinated control of an industrial robot, TUSUR, 1(25), part 2 (2012)
3. Tang, L., Landers, R.G.: Multiaxis contour control – the state of the art. Proc. IEEE Trans. Control Syst. Technol. (2012). https://doi.org/10.1109/tcst.2012.2235179
4. Huo, F., Poo, A.: Precision contouring control of machine tools. Int. J. Adv. Manuf. Technol. (2013). https://doi.org/10.1007/s00170-012-4015-5
5. Huan, J.: Bahnregelung zur Bahnerzeugung an numerisch gesteuerten Werkzeugmaschinen. Dr.-Ing dissertation, Stuttgart Universitaet (1982). https://doi.org/10.1007/978-3-642-45540-7
6. Chen, J., Tsai, H.: A path algorithm for robotic machining. Robot. Comput. Integr. Manuf. 10(3) (1993). https://doi.org/10.1016/0736-5845(93)90054-n
7. Chin, J., Lin, S.: The path precompensation method for flexible arm robot. Robot. Comput.-Integr. Manuf. 13(3) (1997). https://doi.org/10.1016/s0736-5845(97)00003-3
8. Koren, Y.: Cross-coupled biaxial computer control for manufacturing systems. ASME J. Dyn. Syst. Meas. Control 112(2) (1990). https://doi.org/10.1115/1.3149612
9. Chin, J., Cheng, Y., Lin, J.: Improving contour accuracy by Fuzzy-logic enhanced cross-coupled precompensation method. Robot. Comput.-Integr. Manuf. 20 (1) (2004). https://doi.org/10.1016/j.rcim.2003.06.001
10. Li, P.Y., Horowitz, R.: Passive velocity field control (PVFC): part II—application to contour following. IEEE Trans. Autom. Control 46(9) (2001). https://doi.org/10.1109/9.948464
11. Chen, C., Cheng, M., Wu, J., Su, K.: Passivity-based contour following control design with virtual plant disturbance compensation. In: Proceedings of the 11th IEEE International Workshop on Advanced Motion Control (2010). https://doi.org/10.1109/amc.2010.5464041
12. Lyubich, V.K., Kurmashev, A.D.: Soglasovannoe upravlenie programmnym dvizheniem manipulyatora po traektorii, zadannoj kubicheskim splajnom [The coordinated control of the manipulator program movement along a cubic spline trajectory]. In: Proceedings of the Week of Science 2017, ICST, St. Petersburg (2017). (in Russian)
13. Zeng, D., Gao, Y., Hu, Y., Liu, J.: Optimization control for the coordinated system of an ultra-supercritical unit based on stair-like predictive control algorithm. Control Eng. Pract. 82 (2019). https://doi.org/10.1016/j.conengprac.2018.10.001

14. Boichuk, L.M.: Metody strukturnogo sinteza nelineinyh sistem avtomaticheskogo upravleniya [Structure synthesis methods for nonlinear control systems], Energia, Moscow (1971). (in Russian)
15. Galiullin, A.S.: Postroenie sistem programmnogo dvizheniya [Synthesis of program movement systems], Nauka, Moscow (1971). (in Russian)
16. Spur, G., Duelen, G., Wendt, W.: Improvement of the dynamic accuracy of the continuous control paths. Robot. Comput.-Integr. Manuf. **1**(1) (1984). https://doi.org/10.1016/0736-5845(84)90082-6
17. Kurmashev, A.D.: Povyshenie tochnosti i skorosti vosproizvedeniya programmnyh dvizhenij promyshlennymi robotami [An industrial robot program movement accuracy increase], Dissertation, Leningrad Polytechnic Institute, Leningrad (1990). (in Russian)
18. Chen, S., Wu, K.: Contouring control of smooth paths for multiaxis motion systems based on equivalent errors. IEEE Trans. Control Syst. Technol. **15**(6) (2007). https://doi.org/10.1109/tcst.2007.899719
19. Lyubich, V.K.: Razrabotka algoritmov programmnogo upravleniya elektroprivodami dvuhzvennogo manipulyatora s uprugimi swyazyami [Synthesis of a program control for the flexible biaxial manipulator], Master thesis, SPbTU, St. Petersburg (2016). (in Russian)

# Multiple Parameters Estimation of Ship Roll Motion Model with Gyrostabilizer

Mikhail A. Kakanov[1(✉)], Fatimat B. Karashaeva[1], Oleg I. Borisov[1,2], and Vladislav S. Gromov[1,2]

[1] Faculty of Control Systems and Robotics, ITMO University,
St. Petersburg, Russia
makakanov@itmo.ru
[2] Center for Technologies in Robotics and Mechatronics Components,
Innopolis University, Innopolis, Russia

**Abstract.** This paper addresses a problem of parametric identification in marine applications, namely, surface vessels equipped with gyrostabilizers. In this work three approaches are considered: the gradient method, Kalman filter-based method and dynamic regressor extension and mixing (DREM) method. All of them are applied to a ship roll motion model with unknown parameters. For simulation of the system, two types of input signals are injected to test sensitivity of the each approach to the persistency of excitation conditions. Simulation results show advantages of the DREM method in the considered application, in particular, with regard to the convergence rate of the multiple parameters estimation.

**Keywords:** Marine applications · Gyrostabilizers · Gradient method · Kalman filter-based method · Dynamic regressor extension and mixing method · DREM · Ship roll motion model · Simulating

## 1 Introduction

One of the crucial control problems in marine applications is stabilization of the roll motion. External disturbances due to waves and current affecting the surface vessel might cause discomfort to passengers, reduce effectiveness of the crew, and also damage the cargo. Cancellation of these disturbances and stabilization of the roll motion allows not only to improve on-board conditions, but also to enhance the course keeping and the fuel efficiency. One of the approaches to the control of the ship roll motion is to use such devices as gyrostabilizers [1, 2]. Their implementation does not cause any additional hydrodynamic effects and that creates the convenience of such solution.

In order to cancel external disturbances and stabilize the roll motion, one needs to identify the unknown parameters of the relevant dynamical equations describing this process. The parametric identification problem in the case of offshore vessels is typically solved by using the Kalman filter [3]. This method, however, requires the persistency of excitation (PE) condition, which means that the given bounded function $\omega$ satisfies this condition, if at any time there exist $T > 0$, $\gamma > 0$ such that

© Springer Nature Switzerland AG 2020
D. G. Arseniev et al. (Eds.): CPS&C 2019, LNNS 95, pp. 326–334, 2020.
https://doi.org/10.1007/978-3-030-34983-7_31

$$\int_{t}^{t+T} \omega(\tau)\omega(\tau)^T d\tau \geq \gamma\, I, \tag{1}$$

where $I$ is the identity matrix.

This restriction can be mitigated by means of the dynamic regressor extension and mixing (DREM) method proposed in references [4] and [5] and used in particular applications of chaotic oscillators in reference [6] and a robotic arm in [7]. To complete the picture, the classical gradient-based approach is also considered in this work. The paper provides a comprehensive comparison of these three identification methods applied to the ship roll motion model with gyrostabilizer.

In this paper, the addressed problem is formulated in Sect. 2. The considered plant model is parametrized in Sect. 3. All three identification methods, which are the gradient-based approach, advanced Kalman filter and DREM, are outlined in Sect. 4. Their application to a parametrized model of the ship roll motion with a gyrostabilizer is illustrated in Sect. 5. Finally, there is a conclusion with a summary of results.

## 2    Problem Formulation

Let's consider a linearized model of the ship roll motion with a gyrostabilizer [8]

$$I_\phi \ddot{\phi} + B_\phi \dot{\phi} + C_\phi \phi = \tau_w - nK_g \dot{\alpha}, \tag{2}$$

$$I_\alpha \ddot{\alpha} + B_\alpha \dot{\alpha} + C_\alpha \alpha = \tau_p + K_g \dot{\phi}, \tag{3}$$

where $\phi$ is the roll angle, $\alpha$ is the precession angle, $\tau_w$ is the wave-induced roll torque, $\tau_p$ is the precession control torque; $I_{(\cdot)}$, $B_{(\cdot)}$, $C_{(\cdot)}$ are physical parameters of the roll and precession dynamics, namely, the total moment of inertia, damping and restoring coefficients, respectively; $K_g$ is the angular momentum and $n$ is the number of spinning wheels. This linearized model is valid as long as the precession angle satisfies $|\alpha| \leq 1$.

The goal of this paper is to design three parameter estimators, each based on a specific method, namely, the gradient-based approach, advanced Kalman filter and DREM, and ensure the convergence of the estimation error to zero as

$$\lim_{t\to\infty} \left\| \theta - \hat{\theta}(t) \right\| = 0, \tag{4}$$

where $\theta$ is a vector of unknown parameters, $\hat{\theta}(t)$ is a vector of the estimates.

The available signals are the roll angle $\phi$ and precession angle $\alpha$. All the parameters of the model (2)–(3) are considered as unknown.

## 3   Model Parametrization

At the first step, we need to parametrize the original model and derive a regression model comprised of available signals and unknown constant parameters; this is carried out in this section.

Applying the Laplace transform to Eqs. (2) and (3) we obtain

$$I_\phi s^2 \phi(s) + B_\phi s\dot{\phi}(s) + C_\phi \phi(s) = \tau_w(s) - nK_g s\alpha(s), \tag{5}$$

$$I_\alpha s^2 \alpha(s) + B_\alpha s\alpha(s) + C_\alpha \alpha(s) = \tau_p(s) + K_g s\phi(s). \tag{6}$$

Besides the control torque $\tau_p$, the only measurable signals are $\phi$ and $\alpha$. Their first and second derivatives needed for the estimation procedure are not available. Therefore, we introduce an auxiliary linear filter of the form

$$\xi(s) = \frac{\lambda^k}{(s+\lambda)^k}, \tag{7}$$

with $k = 2$ and, multiplying both sides of the Eqs. (5), (6), we obtain

$$\frac{\lambda^2}{(s+\lambda)^2}\left[I_\phi s^2 \phi(s) + B_\phi s\dot{\phi}(s) + C_\phi \phi(s)\right] = \frac{\lambda^2}{(s+\lambda)^2}\left[\tau_w(s) - nK_g s\alpha(s)\right],$$

$$\frac{\lambda^2}{(s+\lambda)^2}\left[I_\alpha s^2 \alpha(s) + B_\alpha s\alpha(s) + C_\alpha \alpha(s)\right] = \frac{\lambda^2}{(s+\lambda)^2}\left[\tau_p(s) + K_g s\phi(s)\right],$$

which we rewrite in a compact form

$$I_\phi s^2 \xi_\phi(s) + B_\phi s\xi_\phi(s) + C_\phi \xi_\phi(s) = \xi_{\tau_w}(s) - nK_g s\xi_\alpha(s),$$

$$I_\alpha s^2 \xi_\alpha(s) + B_\alpha s\xi_\alpha(s) + C_\alpha \xi_\alpha(s) = \xi_{\tau_p}(s) + K_g s\xi_\phi(s),$$

with new notations given as

$$\xi_\phi(s) = \frac{\lambda^2}{(s+\lambda)^2}\phi(s), \quad \xi_\alpha(s) = \frac{\lambda^2}{(s+\lambda)^2}\alpha(s),$$

$$\xi_{\tau_p}(s) = \frac{\lambda^2}{(s+\lambda)^2}\tau_p(s), \quad \xi_{\tau_w}(s) = \frac{\lambda^2}{(s+\lambda)^2}\tau_w(s).$$

Performing the inverse Laplace transform and merging all the terms into the matrix representation, we derive the following linear regression model

$$\ddot{\psi} = \omega^T \theta, \tag{8}$$

where

$$\psi = \begin{bmatrix} \xi_\phi \\ \xi_\alpha \end{bmatrix}, \quad \omega = \begin{bmatrix} \omega_\phi \\ \omega_\alpha \end{bmatrix}, \quad \theta = \begin{bmatrix} \theta_\phi \\ \theta_\alpha \end{bmatrix},$$

$$\omega_\phi = \begin{bmatrix} \xi_\phi \ \dot{\xi}_\phi \ \dot{\xi}_\alpha \ \xi_{\tau_w} \end{bmatrix}, \quad \theta_\phi^T = \begin{bmatrix} -\dfrac{C_\phi}{I_\phi} \ \dfrac{B_\phi}{I_\phi} - \dfrac{nK_g}{I_\phi} \ \dfrac{1}{I_\phi} \end{bmatrix},$$

$$\omega_\alpha = \begin{bmatrix} \xi_\alpha \ \dot{\xi}_\alpha \ \dot{\xi}_\phi \ \xi_{\tau_p} \end{bmatrix}, \quad \theta_\alpha^T = \begin{bmatrix} -\dfrac{C_\alpha}{I_\alpha} \ \dfrac{B_\alpha}{I_\alpha} \ \dfrac{K_g}{I_\alpha} \ \dfrac{1}{I_\alpha} \end{bmatrix}.$$

All signals included in the regressor $\omega$ of the model (8) are available, except for the wave-induced roll moment $\tau_w$. Let's assume that the estimation procedure is carried out under disturbance-free conditions, when $\tau_w = 0$. The next section provides three identification approaches addressed within this study in application to the derived regression model (8).

## 4    Parameter Estimators Design

To estimate the unknown plant parameters by measuring only the input and output signals without performing differentiation operation, we analyzed gradient algorithm, advanced Kalman filter and DREM method.

### 4.1    Gradient Method

Let's consider the quadratic criterion in the continuous form

$$J\left(t, \hat{\theta}\right) = \frac{1}{2}\left(\psi(t) - \omega^T(t)\hat{\theta}(t)\right)^2. \tag{9}$$

The idea of the gradient method is to move along the direction opposite to gradient $\nabla_{\hat{\theta}} J$ minimizing the criterion (9). Then the parameter estimator based on this approach can be defined as

$$\dot{\hat{\theta}} = \gamma\omega(t)\left(\psi(t) - \omega^T(t)\hat{\theta}(t)\right), \tag{10}$$

Where $\gamma$ is the scalar gain coefficient.

The regressor bounded by the estimation law (10) provides asymptotic convergence of the error signal to zero. However, it does not guarantee the convergence of the estimates to the true values of the parameters. That, in its turn, could be ensured if the corresponding regressor satisfies the PE condition (1).

## 4.2   Advanced Kalman Filter

The Kalman filter is a well-known tool in a wide range of control applications. It is often used for the state estimation in vessels. This approach has gained wide popularity due to its ability to filter noise, predict and restore unmeasured states. On the basis of this approach the classical gradient method is modified to advance the parametric identification.

Let's consider the following representation of the estimator based on the advanced Kalman filter

$$\dot{\hat{\theta}} = -\hat{\Gamma}\omega(\ddot{\xi} - \omega^T\hat{\theta}), \tag{11}$$

$$\dot{\hat{\Gamma}} = -\hat{\Gamma}\omega\omega^T\hat{\Gamma} + \beta\hat{\Gamma}, \tag{12}$$

$$k_0 I \geq \hat{\Gamma}(0) = \hat{\Gamma}(0)^T > 0, \tag{13}$$

where $\hat{\Gamma}$ is the matrix of the gain coefficients; $\beta > 0$ is the regularization coefficient; $k_0 > 0$ is the initial value for the matrix of the gain coefficients.

There also are other variations of estimators based on this approach. The particular version given above ensures a relatively high rate of the convergence (see details in references [9] and 10]). Advantages of this algorithm are the simplicity of the initial setup and noise immunity.

## 4.3   DREM

Let's introduce additional $m - 1$ linear $L_\infty$ - stable delay operators $d_m$ with $m$ being the dimension of the regressor $\omega$. This is needed to expand the initial regressor (8). As a result, we have filtered signals

$$\omega_{dm}(t) = \omega_m(t - d_m), \quad \ddot{\xi}_{dm}(t) = \ddot{\xi}_m(t - d_m),$$

which could be combined as

$$\Omega(t) = \begin{bmatrix} \omega^T(t) \\ \omega_{d1}^T(t) \\ \omega_{d2}^T(t) \\ \vdots \\ \omega_{d(m-1)}^T(t) \end{bmatrix}, \quad \Xi(t) = \begin{bmatrix} \ddot{\xi}(t) \\ \ddot{\xi}_{d1}(t) \\ \ddot{\xi}_{d2}(t) \\ \vdots \\ \ddot{\xi}_{d1(m-1)}(t) \end{bmatrix},$$

The estimation law based on the DREM method is of the form

$$\dot{\hat{\theta}} = K \det\{\Omega\}(\text{adj}\{\Omega\}\Xi - \det\{\Omega\}\hat{\theta}), \tag{14}$$

where $\Omega$ is the extended regressor, $\Xi$ is the extended vector of the variables $\ddot{\xi}$, $K > 0$ is the scalar gain coefficient.

This estimator ensures the non-strictly monotonic convergence of the estimates to the true values of the parameters according to Eq. (4). In addition, if the regressor satisfies the PE condition (1), this convergency is exponentially fast.

In the next section, we will proceed with the application of the estimators presented above to the considered ship roll motion model and then compare their performance.

## 5 Application to Ship Roll Motion Model with Gyrostabilizer

Let us show how all the presented estimators work in the application to the ship roll motion with a gyrostabilizer. Let's consider the model (2)–(3) with the parameters

$$I_\phi = 3426300, \ C_\phi = 2.9634e + 06, \ B_\phi = 6.3729e + 05,$$

$$I_\alpha = 548000, \ C_\alpha = 472000, \ B_\alpha = 337000, \ K_g = 520600, \ n = 2.$$

The simulation is carried out without taking into account the wave roll, i.e., $\tau_w = 0$ Actuation of the system is made by means of the gyrostabilizer. Two types of input signals for $\tau_p$, namely, a sine wave and a step, are injected in order to show the sensitivity of the each estimator to the PE condition.

The adaptation coefficient of the gradient-based estimator (10) is chosen as $\gamma = 0.5e + 12$. For the estimator based on the advanced Kalman filter (11)–(13), the following initial parameters are chosen as

$$\beta = 1, \ k_0 = 1.$$

For the DREM-based estimator (14), the following delay operators are chosen as

$$d_1 = 0.1, \ d_2 = 0.2, \ d_3 = 0.3.$$

The gain is set as $K = 10$. As the simulation shows, the filter output has a fairly low magnitude, which complicates the estimation. Therefore, these signals are gained by an additional factor $k_\xi = 8e + 7$ in order to provide a magnitude suitable for the transient process.

### 5.1 Sine Wave Input

For the first run we feed the input with a sinusoidal signal $\tau_p = \sin(t) + \cos(t)$ that meets the PE condition (1). The simulation results are shown in Figs. 1 and 2.

### 5.2 Step Input

For the second run we feed the input with a step signal with the initial time $t = 0.3$ and the final value $\tau_{pf} = 1$. The simulation results are shown in Figs. 3 and 4.

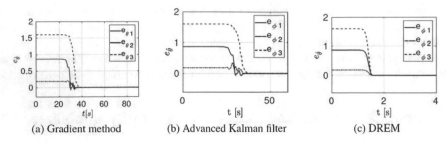

(a) Gradient method    (b) Advanced Kalman filter    (c) DREM

**Fig. 1.** The error signals of the vessel parameters estimation with the PE condition

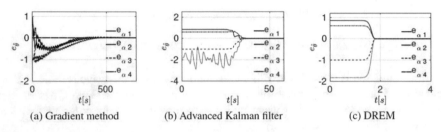

(a) Gradient method    (b) Advanced Kalman filter    (c) DREM

**Fig. 2.** The error signals of the gyrostabilizer parameters estimation with the PE

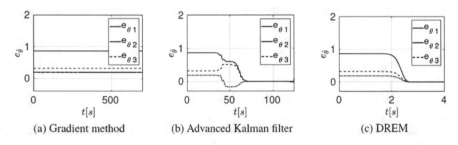

(a) Gradient method    (b) Advanced Kalman filter    (c) DREM

**Fig. 3.** The error signals of the vessel parameters estimation without the PE condition

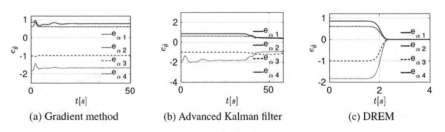

(a) Gradient method    (b) Advanced Kalman filter    (c) DREM

**Fig. 4.** The error signals of the gyrostabilizer parameters estimation without the PE condition

**Table 1.** Evaluation results

| Signal | Standard deviation | | | Settling time | | |
|---|---|---|---|---|---|---|
| | Grad. M. | Adv. Kalman F. | DREM | Grad. M. | Adv. Kalman F. | DREM |
| $\hat{\theta}_{\phi 1}$ | 0.3650 | 0.1113 | 0.0950 | 242.44 | 33.79 | 1.49 |
| $\hat{\theta}_{\phi 2}$ | 0.0865 | 0.0552 | 0.0204 | 314.03 | 34.29 | 1.49 |
| $\hat{\theta}_{\phi 3}$ | 0.7043 | 0.3653 | 0.1756 | 344.12 | 35.97 | 1.49 |
| $\hat{\theta}_{\alpha 1}$ | 0.2193 | 0.4049 | 0.0499 | 337.44 | 32.82 | 1.73 |
| $\hat{\theta}_{\alpha 2}$ | 0.2571 | 0.2991 | 0.0356 | 415.46 | 36.01 | 1.73 |
| $\hat{\theta}_{\alpha 3}$ | 0.3646 | 0.4734 | 0.0579 | 307.22 | 34.14 | 1.73 |
| $\hat{\theta}_{\alpha 4}$ | 4.8020e−07 | 9.6958e−07 | 1.0563e−07 | 365.34 | 34.37 | 1.73 |

## 5.3   Discussion

The results of the simulation runs are shown in Figs. 1, 2, 3 and 4. The quantitative evaluation is given in Table 1. As expected, the DREM method provides plots of the multiple parameters estimation smoother and faster than other methods. The gradient method reveals its dependence on the PE conditions. The advanced Kalman filter ensures the convergence of the estimates to zero in the absence of the PE condition; however, it ultimately loses stability. A detailed proof of the convergence of the estimates to zero by the advanced Kalman filter with PE condition one can find in reference [9].

## 6   Conclusion

Three estimators based on different approaches, namely, the gradient method, Kalman filter-based method and dynamic regressor extension and mixing (DREM) method, are applied to the ship roll motion model with gyrostabilizer. Comprehensive simulation runs with two types of the input signals are carried out to reveal the performance of the each approach. The obtained results are discussed together with the quantitative evaluation for all three approaches provided in the paper. As expected, the DREM method has shown its advantages in the multiple parameter estimation with respect to the closest analogues in the particular application to the ship roll motion with gyrostabilizer.

**Acknowledgments.** This work is supported by the Russian Science Foundation (project 19-19-00403).

## References

1. Townsend, N.C., Shenoi, R.A.: Control strategies for marine gyrostabilizers. IEEE J. Oceanic Eng. **39**(2), 243–255 (2014)

2. Haghighi, H., Jahed-Motlagh, M.R.: Ship roll stabilization via sliding mode control and gyrostabilizer. Bul. Inst. Polit. Iasi, LVIII (2012)
3. Fossen, T.I.: Handbook of Marine Craft Hydrodynamics and Motion Control. Wiley, Hoboken (2011)
4. Aranovskiy, S., Bobtsov, A., Ortega, R., Pyrkin, A.: Performance enhancement of parameter estimators via dynamic regressor extension and mixing. IEEE Trans. Autom. Control **62**(7), 3546–3550 (2017). https://doi.org/10.1109/TAC.2016.2614889
5. Aranovskiy, S., Bobtsov, A., Ortega, R., Pyrkin, A.: Improved transients in multiple frequencies estimation via dynamic regressor extension and mixing. J. IFAC-PapersOnLine **49**(13), 99–104 (2016)
6. Gromov, V.S., et al.: The drem approach for chaotic oscillators parameter estimation with improved performance. J. IFAC-PapersOnLine, **50**(1), 7027–7031 (2017). http://www.sciencedirect.com/science/article/pii/S2405896317318815
7. Borisov, O.I., et al.: Adaptive tracking of a multi-sinusoidal signal with drem-based parameters estimation. J. IFAC-PapersOnLine **50**(1), 4282–4287 (2017)
8. Perez, T., Steinmann, P.: Analysis of ship roll gyro stabiliser control. IFAC Proc. Volumes **42**, 310–315 (2009)
9. Andrievsky, B., Fradkov, A., Stotsky, A.: Shunt compensation for indirect sliding-mode adaptive control. IFAC Proc. **29**(1), 5132–5137 (1996)
10. Gawthrop, P.: Continuous-time self-tuning control – a unified approach. IFAC Proc. **20**(2), 19–24 (1987)

# Comparison of Analytical BP-FBP and Algebraic SART-SIRT Image Reconstruction Methods in Computed Tomography for the Oil Measurement System

Lotfi Zarour[1]($\boxtimes$) and Galina F. Malykhina[1,2]

[1] Peter the Great St. Petersburg Polytechnic University, St. Petersburg, Russia
lotfi.zarour.92@gmail.com, g_f_malychina@mail.ru
[2] Russian State Scientific Center for Robotics and Technical Cybernetics,
St. Petersburg, Russia

**Abstract.** An imbalance between the produced oil entering the pipeline and the oil received by consumers is a real problem. To solve the problem, we propose to use methods of computed tomography. The article is devoted to investigating methods for reconstructing a section of a pipeline to determine time intervals over which no gas inclusions in the oil flow occur. Computed tomography is based on image reconstruction methods. We made comparison between analytical reconstruction techniques: Back Projection (BP) and Filtered Back Projection (FBP) and iterative reconstruction techniques: Simultaneous Algebraic Reconstruction Technique (SART) and Simultaneous Iterative Reconstruction Technique (SIRT); the simulation was performed using Astratoolbox, an open source image reconstruction tool for tomography, and then the reconstructed images were compared using the relative root mean square error and a conclusion was achieved. The results demonstrate that the SIRT and SART method have given the closest to each other reconstructed images.

**Keywords:** Oil imbalance · Industry computed tomography · Simultaneous algebraic reconstruction · Simultaneous iterative reconstruction

## 1 Introduction

Tomography is a standard tool for medical analysis and diagnosis; however, nowadays, tomography is used in industrial applications [5], especially if the interstitial part of the object under analysis is inaccessible [14]. Such are the internal flow structures in petroleum industries.

Computed tomography (CT) based on gamma rays and x-rays consists of two essential steps: at the first step, the object is scanned with gamma and x rays, while which the rays undergo an attenuation due to the absorption of the material [12] of the scanned object; then this attenuation will be detected by a sensor: this is called information collection. The second step consists of reconstructing the structure of the object with the means of suitable reconstruction techniques. Image reconstruction from the data collected during the scan can be done using several techniques, including

© Springer Nature Switzerland AG 2020
D. G. Arseniev et al. (Eds.): CPS&C 2019, LNNS 95, pp. 335–343, 2020.
https://doi.org/10.1007/978-3-030-34983-7_32

analytical ones, such as Back Projection (BP) and Filtered Back Projection (FBP), as well as algebraic techniques, such as Simultaneous Algebraic Reconstruction Technique (SART) and Simultaneous Iterative Reconstruction Technique (SIRT).

In this paper, we will present the results of a comparison among these different tomographic reconstruction techniques.

## 2  Analytic Reconstruction Methods

Analytical reconstruction consists of obtaining an image by inverting the Radon transform [13]. Operator $R_f$ which expresses the projections of $f$ is called Radon transform. In a precise way, at point $(u, \theta)$ it is expressed by the equation:

$$R_f(u, \theta) = p(u, \theta) \tag{1}$$

with:

$$p(u, \theta) = \int_{D_\theta} f(u \cos \theta - v \sin \theta, u \sin \theta + v \cos \theta) \mathrm{d}v \tag{2}$$

If we present the values of $p_\theta(u)$ in an axial plane and $u$, for example, by grayscale, we obtain a sinogram [8]. This sinogram, i.e., a set of sinusoids, is not the desired image, and it is not visually interpretable by a human. The desired images are calculated from the projections. It is, therefore, a reconstruction problem that has to be solved.

### 2.1  Back-Projection

The principle of back-projection consists in assigning the value $p_\theta(u)$ to any point placed on the projection ray that gave this value, then summing over all contributions from all projections. The back-projection of all projections is defined by operator $B_p$, known as the back-projection operator, obtained by summing over all angles:

$$B_p(x, y) = \int_0^\pi (x \cos \theta + y \sin \theta) \mathrm{d}\theta \tag{3}$$

In this case, the resulting image is not the desired image, but a blurred version of this image, as is shown in the results and discussion section.

## 2.2 Filtered Back-Projection

Filtered back-projection is based on the following mathematical expressions:

$$P(\Phi, \theta) = \int\limits_{-\infty}^{+\infty} p(u, \theta) \exp(-j2\pi\Phi u)\mathrm{d}u \tag{4}$$

$$f(x, y) = \int\limits_{0}^{\pi} \left( \int\limits_{-\infty}^{+\infty} P(\Phi, \theta)|\Phi| \exp(j2\pi\Phi u)\mathrm{d}\Phi \right) (x\cos\theta + y\sin\theta)\mathrm{d}\theta \tag{5}$$

These equations express f as the back-projection of the projections filtered by the ramp filter $|\Phi|$. This result is obtained by representing f as an inverse Fourier transform of its Fourier transform $P(\Phi, \theta)$.

# 3 Iterative Reconstruction Methods

## 3.1 Simultaneous Algebraic Reconstruction Technique (SART)

SART is a basic iterative method designed to solve the linear system in the image reconstruction [9]. The algorithm starts with an arbitrary $f^0$, and then begins to converge through the iteration [1].

$$f_j^{k+1} = f_j^k + \frac{\lambda}{\sum\limits_{i=1}^{M} a_{i,j}} \sum\limits_{j=1}^{M} a_{i,j} \frac{b_i - \sum\limits_{l=1}^{M} a_{i,l} f_l^k}{\sum\limits_{l=1}^{M} a_{i,l}}. \tag{6}$$

Where $k$ is the iteration number, $a_{i,j}$ is the probability of a photon emitted in the pixel of image $j$ to be detected by detector $i$ [6]. The value of the detection probability $a_{i,j}$ is under direction of physical phenomena, such as geometric efficiency of the collimator, the attenuation provided by the material surrounding the radioactive pixel $j$ and detection efficiency of the imaging system [2].

## 3.2 Simultaneous Iterative Reconstruction Technique (SIRT)

SIRT is a commonly used reconstruction algorithm that tends to be more stable in case of noisy projections [7]. It consists of correcting each $f_i$ by using all the rays passing through pixel $x_i$; the equation for estimating $f^{(k)}$ by correcting $f^{(k-1)}$ is as follows:

$$f_i^k = f_i^{k-1} + \frac{\sum\limits_j P_j}{\sum\limits_j \sum\limits_i R_{i,j}} \frac{\sum\limits_j R_i f^{k-1}}{\sum\limits_j \|R_j\|^2} \tag{7}$$

The summation relating to all the indices $j$ that the beam $j$ crosses pixel $x_i$. Normalization coefficient $\|R_j\|^2$ is the norm of the line of the matrix $R$ corresponding to $j$, i.e., in the simplest case, it is equal to the number of pixels crossed by the beam $j$ [11]. In brief, the beam $j$ allows to correct all pixels it passes through.

## 4 Tools and Evaluation Method

A phantom was created containing different shapes and grey scale intensities, representing horizontal section of a multiphasic tube containing liquid with gas bubbles. The phantom image of size $256 \times 256$ pixels is projected into radon space that we saw in the second section. Phantom image is used as input for the test simulation (see Fig. 1).

**Fig. 1.** Phantom image used as input for the test simulation

In order to evaluate the image reconstruction algorithms, a relative root mean squared error (RRMSE) criterion was calculated between the original and the reconstructed images. RRMSE ($\Delta$) is the square root of average of the square root of all the errors divided by the average value of measured data; it provides the mean magnitude of the error [10].

$$\Delta = \sqrt{\frac{1}{n} \frac{\sum\limits_{i=1}^{n} (f_i - \hat{f}_i)}{\sum\limits_{i=1}^{n} \hat{f}_i}}, \tag{8}$$

where $f, \hat{f}$ are the original and restored images respectively.

The algorithms were implemented and compared in the MATLAB environment using the ASTRA Toolbox package [3, 4].

# 5 Results and Discussion

This study was set out to compare different tomographic image reconstruction techniques, such as BP, FBP, SART and SIRT; the results are presented in this section. For the different algorithms mentioned above, the simulation was performed with projection numbers of 16, 32, 72 and 180 for a coverage angle ranging from 0° to 180°, while the remaining 180° were only a mirror of the other half, since the direction in which the rays pass through the material does not matter.

The results obtained revealed that for the BP technique, the reconstructed images are highly blurred, although the improvement in the quality of the reconstructed image depends on the number of projections as it is shown in Fig. 2.

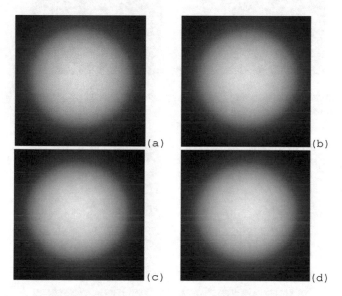

**Fig. 2.** Images reconstructed using BP (a: 16 projections, b: 32 projections, c: 72 projections, d: 180 projections)

This blur is due to the overlapping of the Fourier transform of images in low-frequency segments.

To fix the blur issue, a filter is used; the choice of this filter will not be discussed in this article.

In Fig. 3 the results of the reconstruction using the FBP method can be seen; we can notice at once that images reconstructed from a low projection number are affected by artefacts that decrease the quality of reconstruction; however, we can claim that the quality from 72 projections is acceptable.

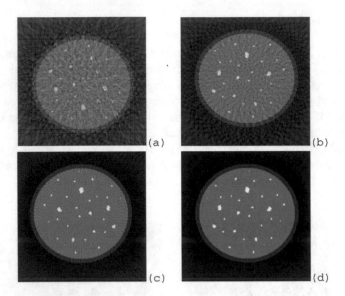

**Fig. 3.** Images reconstructed using FBP (a: 16 projections, b: 32 projections, c: 72 projections, d: 180 projections)

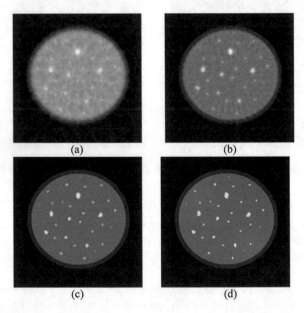

**Fig. 4.** Images reconstructed using SART (a: 16 projections, b: 32 projections, c: 72 projections, d: 180 projections)

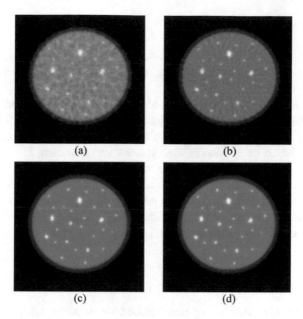

(a)                              (b)

(c)                              (d)

**Fig. 5.** Images reconstructed using SIRT (a: 16 projections, b: 32 projections, c: 72 projections, d: 180 projections)

**Table 1.** Projection number and RRMSE

| Projection number | RRMS error | | |
|---|---|---|---|
| | FBP | SART | SIRT |
| 16 | 0.85 | 0.19 | 0.21 |
| 32 | 0.81 | 0.16 | 0.17 |
| 72 | 0.59 | 012 | 0.12 |
| 180 | 0.39 | 0.11 | 0.099 |

In algebraic techniques, the number of iterations was set at 200 for SART and SIRT in order to focus on the effect of the projection numbers in these techniques; the simulation revealed a major observation that despite the lower number of projections (16 and 32), the images reconstructed by SART and SIRT, as can be seen in "Figs. 4" and "5" respectively, are clearly good; we also notice the absence of artifacts generated when using analytical methods (FBP).

Quality evaluation presented in Table 1 consolidates our previous analysis where the smaller the RRMSE was, the closer the reconstructed image was to the original image.

# 6 Conclusion

The application of industrial computed tomography in the measurement of two-phase oil and gas flows allows to get accurate oil and gas flow measurement results without direct access to the flow.

The study of tomography algorithms showed that the SART and SIRT methods allow us to obtain a smaller RRMSE in image reconstruction using fewer projections than in the BP-FBP analytical methods. This allows us to conclude that the SART and SIRT methods should be applied in the systems for measuring the production of oil wells tomography systems, which will be our next work.

To improve the performance of measuring instrument, we can use a smaller number of projections with an acceptable deterioration in the quality of restoration. With a small number of projections, the SART and SIRT methods have a more significant advantage over the BP and FBP methods.

# References

1. Andersen, A.: Simultaneous algebraic reconstruction technique (SART): a superior implementation of the ART algorithm. Ultrason. Imaging **6**(1), 81–94 (1984)
2. Boudjelal, A., Messali, Z., Elmoataz, A., Attallah, B.: Improved simultaneous algebraic reconstruction technique algorithm for positron-emission tomography image reconstruction via minimizing the fast total variation. J. Med. Imaging Radiat. Sci. **48**(4), 385–393 (2017)
3. Aarle, W.V., Palenstijn, W.J., Beenhouwer, J.D., Altantzis, T., Bals, S., Batenburg, K.J., Sijbers, J.: The ASTRA toolbox: a platform for advanced algorithm development in electron tomography. Ultramicroscopy **157**, 35–47 (2015)
4. Aarle, W.V., Palenstijn, W.J., Cant, J., Janssens, E., Bleichrodt, F., Dabravolski, A., Beenhouwer, J.D., Batenburg, K.J., Sijbers, J.: Fast and flexible X-ray tomography using the ASTRA toolbox. Opt. Express **24**(22), 25129–25147 (2016). https://doi.org/10.1364/OE.24.025129
5. Kalaga, D.V., Kulkarni, A.V., Acharya, R., Kumar, U., Singh, G., Joshi, J.B.: Some industrial applications of gamma-ray tomography. J. Taiwan Inst. Chem. Eng. **40**(6), 602–612 (2009)
6. Abdullah, J., Cassanello, M.C.F., Dudukovic, M.P., Dyakowski, T., Hamada, M.M., Jin, J. H.: Industrial Process Gamma Tomography, IAEA, Vienna, Austria (2008)
7. Banjak, H., Grenier, T., Epicier, T., Koneti, S., Roiban, L., Gay, A.-S., Magnin, I., Peyrin, F., Maxim, V.: Evaluation of noise and blur effects with SIRT-FISTA-TV reconstruction algorithm: application to fast environmental transmission electron tomography. Ultramicroscopy **189**, 109–123 (2018)
8. Chetih, N., Messali, Z.: Tomographic image reconstruction using filtered back projection (FBP) and algebraic reconstruction technique (ART). In: Proceedings of the 2015 3rd International Conference on Control, Engineering & Information Technology (CEIT) (2015)
9. Rit, S., Sarrut, D., Desbat, L.: Comparison of analytic and algebraic methods for motion-compensated cone-beam CT reconstruction of the thorax. Proc. IEEE Trans. Med. Imaging **28**(10), 1513–1525 (2009)
10. Vijayalakshmi, G., Vindhya, P.: Comparison of algebraic reconstruction methods in computed tomography. Int. J. Comput. Sci. Inf. Technol. **5**, 6007–6009 (2014)

11. Aarle, W.V., Batenburg, K.J., Gompel, G.V., Casteele, E.V.D., Sijbers, J.: Super-resolution for computed tomography based on discrete tomography. IEEE Trans. Image Process. **23**(3), 1181–1193 (2014)
12. Johansen, G.: Gamma-ray tomography. Ind. Tomogr. 197–222 (2015)
13. Askari, M., Taheri, A., Larijani, M.M., Movafeghi, A.: Industrial gamma computed tomography using high aspect ratio scintillator detectors (A Geant4 simulation). Nucl. Instrum. Methods Phys. Res. Sect. A **923**, 109–117 (2019)
14. Mesquita, C.H.D., Carvalho, D.V.D.S., Kirita, R., Vasquez, P.A.S., Hamada, M.M.: Gas–liquid distribution in a bubble column using industrial gamma-ray computed tomography. Radiat. Phys. Chem. **95**, 396–400 (2014)

# Control of Solar PV/Wind Hybrid Energy System in Grid-Connected and Islanded Mode of Operation

Anatoli L. Loginov[1], Bekbol Mukhambedyarov[2(✉)],
Dmitry V. Lukichev[2], and Nikolay L. Polyuga[2]

[1] Peter the Great St. Petersburg Polytechnic University, St. Petersburg, Russia
[2] ITMO University, St. Petersburg, Russia
spase_line93@mail.ru

**Abstract.** Hybrid energy system which includes photovoltaic (PV) arrays and wind turbine with synchronous generator (WT/SG) is considered in this paper. The structure of the system was designed according to the most popular and efficient scheme. It includes generation sources, a DC bus, power converters, a storage battery, and a load and stiff grid. Maximum power point tracking algorithms (MPPT) were developed to increase power generation of the PV array and wind turbine. MPPT algorithm based on fuzzy logic controller shows efficient performance for PV arrays. A eprturb and observe algorithm gives an opportunity to achieve maximum power from the wind turbine. Boost converter and active rectifier (AR) were used for power conversion with the PV arrays and wind turbine accordingly. Novel control strategy for active rectifier was introduced in the article. Here, an active rectifier was implemented with synchronous generator behavior, and this approach is called a virtual synchronous machine (VSM). The concept of virtual synchronous machine is an alternative method of grid feeding. The model of a VSM can emulate properties, such as damping and inertia. The virtual synchronous machine allows smooth synchronization in a grid-tied mode and shows high operation speed and accuracy. In this paper, an energy system is designed for 10 kW and all essential points, such as reliability, optimal control strategy and high efficiency are inherited in our system. Modelling was held in MATLAB/Simulink software package.

**Keywords:** Power converter · Active rectifier · PV array · Wind turbine · Hybrid system · Control system · Power plant · MPPT algorithm · Fuzzy logic · Distributed generation · Microgrid · Power electronics

## 1 Introduction

Nowadays development of renewable energy sources increases, and this trend carries a great influence on the implementation and spread of such types of energy sources in industry and everyday life [1]. For instance, the European Union developed a plan which states that 20% of all generated energy will come from renewable sources by 2020 [2]. Today, many countries begin modifying their usual energy system to a distributed energy system structure [3]. Microgrid implementation between small

© Springer Nature Switzerland AG 2020
D. G. Arseniev et al. (Eds.): CPS&C 2019, LNNS 95, pp. 344–359, 2020.
https://doi.org/10.1007/978-3-030-34983-7_33

sources, load and grid with a reliable control strategy is one of the new technologies in the intelligent energy generation [4].

There is a lot of people who have no access to centralized generation sources in Russia and the Commonwealth of Independent States. Such areas require tons of fuel resources like diesel or gasoline because people use local fuel power plants. That's why is the demand to enable people with high-quality and efficient energy generated by renewable sources. Taking advantage of renewable energy sources and power electronics, we can design hybrid power energy system to meet the demands.

Hybrid energy system is the system which includes a lot of power devices for energy converting to meet customers' needs. Often such systems are used in autonomous mode and usually have one renewable energy source at least. Also, hybrid energy systems may operate as an energy distribution system. Up to now, distributed generation is developing very rapidly due to the microgrid and smart grid concepts [5]. The concept of microgrid gives an opportunity to generate high-efficient and available energy due to operation modes and control strategies. It has many beneficial aspects of using distributed generation. Usually, a microgrid acts in a grid-tied mode, but it can operate in an autonomous islanded mode [6]. One of the essential points is that when a stiff grid is disconnected because of a failure, the crash and emergency microgrid switches in an islanded mode and provides consumers with accumulated energy. The main advantages of a microgird are its redundancy, reliability and modularity.

There are some schemes of hybrid energy system discussed in [7] and [8]. The most popular schemes are with an AC bus and with a DC link bus. Using AC buses leads to more complicated control strategies and higher financial expenses though it does not provide for higher efficiency [9]. Systems with generation concentrated in the DC bus are discussed in [10] and [11]. In this case, a storage battery is used for energy accumulation. Micro sources are connected with power converters to interact with the whole system. Power converters play an important role in energy management and control strategy [12]. One of the efficient power converters for PV arrays is DC/DC boost converter which is discussed in [13]. Active rectifier is used for power conversion from the wind turbine and synchronous generator to the DC link bus. A new approach to the DC/AC converter control strategy is called virtual synchronous machine [14].

In this paper, authors consider energy management and control strategy of a hybrid energy system based on solar PV and wind turbines. The paper discusses design of the whole system and each component in particular. The main purpose is to achieve a smooth synchronization with the grid, energy balance abidance, short transient time and accuracy.

## 2 System Description and Modelling

A schematic of a hybrid power plant is shown in Fig. 1. All units in this system can be controlled separately, which means that these units can work independently and we can configure them, so that it gives the opportunity to achieve the optimal operation mode. In this case, the load has supply from a common active rectifier. The ability to control the active rectifier allows to transmit high-quality and reliable energy to the consumer. Additional units can be added to the system and this is one of the main advantages of this topology. Each unit consists of a power source and power converter which are connected to a DC bus. The power flows to the load through an active rectifier that converts the DC to AC. PV arrays and a wind turbine are used in our case as power sources. Also, a buffer energy storage, such as a battery, is connected to the DC bus for energy accumulation. On the other hand, the load can consume energy from the stiff grid. The advantages of this schematic are: control of power flow, optimal control strategies of power converters, modularity and software support.

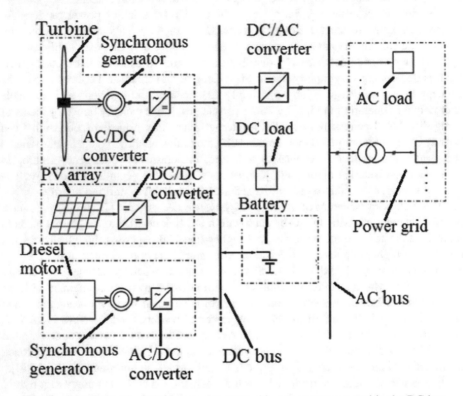

**Fig. 1.** Schematic of a hybrid energy system with energy concentrated in the DC bus

## 2.1    Modelling of a Photovoltaic Module

A photovoltaic cell is a semiconductor heterostructure, which has one p-n junction that appears on the boundaries of two semiconductor materials of p and n types accordingly. This device works on the principle of photoelectric effect. When the light falls onto PV cell photons, it "knocks out" electrons from the crystal lattice, so an electron-hole pair is formed. The charge carries are moving freely under the influence of the electric field and the electromotive force appears on terminals of a PV cell. A solar panel equivalent circuit is shown in Fig. 2.

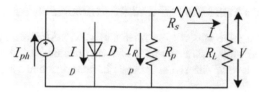

**Fig. 2.** A solar panel equivalent circuit

The expression describing the output current and voltage according to the scheme [15] is:

$$I = I_{ph} - I_{sat}\left\{\exp\left[\frac{q(V + IR_s)}{AkT}\right] - 1\right\} - \frac{V + IR_S}{R_p},$$

where $I_{ph}$ is the photocurrent, $I_{sat}$ is the diode saturation current, $R_s$ is the series resistance, $R_p$ is the shunt resistance, $V$ is the output voltage, $I$ is the output current, $q$ is the electron charge, $A$ is the ideality factor, $k$ is the Boltzmann's constant, $T$ is the ambient temperature.

The PV cell's photocurrent depends on the temperature and solar irradiance described in [16]. Therefore, according to the expression above, the output current and voltage depend on the solar irradiation and temperature. In addition, we can find relationship between the output current and voltage, and consequently, between the power and voltage (see Fig. 3) of the solar panel. We consider 5 kW power generation at 600 W/m$^2$ irradiation from a PV array.

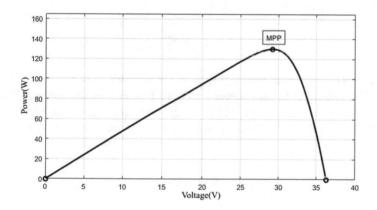

**Fig. 3.** P-V characteristics

It can be seen from the characteristics that the maximum power is reached at a certain voltage, and this point is the called maximum power point (MPP) [17]. Further on, we can use it for the maximum power borrowing from the PV array. A model of photovoltaic cell was chosen from the MATLAB/Simulink standard library.

## 2.2    Modelling of a Wind Turbine and Synchronous Generator

A model of a wind turbine was chosen from the MATLAB/Simulink standard library. Nominal mechanical output power was 7 kW at the rated wind speed of 5 m/s and rated rotation frequency of 785 rad/s. Power characteristics are shown in Fig. 4.

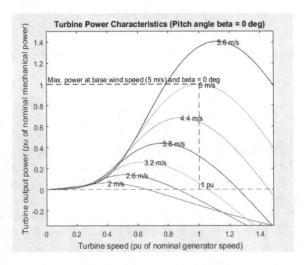

**Fig. 4.** Power characteristics

As we can see, a certain wind speed causes the maximum power generation. This gives the opportunity to implement the maximum power point tracking algorithm. The power of the wind turbine is descripted as

$$P_m = c_p(\lambda, \beta) \frac{\rho A}{2} v_{wind}^3,$$

where $P_m$ is the mechanical output power of the turbine, $c_p$ is the performance coefficient of the turbine, $\lambda$ is the tip speed ratio of the rotor blade, $\beta$ is the blade pitch angle, $\rho$ is the air density, A is the turbine swept area, and $v_{wind}$ is the wind speed. The wind turbine reaches its maximum power when $\lambda$ is optimal. The rotation speed is given as

$$\omega_t = \lambda_{opt} \frac{V}{R}.$$

It is obvious that the optimal rotation speed depends on the wind speed and optimal coefficient. Figure 5 shows a graph of the rotation speed versus the wind speed.

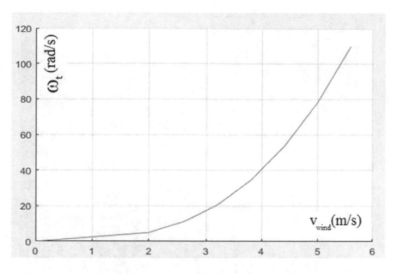

**Fig. 5.** Rotation speed versus wind speed

## 2.3    Power Control System

### 2.3.1    Photovoltaic Control System

A power unit consists of a PV array and a boost converter [18]. The maximum power point is reached at a certain voltage, and we need to adjust the boost converter output voltage to achieve the maximum generation. We suggest a fuzzy logic controller (FLC) to control the output voltage. The FLC has two input and one output variables in our case. The input variables are the error and change in error which are calculated as follows:

$$E(k) = \frac{P(k) - P(k-1)}{V(k) - V(k-1)},$$

$$\Delta E = E(k) - E(k-1),$$

where $P(k)$, $P(k-1)$ are the current power and power on the previous clock cycle of the PV panel,

$V(k)$, $V(k-1)$ are the current voltage and voltage on the previous clock cycle of the PV panel,

$E(k)$, $E(k-1)$ are the current error and error on the previous clock cycle.

Expression $\Delta P/\Delta V$ [19] was chosen because using it we can determine location of the operating point, whether it is on left from MPP or on the right. When $\Delta P/\Delta V = 0$ MPP is reached, the change in the error gives information about the direction of the operating point moving along the curve in Fig. 3. The duty cycle is the output variable. Each of input and output variables have membership functions (see Fig. 6).

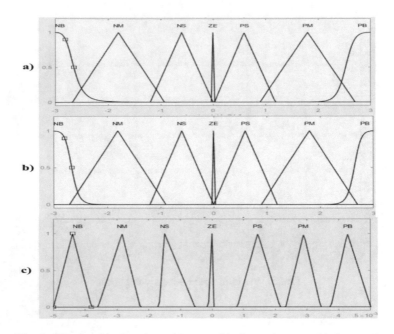

**Fig. 6.** Membership functions: (a) error, (b) change in error, (c) duty cycle

Model of photovoltaic station is showed in Fig. 7.

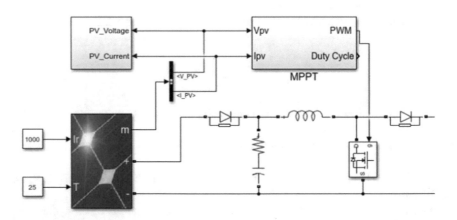

**Fig. 7.** Photovoltaic power unit model

### 2.3.2    Wind Turbines and Synchronous Generator Control System

We need to know the wind speed to reach the maximum power production point, and we use am anemometer for this purpose. If we know the wind speed, then it is easy to determine the optimal rotation speed when the maximum power generation is reached. A power unit consists of a wind turbine, synchronous generator and active rectifier. The control system of an active rectifier has two control circuits. The first circuit is the speed circuit, and the second one is the current circuit (see Fig. 8). The input of the speed controller is the optimal demanded speed, and we have the required current as the output. Driving pulses for the active rectifier are the output of the current controller. A model of a wind turbine with synchronous generator is shown in Fig. 9.

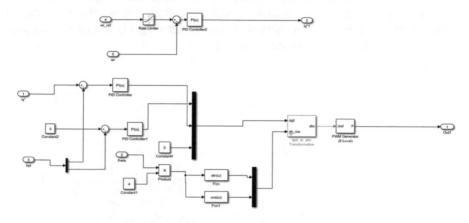

**Fig. 8.** Speed and current controllers

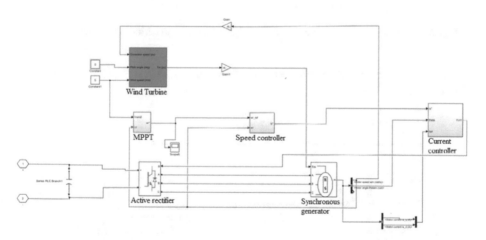

**Fig. 9.** Model of a wind turbine with synchronous generator

### 2.3.3 Virtual Synchronous Machine

Active rectifier is used as a DC/AC converter for power conversion from the DC bus to the grid and load. We use a relatively new approach which is called a virtual synchronous machine. The virtual synchronous machine is an alternative method of power control in electronic converters [20]. Control of a three-phase active rectifier is carried out, so that it imitates the synchronous generator behavior [21]. This gives some advantages, due to that the synchronous generator has the same inertia due to the rotating masses, the damping effect due to the damper winding, and the speed-droop characteristics for the load sharing. The properties are described by dynamics:

$$J \frac{d\omega}{dt} = T_m - T_e - D\Delta\omega,$$

where $J$ is the moment of inertia of the rotating masses; $\omega$ is the angular speed; $T_m$ is the mechanical torque; $T_e$ is the countering electromagnetic torque; $D$ is the damping torque coefficient. The proposed control structure including the virtual synchronous machine for an active rectifier is shown in Fig. 10.

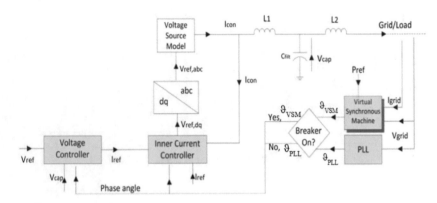

**Fig. 10.** Control structure including a virtual synchronous machine for an active rectifier

A model of a virtual synchronous machine in MATLAB/Simulink is shown in Fig. 11.

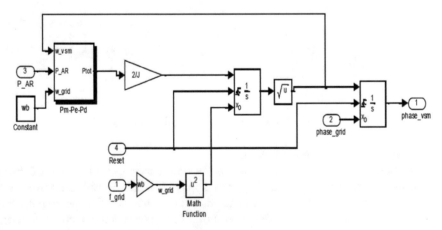

**Fig. 11.** Model of virtual synchronous machine

Actually, according to the formula and model, the output of a VSM is the phase angle which gives smooth and fast synchronization with the grid and a quick and qualitative transient response.

## 3 Simulation Results and Discussion

A model of the whole system is shown in Fig. 12.

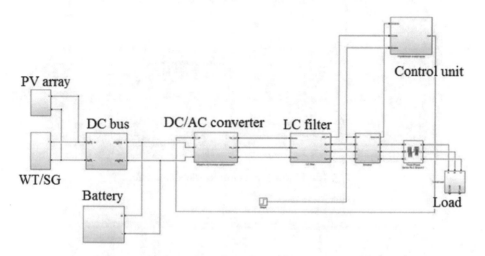

**Fig. 12.** Model of total system

10 kW power station with phase voltage 220 V is considered. Power generation in a islanded mode is shown in Fig. 13. We can see that the load demands are satisfied. Power generation increases under the changing weather circumstances and reaches its maximum. The storage battery in this case is charging. If power consumption and generation are equal, power does not flow into the battery. The grid frequency (freq.) in the isolated mode is shown in Fig. 14. The frequency reaches the steady state rather quickly, in about 250 ms, but it has a drawdown of about 2%.

Let's connect the station to the grid. The power flow in the grid-tied mode is shown in Fig. 15. Here, the power generation of our station coverts the load demand completely; that is why some power flows go into the stiff grid. The grid frequency is shown in Fig. 16. It must be noted that we have got a fast synchronization and the process almost without any oscillations.

Results with a sharp grid breakage at the loading and further operation of the power plant on alternative sources are shown in Fig. 17. Here we can see when the stiff-grid crash station quickly turned on and supported loading. The transient time is about 20 ms.

Power generation of a hybrid energy system with a changing damping coefficient is shown in Fig. 18.

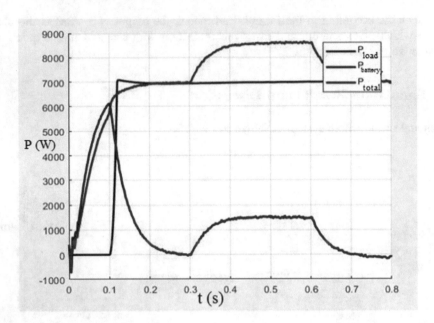

**Fig. 13.** Power generation in islanded mode

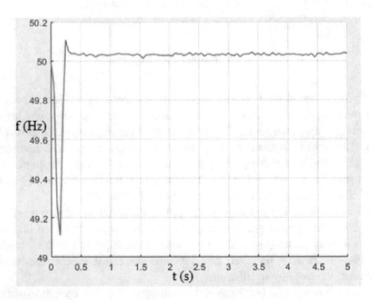

**Fig. 14.** Grid frequency in an isolated mode

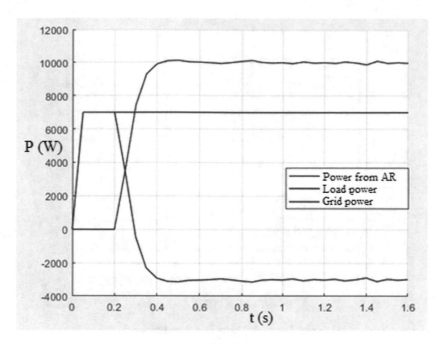

**Fig. 15.** Power generation in a grid-tied mode

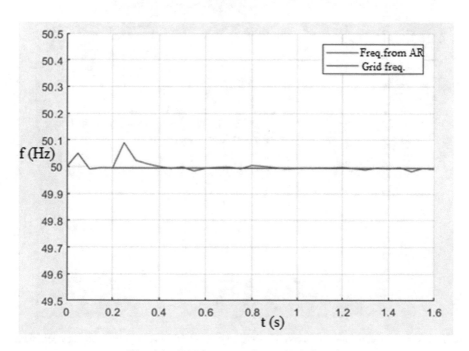

**Fig. 16.** Grid frequency in a grid-tied mode

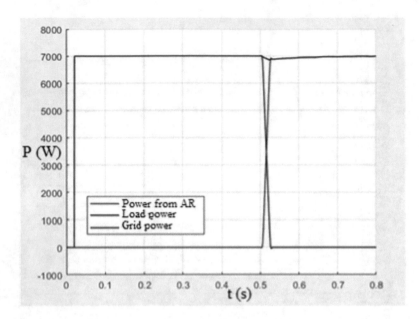

**Fig. 17.** Power flow in a grid breakage

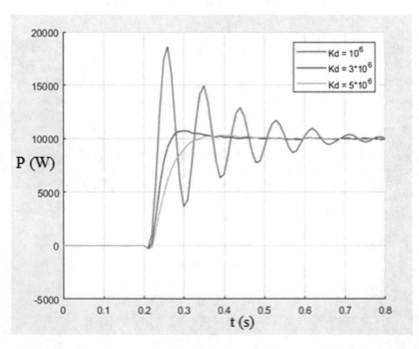

**Fig. 18.** Power generation of a hybrid energy system with a changing damping coefficient

As the damping factor decreases, the transient process becomes oscillatory. The best option is in the case when the value of damping factor 5 · 106.

Power generation of a hybrid energy system with changing inertia is shown in Fig. 19. The transient response when the moment of inertia changes seems to be optimal at $J = 5$ kg · m$^2$, which is not the highest value among the compared ones. The physical reason for this behavior is that when J increases, the kinetic energy (virtual) increases proportionally.

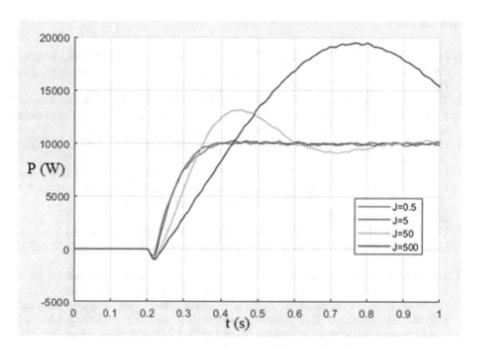

**Fig. 19.** Power generation of a hybrid energy system with changing inertia

As a result, it may take a longer time for the rotating system to stop or go back into the synchronism. When J is too low, the system is oscillatory.

## 4   Conclusions

The structure and the corresponding model of the power supply system with distributed generation based on a joint use of wind and solar power plants are considered.

A model of an active three-phase rectifier for a wind power station was built, and its control system is implemented using the vector control.

DC/DC boost converter was chosen and built as a power converter for solar panel.

The maximum power production point tracking algorithms for renewable energy management systems were implemented. Also, a model of a virtual synchronous machine based on an active rectifier for the use in MicroGrid power networks is discussed. The device has the property of universality, so it can be used for different types of power plants in distributed networks.

An active rectifier maintains the required grid frequency and produces the required power in the isolated mode. In a grid-tied mode, the active rectifier quickly finds the synchronism, as it can be seen from the graph of the grid frequency. The output power of an active rectifier corresponds to the required value.

# References

1. Lamell, J.O., Trumbo, T., Nestli, T.F.: Offshore platform powered with new electrical motor drive system. In: Petroleum and Chemical Industry Conference, Basle, Switzerland, 26–28 October 2005, pp. 259–266 (2005)
2. T. E. P. A. T. C. O. T. E. UNION, DIRECTIVE 2009/28/EC of the European Parliament and of the Council on the promotion of the use of energy from renewable sources and amending and subsequently repealing Directives 2001/77/EC and 2003/30/EC (2009)
3. Majumder, R.: Some aspects of stability in microgrids. IEEE Trans. Power Syst. **28**(3), 3243–3252 (2013)
4. Shuhui, L., Proano, J., Dong, Z.: Microgrid power flow study in grid-connected and islanding modes under different converter control strategies. In: Proceedings of the 2012 IEEE Power and Energy Society General Meeting, Manchester Grand Hyatt, San Diego, 22–26 July 2012, pp. 1–8 (2012)
5. Zhong, Q.-C., Weiss, G.: Synchronverters: inverters that mimic synchronous generators. IEEE Trans. Ind. Electron. **58**(4), 1259–1267 (2011). https://doi.org/10.1109/TIE.2010.2048839
6. Lopes, J.A.P., Moreira, C.L., Resende, F.O.: Microgrids black start and islanded operation. In: Proceedings of the 15th PSCC, Liege, Belgium, 22–26 August 2005 (2005)
7. Obukhov, S.G., Plotnikov, I.A.: Sravnitelny alaliz skhem postroeniya avtonomnykh elektrostantsy, ispolzuyuschikh ustanovki vozobnovlyaemoi energetiki [Comparative analysis of construction schemes of autonomous power plants using renewable energy sources]. J. Promyshlennaya Energetika **7**, 46–51 (2012)
8. Hatziargyriou, N. (ed.): Microgrids: Architectures and Control. IEEE Press, Wiley (2013)
9. Justo, J.J.: AC-microgrids versus DC-microgrids with distributed energy resources: a review. Renew. Sustain. Energy Rev. **24**, 387–405 (2013). https://doi.org/10.1016/j.rser.2013.03.067
10. Ahamed, F., et al.: Modelling and simulation of a solar PV and battery based DC microgrid system. In: Proceedings of the 2016 International Conference on Electrical, Electronics, and Optimization Techniques, DMI College of Engineering, Chennai, 3–5 March 2016 (2016). https://doi.org/10.1109/iceeot.2016.7754977
11. Dede, A., et al.: A smart PV module with integrated electrical storage for smart grid applications. In: 2016 International Symposium on Power Electronics, Electrical Drives, Automation and Motion, Sorrento, Italy, 20–22 June 2016 (2016). https://doi.org/10.1109/speedam.2016.7525997
12. Zamora, R., Srivastava, A.K.: Energy management and control algorithms for integration of energy storage within microgrid. In: Proceedings of the 2014 IEEE 23rd International Symposium on Industrial Electronics, Grand Cevahir Hotel and Convention Center, Istanbul, 1–4 June 2014, pp. 1805–1810 (2014). https://doi.org/10.1109/isie.2014.6864889

13. Hariharan, S., Kumar, V.N.: Desigh of a DC-DC converter for a PV array. In: Proceedings of the 2010 International Conference on Industrial Electronics, Control and Robotics, Rourkela, India, pp. 79–84 (2010). https://doi.org/10.1109/iecr.2010.5720154

14. Visscher, K., De Haan, S.W.H.: Virtual synchronous machines for frequency stabilisation in future grids with a significant share of decentralized generation. In: SmartGrids for Distribution, CIRED Seminar, pp. 1–4 (2008). https://doi.org/10.1049/ic:20080487

15. Wang, P., Zhao, J., Li, F.: A new adaptive duty cycle perturbation algorithm for peak power tracking. In: Proceedings of the 2nd International Asia Conference on Informatics in Control, Automation and Robotics, Wuhan, China, pp. 298–301 (2010). https://doi.org/10.1109/car.2010.5456543

16. Khatoon, S., Ibraheem, Jalil, M.F.: Analysis of solar photovoltaic arrays under partial shading conditions for different array configrations. In: Proceedings of the 2014 Innovative Applications of Computational Intelligence on Power, Energy and Controls with their impact on Humanity, Ghaziabad, India, 28 29 November 2014, pp. 452–456 (2014). https://doi.org/10.1109/cipech.2014.7019127

17. Vinifa, R., Kavitha, A.: Maximum power point tracking of boost converter on a PV system using fuzzy logic. Int. J. Mech. Eng. Technol. **8**(12), 583–593 (2017)

18. Chan, P.-W., Masri, S.: DC-DC boost converter with constant output voltage for grid connected photovoltaic application systems, pp. 1–5 (2010)

19. Gupta, A., Pachauri, K.P., Chauhan, Y.K.: Performance analysis of neural network and fuzzy logic based MPPT techniques for solar PV systems. In: 2014 6th IEEE Power India International Conference, pp. 1–6 (2014). https://doi.org/10.1109/poweri.2014.7117722

20. Hochgraf, C., et al.: Comparison of multilevel Inverters for Static Var Compensation, pp. 1–8 (1994)

21. Martins, F.J.N.: Virtual Synchronous Machine, pp. 1–6 (2016)

# On the Issue of the Green Energy Markets Development

Yury R. Nurulin[(⊠)], Inga Skvortsova, and Elena Vinogradova

Peter the Great St. Petersburg State Polytechnic University,
St. Petersburg, Russia
yury.nurulin@gmail.com, ingaskvor@list.ru,
vinogradova@spbstu.ru

**Abstract.** The article is devoted to the analysis of the current state and pro-spects for the development of renewable (green) energy market. The achieved level of technologies and equipment for production of renewable energy and the main trends in the development of these issues is analyzed. The demand of green energy is analyzed from the point of view of free market niches for this type of energy. The main focus in the study is made on the organizational and economic mechanisms for connecting new suppliers to existing grids. The accessibility of energy systems is considered as a key factor for the development of the green energy market. Technical, organizational and economic issues of ensuring the accessibility of energy systems are analyzed. The barriers that prevent reaching the level of grids accessibility necessary for the formation of an effective market for green energy, are analyzed. Getting electricity index as a component of the Doing Business ranking is used for comparative analysis of availability of electricity grids in the world economies and Russia. The Smart Grid concept is analyzed from the point of view of the development of grid's availability.

**Keywords:** Renewables · Green electricity · Smart grids · Energy consumption · Grids accessibility

## 1 Introduction

The necessary conditions for the existence of any market are the presence of product producers, the presence of product consumers and the presence of mutually beneficial conditions for the sale/purchase of a product. The key factor for the first component is the level of technological development. Over the past decade, we have seen a rapid growth of technical and technological solutions that underlie successful innovative projects in green energy sources [1]. This process is fully consistent with the "tech-nology push" model that reflects the importance of the technology for innovation [2]. As the technology develops, it stimulates the growth of market demand for products based on the technology [3]. For the green energy, the meaning of the term of "market demand" should be clarified. The classic market, which is based on the principles of free mutually beneficial interaction of suppliers and consumers, in the field of green energy acquires features that are more inherent to the planned economy. In addition to economic mechanisms, political and environmental factors play a significant role in the

© Springer Nature Switzerland AG 2020
D. G. Arseniev et al. (Eds.): CPS&C 2019, LNNS 95, pp. 360–367, 2020.
https://doi.org/10.1007/978-3-030-34983-7_34

green energy market. Formally not being economic, these factors have a serious impact on the economy of green energy through subsidies and tax preferences on the one hand, and restrictions and penalties on the other [5]. Considering the above, the article is structured as follows. The first section contains a literature review of the current state of the green energy market components. The next section is devoted to grids which are one of key elements of energy system. The final sections provide analysis of current procedures for connecting to existing grids and present the general conclusion of the study.

## 2  Literature Review

### 2.1  Renewable Energy Sources

Global trends in renewable energy development are in the focus of expert assessments around the world [6]. All experts point out that at the moment, renewable energy shows the highest growth rates among all energy carriers. While a general increase in energy demand is forecast at the level of 47% by 2040, the consumption of renewable energy sources (RESs) will increase by 93% within the same period [7]. While this trend is global in its nature, the specific values of growth substantially depend on the region. In developed countries, the consumption of renewable energy will increase by about 130% in the period up to 2040, especially due to new renewable sources (sun, wind, etc.), which are expected to have a nearly sevenfold increase. In developing countries, consumption of renewable energy will increase by 80%, an impressive almost tenfold increase will also be demonstrated by new renewable energy, which will increase its share in the structure of renewable energy consumption to 21%.

The largest increase in consumption of renewable energy will be in the developing countries of Asia (about 31%), while the Middle East and the developed countries of Asia will demonstrate the highest growth rates, which is largely due to their low baseline indicators. Of the largest consumers, more than a double growth is forecast for Europe and North America (see Fig. 1).

Increasing consumer demand is the main driving force for renewables [8]. It consists of two independent components. The first is the increase of energy consumption in a whole and the second is environmental protection issues. These components are independent and often contradictory from the economic point of view. The second component has a non-market nature and influences the market indirectly by non-economical mechanisms (policy, stakeholder behavior, etc.).

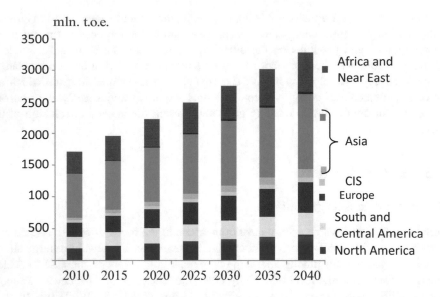

**Fig. 1.** Forecast of consumption of renewable energy sources by region [7]

The economic attractiveness of renewable energy is in the analysis of world experts. While some experts state that "renewables are reaching price and performance parity in the grid and at the socket" [9], other underline that high rates of growth in production of renewable energy are largely due to the significant state support [3]. In a number of countries it allows to make the renewable energy attractive, even in the cases where the initial economic indicators (excluding support mechanisms, taxation, etc.) are more than 50% worse than when using fossil fuels [10]. It is forecast that until 2040, many types of renewable energy will continue to need such support, despite the expected technological improvements [11].

## 2.2 Technological Improvement

The main trend in improvements of renewable energy technologies is reducing costs while increasing capacity and reliability of solar and wind generators [12]. According to BNEF, there has been a 28,5% drop in the price for crystalline silicon PV modules for every doubling of cumulative capacity since 1976 [13].

The rapid increase in the use of new RESs in the electric power industry poses additional tasks for the entire energy sector, associated with the need to reserve capacity and storage to ensure flexibility of the power system. In addition to the daily irregularity of electricity production at renewable energy facilities, the seasonal component, which is manifested due to changes in the intensity of solar radiation and wind power in different periods of the year should be noted. Because of this, the average monthly capacity utilization figures may differ in several times. In fact, this sets an additional hidden investment premium for the energy sector, which is often not taken into account when direct comparing the costs of electricity generation from various sources. Batteries, which are necessary to replenish energy in the dark or calm weather, have become one of the limiting factors to increase efficiency of RESs. Their production and disposal can have a negative impact on the environment, and the limited service life and the need for replacement cause additional costs during the RESs operation.

The effectiveness of batteries is increasing and the price is decreasing because of technology development and increasing demand. A significant impact on this process have both RESs and electric vehicles (EV) in which batteries are one of key elements as well. BNEF expects EV sales raise from 1.1 million at the moment to 30 million by 2030; that will cause essential cost decline for battery packs [14].

The most important trend in the development of world energy is the increase of the share of electricity in the final energy consumption. Electricity is the most convenient form of energy to use and will crowd out all other ones. Therefore, the demand for electricity is growing in all countries of the world, without exception, even in those OECD countries that stabilize their primary energy consumption. The electric power industry has a pronounced regional character and, in the absence of cheap long-distance transmission methods, power is mainly produced in the regions of consumption. Accordingly, the main increase in electricity production in the world (87%) will be provided by developing countries.

The electric power industry is the sector in which the main competition between all types of fuel occurs, and this competition will intensify. The improvement of new technologies leads to a decrease in the specific cost of renewable energy, and the rise in prices of traditional energy resources; on the contrary, will push up the costs of gas and coal generation, bringing all technologies into a rather narrow range of competition. Wind and solar power production have reached grid price parity and are moving closer to the performance parity with conventional sources [15].

## 3 Methodology

In the energy market in general, and in the green electricity market in particular, grids which act as equal and some time, even as the most important players play an essential role. Like any commodity, energy is needed not where it is produced, but where it is consumed. When delivering goods, the consumer is an active participant in the process, choosing the place where they buy the goods and moving it at their discretion. When organizing the supply of energy, the consumer participates in the process only at the initial stage, when the connection to the grid occurs. Further on, their participation in the supply is limited to the payment of the supplied energy, and a change of the supplier is practically impossible, or at least very difficult. Thus, the internal properties of the grid (technical parameters, business processes, regulatory framework, etc.) have a significant impact on the efficiency (or even on the very existence) of the energy market. Substantial development of these properties is provided in the framework of the Smart Grid concept which has the following attributes [16]:

- Flexibility. The grid must adapt to the needs of electricity consumers.
- Reliability. The grid must ensure the security and quality of the electricity supply in accordance with the requirements of the digital society.

- Cost effective. A synergistic effect should be provided by a combination of innovative technologies for building the grid together with effective business models and management of the grid functioning
- Availability. The grid should be accessible to new users, and new users could be generating sources, including RES with zero or reduced $CO_2$ emissions.

The last parameter (availability) will be in the focus of further analysis which is based on the Doing Business rating developed by the World Bank [17]. The subject for analysis will be Getting Electricity indicator for the city of St. Petersburg Russian Federation,

## 4  Discussion

During the last 7 ears the Russian Federation conducted a series of reforms to rise availability of existing grids for new customers (see Table 1).

As a result, the Russian Federation rank according to the Doing Business essentially increased and in 2019 the only one indicator (the time for connection) indicator is far from the desired value (see Table 2).

**Table 1.** Changes in getting electricity procedures in Russia [18]

| Year | Reform |
|------|--------|
| 2012 | Getting electricity became cheaper due to lower tariffs for connection |
| 2014 | Getting electricity became simpler and less costly due to setting standard connection tariffs and eliminating many procedures previously required |
| 2016 | The process of obtaining an electricity connection became simpler, faster and less costly due to eliminating a meter inspection by electricity providers and revising connection tariffs |
| 2018 | The number of procedures necessary for getting electricity and the time to complete were reduced |

**Table 2.** Getting Electricity – St. Petersburg, standardized connection [18]

| Indicator | St. Petersburg | Europe & Central Asia | OECD high income | Best Regulatory Performance |
|-----------|----------------|-----------------------|------------------|-----------------------------|
| Procedures (number) | 2 | 5,3 | 4,5 | 3 (25 Economies) |
| Time (days) | 80 | 110,3 | 77,2 | 18 (3 Economies) |
| Cost (% of income per capita) | 6,5 | 325.1 | 64.2 | 0.0 (3 Economies) |
| Reliability of supply and transparency of tariff index (0–8) | 8 | 5,5 | 7,5 | 8.0 (27 Economies) |

Obtaining an electricity connection is essential to enable a business to conduct its most basic operations. Whether electricity is available or not, the first step for a customer is always to gain access by obtaining a connection to electric grid. In many economies the connection process is complicated by the multiple laws and regulations involved – covering service quality, general safety, technical standards, procurement practices and internal wiring installations. Doing Business rating provides a common average picture of the problem that a consumer faces when connecting in a standard way. In reality, the consumer may face unexpected problems for him, which can significantly increase the cost and time to connect.

A survey of 43 organizations that implemented connection projects in 2017–2019, conducted in St. Petersburg, showed that these problems can appear even before the first formal procedure (Submitting application to the responsible organization and waiting for technical conditions and a connection contract). More than half of the companies surveyed noted the difficulty of collecting the documents necessary for technological connection. Upon having the connection contract signed by the customer, the responsible company prepares project design for building a network for the connection and obtain all approvals for construction work (such as a permit required for the laying of underground or overhead lines, etc.), and carry out all external works according to the contract and technical conditions. In an urban setting, another typical difficulty identified by the companies surveyed is the presence of a large number of other communications, which entails a search for an extraordinary design solution for the passage of intersections with heating and water supply lines, telephony, electricity, sewers, highways. As a result, the expected typical duration and cost may be exceeded in several times.

The second formal procedure (Receiving external works and final connections) may also cause unexpected difficulties in the absence of sufficient capacity stipulated by that need for the construction of new or modernization of existing substations. The second procedure may also cause unexpected difficulties in the absence of sufficient capacity and the resulting need for the construction of new or modernization of existing substations.

60% of surveyed companies used a temporary connection for the period of work so that the business would not stand idle while waiting for the technological connection. Most of them rented traditional mobile diesel power plants and only 3 used wind and biomass generators. All respondents from this group noted on the difficulties in choosing a specific option for implementing a temporary connection due to the lack of necessary information.

Formally, the conducted survey is not related to the green energy market because it concerns only electricity consumers who are generally indifferent to the type of energy source: the connected electricity must be cheap, high-quality and reliable. At the same time, it clearly shows that possible green energy suppliers will have similar or even much more acute difficulties while connecting to existing grids to sell electricity. By analogy with the problems of connecting new consumers, it can be assumed that most of the problems for green energy suppliers will be of organizational nature.

## 5 Conclusions

At the moment, the availability of existing grids for new customers and especially for new suppliers could be a restraining factor for the development of green energy markets. To counter this, targeted organizational measures are needed to develop grids accessibility. Best practices in this sphere should be studied and benchmarked.

The Doing Business ranking approach that has proved its effectiveness for the analysis of different aspects of business, should be extended to accessibility of existing grids for green electricity suppliers.

**Acknowledgments.** The article is prepared in the frame and with financial support of the KS1024 project of the CBS ENI program. Conflicts of interest: the authors declare no conflict of interest.

**Author Contributions.** Yury R. Nurulin developed the main conceptual idea and has edited the text of the paper, Inga Skvortsova designed the methodological framework and contributed to the analysis of the background, and Elena Vinogradova contributed to the analysis of the context and background. All authors provided critical feedback and contributed to the editing final version of the paper.

## References

1. Office of Energy Efficiency & Renewable Energy. Renewable Electricity Generation. https://www.energy.gov/eere/office-energy-efficiency-renewable-energy. Accessed 04 May 2019
2. Freeman, C.: The 'National System of Innovation' in historic perspective. Camb. J. Econ. **19**, 5–24 (1995)
3. Kline, S.J., Rosenberg, N.: An overview of innovation. In: Landau, R., Rosenberg, N. (eds.) The Positive Sum Strategy, pp. 275–305. National Academy Press, Washington (1986)
4. Verbeke, S.: Realising the clean energy revolution in the existing building stock. EEI 28–31 (2018)
5. Islam, S.: Key European solar markets trends from the installers' perspective. EEI **3**, 24–25 (2018)
6. REN21. Renewables 2018 Global Status Report (2018). http://www.ren21.net/gsr-2018/. Accessed 04 May 2019
7. Makarov, A.A., Grigoriev, L.M., Mitrova, T.A. (eds.): Prognoz razvitiya energetiki mira i Rossii, INEI RAN pri Pravitelstve RF, Moscow, 200 p. (2016). (in Russian)
8. Regulatory and Business Model Reform. https://rmi.org/our-work/electricity/regulatory-business-model-reform/. Accessed 04 May 2019
9. Motyka, M., Slaughter, A., Amon, C.: Global renewable energy trends (2018). https://www2.deloitte.com/insights/us/en/industry/power-and-utilities/global-renewable-energy-trends.html. Accessed 04 May 2019
10. Bloomberg, N.E.F.: Corporations already purchased record clean energy volumes in 2018, and it's not an anomaly (2018). https://www.bloomberg.com/professional/blog/corporations-already-purchased-record-clean-energy-volumes-2018-not-anomaly/. Accessed 04 May 2019
11. Giannakopoulou, E.: The Power Transition – Trends and the Future, Bloomberg New Energy Outlook (2017)

12. Jäger-Waldau, A.: Rooftop PV and self consumption of electricity in Europe Benefits for the climate and local economies. EEI **3**, 16–20 (2018)
13. Cheung, A.: Power Markets Today, BNEF (2018)
14. Bloomberg New Energy Outlook 2018 (2018). https://bnef.turtl.co/story/neo2018?src=pressrelease&utm_source=pressrelease. Accessed 04 May 2019
15. Motyka, M., Slaughter, A., Amon, C.: Global renewable energy trends (2019). https://www2.deloitte.com/insights/us/en/industry/power-and-utilities/global-renewable-energy-trends.html. Accessed 04 May 2019
16. European SmartGrids Technology Platform. Vision and Strategy for Europe's Electricity Networks of the Future, Luxembourg: Office for Official Publications of the European Communities (2006)
17. Doing Business (2018). http://www.doingbusiness.org/en/reports/. Accessed 04 May 2019
18. Economy Profile of the Russian Federation (2018). http://russian.doingbusiness.org/ru/reports/global-reports/doing-business-2019. Accessed 04 May 2019

# Digital Twin Analytic Predictive Applications in Cyber-Physical Systems

Anton P. Alekseev[1], Vladislav V. Efremov[1],
Vyacheslav V. Potekhin[1(✉)], Yanan Zhao[2], and Hongwang Du[3]

[1] Peter the Great St. Petersburg Polytechnic University, St. Petersburg, Russia
`alekseev.ap@google.com`, `vladefr97@yandex.ru`,
`Slava.Potekhin@gmail.com`
[2] College of Mechanical and Electrical Engineering, Harbin Engineering
University, Harbin, China
`zhao.nan@yandex.ru`
[3] College of Automation, Harbin Engineering University, Harbin, China
`duhongwang@hrbeu.edu.cn`

**Abstract.** The article shows the relevance of the use of predictive models of digital counterparts for the formation and analysis of time trends obtained from the sensors of an automated control system. The requirements for the predictive model are shown; machine learning algorithms, regressions for time series forecasting are described; analysis and comparison of algorithms based on RMSE, MAE, R2 error readings are presented. Also, the article shows methods of automatic determination of emissions and novelty in time series and methods of detection of dependencies between parameters are brought. The authors give an example of integration of the predictive model into the infrastructure of a digital double, describe the life cycle and full functionality of such a system. In conclusion, the prospects of using the predictive model in systems where it is difficult to read the necessary parameters with low frequency are shown.

**Keywords:** Industry 4.0 · Digital twin · Intelligent control system ·
Automation · Machine learning · Regression · Neural network

## 1 Introduction

Industry 4.0 has become the most discussed concept of industrial business lately. One of the most important components of this concept is the digital production model. Automated control systems based on the digital twin principle make it possible to achieve efficiency in the process of measuring and optimizing production. Production will become more reliable as monitoring systems will be alerted to emerging issues in participation. Thus, we can get highly efficient production.

- Increased production flexibility is achieved by abandoning hard "conveyor" solutions, which leads to an increase in mass production of individual products and production scaling through the introduction of new solutions
- Due to functioning on a single technological platform and control at each level, there is an improvement of production setup

© Springer Nature Switzerland AG 2020
D. G. Arseniev et al. (Eds.): CPS&C 2019, LNNS 95, pp. 368–377, 2020.
https://doi.org/10.1007/978-3-030-34983-7_35

- Risks and costs that previously arose due to the human factor are reduced to a minimum
- The digital production model provides all the information necessary for analysis in real time and is also able to predict the future states of the system and make the most effective decisions.

In many automated systems in industrial enterprises, data collection is carried out at the lower level of the modern concept of a multi-level structure of automatic control systems, where it is not always possible to obtain information about the quality of the product in real time. Sometimes measurements are made once a day or once a week, which does not provide full information to confirm the quality of the product. To ensure a high level of observability of the object, predictive analytical applications built on machine learning are used.

For example, continuous data collection and condition monitoring is necessary when using industrial robots. Industrial robots are highly integrated products of mechatronics. It relates to many disciplines, such as computers, artificial intelligence and mechanical design, and has a high scientific and technological value [3]. Cutting robots are a high proportion of industrial robot applications. The use of CNC cutting robots can improve product quality, increase production efficiency, reduce production costs, and complete more complex cutting fields, such as 3D curved cutting and flexible cutting, becoming an indispensable key technology in the petrochemical industry and shipbuilding industry [4, 5]. CNC cutting robots can also be an integral part of flexible manufacturing systems, intelligent distributed manufacturing systems, and computer-integrated manufacturing systems [6, 7].

In this article, we will develop a predictive analytical application based on the digital twin in cyber-physical systems to restore the gas content to 1-minute discretization, as well as predicting the parameters for the next hour.

The article also provides an example of the integration of the cutting line based on the robotic complex RZV-016 in the production cycle of an enterprise, where it would be possible to use a digital twin.

## 2  Dataset for Analysis

To show an example of how the predictive analytical applications work, we used the open data bank of time series. This article discusses a data set that contains the obtained time series from 16 chemical sensors exposed to gas mixtures with different levels of concentration. In particular, there are two gas mixtures: ethylene and methane in the air and ethylene and CO in the air.

The data set was assembled in a gas delivery platform installation at the Chemical Signal Laboratory at the Bioschem Institute, University of California San Diego. The measurement system platform provides versatility to obtain the required concentrations of interest chemicals with high accuracy and in a highly reproducible manner.

The sensor array included 16 chemical sensors (Figaro Inc., USA) of 4 different types: TGS-2600, TGS-2602, TGS-2610, and TGS-2620 (4 blocks of each type). Sensors were integrated with signal conditioning and control electronics. The operating voltage of the sensors that control the operating temperature of the sensors was kept constant at 5 V throughout the duration of the experiments. Sensor conductances were obtained continuously at a sampling rate of 100 Hz. The sensor array was placed in a measuring chamber with a volume of 60 ml, into which a gas sample was injected with a constant flow rate of 300 ml/min [1].

# 3   Architecture of an Analytics Predictive Application

In order to develop a digital object model, you must first think through all the steps from the choice of algorithms to the software implementation of the product. To analyze and prepare a digital model for any dataset with time series, you need to:

- Determine the nature of the process: stationary or not, whether seasonality, trend
- Determine outliers, exclude them from the sample, interpolate empty values
- Find the dependencies between the parameters, determine the correlation of each indicator
- Create a digital predictive or interpolation model, taking into account dependent parameters.

Thus, following these steps, you can develop an application for data analysis and decision making.

# 4   Analysis and Pre-processing of the Source Dataset

In most cases, the initial data set contains unwanted outliers and trends, which may occur due to measuring equipment errors, random errors, as well as a number of other non-obvious factors. Such data distortions should be eliminated.

## 4.1   Time Series Anomaly Detection

There are two areas in data analysis that look for anomalies: Outlier Detection and Novelty Detection. Like the ejection "new object" is an object that differs in its properties from the objects of the (training) sample. But unlike the outlier, it is not yet in the sample itself (it will appear after a while, and the task is precisely to detect it when it appears). For example, if you analyze temperature measurements and drop the abnormally large or small ones, then you are struggling with emissions. And if you create an algorithm that for each new measurement evaluates how much it looks like the past, and throws out abnormal ones – you are "struggling with novelty" [2].

### 4.1.1  Box-and-Whiskers Diagram and IQR Score

In order to visually show the distribution of values, the authors use the box-and-whiskers diagram. Emissions will be indicated as points behind the vertical lines – extremes. To automatically determine black box emissions, the IQR score algorithm is used. In descriptive statistics, the interquartile range (IQR) (also called the average or average 50%, or technically H-spread) is a measure of statistical dispersion equal to the difference between the 75th and 25th percentiles or between the upper and lower quartiles: IQR = Q3 − Q1. In other words, the IQR is the first quartile deducted from the third quartile; these quartiles are clearly visible on the chart. It is a truncated estimator, defined as a 25% truncated range, and is a widely used and reliable measure of scale.

### 4.1.2  Z-Value and the Distance of Mahalanobis

Z-value is a measure of the relative spread of the observed or measured value, which shows how many standard deviations make the spread of the relative average value. This is a dimensionless statistics used to compare values of different dimensions or scale measurements. Also, to identify anomalies, the Mahalanobis distance is used, i.e., a measure of the distance between vectors of random variables, which generalizes the concept of Euclidean distance.

The simplest approach is to calculate the root-mean-square deviation of the set points from the center of mass. If the distance between a given point and the center of mass is less than the standard deviation, then we can conclude that the probability that a point belongs to a set is high. The farther the point is, the greater is the likelihood that it does not belong to the set.

This intuitive approach can be determined mathematically through the distance between a given point and a set using an equation:

$$d(\overrightarrow{x}, \overrightarrow{y}) = \sqrt{\sum_{i=1}^{N} \frac{(x_i - y_i)^2}{\sigma_i^2}} \tag{1}$$

By substituting this value into the normal distribution, one can find the probability of the point's belonging to a set.

In Fig. 1a the graph shows the calculated Mahalanobis distances for each point, and the anomalous ones are those where the distance exceeds a sufficiently large value. Figure 1b shows the calculated Z-values for each point. In Fig. 1c you can see a box with a "mustache", where the values of abnormal points are immediately seen. Figure 1d shows all found anomalies (the methods considered found the same amounts of emissions).

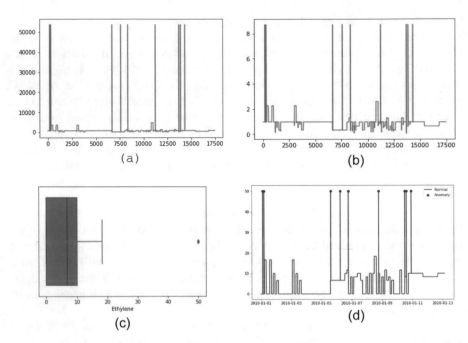

**Fig. 1.** (a) Mahalanobis distances; (b) Z-values; (c) a box with a "mustache"; (d) found anomalies

## 4.2    Data Smoothing with Interpolation Polynomials

Often a characteristic feature of cyber-physical systems is the high sampling rate of measured parameters in the studied data sets. As a result, the parameter values acquire a stepped character with the presence of pronounced thresholds (see Fig. 2).

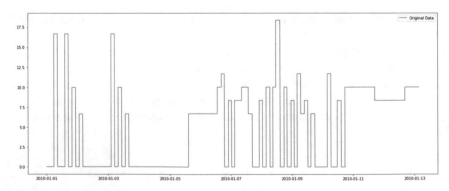

**Fig. 2.** The graph shows the stepwise character of the data

It can be assumed that the concept of building predictive models based on data with a high sampling rate might be redundant.

As a way to reduce the high discreteness of the source data, it is proposed to use the construction of a cubic spline interpolation on average values and the subsequent training of predictive models on the interpolated data.

As a concrete solution to this problem, it is proposed to use spline interpolation with cubic polynomials (see Fig. 3).

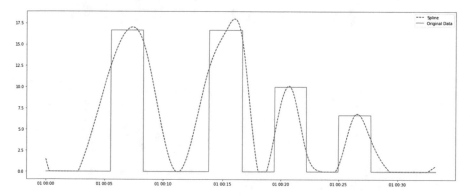

**Fig. 3.** The graph shows the interpolation of the original step data using a cubic polynomial

In order to interpolate the data, points located on the same threshold are used. The group of such points is replaced by a single value – their arithmetic average (shown in the Figure by red dots). These are arithmetic mean values for each threshold.

Pre-processing of input data by spline interpolation allows to reduce the pronounced degree of input data and thus help reducing the error in the prediction.

## 5  Dependence Determination

To identify the most significant parameters, it is necessary to find the Pearson's correlation coefficients. Based on this, it is possible to choose for training only those parameters that have a coefficient greater than 0.5 (see Fig. 4).

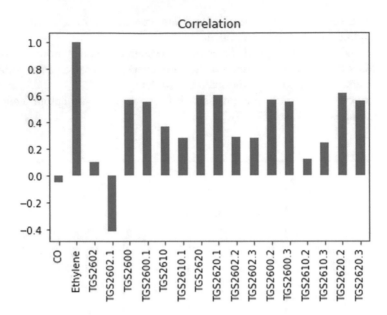

**Fig. 4.** The graph shows Pearson's correlation of all parameters relative to Ethylene. The most significant parameters have the 0.5 coefficient

## 6   Time Series Interpolation and Forecasting

The RMSE is used as metrics to identify the most efficient digital model. The standard deviation (RMSE) $d_{1j}$ of the forecast based on the results of laboratory analysis for the time interval with $n$ laboratory analysis results available, are calculated using an equation:

$$d_{1j} = \sqrt{\sum_{i=1}^{n} \frac{(y_i - \hat{y}_i)^2}{n}} \tag{2}$$

In this article, a random forest was chosen as a learning model—this is a set of decisive trees. In the regression problem, their answers are averaged, in the classification problem the decision is made by a majority vote. All trees are built independently according to the following scheme.

To build each splitting in a tree, we review max_features of random features (for each new splitting, its own random features).

We choose the best signs and splitting on it (by a predetermined criterion). A tree is built, as a rule, until the sample is exhausted (until representatives of only one class remain in the leaves), but in modern implementations, there are parameters that limit the height of the tree, the number of objects in the leaves, and the number of objects in the subsample at which the splitting is performed.

In Fig. 5, there is a chart that illustrates the result of the Etinol Parameter forecast on the most dependent indicators for 2 h ahead of our trained model. For example, if real measurements are made every ten minutes, we can thus use a digital model to obtain information with a discretization of 1 min, which will significantly reduce the risks and increase the efficiency of monitoring processes.

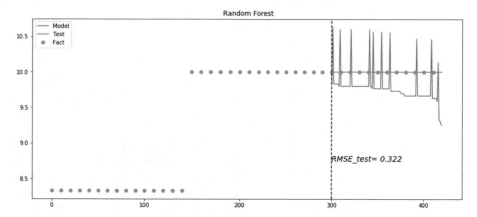

**Fig. 5.** The result of the Etinol Parameter forecast on the most dependent indicators for 2 h ahead of our trained model

## 7   Function and System Plan of CNC Cutting Robot

This CNC cutting robot is mainly used for cutting a hole in the pressure vessel in the boiler in chemical industry and other industries. The types of open-hole containers include: a cylinder, a cone shell, an ellipsoid head, a ball head, an opening of a pipe of equal diameter, etc., as shown in Fig. 6.

The types of groove are shown in Fig. 7, including a straight mouth, single-sided outer groove, single-sided inner groove, double-sided groove; the groove angle is variable. This is suitable for the cutting of carbon steel pipe with the diameter Ø30–800 mm.

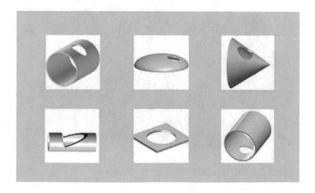

**Fig. 6.** Type of hole groove

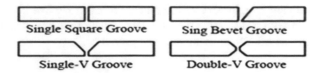

**Fig. 7.** Type of groove

The system structure scheme adopts a series of multi-joint mechanical arm type, as shown in Fig. 8. It has four degrees of freedom, three of which are joints and one of which is a swing joint. The latter consists of a fixed base, a boom, a small arm, a rotating joint, and a swing joint shaft. The fixed base is connected to the boom by a rotary joint U and has a degree of freedom of rotation; the boom and the arm are also connected by a rotary joint V, and also have a degree of freedom of rotation; the small joint position can be determined to stay small. The spatial position of the arm and end effector are relative to the base. Between the arm and the end effector headstock, there is a rotary joint W and a swing joint R. The two joints are linked to implement the attitude adjustment of the end effector relative to the base; when the position is adjusted, according to the operation requirements, the end effector holds different work tools for specific operations on the pipe end. In addition, a fine adjustment can be achieved by the robotic arm adjustment mechanism to ensure the positioning accuracy of the end effector.

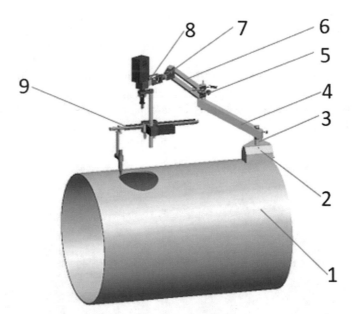

**Fig. 8.** Frames model of a CNC cutting robot: 1 - steel pipeline; 2 - base frame; 3 - rotating joint U; 4 - link 1; 5 - rotating joint V; 6 - link 2; 7 - rotating joint W; 8 - pitch joint R; 9 - ending assembly with a cutting torch

# 8   Conclusion

Having considered a real example of production, we can say with confidence that it is now necessary to develop such preliminary analytical applications to improve the efficiency of obtaining information about the processes in real time. We have developed a digital model that was able to interpolate hard-to-measure indicators and predict the future timing with an RMSE = 0.322, that speaks about the effectiveness of this model.

**Acknowledgments.** The article is published with the support of the project Erasmus+ 573545-EPP-1-2016-DE-EPPKA2-CBHEJP Applied curricula in space exploration and intelligent robotic systems (APPLE) and describes a part of the project conducted by SPbPU.

# References

1. Fonollosa, J., et al.: Reservoir compensates for continuous monitoring. Sens. Actuators B **215**, 618–629 (2015)
2. Shipmon, D.T., Gurevitch, J.M., Piselli, P.M., Edwards, S.T.: Time series anomaly detection; detection of anomalous drops with limited features and sparse examples in noisy highly periodic data. Google, Inc., Cambridge (2017)
3. Du, H.W., Wang, Z.Y., Liu, S.G., Zhao, Y.: Kinematics and track amendments of intersecting pipe welding robot. Trans. China Weld. Inst. **30**(7), 45–48 (2009)
4. Keraita, J.N., Kim, K.H.: PC-based low-cost CNC automation of plasma profile cutting of pipes. ARPN J. Eng. Appl. Sci. **2**(5), 1–7 (2007)
5. Rashid, T.H.: Improvement of cutting conditions using oxy-propane flame through CNC cutting machine. J. Iraqi J. Mech. Mater. Eng. **11**(1), 92–103 (2011)
6. Liu, Y., Liu, Y., Tian, X.: Trajectory and velocity planning of the robot for sphere-pipe intersection hole cutting with single-Y welding groove. J. Robot. Comput. Integr. Manuf. **56**, 244–253 (2019)
7. Fedorov, A., Shkodyrev, V., Zobnin, S.: Knowledge based planning framework for intelligent distributed manufacturing systems. In: Lecture Notes in Computer Science (including subseries Lecture Notes in Artificial Intelligence and Lecture Notes in Bioinformatics), vol. 9141, pp. 300–307. Springer (2015)

# Power Management in Autonomous Optical Sensor Nodes

Uliana Dudko$^{(\boxtimes)}$ and Ludger Overmeyer

Institute for Transport and Automation Technology,
Leibniz University Hannover, Hannover, Germany
uliana.dudko@ita.uni-hannover.de

**Abstract.** Solar cells are the most common energy scavengers due to their ease of use, reliability and wide range of applications wherever the light is present. One of such applications is a wireless autonomous sensor network, where each sensor node tends to be as small as possible. In this paper, we describe a way to optimize the optical energy harvesting process using as an example a power management IC BQ25570. We introduce an approach to calculate the minimal solar cell area and storage capacitance, which are necessary to supply the module in the active mode for a specified time period and to minimize the charging time.

**Keywords:** Energy harvesting · Solar cell · Autonomous sensor node · Visible light combination · Power management

## 1 Introduction

The past decade saw a rapid transition in low-power electronics from primary batteries to an energy harvesting concept. All wearable devices, the Internet of Things (IoT) gadgets and smart sensors tend to be wireless and autonomous to provide greater freedom to their users. Low prices and the variety of energy scavengers with off-the-shelf supportive electronics facilitate the design process for developers and broadens the range of energy harvesting applications. Industrial wireless sensor networks belong to such applications [1]. Wireless sensor nodes spend most of the time in standby mode and are activated periodically by an external or internal signal to send the measurement data. In our study, we develop a novel design of a miniature optical sensor node ($30 \times 10 \times 10$ mm), which can send the measurements using visible light and can autonomously charge itself from a solar cell [2, 3]. The sensor node is activated optically from a solar cell by an external optical signal. Solar cells are the most widely used types of ambient energy sources due to their long lifetime and ease of use wherever light is present. The size of a solar cell is the main limiting factor for node miniaturization.

The straightforward approach of harvesting energy from a solar cell implies direct electrical connection of a solar cell with an energy storage through a diode (direct-coupled circuit). This method has proven its efficiency only for high irradiances exceeding $100$ W/m$^2$, which makes it suitable only for outdoor use [4]. The optical power generated by a solar cell indoors must be properly managed in order to avoid

D. G. Arseniev et al. (Eds.): CPS&C 2019, LNNS 95, pp. 378–387, 2020.
https://doi.org/10.1007/978-3-030-34983-7_36

energy leakage and to maximize its efficiency. Along with energy harvesting and storage some other important functions must be implemented, such as overvoltage protection, constant output voltage for the load (e.g., a microcontroller) and maximum power point tracking (MPPT) of the solar cell under different illumination conditions. The general diagram of an energy harvesting unit as part of an optical sensor node is presented in Fig. 1.

The typical amount of power harvested from a solar cell in indoor light conditions is in the range of microwatts. This is the main challenge in the design of an optical autonomous sensor node, since the power required for a LED to transmit a data signal is typically rated in milliwatts. In this paper, we present an approach for overcoming this challenge using a power management integrated circuit (PMIC) BQ25570 and a mathematical model for calculating the optimized power parameters of an optical autonomous sensor node. These parameters are the instantaneous power amount required for the solar cell and the minimal required solar cell area. Precise definition of such system's attributes is very important for the sensor node optimization and for efficient energy harvesting.

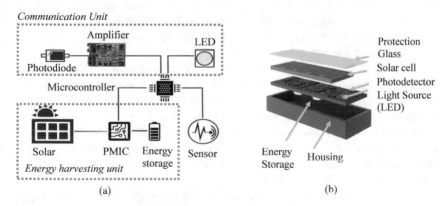

**Fig. 1.** a. Schematic representation of the optical autonomous sensor node. b. Conceptual design.

Section 2 explores the related work and existing designs of power management in autonomous sensor nodes. Section 3 provides a comparison of existing PMICs in the market. Section 4 suggests a mathematical model and computation of optimized power parameters for the optical autonomous sensor node. In Sect. 5, design of the energy harvesting unit is presented and discussed.

## 2 Related Work

There is a relatively small body of literature that is concerned with the combination of energy harvesting and optical wireless communication in sensor nodes. The main reason for that is the relatively high power requirement for an optical transmitter in comparison with the energy scavenged from the ambient. Authors of [5] tried to overcome this challenge using a retro-reflector instead of a power-intensive LED for the uplink information transmission. The retro-reflector uses less energy harvested from a solar cell due to backscattering the incoming light from the externally powered illuminating the LED lamp. The disadvantage of such design is that the sensor nodes have to rely on the LED lamp and they are unable to communicate with one another. Another design proposed by [6] employs a conventional LED integrated in the sensor node for communication with a smartphone camera. Here, the authors power the node by a 3.7 V polymer lithium battery rechargeable through a USB without employment of the energy harvesting concept.

Different examples exist in the literature regarding the use of solar cells in visible light communication not only as energy sources, but also as light signal receivers [7–9]. Unlike common photodiode detectors, whose circuitry requires additional energy, the solar cells generate significant amount of power along with signal detection. In this case, the frequency response of solar cells, which affects the data rate, is limited to hundreds of kilohertz due to their physical properties. Only the direct-coupled circuit for energy harvesting was investigated in the mentioned studies.

To date, a number of studies have examined various off-the-shelf PMICs for energy harvesting in sensor nodes with radio-frequency (RF) communication. These PMICs integrate in small-sized packages (e.g. $3 \times 3$ mm) the most important functions of energy harvesting: battery over-/under voltage protection, MPPT for solar energy sources, and buck-boost DC/DC conversion. Typically, PMICs require nanowatts of power. In [10] the authors successfully implemented a fault-tolerant solar harvesting system based on LTC3330 IC for sensor nodes which ensures goods traceability in sea ports and airports using RF communication. In the Ref. [11] the team managed to design a miniature autonomous inertial sensor node, which employs BQ25504 IC for power management from two small-size ($6 \times 4 \times 1.5$ mm) monolithic solar cell strings. The node communicates using the Bluetooth Low Energy protocol and can be utilized only under sunlight conditions.

However, none of the solutions discussed above investigate the power management of energy harvesting in sensor nodes with optical wireless communication, which employs an active LED for data transmission. In this study, we report the design of the energy harvesting system for a miniature optical autonomous sensor node with 0.08% duty cycle in indoor environment.

## 3   Comparison of PMICs

In order to analyze the PMICs existing on the market, it is important to define the criteria and limitations of the miniature autonomous optical sensor nodes. The solar cell has to generate enough energy to supply the node and at the same time, its sensitive area has to be as small as possible. For solar cells with area less than 30 cm$^2$, the typical open circuit voltage (input voltage for PMIC) is in the range from 1 V to 5 V.

A microcontroller of the sensor node requires a regulated output in the range from 1.8 to 3.3 V. This output voltage should be implemented via a step-down voltage regulator. Out of two common types of voltage regulators, linear dropout regulators

(LDOs) have the simplest circuit configuration; however, their efficiency is relatively weak. In contrast, buck converters, another type of regulators, has high efficiency and low heat generation. Therefore, they are preferable in the design of an energy harvesting unit of the optical sensor module.

Another criteria is existence of an embedded Maximum Power Point Tracking circuit to maximize power extraction under any conditions. This circuit for a solar harvesters is essential, since the maximum power that can be harvested significantly depends on the constantly changing light conditions.

Additionally, the cold-start voltage also is a significant parameter, which defines the minimal voltage level required to start the boost charger. The quiescent current is the current drawn by the PMIC when it is not supplying a microcontroller but only charging an energy storage. Quiescent current must be as low as possible to reduce the waste of energy. Other important functions are overvoltage and undervoltage protection of the energy storage. In Table 1 we described the latest and most prevalent PMICs used in wireless sensor nodes that are available on the market.

All PMICs presented in Table 1 have an input voltage range suitable for operating in combination with miniature solar cells, except for LTC3330, which requires for functioning at least 3 V. The ICs ADP5090, MAX17710 and BQ25504 have one unregulated output. In order to supply a microcontroller with constant voltage, the output has to be converted to a lower voltage level using additional electronic components. Therefore, the complexity and the size of the resulting electric circuit would increase. The chip MAX17710 has an additional regulated LDO pin with selectable voltage levels. If one of these levels suits the application, the LDO can be used for output without any additional step-down circuit.

**Table 1.** The latest PMICs available on the market.

| PMIC | BQ25570 | ADP5090 | MAX17710 | BQ25504 | LTC3330 |
|---|---|---|---|---|---|
| Manufacturer | TI | Analog devices | Maxim | TI | Linear technologies |
| Input voltage [V] | 0.1–5.1 | 0.08–3.3 | 0.75–5.3 | 0.13–5.5 | 3.0–19 |
| Input power [µW] | 5–500 000 | 16–200 000 | 1–100 000 | 15–300 000 | [-] |
| Cold-start voltage [mV] | 330 | 380 | 750 | 600 | [-] |
| Quiescent current [nA] | 488 | 320 | 625 | 300 | 750 |
| Outputs | 1 regulated | 1 un-regulated | 1 unreg., 1 LDO | 1 un-regulated | 1 unreg., LDO |
| Output voltage [V] | 2.0–5.5 | 2.0–5.2 | Selectable: 1.8, 2.3, 3.3 | [-] | 1.8–5.0 |
| Over-/under voltage protection | Yes | Yes | Yes | Yes | No |
| MPP-tracking | Yes | Yes | Yes | Yes | No |
| Boost charger | Yes | Yes | Yes | Yes | Yes |
| Buck converter | Yes | No | No | No | Yes |
| Size | 3 × 3 mm | 3 × 3 mm | 3 × 3 mm | 3 × 3 mm | 5 × 5 mm |

However, linear regulators are less efficient in comparison to buck converters. The PMIC BQ25570 is the only one IC with an embedded buck converter and programmable output voltage, which makes it the most compact and efficient IC available on the market. Moreover, BQ25570 requires the least cold start voltage to drive the boost charger. This gives the sensor an opportunity to work in extremely poor light conditions.

All mentioned advantages of BQ25570 make this IC the most beneficial for employment in autonomous optical sensor nodes.

In the further sections we will exploit the BQ25570 IC in the design of an energy harvesting unit of the optical sensor node. In Sect. 4, along with mathematical model, we provide an example of calculation of optimized power parameters based on BQ25570 IC.

# 4  Mathematical Model of Power Parameters

In this section, we describe the mathematical model of the power parameters for the optimal energy harvesting configuration. We also provide an example of calculation of a power unit oriented towards the design of the autonomous optical sensor module.

## 4.1  Mathematical Model

In order to calculate the power parameters, it is necessary to analyze the system loads and power consumption of an autonomous sensor node. The sensor node has several energy modes: one or two active modes (e.g., for data transmission and data reception) and a standby mode, when the module waits for an activation signal. The consumed current and power per day in active modes can be calculated using the following equation:

$$mAh\ per\ Day = I_{stor} \cdot Mode\ hours\ per\ Day; \tag{1}$$

$$mWh\ per\ Day = V_{stor} \cdot mAh\ per\ Day; \tag{2}$$

where $V_{stor}$ is the predefined voltage level of an energy storage, and $I_{stor}$ is the current supplied to the load (microcontroller).

$I_{stor}$ can be defined as follows:

$$I_{stor}(mA) = \frac{V_{out} \cdot I_{out}}{V_{stor} \cdot Effcy}, \tag{3}$$

where $V_{out}$ and $I_{out}$ are the predefined output voltage and current, which sensor node requires in the active mode. The $Effcy$ parameter stands for the efficiency of the PMIC and is given by its data sheet. When the duration and period of the active mode is defined, the amount of its operation hours per day can be calculated as follows:

$$Mode\ hours\ per\ Day = \frac{24\ hours \cdot Duration}{Period}. \tag{4}$$

Since for the rest of the daytime the sensor node operates in the standby mode, the standby mode hours per day are defined as subtraction of the operation hours of n active modes from 24 h:

$$Standby\ mode\ hours\ per\ Day = 24 - \sum_{i=1}^{n} Mode_i\ hours\ per\ Day. \qquad (5)$$

The current and power consumed in the standby mode can be defined using Eqs. 1 and 2 respectively. Additionally, it is important to take into account the leakage current of the storage element $I_{stor\ leak}$ and quiescent current of the PMIC $I_{quiescent}$:

$$mAh\ per\ Day_{leak} = 24 \cdot (I_{stor\ leak} + I_{quiescent}); \qquad (6)$$

$$mWh\ per\ Day_{leak} = V_{stor} \cdot mAh\ per\ Day_{leak}. \qquad (7)$$

The instantaneous power required from the solar cell can be calculated as ratio of the power required from the solar cell per day during charging days $P_{req}$ to the efficiency of PMIC:

$$P_{req\ min}(mW) = \frac{P_{req}}{Effcy}. \qquad (8)$$

$P_{req}$ can be defined as follows:

$$P_{req} = \frac{Total\ mWh\ per\ Day}{CHD}. \qquad (9)$$

CHD represents charging hours per day and should be predefined for a particular application.

The minimum required solar cell area can be defined using Eq. (10):

$$A_{sc} = \frac{P_{req\ min}}{P_{sc}}, \qquad (10)$$

where $P_{sc}$ is the solar cell power density from the lowest expected light level. It can be taken either from the datasheet of the employed solar cell or, preferably, from preliminary experiments.

## 4.2 Calculation Example for Autonomous Optical Sensor Module

In this subsection we demonstrate the calculation of the optimized power parameters for an autonomous optical sensor node using the mathematical model described in Sect. 4.1. Table 2 presents the known characteristics of the sensor module design and the given application conditions.

We define the "Active mode 1" when the sensor node operates as a transmitter. This mode is the most power consuming due to the active LED. The "Active mode 2" stands for the operation as a receiver. This mode requires power for the functioning of the amplifier circuit of the photodiode. The duration time of transmission mode we set to 240 ms. This time is enough to send measurement data from the sensor node according to the designed modulation scheme [3] for 10 times.

**Table 2.** Given characteristics of the optical sensor node design and environmental conditions

| Modes | Vout | Iout | Effcy | Duration | Period |
|---|---|---|---|---|---|
| Active mode 1 | 2.8 V | 10 mA | 0.9 | 0.24 s | 300 s |
| Active mode 2 | 2.8 V | 1 mA | 0.9 | 0.48 s | 300 s |
| Vstor | 4.3 V | | | | |
| BQ25570 Iquiescent | 488 nA | | | | |
| Storage leakage Istor leak | 0.5 µA | | | | |
| Charging hours per day CHD | 24 | | | | |
| Solar cell power density | 0.016 mW/cm$^2$ | | | | |

The reception duration is doubled. We assume that the node is able to operate in active modes every 5 min (300 s). The designed optical wake-up system of a solar cell can activate the node from fully off-state. Therefore, the node does not require any power during the standby mode. For this reason, we exclude the standby mode from calculations. We also assume that in the industrial environment, where the shift work is quite common (at night as well), the light in the hall is always present. Thus, we set CHD to 24 h. Amorphous silicon solar cells have shown the best efficiency under the indoor illumination. Therefore, we take the measurement results of this material taken at 500 lx for the solar cell power density parameter. Table 3 demonstrates the calculation results.

**Table 3.** Calculated optimal power parameters for the optical autonomous sensor module

| Modes | Istor [mA] | Mode hours per day | mAh per day | mWh per day |
|---|---|---|---|---|
| Active mode 1 | 7.33 (Eq. 3) | 0.019 (Eq. 4) | 0.139 (Eq. 1) | 0.597 (Eq. 2) |
| Active mode 2 | 0.72 (Eq. 3) | 0.038 (Eq. 4) | 0.028 (Eq. 1) | 0.119 (Eq. 2) |
| Leakage | 0.001 (Istor + Iquiescent) | 24 | 0.024 (Eq. 6) | 0.103 (Eq. 7) |
| Total mWh per day | 0.819 mWh/day | | | |
| Preq | 0.034 mW (Eq. 9) | | | |
| Preq min | 0.038 mW (Eq. 8) | | | |
| Solar cell area | 2.411 cm$^2$ (Eq. 10) | | | |

The results show that for the defined operation conditions the node requires 38 μW of instantaneous power from the solar cell. Therefore, the solar cell area should be no less than 2.411 cm$^2$. The solar cell area is the main limiting factor in the sensor node miniaturization. With the help of precise calculation, we are able to design the miniature sensor node with the area less than a one-euro coin.

## 5   Design of the Energy Harvesting Unit

In the design of the optical autonomous sensor node we base our choice of the solar cell on the results obtained in the calculations from the previous section. We selected two amorphous silicon solar cells from the Amorton catalog (Panasonic) with different dimensions: AM-1819 and AM-1456 (Fig. 2b). The parameters of the BQ25570 PMIC were set according to the Fig. 2a.

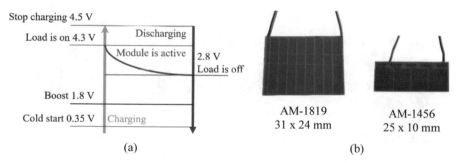

**Fig. 2.**  a. Principal of BQ25570 PMIC operation, b. Amorphous silicon solar cells

We conducted experiments to investigate charging and discharging times of the designed energy harvesting unit for the selected solar cell types in different illumination conditions. On the one hand, we measured the charging time under ambient light illumination of around 450 lx. On the other hand, we used a smartphone flashlight to speed up the charging. Table 4 shows that a small solar cell with the area of 2.5 cm$^2$ can be charged within 5 min under ambient light and would require only 13.6 s for charging with the help of a smartphone flashlight. The discharging time for both solar cells is 240 ms. Larger solar cells would also suit the application. However, since the miniature design is preferable, we selected the smaller one, AM - 1456, for the further node implementation.

**Table 4.** Experiments on the charging and discharging times of the energy harvesting unit

| Solar cell | Area | Charging | | Discharging | Energy storage |
|---|---|---|---|---|---|
| | | Ambient light (450 ± 20 lx) | Smartphone (17 000 lx) | | |
| AM - 1819 | 31 × 24 mm | 2 m 31 s | 6.6 s | 0.24 s | 1 mF |
| AM - 1456 | 25 × 10 mm | 4 m 26 s | 13.6 s | 0.24 s | 1 mF |

Figure 3 demonstrates the evaluation board with the main electric circuit in the middle; it is designed according to Fig. 1. The main circuit area corresponds to the dimensions of the smallest solar cell. The circuit comprises both the communication unit and the energy harvesting unit. The LED and the photodiode are located on the side of the evaluation board. In the designed prototype, they are placed close to the main circuit, extending the node length up to 32 mm. The electronic components of the main circuit are the smallest available on the market.

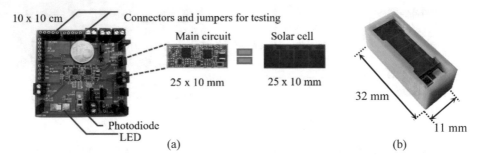

**Fig. 3.** a. Evaluation board for optical autonomous sensor node. b. Designed prototype

The measurement results are consistent with the parameters obtained from mathematical modeling and show the appropriateness of the designed energy harvesting system for the application.

## 6   Conclusion

The aim of the present research was to examine the use of power management ICs in order to achieve efficient energy harvesting for autonomous optical sensor nodes. This study has shown the possibility of combining the energy demanding communication with a LED with energy harvesting from a miniature solar cell. The novel design of the energy harvesting unit in the optical sensor module provides for optical wireless communication for 240 ms by sending 10 measurements, i.e., every 5 min under the indoor ambient light conditions. The resulting node dimensions are 32 × 11 × 10 mm. The relevance of the proposed mathematical model was clearly supported by the experimental results.

**Acknowledgments.** This work was financially supported by the Lower Saxony Ministry for Science and Culture, Germany, within the framework of "Tailored Light" project.

# References

1. Shaikh, F.K., Zeadally, S.: Energy harvesting in wireless sensor networks. A comprehensive review. Renew. Sustain. Energy Rev. **55**, 1041–1054 (2016)
2. Dudko, U., Overmeyer, L.: Intelligent photoelectric sensor module utilizing light for communication and energy harvesting. In: DGaO Proceedings (2017)
3. Dudko, U., Pflieger, K., Overmeyer, L.: Optical autonomous sensor module communicating with a smartphone using its camera. In: Proceedings of the Smart Photonic and Optoelectronic Integrated Circuits XXI (SPIE OPTO), San Francisco, CA, USA, vol. 10922, p. 1092201 (2019). https://doi.org/10.1117/12.2506777
4. Penella-López, M.T., Gasulla-Forner, M. (eds.): Powering Autonomous Sensors: An Integral Approach with Focus on Solar and RF Energy Harvesting. Springer, Heidelberg (2011)
5. Li, J., Liu, A., Shen, G., Li, L., Sun, C., Zhao, F.: Retro-VLC. In: Proceedings of the 16th International Workshop on Mobile Computing Systems and Applications, pp. 21–26 (2015)
6. Duquel, A., Stanica, R., Rivano, H., Desportes, A.: Decoding methods in LED-to-smartphone bidirectional communication for the IoT. In: Global LIFI Congress (GLC), pp. 1–6 (2018). https://doi.org/10.23919/glc.2018.8319118
7. Wang, Z., Tsonev, D., Videv, S., Haas, H.: Towards self-powered solar panel receiver for optical wireless communication. In: Proceedings of the 2014 IEEE International Conference on Communications (ICC 2014), pp. 3348–3353. Institute of Electrical and Electronics Engineers (IEEE) (2014). https://doi.org/10.1109/icc.2014.6883838
8. Rakia, T., Yang, H.-C., Gebali, F., Alouini, M.-S.: Optimal design of dual-hop VLC/RF communication system with energy harvesting. IEEE Commun. Lett. **20**(10), 1979–1982 (2016)
9. Shin, W.-H., Yang, S.-H., Kwon, D.-H., Han, S.-K.: Self-reverse-biased solar panel optical receiver for simultaneous visible light communication and energy harvesting. Opt Express **24**(22), A1300–A1305 (2016). https://doi.org/10.1364/oe.24.0a1300
10. Visconti, P., Ferri, R., Pucciarelli, M., Venere, E.: Development and characterization of a solar-based energy harvesting and power management system for a WSN node applied to optimized goods transport and storage. Int. J. Smart Sens. Intell. Syst. **4**(9), 1637–1667 (2016)
11. Lee, C.-T., Liang, Y.-H., Chou, P.H., Gorji, A.H., Safavi, S.M., Shih, W.-C., Chen, W.-T.: EcoMicro. Energy Convers. Manag. **80**, 1–6 (2014)

# The Model of a Cyber-Physical System for Hybrid Renewable Energy Station Control

Dmitry G. Arseniev, Vyacheslav P. Shkodyrev,
and Kamil I. Yagafarov[(✉)]

Peter the Great St. Petersburg Polytechnic University, Saint Petersburg, Russia
kem4lk@gmail.com

**Abstract.** Cyber-Physical Systems (CPSs) are a modern engineering system class based on the synergy of software and hardware components. In CPSs, embedded computers and networks monitor and control physical processes using feedback loops where physical processes have impact on calculations and vice versa. The present paper is devoted to the CPSs development for the control of the energy production process.

**Keywords:** Control system · Cyber-physical system · Intelligent automation · Renewable energy production

## 1 Introduction

Several modern computing systems are represented as a combination of cyber and physical systems, which are independently developed. That resulted in the appearance of a new approach where cyber and physical components are connected at all levels. Various sources describe cyber-physical systems (CPSs) as in the following.

A CPS is an integration of computation with physical processes whose behavior is defined by both cyber and physical parts of the system. Embedded computers and networks monitor and control the physical processes, usually with feedback loops where physical processes affect computations and vice versa. As an intellectual challenge, a CPS is about an intersection, not a union, of the physical and the cyber [1].

CPSs are defined as systems which offer integration of computation, networking, and physical processes, or, in other words, as such systems where physical and software components are deeply intertwined, each operating on different spatial and temporal scales, exhibiting multiple and distinct behavioral modalities, and interacting with each other in a myriad of ways that change with the context. Some of the defining characteristics of CPSs include:

- Cyber capability in every physical unit
- High-degree of automation
- Networking at multiple scales
- Integration at multiple temporal and spatial scales
- Reorganizing/reconfiguring dynamics [2].

D. G. Arseniev et al. (Eds.): CPS&C 2019, LNNS 95, pp. 388–397, 2020.
https://doi.org/10.1007/978-3-030-34983-7_37

The term of a cyber-physical system refers to the tight conjoining of and coordination amid computational and physical resources. Research advances in cyber-physical systems promise to transform our world with the systems that can respond more quickly and be more precise, work in dangerous or inaccessible environments, provide large-scale distributed coordination, are highly efficient, augment human capabilities, and enhance societal well-being. These capabilities will be implemented by deeply embedded computational intelligence, communication, control, and new mechanisms for sensing and actuation of and adaptation to physical systems with active and reconfigurable components [3].

Therefore, we can summarize and define CPSs as a modern engineering system class based on the synergy of software and hardware units. This includes such concepts as:

- Intellectualization
- Network hierarchy
- Self-organization and development [4].

The concept of intellectualization is associated with the ability of a cyber-physical system to extract knowledge from the environment through multiple sensors and channels, to accumulate knowledge in special storages (knowledge bases), and to use it for decision-making. The idea is to implement artificial intelligence approaches, such as machine learning based on knowledge graphs [5], for data processing.

The network hierarchy implies the use of distributed management principles, i.e., management of a set of distributed objects. It is necessary to take into account that the objects (also known as agents) are able to interact not only with each other but also with the environment. Therefore, it is required to use technologies that allow continuous information exchange and agents' behavior correction under environmental condition variations. For instance, this includes swarm control in unmanned aerial vehicles [6, 7].

Finally, the third concept includes principles of self-organization and development of the management environment on the basis of unsupervised, semi-supervised or supervised learning. Such approaches are often implemented in robotic systems in order to achieve autonomy [8].

The work of a modern CPS can be evaluated and optimized by several key parameters. The relevance of this study lies in the need of efficiency improvement of such systems which are shown through the example of a hybrid renewable energy station (HRES). The first step of the research was aimed at preparing the initial data (dimension reduction, normalization, feature selection). The next step was devoted to the analysis of data dependences, i.e., evaluation of how the input parameters (rated equipment power, weather condition, equipment condition, resource potential) affect the output ones (amount of energy produced, resource substitution, maintenance cost, etc.). The complexity is aggravated by the fact that improvement of some indicators leads to degradation of others. For example, the energy production increase results in the growth of resource consumption, equipment deterioration, etc. Thus, the problem of multi-criteria optimization should be solved. It is necessary to find the best input parameter combination to obtain the best performance at the output.

# 2   Intelligent Data Analysis Using Cognitive Models

## 2.1   Problem Formulation

A technological process is a sequence of actions aimed at creating the final product with the required characteristics. Materials, energy and information are transferred and/or transformed at each stage of this process. In terms of automation and control, these processes are defined as a controlled object, which is managed by an automated control system.

Let us consider a certain technological process of energy production by an HRES [9]. The input data is determined by a set of signals $X(t) = \{x_1(t), x_2(t), \ldots, x_n(t)\}$ collected from different sensors. Other inputs are environmental condition $E(t) = \{e_1(t), e_2(t), \ldots, e_m(t)\}$ that have an influence on the technological process. For instance, those can be weather condition (solar irradiation, wind potential power, etc.). The final input consists of control signals $U(t) = \{u_1(t), u_2(t), \ldots, u_k(t)\}$. The output can be divided into two categories. There are physical products $Y(t) = \{y_1(t), y_2(t), \ldots, y_i(t)\}$ as a result of the technological process and information signals $G(t) = \{g_1(t), g_2(t), \ldots, g_j(t)\}$ that can be considered as key performance indicators (KPIs). KPIs correspond to the technological process, so there is a great amount of alternatives, like product quality, productive efficiency, resource cost, profitability, safety, etc. We will take a close look at them in Sect. 3.1. A simplified model of the technological process as a controlled object is presented in Fig. 1.

As described above, the problem lies in solving a multi-criteria optimization problem, where the criteria are represented as key performance indicators. It is necessary to define a set of input data that would allow to get the most effective output. It can be formulated as

$$G = \min(f_1(x), f_2(x), \ldots, f_k(x)), \ x \in X, \tag{1}$$

where $k \geq 2$ is the number of the criteria and $X$ is a feasible set of decision vectors. Pareto optimal solution [10] is one of the possible ways to handle this task. Formally, a feasible solution $x_1 \in X$ is more preferable than another solution $x_2 \in X$, if

$$f_i(x_1) \leq f_i(x_2) \text{ for all indices } i \in \{1, 2, \ldots, k\}, \tag{2}$$

$$\text{and } f_j(x_1) < f_j(x_2) \text{ for at least one index } j \in \{1, 2, \ldots, k\}. \tag{3}$$

In this case, the solution $x^* \in X$ is called Pareto-optimal [11].

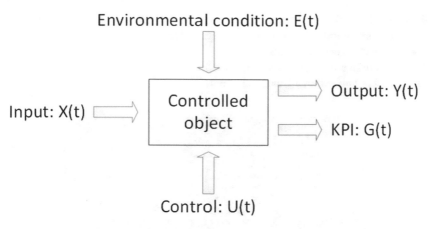

**Fig. 1.** The model of a technological process as a controlled object

## 2.2 Controlled Object

The technological process of energy production from renewable sources and its conversion is considered as the controlled object. Wind and solar activity potential is widespread and common, but it has disadvantages, such as the dependence on geographical location and weather condition. E.g., solar irradiation is only 1 kW/m² on a clear day and its annual average value does not exceed 250 W/m²; the average wind flow power density is approximately 250 W/m² in the case of 10 m/s wind speed, but usually it is less; the average water flow power density is the same in the case of 1 m/s water flow speed [12]. Consequently, provision of a sufficient amount of energy requires the usage of an energy storage system and additional sources like diesel generators. In that case, renewable energy sources allow to save 50–60% of diesel fuel. It is significant for remote locations that are not connected to power grids and have to burn fuel for electricity production. At the same time, a control system is required [13]. As a result, the power plant is able to provide the consumer with sufficient amount of electricity at any time. If weather condition do not allow to produce electricity from renewable sources or the accumulators are discharged, the control system will use additional sources, e.g., diesel generators. So, the energy production can be represented as a graph showing objects and processes and cause-and-effect relations between them (see Fig. 2).

Environmental conditions are characterized by geographical location that has an influence on climatic conditions. Another factor is the amount of energy required by the consumer at each moment of time. Input parameters consist of equipment characteristics, e.g., the nominal power. The output channel presents results of the HRES operation expressed in the amount of produced electricity, its quality, etc. The final block contains various control actions as follows:

- Switch on/off solar plants and corresponding DC-AC converters, e.g., in the case of malfunction or maintenance

- Switch on/off wind generators and corresponding DC-AC converters, e.g., in the case of storm wind that can damage impeller blades
- Switch on/off diesel generators
- Switch batteries to charge/discharge mode
- Battery charge/discharge mode selection.

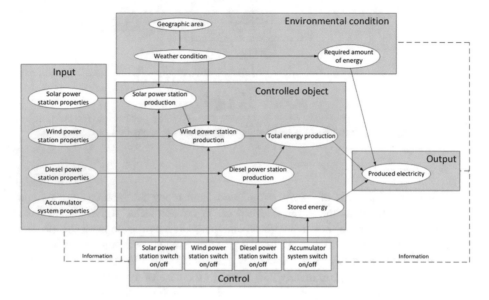

**Fig. 2.** The model of a technological process as a controlled object (detailed)

## 3    Simulation Results

### 3.1    Dataset

The data obtained from a certain HRES were used as the experimental dataset. This power plant uses various energy sources, such as the solar array, diesel driven generators and a storage system to provide the nearby villages with electricity. Through the data analysis, we have selected four KPIs to evaluate the efficiency of the energy production process. The time series consists of the following technological parameters (see Fig. 3):

- Fuel consumption rate that should be minimized
- Output active power
- Output frequency and voltage that should be equal to 50 Hz and 400 V.

As it could be seen, these parameters can deviate from their nominal values. In some situations, this is connected with specific electrical appliances used in the villages, e.g., pumps or bakery equipment. This condition could not be changed. In other cases, KPIs can be improved by the control system that should identify the reason of deviation and remove or minimize it.

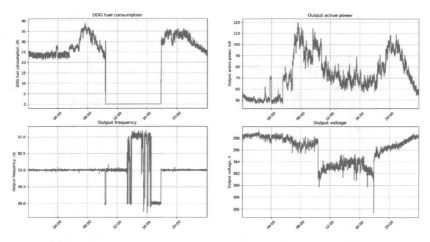

**Fig. 3.** Key performance indicators in the HRES energy production

## 3.2 Intelligent Data Analysis

We propose to use different approaches in the model of a cyber-physical control system.

### 3.2.1 Principle Component Analysis

The principal component analysis (PCA) is one of the approaches to reduce the data dimension and find correlation between the parameters [14, 15]. It is used in many areas, such as pattern recognition, computer vision, data compression, and so on.

A PCA plot in Fig. 4 demonstrates that accumulated in the storage energy and energy from solar panels are similar in the way they impact the output frequency. The control system should pay attention to their work to monitor the output frequency that must be equal to 50 Hz. At the same time, the diesel generator has a minor influence on the considered KPI.

### 3.2.2 Cluster Analysis

Cluster analysis is a multidimensional statistical procedure focused on data collection and separation into groups (called clusters) [16]. The clustering task relates to statistical data processing, as well as to unsupervised learning. It usually is implemented for solving the following tasks [17]:

- Typology or classification development
- Hypothesis generation based on data research
- Hypothesis testing.

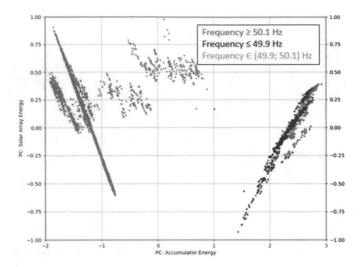

**Fig. 4.** Principle component analysis of the output frequency

Figure 5 presents plots where the data were separated into two clusters. The first, circumscribed with the green line, represents the correct behavior of the power plant, i.e., the interface voltage should be approximately 400 V. The second one, circumscribed in red, shows cases where the output voltage does not correspond to the required value, so the control system should take appropriate measures to prevent these situations.

### 3.2.3    Trend and Predictive Analytics

Trend analysis is a popular approach to accumulate data and spot patterns [18]. Although trend analysis is often used to predict future events, it could be used to estimate uncertain events in the past. For instance, Fig. 6 demonstrates predicted behavior of the system based on its current behavior.

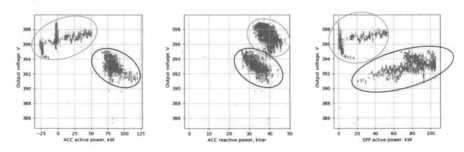

**Fig. 5.** Output frequency cluster analysis

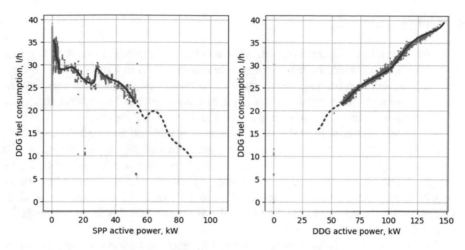

**Fig. 6.** Fuel consumption prediction analysis

Predictive analytics is a class of data analysis methods that focuses on predicting the future behavior of objects and subjects in order to make optimal decisions [19]. Predictive analytics uses statistical and data mining methods, game theory; it analyzes current and historical data to make predictions about future events. Models represent relationships among many factors to evaluate risks and make decisions.

### 3.2.4  Blind Source Separation

The described method is presented as separation of a signal set from a variety of mixed signals with no information about their sources [20]. Another approach, quite close to the blind source separation, is independent component analysis. It is a statistical and computational technique for revealing hidden factors from raw data [21]. Figure 7 shows an example where the output active power is presented as a combination of the active power of the accumulator, solar panels and diesel generator.

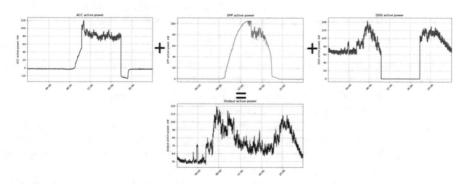

**Fig. 7.** Independent component analysis of output active power

## 4    Conclusions

This paper is devoted to the cyber-physical system development to control a hybrid renewable energy station. Its efficiency can be estimated via key performance indicators, such as fuel consumption rate, amount and quality of produced electricity, etc. Therefore, we propose to use the following approaches: principal component analysis, cluster analysis, blind source separation, trend and predictive analysis.

As a result, the technological process of energy generation was analyzed, real data experiments were made to demonstrate the possibilities of the chosen algorithms. It allows us to produce recommendations for the optimal usage of the HRES. For example, electrical circuits from energy storage system should be checked because of the grid energy losses; diesel generator should be loaded by 40–80% to produce electricity with the required quality; solar DC-AC converters should be set properly, because they perturb the grid, etc.

The experimental results demonstrate the effectiveness of the proposed approach and show the necessity of further research. More experiments with real data will be performed and the prototype CPS will be developed to control the process of energy generation on a real power station.

**Acknowledgments.**    We thank the Siemens Company for its support. The research was covered by the donation agreement No. 1 CT-SPbPU-2016 as of 29.09.2016 to Peter the Great St. Petersburg Polytechnic University.

## References

1. Lee, E.A., Seshia, S.A.: Introduction to Embedded Systems – A Cyber-Physical Systems Approach, 2nd edn, 564 p. MIT Press, Cambridge (2017)
2. Khaitan, S.K., McCalley, J.D.: Design techniques and applications of cyber physical systems: a survey. IEEE Syst. J. **9**(2), 350–365 (2015)
3. US National Science Foundation (2010). https://www.nsf.gov/pubs/2010/nsf10515/nsf10515.htm
4. Shkodyrev, V.P.: Technical systems control: from mechatronics to cyber-physical systems, studies in systems, decision and control. In: Gorodetskiy, A.E. (ed.) Smart Electromechanical Systems, vol. 49, pp. 3–6. Springer, Cham (2016)
5. Xie, R., Liu, Z., Sun, M.: Representation learning of knowledge graphs with hierarchical types. In: Proceedings of the 25th International Joint Conference on Artificial Intelligence, New York, USA, 9–15 July 2016, pp. 2965–2971 (2016)
6. McCune, R.R., Madey, G.R.: Swarm control of UAVs for cooperative hunting with DDDAS. Procedia Comput. Sci. **18**, 2537–2544 (2013)
7. Trinh, M.H., Zhao, S., Sun, Z., et al.: Bearing-based formation control of a group of agents with leader-first follower structure. IEEE Trans. Autom. Control **64**(2), 598–613 (2018)
8. Silva, F., Correia, L., Christensen, A.L.: Evolutionary online behavior learning and adaptation in real robots. Roy. Soc. Open Sci. **4**(7), 15 (2017)
9. Wang, Y., Liu, D., Sun, C.: A cyber physical model based on a hybrid system for flexible load control in an active distribution network. Energies **10**(3), 267–286 (2017)

10. Tomoiagă, B., Chindriş, M., Sumper, A., et al.: Pareto optimal reconfiguration of power distribution systems using a genetic algorithm based on NSGA-II. Energies **6**, 1439–1455 (2013)
11. Mornati, F.: Pareto optimality in the work of pareto. Eur. J. Soc. Sci. **51**(2), 65–82 (2013)
12. Fortov, V.E., Popel, O.S.: Renewable energy sources in the world and Russia [Vozobnovlyayemyye istochniki energii v mire i v Rossii]. In: Popel, O.S. (ed.) Proceedings of the 1st International Forum "Vozobnovlyaemaya energetika. Puti povysheniya energeticheskoy i ekonomicheskoy effektivnosti" (REENFOR-2013), Moscow, Russia, 22–23 October 2013, pp. 12–22. OIVT RAN Publ., Moscow (2013). (in Russian)
13. Elistratov, V.V.: Renewable Energy [Vozobnovlyaemaya energetica]. Alternativnaya energetica i ecologiya **1**, 142–143 (2017). (in Russian)
14. Richardson, M.: Principal Component Analysis, pp. 1–23 (2009)
15. Jolliffe, I.T., Cadima, J.: Principal component analysis: a review and recent developments. Philos. Trans. Ser. A Math. Phys. Eng. Sci. **374**(2065), 16 p. (2016)
16. Mythili, S., Madhiya, E.: An analysis of clustering algorithms in data mining. Int. J. Comput. Sci. Mob. Comput. **3**(1), 334–340 (2014)
17. Battaglia, O.R., Paola, B.D., Fazio, C.: Cluster Analysis of Educational Data: an Example of Quantitative Study on the Answers to an Open-Ended Questionnaire, 30 p. (2015)
18. Jain, A., Gupta, A., Gupta, A., et al.: Trend-based on networking driven by big data telemetry for SDN and traditional networks. Int. J. Next-Gener. Netw. **11**(1), 1–15 (2019)
19. Selvaraj, P., Marudappa, P.: A survey of predictive analytics using big data with data mining. Int. J. Bioinform. Res. Appl. **14**(3), 269–282 (2018)
20. Pal, M., Roy, R., Basu, J., Bepari, M.S.: Blind source separation: a review and analysis. In: Proceedings of the International Conference Oriental COCOSDA held jointly with 2013 Conference on Asian Spoken Language Research and Evaluation (O-COCOSDA/CASLRE), 25–27 November 2013 (2013)
21. Naik, G.R., Kumar, D.K.: An overview of independent component analysis and its applications. Informatica **35**, 63–81 (2011)

# Linear Direct Drive for Light Conveyor Belts to Reduce Tensile Forces

Malte Kanus[1], Alexander Hoffmann[2], Ludger Overmeyer[1(✉)], and Bernd Ponick[2]

[1] Institute of Transport and Automation Technology, Leibniz University Hannover, Hanover, Germany
{malte.kanus,ludger.overmeyer}@ita.uni-hannover.de
[2] Institute for Drive Systems and Power Electronics, Leibniz University Hannover, Hanover, Germany
alexander.hoffmann@ial.uni-hannover.de

**Abstract.** Due to increasing demands of the mass flow and transport lengths, the use of intermediate drives for continuous conveyors for both packaged and bulk materials is constantly growing. Intermediate drives allow the transmission of drive forces along the conveyor and thus lead to a reduction in the maximum belt tensile force. This paper presents a new drive concept for light conveyor belts. To reduce the belt tensile force, intermediate drives in form of linear direct drives are allocated along the transport distance. In the first part, a new belt design is presented which enables the implementation of the linear direct drive runner elements. The conveyer belt is characterized by low additional weights of the runner elements and has only a slightly higher bending stiffness compared to conventional conveyor belts, whereby small pulley diameters can be achieved. The second part explains the drive concept in the form of an Integrated Linear Flux Modulating Motor in more details. In particular, possible problems and the developed solutions, which were implemented and verified in a demonstrator, are presented. The results of the research show the high potential of the new drive technology.

**Keywords:** Linear direct drive · Light conveyor belts · Integrated Linear Flux Modulating Motor · Reducing tensile forces

## 1 Introduction

In conventional belt conveyors, the driving force is transmitted into the system via drive pulleys. Friction between the pulley and the belt results in a friction-locked power transmission. For a high force transmission with a low coefficient of friction µ, a high normal force must be applied. For this purpose, the belt must be subjected to a correspondingly high pretension. This is shown schematically in Fig. 1(left). The total belt tensile force (yellow) is applied perpendicular to the direction of action, which is divided into the constant pretension (red) and the position-dependent belt tensile force. The belt tensile force is at its greatest when it runs onto the drive pulley and is reduced

© Springer Nature Switzerland AG 2020
D. G. Arseniev et al. (Eds.): CPS&C 2019, LNNS 95, pp. 398–406, 2020.
https://doi.org/10.1007/978-3-030-34983-7_38

to the level of the pretension until it runs out. This force progression can be described by Euler-Eytelwein's formula [1]:

$$F_1 \leq F_2 \cdot e^{\mu\alpha}. \tag{1}$$

$F_1$ represents the preload tension and $F_2$ the maximum belt tensile force right before the drive pulley. The maximum belt tensile force depends on the length of the conveyor and on the load condition, which is limited by the nominal strength of the belt. In contrast, Fig. 1(right) shows the schematic distribution of the belt tensile force using two distributed intermediate drives. The force is no longer transmitted via the frictional connection to the drive pulley, so that the preload is significantly reduced. The distributed transmission of drive force reduces the belt tensile force (yellow) along the upper strand, whereas the peak force is significantly lowered compared to conveyors with a conventional drive pulley.

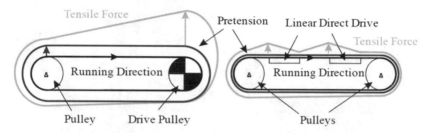

**Fig. 1.** Schematic diagram of the belt tensile force for a belt conveyor; conventional head drive (left), linear direct drives (right) [2]

The main objective of this research work is to reduce the maximum belt tensile forces by distributing the driving forces along the conveyor section. This was achieved by using linear direct drives as intermediate drives. Another objective is to point up the dynamic advantages of a linear direct drive in comparison to conventional drive pulleys. The direct force transmission into the belt enables significantly higher dynamics, resulting in new application scenarios. For this purpose, several different approaches were made in the last decades. Patent [3] describes the use of linear drives for conveyors. Therefore, metallic inlets are placed in the belt edges before vulcanisation; they are used as moving runner elements. A similar method describes the patent in reference [4] where flat metallic inlets are implemented perpendicular to the running direction over the width of the belt. By coupling the elements, they built a loop and can be used as runner elements. Patent [5] shows the use of a conveying device as a runner element for a linear drive, where the perpendicular parts are coupled with a flexible wire mesh. A similar approach with meshed electrically conductive inlets in the belt describes the patent [6]. In addition, the patents [7] and [8] describe related usable inventions. Patent [9] shows an expensive solution where flexible rare earth strips are embedded in the belt as passive runner elements. The US patent [10] describes a conveyor system with a 3-phase linear hybrid motor. It uses metallic runner elements on the conveyor belt for the horizontal movement.

## 2   Linear Direct Drive for Light Conveyor Belts

The linear direct drives used as intermediate drives operate according to the Integrated Linear Flux Modulating Motor principle [11]. This enables the use of soft magnetic materials for the runner elements. This is especially good for long distances, as no expensive permanent magnets have to be integrated along the entire length. Instead, these are accommodated in the stator, which is short relative to the conveying length. Figure 2 shows two different technical designs, developed by the Institute of Transport and Automation Technology (ITA) and the Institute for Drive Systems and Power Electronics (IAL). The left side shows a version where a complete linear direct drive was set between the upper and lower strands. The runner elements are attached to the running side of the belt. Due to these restrictions, only large end diameters are feasible.

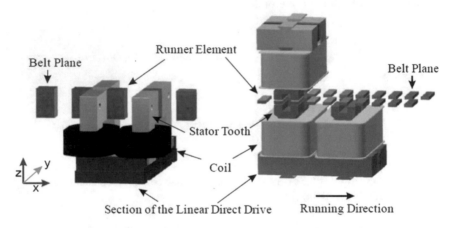

**Fig. 2.** Motor concept with attached runner elements at the running side (left) [11], motor concept with runner elements for integration into the belt (right)

For the implementation of smaller end diameters, ITA and IAL developed a new concept of the linear direct drive. Figure 2 (right) shows the smaller dimensions of the runner elements. This enables the full integration of the runner elements into the belt. The configuration of the stator at the right side is divided into an upper and a lower part. Therefore, only one stator part is located between the upper and the lower strand. This allows to reduce the space between the upper and the lower strands and thus the diameter of the pulleys.

The Integrated Linear Flux Modulating Motor enables operating in a stepper mode [11]. Figure 3 shows the operating principal of a two-phase Integrated Linear Flux Modulating Motor in full step mode.

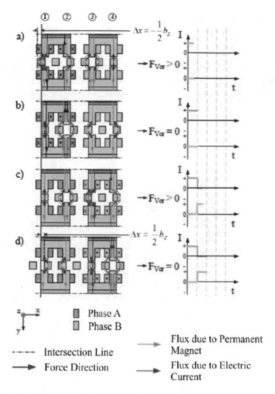

**Fig. 3.** Operating principle of a double-phase Integrated Linear Flux Modulating Motor [11]

In this scenario, only one phase per step carries current, which results in different magnetic flux distributions in the stator teeth 1 to 4 in Fig. 3. The perpendicular parts of the drive force (y-direction) eliminate themselves in case of symmetry. In the first step (a), the current flows as shown in phase A (orange). The magnetic flux is enlarged in the first tooth of the stator and reduced in the second one. This leads to a driving force on the left runner element, which moves the element in positive x-direction into the position shown in Fig. 3 step (b). For the following step (c), the current flows through phase B (turquoise). This leads to the movement to the position (d) similar to the previous step (b). This sequence corresponds to a movement of one tooth pitch $\tau_T$, which corresponds to one electrical period. The repetition of the whole sequence leads to a continuous movement of the runner elements. The velocity $v$ of the movement corresponds to the electrical frequency $f$ multiplied with the tooth pitch:

$$v_1 = \tau_T \cdot f_1. \tag{2}$$

As mentioned, the perpendicular parts of the drive force (y-direction) eliminate themselves in perfect symmetry. In case the runner elements deflect from this position in y-direction, the sum of the forces deviates from zero. Figure 4 shows the simulated normal force $F_N$ in dependence of the displacement y. The data presented here begins with the right side of the runner element, whereas 7.5 mm is the symmetric position. The force increases until the runner element contacts the stator at 8.02 mm. The normal force is about ten times higher than the driving force of the motor. Deviating from the unstable symmetry position, the runner element moves to the stator, which ends up in frictional resistance between the parts. The prevention of this high normal force is one of the challenges that need to be solved for the functional efficiency of the linear direct drive.

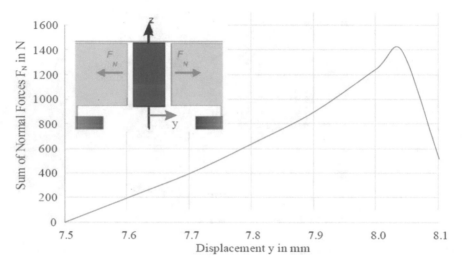

**Fig. 4.** Normal force on the runner element in dependence to its displacement from symmetry [12]

## 3   Belt Structure

There are two main objectives for the belt structure. For the transmission of a continuous driving force, the runner elements must be coupled along the belt. Another challenge of a linear direct drive is the constant spacing between the runner elements. This must be given by the belt structure for the functional capability.

Furthermore, the belt structure must not be too stiff, as otherwise the overcoming of the flexural and bending resistances at the pulleys would require a large portion of the drive forces. In the novel approach presented here, the researchers integrated the runner elements into a conventional conveyer belt. For this purpose, the height of the runner elements was chosen in relation to the belt thickness. For the durability, long-term tests with different heights of the runner elements were performed in the development phase. These tests have shown that a height ratio between a runner element and the belt smaller than one is the best choice. For a low-energy deflection at the pulleys, the length in the running direction of the runner elements has to be minimal. Otherwise, it leads to a polygon-effect known from load chains and plastic modular belts [13]. Small dimensions are contradictory to a high driving force. For a high force transmission, a runner element with a large volume is advantageous. As a result, the runner elements are limited in height and length, so that the width is the only free parameter for a high volume. This leads to a high width transversely to the running direction of the belt. The selected geometry of the runner elements only leads to a small increase of the belt stiffness. This advantage allows choosing small pulley diameters for the demonstrator.

For the integration of the runner elements into the belt, suitable recesses were cut. The belt itself ensures the constant distance between the elements. The belt is covered with a low-friction material over the entire width of the Linear Direct Drive to fix the elements and reduce frictional resistance (see Fig. 5). Due to the motor power, the runner elements push the belt in front of them. By using more than one intermediate drive, the belt tensile force along the conveyor section can be significantly reduced and sufficient drive force is available. The properties of the belt with regard to its transverse stiffness (perpendicular to the running direction), despite the introduction of defects, are sufficient for the design as a secondary part of the motor. This also allows a one-sided drive, which saves a lot of space and weight compared to the two-sided drive.

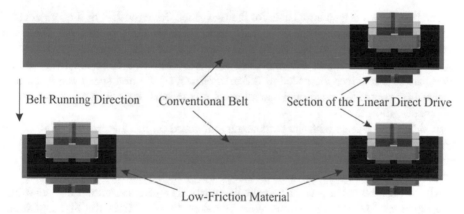

**Fig. 5.** View on a part of the belt with a one-sided drive (top) and a two-sided drive (bottom)

## 4   Demonstrator

The demonstrator combines an Integrated Linear Flux Modulating Motor with the runner elements in the belt. For this purpose, a conventional light conveyor was modified to be used with linear direct drives. Figure 6 shows the demonstrator with a one-sided drive arrangement.

Light Conveyor Belt with Runner Elements — Linear Direct Drive

Deflection Pulley

Deflection Pulley

Control Unit

**Fig. 6.** Demonstrator of linear direct drives for light conveyor belts at the Institute of Transport and Automation Technology

The belt shown is also applicable for the two-sided drive mode. For the guidance of the runner elements through the drive, adjustable bearings are implemented. These are located close to the primary part of the motor and brace single runner elements against the acting normal force. In the one-sided scenario with two allocated linear direct drives, the belt achieves the speed of 0.7 m/s. However, the belt speed can be further increased by controlled positional operations. Conventional logistic conveyors operate with speeds up to 2 m/s.

First long-term tests with the demonstrator focused on the thermal warming of the motor. For the test under laboratory conditions, the current was set at its rated value. The belt speed was set to 0.5 m/s. Figure 7 shows an increase of the temperature. In the first half hour, the temperature was measured every five minutes.

Afterwards, the value was measured half-hourly. The test was started with an initial temperature of 25.6 °C. The temperature increases to 46.7 °C in the first half hour. After two hours, the temperature reaches a consistent value of about 58 °C. With regard to the thermal durability of the single components, this temperature proves the stability of the Linear Direct Drive.

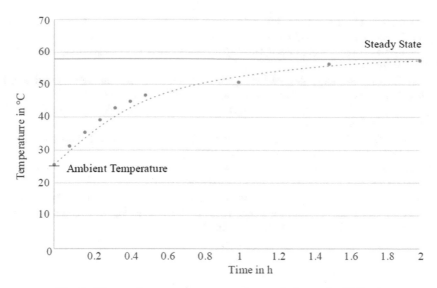

**Fig. 7.** Measured motor temperature rise at a belt speed of 0.5 m/s

## 5   Conclusion

The demonstrator build by ITA and IAL shows the general functionality of Linear Direct Drives for light conveyors. In case of light conveyors, where the belt stiffness is not high enough for the guidance of the runner elements, an additional bearing guidance was developed. This allows reducing the normal forces on the runner elements. The small dimensions of the runner elements in the conveying system increase the belt stiffness only by a small amount. The design and arrangement of the motor allows the use of small pulleys. This enables the transport of packaged goods from preceding and further conveyors without obstructive gaps. Integration of the runner elements into the belt keeps the complexity of the entire system to a minimum.

## References

1. Grote, K.-H., Feldhusen, J.: DUBBEL: Taschenbuch für den Maschinenbau, 23rd edn. Springer, Heidelberg (2011). ISBN 978-3642-17305-9
2. Alles, R.: Zum Zwischenantrieb von Gurtförderern mittels angetriebener Tragrollen und Linearmotoren. Dissertation, Universität Hannover (1976)
3. Gebhardt, R., Heidolf, W.: Fördergurt für den Antrieb von Bandanlagen durch Linearmotoren. Patentschrift 107 882 (1974)
4. Noack, A.: VEB Braunkohlewerk Oberlausitz, Sekundaerteil eines Drehstromlinearmotors fuer Bandantriebe, Patentschrift 201 133 (1983). ISSN 0433-6461
5. Joug, R., Ragout, B.: Pneumatiques, Caoutchouc Manufacture et Plastic Kleber Colombes S. A. Fördervorrichtung, Offenlegungsschrift 2032856 (1970)
6. Keuth, H., Traupe, W., Siemens, A.G.: Fördereinrichtung mit elektrisch durch Wanderfeld angetriebenem endlosen Förderband, Offenlegungsschrift 2134055 (1973)

7. Woodridge, C., Lawson, C.: Brockway Engineering Co. Ltd., Band, insbesondere Förderband, zum Antrieb durch einen Linearmotor, DE 3214811 A1 (1981)
8. Kletterer, H., Kletterer, K., Lewin, H.-U.: Gurtbandförderer mit Linearantrieb, DE 37 41 054 C2 (1997)
9. Lee, R.: Belt including flexible rare earth magnetic strip and conveyor utilizing a belt including a flexible rare earth magnetic strip, US 7597190 B2 (2009)
10. Lucchi, M., Simmendinger, S., Elsner, D.: Habasit AG, hybrid modular belt, US 2017/0088356 A1 (2017)
11. Jaztrembzki, J.-P.: Synchrone Linear-Direktantriebe für Förderbänder, Fortschritt-Berichte VDI Düsseldorf (2014). ISBN 978-3-18-341221-1
12. Radosavac, M.: Untersuchungen zum Abtrieb von Gurtförderern mittels Linearmotor, Berichte aus dem ITA (2018). ISBN 978-3-95900-238-7
13. Katkow, A., Wehking, K.-H.: Kraftübertragung zwischen Lastkette (Flyerkette) und Umlenkrolle. Log. J. Proc. **2016**(10) (2016). ISSN 2192-9084. https://www.logistics-journal.de/proceedings/2016/fachkolloquium2016/4447

# A Cyber-Physical System for Monitoring the Technical Condition of Heat Networks

Gennady I. Korshunov[1]([⊠]), Alexander A. Aleksandrov[2],
and Artur R. Tamvilius[3]

[1] Peter the Great St. Petersburg Polytechnic University, St. Petersburg, Russia
kgi@pantes.ru
[2] Russian Monitoring Systems LLC, St. Petersburg, Russia
[3] Teploelektroproekt St. Petersburg LLC, St. Petersburg, Russia

**Abstract.** Issues of creating a new generation of heat pipe monitoring systems using the method of pulse reflectometry are considered. The control conductors built into the thermal insulation layer along the thermal conductor ensure the detection and localization of an ongoing or forecast accident during the thermal insulation dampening. The introduction of technological innovations in the primary basic structure has led to a reduction in equipment and an increase in system reliability. New hardware and software tools have created the conditions for in-depth analysis of the technical condition of heating networks and solving new control tasks. The proposed approach and its development correspond to the concept of cyber-physical systems. The physical process of wetting insulation is an indirect but significant accident factor. Control of wetting in the prescribed tolerances fixes a sign of an accident when going beyond the tolerance field. However, the value of the parameters of a pulse reflected from the place of moistening, the dynamics of its changes contain important additional information for a possible forecast of the technical condition and energy efficiency of pipelines and heat networks in general. The formation, transmission, processing of information from many objects of heat networks using knowledge bases, the Internet of Things and other elements of the Industry 4.0 concept ensures the generation of near-optimal solutions and the achievement of high energy efficiency.

**Keywords:** Heat networks · Pulse reflectometry · Technological innovations · Energy efficiency · Estimation criterion

## 1 Introduction

The basic structure of the system was created by Karsten Moller at EMS A/S (Denmark) [1] on the basis of locators and related components. The system based on the EMS 4000 locator was adapted in collaboration with EMS A/S for the Russian heat pipelines [2] and introduced by the authors in 2005. The development of cybernetics tools and increased requirements for energy efficiency of heat networks has led to the creation of a new generation of heat conduction monitoring systems using pulsed reflectometry. The control conductors built into the thermal insulation layer along the heat conductor ensure the detection and localization of an ongoing or forecast accident

© Springer Nature Switzerland AG 2020
D. G. Arseniev et al. (Eds.): CPS&C 2019, LNNS 95, pp. 407–412, 2020.
https://doi.org/10.1007/978-3-030-34983-7_39

during the insulation wetting. The physical process of insulation wetting is an indirect but significant accident factor. Control of wetting in the prescribed tolerances fixes a sign of an accident when going beyond the tolerance field. However, the value of the parameters of a pulse reflected from the place of moistening, the dynamics of its changes contain important additional information for a possible forecast of the technical condition and energy efficiency of pipelines and heat networks in general.

## 2   Adaptation of the Basic Structure and the Introduction of Innovations in the Hardware

The EMS 4000 [1] stationary reflectometer software was built on the basis of the MODBUS redundant system, and communication with the control conductors was carried out via matching devices. Monitoring of the heat network without defects that appear or are predicted is described by matched lines with an impedance value of $Z_0 = 200\ \Omega$. The dependence of the impedance of the polyurethane foam insulated pipeline on the distance h from the signal conductor with diameter d to the surface of the pipe is

$$z_0(h, d, z_r) = \frac{\sqrt{\mu_0 \frac{\mu_r}{r_0} a \cosh\left(\frac{2h}{d}\right)}}{2x\sqrt{z_r}}. \tag{1}$$

The $h$ value in European heat pipelines is 18 mm, and in Russian it amounts to 20 mm. The appearance of wetting or the test conductor breaking through which the probe pulse propagates is characterized by alarm thresholds "$-200\ \Omega$", "$+300\ \Omega$".

Figure 1 shows the reflectogram of the EMS 4000 locator with a detected pipe defect (wetting of the insulation) at a distance of 350 m from the instrument. The "Low limit" curve determines the tolerance value and its shape takes into account the

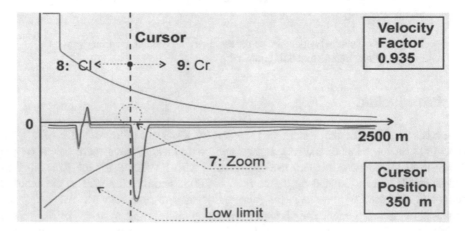

**Fig. 1.** Reflectogram with a detected pipe defect (insulation wetting) at a distance of 350 m from the instrument

attenuation of the pulse in a long line. The negative impulse crosses the tolerance line. Other details of the technology are known and not presented here.

Operating practices and further research led to the need for technological innovation and the creation of a new generation of hardware (see Fig. 2). The task of equipment modernizing built on the basis of EMS 4000 [1] included a significant improvement in the performance of the reflectometer. It was envisaged the elimination of bulky and technologically difficult for embedding matching devices and the combination of control and transmission functions in one processor. It was proposed to replace the redundant MODBUS-based software with specialized software that provides more in-depth monitoring and forecasting of the state of heat pipes in real time. In the considered system, the stationary reflectometer is several times smaller in size and mass, and specialized software was developed for it. A special flat cable with an impedance value of $Z_0 = 200 \ \Omega$ was developed and patented for direct connection to the heat pipe without the use of matching devices. The low-level processor for local control is combined with a radio modem for packet transmission of pulse parameters (previously, these were separate devices). A GPRS communication channel with a top-level computer was introduced. These technical innovations allowed us to simplify the hardware and software and increase the reliability of the system. Thus, the possibility for an in-depth analysis of the primary processes of thermal insulation wetting, their digitalization, transmission for processing, accumulation of databases and knowledge, and support for management decisions on the resources distribution appeared.

**Fig. 2.** Pilot model of a new generation monitoring system

## 3   Creation of the Cyber-Physical System for Monitoring the Technical Condition of Heat Networks

A top-level computer provides for the organization of tens/hundreds of monitoring systems for heat pipelines across the region, the operation of databases and knowledge bases for storage, cognitive processing of information, and support for managerial decision-making. The direct penetration of methods and means of identification and evaluation into the physical processes of wetting insulation allows minimizing the influence of the "human factor" and generating options for management decisions to achieve near-optimal energy efficiency values.

The new tasks now are the study and assessment of the dynamics of changes in the pulse in the tolerance field and prediction of an emergency condition. For this purpose, fuzzy models and algorithms for estimating the behavior of parameters within the tolerance are used. The approach proposed in references [3, 4] includes an assessment of the parameter change and is based on the application of six local criteria and a complex criterion for assessing the level of system performance. The criteria are formed on the basis of time series, linguistic variables, membership functions and the proposed fuzzy classifier algorithm.

The criteria system includes:

- Operating margin on the admission of the parameter
- Stability of the trend of the parameter change
- The value of the translational change of the parameter
- Rate of change of the parameter
- Acceleration of the parameter change.

"The value of the reverse parameter change" criterion given in reference [4] makes no sense if the insulation becomes wet and is not considered.

The convolution of criteria is presented as an integral "level of efficiency" criterion and can be interpreted as an energy efficiency criterion for the heat conductor. Convolution is performed using the Mamdani fuzzy inference algorithm as a procedure for determining the quantitative value of the output linguistic variable in the form of a real number [4]. The process of transition from the membership function of the output linguistic variable $\mu_{acc}(y)$ to its precise value $y^*$ is performed by the method of the left modal value. The output variable value is defined as the modal value of a fuzzy set for the corresponding output linguistic variable or the smallest modal value (leftmost) if the fuzzy set has several modal values

$$y^* = \min\{y_m\} = \min\{\arg\ \max\{\mu_{acc}(y)\}\},  \tag{2}$$

where $y_m$ is the modal value of the fuzzy set with $\mu_{acc}(y)$.

Another option for convolution of criteria is the center of gravity method [5]. The center of gravity of the complex indicator is the set of the centers of gravity of the fuzzy sets entering this state. The distance between two states is represented by the Euclidean distance between their centers of gravity. It is assumed that centers of gravity of any two fuzzy sets do not match.

The simplest way to calculate $K$ is linear convolution modified to reflect the fuzzy-multiple nature of the exponents. In this case:

$$K_{ijklm} = q_1\overline{Z}_i + q_2\overline{Z}_j + q_3\overline{Z}_k + q_4\overline{Z}_l + q_5\overline{Z}_m;$$ (3)

$$\overline{Z}_i = \frac{\int_0^1 Z\mu_i dZ}{\int_0^1 \mu_i dZ}.$$ (4)

where $q_1,...,q_5$ are the weights of the local indicators, $\overline{Z}_i$ is a clear value of the local indicator $Z_i$ after dephasing it by the center of gravity method. In addition, each of the indices $i, j, k, l, m$ can take values from 1 to 5.

An in-depth study of the technical condition and energy efficiency of heat networks leads to the need to create an interactive map of the territory in question with indication of the necessary investments.

It becomes relevant and possible to create a managed database on the status of heating networks, determine service priorities, and allocate resources for energy efficiency.

## 4   Conclusion

The formulation and solution of the above-mentioned and some other tasks makes it possible to classify the monitoring system of the technical state of heating networks as cyber-physical, and to further their development in terms of the "Industry 4.0" concept. Such solutions include determining priorities for the distribution of technical and economic resources, network reconfiguration, assessment of the energy efficiency of heating networks and energy saving of thermal energy. The given mathematical models are used to interpret behavior of the primary parameters of moistening at the level of each monitoring system for the formation and selection of management decisions. The planned implementation of self-learning and adaptation algorithms will reduce the unreliable influence of the "human factor" and improve the cyber-physical system based on the operating experience gained.

## References

1. European monitoring systems. http://www.ems-as.dk
2. Moller, K., Korshunov, G.: Development and adaptation to the Russian conditions of the technology of monitoring of heat networks with polyurethane insulation based on stationary reflectometers. In: Proceedings of the IEHS 2007, pp. 253–259. SUAI, St. Petersburg (2007). (in Russian)

3. Korshunov, G., et al.: Fuzzy models and system technical condition estimation criteria. In: Proceedings of ICICT 2019 – Fourth Conference in the Series International Congress on Information and Communication Technology, 25–26 February 2019. Brunel University, London (2019)
4. Korshunov, G., Smirnov, V., Milova, V.: Multi-criteria fuzzy model for system technical condition estimation at the life cycle stages. In: International Workshop "Advanced Technologies in Material Science, Mechanical and Automation Engineering" (MIP: Engineering-2019), Krasnoyarsk, Russia, 4–6 April 2019, Within the Framework of XXIV International Scientific and Research Open Conference "Modern Informatization Problems", Yelm, WA, USA, p. 131 (2019)
5. Milova, V., Milova, N.: The approaches to the formalizing of the uncertainty for the fuzzy defined objects and systems description and researching. J. Voprosy radioelektroniki **7** (2013). (in Russian)

# Enhancing the Performance of Reservation Systems Using Data Mining

Elena N. Desyatirikova[1], Alkaadi Osama[2], Vladimir E. Mager[3($\boxtimes$)],
Liudmila V. Chernenkaya[3], and Ahmad Saker Ahmad[4]

[1] Voronezh State Technical University, Voronezh, Russia
science2000@ya.ru
[2] Voronezh State University, Voronezh, Russia
oalkadee@gmail.com
[3] Peter the Great St. Petersburg State Polytechnic University,
Saint Petersburg, Russia
mv@qmd.spbstu.ru
[4] Tishreen University, Latakia, Syria
dr-ahmad@scs-net.org

**Abstract.** This paper is dedicated to applying data mining techniques to obtain knowledge from large databases of online resource reservation systems, such as air travel, post office, hotels, hospitals, and many more. The acquired knowledge is used to predict customers' behavior and improve resource planning through improved overbooking management. Overbooking is a common trick, for example, in the area of tourism or hotels, where the consumer is completely expected to be denied services that have been pre-ordered. In other terms, such cases are referred to as "non-admittance".

**Keywords:** Neural networks · Data mining · Overbooking

## 1 Introduction

Companies that provide resources are interested in increasing the number of applications to ensure the financial success of the company even in the face of multiple returns and customer failure. Improving overbooking is possible for the following reasons. On the one hand, if the actual non-attendance amount is reassessed, some customers will be moved to the "queue" and will be able to claim a certain compensation, which entails additional financial costs for the company, as well as loss of customers' confidence. On the other hand, reducing non-attendance can result in a loss of potential profits. Thus, the challenge is to strike a balance between service delivery costs and losses with lost profits.

This paper discusses ways to predict the probability of not appearing based on data extraction [1], including decision trees, regression models, and neural networks.

© Springer Nature Switzerland AG 2020
D. G. Arseniev et al. (Eds.): CPS&C 2019, LNNS 95, pp. 413–421, 2020.
https://doi.org/10.1007/978-3-030-34983-7_40

## 2 Conventional Methods of Forecasting

### 2.1 Weighted Average Method

For no-show forecasting, the usual weighted mean (WM) [2] method is used, which uses historical information about customers. The calculation is based on an estimate of the average historical rate of orders (SUR – show-up rate) [3] multiplied by the number of final orders:

$$SUR = BO/FB,$$

where $BO$ (Booking Out) is the number of final used reservations (or shows), and $FB$ (Final Bookings) is the total number of reservations, including overbooking.

To improve the results of the WM model, the method of the average absolute deviation MAD (Mean Average Deviation) [4] can be applied, which has some advantages: it is simpler and based on using a small amount $N$ of data available at any time

$$SUR = \frac{\sum_{i=1}^{N} \frac{BO_i}{FB_i}}{N}.$$

However, MAD, as a rule, shows rather a high error in the estimation of the forecast [5].

### 2.2 The Linear Regression Model

The regression equation on a standardized scale has the form: $t_y = \beta_1 t_{x_1} + \beta_2 t_{x_2} + \ldots + \beta_p t_{x_p} + \varepsilon$ where $t_y, t_{x_1}, t_{x_2}, \ldots, t_{x_p}$ are standardized variables: $t_y = \frac{y-\bar{y}}{\sigma_y}$; $t_{x_j} = \frac{x_j-\bar{x}_j}{\sigma_{x_j}}$, $j = \overline{1,n}$, for which the average value is zero: $\bar{t}_y = \bar{t}_{x_1} = \bar{t}_{x_2} = \ldots = \bar{t}_{x_p} = 0$, and the mean square deviation is equal to one: $\sigma_y = \sigma_{t_{x_j}} = 1, j = \overline{1,n}$; $\beta_j$ are standardized regression coefficients. Applying the least squares method (LSM) to the equation, after corresponding transformations we obtain a system of normal equations:

$$\begin{cases} \beta_1 & + \beta_2 r_{x_2 x_1} & + \beta_3 r_{x_3 x_1} & + \beta_p r_{x_p x_1} & = r_{yx_1} \\ \beta_1 r_{x_1 x_2} & + \beta_2 & + \beta_3 r_{x_3 x_2} & + \beta_p r_{x_p x_2} & = r_{yx_2} \\ \ldots & \ldots & \ldots & \ldots & \ldots \\ \beta_1 r_{x_1 x_p} & + \beta_2 r_{x_2 x_p} & + \beta_3 r_{x_3 x_p} & + \beta_p & = r_{yx_p} \end{cases}$$

In this system $r_{yx_j}$, $r_{x_i x_j}$, $j, k = \overline{1,p}$ are elements of the expanded matrix of paired correlation coefficients or, in other words, coefficients of correlation pair between different factors or between the factors and the resultant characteristics.

The system determines β – coefficients. These coefficients show how many values of LSM in the average result is changed if the corresponding factor $x_j$ changes by one LSM with the same average level of other factors. Since all variables are given as centered, β coefficients are comparable.

## 3   Using Data Mining Methods

In the case of a large amount of data, many applications use knowledge management and data mining techniques. Given that the accuracy of the forecast for overbooking determines the adoption of decisions on profit management, the choice of adequate methods is critical in our work. Factors affecting the success of these methods are the following: data collection and preparation; choice of method of data reduction; and choice of the prediction method.

This work is aimed at collecting data from various sources that represent records from multiple fields to which the analysis and reduction method based on the decision tree methods can be applied. The goal is to discover the relationship between the data and the allocation of patterns. This will allow to select only those variables that are expected to have the greatest impact on the model being constructed. An analysis based on the decision tree will always give a result, even if there are very weak relations between the input and output variables.

In the case of weak relationships or insufficient data, it may be useful to construct a random complement. In this project, the CHAID method proposed in [6] is used as a variant of the decision tree method. The Logit model, which is a linear regression model, is also well suited for estimating conditional probabilities.

Note that the constructed model will have to adapt in the next stages of evaluation and verification. In this connection, attention is drawn to modeling based on neural networks, which in applications for data mining came from the field of machine learning. In this paper, neural networks of backward propagation (BPNN) and common regression neural networks (GRNN) [7] are used to build a model for forecasting customer failures and improve forecasts to reduce financial losses.

An example of the structure of a probabilistic neural network for solving the problem of classifying p-component input vectors $x$ into $M$ classes is depicted on the input layer of the calculation network and serves to receive and divide the character-istics of the input vector. The number of neurons of the input layer is determined by the number of signs of the vector x. The sample layer contains one neuron for each sample of the input vector from the training sample. That is, for a total training sample size that contains $L$ samples, the sample layer must have $L$ neurons. The summation layer contains the number of neurons equal to the number of classes into which input images are divided. Each neuron of the summation layer has connections only with the neurons of the layer of samples that belong to the corresponding class. All weights of the links of the summation layer in the traditional probabilistic neural network are equated to ones. The initial neuron functions as a discriminator of the threshold value. It indicates which neuron of the summation layer has the maximum output signal. The weights of the neuron connections of the outgoing layer are set so that the neuron of the summation layer with the highest activity value is identified at its output. The dimension $N$ of the training sample vectors $X_i, i = 1, \ldots, L$ determines the number of neurons and the structure of the incoming layer of the probabilistic neural network. The total size $L$ of the training sample $X_i, i = 1, \ldots, L$ corresponds to the total number of neurons in the sample layer. The presentation of the network of each of the $L$ samples is accompanied by an indication from the teacher of the number of the $k$-th class to which the incoming sample belongs. The sequence of presentations of training samples can be arbitrary.

After the presentation of all $L$ vectors of the learning sample, the network structure is formed, and the network parameters in the form of a matrix are determined. The activity function of the $k$-th summation neuron determines the value of the probability density for the entire $k$-th class. In general, it is calculated by the formula:

$$Y^k(X) = \frac{1}{N(2\pi)^{\frac{p}{2}}\sigma^p} \sum_{j=1}^{L_k} \exp\left(-\frac{(x - x_{kj})^T (x - x_{kj})}{2\sigma^p}\right), \; k = \overline{1, M}.$$

The output is associated with the center closest to the incoming vector as the most suitable output.

The model of such a network is:

$$\hat{y}(x) = \frac{\sum\limits_{i=1}^{N} Z_i y_i \exp\left(-\frac{(x-c_i)^2}{2\sigma^2}\right)}{\sum\limits_{i=1}^{N} Z_i \exp\left(-\frac{(x-c_i)^2}{2\sigma^2}\right)}.$$

where $c_i$ – the center of a vector for a class in the input space, $X_i$ – the number of input vectors that are associated with the center $i$.

The use of specialized software *RapidMiner* is quite promising In the *RapidMiner*, all operators receive certain data at the input, after which corresponding actions take place on these data, and at the output the operator produces some result. Thus, three things are always interesting for operators: input data, actions performed over them, output data. Actions performed on the input data for each algorithm are different, but what the algorithm receives at the output (outputs) at the output can be generalized. We build a model and add blocks of linear regression and neural networks, after which we get numerical values equal.

Figures 1 and 2 show the processing of an array of values through linear regression and neural networks.

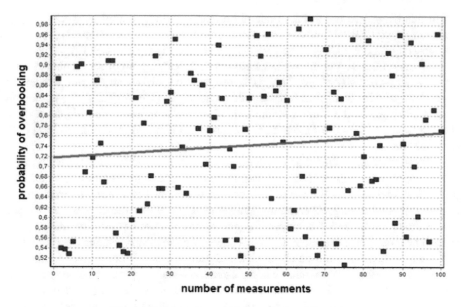

**Fig. 1.** Results of regression analysis of the array

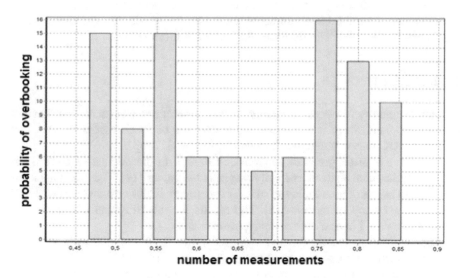

**Fig. 2.** Result of the array analysis

## 4  Artificial Neural Networks (ANN)

ANN is a system of connected and interacting simple processors (artificial neurons). Such processors are usually quite ordinary (especially in comparison with processors used in personal computers).

In this paper, we use artificial neural networks, namely neural networks of backward propagation (BPN) and common regression neural networks (GRNN) [8] to build a forecasting model and improve forecasts to reduce financial losses [9].

The most famous variant of the neural network learning algorithm is the so-called back propagation algorithm. The back propagation algorithm is the simplest to understand, and in some cases it has certain advantages [10]: for back propagation, the error surface gradient vector is calculated. This vector indicates the direction of the shortest descent along the surface from this point, so if we "move" a little along it, the error will decrease.

Another method is *Generalized regression neural network* (GRNN). GRNN copies init all the training observations and uses them to evaluate the response at an arbitrary point. The final output estimate of the network is obtained as a weighted mean of the outputs for all training observations, where weights reflect the distance from these observations to the point at which the estimation is made (and thus closer points contribute more to the assessment). In this paper, we use the reservation data for 14,000 customers, taken from the database of the Syrian Airline. 9 consumer variables are taken into account: day, month, year, age, gender, education, marital status, cancellation history, and the last cancellation number. In this case, there are 9 neurons in the input layer and one neuron in the output layer (the output values are zero or one, 0 means complete reservation, 1 means cancel) [11, 12].

The study will analyze the capabilities of the predictive model, as well as the identification of its capabilities in accordance with the sensitivity, specificity and the ROC curve which is sensitivity depending on the fall-out [13].

# 5    ROC Curve

A graph that allows to assess the quality of separation of two classes. In addition to the visual component, there is a numerical characteristic of ROC, that is the AUC area under the ROC curve; the higher its value is, the better (Fig. 3).

The curve is just a graph of two values: the ratio of the number of correctly and incorrectly classified characteristics of some selected values [14]. The original data was divided into two groups as follows: 70% for training and 30% or 4,200 sets of data for testing the accuracy of models [15]. The results for the BPN and GRNN tests are shown in Table 1 (*1: cancel/0: complete reservation*).

The sensitivity, specificity, and area under the ROC curve shown in Table 2 indicate that both models give good results for classification [16, 17].

The ideal test will show the points in the upper left corner, with 100% specificity and 100% sensitivity. The network performs this task very well.

The ROC curve shown in Fig. 4 indicates that BPN has more accurate classification results than GRNN [18].

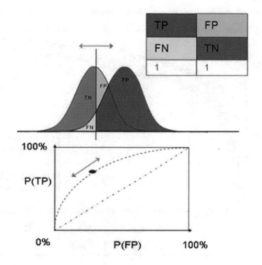

**Fig. 3.** ROC curve

**Table 1.** Test results

|  | BPN | | | GRNN | | |
|---|---|---|---|---|---|---|
|  | Actual value 1 | Actual value 7 | Total | Actual value 1 | Actual value 0 | Total |
| Prediction value 1 | 560 | 880 | 1440 | 610 | 1070 | 1680 |
| Prediction value 0 | 140 | 2620 | 2760 | 90 | 2430 | 2520 |
| Total | 700 | 3500 | 4200 | 700 | 3500 | 4200 |

**Table 2.** Test Results for BPN and GRNN

|  | Sensitivity | Specificity | Square under curve |
|---|---|---|---|
| BPN | 80% | 75% | 80.87% |
| GRNN | 87.5% | 69.5% | 75.34% |

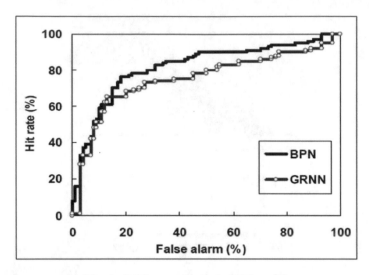

**Fig. 4.** ROC curve for both ANN models

## 6 Conclusion

In the past, the success of various companies depended on the experience of managers capable of guessing the number of cancellations. Intelligent analysis of these technologies can improve forecasting and accuracy of judgment [19, 20]. Both models give good classification results. They can help managers when assessing whether customers will cancel reservations, and can also help in planning the dynamic capacity of the service.

## References

1. Shirley, C., Andera, S.: A Practical Guide to Data Mining for Business and Industry, 1st edn., Pondicherry, Minion, 303 p. (2014)
2. Mark, G.: GNU Scientific Library. Free Software Foundation, Edition 2.1, Invariant, Boston, 589 p. (2015)
3. Janakiram, S., Shaler, S., Conrad, L.: Airline yield management with overbooking, cancellations, and no-shows. J. Transp. Sci. **33**(2), 147–167 (1999)
4. Elsayed, A.: Mean absolute deviation about median as a tool of explanatory data analysis. In: International Association of Engineers: Collection of the Materials of Scientific Conference "Proceedings of the World Congress on Engineering 2012", London, Great Britain, vol. I, pp. 324–329 (2012)
5. Francesco, V.: An application of data mining methods to airline overbooking optimization. In: International Fuzzy Systems Associations (IFSA): Collection of the Materials of Scientific Conference, Fifth International Conference on Application of Fuzzy Systems and Soft Computing, Milan, Italy, pp. 88–93 (2002)

6. Antipov, E., Pokryshevskaya, E.: Applying CHAID for logistic regression diagnostics and classification accuracy improvement. J. Target. Meas. Anal. Mark. **18**, 109–117 (2010)
7. Hilbe, M.: Practical Guide to Logistic Regression, 1st edn. Taylor & Francis Group, CRC Press, London, Great Britain, 174 p. (2015)
8. Merry, C., Paul, S.: Neural network based ACC for optimized safety and comfort. Int. J. Comput. Appl. **4**, 1–4 (2012)
9. Griewank, A.: Who invented the reverse mode of differentiation? Optimization Stories, Documenta Matematica, Extra Volume ISMP, 389–400 (2012)
10. Negnevitsky, M.: Artificial Intelligence: A Guide to Intelligent Systems. Pearson Education Limited (2005)
11. Tahmasebi, P., Hezarkhani, A.: A hybrid neural networks-fuzzy logic-genetic algorithm for grade estimation. Comput. Geosci. **42**, 18–27 (2012)
12. Desyatirikova, E.N., Khodar, A., Osama, A.: Scheduling approach for virtual machine resources in a cloud computing. In: XIX International Conference "Informatics: Problems, Methodology, Technology" (IPMT-2019), Voronezh, Russia (2019)
13. Hernandez-Orallo, J.: ROC curves for regression. Pattern Recognit. **46**(12), 3395–3411 (2013). https://doi.org/10.1016/j.patcog.2013.06.014
14. Ivanov, S.: Hotel Revenue Management: From Theory to Practice. Zangador, 205 p. (2014)
15. Wojtek, J., David, J.: ROC Curves for Continuous Data. Chapman and Hall, CRC Press, London, Great Britain, 256 p. (2009)
16. Volkova, V.N., Chernenkaya, L.V., Desyatirikova, E.N., Hajali, M., Khodar, A., Osama, A.: Load balancing in cloud computing. In: Proceedings of the 2018 IEEE Conference of Russian Young Researchers in Electrical and Electronic Engineering (2018 ElConRus), St. Petersburg and Moscow, Russia, pp. 387–390 (2018). https://doi.org/10.1109/eiconrus.2018.8317113
17. Alkaadi, O., Vlasov, S.V.: Adaptation in reservation systems. In: Informatics: Problems, Methodology, Technology: Collection of Materials of the XVI International Scientific and Methodological Conference, Voronezh, Russia, pp. 12–15 (2016)
18. Alkaadi, O., Desyatirikova, E.N., Algazinov, E.K., Gubkin, I.M.: Data Mining in Booking System Optimization, Technological Perspective in the Framework of Eurasian Space: New Markets and Points of Economic Growth, pp. 328–334. Publisher "Asterion", St. Petersburg, Russia (2019)
19. Desyatirikova, E.N., Hajali, M., Khodar, A., Osama, A.: Load balancing in cloud computing. In: Vestnik Voronezh State University. System Analysis and Information Technology, vol. 3, pp. 103–109 (2017)
20. Desyatirikova, E.N., Osama, A., Khodar, A.: The improvement of results in online reservation systems using data mining. In: XVIII International Scientific and Methodological Conference "Informatics: Problems, Methodology, Technology" (2018)

# Cognitive Monitoring of Cyber-Physical Systems in Agriculture

Igor A. Katsko[✉] and Elena V. Kremyanskaya

Kuban State Agrarian University named after I.T. Trubilin,
Kalinina Str., 13, 350044 Krasnodar, Russia
ingward@mail.ru

**Abstract.** The article contains the rationale for using cognitive ideology to monitor the cyber-physical systems adaptation process in agricultural production. The authors give an example related to the milk production process. It is proposed to use this ideology for building cognitive maps when studying the problems of subsystems of individual commodity markets and embedding cyber-physical systems in them.

**Keywords:** Cognitive monitoring · Cyber-physical system · Cognitive analysis · Cybernetics 2.0 · Panelregression · Market · Milk and dairy products

## 1 Introduction

The recent decades were characterized by the development of a cyber-space with modern means of communication (Internet, cloud services, social networks) [16, 17]. An attempt to unite the real (analog) world and cyberspace became known as cyber-physical systems (cyber-physical systems, CPS) [21, 23]. Today, CPS appear as a regular theory of the "grand unification", based on the ideology of cybernetics 2.0, and physical and computational elements. An important feature of these systems [2–4] is their openness, determined by the continuous data acquisition from the environment to ensure effective management. CPS are considered to be the next step compared to the Internet of Things (Internet of Things, IoT), similar to it in architecture. The difference is that the CPS have capacities in adapting to changing environmental conditions, which may impose a change in the initial goals.

Modern technologies of the Internet of Things are used in precision farming, dairy cattle breeding, processing agricultural products, etc. The cybernetics ideas are broadly used: a feedback loop for machine data (cybernetics 1.0), an external feedback loop for control information accessible to a human being (control) (cybernetics 2.0). Thus, the basis for the formation of CPS in agricultural production may be the Boyd cycle (1995), OODA [22]:

- Observation
- Orientation
- Decision
- Action.

D. G. Arseniev et al. (Eds.): CPS&C 2019, LNNS 95, pp. 422–430, 2020.
https://doi.org/10.1007/978-3-030-34983-7_41

Using the OODA cycle should form CPS in agricultural production with reduced human influence. However, the choice of goals, indicators, quality criteria and their regular monitoring should be performed under human control, because mass production of low-quality products does not make any sense. Cognitive data analysis based on building a domain model in the form of a directed graph [7–9] can provide the foundations for the formation of CPS and their monitoring. The article discusses the application of the proposed ideology using an example of the milk and dairy products market.

## 2   A Meta-Set of Models as the Basis for a Monitoring System

The complexity and high degree of uncertainty in the decision-making process in the milk and dairy products market require synthesizing procedures for formalizing research problems, which leads to the need to build a model of a meta-set (sets) of this system:

$$M = \left\{ M_O(Y, U, P), M_O(X), M_{YS\chi}, M_D(Q), M_{MO}, M_{ME}, M_U, A, M_H \right\} \tag{1}$$

where $M_O(Y, U, P)$ is an identifying model of the system (regional milk and dairy products market), in which vector $Y$ is the vector of endogenous variables $y \in Y \subseteq E^m$, characterizing the phase state of an object (for example, demand volumes, supply volumes, investments, level of integration processes development, production concentration and specialization, etc.); $U$ is a vector of controlled variables $u \in U \subseteq E^r$ (for example, the number of cows, the costs of feed and veterinary services, renovation of equipment and production facilities, development of production capacities for dairy processing, etc.); $P$ is the vector of available resources $p \in P \subseteq E^s$ (for example, fixed and current assets, net profit of organizations producing raw milk and its products, investments);

$M_O(X)$ is an environmental model (regional socio-economic system), in which $X$ is the vector of exogenous values (for example, the natural environment, inter-sectoral, interregional and foreign economic market exchange);

$M_{YS\chi} = \left\{ M_{S\chi}, M_{YS} \right\}$ is the model of interaction between the object and the environment ($M_{S\chi}$ – input models of connections with the environment, $M_{YS}$ – system models with connections with the environment at the output);

$M_D(Q)$ is the system behavior model, where $Q$ are disturbing influences;

$M_{MO}$ and $M_{ME}$ are models of measuring the state of the system and the environment (e.g., programs, data acquisition plans, organization of measurements);

$M_U$ is the model of the governing (regulatory) system (for example, regional and federal regulatory systems) not included in the meta-set if only the object research tasks are being solved;

$A$ is the selection rule of the object change processes;

$M_H$ is the "observer" (researcher, expert, cognitive engineer) model. The presence of the "observer" in (1) – the meta-set M enables the account to be taken of the changes in understanding (cognition) of the object under the study (makes it possible to take into account managerialt information) and synthesize the methodology of research and decision-making, corresponding to the ideology of Cybernetics 2.0 [12, 14, 20].

Internal models are statistical and econometric models based on retrospective data on the system functioning; external models of the system, environment and their interaction with internal models are cognitive models; behavioral models are impulse processes on cognitive maps and development scenarios, obtained by means of simulation modeling. From our point of view, the cognitive approach has the most generic character due to the fact that it makes it possible to link internal and external models, thus eliminating the criticized use of historical data for management and decision-making processes.

## 3   Cognitive Map of the Regional Dairy Market

One of the most successful tools for studying and analyzing the market is cognitive modeling, which involves building cognitive models (primarily cognitive maps), their expert analysis, forecasting the development of situations on models, creating and

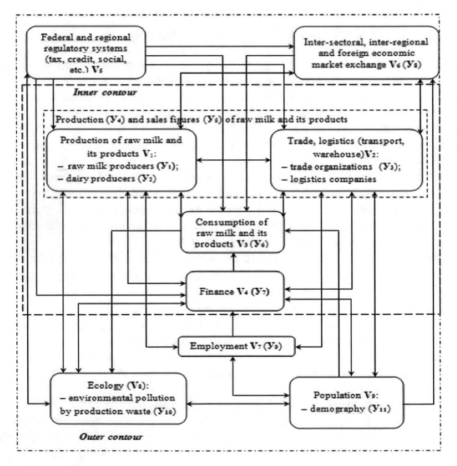

**Fig. 1.** Integrated linkage scheme in the regional milk and dairy products market (adapted by the authors)

substantiating the algorithm of practical actions [8]. In the study, the authors developed a scheme for the milk and dairy products market (see Fig. 1), demonstrating the multiplicity of its interconnected main blocks (vertex $V_i$, $i = \overline{1,9}$).

It can be called a cognitive map in the simplest form, i.e., in the form of a directed graph. The design of the scheme was based on an integrated model of a regional economic mechanism proposed by Granberg [10]. In a highly volatile economy, the essence and strength of the links between the structural elements of the presented scheme are constantly changing. In this regard, such visualization helps to envision the unification of a number of reproductive processes, the continuity of which is provided by the market: the reproduction of products, capital, labor, the environment and natural resources. The scheme also reflects material and financial flows between the main market participants: producers, trade organizations, households, government agencies, etc. [10].

Such a look at the market structure allows us to link social and economic indicators of various levels (micro-, meso- and macro- levels) with each other, which has become particularly relevant in recent decades, due to the increasing complexity of the interaction of various market sectors between them and with the external environment. It should be added that the content of the blocks of the internal and external contours may vary depending on the specific economic situation in the region and the country as a whole.

For the purpose of a comprehensive, objective assessment of the regional milk and dairy products market, it is important to study the situation in the designated blocks simultaneously at the current or forecasted moment (period) of time. This is the requirement of a systemic approach, according to which the interrelation of system elements and their simultaneous analysis are necessary [1, 6]. For its implementation, were selected those indicators (parameter vectors), the consideration of which is most relevant for diagnosing the state of the market and processes occurring in it, and the following coding was introduced: $Y1$ – indicators of raw milk producers; $Y2$ – indicators of dairy products manufacturers; $Y3$ – indicators of trade organizations; $Y4$ – indicators of raw milk and its products production; $Y5$ – sales figures of raw milk and its products indicators; $Y6$ – indicators of raw milk and its products consumption; $Y7$ – financial indicators; $Y8$ – indicators of inter-sectoral, interregional and foreign economic market exchange; $Y9$ – employment indicators; $Y10$ – indicators environmental pollution by production waste; $Y11$ – demographic indicators.

The introduced scheme will provide an opportunity to link the milk and dairy products market with other markets and the external socio-economic environment, as well as build a set of cognitive maps for its individual subsystems. In its turn, an analysis of the latter will help improve management and decision-making processes by reducing the uncertainty caused by external factors.

On the basis of the enlarged scheme, the cognitive G0 map "Regional market of milk and dairy products" was developed (see Fig. 2).

Relations between its concepts were established according to the results of expert assessments and were corrected in accordance with the dialectics of the interconnection between economics and politics.

Since the market is imperfect in its nature, the direct or indirect impact of the policies on the market to a certain degree is always justified. The instability of the economic situation in modern Russia owes much to the imperfection of the long-term state development program of the country. Meanwhile, a competent government policy that could contribute to successful confronting various kinds of risks that impede the development of the market environment.

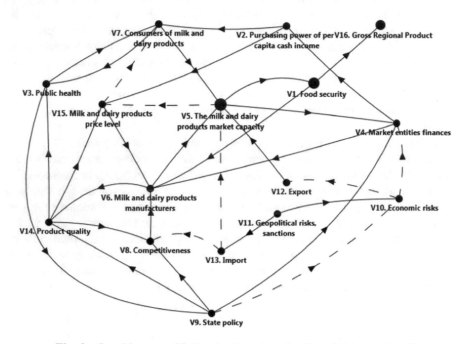

**Fig. 2.**  Cognitive map G0 "Regional market of milk and dairy products"

The analysis of the paths of the cognitive map allowed us to confirm the absence of any contradictions between the developed model and the real system of "The market of milk and dairy products". Since the main goal of improving the market mechanism is to strengthen the country's food security, the main focus of the research was on studying all possible paths from peak V9 "State policy" to the target peak V1 "Food security". The total number of paths is 22. The longest positive one of them is:

$$V_9 \rightarrow V_{10} \rightarrow V_4 \rightarrow V_2 \rightarrow V_{15} \rightarrow V_6 \rightarrow V_{14} \rightarrow V_3 \rightarrow V_7 \rightarrow V_5 \rightarrow V_1$$

Interpreting the result obtained, it can be noted that the improvement of the pricing and financial and crediting policy of the state leads to a decrease in economic risks. This, in its turn, contributes to strengthening the financial condition of market entities, increasing their purchasing power and, as a result, to establishing the most reasonable prices for milk and dairy products, allowing producers to operate in expanded repro-duction mode, saturating domestic market with high-quality domestic goods. Con-sumption of the latter entails strengthening the health of the nation, increasing the market size and, ultimately, increasing the level of food security in the relevant market segment.

The main consistently important players in the milk and dairy products market are producers of raw milk, and an important condition for increasing the efficiency of their operation is optimization of the production resources use. A classic tool for analysis is the econometric modeling method. The expediency of its application is necessitated by the scarcity of available information, the possibility of a formalized description of complex systems and their presentation in a compact form.

## 4   Monitoring Task and Panel Regression

With the structured data, the task of monitoring the condition of a real system managed by a cyber-physical system can be solved using econometric models [11, 15, 18, 19]. In the case under consideration, the array of panel data formed on the basis of annual reporting of 134 agricultural organizations of the Krasnodar Krai for 2015–2017 served as the initial information for the simulation. The analysis was performed using the STATA 14 package. In the course of solving the problem, a regression model *"within"* was constructed. The productive attribute ($Y$) in this research was the cost of milk produced on average per cow. In contrast to the natural indicator of productivity, this indicator allowed us to take into account the quantity and quality of raw materials obtained. Of all the many possible factors, the most significant ones are included in the final model:

- The average monthly wage of one milking machine operator, rubles
- Concentration of livestock of cows per 100 hectares of agricultural land, heads
- Production costs per 100 hectares of agricultural land, thousand rubles
- Calf yield per 100 cows and heifers, heads
- Production costs per cow, thousand rubles
- $d2017$ is a dummy variable that takes the value of 1 if 2017 is considered, and the value of 0 otherwise.

In addition, a dummy variable (subsidies) was introduced, taking a zero value in the absence of subsidies for reimbursement of part of the cost of 1 L of marketable milk sold, and a value of 1 provided that subsidies are available. This action is caused by the desire to identify the significance of state support measures for the efficiency of raw milk producers functioning.

To level the existing asymmetry in the distribution of the data, the logarithm of the information in the initial array was calculated. Table 1 illustrates the parameters of the econometric estimation of the model.

Quite a high value of *Rsq, within* = 0.7478, characterizing the quality of the regression fitting, indicates that in the framework of the model constructed, the inter-individual differences were stronger pronounced than the dynamic ones.

The studies have shown that the most significant factor affecting the cost of milk production per cow is the level of production intensification, expressed in production costs per head. With the increase of this factor by 1%, the cost of milk obtained from one cow grew up by 0.876% on average, as is evidenced by the corresponding coefficient of elasticity. With an increase in the average monthly wage of one milking machine operator by 1%, the effective index increased by 0.117%. This proves the importance of adequate material incentives for dairy cattle workers, since, as practice shows, with all other things being equal, the higher is the level of material incentives, the higher is the productivity of staff and their interest in the results of their work.

Increasing the concentration of livestock per 100 hectares of agricultural land by 1% entailed an increase in the cost of milk obtained from one cow by 0.088%. Therefore, in modern conditions, the highest productivity of animals and the quality of raw milk are demonstrated by agricultural organizations with a large number of cows and a highly intensive dairy cattle breeding organization. An increase in calves output per 100 cows and heifers by 1% resulted in an increase in the cost of milk, obtained on average from one cow, by 0.124%.

**Table 1.** Econometric estimation results of the "within" regression model for lnY

| Variable | Elasticity coefficient | Standard error |
|---|---|---|
| lnX2 (average monthly wage of one milking machine operator, rub.) | 0.117*** | 0.0427 |
| lnX4 (concentration of livestock of cows per 100 hectares of agricultural land, heads) | 0.088** | 0.0377 |
| lnX5 (production costs per 100 hectares of agricultural land, thousand rubles) | − 0.108*** | 0.0302 |
| lnX9 (calf yield per 100 cows and heifers, heads) | 0.124 * | 0.0652 |
| lnX11 (production costs per cow, thousand rubles) | 0.876*** | 0.0468 |
| Subsidies | 0.043 | 0.0287 |
| d2017 | 0.382*** | 0.0623 |
| Const | − 0.610 | 0.467 |

R-sq, within = 0.7478

***, **, * — significance levels 1%, 5% and 10% respectively

This circumstance is objectively due to the fact that after calving the milk yield increases, reaching a maximum in the first or second month of lactation. For this reason, the organization of year-round calving makes it possible to get more milk per head. It should also be noted that the increase in the level of subsidies for dairy cattle breeding by 1% led to an increase in the effective index by 0.043%. This proves the correctness and timeliness of government measures aimed at supporting the sub-sector, and also demonstrates the feasibility of attracting investment funds to its further development. The reduction by 0.108% of the productive indicator with an increase of 1% in production costs per 100 hectares of agricultural land is probably due to the fact that agricultural organizations with significant use land are currently very strongly focused on the development of crop production; more precisely, on grain production as basis of the export potential in the Krasnodar Krai. The development of dairy cattle at such farms is not given the due importance.

# 5   Conclusion

The research proved the expediency of further production intensification at farms providing the regional market for milk and dairy products with raw materials, as well as improving state policy in terms of subsidizing dairy cattle breeding.

Synthesis of cognitive and econometric modeling has good prospects not only for identifying efficiency of the resources in dairy cattle breeding, but also for predicting gross milk production for a given amount of resources, assessing the reality of meeting the targets for the development of the sub-industry, and the regional market for milk and dairy products in general. All this shows the need and usefulness of integrating cognitive monitoring tools into cyber-physical systems.

# References

1. Arnold, V.I.: "Hard" and "soft" mathematical models. MCNMO, 32 p. (2008)
2. Volkova, V.N., Denisov, A.A.: Fundamentals of the theory of systems and systems analysis [Osnovy Teorii System I Systemnogo Analiza]. Publ. St. Petersburg State Technical University, Russia, 510 p. (1997). (in Russian)
3. Volkova, V.N.: Gradual formalization of decision-making models. Publishing House of Polytechnic University, St. Petersburg, Russia, 120 p. (2006). (in Russian)
4. Volkova, V.N.: Open Systems. Rethinking L. von Bertalanffy: monograph. Politeh-Press, St. Petersburg, 440 p. (2019). (in Russian)
5. Vyugin, V.V.: Mathematical Foundations of Machine Learning and Forecasting, 2nd edn. MCCME, Moscow, Russia, 384 p. (2018)
6. Garaedagi, J.: Systems Thinking: How to Manage Chaos and Complex Processes: A Platform for Modeling Business Architecture. Grevtsov Books, Russia, 480 p. (2010)
7. Gorelova, G.V., Zakharova, E.N., Ginis, L.A.: Cognitive analysis and modeling of sustainable development of socio-economic systems. Publ. House of Rostov University, Rostov on Don, Russia, 288 p. (2005). (in Russian)
8. Gorelova, G.V., Zakharova, E.N., Radchenko, S.A.: Investigation of semi-structured problems of socio-economic systems: a cognitive approach. Publishing house of RSU, Rostov on Don, 332 p. (2006). (in Russian)
9. Gorelova, G.V., Maslennikova, A.V.: Simulation modeling based on cognitive methodology and system dynamics, analysis of the system South of Russia, System analysis in the economy – 2012: collection of scientific articles of the scientific-practical, Conf., Section 2, pp. 50–65. CEMI RAS, Moscow, Russia (2012). (in Russian)
10. Granberg, A.G.: Fundamentals of Regional Economics: A Textbook for Universities, 4th edn. Publishing House of the Higher School of Economics, Moscow, Russia, 495 p. (2004). (in Russian)
11. Doych, D.: Beginning of infinity. Explanations that change the world. Transl. from English, Alpina non-fiction, Russia, 582 p. (2016)
12. Gorelova, G.V., Pankratova, N.D., Bondarenko, V.L., et al.: Innovative development of socio-economic systems based on foresight and cognitive modeling methodologies. Naukova Dumka, Kiev, Ukraine, 464 p. (2015)
13. Katsko, I.A.: Information support of the socio-economic systems management process. In: Gorelova, G.V. (ed.) KubSAU, Krasnodar, Russia, 421 p. (2008)

14. Novikov, D.A.: Cybernetics: navigator. The history of cybernetics, current state, development prospects [Kibernetika: Navigator]. LENAND, Moscow, Russia, 160 p. (2016)
15. Zagoruyko, N.G.: Applied methods of data and knowledge analysis. IM SB RAS, Russia, 270 p. (1999)
16. Karr, N.: The Great Transition: What the Cloud Technology Revolution Prepares. Transl. from English by A. Baranov, Mann, Ivanov and Farber, Moscow, Russia, 272 p. (2014)
17. Franks, B.: The taming of Big Data: How to Extract Knowledge from Large Arrays of Information with the help of In-depth Analytics. Transl. from English by A. Baranov, Mann, Ivanov and Farber, Moscow, Russia, 352 p. (2014)
18. Franks, B.: Revolution in analytics: how to improve your business in the era of Big Data with the help of operational analytics. Transl. from English, Alpina Publisher, Moscow, Russia, 316 p. (2016)
19. Khabbard, D.: How to measure anything. Valuation of intangible in business. Transl. from English by E. Pestereva, Olymp-Business, Moscow, Russia, 320 p. (2009)
20. Hammond, J., Keene, R., Rife, G.: The Right Choice in Decision Making. Transl. from English by V. N. Egorov, Binom Laboratoriya znaniy, Moscow, Russia, 254 p. (2014)
21. Cyber-physical systems in the modern world. https://habr.com/ru/company/toshibarus/blog/438262/
22. Boyd's loop and second order cybernetics. https://www.osp.ru/os/2013/07/13037357/
23. Lee, E.A.: Cyber-Physical Systems – Are Computing Foundations Adequate? (2006). https://ptolemy.berkeley.edu/publications/papers/06/CPSPositionPaper/Lee_CPS_PositionPaper.pdf

# Transformation PLM-Systems into the Cyber-Physical Systems for the Information Provision for Enterprise Management

Alla E. Leonova[1], Valery I. Karpov[1], Yury Yu. Chernyy[2],
and Elena V. Romanova[3(✉)]

[1] "NICEVT", Moscow, Russia
alla.leonova@nicevt.ru, vikarp@mail.ru
[2] Institute of Scientific Information for Social Sciences of the Russian Academy
of Sciences, Moscow, Russia
yuri.chiorny@mail.ru
[3] MIREA – Russian Technological University, Moscow, Russia
porabot@inbox.ru

**Abstract.** The article proposes a concept of transforming the existing Product Lifecycle Management System (PLM-System) into a Cyber-Physical System for the production and management of an enterprise (in an extended meaning of this term). The concept is based on the study of interaction of product development systems with decision making about the feasibility of transferring relevant processes into an automated digitalization mode. When implementing transformation of a PLM-System into a Cyber-Physical System, it is necessary to investigate the business-processes of the enterprise by developing appropriate models briefly described in the paper. Such transformation, based on introduction of emerging technologies, can be used as the basis for creating a cyber-physical system for managing the entire process lifecycle of the enterprise from a job order and organization of production process to product delivery and support.

**Keywords:** Innovative technologies · Integrated information system · Cyber-physical system · Digitalization · Enterprise · Management · PLM-System

## 1 Introduction

In the paper, based on the analysis of the development of production management automation, digitalization of production and management processes elaborated as system conversion of the existing automation systems is proposed, which is possible when creating a cyber-physical system (in the extended meaning of the term) as transformation of the systems supporting the product life cycle (Product Data Management, PDM) and systems supporting the life cycle of an enterprise as a whole (PLM-type systems – Product Lifecycle Management).

In such systems, sensors, controllers and procedure-oriented information systems are integrated into a unified network throughout the entire product life cycle. For transformation of the existing integrated information systems, of the PDM and PLM

© Springer Nature Switzerland AG 2020
D. G. Arseniev et al. (Eds.): CPS&C 2019, LNNS 95, pp. 431–439, 2020.
https://doi.org/10.1007/978-3-030-34983-7_42

types into systems like CPSs, it is necessary to investigate businessprocesses of the enterprise by developing appropriate models.

## 2  Analysis of Enterprise Information Systems Development

In the course of development of the automation of information support of enterprise management, with the emergence of new technologies, the terminology, types of information systems, standards governing their development and functioning were changing.

In the initial period of automation back in the USSR, the following terms were adopted at enterprises: automated systems of technological processes of production (ASUTP) and automated systems of enterprise management in general (ASUP).

Further, as technologies and automation were improving, more and more functions that humans had previously performed began to be carried out by information systems. Such IT solutions as automated control systems for technical preparation of production (ASTPP), design automation systems (SAPR), automated control systems for scientific research (ASNI), automated systems for scientific and technical information (ASNTI) and other systems has been developed.

As information technologies developed, information systems that provided for the ability to automate various types of organization activities appeared; such systems were called integrated information systems (IISs) or corporate information systems (CISs).

The most developed information systems that ensure the internal activity of mass production enterprises include:

$$IIS \ = \ <ERP, MRP, MRPII, MES, PDM>, \qquad (1)$$

where MES means Manufacturing Execution System;

MRP means Material Requirement Planning;

MRPII means Manufacturing Resource Planning;

ERP means Enterprise Resource Planning;

PDM is Product Data Managerment providing data management for a specific technical product.

In the late 1990s, the CRM (Customer Relationship Managerment) standard was developed, and then appeared the CSRP (Customer Synchronized Resource Planning) approach which covers the full life cycle of a product, from the design to customers' requirements imposed on the warranty and aftersales service. That can be indicated as follows:

$$CSRP = \ ERP \ + \ CRM.$$

At present, research and production enterprises, which better adapt to specific customers' orders, implement complex and unique orders and are geared to individualized production, use information systems corresponding with the PDM concept. Such enterprises use PDM and PLM, combinations of these types of systems (i.e., PDM/PLM modules). PDM and PLM systems seem to be very similar. However, PDM and PLM are fundamentally different. In many ways, it is essential to take into account these differences when choosing a management module.

A PDM system is a product management system that provides management of comprehensive product information. A PDM system provides a means to establish interaction between users, to control large flows of engineering information, and to obtain partial access to data at any stage of product development/manufacturing. Among the main functions of a PDM system is management of product documentation including storage, processing; engineering and technical data; visual, graphic and other information about specific products; product structure; workflows; authorization mechanisms, automated reporting, etc. PDM systems form a good basis for transformation into PLM systems that control the product life cycle as a whole.

Relatively to PDM systems, PLM systems provide many additional options. Among these options are, for example, project planning, creation of production waste disposal schemes, market research conduction, designing and creating of products and workflows, purchases of raw materials, production and inspection of products; packaging, storage, and sales; technical and operational support; ensuring interaction between different systems, their integration into the general information field; recycling, and so on.

All stages of a full product life cycle taken into account, it allows to reduce costs, combine complex processes, track each item of the product released, and consider various requirements.

All product lifecycle processes (order placement, planing, design, logistics, production, operation, recycling, and other related processes) occur in the physical environment. At the same time, in a PLM system, these processes correspond to processes occurring in the information space (in computer systems).

In fact, PLM systems include PDM systems. Product management is a key function of a PLM system; however, it is not only the availability of a Product Lifecycle Management unit and the wide range of functions that allow speaking about the fundamental differences between the PDM and PLM systems.

A PLM system represents a complex of domain-specific application software designed to structure data sets and automate the management of physical and information processes throughout the product life cycle, i.e.,– from a customer's order to the moment of product delivery to the customer or releasing it to the market. With further digitalization, the introduction of emergent information technologies of the Fourth industrial revolution, PLM systems can be transformed into the systems that can be attributed to the class of cyber-physical systems in the modern expanded understanding of this concept.

## 3   CPS Concept

Currently, there is no unanimously accepted definition of a cyber-physical system (Cyber-Physical System, CPS). In various papers (see References [1–7], and others), a CPS is interpreted as integration of computing resources into physical processes; in the applied aspects, it is a set of automatic systems for collecting, processing, and analyzing data, including geographically distributed data. CPSs include autonomous vehicle systems, process control systems, distributed robotics, and avionics systems. CPS applications also include sensor-based systems, data acquisition systems, wireless sensor networks that control a certain aspect of the environment and transmit processed information to the central node. Some researchers associate Socio-Cyberspace with physical processes of the CPS class [8–10]. However, all definitions of CPSs have one thing in common: that is the fact that CPSs use "phenomenal" (as they are called by Lee and Seshia [1, 6]) information technologies.

Examples of individual innovative technologies application at industrial enterprises are given in many contemporary scientific and practical works, while the ideologist of the Fourth industrial revolution Klaus Schwab believes that the most unpredictable result will come out with the complex introduction of interacting technologies [11, 12].

The term of CPS is proposed for the study of complexes consisting of various natural objects, artificial systems and controllers that are integrated into a single whole and include embedded real-time systems, distributed computing systems, automated control systems for technical processes and units, and wireless sensor networks. Actually, such complexes are automated systems, but they are larger and more complex than the existing ones, and in CPSs computers are integrated or built into certain physical devices or systems. The study of such complexes requires the combined use of two types of models. On the one hand, these are traditional engineering models (models from the field of construction, mechanical, electrical, biological, chemical models, etc.), and on the other hand, these are computer models.

The theory of CPSs has been developing as an interdisciplinary scientific theory, claiming to unite two scientific schools that have been developing independently. First of them is the Computer Science school, based on mathematical linguistics and theory of algorithms, and the second one is formed by schools based on the automatic control theory, where integro-differential equations are used to simulate dynamic processes.

Thus, the concept of CPSs should be considered as a new interdisciplinary area, which should combine models of computational and physical processes. In this case, physical processes can be understood not only as production processes, but also processes of enterprise management in general.

## 4   Understanding the Enterprise CPS

Understanding the role of CPSs at an industrial enterprise is enabled by the definition offered by Lee and Seshia [1]. As an example, the structure of a CPS network is given. It combines: (1) the "physical" part of a CPS that is not implemented using computers or digital networks and includes mechanical parts, biological or chemical processes, human operators, etc.; (2) one or more computing platforms, sensors, actuators, computers, and, possibly, one or more operating systems; (3) a network structure that provides mechanisms for computer communication. Together, the platform and network form a technical or "cyber" part of the CPS.

A similar interpretation of CPSs was given by the Deputy General Director of the Ruselprom Concern Masyutin [15]. In his definition, a CPS is understood as a cyber-physical network, in which "… sensors, controllers and information systems are combined into a single network throughout the product life cycle", and which "… may be both within a single enterprise and within a dynamic business model consisting of several companies. Operations throughout the life cycle interact with each other using standard Internet protocols for managing, planning, self-tuning and adapting to changes. A CPS is the infrastructure for the Internet of Things" [15].

Based on the above definitions, it is natural to assume, that the CPS concept should be based on the systems theory and should use methods and models of system analysis.

A cyber-physical system for an industrial enterprise may include the following set of technologies:

$$CPS \equiv \;< CAD,\; CAE,\; IR,\; PLM,\; CV,\; 3D,\; AR,\; VR, ABD,\; IIoT,\; CRM,\; M2M,\; \ldots >$$

$$(2)$$

where CAD/CAE are computer-aided design systems;

IR are industrial robots and networks coordinating their interaction;

CV are computer vision systems (image processing, machine vision, visualization, etc.),

3D are 3D technologies (3D modeling, 3D printing for prototyping, etc.);

AR are the augmented reality technologies that help to create visual instructions/ "tips" at a workplace;

VR are Virtual reality technologies for creating physical models, advertising and promoting product sales;

ABD are Big Data analysis technologies (Big Data) for the online decision support;

IIoT is the Industrial Internet of Things;

CRM is an automated customer ("supplier - customer") relationship management syste, as well as its integration into the loop of end-to-end business processes and data exchange management;

M2M (Machine-to-Machine) is a set of technologies that allows machines to exchange information with each other, or transmit it unilaterally; these can be wired and wireless systems for sensors monitoring or for control of any device parameter (e.g., temperature, location, etc.); and other technologies (for example, [18, 19]).

Technological trends underlying CPSs are already separately used in various areas, but being integrated into a single system, they could change the existing relationships between manufacturers, suppliers and customers, as well as between men and engines.

Application of new technologies will make it possible to obtain a new quality of production processes and automate management of the entire life cycle of an enterprise, expanding the capabilities of PLM systems.

## 5 Principles of Transformation a PLM-System into a Cyber-Physical System

In making decisions on the need to employ innovative technologies and transform information systems, such as PLM systems, into a system with the CPS philosophy, one should proceed from the features of a particular production, types of products manufactured by an enterprise, and the need for the lifecycle of an enterprise prompt updating.

The most promising form ensuring the development of an enterprise is the engineering system in the original understanding of this term, which arose in Europe in the 16th century (engineering from Latin "ingenium" that means ingenuity, invention, knowledge), i.e., not only computer engineering (software for engineering analysis and design) [10], but first of all, the use of scientific and technical knowledge of how to create systems, devices, materials, organize and control production processes and implement enterprise management in general.

If compared to other engineering systems, a fundamentally new fact in CPSs is that such systems should provide for close communication and coordination between digital and physical resources. Quick response to emerging cyber-physical effects is very important; such emerging effects are primarily associated with the mutual influence of physical and computational processes on each other. Computers monitor and control physical processes using the feedback through which physical systems are influenced by the events that take place in digital systems and vice versa.

For efficient use of IT technologies, it is necessary to transform the processes occurring in the physical space into the tasks in the information sphere, as well as to have an ability to reverse transformations of information processes into physical ones. Such transformations are considered as a problem of adequate modeling of a product life cycle algorithm, i.e., establishing a one-to-one correspondence between the physical and information spaces.

When transforming PLMs into CPSs, it is necessary to explore the interaction of product development and decision-making systems and the feasibility of transferring relevant processes to automatic digital modes. It is important to research the technical and logical design support using the PLM system and to provide information support for developing preliminary strategies (long-term and medium-term) and plans for the transition to digital technologies and for the development of enterprises along with the support of introduced digitat technologies.

For CPSs development, methods and means of information systems design based on various methods of generating graphs can be used. The most common methods for modeling production processes are methods developed in the theory of object-oriented modeling based on the formation of hybrid or complex dynamic systems, which are hierarchical, event-driven systems of variable structure.

For the study of organizational management models, automated procedures (computer programs) that help to design and analyze business processes are used. Previously, in the practice of large industrial enterprises, special management graphs were used; they were called organizational and technological procedures and described various organizational processes. For example, organizational and technological procedures were used in the organization of production preparation; they could be presented as basic production models, technological models, algorithms for the implementation of the technological processes, and flow charts. The types of these models depend on particular characteristics of specific productions.

When adjusting organizational structure of an enterprise, the models are determined by a general methodology which should include the stages of forming the structure of goals and functions of the enterprise, function evaluating, and distribution of functions among the implementors. To accomplish these steps, methods for structuring goals and functions, methods of complicated expertise are used; for all those appropriate auto-mated procedures were developed (for example, [22–30]).

The use of emerging technologies allows to obtain a new quality of production processes and expand the possibilities of managing the entire life cycle of an enterprise expanding the capabilities of PLM systems.

When developing a PLM system, it is necessary to improve the management of enterprise sustainable development (e.g., [31, 32]) and constantly maintane its sus-tainable functioning; in this field, the concept of PLM plays a big role, which must be preserved when transforming PLM systems into cyber-physical ones on the basis of introducing new intellectual technologies into it. In other words, the development of a PLM system in a cyber-physical system should be carried out under the control of ideologists of the PLM concept and administrators of integrated information systems that provide management of all processes of company's activities.

## 6   Conclusion

The paper proposes the concept of transforming the existing PLM system into a cyber-physical system (in the broad understanding of this term) for the production process and enterprise management.

Such transformation, based on the introduction of emergent technologies, can be the basis for creating a cyber-physical system for managing all life cycle processes of a product – from order placement and organization of the production to the delivery of the product to the customer and product support.

More and more functions that previously only people could perform are to be transferred to artificial intelligence systems. At the same time, there may appear unforeseen consequences that can negatively affect the development of the enterprise, or even be dangerous. Therefore, it is important to control the interoperability of the technologies; it is necessary to develop models for managing the sustainable devel-opment of enterprises.

## References

1. Lee, E.A., Seshia, S.A.: Introduction to Embedded Systems. A Cyber-Physical Systems Approach. LeeSeshia.org (2011)
2. Colombo, A., Bangemann, T.: Industrial Cloud-Based Cyber-Physical Systems: The IMC-AESOP Approach, 245 p. Springer, Cham (2014)
3. Kupriyanovskiy, V.P., Namiot, D.E., Sinyagov, S.A.: Cyber-physical systems as the basis of the digital economy [Kiberfizicheskiye sistemy kak osnova tsifrovoy ekonomiki]. Int. J. Open Inf. Technol. 4(2), 18–25 (2016). ISSN: 2307-8162. (in Russian)
4. Dobrynin, A.P., et al.: Digital economy - various ways to efficiently apply technology. (BIM, PLM, CAD, IOT, Smart City, BIG DATA etc.) [Tsifrovaya ekonomika - razlichnyye puti k effektivnomu primeneniyu tekhnologiy (BIM, PLM, CAD, IOT, Smart City, BIG DATA i dr.)]. Int. J. Open Inf. Technol. 4(1), 4–11 (2016). (in Russian)
5. Namiot, D.: On Big Data stream processing. Int. J. Open Inf. Technol. 3(8), 48–51 (2015)

6. Lee, E.A.: The Past, Present and Future of Cyber-Physical Systems: A Focus on Models. https://citeweb.info/20150013436
7. Lee, E.: Cyber physical systems: design challenges, University of California, Berkeley, Technical Report No. UCB/EECS-2008-8, 23 January 2008 (2008). https://www2.eecs. berkeley.edu/Pubs/TechRpts/2008/EECS-2008-8.pdf
8. Zhuge, H.: Cyber-physical society. In: Materials of the 1st Workshop on Cyber Physical Society, in conjunction with the 6th International Conference on Semantics, Knowledge and Grids, Ningbo, China (2010)
9. The Cyber-Physical Society website. http://www.knowledgegrid.net/ ~ H.Zhuge/CPS.htm
10. Zhuge, H.: Socio-natural thought semantic link network: a method of semantic networking in the cyber physical Society. Keynote at IEEE AINA 2010, Proceedings of the 24th IEEE Advanced Information Networking and Applications, Perth, Australia, 20–23 April 2010, pp. 19–26 (2010)
11. Schwab, K.: The Fourth Industrial Revolution. Portfolio/Penguin, 184 p. (2017)
12. Schwab, K., Devis, T.: Shaping The Fourth Industrial Revolution [Tekhnologii chetvertoy prpomyshlennoy revolyutsii]. "E" Publ., Moscow, 320 p. (2018). (in Russian)
13. Chernyak, L.: The Internet of Things: new challenges and new technologies. Open Syst. J. DBMS **4**, 14–18 (2013). http://www.osp.ru/os/2013/04/13035551
14. Questions of philosophy [Voprosy philosofii] **6**, 62–78 (1980). (in Russian)
15. Masyutin, S.A.: Enterprise strategy in the transition to "Industry 4.0" [Strategiya predpriyatiya pri perekhode k "Industrii 4.0"]. PLM.pw›2016/09/The-6-Factors-of-Industry-4.0.html (2016). (in Russian)
16. Emerging Technologies. HuffPost/huffingtonpost.com›topic/emerging-technologies
17. Emerging Technologies, Deloitte Insights. https://www2.deloitte.com/insights/us/en/tags/ emerging-technologies.html
18. L'vov, D.S., Glaz'ev, S.Yu.: Theoretical and applied aspects of scientific and technological progress management [Teoreticheskie i prikladnye aspekty upravleniya NTP]. J. Econ. Math. Methods [Ekonomika i matematicheskie metody] **5**, 793–804 (1986). (in Russian)
19. Rifkin, J.: The Third Industrial Revolution: How Lateral Power Is Transforming Energy, the Economy, and the World, 304 p. St. Martin's Press, N. Y., USA (2011)
20. Volkova, V.N., Vasiliev, A.Y., Efremov, A.A., Loginova, A.V.: information technologies to support decision-making in the engineering and control. In: Proceedings of 2017 20th IEEE International Conference on Soft Computing and Measurements (SCM 2017), St. Petersburg, Russia, pp. 727–730 (2017)
21. Volkova, V.N., Efremov, A.A. (eds.): Information Technology In Control Systems. [Informatsionnyye tekhnologii v sisteme upravleniya.], 408 p. St. Petersburg Polytechnic University Publishing House (2017). (in Russian)
22. Volkova, V.N., Kozlov, V.N. (eds.): Modeling systems and processes [Modelirovanie sistem i protsessov], 450 p. "Yurait" Publishing House, Moscow (2016). (in Russian)
23. Volkova, V.N., Gorelova, G.V., Efremov, A.A., et al.: Modeling systems and processes: practicum [Modelirovanie sistem i protsessov. Praktikum]. In: Volkova, V.N. (ed.) "Yurait" Publishing House, Moscow, 295 p. (2016). (in Russian)
24. Volkova, V.N., Loginova, A.V., Yakovleva Ye., A., et al.: Models of innovation management of enterprises and organizations [Modeli upravleniya innovatsionnoy deyatel'nost'yu predpriyatiy i organizatsiy], 246 p. St. Petersburg Polytechnic University Publishing House (2014). (in Russian)

25. Volkova, V.N., Loginova, A.V., Leonova, A.Ye., Chernyy, Yu.Yu.: Approach to comparative analysis and selection of technological innovations of the third and fourth industrial revolutions [Podkhod k sravnitel'nomu analizu i vyboru tekhnologicheskikh innovatsiy tret'yey i chetvertoy promyshlennykh revolyutsiy]. In: Proceedings of 2018 21st IEEE International Conference on Soft Computing and Measurements (SCM 2018), 23–25 May 2018, St. Petersburg, Russia, pp. 373–376 (2018). (in Russian)
26. Volkova, V.N., Kudryavtseva, A.S.: Models for managing the innovation activities of industrial enterprises [Modeli dlya upravleniya innovatsionnoy deyatel'nost'yu promyshlennogo predpriyatiya]. Open Educ. J. [Otkrytoye obrazovaniye] **22**(4), 64–73 (2018). https://doi.org/10.21686/1818-4243-2018-4-64-73. (in Russian)
27. Kudryavtseva, A.S.: Models for the management of industrial enterprises in the implementation of technological innovations [Modeli dlya upravleniya innovatsionnoy deyatel'nost'yu promyshlennogo predpriyatiya prI vnedreniyi tekhnologicheskikh innovatsiy]. In: Proceedings of 22th International Scientific Practical Conference on System Analysis in Engineering and Control, Part 1, pp. 389–398. Polytechnic University Publishing House, St. Petersburg (2018). (in Russian)
28. Volkova, V.N., Kozlov, V.N., Karlik, A.E., Iakovleva, E.A.: The impact of NBIC-technology development on engineering and management personnel training. Strat. Partn. Ship Univ. Enterp. Hi-Tech Branches. (Sci., Educ. Innov.) **6**, 51–54 (2018)
29. Volkova, V.N., Kozlov V.N., Mager, V.E., Chernenkaya, L.V.: Classification of methods and models in system analysis. In: Proceedings of the 20th IEEE International Conference on Soft Computing and Measurements (SCM 2017), St. Petersburg, Russia, pp. 183–186 (2017)
30. Volkova, V.N., Denisov, A.A.: Methods of the procedures of complicated expertise organization [Metody organizatsii slozhnoy ekspertizy.], 128 p. Publishing House of St. Petersburg Polytechnic University (2010). (in Russian)
31. Volkova, V.N., et al.: Problems of sustainable development of socio-economic systems in the implementation of innovations. In: Proceedings of the 3rd International Conference on Human Factors in Complex Technical Systems and Environments (Ergo 2018), St. Petersburg, Russia, pp. 53–56 (2018)
32. Volkova, V.N., Lankin, V.Ye.: The problem of the sustainability of the socio-economic system in the context of innovation of the Fourth industrial revolution [Problema ustoychivosti sotsial'no-ekonomicheskoy sistemy v usloviyakh vnedreniya innovatsiy chetvertoy promyshlennoy revolyutsii]. Sci. Pract. J Econ. Manag. Probl. Solut. [Nauchno-prakticheskiy zhurnal "Ekonomika i upravleniye: problemy i resheniya"] **7**(77), 25–29 (2018). (in Russian)

# Criterion of Stability of a Linear System with One Harmonic Time-Varying Coefficient Based on a Formalized Filter Hypothesis

Anton Mandrik[(✉)]

Peter the Great St. Petersburg Polytechnic University, Saint Petersburg, Russia
lasthero1987@mail.ru

**Abstract.** Stability criterion for a linear time-varying (LTV) system with one harmonic time-varying coefficient in feedback is suggested. The found criterion is based on the hypothesis that the linear time-invariant (LTI) part of the system is a low-frequency filter. The criterion is simple and suitable for calculation of stability borders for LTV systems. The suggested criterion is compared with a numerical experiment, Bonjiorno criterion, stationarization method.

**Keywords:** Stability criterion · Linear time-varying system · Stability borders · Numerical experiment · Bonjiorno criterion · Stationarization method

## 1 Introduction

The theory of linear time-varying (LTV) systems was intensively developing in the last 20 years [5, 7, 12, 14, 15]. Usually, stability of such systems is estimated with the usage of the second Lyapunov method which is concentrated on searching for an appropriate Lyapunov function [2]. There are no strict algorithms for building such functions for time-varying systems. This significantly complicates programming control systems which are usually built on a microcontroller platform which has no pre-installed algorithms for Lyapunov functions search.

There also are some frequency domain analysis methods which can help building algebraic stability criteria for linear time-varying systems. One of the known frequency domain methods is represented by spectral analysis theory [1, 3]. A spectral method [3] allows building a stability criterion by analysis of an infinite dimensional matrix. It is possible to evaluate the determinant of this matrix in the form of a corresponding approximation matrix with finite dimensions [3].

Due to the mentioned obstacles, the estimation of LTV system stability remains a difficult problem for researchers. In reference [14] the control system design for crane payload oscillations is based on a particle swarm optimization method which means a simple brute force method. A special control system for active control damping systems [9] requires frequency adjusting of oscillations of a time-varying mass in the end of every period of oscillations, and no stability criterion is used. Meanwhile, the importance of stability estimation of gain-scheduled and parameter-varying systems from the parametric resonance point of view is considered in reference [4]. Similar stability estimation problems appear in the reset systems given by special integrators with

© Springer Nature Switzerland AG 2020
D. G. Arseniev et al. (Eds.): CPS&C 2019, LNNS 95, pp. 440–448, 2020.
https://doi.org/10.1007/978-3-030-34983-7_43

zeroing integrator value after zero-crossing [10]. LTV system stability concerns different mechanical systems, such as mathematical pendulums, crane payloads or springloaded masses [11, 13, 16, 18]. Time-varying parameters can be used for the control of linear time-invariant (LTI) or LTV systems [19] but the method of control is rather empirical. Frequency analysis methods are used in [17, 20] and allow to receive a new design of parametric control systems.

In this article, we are going to receive a method of stability evaluation of an LTV system with one harmonic time-varying coefficient. The first part of the article shows the problem of stability evaluation of LTV systems with one harmonic time-varying coefficient. The second part of the article reveals a new stability criterion which requires special type of the transfer function of the LTI part of an LTV system.

## 2   Model of Harmonic Signals Generation in a Linear Time-Varying System with One Harmonic Time-Varying Coefficient

We are going to discuss the following linear time-varying system:

$$
\begin{aligned}
a_n x^{(n)} + a_{n-1} x^{(n-1)} + \ldots + a_1 x' + a_0 x &= u(t), \\
u(t) &= -k(t)x,
\end{aligned}
\tag{1}
$$

where $a_n, a_{n-1}, \ldots, a_1, a_0$ are real coefficients, $x = x(t)$ is the output of the system, $x^{(n)}, x^{(n-1)}, \ldots, x'$ are the derivatives of the system output, $u(t)$ is the feedback signal, $k(t)$ is the time-varying coefficient represented by a harmonic function of time $t$.

Although the system of Eq. (1) looks quite simple, its solution may be found only using a Fourier series representation which is discussed in Floquet theory.

The first equation of system (1) represents the LTI part of the system; the second equation of system (1) represents the LTV part of the system. The LTI part of the system may be described by a transfer function:

$$
W(p) = \frac{x(s)}{u(s)} = \frac{1}{a_n s^{(n)} + a_{n-1} s^{(n-1)} + \ldots + a_1 s' + a_0},
$$

where $s = j\omega$ is the Laplace transform complex variable. The magnitude of the LTI part for frequency $\omega$ will be as follows:

$$
|W(j\omega)| = \left| \frac{1}{a_n (j\omega)^{(n)} + a_{n-1} (j\omega)^{(n-1)} + \ldots + a_1 (j\omega) + a_0} \right|.
$$

The Laplace transform is needed to evaluate the magnitudes for Fourier series components of the solution. The LTV part is represented as feedback with a time-varying coefficient:

$$k(t) = K \sin(2\omega_0 t + \varphi), \tag{2}$$

where $\omega_0$ is an Eigen frequency of the LTI part of system (1). According to Floquet theory [2], it is possible to represent the solution of system (1) in the following form:

$$x^* = \sum_{i=1}^{\infty} A_i \exp(\lambda_i t) \sin(i\omega_0 t + \varphi_i), \tag{3}$$

where $A_i$, $\lambda_i$ $\varphi_i$ are real values. We insert Eq. (3) into the second equation of system (1) and receive a feedback signal:

$$
\begin{aligned}
u(t) = -k(t)x^* &= K \sin(2\omega_0 t + \varphi) \sum_{i=1}^{\infty} A_i \exp(\lambda_i t) \sin(i\omega_0 t + \varphi_i) \\
&= \sum_{i=1}^{\infty} A_i' \exp(\lambda_i' t) \sin(i\omega_0 t + \varphi_i'),
\end{aligned}
\tag{4}
$$

where $A_i'$, $\lambda_i'$ $\varphi_i'$ are real values. So, the feedback of system (1) with coefficient (2) transfers signal (3) keeping the same spectrum of frequencies. As long as signal $x^*$ is a solution of the system, the reaction of the LTI part of system (1) to signal (4) will be equal (3). Providing that the components of $x^*$ given in Eq. (3) have several different frequencies, it is possible to say that stability of all components of $x^*$ means that the solution $x^*$ is stable.

The components of $x^*$ in Eq. (3) are harmonic functions which turn into harmonic functions with another frequency after passing through the linear time-varying feedback. The most dangerous for stability case is when the frequency of variation of the feedback equals to doubled Eigen frequency of the LTI system. We will build a graph of interferences between the harmonic functions of the system output while they are passing through the feedback several times (see Fig. 1).

In Fig. 1 each bone of the graph defines a transfer between the harmonic function of one frequency and the next generated harmonic function of another frequency. The weight of each bone may be represented with the transfer function.

Phase analysis is quite difficult, so we will analyze the absolute values of the transfer function. Now, we will write down the absolute value of the transfer function between the harmonic functions with frequencies $\omega$ and $-\omega$ given by one passing through the feedback:

$$|W_1(\omega_0, -\omega_0)| = |W_1(-\omega_0, \omega_0)| = 0{,}5K|W(j\omega_0)|. \tag{5}$$

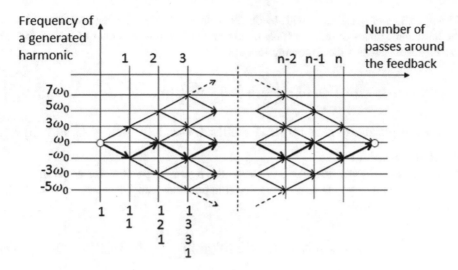

**Fig. 1.** Graph of transfers between the harmonic signals while passing through the time-varying feedback for several times

Providing that we have defined $\omega_0$ as the Eigen frequency of the LTI part of system (1), the absolute values of transfer functions between the signals of other frequencies given by one passing through the feedback will be:

$$|W_1(\omega_n, \omega_0)| = 0{,}5K|W(j\omega_0)|,$$
$$|W_1(\omega_0, \omega_n)| < 0{,}5K|W_1(j3\omega_0)|, \qquad (6)$$
$$|W_1(\omega_n, \omega_m)| < 0{,}5K|W_1(j3\omega_0)|,$$

where $\omega_n = n\omega_0$, $\omega_m = m\omega_0$, $n$ and $m$ are natural values with exceptions: $n \neq 1$ and $m \neq 1$.

## 3   Stability Criterion

According to the graph in Fig. 1, it is possible to suggest the following definition:

*For stability of the signal with frequency $\omega_0$ in system* (1) *described by* Eqs. (1) *and* (2)*, the signal must have descending amplitude after infinite number of passes around the feedback.*

This requirement is an equivalent of the Nyquist criterion, assuming the difference represented by generation of a variety of harmonic functions with different frequencies. According to the graph on Fig. 1, it is possible to build a transfer function $W_k(\omega_0, \omega_0)$ between the signal before passing through the feedback and the signal after $k$ passes through the feedback. For an infinite number of passes through the feedback, we will have the transfer function defined as $W_\infty(\omega_0, \omega_0) = \lim\limits_{k \to \infty} W_k(\omega_0, \omega_0)$.

**Theorem 1.** For system (1) with feedback of form (2), the transfer function between the harmonic signals of frequency $\omega_0$ before and after infinite number of passes around the feedback satisfies the following inequality:

$$|W_\infty(\omega,\omega)| < \left(\frac{K}{2}|W(\omega)| + \sqrt{3}\frac{K}{2}|W(3\omega)|\right)^\infty.$$

So, it is possible to note that $|W_\infty(\omega,\omega)| < 1$ when $\frac{K}{2}\left(|W(\omega)| + \sqrt{3}|W(3\omega)|\right) < 1$.

**Proof of Theorem 1**

There are several routs from the leftmost to the rightmost point of the graph in Fig. 1. Each route with $k$ points has a transfer function that can be represented with the following expression:

$$\left|W_{\text{single\_rout}}(\omega_0,\omega_0)\right| = \left[\frac{K}{2}|W(\omega_0)|\right]^{k-i}\left[\frac{K}{2}|W(3\omega_0)|\right]^i, \tag{7}$$

where $i$ is the number of points with frequencies different from $\omega_0$ or $-\omega_0$ passed along the route.

The number of combinations (sequences) with $(k-i)$ points of $\omega_0$ or $-\omega_0$ and $i$ points of another type can be limited by a binomial coefficient:

$$C_i^k = \frac{(k)!}{(k-i)!i!}. \tag{8}$$

It is important to say that such estimation of combinations is larger than the real number of combinations. For example, it is impossible to change the frequency along the chain $\omega_0 - 3\omega_0 - 5\omega_0 - \omega_0$; however, Eq. (8) takes this case into account. As we build the upper margin for the transfer function, this estimation is quite convenient. Also, it is important to note that the last expression defines the number of combinations but not the number of routes. Passing through $i$ consequent points with frequencies different from $\omega_0$ and $-\omega_0$ allows to have a variety of routes that have triangle form (see Fig. 2).

The size of a triangle can be measured by the number of points between the end points of the route. Each triangle can be built from three triangles of smaller size.

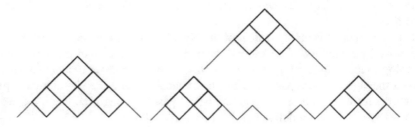

**Fig. 2.** Illustration for calculation of the number of ways from the left point to the right point of the triangle

It is possible to write down the following expression for the number of the routes in a triangle:

$$P\Delta_i = 3P\Delta_{i-2} - 1, \tag{9}$$

where $P\Delta_i$, $P\Delta_{i-2}$ are numbers of routes in the triangles of sizes $i$ and $(i-2)$.

The right part of Eq. (9) contains a minus one because the lowest route is counted two times for both one bottom left small triangle and one bottomright small triangle. Now, we can write down a few consequences of (9):

$$P\Delta_{i-2} = 3P\Delta_{i-4} - 1,$$
$$P\Delta_i = 3(3P\Delta_{i-4} - 1) - 1 = 3^2 P\Delta_{i-4} - 3 - 1$$
$$= 3^2(3P\Delta_{i-6} - 1) - 3 - 1 = 3^3 P\Delta_{i-6} - 3^2 - 3 - 1 = \ldots,$$
$$P\Delta_i = 3^{\frac{i-1}{2}} - \frac{3^{\frac{i-1}{2}} - 1}{3 - 1} = \frac{3^{\frac{i-1}{2}}}{2} + \frac{1}{2}.$$

Thus, for each triangle of size $i$, the following expression is correct:

$$P\Delta_i < 3^{i/2} 3^{-1/2} 2^{-1}. \tag{10}$$

Taking into account Eqs. (7), (8) and (10), we receive the upper margin for the transfer function between the signals of frequency $\omega_0$ before and after infinite passes through the time-varying feedback and the LTI part of system (1):

$$|W_\infty(\omega_0, \omega_0)|$$
$$= \lim_{k \to \infty} \left( \sum_{i=0}^{i=k} \frac{(k)!}{(k-i)!i!} \left[ \frac{K}{2} |W(\omega_0)| \right]^{k-i} \left[ \frac{K}{2} |W(3\omega_0)| \right]^i 3^{i/2} 3^{-3/2} 2^{-1} \right)$$
$$= 3^{3/2} 2^{-1} \lim_{k \to \infty} \left( \frac{K}{2} |W(\omega_0)| + \frac{K}{2} 3^{1/2} |W(3\omega_0)| \right)^k.$$

Now we can apply the infinite route function and eliminate the infinity:

$$\lim_{k \to \infty} \sqrt[k]{|W_\infty(\omega_0, \omega_0)|} = \frac{K}{2} \left( |W(\omega_0)| + 3^{1/2} |W(3\omega_0)| \right).$$

The expression $|W_\infty(\omega_0, \omega_0)| < 1$ will be satisfied when the following expression is correct:

$$\frac{K}{2} \left( |W(\omega_0)| + 3^{1/2} |W(3\omega_0)| \right) < 1.$$

*The theorem is proved.*

## 4  Comparing the Criteria

Let us assume the following LTV system:

$$x^{(3)} + 2x^{(2)} + 2x^{(1)} + 2x = u,$$
$$u = -K\sin(2\omega t)x,$$

where $x$ is the output value, $x^{(1)}$, $x^{(2)}$, $x^{(3)}$ are the output derivatives. The transfer function of the LTI part of the system will be as follows:

$$W(j\omega) = (-j\omega^3 - 2\omega^2 + 2j\omega + 2)^{-1}.$$

Figure 3 illustrates calculated stability borders and the real stability border received by a numerical experiment.

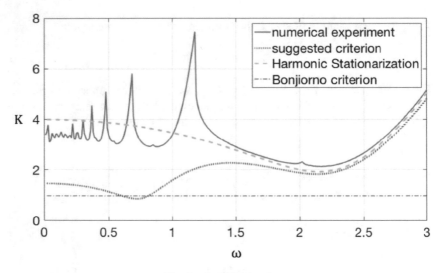

**Fig. 3.**  Stability borders

Three stability criteria are compared. First is the criterion suggested in Theorem 1:

$$\frac{K}{2}\left(|W(\omega_0)| + 3^{1/2}|W(3\omega_0)|\right) < 1.$$

Frequency domain method of stationarization uses a single frequency approximation and provides for the following criterion:

$$0{,}5K|W(j\omega)| < 1.$$

According to Bonjiorno criterion, stability is guaranteed when the following inequality is correct:

$$K|W(j\omega)| < 1.$$

The criterion suggested in Theorem 1 allows to build a more precise stability border than in the Harmonic Stationarization method. For lower frequencies the accuracy of the suggested criterion is worse, however for higher frequencies the experimental and calculated boundaries almost coincide. The main advantage of the suggested criterion is the estimation of higher frequencies generated in the LTV system.

## 5 Conclusion

The result presented in this article is a new sufficient criterion of stability that can be used for LTV systems; it is showing more precise stability calculation. Obviously, the calculation of transfer functions between the graph points from Fig. 2 can help to build stability criteria. The further research may be concerned with the application of the described approach for non-linear systems.

## References

1. Taft, V.A.: On the analysis of stability of periodic modes of operation in non-linear automatic control systems. J. Avtomat. i telemek **9** (1959)
2. Hartman, P.: Ordinary Differential Equations. Wiley, New York (1964)
3. Taft, V.A.: Electrical Circuits with Variable Parameters including Pulsed Control Systems. Pergamon, Oxford (translation from Russian) (1964)
4. Shamma, J.S., Athans, M.: Gain scheduling: potential hazards and possible remedies. IEEE Control Syst. Mag. **12**, 101–107 (1992)
5. Chechurin, S.L., Chechurin, L.S.: Elements of physical oscillation and control theory. In: Proceedings of the IEEE International Conference Physics and Control, St. Petersburg, vol. 2, pp. 589–594 (2003)
6. Bruzelius, F.: Linear Parameter-Varying Systems: An Approach to Gain Scheduling. Chalmers Univ. Technol. (2004)
7. Insperger, T., Stépán, G.: Optimization of digital control with delay by periodic variation of the gain parameters. In: Proceedings of IFAC Workshop on Adaptation and Learning in Control and Signal Processing, and IFAC Workshop on Periodic Control Systems, Yokohama, Japan, pp. 145–150 (2004)
8. Allwright, J.C., Alessandro, A., Wong, H.P.: A note on asymptotic stabilization of linear systems by periodic, piecewise constant, output feedback. Automatica **41**, 339–344 (2005)
9. Szyszkowski, W., Stilling, D.S.D.: On damping properties of a frictionless physical pendulum with a moving mass. Int. J. Non Linear Mech. **40**(5), 669–681 (2005)
10. Baños, A., Barreiro, A., Beker, O.: Stability analysis of reset control systems with reset band. IFAC Proc. Vol. **42**(17), 180–185 (2009)
11. Mailybaev, A.A., Seyranian, A.P.: Stabilization of statically unstable systems by parametric excitation. J. Sound Vib. **323**, 1016–1031 (2009)

12. Eissa, M., Kamel, M., El-Sayed, A.T.: Vibration reduction of a nonlinear spring pendulum under multi external and parametric excitations via a longitudinal absorber. Mechanica **46**, 325–340 (2011)
13. Arkhipova, I.M., Luongo, A., Seyranian, A.P.: Vibrational stabilization of the upright statically unstable position of a double pendulum. J. Sound Vib. **331**(2), 457–469 (2012)
14. Abe, A.: Non-linear control technique of a pendulum via cable length manipulation: application of particle swarm optimization to controller design. FME Trans. **41**(4), 265–270 (2013)
15. Amer, Y.A., Ahmed, E.E.: Vibration control of a nonlinear dynamical system with time varying stiffness subjected to multi external forces. Int. J. Eng. Appl. Sci. (IJEAS) **5**(4), 50–64 (2014)
16. Arkhipova, I.M., Luongo, A.: Stabilization via parametric excitation of multi-dof statically unstable systems. Commun. Nonlinear Sci. Numer. Simul. **19**(10), 3913–3926 (2014)
17. Mandrik, A.V., Chechurin, L.S., Chechurin, S.L.: Frequency analysis of parametrically controlled oscillating systems. In: Proceedings of the 1st IFAC Conference on Modelling, Identification and Control of Nonlinear Systems (MICNON 2015), St. Petersburg, Russia, 24–26 June 2015. IFAC-PapersOnLine, **48**(11), 651–655 (2015)
18. Reguera, F., Dotti, F.E., Machado, S.P.: Rotation control of a parametrically excited pendulum by adjusting its length. Mech. Res. Commun. **72**, 74–80 (2016)
19. Scapolan, M., Tehrani, M.G., Bonisoli, E.: Energy harvesting using a parametric resonant system due to time-varying damping. Mech. Syst. Signal Process., 1–17 (2016)
20. Mandrik, A.V.: Estimation of stability of oscillations of linear time-varying systems with one time-varying parameter with calculation of influence of higher frequency motions. Autom. Control. Comput. Sci. **51**(3), 141–148 (2017)

# Performance Analysis of Available Service Broker Algorithms in Cloud Analyst

Elena N. Desyatirikova[1], Almothana Khodar[1],
Alexander V. Rechinskiy[2], Liudmila V. Chernenkaya[2(✉)],
and Iyad Alkhayat[3]

[1] Voronezh State Technical University, Voronezh, Russia
science2000@ya.ru, mothana-sy@hotmail.com
[2] Peter the Great St. Petersburg State Polytechnic University,
Saint Petersburg, Russia
{mv,liudmila}@qmd.spbstu.ru
[3] Damascus University, Damascus, Syria
iyad_alkhayat@hotmail.com

**Abstract.** Cloud computing is a reasonably distributed computing with heterogeneous computational resources. Resource sharing, convenience of the resources that correspond to the needs and a lot of box-management issues facing the cloud are considered. In this paper, analysis and comparison of various existing algorithms of service brokers and load equalizing algorithms in cloud computing is presented. To check the implementation of existing algorithms, various modeling tools were developed, for instance, the Cloud Analyst. Performance of the present policies of service brokers is compared by considering the total response time and execution time in a data center.

**Keywords:** Cloud computing · Cloud analyst · Infrastructure performance · Service broker · Load balancing algorithms

## 1 Introduction

Cloud computing is a technology that helps to exchange data and provide many resources to users. Users pay only for the resources they used [1].

The main feature that characterizes cloud computing is the use of virtualization. In a broad sense, the concept of virtualization is concealment of the present implementation of any process or object from its true representation for the one who uses it [2]. In other words, there is a separation of the representation from the implementation of something. In cloud computing, the term "virtualization" usually refers to abstraction of computational resources and provision of a system to a user that "encapsulates" (hides in itself) its own implementation [3, 4]. Virtual machines are run over the available hardware to address the user's needs. Selection of virtual machines whenever the workload is encountered is done by the load balancer, whose aim is to distribute the load in such a way that no virtual machine is flooded by requests at one time, while

© Springer Nature Switzerland AG 2020
D. G. Arseniev et al. (Eds.): CPS&C 2019, LNNS 95, pp. 449–457, 2020.
https://doi.org/10.1007/978-3-030-34983-7_44

remaining idle at other times [5]. Above this level lies another abstraction called the service broker, which is the intermediary between the users of the cloud and the cloud service providers. It makes the use of the existing service broker policies in order to route the user's request to the most appropriate data centre. Therefore, the selection of the best policy determines the response time of a particular request and the efficiency of utilization of the data centre [6, 7].

However, measuring the performance of internet-based applications using a real cloud platform is very difficult. Therefore, simulation-based approaches are provided to solve such issues virtually and free of charge in a stable and controllable environment; an extensible toolkit for modeling and simulating cloud computing systems called CloudAnalyst is proposed in [8]. The Cloud Analyst provides a set of components that provide the base for cloud computing, including Virtual Machines (VM), Cloudlets (Jobs and user's request will be used interchangeably), datacenters (DCs), Service broker and hosts [9, 10]. Each of them has its own characteristics and functionality, which we will review in our paper.

## 2  Routing of User Request in Cloud Computing

In the cloud, from a user's point of view, the main factors are cost optimization and a provider that provides efficacy for the user's needs. Thus, routing a user's request is a very important aspect in the cloud. Figure 1 shows the routing of a user's request in one of the Cloud Analyst simulation tools [11].

- The user base creates an Internet Cloudlet which includes both the application identifier ID and the name of the user base in order to route back the RESPONSE to the user.
- REQUEST is sent to the Internet without delay.
- The data center is chosen by the Internet after consulting the service broker. The policy of the service broker is determined based on the information in the REQUEST
- Service broker sends information about the chosen data center controller to the Internet.
- The Internet adds appropriate network delay to the REQUEST; then it will be sent to the chosen data center controller.
- The selected data center controller uses a suitable policy for load balancing of the virtual machines.
- The load balancing of the virtual machines allocates the virtual machine to the user's request.
- Selected data center sends the RESPONSE to the Internet after handing out the REQUEST.
- Internet uses the originator field of the Cloudlet information and adds an appropriate network delay with the RESPONSE; then it sends all this info to the user base.

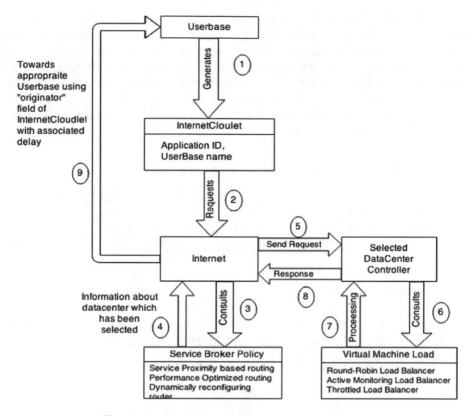

**Fig. 1.** Routing of the user's request in Cloud Analyst

## 3 Existing Service Broker Algorithms

The most common standard algorithms [12, 13] are as the following.

*Closest data centre policy:* This policy makes use of the construct of region proximity for choosing the info center that processes client's requests. The proximity list is maintained by using of the "lowest network latency first" criteria to put the order of incidence of information centers within the list. The info centre located 1st within the list, i.e., the nearest information centre, is chosen to satisfy the request. For just in case, if one information center with constant latency square measure is offered, a random choice of info centers is created. This policy is so helpful in the case if the request may be satisfied by a relatively close information center or a center in a permanent region [14].

*Optimal response time policy:* the first step in this policy is identification of the nearest data center using the network delay constraint, as in the earlier described policy. Then, for each of those, the current response time is predictable. If the estimated response time of the nearest data center is the shortest response time, the nearest data center will be chosen. Otherwise, the closest data center or data center with the lowest response time will be chosen with the probability of 50:50 [15].

*Dynamically reconfigurable routing with load balancing:* This broker policy makes use of the execution load to scale the applying preparation. It will additionally increase or decrease the amount of consequent virtual machines. The router has to hold an extra responsibility of scaling the applied preparation. This can be done supporting the load that it's presently facing. In this policy, scaling is completed by considering these time intervals and, therefore, the best process time ever is achieved [16].

From the above, we identified some shared characteristics among the three service broker policies summarized in Table 1.

**Table 1.** Shared characteristics among the three service broker policies

| Policy name | Available BW | Latency | DC current load |
|---|---|---|---|
| Closest data centre | No | Yes | No |
| Optimal response time | Yes | Yes | Yes |
| Dynamically reconfigurable routing | No | Yes | Yes |

## 4  Simulation Tools

Cloud-Analyst is one of the most essential simulators, it is an open-source toolkit [4] that helps us model and evaluate the performance of cloud services. It is built on top of Cloudsim, as is shown in Fig. 2.

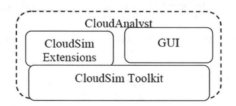

**Fig. 2.**  Cloud-Analyst architecture

The main components in Cloud Analyst are:

*Region*: World is divided to six regions according to the continents by the Cloud-Analyst.

*User Base*: This component generates traffic which represents the users. Thousands of users can be considered as a single user base which is configured as a single unit.

*Cloudlet*: This component represents a group of users' requests. The Cloudlet has all the information about the number of requests, identifier of the sender and target application used for routing over the Internet [17].

*Virtual Machines*: A completely isolated container in which the operating system and various applications can be launched.

*VM Load Balancer*: It is responsible for specifying the virtual machine that will process the incoming from the user request according to its scheduling algorithms [18].

*Cloud Application Service Broker*: The service broker is responsible for scheduling at the data center level, which is higher than the level of virtual machines, specifying the data center that will receive the request sent by the user. Thus, the service broker controls the routing of the traffic between user databases and data centers [19, 20].

# 5  Experimentation

For the experiments, we applied this modeling to the Western Union Bank, which has the largest number of users in the world and mass bank transfers every hour. Different scenarios are considered with the three data centers. Each data center uses one physical computer with the number of processors equal to 4, with time-shared policy for virtual machine allocation. Main configuration parameters for user bases are shown in Table 2. We assume that all users in one continent are grouped in one base, the peak period for users is only two hours per day, and 10% of customers make structural transfers at non-peak times.

We have also identified the characteristics of the datacenter, in terms of the place, processor architecture, number of processing units, operating system, and cost of use over a given period of time as shown in Table 3.

**Table 2.** Parameters for the user bases

| Base | Region | Requests per User per Hr | Data size per request (bytes) | Peak hours start (GMT) | Peak hours end (GMT) | Users online during peak hours | Users online in non-peak hours |
|------|--------|------|------|------|------|------|------|
| UB1 | 0 | 12 | 100 | 12 | 14 | 4000 | 400 |
| UB2 | 1 | 12 | 100 | 14 | 16 | 1500 | 150 |
| UB3 | 2 | 12 | 100 | 19 | 21 | 3000 | 300 |
| UB4 | 3 | 12 | 100 | 00 | 02 | 2000 | 200 |
| UB5 | 4 | 12 | 100 | 20 | 22 | 500 | 50 |
| UB6 | 5 | 12 | 100 | 10 | 12 | 800 | 80 |

**Table 3.** Data center configuration

| Name | Region | Arch | OS | Cost per VM $/Hr | Physical HW units |
|------|--------|------|------|------|------|
| Dc1 | 0 | x86 | Linux | 0.1 | 1 |
| Dc2 | 1 | x86 | Linux | 0.1 | 1 |
| Dc3 | 2 | x86 | Linux | 0.1 | 1 |

Configuration for the deployment of the application contains virtual machine properties, (i.e., the number of virtual machine-associated data centers, etc.) shown in Table 4.

**Table 4.** Application deployment configuration

| Data centre | No of VM | Image size | Memory | B.W. |
|---|---|---|---|---|
| DC1 | 25 | 10000 | 512 | 1000 |
| DC2 | 25 | 10000 | 512 | 1000 |
| DC3 | 25 | 10000 | 512 | 1000 |

In the simulation we choose, for the advanced configuration parameters, to include the user grouping factor equal to 500, the request grouping factor is chosen to be 300, while the executable instruction length is 200 Byte per request, and to balance the load across the virtual machine in a single data centre, the round-robin policy is used. Simulation duration has been limited to 24 h.

## 6 Simulation Results

Two tests were carried out to demonstrate two different scenarios. The first scenario has three data centers in the same region (North America). The second scenario has three data centers in different regions (North America, South America and Europe). In each scenario we used the same previous settings but we changed the service broker algorithm. We compared the three service broker policies by considering the overall response time and data center processing time as shown in Table 5.

As shown above, Fig. 3 presents the comparison between the three service broker algorithms with the three data centers in the same region. The Optimized Response Time algorithm provides the best result. The second one is Closest Data Center algorithm. The Reconfigure Dynamically algorithm required the longest time to execute users' requests, and the data center load was the largest.

**Table 5.** Simulation results for two scenarios with three service broker algorithms

| Scenario | Service broker algorithms | Overall response time in ms | DC processing time in ms |
|---|---|---|---|
| Three data centers in the same region | Closest data center | 257.83 | 17.85 |
| | Optimized response time | 257.78 | 17.84 |
| | Reconfigure dynamically | 710.62 | 388.99 |
| Three data centers in different regions | Closest data center | 133.02 | 17.68 |
| | Optimized response time | 133.15 | 17.67 |
| | Reconfigure dynamically | 451.55 | 266.43 |

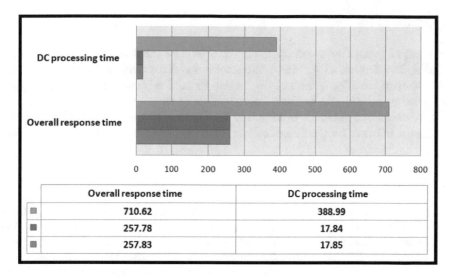

| | Overall response time | DC processing time |
|---|---|---|
| ▪ | 710.62 | 388.99 |
| ▪ | 257.78 | 17.84 |
| ▪ | 257.83 | 17.85 |

**Fig. 3.** Results for first scenario with three data centers in the same region

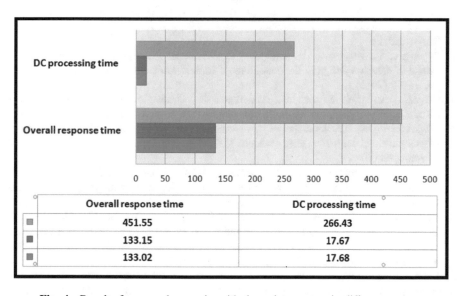

| | Overall response time | DC processing time |
|---|---|---|
| ▪ | 451.55 | 266.43 |
| ▪ | 133.15 | 17.67 |
| ▪ | 133.02 | 17.68 |

**Fig. 4.** Results for second scenario with three data centers in different regions

Figure 4 shows the comparison between the three service broker algorithms when data centers are placed in three different regions. Clearly, there is a large decrease in the Overall response time compared to the first scenario due to the distribution of processing facilities near the places of request. The best service broker algorithm here is the Closest Data Center algorithm, and the second one is the Optimized Response Time algorithm

# 7 Conclusion

The main advantage of cloud computing is that customers are freed from concerns regarding learning about the basic instrumentality that caters to their requests. This paper assesses the effectiveness of the 3 existing service broker policies with constant load leveling policy. The results of our work have shown that once the information centers exist in one place, it's better to use the Optimal Response time policy, whereas the Closest Information Center policy has higher performance in distribution centers.

# References

1. Calherios, R.N., Wickremasinghe, B.: Cloud Analyst: a cloud-sim-based visual modeler for analyzing cloud computing environments and applications. In: Proceedings of IEEE International Conference on Advance Information Networking and Applications (2010)
2. Desyatirikova, E.N., Almothana, K., Osama, A.: Load balancing in cloud computing using genetic algorithm. In: XVIII International Scientific and Methodological Conference "Informatics: Problems, Methodology, Technology", 8–9 February 2018, Voronezh, Russia (2018)
3. Volkova, V.N., Loginova, A.V., Shirokova, S.V., Iakovleva, E.A.: Models for the study of the priorities of innovative companies. In: Proceedings of the 19th International Conference on Soft Computing and Measurements, SCM 2016, pp. 515–517. IEEE, St. Petersburg, Russia (2016)
4. Klochkov, Y., Klochkova, E., Antipova, O., Kiyatkina, E., Vasilieva, I., Knyazkina, E.: Model of database design in the conditions of limited resources. In: ICRITO, Noida, India, pp. 64–66 (2016)
5. Desyatirikova, E.N., Moussa, H., Almothana, K., Osama, A.: Load balancing in cloud computing. Syst. Anal. Inf. Tech. 3, 103–109 (2017). Vestnik Voronezh State University
6. Wickremasinghe, B.: CloudAnalyst: a CloudSim-based tool for modelling and analysis of large scale cloud computing environments. MEDC Project Report, University of Melbourne, Australia (2009)
7. Tordsson, J., Montero, R.S., Vozmediano, R.M., Llorente, I.M.: Cloud brokering mechanisms for optimized placement of virtual machines across multiple providers. Future Gener. Comput. Syst. 28(2), 358–367 (2011)
8. Volkova, V.N., Chernenkaya, L.V., Desyatirikova, E.N., Moussa, H., Almothana, K., Osama, A.: Load balancing in cloud computing. In: Proceedings of 2018 IEEE Conference of Russian Young Researchers in Electrical and Electronic Engineering (2018 ElConRus), St. Petersburg and Moscow, Russia, pp. 397–400 (2018)
9. Dash, M., Mahapatra, A., Ranjan, N.: Cost effective selection of data center in cloud environment. Int. J. Adv. Comput. Theory Eng. (IJACTE) (2013). ISSSN 2319–2526
10. Algazinov, E.K., Desyatirikova, E.N., Almothana, K.: Performance analysis of the cloud service broker algorithm. In: International Scientific Conference, St. Petersburg, Russia (2018)
11. Zhao, W., Peng, Y., Xie, F., Dai, Z.: Modeling and simulation of cloud computing: a review. In: IEEE Asia Pacific Cloud Computing Congress (APCloudCC), pp. 20–24 (2012)
12. Limbani, D., Oza, B.: A proposed service broker strategy in cloud analyst for cost-effective data center selection. Int. J. Eng. Res. 2(1), 793–797 (2012)

13. Desyatirikova, E.N., Almothana, K., Osama, A.: Scheduling approach for virtual machine resources in a cloud computing. In: XIX International Conference "Informatics: Problems, Methodology, Technology" (IPMT-2019), Voronezh, Russia (2019)
14. Shah, M.M.D., Kariyani, M.A.A., Agrawal, M.D.L.: Allocation of virtual machines in cloud computing using load balancing algorithm. Int. J. Comput. Sci. Inf. Technol. Secur. (IJCSITS) (2013)
15. Dinh, H.T., Lee, C., Niyato, D., Wang, P.: A survey of mobile cloud computing: architecture, applications, and approaches. Wirel. Commun. Mob. Comput. **13**(18), 1587–1611 (2013)
16. Rani, P., Chauhan, R., Chauhan, R.: An enhancement in service broker policy for cloud-analyst. Int. J. Comput. Appl. **115**(12), 5–8 (2015)
17. Rekha, P.M., Dakshayini, M.: Cost based data center selection policy for large scale networks. In: ICCPEIC, 2014 International Conference (2014)
18. Arora, V., Tyagi, S.S.: Performance evaluation of load balancing policies across virtual machines in a data center. In: ICROIT, 2014 International Conference on the Optimization, Reliability, and Information Technology (2014)
19. Ranbhise, S.M., Joshi, K.K.: Simulation and analysis of cloud environment. IJARCST **2**(4), 2347–9817 (2014)
20. Mishra, R., Jaiswal, A.: Ant colony optimization: a solution of load balancing in cloud. Int. J. Web Semant. Technol. (IJWesT) **3**(2), 33–50 (2012)

# Cyber-Physical System Control Based on Brain-Computer Interface

Filipp Gundelakh[1], Lev Stankevich[2], Nikolay V. Kapralov[1(✉)], and Jaroslav V. Ekimovskii[1]

[1] Peter the Great St. Petersburg Polytechnic University, Saint Petersburg, Russia
nikolay.kapralov@gmail.com
[2] St. Petersburg Institute of Informatics and Automatics of RAS, Saint Petersburg, Russia

**Abstract.** The study describes approaches of direct and supervisory control of cyber-physical systems based on a brain-computer interface. The interface is the main component of the control system, performing electroencephalographic signal decoding, which includes several steps: filtering, artefact detection, feature extraction, and classification. In this study, a classifier based on deep neural networks was developed and applied. Description of the classifiers based on convolutional neural network is given. The developed classifier demonstrated accuracy $73 \pm 5\%$ of decoding four classes of imaginary movements. Prospects of using non-invasive brain-computer interface for control of cyber-physical systems, in particular, mobile robots for maintenance of immobilized patients and devices for rehabilitation of post-stroke patients are discussed.

**Keywords:** Brain-computer interface · Neural networks · Classifying

## 1 Introduction

Recently, a new informational-technological concept named *cyber-physical system* received great attention. This concept concerns integration of computing resources and physical processes. In this connections, it is possible to consider as cyber-physical systems the complexes that consist of objects of various natures merged with computer control means and working as a whole. These can be complexes of real-time embedded systems solving problems of "smart homes" or "internet of things", distributed computing and controlling systems, automated control systems of industry, sensor networks, and so forth.

A special place is occupied by *assistive complexes* helping people with restricted mobility and complexes for rehabilitation of patients after strokes and traumas. Such complexes generally include various *robotized means*. These can be arm and leg artificial limbs, exoskeletons, and mobile robots with direct or remote control. Such mobile robots as wheel robots with manipulators or anthropomorphic robots, whose shape is similar to the human's one, can be basic robotized means for assistive and rehabilitative complexes. In particular, according to forecasts, in the near future, anthropomorphic robots will be widely used as assistants to people, especially when working in extreme areas and service sectors, including helping immobilized people in hospitals and at home.

© Springer Nature Switzerland AG 2020
D. G. Arseniev et al. (Eds.): CPS&C 2019, LNNS 95, pp. 458–469, 2020.
https://doi.org/10.1007/978-3-030-34983-7_45

*Mobile robots* can be implemented in various ways, but for working with immobilized people, contact-free means of communication between the person and the robot are required.

The robots are controlled by computers. A promising way to realize communication between a person with dexterity impairment and a computer that controls the robot is the use of so-called *brain-computer interface* (BCI). BCI is a modern technology that can provide human interaction with external electronic and electromechanical devices based on the registration and decoding of signals of brain electrical activity [1]. It is shown that for immobilized people who need to control the service robotic devices themselves, a BCI, based on the recognition of imaginary movements can be very helpful [2].

As a means of obtaining information about the bioelectric activity of the brain corresponding to motor commands, *electroencephalography* (EEG) is often used. The EEG is a non-invasive and relatively inexpensive technology that has significant potential for the creation of BCIs. However, the factor limiting practical application of BCIs based on EEG signals is the complexity of reliable and reproducible interpretation (decoding) of brain signals. Another factor is the difficulty in classifying the EEG patterns of imaginary movements in real time.

Several Russian scientific groups and organizations carried out researches related to the development of BCI, of which some are known internationally [3, 4]. Thus, efforts to create a *human-robot interface* using BCI based on EEG signals have been performed. For example, such tools allow detecting certain mental commands of a person using P300 signals [4]. However, such interfaces are limited by the number of degrees of freedom and are only applicable for simple control of a robot.

A review of the prospects of brain and neural computer interfaces (BNCIs) is presented in the study [5].

The aim of this work is to develop the robotized control system means based on BCI performing classification of imaginary movements using EEG signals. The robotized means with such control systems can be used, for example, for prosthetics, helping people with limited motor abilities, as well as for rehabilitation in clinics and at home. A variant of EEG imagery motor pattern classifier based on a convolutional neural network is discussed. Examples of application of a BCI to the control of an intellectual sensor platform and an anthropomorphic robot are considered.

## 2  Robotized Means

Variants of robotized means for the assistance to people with the limited mobility and of rehabilitation complexes for stroke and post-trauma patients include arm or leg artificial limbs, exoskeletons, and mobile robots. Artificial limbs of the arms and legs, in essence, are manipulators operated based on signals from nervous system. The use of bioelectric signals of brain's activity for their control also is possible but demands special interfaces detecting and decoding matching motor commands of the brain. Exoskeletons can be of various designs and mission. In particular, for the rehabilitation of stroke patients are used exoskeleton brushes of the arm which can be simple mechanical benders and straighteners of fingers or complicated ones, operating using special computer interfaces [6]. Mobile robots appeared in connection with the need to expand the working area of manipulators, as well as to perform transport operations.

The modern assistive mobile robots have various means for movement and manipulation, as well as computer control systems [7]. In particular, to make a movement, they use a variety of vehicles, including wheeled and walking platforms. For working with objects they use simple manipulators or anthropomorphic arms.

Wheeled robots, as a rule, have many sensors for the perception of environment. Such robots can perform service tasks, for example, bring water or medicines, call for medical personnel and so on. For contactless control of the robots, remote direct or supervisory control systems are required.

Walking type of movements have higher maneuverability in rugged terrains as compared to wheeled or tracked ones, and are more economical in these conditions because interaction with the ground occurs only at the places of the impression of the foot, while the wheeled and crawled types of movements leave a continuous track behind machines. Since the main purpose of walking machines is moving across imperfect surfaces, their control systems must be adaptive.

Anthropomorphic robotics engaged in researching and creating humanoid robot-shas been actively developed in the recent years. Such robots are oriented to work in a poorly determined environment and are positioned as universal human assistants. Synthesis of an anthropomorphic robot control system is a complex problem. The rapid development of this area is associated with successes in the field of artificial intelligence.

## 3 Remote Control

Remote control of a mobile robot can be implemented by acting on different levels of the control system, among which may be singled out the direct control of individual robot drives (direct control) and supervisory control using a set of high-level commands.

Direct control provides sufficient freedom of control but it is very laborious and requires a lot of skill. Time lag in the communication channel makes it even more difficult to work in this mode.

In its turn, supervisory control provides a limited set of commands, but these commands can run complex behavior algorithms.

EEG-based BCIs are mainly targeted on disabled people thanks to a rich variety of applications. There are several low-cost realizations of such BCIs, for example, based on Emotive Epoc headset and P-300 method [8]. It could use a mouse to control a web browser or to spell words mentally. Other main research areas are the study of motor substitution or motor rehabilitation; the main applications are hand-grasping [9] and wheelchair control [10]. For healthy end-users, BCIs have also been used to augment interactivity in games by using multimodality from the EEG signals and standard control [11, 12].

# 4  Human-Robot Interfaces

At present, the development of neurophysiology, psychology and artificial intelligence, as well as methods and devices for recording bioelectric signals, has led to the possibility of creating new means of human-robot interfaces (HRIs) based on direct decoding of the nervous system signals and using them for robot control.

Currently, HRIs are implemented on non-invasive BCIs based on EEG. Such BCIs can recognize imaginary and real movements of body parts, e.g., hands. However, for robot control, such BCIs must be taught decoding fine movements, for example, the movements of the fingers of one hand. This problem is difficult to solve because of the anatomical proximity of the brain structures participating in the implementation of imaginary movements and the slight differences in EEG signals imagining movements in small parts: it requires large computational resources for data analysis. Furthermore, HRIs should operate in real time, that is, generate control signals with a minimum delay determined by the operation speed of external devices.

There are many studies which highlighted the importance of EEG-BCIs for controlling external devices. In [13] it was shown that people with severe Neuromuscular Disorder can acquire and maintain control over detectable patterns of brain signals and use that in order to control devices. Subjects were asked to execute or to imagine movements of their hands or feet in response to the appearance of a respective target. This study showed, for the first time, that an EEG-based BCI system can be integrated into an environmental control system. If the user was not able to master any of the devices, a training procedure followed.

Users needed to learn to modulate their Sensorimotor Rhythm to achieve a more robust control than a mere imagination of the movements a limb can produce. After 10 sessions of training, the subjects acquired brain control with an average classification higher than 75% in a binary selection task. Nevertheless, it should be stressed that only 4 out of 14 participants managed to achieve the above-mentioned accuracy.

In another study, the potential use of SMR for the control of external domestic devices was investigated. As far as the experimental procedure is concerned, initially, a rectangular target appeared on the right side of the screen (upright or downright). Afterward, the cursor appeared in the middle of the left side of the screen and moved horizontally with a constant speed to the right. During the training sessions, subjects were asked to imagine the same kinesthetic movement during the visual session. Four out of six participants with Discrete Muscular Dystrophy were able to control several electronic devices in the domestic context with the BCI system, with a percentage of correct responses averaging over 63%, whereas healthy subjects achieved the rates of 70–80%.

Another study [14] introduced the one-dimensional feedback task, i.e., moving a cursor from the center of a monitor to a randomly indicated horizontal direction. In the second stage of the experiment, the feedback signal was provided only at the end of each trial. Four out of seven subjects were able to operate the BCI system via attempted (not imagined) movements with their impaired limbs (both foot and hand) with the classification accuracy (CA) up to 84%. Moreover, it was highlighted that the foot movements were easier discriminated than hand movements for the majority of subjects.

Furthermore, a recent study demonstrated the potential use of a telepresence robot, which was remotely controlled by a BCI system in the course of a navigation task. In this task, the participants with motor impairment achieved similar CAs as ten able-bodied participants who had already been familiar with the environment [15]. The robot starts from a specific position, and there were four target positions to be reached. The user's task was to drive the robot along one of three possible paths and then back to the starting position. Some end-users were able to press specific buttons on a modified keyboard, while others were using "head switches" by imagining left hand, right hand and feet movements during the calibration recordings. However, people with motor impairment needed more time in comparison with healthy participants to complete the path. It is noteworthy that the environment of the experiment contained natural obstacles as in a real-life environment.

## 5   Human-Robot Interface Structure

While controlling robotic devices, such as a wheelchair, robotic manipulator, pros-thesis, exoskeleton, etc., a paralyzed person should be able to give commands that allow these devices to perform specific tasks for human's movement, object manipu-lation, or prosthesis control. In this paper, it is proposed to do this using a non-invasive BCI decoding brain activity signals, and form supervisory commands to external devices. For the practical implementation of such a system, it is necessary to use biological feedback through the vision, which allows the user to learn how to control the system by amplifying EEG signals corresponding to imaginary commands [3]. Delays in calculating the response must be small enough, so that in the user's mind these events would be clearly associated.

It is possible to formulate the following goals for constructing systems that would allow a person to control robotic devices through a BCI: (1) achieving high accuracy and speed of classification of EEG patterns to levels acceptable for the use of BCI for the control of robotic devices; (2) ensuring sufficient BCIs degrees of freedom, that is, of at least 4 recognized imaginary commands; (3) optimization of computational resources for the reduction of the run-time calculations of the BCI while maintaining the recognition accuracy, including the classification of imaginary commands by single trial of imagination.

EEG-based BCIs include a stage of recording and preprocessing EEG signals, extracting features in spectral or temporal domains, classifying EEG patterns of imaginary commands, and forming a final decision for the class of the recognized command (Fig. 1).

To study the possibility of implementing a human-computer interaction system based on a non-invasive BCI, the classifier of imaginary finger movements of one hand, developed by the authors earlier [16, 17], was used.

The adaptation of this version of the BCI to the classification of imaginary com-mands was carried out using a specially developed program and methodology. According to the method, subjects were asked to consistently imagine 4 types of imaginary movements with the fingers of one hand (thumb, index, middle finger and little finger) in accordance with the commands assigned to them to control the behavior of the robotic device. During the training, the subjects had first to press the computer mouse button with a finger assigned to control the necessary behavior, in the rhythm defined by the sounds, and after the sound was turned off, to continue imagining the movement with the same finger.

Training was focused on the imagination of movements on the kinesthetic sensations of the subject. The test series of real and imaginary movements were repeated several times. As a result, the subject had to perform no less than one hundred real and one hundred imaginary movements in a given rhythm in one block of the test. The number of test blocks should correspond to the number of types of performed movements (real/imaginary finger movements). Thus, the control of the robotic devices with the command buttons is simulated.

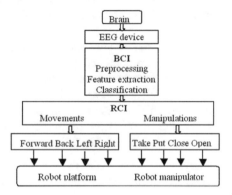

**Fig. 1.** Structure of HRI based on non-invasive BCI (RCI stands for Robot Control Interface)

The EEG signals were recorded using a 32-channel digital electroencephalograph of the Mitsar Company. We used 19 electrodes, which were located using the 10–20 system. EEG signals with a duration of 600 ms were recorded in the frequency band of 0.53 Hz–30 Hz. The sampling rate was 500 Hz. When analyzing EEG recordings, artifacts such as eye movements, slow and fast waves, fragments of EEG signals with an amplitude of more than 100 µV were excluded. For the analysis and subsequent classification, EEG signal records without artifacts registered from the sensorimotor region of the cerebral cortex were used: C3, Cz, F3 sites.

Analysis of EEG signals was carried out in the time and frequency domain. To ensure the necessary speed of calculating the function of the system, analysis of each individual test without data accumulation was made. An algorithm utilizing a combination of two feature spaces was used: the length of the curve and the cepstral coefficients calculated in the signal segment. These functions were calculated in a sliding window, which allowed selecting the most informative windows.

Choosing the most optimal size of the analysis window is important for increasing the accuracy of classification. Previously, it was shown that the selection of optimal values of the analysis window width can significantly improve the accuracy of the classification of imaginary movements. On the basis of the earlier studies [17], a 100 ms analysis window width with a shift of 50% was chosen.

Classification of the EEG patterns corresponding to imaginary movements was carried out using the calculated features as inputs of neural network and support vector machine classifiers.

# 6  Classification of the Imaginary Movements

Classification of the imagery movement of fingers is implemented using a convolution neural network (CNN) [18]. It is possible to use a CNN of the Deep ConvNet and Shallow ConvNet types for raw EEG signal classification. Studies have shown that they are effective classification means, including that for classification of the imaginary movements.

In this work, Shallow ConvNet neural network was implemented as a CNN with four layers: two convolution layers (temporal and spatial), one mean pooling layer, and an output softmax layer. As activating functions in the hidden layers and an output layer, exponential linear functions (ELUs) were used.

At the input of the CNN, raw EEG signals were introduced. On the basis of these signals, the network carried out the classification of imaginary movement of fingers for one hand (Fig. 2). The layers of the network are described as following:

- *Temporal Convolution Layer:* the input signal shape is $21 \times 300$ (21 channels each consisting of 300 samples) and the kernel size is $1 \times 25$ (convolution on time), and 40 filters are used.
- *Spatial Convolution Layer:* the size of input tensor is 40 (filters) $\times$ 21 (channels) $\times$ 276 (time points after the convolution on time), the kernel size is $40 \times 21$; again, 40 filters were used.
- *Square:* all values of the matrixes are squared as elements.
- *Mean Pooling:* the shape of the input matrix is $40 \times 276$, the kernel size is $1 \times 75$.
- *Log:* each element of the matrix is transformed by natural logarithm.
- *Classification:* the shape of the input matrix after the *Mean Pooling* is $40 \times 14$; the classification is performed by a combination of fully connected and softmax layers.

During the operation of the BCI in real time, it is assumed that the EEG data is being read continuously [19]. To ensure the continuity of receiving input data and their parallel processing, the multithreaded programming method was used. The processing scheme is as follows: as soon as the trial N ends and the trial N + 1 begins, thread 2 reads the data related to the trial N, pre-processes data, extracts features, classifies, and returns the result of the classification (Fig. 3).

The conducted studies of the BCI prototype have shown that its characteristics correspond to real-time requirements. The time delays needed to obtain a response from the classifier are within 150 ms, which is acceptable for the use of the BCI with biological feedback. According to the subjective self-report of the subjects, they really connect the imaginary act with the received response of the system.

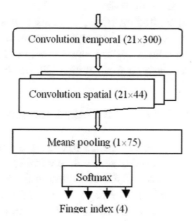

**Fig. 2.**  Scheme of the CNN classifier

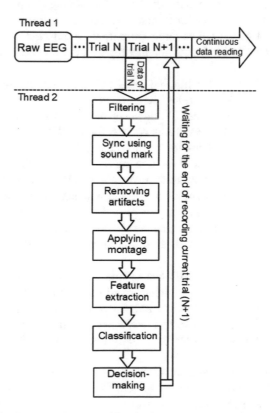

**Fig. 3.**  Scheme of the classification system operating in real time

The following results were obtained in a BCI with the CNN classifier studies. In autonomous sessions, when classifying EEG patterns of imaginary commands without accumulating trials, the probability of successful classification of four types of imaginary commands exceeded the random threshold of 25% and averaged at $58 \pm 5\%$ for channels F3, C3. Using accumulating trials, the classifier demonstrated accuracy of $73 \pm 5\%$ of decoding four classes of imaginary movements, if the decision was accepted at 3 successive decodings. At the same time, studies have shown that some subjects are not able to learn how to work with BCIs on the basis of EEG signals.

# 7 Application Examples of BCI-Based Control Systems

An operator should be able to give commands to change the behavior of robotic means. As part of the study, it was proposed to remotely control service mobile robots through a BCI. A control system based on a BCI can allow immobilized people both to self-service using a robot and to train the damaged motor functions.

Experiments were conducted to control a robotic cart and an anthropomorphic robot using the noninvasive BCI described in this work with the classification system of the EEG –patterns of imaginary movements based on a CNN.

The robotic cart has 4 controlled drives (2 for moving the platform and 2 for the sensor unit orientation for directional stereo vision) and can be used, for example, to service immobilized patients. The anthropomorphic robot NAO with 23° of mobility can be used as a mobile device for interacting with a patient and performing simple actions.

These mobile robots were controlled using a BCI by issuing commands directly to the drives (direct control) or high-level command signals (supervisory control). Using supervisory control by simulating the pressing of control keys with different fingers, four classes of commands were formed: "Forward", "Stop", "Right", "Left". The experiments have shown the possibility of training the BCI classifier for these commands and controlling a cart and a robot using them. Since both the cart and the robot had vision systems and their own systems for recognizing and avoidance of obstacles, such supervisory control was sufficient for the practical application of these robotic devices. Thus, a robotic cart is capable of delivering medicines and water by the operator's commands and transporting away the used flatware. In the future, it is planned to equip the platform with an artificial hand for loading and unloading the required items.

An anthropomorphic robot NAO has an intelligent control system which allows it to perform a wide range of actions, such as hand and foot movements, steps, turns, etc. Such a man-like robot equipped with a BCI-based control system can become a novel tool for rehabilitation of patients with severe movement disorders caused by stroke and brain injuries and assistance robots for immobilized people. There might be two use cases:

The user imagines a movement of an immobilized limb, the BCI decodes this intention and a robot performs this movement (imitating human movements);

The User controls the robot by supervisory commands ("Forward", "Stop", "Right", "Left"), and the robot passes through a complex path performing them.

In the first case, the user receives feedback from the robot, which significantly increases the user's motivation and enables binding mental commands and real movements. These results are the necessary conditions for the facilitation of rehabilitation process [20, 21].

The second scenario aims at objective tracking of the rehabilitation process in the absence of real movements, in a period when brain training based on brain plasticity has already started but the ability to control the user's own movements has not yet appeared. In this case, the path of the robot and the required time might be used as objective markers to demonstrate the day-to-day dynamics.

# 8  Conclusion

Direct and supervisory control of cyber-physical systems can be performed using BCIs with EEG pattern classifiers. For example, such classifiers can be effectively used for remote control of mobile robots at performing service tasks for immobilized people. For this case, the remote control is proposed to implement by imitation of the imagery movement of fingers pushing control switches (4 classes of motor commands) or compressing right and left hands (2 classes of motor commands). It is implemented using a convolution neural network of the type Shallow ConvNet for the raw EEG signal classification. This classification system is capable of learning using biological feedback for recognizingthe imaginary motor commands. It is shown that such an EEG pattern classification system can provide acceptable accuracy of imagery motor command classification. This allows it to be used to control various robotic means, such as a robotic cart with a computer-vision system or an anthropomorphic robot.

Investigations of the developed BCI have shown that it meets the requirements of practical application in real time. Thus, the maximum delay period for generation of control signals should not exceed 150 ms, which makes it possible to use biological feedback. The multithreading implemented in the system allows to record and process simultaneously a EEG signal without data loss and to perform a reliable online classification.

The application examples of BCI-based control systems of robotic cart with computer-vision system or an anthropomorphic robot have shown that for the development of effective systems for remote control based on a noninvasive BCI, it is required to solve the problems of increasing the accuracy and speed of classification of imaginary movements, increasing the degree of freedom of a BCI and optimizing the computational resources for implementing the algorithms of the BCI functioning. Solving these problems will create effective BCIs that will provide people with limited motor functions the ability to control robotic devices using brain signals, which can significantly improve the quality of their lives in the world of healthy people.

**Acknowledgements.** The work was financially supported by RFBR grant 16-29-08296.

# References

1. Wolpaw, J.R., Wolpaw, E.W.: Brain-Computer Interfaces: Principles and Practice. Oxford University Press, New York (2012)
2. Daly, I., Billinger, M., Laparra-Hernández, J., Aloise, F., García, M.L., Faller, J., Scherer, R., Müller-Putz, G.: On the control of brain-computer interfaces by users with cerebral palsy. Clin. Neurophysiol. **124**, 1787–1797 (2013)
3. Frolov, A.A., Roshin, V.U.: Brain-computer interface. Reality and perspectives. In: Scientific Conference on Neuroinformatic MIFI 2008. Lections on Neuroinformatic (2008). (in Russian). http://neurolectures.narod.ru/2008/Frolov-2008.pdf
4. Kaplan, A.Ya., Kochetkov, A.G., Shishkin, S.L., et al.: Experimental-theoretic bases and practical realizations of technology "Brain-computer interface". Sibir. Med. Bull. **12**(2), 21–29 (2013). (in Russian)
5. Brunner, C., et al.: BNCI Horizon 2020: towards a roadmap for BCI community / Brain-Computer Interfaces. Taylor&Fransis Group (2015). (https://doi.org/10.1080/232623x.2015.1008956)
6. Frolov, A.A., et al.: Principles of motor recovery in post-stroke patients using hand exoskeleton controlled by the brain-computer interfaces based on motor imagery. CNU FIS (2017). (https://doi.org/10.14311/nnw.2017.27.006)
7. Cherubini, A., et al.: A multimode navigation system for an assistive robotics project. Auton Robot **25**, 384–404 (2008). https://doi.org/10.1007/s10514-008-9102-y
8. Duvinage, M., et al.: Performances of Emotive Epoc headset for P-300 based applications. BioMed. Eng. Online **12**, 56 (2013). (https://www.biomedical-engineering-online.com/content/12/1/56)
9. Muller, P.G., et al.: "Thought" – control of functional electrical stimulation to restore hand grasp in a patient with tetraplegia. Neurosci. Lett. **351**, 33–56 (2003)
10. Guan, R.B., et al.: A brain-controlled wheelchair to navigate in familiar environments. IEEE Trans. Neural Syst. Rehabil. Eng. **18**(6), 590–598 (2010)
11. Nijkolt, A., et al.: Turning shortcomings into challenges DCIs for games. Entertainment Comput. **1**(2), 85–94 (2009)
12. Muhi, C., et al.: Bacteria Hunt. J. Multimodal User. Interfaces **4**, 11–25 (2010)
13. Cincotti, F., Mattia, D., Aloise, F., Bufalari, S., Schalk, G., Oriolo, G.: Non-invasive brain-computer interface system: towards its application as assistive technology. Brain Res. Bull. **75**, 796–803 (2008). https://doi.org/10.1016/j.brainresbull.2008.01.007
14. Conradi, J., Blankertz, B., Tangermann, M., Kunzmann, V., Curio, G.: Brain-computer interfacing in tetraplegic patients with high spinal cord injury. Int. J. Bioelectromagn. **11**, 65–68 (2009)
15. Leeb, R., Tonin, L., Rohm, M., Carlson, T., Millan, J.D.R.: Towards independence: a BCI telepresence robot for people with severe motor disabilities. Proc. IEEE **103**, 969–982 (2015). https://doi.org/10.1109/JPROC.2015.2419736
16. Sonkin, K.M., Stankevich, L.A., Khomenko, Ju.G., Nagornova, Zh.V., Shemyakina, N.V.: Development of electroencephalographic pattern classifiers for real and imaginary thumb and index finger movements of one hand. Art. Intell. Med. **63**(2), 107–115 (2015)
17. Stankevich, L.A., Sonkin, K.M., Shemyakina, N.V., Nagornova, Zh.V., Khomenko, Ju.G., Perts, D.S., Koval, A.V.: Pattern decoding of rhythmic individual finger imaginary movements of one Hand. Hum. Physiol. **42**(1), 32–42 (2016)

18. Ball, T., et al.: Deep Learning with convolution neural networks for brain mapping and decoding of movement-related information from the human EEG. Preprint arXiv:1703.05051v5 [cs.LG] 8 June 2018
19. Stankevich, L.A., Sonkin, K.M., Shemyakina, N.V., Nagornova, Zh.V., Khomenko, Ju.G., Perts, D.S., Koval, A.V.: Development of the real-time brain-computer interface based on the neurological committee of EEG signal classifiers. In: Proceedings of the XVIII International Conference "Neuroinformatics-2016", Moscow, 25–29th April 2016, vol. 3, pp. 172–183 (2016). (in Rus)
20. Sonkin, K.M., Stankevich, L.A., Khomenko, Ju.G., Nagornova, Zh.V., Shemyakina, N.V., Koval, A.V., Perets, D.S.: Neurological classifier committee based on artificial neural networks and support vector machine for single-trial eeg signal decoding. In: Advances in Neural Networks, pp. 100–107. ISNN-2016 (2016)
21. Stankevich, L., Sonkin, K.: Human-robot interaction using brain-computer interface based on EEG signal decoding. In: First International Conference, ICR-2016. Budapest, Hungary, 24–26 April, 2016. Lecture Notes on AI, vol. 9812, pp. 99–106. Springer (2016)

# Modern Approaches to the Language Data Analysis. Using Language Analysis Methods for Management and Planning Tasks

Andrei N. Vinogradov[1]([✉]), Natalia Vlasova[2], Evgeny P. Kurshev[2], and Alexey Podobryaev[2]

[1] Peoples' Friendship University of Russia (RUDN University),
Moscow, Russia
vinogradov-an@rudn.ru
[2] Ailamazyan Program Systems Institute of RAS (PSI RAS),
Pereslavl-Zalessky, Yaroslavl Region, Russia
vlassova@gmail.com, epk@epk.botik.ru,
alex@alex.botik.ru

**Abstract.** The article discusses promising directions for the use of modern automatic methods for analyzing natural language data for solving a wide range of practical problems. The technology of creating electronic corpora (collections) of texts is considered as a tool for the transition from model linguistics to tagged data linguistics. The principles of the creation of marked corpora of texts, the possibilities and limitations of their use are considered. Creation of a marked corpus of texts in which language data that is downloaded from the Internet is processed sequentially before issuing the results to users is described. The conveyor consists of the following steps: uploading data from the Internet; definition of the language in which the text is written; unloading metadata; splitting the texts into paragraphs and sentences; deduplication; tokenization; automatic language markup; uploading cleared and marked data to the network. The prospects for the development of language data analysis systems are presented. Requirements for the creation of corpora for solving problems of public administration and strategic planning are developed. Properties that should have such bodies are considered. Those include: corpus format, corpus volume, the degree of the linguistic analysis depth, corpus-manager structure. A description of the marked corpora of texts developed at the Artificial Intelligence Research Center (AIReC) of Ailamazyan Program Systems Institute of the Russian Academy of Sciences, with a reference to the tasks of extracting information about persons, events and situations from the texts of news reports is presented. A retrospective review of the development of systems for automatic processing of natural language texts in the areas of machine translation and human-machine interaction is given.

**Keywords:** Artificial intelligence · Machine translation · Natural language processing · Strategic management · Digital economy

© Springer Nature Switzerland AG 2020
D. G. Arseniev et al. (Eds.): CPS&C 2019, LNNS 95, pp. 470–481, 2020.
https://doi.org/10.1007/978-3-030-34983-7_46

# 1 Introduction

In the modern world, automated control systems (ACS) are used to solve increasingly sophisticated problems, which, in their turn, require the use of new effective approaches for solving them. Previously, ACS, as a rule, were not very complex in the database structure; the management task in those was to build mathematical models based on the available numerical input data, control the logical consistency of the database, and perform a simple set of control actions. Most of the input data in such systems were strictly structured and fairly easy to process. At present, the tasks assigned to information management systems are becoming more and more complex, the subject areas covered by them are becoming more far-reaching. The input data streams requiring processing, analysis and making adequate management decisions based on them, have ever-increasing volumes. Along with the increase in the volume of information processed, the following problem arises: a vast majority of information is contained in the raw, unstructured form [1]. For systems, the level of operational monitoring of the situation is, as a rule, video and audio information, as well as data streams from sensors, but in the case of strategic management systems, most often, the input information is presented as arrays of unstructured natural language texts. The problem of extracting suitable for further analysis data, facts and knowledge involved in computational linguistics is discussed.

# 2 Natural Language Processing Tools Review

A very important role in the development of computational linguistics played the creation of machine translation systems (MTs), which started almost at the same time with the birth of the first computers (the first ENIAC computer appeared in 1945, and in 1947, the first attempt to create a similar program was made). Generally speaking, MTs were the first attempt in using computers for solving non-computational problems. Held in January 1954, The Georgetown experiment (a joint project of Georgetown University and IBM) set as its main goal to publicly demonstrate the feasibility of automatic translation of the text. The translation of 49 carefully selected Russian sentences into English using a dictionary of 250 words and only 6 rules using the IBM 701 computer, although had little scientific significance, nevertheless caused the rapid development of working on MTs around the world. An interesting fact is that one of the world's leading IT specialists, Charles Anthony Richard Hoar (known as the developer of the quick sorting algorithm and Hoar's logic), at the beginning of his scientific career, then a graduate student at Moscow State University, was engaged in the research on machine translation tasks. One of the first commercial MT systems is the SYSTRAN system [2] was developed in 1968 by Latsec under direction of Peter Toma and ordered by the United States Air Force to translate documents from Russian into English. It was first installed in 1970 at the Department of Foreign Technology of the U.S. Air Force, replacing the IBM Mark-II system. In 1974, the SYSTRAN system was selected by NASA to translate the documentation for the Apollo-Soyuz project. In 1976, General Motors of Canada purchased the SYSTRAN system for the English-French translation of documentation (and later, in 1982, and the English-Spanish translation). In addition to the active use of SYSTRAN for translation of its own documentation, Xerox has developed a special subset of the English language, the Multinational Customized English, which has been successfully used to create and translate multilingual technical documentation. Also, in 1976, the European Commission purchased the English-to-French version of SYSTRAN for its evaluation and potential use for translating community documentation. Up to 1978, various versions of SYSTRAN were investigated for this purpose, but in 1978 the European Economic Community announced the creation of a European EUROTRA program (1978–1992) as part of the EURONET-DIANE (Direct Information Access Network for Europe) project, which aimed to develop a computer translation system for all European languages. And although in the course of work on EUROTRA, a working MT system has never been created, the approaches and technologies developed during the project were later on used in a lot of commercial MT systems.

Most of the machine translation systems at that time were developed to translate technical documentation; similar work was carried out in the USSR: for example, as a result of the developments in the field of MT started in the 1970s at VINITI by the group of Professor Belonogov, the first Russian industrial version, the RETRANS system [3] was created here in 1993.

Also, an important contribution to the development of computational linguistics and machine translation was made by LUNAR and LIFER/LADDER systems (Language Interface Facility which Ellipsis and Recursion), created in the 1970s as natural-language interfaces to databases (samples of lunar minerals and U.S. Navy ships database), which used semantic grammar [4]. This area was developed in the 1980s in the United States, where a number of systems with natural language interfaces were developed: Q&A (Symantec), Language Craft (Carnegie group). Further on, with the appearance and broad use of personal computers, the number of commercial dictionaries and machine translation systems began to grow. Translation memory technology [5] has become another technology from the 1970s and 1980s, based on the accumulation of ready-made translations of text fragments in the form of a linguistic database, examples of which are actively used in the translation of new texts. This technology is now actively used by both machine translation systems and human translators.

Currently, there are two main directions of the MT. The first one is Statistical Machine Translation (SMT) based on the comparison of a huge number of already existing "correct" translations. Obviously, for such an approach to work effectively, you need large corpora of multilingual texts, so the main companies developing this area are Internet search engines with products such as Google Translate, YandexTranslator and Bing Translator from Microsoft. Within this direction, artificial neural network technologies and word2vec (word to vector) and sec2sec (sequence to sequence) approaches are actively used [6]. The second direction is rule-based Machine Translation (RBMT) [7] using the linguistic model bases created by linguistic specialists. The working quality of this approach depends entirely on how fully linguists can describe the constructions of a natural language. The advantage of this direction is the relatively low computing power required for MT systems. Examples of such systems are Multillect, Linguatec and PROMT. There is also a Hybrid Machine

Translation (HMT) [8], combining the advantages of each of these two approaches (used in the latest versions of the SYSTRAN software system).

Nevertheless, each of these approaches requires effective and high-quality work and the availability of a large number of pre-marked multilingual texts in natural languages.

## 3   Current State and Prospects of Natural Language Processing Systems Development

The development of information technologies, such as neural networks and BigData, allows us to solve new problems in the field of word processing. Big Data is such a large and complex data collection that it becomes difficult to process it using standard database management tools or traditional data processing applications.

On the one hand, for the high-quality machine learning, labeled corpora of texts are necessary; on the other hand, processing with the help of BigData technologies of unmarked corpora allows to reveal some regularities for different classes of texts, although errors are quite possible. Combining machine learning results on large corpora of unmarked texts and on smaller but labeled cases allows us to more accurately solve certain classes of word processing tasks that were previously difficult to solve.

Such tasks, in particular, include a whole class of tasks related to the technology of creating bots and using them.

The term bot is an abridged form of the word "robot." A bot is a program that operates autonomously on the Internet.

Bots are used in a broad variety of fields, such as customer service, virtual help, answers to frequently asked questions, etc.

As it often happens, advances in scientific and technical areas can be used both for the benefit of man and for harm.

In this case, the activity of the so-called social bots (special programs created to simulate people's behavior in social networks) is mainly destructive.

The main areas of application of social bots are distribution of fake news, attempts to discredit politicians and companies, blocking websites by massive attacks, the spread of propaganda including in the interests of terrorist organizations, etc. [9, 10].

Since such bots are becoming more common, the challenge is to identify them.

Various ways to solve this problem have been described. They use a variety of features, such as activity time, sequence of actions, data on the account of a potential bot, etc. [11, 12].

At the same time, one of the most important features is the text of the bot message: its character, style, and the dictionary.

Virtually all bots use natural language processing methods; otherwise the dialogue would look unnatural and, besides, it would be difficult for bots to interpret the variations in the expression of particular statements. Nevertheless, bot text messages have their own characteristics by which you can determine whether the text belongs to a bot.

To identify such features through machine learning, you must have a marked corpus of texts generated by bots.

As another example of the use of combined technologies, one can cite the task of identifying social tensions in regions based on the analysis of messages in social networks.

One of the signs of social tension is verbal aggression in the texts of social networks [13]. Detection of texts with speech aggression can be done using BigData processing methods. But, by no means, all texts with speech aggression characterize the growth of social tension. And to reduce errors in the determination of social tension, one should use labeled corpora of texts. In this case, a more precise definition of the object or situation leading to tension is necessary. On the other hand, the use of machine learning methods for identifying social tensions based on only marked corpora can lead to omissions, inability to reveal the situation due to the relatively small size of the text bodies. And only a combination of the two approaches (using BigData and labeled corpora of texts) would provide for solving the problem with sufficient accuracy and completeness.

## 4  Language Corpora as an Information Tool

With the development of computer-aided analysis of natural languages, researchers faced the problem of presenting language data in electronic form. A need to create corpora (collections) of texts for solving problems that could be formulated in connection with the emergence of new methods and approaches to language analysis appeared. A lot of data for processing was needed; those should be representative, with the format and specificity corresponding to the goals and objectives of the planned research. Thus, in the 1960s, a separate direction in linguistics appeared, i.e., corpus linguistics, which actively developed over the past decades, in parallel with technical progress; as a result of which more and more powerful tools of automatic language analysis appeared. In addition, the development of corpus linguistics is also related to the fact that more and more tasks appear at the intersection of disciplines, as well as applied tasks not related to theoretical and applied linguistics, which require data analysis of the linguistic material. The latter includes management and strategic planning tasks.

The first language corpora were collections of texts, hand-picked and processed. One of the first cases was the corpus of texts in English created in 1963 (Brown University, USA, by W. Francis and G. Kucera). In the future, many corpora were created for different languages. The basic principles of the compilation of such corpora are:

- accessibility in electronic form (are open to all, available by subscription, limited available and commercial);
- representativeness (the texts that make the corpus should represent various linguistic phenomena required for research);
- balance (there should be no bias towards any given phenomena, topics, or features of the language);
- easiness of use;
- possibility of a smart search (appropriate tools must be provided inside the case).

For the Russian language, one of the first, still developing and not losing relevance for corpora researchers, is the National Corpus of the Russian language (NCRF) [14]. The work on its creation began in the early 2000s; linguists from various organizations took part in those. Currently, the volume of the main body of the NCRF exceeds 280 million word usage. Corpus word forms are equipped with automatic morphological and semantic markup. Approximately one and a half percent of texts are checked manually, the morphological and semantic homonymy is removed in them. Of the minuses of working with the NCRC, it is to be noted that it is closed (you can freely use only the search and a small fragment for research purposes); the corpus manager configured for a certain range of tasks; limited set of genres of the submitted texts.

For researches, both theoretical and applied, the language corpus is a language model, because at present, the work of automatic language analysis tools (algorithms based on machine learning, neural networks) is carried out only on the material of the corpora, thus only that would fall into the case. As a result, the principles of creating cases are very important, as well as the information that is supplied (automatically or manually) with the language units of the body - the so-called markup and meta-information. Experts note that the vector of development of theoretical linguistics is shifting away from corpus linguistic models to deep development of the principles of data markup (in its turn, markup is used primarily in the creation of language corpora). There is a transition from linguistic models to linguistic labeled data [15]. Therefore, in recent years, scientists, as well as representatives of commercial companies engaged in the analysis of language for the production of the product to sell, are moving to the creation of language Internet corpora. The peculiarity of the Internet corpora in comparison with the conventional, traditional corpora is a much larger volume (hundreds of millions of texts, billions of tokens - word forms), which makes the corpora much more representative and better reflecting the natural language. The source of texts and other language units of such corpora is the Internet information and telecommunication network. Of course, a body of this size can only be created and marked automatically. Accordingly, the technologies for creating Internet cases have been developed and are constantly being improved. These technologies are universal and do not depend on a specific language. Often, entire bunches of corpora of different languages are created in the same way. Corpus linguistics has long gone beyond the limits of the space of one language, although corpora created for the needs of researchers in a single specific language are also not uncommon. In such a corpus, as we will see below in the review, can be different tasks.

Consider the usual sequence of creating an Internet corpus:

- Uploading data from the Internet, extracting text information, correcting encoding errors. There are two possible ways to bypass the Internet to obtain the required data. One way is to set words from a predetermined set of words of high frequency in the search string. The second one, the so-called crawling, is based on the use of a special program (crawler), which bypasses the sites specified during the initial search, and then follows the links found on the previous sites. The crawler version is more often used to create truly voluminous language corpora, while the first method is convenient for creating not very large but specific collections of texts or sentences.

- Determining the language in which the text is written, deleting the text in case the language is "wrong".
- Uploading metadata (for example, the date and time of creating the text, the author of the text, the gender of the author of the text, the place of creation of the text, the genre (if known)).
- Breaking up the received texts into paragraphs and sentences.
- Deduplication - removing duplicate texts and text fragments.
- Tokenization - defining the boundaries of word forms, i.e., the basic language units.
- Automatic language markup (traditionally includes morphological markup, sometimes syntactic and semantic).
- Uploading cleared and tagged data to the network. The data becomes available through the case manager, allowing users to work with the case.

The described technology is a pipeline (English pipeline), on which language data downloaded from the Internet, are consistently processed before issuing the result to users. Accordingly, this hull technology allows for a flexible configuration of the hull for a specific task. Recent trends in the construction of language Internet corpora are such that researchers are striving to make their corpora accessible to all users.

Currently, several Internet corpora have been created or are in the process of creating in the Russian language [16]. Let's consider them.

- Project Open Corpora [17]. Start of creation: 2010–2011. The corpus emerged as an alternative to the National Corpus of the Russian language, so that it could be freely used without restrictions related to copyright. It undertook only such texts that were distributed under the CC-BY-SA license. This corpus is not a fully internet corpus due to its small size, as well as manual marking and manual text selection at the initial stage of creation. The case has not been updated since 2015.
- The General Corpus of the Russian Language (GKRYa) [18]. This corpus is created exclusively for the Russian language for solving linguistic and, in particular, sociolinguistic, lexicographical and semantic tasks. The project involved the ABBYY, RSUH, MIPT, MSU companies. The corpus is created completely automatically; currently it mostly contains texts from social networks and the blogosphere. Due to the specifics of the tasks for which the GKRYa is created, each text is supplied with detailed meta-information and automatic morphological marking.
- The Aranea family of multilingual enclosures [19, 20]. Within the framework of a large multilingual project for the creation of representative Internet corpora, the Russian corpus Aranea Russicum is created. The goal of the project is to prepare large and extra-large corpora for different languages (currently, there are about 20 corpora in the Aranea family). Shells are created according to the same principles, with the possibility of comparing and applying to them common search and analysis algorithms.
- The TenTen multilingual enclosures family with the Russian enclosure RuTenTen [21]. There is also a multilingual project, in which the Russian language corpus is being created and developed. One of the goals of the project is the possibility of compiling lists of frequency phrases distributed by syntactic functions (word sketches).

- Internet corpus with TAIGA syntax markup [22]. A completely new Internet corpus, which is different from the corpora listed above due to the presence of morphological and syntactic markup in the UD format (universal dependencies) [15, 23]. The enclosure will be open for use. At the moment, filling the body continues.

As you can see, language Internet corpora are created to solve various problems. Fully automatic conveyor technology is universal for any language and allows to vary the design parameters of the body. This allows to quickly create a new or adapt an existing case for new tasks.

## 5  Language Corpus for Solving Problems of Strategic Planning. Perspectives of Creation

The language Internet corpus, originally created exclusively for the needs of linguistics, is currently used to solve not only theoretical and applied problems of linguistics. Now any professionally built language corpus is a powerful information tool, indispensable in those areas where expert analysis of natural languages is required. The scope of application of language corpora is constantly expanding. So, for example, they are widely used in teaching translators: these cases allow students to see the full use of one or another word or phrase, to make the right choice when translating. A language corpus with a properly organized system of searching and issuing the required information opens up completely unlimited possibilities for the translator to analyze words and expressions of the language from or into which the translation is being made [24]. The article by Matveychuk [25] sets the task of creating a language corpus for the needs of game management - in particular, the corpus would allow to solve the practical problem of the legal and technical registration of state expertise acts. As the author notes in her article, the existing language classes of the Russian language (listed in the previous section) are not suitable for solving strategic problems in hunting management. Another example of the potential use of language corpora is the study of so-called conflict texts. Conflict texts, of which court documents are a prime example, represent a certain text invariant that objectively allows for a number of interpretative options. A language corpus containing conflicting texts would be very relevant in the field of resolving conflicts of different types of text [26].

Accordingly, many experts in areas not directly related to linguistics think about using such powerful tools as language corpora for obtaining information for their fields of research. It is safe to believe that a specially constructed and streamlined language corpus can serve as a basis for solving many problems of public administration and strategic planning, since the language data are most often used in the analysis and forecasting.

What properties should such a housing have and whether it is possible to adapt existing corpora for tasks of this kind? First of all, it is necessary to understand what data for state administration and strategic planning of work of organizations and enterprises can be obtained using the language corpus. Based on this, we can suggest:

1. Corpus format: which language units will be included in it. If these are texts, from which sources they will be taken, what meta information should be stored in the corpus.
2. The volume of the case. It is important to decide on the basis of the task, what amount of text or sentences will be sufficient for effective work with the corpus. For different tasks, it is possible to create subcorpora. For example, for the analysis of opinions, it is important to collect as large a sample of subject texts as possible. If the task is to analyze the state and municipal documents in order to build development strategies, find inconsistencies and irregularities, the set and number of texts, of course, will be completely different.
3. The degree of the depth of linguistic analysis. Since the tasks of strategic planning and public administration are high-level tasks that require in-depth data analysis, it can be assumed that the prospective corpus will need to provide for the highest possible level of linguistic study.
4. The structure of the corpus manager, which would allow the expert to obtain the required information. The corpus should include tools that allow you to formulate a request for language data and get an answer in a convenient for further processing form.

Thus, we see that to create a promising language corpus for the tasks of strategic planning and public administration, it is necessary to combine the efforts of specialists from different areas. The automatic case-building technologies developed by corpus linguistics make it possible to create large-scale collections of texts in a short time; however, fine tuning of these technologies is necessary. Such adjustment is a task for experts in linguistics, as well as in the field of strategic management and state planning.

In the research center of artificial intelligence of the A.K. Aylamazyan Institute of Software Systems of RAS, several marked text corpora designed to solve applied problems associated with the automatic extraction of information from news texts (Situations-1000, Relations-1000, Persons-1000, Persons-1111-F) were developed; in particular, they used to assess the accuracy and completeness of the developed algorithms for extracting information from texts.

The "Situations-1000" corpus contains markup of appointment and resignation events and is used in tasks that require determining the type of an event, the place where the event is mentioned in the text, as well as identifying participants of the event in the form of a text fragment (string). The "Situations-1000" case is close to the ACE in terms of marking principles but has some differences (the search area for information about the participants is a clause containing a marker structure). Annotation type enclosures (with reference to markup to text), in contrast to the markup type MUC, despite the complexity of their creation, have some advantages. In particular, such markup allows to reliably assess the quality of the extraction algorithms, referring to the text fragment containing the selected information.

The "Relations-1000" corpus contains the markup of role-to-face and role-by-relation relations. Relations of the first type express the connection of a person with his role or aspect. For example, director Ivan Smirnov (role), positivist Fedor Ivanov (aspect). The second type of relationship indicates what the person plays this role relative to. For example, the shareholder of the company, head of the department, Boris's nephew.

The "Persons-1000" and "Persons-1111-F" cases are marked for the task of extracting personal names from news texts. The task of extracting involves determining the place of mentioning persons in a text, as well as bringing this mention to a given canonical form. Typing in "Persons-1111-F" is different from "Persons-1000" and covers news from Southeast Asia, the Central Asian region and the Arab world.

In all marked packages, the markup is stored in .xml files. The root tag is called markup. It can contain an arbitrary number of elements with an entry tag, each corresponding to a single reference to the marked object in the text.

An example of an entry element for a "Situations-1000" case is shown below.

```xml
<entry>
<id>1</id>
<descr> assigned </descr>
<class> appointment/entry into office </class>
<ancor>
<offset>1266</offset>
<length>8</length>
</ancor>
<slot>
<name> occupying </name>
<position>
<offset>1231</offset>
<length>30</length>
</position>
</slot>
<slot>
<name> position </name>
<position>
<offset>1275</offset>
<length>20</length>
</position>
</slot>
</entry>
```

# 6    Conclusion

Modern automatic methods for the analysis of natural language data provide ample opportunities for use in other areas other than linguistics: not only the results of the work of language modules, but also for the independent formulation of problems and the selection of technologies for analyzing language data for each specific task. Of particular relevance in recent years are large language enclosures, which are created as a result of the consistent work of automatic tools for searching, cleaning and marking text data. Unlike traditional hand-made cases, automatic case filling technologies make it possible to quickly find the texts needed for solving a specific task and provide them with the necessary markings. The scope of application of large language corpora is constantly expanding. The creation of a corpus of the Russian language for solving problems of state administration and strategic planning seems to be very promising. Language data contained in a huge number of digitized documents and texts are one of the most important sources of information for solving such problems. Units in a promising language corpus should be whole texts corresponding to a specific subject matter, provided with detailed meta-information and time and place of creation, authorship and links to other documents. In addition, for a full analysis, the texts in the proposed corpus should be equipped with modern reliable linguistic markup - this will allow you to use the achievements of linguistics to search and analyze relevant information.

**Acknowledgments.** The publication was prepared with the support of the state program AAAA-A19-119020690042-2 «Research and development of data mining methods».

# References

1. Isaksson, A.J., Harjunkoski, I., Sand, G.: The impact of digitalization on the future of control and operations. Comput. Chem. Eng. **114**, 122–129 (2018). https://doi.org/10.1016/j.compchemeng.2017.10.037
2. Comparin, L.: Quality in machine translation and human post-editing: error annotation and specifications. Diss. (2017)
3. Belonogov, G.G.: Systems of phraseological machine translation of polythematic texts from Russian into English and from English into Russian (RETRANS and ERTRANS Systems). Int. Forum Inf. Documentation **20**(2), 29–35 (1995)
4. Sowah, E.: Natural language processing in cooperative query answering databases (NLPICQA) (2018)
5. Schneider, D., Zampieri, M., van Genabith, J.: Translation memories and the translator: a report on a user survey. Babel. **64**(5–6), 734–762 (2018)
6. Johnson, M., Schuster, M., Le, Q.V., Krikun, M., Yonghui, W., Chen, Z., Thorat, N., Viegas, F., Wattenberg, M., Corrado, G., Hughes, M., Dean, J.: Google's multilingual neural machine translation system: enabling zero-shot translation. Trans. Assoc. Comput. Linguist. **5**, 339–351 (2017)
7. Macketanz, V., Avramidis, E., Burchardt, A., Helc, J., Srivastava, A.: Machine translation: phrase-based, rule-based and neural approaches with linguistic evaluation. Cybern. Inf. Technol. **17**(2), 28–43 (2017)
8. Costa-jussa, M.R., Fonollosa, J.A.R.: Latest trends in hybrid machine translation and its applications. Comput. Speech Lang. **32**(1), 3–10 (2015)
9. Ferrara, E., Varol, O., Davis, C., Menczer, F., Flammini, A.: The rise of social bots. Commun. ACM **59**(7), 96–104 (2016)
10. Oberer, B., Erkollar, A., Stein, A.: Social bots – act like a human. In: Think Like a Bot. Stumpf, M. (eds.) Digitalisierung und Kommunikation. Europäische Kulturen in der Wirtschaftskommunikation, vol. 31, pp. 311–327. Springer VS, Wiesbaden (2019)
11. Shi, P., Zhang, Z., Choo, Raymond, K.K.: Detecting malicious social bots based on clickstream sequences. IEEE Access. **1**, 1 (2019)
12. Davis, C., Varol, O., Ferrara, E., Flammini, A., Menczer, F.: BotOrNot: a system to evaluate social bots. arXiv preprint:1602.00975 (2016)

13. Antropova, V.V.: Speech aggression in the texts of social networks: the communicative aspect. Vestnik VGU. Serija: Filologija. Zhurnalistika. [VSU Herald. Series: Philology. Journalism]. (3), 123–127 (2015). (in Russian)
14. http://www.ruscorpora.ru/
15. Lyashevskaya, O.N., Toldova, S.A.: Modern problems and trends in computational linguistics. Voprosy jazykoznanija [Questions of linguistics] 1, 120–145 (2014). (in Russian)
16. Kozlova, N.V.: Linguistic corpora. Definition of basic concepts and typology. Vestnik NGU Serija: Lingvistika i mezhkul'turnaja kommunikacija [NSU Herald, Series: Linguistics and Intercultural Communication], 11 (1), 79–88 (2013). (in Russian)
17. Granovsky, D.V., Bocharov, V.V., Bichineva, S.V.: Open corpus: principles of work and prospects. Kompjuternaja lingvistika i razvitie semanticheskogo poiska v Internete: Trudy nauchnogo seminara XIII Vserossijskoj ob'edinennoj konferencii "Internet i sovremennoe obshhestvo" [Computational linguistics and the development of semantic search on the Internet: Proceedings of the scientific seminar of the XIII All Russian Joint Conference "The Internet and Modern Society"]. St. Petersburg, October 19–22, 2010, Ed. by V.Sh. Rubashkin, 94 p. (2010). (in Russian)
18. Belikov, V., Kopylov, N., Piperski, A., Selegey, V., Sharoff, S.: Big and diverse is beautiful: a large corpus of Russian to study linguistic variation. In: Proceedings of the 8th Web as Corpus Workshop (WAC-8) @Corpus Linguistics 2013, 24–29 (2013). http://www.webcorpora.ru/wp-content/uploads/2015/10/wac8-proceedings.pdf
19. Benko, V., Zakharov, V.P.: Very large Russian corpora: new opportunities and new challenges. DIALOG-2016 (2016). http://www.dialog-21.ru/media/3383/benkovzakharovvp.pdf
20. Benko, V.: Yet another family of (comparable) Web corpora. In: Conference Text, Speech and Dialogue. 17th International Conference, at Brno, Czech Republic (2014). https://www.researchgate.net/profile/Vladimir_Benko/publication/313904118_Aranea_Yet_Another_Family_of_Comparable_Web_Corpora/links/58c675fdaca272e36dde59c6/Aranea-Yet-Another-Family-of-Comparable-Web-Corpora.pdf
21. Jakubíček, M., Kilgarriff, A., Kovář, V., Rychlý, P., Suchomel, V.: The TenTen corpus family. In: Proceedings of the 7th International Corpus Linguistics Conference. Lancaster, pp. 125–127 (2013)
22. Shavrina, T.O., Shapovalova, O.A.: To the methodology of corpus construction for machine learning: TAIGA syntax tree corpus and parser. Trudy mezhdunarodnoj konferencii "Korpusnaja lingvistika-2017". In: Proceedings of the International Conference "Corpus Linguistics-2017", Saint-Petersburg, Ch.13, pp. 78–84 (2017)
23. Lyashevskaya, O., Droganova, K., Zeman, D., Alexeeva, M., Gavrilova, T., Mustafina, N., Shakurova, E.: Universal Dependencies for Russian: a New Syntactic Dependencies Tagset. Basic reaearch program. Working papers (2016). http://olesar.narod.ru/papers/44LNG2016.pdf
24. Osipova, E.S., Tarnaeva, L.P.: Using of corpus linguistic resources in the preparation of translators in the field of professional communication. Filologicheskie nauki. Voprosy teorii i praktiki [Philology. Theory and practice], 63(9), 205–209 (2015). (in Russian)
25. Matveychuk, S.P.: Prospects for the use of text (linguistic) corpora in hunting research. Gumanitarnye aspekty ohoty i ohotnich'ego hozjajstva, trudy konferencii [Humanitarian aspects of hunting and hunting, conference proceedings], pp. 29–35 (2015). (in Russian)
26. Kovalchuk, A.N.: The relevance of the creation of specialized linguistic corpora for solving practical problems of legal linguistics. Intellektual'nyj potencial XXI veka. Stupeni poznanija [XXI century Intellectual potential. Steps of knowledge] 21, 142–146 (2014). (in Russian)

# Dynamic Container Virtualization as a Method of IoT Infrastructure Security Provision

Andrey Iskhakov[✉], Anastasia Iskhakova,
and Roman Meshcheryakov

V. A. Trapeznikov Institute of Control Sciences of Russian Academy
of Sciences, Moscow, Russia
iskhakovandrey@gmail.com, shumskaya.ao@gmail.com,
mrv@ieee.org

**Abstract.** This article proposes an approach to security provision of one of the key bases of digital transformation – the technology of the Internet of things (IoT). The effective technology of carrying out the information security audit with application of Honeypot systems is the cornerstone of the article and the offered method. The main advantages of the use of container virtualization, unlike application of traps on the basis of virtual machines, are formulated by the authors. The method of protection of a similar infrastructure by means of integration of dynamic container virtualization of network traps is considered. The article contains information on implementation of the offered method, comparison of results with existing solutions, and a summary table with the actual results of an experiment. A detailed flowchart of functioning of the offered method is also provided in the work. The proposed solutions allow to increase efficiency of the malefactor's actions analysis. The administrator of IoT devices network can obtain information about the priority purposes, the used by malefactor means, and vulnerabilities of various elements of network. These circumstances give an opportunity to quickly take measures for increase in security of network and to avoid its compromise.

**Keywords:** Internet of Things · Container virtualization · HoneyPot · Information security · Audit of network infrastructure

## 1 Introduction

Today, there is a conclusive fact that one of the bases for new digital economy is the universal development of the Internet of Things. Active development of the given concept supported by hi-tech manufacture on a global scale causes the growth of the quantity of final "smart" devices and technologies to connect them. All of this forms the basis for positive induction in the sphere of new cyber threats and vulnerabilities. In this connection, according to the research of the Symatec Company by Cleary et al. [4], in 2018, the number of attacks on IoT-devices increased approximately sixfold. The statistics published by Kuskov et al. [9], employees of Russian manufacturers of anti-virus protection products, confirms this tendency.

© Springer Nature Switzerland AG 2020
D. G. Arseniev et al. (Eds.): CPS&C 2019, LNNS 95, pp. 482–490, 2020.
https://doi.org/10.1007/978-3-030-34983-7_47

For example, in May 2017, the Kaspersky Lab database of contained more than 7000 various samples of malicious software for the IoT infrastructure and about a half of that was registered in 2017. It is important to note that the mentioned smart devices are frequently applied in crucial systems. For example, elements can be connected in an interconnected network of monitoring and management of the city firefighting system. Public services quite often consolidate the data collected from various IoTsensors and provide citizens with public access to it including the data on the level of environmental pollution or fire security.

Thus, the development and adaptation of methods and products of information security provision for their application in the security systems of IoT devices is of interest [7]. One of the methods showing high efficiency in the problem of construction of the model of a malefactor and allowing one to correct the parameters of security functions is a constant audit of an IoT infrastructure by means of applying Honeypot traps considered in [13].

## 2    Trap Devices and Container Virtualization

In general, Honeypots are widely used for studying adversary behaviour. More information about this can be found in articles [1, 11, 12, 14, 19]. For example, the authors of [2] proposed two solutions for monitoring and analytics implemented on different honeypot systems. The first solution applies data mining techniques on gathered from a honeypot data for the automatic discovery of patterns. It does not consider monitoring honeypots and is focused on analysis. It only handles connections that reached a honeypot; they consist of such items as source and destination IP address and port and the used protocol. The analysis uses two types of patterns: frequent sets and jumping emerging patterns; their application allows reduction of the amount of data that should be revised by an analyst. There are well-known works [3, 6, 15, 16] in the field of adaptive HoneyPot systems but they focus on reviewing end-agent reconfiguration issues.

The results of the practical experiment referred to in article [8] show the efficiency of the Honeypot technology application for an IoT infrastructure as a tool of the analysis of malefactors' attack vectors. In the discussed case, the used stand represented a highly interactive Honeypot system with an application of hardware agents – separate hosts located in a corporate network. However, not every infrastructure can have an opportunity to select a separate segment of corresponding devices for the realization of similar traps. Considering the prominent features of the investigated infrastructure and the requirement to minimize the load on the system agents, the problem of development of a trap organization method that does not require additional hardware is of interest. After the analysis of possible technologies, the method was based on the principle of a container virtualization dynamic system.

When comparing container virtualization with the use of full-fledged virtual machines, one of the main differences (taking into account the absence of differences in functionality) is its light weight and high speed of creation and launching of containers. A similar idea has already found its reflexion in papers [5, 6, 10], where separate class of Honeypot of systems – dynamic systems was selected. Such systems change their status in the course of interaction with users (including malefactors) who have access to the system. The technologies of container virtualization allow us to operatively develop a multi-agent Honeypot system which a malefactor sees as a fully-fledged corporate network. Besides, the efficiency of such solution regarding resource-intensiveness is ensured at the expense of the lack of necessity for constant support of activity of all elements. In the conditions of the IoT infrastructure segment, the implementation of such approach has a number of advantages.

- Transparency: in the condition of full virtualization, each guest system has an isolated stack of equipment (albeit virtual), whereas all containers work directly using the equipment of the host system.
- Productivity: the productivity of guest systems constructed with the use of container virtualization technology essentially increases. This is directly connected with the fact that when using the containers there is no necessity to emulate the operation of the additional (not used in the given problem) equipment and many other resource-intensive operations of the transmission of system calls of guest systems.
- Compactness: container operation does not require a fully-fledged copy of operational system; and hence, the container occupies much less of the disk space in comparison with an image of a virtual machine.
- Flexibility of control: to provide the interaction between containers and control their operation, open interfaces are provided and there is no need to use various third-party mechanisms.

## 3   Description of the Method

The offered method represents a set of instructions for the dynamic formation of HoneyNet (a set of Honeypot agents connected in a uniform network) for an IoT infrastructure. Figure 1 shows the structure of elements of a necessary infrastructure; the dotted line represents information streams.

Elements $[U1...Un]$ are the users of the IoT infrastructure consisting of devices $[D1...Dn]$. Let us designate a certain user $[U_H]$ as a malefactor using a certain toolkit of automated scanning. The Manager of Connections $[Mcon]$ element can act as a perimeter gateway screen or an intelligent analyzer of network packages. As soon as $[Mcon]$ registers the signature of a potentially dangerous inquiry or an attempt of automated scanning of the network infrastructure, the inquiry is sent to the device $[D1]$ representing a kernel – HoneyNet server. The analytical block of this server – the Manager of Containers $[Mc]$ – either generates a new knot of the HoneyNet network or chooses a necessary element from the previously generated network-trap and returns the necessary address in $[Mcon]$ on the basis of the data on the available infrastructure. Further, $[Mc]$ will transmit all malefactor's packages assigned for one or another HoneyPot to the chosen container.

The expanded methodological description of an organization similar to HoneyNet in an IoT infrastructure is presented below.

Input: the configuration of an IoT infrastructure.

Output: dynamic HoneyNet.

- Step 1: carry out the primary initialization of the medium.
- Step 1.1: adjust the Manager of Connections in such a manner that a certain host with necessary open ports was accessible to a malefactor.
- Step 1.2: prepare the containers containing chosen operational systems and sets of applications. Note: it is recommended to use obviously vulnerable versions of applications and OS with an objective that it was easier for the malefactor to carry out an attack on the infrastructure.
- Step 1.3: configure a necessary HoneyNet structure in the manager of containers: define all hosts and their parameters.
- Step 2: expect an attempt of connection by the malefactor (an inquiry for the access to an element of the IoT infrastructure).
- Step 3: receive the container address.
- Step 4: the Manager of Connections needs to transmit the command to form a new copy of the container to the manager of containers.
- Step 5: the Manager of Containers needs to check the inquiry signatures in the database of the preliminarily prepared containers with Honeypot.
- Step 6: create a new container, if the container does not yet exist in the infrastructure; if it does, then pass to step 7.

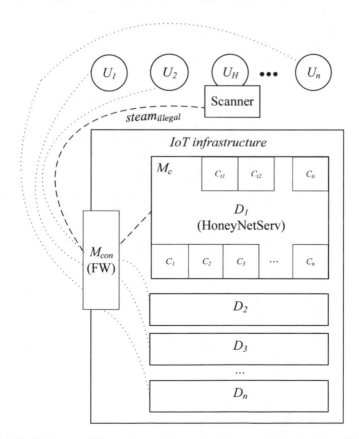

**Fig. 1.** Principal diagram of interaction of elements when organizing the dynamic HoneyNet

- Step 7: start the container, if in the list of active containers there is no necessary element; if the container is started, then pass to step 8.
- Step 8: the Manager of Containers needs to forward all packages of the malefactor assigned to Honeypot in the chosen container.
- Step 9: carry out connection dump if packages stopped coming within the determined limit of time.

In Fig. 2 the flowchart showing the maintenance of the presented method is submitted.

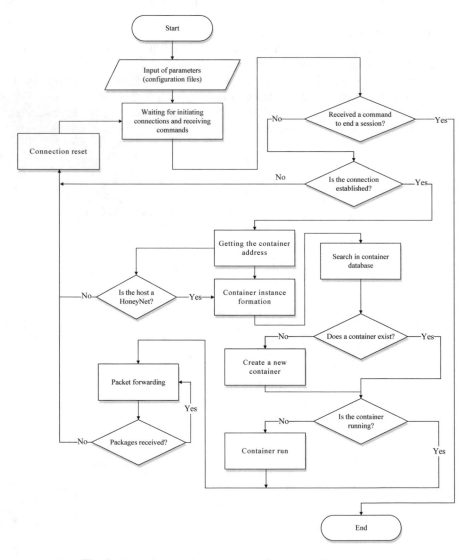

**Fig. 2.** Flowchart of dynamic generation of the Honeypot traps

Thus, at step-by-step fulfilment of the scenario realized above, a dynamic HoneyNet will be generated, allowing one to give the malefactor an illusion of operation with a corporate network of any size and complexity, using a reasonable amount of computing and memory resources.

## 4 Implementation

The implementation of the presented above method does not require a high-efficiency infrastructure at the site of implementation, as the deployment of a separate environment of virtualization is not required. This fact characterizes its high degree of adaptation for the IoT segment. A complex of programs for Linux OS family, implementing the given algorithm, was developed. The manager of containers and the manager of sessions were developed with the C++ and Python programming languages. The network interaction in the program was made with sockets. Additional functionality (subsystem of clearing, gateway screen) were developed by means of an operational system with the application of sh-scripts.

The results presented in Table 1 show the main ports used by the malefactor when checking network points on existence of vulnerabilities. Figure 3 shows the distribution of attack categories by experiment day.

**Table 1.** The most attackable ports

| Service | Port | Number of the attacks | Percentage of total attacks, % |
|---|---|---|---|
| Secure Shell (SSH) | 22 | 3459 | 45.59 |
| Web services (HTTP) | 80 | 1791 | 23.60 |
| Telnet | 23 | 599 | 7.90 |
| Real Time Streaming Protocol | 554 | 572 | 7.54 |
| MySQL DB System | 3306 | 302 | 3.98 |
| Remote Desktop Protocol | 3389 | 299 | 3.94 |
| Secure web services (HTTPS) | 443 | 235 | 3.10 |
| File Transfer Protocol | 20 | 189 | 2.49 |
| VNC Remote Desktop Protocol | 5900 | 79 | 1.04 |
| Radmin Remote Admin | 4899 | 32 | 0.42 |

The approbation of the program solution based on the presented algorithm with 10 Honeypot knots in the IoT infrastructure was carried out. The comparison of the results with the existing solutions (Dockerpot, HoneyD, Dockerpot, T-Pot) showed the following advantage: the level of load of the system resources decreased (from 7% to 19%) and good indicators of compatibility of LXC technology with such devices as Raspberry PI were registered. When comparing, the decisions presented in the works [10, 17, 18] were also taken into account.

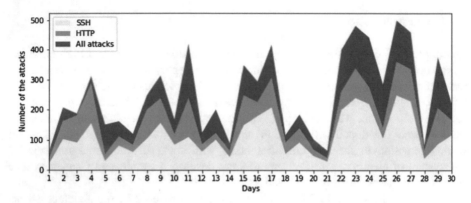

**Fig. 3.** Attacks registered using the dynamic HoneyPot traps

## 5  Conclusion

The presented method of the dynamic container trap virtualization for an IoT infrastructure successfully encompasses the advantages of the container virtualization and the capabilities of the highly interactive Honeypot systems. The basic advantage that creates the conditions for its application in IoT systems is the principle of dynamic generation of containers: the implementing system is capable to change its status depending on the actions of a malefactor, which ensures the increase of efficiency of the use of host system resources. Using the proposed solutions, the administrator of the network obtains information about the priority purposes of the malefactor, means used and vulnerabilities of various elements of the network. Application of this approach allows to maintain security at the required level in the conditions of dynamics of threats set change, taking into account scaling of the IoT-infrastructure.

In comparison with the existing Honeypot solutions, the offered method has a unique capability to automatically control the containers, which allows to use resources more efficiently and also to a create big HoneyNet on the base of an IoT infrastructure which is impossible for the implementation of a fully-fledged virtualization.

**Acknowledgments.** The reported study was partially funded by RFBR according to the research project № 19-01-00767.

# References

1. Alosefer, Y., Rana, O.: Honeyware: a web-based low interaction client honeypot. In: 2010 Third International Conference on Software Testing, Verification, and Validation Workshops (ICSTW), pp. 410–417 (2010)
2. Cabaj, K., Denis, M., Buda, M.: Management and analytical software for data gathered from HoneyPot system. Inf. Syst. Manag. **2** (2013)
3. Chawda, K., Patel, A.D.: Dynamic & hybrid honeypot model for scalable network monitoring. In: International Conference on Information Communication and Embedded Systems (ICICES2014), Chennai, pp. 1–5 (2014)
4. Cleary, M., Corpin, M., Cox, O., Lau, H., Nahorney, B., O'Brien, D., O'Gorman, B., Power, J.-P., Wallace, S., Wood, P., Wueest, C.: ISTR. Internet Security Threat Report (Symantec), vol. 23. Symantec Corporation, Mountain View, USA (2018)
5. Eftimie, S., Răcuciu, C.: Honeypot system based on software containers Mircea cel Batran. Naval Acad. Sci. Bull. **19**(2), 415–418 (2016)
6. Fraunholz, D., Zimmermann, M., Schotten, H.D.: An adaptive honeypot configuration, deployment and maintenance strategy. In: 19th International Conference on Advanced Communication Technology (ICACT), pp. 53–57 (2017)
7. Iskhakov, A., Meshcheryakov, R., Ekhlakov, Yu.: The Internet of Things in the security industry. In: Interactive Systems Problems of Human-Computer Interaction: Collection of Scientific Papers, pp. 161–168 (2017)
8. Iskhakova, A., Meshcheryakov, R., Iskhakov, A., Timchenko, S.: Analysis of the vulnerabilities of the embedded information systems of IoT-devices through the honeypot network implementation. In: Proceedings of the IV International Research Conference "Information Technologies in Science, Management, Social Sphere and Medicine" (ITSMSSM 2017), pp. 363–367 (2017)
9. Kuskov, V., Kuzin, M., Shmelev, Ya., Makrushin, D., Grachev, I.: Traps of "Internet of things". The analysis of the data collected on IoT-traps of Kaspersky Lab. SecureList (2017)
10. Kyriakou, A., Sklavos, N.: Container-based honeypot deployment for the analysis of malicious activity. In: 2018 Global Information Infrastructure and Networking Symposium (GIIS), Thessaloniki, Greece, pp. 1–4 (2018)
11. Lihet, M., Dadarlat, P.D.V.: Honeypot in the cloud. Five years of data analysis. In: 17th RoEduNet Conference Networking in Education and Research (RoEduNet), Cluj-Napoca, pp. 1–6 (2018)
12. Lipatnikov, V.A., Shevchenko, A.A., Yatskin, A.D., Semenova, E.G.: Information security management of integrated structure organization based on a dedicated server with container virtualization. Inf. Control Syst. **89**(4), 67–76 (2017). (in Russia)
13. Luo, T., Xu, Z., Kin, X., Jia, Y., Ouyang, X.: IoTCandyJar. Towards an Intelligent-Interaction Honeypot for IoT Devices, pp. 1–11. Blackhat (2017)
14. Pa, Y.M.P., Suzuki, S., Yoshioka, K., Matsumoto, T., Kasama, T., Rossow, C.: IoTPOT analysing the rise of IoT compromises. USENIX WOOT (2015)
15. Pauna, A., Bica, I.: RASSH – Reinforced adaptive SSH honeypot. In: 10th International Conference on Communications (COMM), Bucharest, pp. 1–6 (2014)
16. Sekar, K.R., Gayathri, V., Anisha, G., Ravichandran, K.S., Manikandan, R.: Dynamic honeypot configuration for intrusion detection. In: 2nd International Conference on Trends in Electronics and Informatics (ICOEI), Tirunelveli, pp. 1397–1401 (2018)

17. Sembiring, I.: Implementation of honeypot to detect and prevent distributed denial of service attack. In: 3rd International Conference on Information Technology, Computer, and Electrical Engineering (ICITACEE), Semarang, pp. 345–350 (2016)
18. Sever, D., Kišasondi, T.: Efficiency and security of docker based honeypot systems. In: 41st International Convention on Information and Communication Technology, Electronics and Microelectronics (MIPRO), Opatija, pp. 1167–1173 (2018)
19. Yagi, T., Tanimoto, N., Hariu, T., Itoh, M.: Enhanced attack collection scheme on high-interaction web honeypots. In: The IEEE Symposium on Computers and Communications (ISCC), pp. 81–86 (2010)
20. Yatskin, A.D.: Dynamic honeypot-systems based on container virtualization, diploma work of a specialist. (in Russian). http://elib.spbstu.ru/dl/2/v16-128.pdf/download/v16-128.pdf

# The Analysis of Cybersecurity Problems in Distributed Infocommunication Networks Based on the Active Data Conception

Sergey V. Kuleshov[1]([✉]), Alexey Y. Aksenov[1], Iliya I. Viksnin[2],
Eugeny O. Laskus[2], and Vladislav V. Belyaev

[1] St. Petersburg Institute for Informatics and Automation of RAS,
St. Petersburg, Russia
kuleshov@iias.spb.su
[2] ITMO University, St. Petersburg, Russia

**Abstract.** The paper considers the problems of cybersecurity for distributed infocommunication networks, related to the violation of the access rights of the executable code of active data to network node resources (shared memory, radio channel reconfiguration, motion control functions, onboard node sensor).

**Keywords:** Cybersequrity · Distributed networks · Active data

## 1 Introduction

In order to ensure the platform-independence of active data (AD) technology, as well as to increase its security, it is proposed to apply the technology of virtual machines. A virtual machine can use both the principles of para-virtualization and full virtualization [1–5] because the executable active data code should not receive direct access to hardware resources for security reasons. The role of the software layer is performed by the hypervisor. This software layer controls the provision of resources to virtual machines and decides which instructions must execute directly, and which ones must emulate. In the case of the paravirtualization, the embedding of the code responsible for the virtualization of hardware re-sources in the code of AD is implied. This eliminates the need to determine which instructions are safe and are allowed to run directly on the processor, and which ones are unsafe and require emulation. In the code prepared in this way, the AD will not contain unsafe instructions; instead, they may be called hypervisor application programming interface (API) [6–8].

## 2 Formalization of Active Data Virtual Machine

The principle of digital content separation to the transport (initializing) stream and the generating program [9, 10] enables flexible adaptation of the content to the existing features and limitations of the physical transmission channels. Within the active data conception, the decoding program can be generated on the transmitter side for every data type to be sent and be transmitted before initializing stream. If predetermined

© Springer Nature Switzerland AG 2020
D. G. Arseniev et al. (Eds.): CPS&C 2019, LNNS 95, pp. 491–499, 2020.
https://doi.org/10.1007/978-3-030-34983-7_48

standard data types are to be used (in this case, on the receiving side, there is a set of standard data recovery programs needed) it is possible to transfer only the index of the program required for recovery of the digital information object. The approach of software-defined systems being configured in accordance to demands and specifics of the transmitted active data enables to create flexible virtual communication environment. The single packet of active data can be described as a bit structure containing three components: signature S, program P and initializing stream D. The only mandatory component is the signature which is needed for the active data packet (ADP) identification and the program to be executed on the receiver side. Initializing stream (based on terminology in [9]) is an input data for the program P and is being transmitted to ADP only if it is needed.

To organize data transmission in mobile networks, there are a number of approaches described in [11–13], each of which has its own limitations. The AD concept is the expansion of such approaches, which allows solving the problem of limited controllability [14]. As an example of the scenario, we will give an example of the organization of relaying AD through a network of mobile nodes controlled by AD, by the analogy with [14].

In this paper, we use a formalization of the Active Data Virtual Machine (ADVM). As most of the modern computer systems are based on Turing Machine [15], we can define a computer system as:

$$M = \left(S,\ I,\ \delta,\ s_0,\ S_f\right)$$

where $S$ denotes the set of all the states a machine (computer system) may have; $I$ denotes the set of all the instructions a machine can provide; $\delta: S \times I \to S \times I$ is the execution operator of an instruction of $I$; $s_0$ denotes the initial state of a machine; $S_f$ denotes the set of all the possible final states of a machine.

Furthermore, we define a series of Execution Operator as $\delta(m)$, so we can get:

$$\delta^{(m)}(s_n, i_n) = \underbrace{\delta^\circ \ldots \delta^\circ}_{m}(s_n, i_n) = \delta(\ldots \delta(s_n, i_n)) = (s_{n+m}, i_{n+m}).$$

Accordingly, the active data is defined as:

$$A = \left(\delta^{(m)}, D\right)$$

where $\delta^{(m)}$ is a series of Execution Operator, $D$ denotes the data stream.

Using the principle of homoiconicality (unity of the presentation of instructions and data) allows you to build a combined alphabet $T$:

$$T = I \cup D$$

Comparing to the classic Turing Machine model, we do the following changes to our model: redefine the Input Symbol as the set $T$ (instructions and data), which are actually all the symbols a virtual machine can accept.

Depending on the permissions on the actions of the active data executable code, there are 2 possible variants of the virtual machine implementation, on the choice of which the analysis of potential system vulnerabilities depends:

- Active data are forbidden to modify the processor:

$$\forall \delta \quad \delta : M \to M.$$

- Active data allowed to modify the processor:

$$\forall \delta \quad \delta : M \to M^*.$$

# 3  Information Security of AD

Information security threats in info-communication systems can be divided into two broad classes - internal and external. In this work, attention will be paid to the elimination of internal threats arising from the use of active data, since external threats can be mainly eliminated using the classical methods of ensuring information security.

In data networks built using active data, internal threats can be caused by vulnerabilities in both hardware and software, including those associated with disruptions in the execution of an active data program, since active data can access control of network device components. These threats include [16–20]:

- The threat of unauthorized access to the host network memory. Realization of this threat may entail a violation of the confidentiality of information stored and transmitted on the network, or a violation of its integrity, which, if necessary for the correct functioning of a network device or network, may lead to problems in the operation of the system.
- The threat of access to the network functions of the device, in particular, to the possibility of reconfiguring the communication channel. The implementation of this threat makes it possible to change the properties of a communication channel organized by a compromised network device and the subsequent disruption of communication with it due to the discrepancy between the communication channel parameters of the nodes using it. A by-effect of this threat is a violation of the availability of information on the network.
- The threat of access to the functionality of the components of the node: on-board sensors, mechanical drives, etc. The implementation of this threat may lead to a violation of the correct operation of the node itself and, in consequence, of the network itself. For example, if an intruder gets access to the functionality of the onboard sensors, it can disrupt their work, which in turn will disrupt the node receiving information about the state of the environment. This may lead to the impossibility for the node to implement a number of functions depending on the information received from the sensors.

The threats listed above are related to the possibility of an access violation by the executable code of active data. To eliminate them, a clear distinction is required between the access rights of different types of active data programs to the node functionality depending on the purpose of their execution.

- The threat of selfishness. This threat is peculiar to mobile networks, nodes of which have limited resources, both battery and computing power. When reducing the number of available resources, the node that trying to save them may limit the provision of its own routing services to other nodes, which will lead to a violation of the availability of information in the network.
- Considering these threats, the following attacks on information-communication networks of active data are possible [16–20]:
- Man-in-the-middle attack. An attack in which the intruder changes the connection between the nodes, which in this case continue to assume that they communicate directly with each other. This attack leads to a breach of confidentiality of information, since information transmitted between nodes passes through the node compromised by the intruder. Also, depending on the violator's goals, it may also result in a violation of the integrity or availability of information: the intruder may replace the transmitted data with his own or not transmit partially or completely received data to the receiving node, while the sending node will consider them delivered.
- Insomnia test. An attack on mobile wireless networks that poses a threat of selfishness, in which the intruder increases the power of the target node, which is why the node, trying to save battery power, limits its work, resulting in a violation of information availability in the network.

## 4   Empirical Part

An experiment on the implementation of active data technology will be conducted based on a network of unmanned ground vehicles. The purpose of this experiment is to analyze the vulnerabilities and threats in these networks when using active data and to find ways to eliminate them.

The experimental stand consists of models of unmanned vehicles (MUVs) and an external information center (EIC), which is an intermediary in the exchange of information between the MUVs and is responsible for its storage. Each MUV is a system of physical elements, the interaction of which allows for the analysis of the environment, movement in it, communication within a group of models, as well as various algorithms necessary for the functioning of the system.

The MUV system includes two subsystems - computational and executive. The computational subsystem is responsible for building the route and analyzing deviations from it, developing a list of commands for execution and making communications. The executive subsystem, in turn, implements the commands developed by the computing subsystem and provides for the detection of obstacles to movement. The structure of the IBA system is presented in Fig. 1.

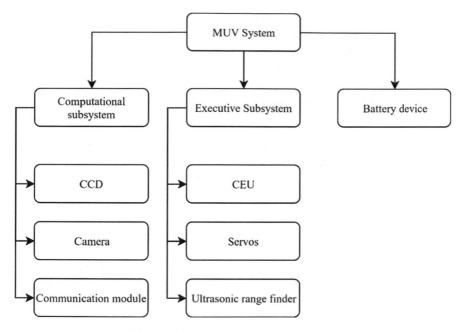

**Fig. 1.** The structure of the MUV system

The computational subsystem consists of three elements - the central computing device (CCD) of the Raspberry Pi, the Raspberry Pi Camera Board camera and the Xbee communication module. The main device of this subsystem is the CCD, which is responsible for processing information from other devices, developing the route and command lists.

The camera is responsible for the computer vision of the MUV - it accepts frames with road markings, thanks to which the CCD is able to determine the distance traveled and the need for route correction.

The communication module exchanges data between the MUV and the EIC, based on which the CCD determines the identification number of the MUV in the group and the priority of movement in disputable situations.

The executive subsystem consists of four elements - the central executive unit (CEU) Arduino Nano, two servo drives and an ultrasonic range finder. CEU is the main device of the subsystem, which analyzes the commands coming from the CCD necessary for execution, the processing of obstacle data from a rangefinder and the control of servo drives. The main task of the CEU is to supply pulses to servos, the change in the magnitude of which allows you to adjust the speed and direction of rotation of the servos. The ultrasonic range finder, based on the time elapsed from the moment of radiation to the moment of returning the signal, determines the distance to the obstacles and, in case of detection of an obstacle in dangerous proximity to the MUV, sends the command of the CEU to stop the servo drives.

The generalized model of information interaction is presented in Fig. 2: the information center includes CCD and CEU - elements of information processing and control of other elements, and elements with limited functionality (i.e., capable of performing only individual actions) are related to the physical level.

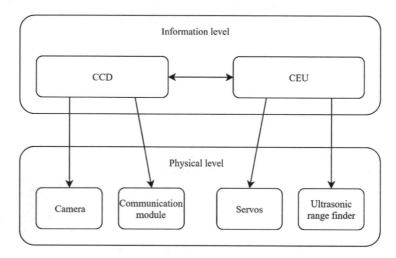

**Fig. 2.** The interaction of the MUV system elements

In this network, active data is a set of commands generated by the CCD of one MUV to execute the CEU of another, with which several system security problems are associated: the possibility of violating the confidentiality of the transmitted data, i.e., execution of the transmitted active data packet by the transit node, and the possibility of unauthorized access of the executable code of the active data to the elements of the physical layer MUV. To eliminate the first MUV threat, node authentication is required when attempting to execute an active data code; for the second - the implementation of the system of differentiation of access rights at the information level MUV.

Scenarios for the implementation of attacks on the MUV network [18, 19]:

- Man-in-the-middle attack. The implementation of this attack is as follows:


1. The offending node C sends a request to establish a connection to node A, pretending to be node B.
2. Node A establishes communication with Node C.
3. Node C sends a request to establish a connection to Node B, appearing as Node A.
4. Node B establishes communication with Node C.
5. In the process of transferring data from A to B and back, node C, when received, performs the necessary manipulations with them and then passes them on.

However, when a connection is already configured between nodes A and B, in order to launch an attack, it is necessary to break the current connection before performing these actions. To do this, you can send a request on behalf of the second one to stop the connection to one of the nodes, or make the channel ineffective for high-quality data transmission (for example, increase the noise level in it), as a result of which the connection through this communication channel is interrupted. In addition, this attack can be implemented in case of violation of the access rights of the executable code of active data to the functions of the communication module, as a result of which it is possible to reconfigure communications in the network in a manner necessary for the intruder.

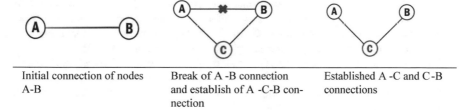

| Initial connection of nodes A-B | Break of A -B connection and establish of A -C-B connection | Established A -C and C-B connections |

- Insomnia test. The implementation of this attack is possible in several ways. The first is sending more traffic through the target node, as a result of which the node will need to direct additional resources to the processing of input data. The second – if there is access to the communication module functionality obtained as a result of an access violation of the active data programs being executed – the intentional non-optimal use of the device components from the point of view of energy consumption.

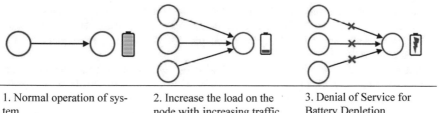

| 1. Normal operation of system | 2. Increase the load on the node with increasing traffic | 3. Denial of Service for Battery Depletion |

# 5   Conclusion

In this paper, an approach to ensure security in active data networks, based on the use of virtual processing, is proposed. The main advantage of this approach is the independence of the solution implementation from the system hardware and network decentralization.

The threats and possible attacks on the active data networks were considered. Experimental stand for the implementation of active data technology was presented.

At the moment it is planned to consider described experiments on a simulator. After testing on a virtual network, to complete the study, this technology will be implemented on an unmanned drones network.

# References

1. Carvalho, A., Silva, V., Afonso, F., Cardoso, P., Cabral, J., Ekpanyapong, M., Montenegro, S., Tavares, A.: Full virtualization on low-end hardware: a case study. In: IECON Proceedings of Industrial Electronics Conference, 21 December 2016, pp. 4784–4789 (2016)
2. Understanding Full Virtualization, Paravirtualization, and Hardware Assist. VMware technical documentation. Accessed 18 Apr 2019. https://www.vmware.com/content/dam/ digitalmarketing/vmware/en/pdf/ techpaper/VMware_paravirtualization.pdf
3. Chen, W., Xu, W., Wang, Zh., Dou, Q., Zhao, B.: A formalization of an emulation based co-designed virtual machine. In: 2011 Fifth International Conference on Innovative Mobile and Internet Services in Ubiquitous Computing (IMIS), pp. 164–168 (2011). https://doi.org/10. 1109/imis.2011.144
4. Craig Iain, D.: Virtual Machines. Springer, London (2006). 269 p. https://doi.org/10.1007/ 978-1-84628-246-1
5. Shi, Y., Gregg, D., Beatty, A., Ertl, M.A.: Virtual machine showdown: stack versus registers. https://www.usenix.org/legacy/events/vee05/full_papers/p153-yunhe.pdf. Accessed 18 Apr 2019
6. Polenov, M., Guzik, V., Lukyanov, V.: Hypervisors comparison and their performance testing. In: Advances in Intelligent Systems and Computing, vol. 763, pp. 148–157. Springer, Cham (2019). https://doi.org/10.1007/978-3-319-91186-1_16
7. Cheng, Y., Chen, W., Wang, Z., Yu, X.: Performance-monitoring-based traffic-aware virtual machine deployment on NUMA systems. IEEE Syst. J. 11(2), 973–982 (2017). https://doi. org/10.1109/JSYST.2015.2469652
8. Rao, J., Wang, K., Zhou, X., Xu, C.: Optimizing virtual machine scheduling in NUMA multicore systems. In: 2013 IEEE 19th International Symposium on High Performance Computer Architecture (HPCA), Shenzhen, pp. 306–317 (2013). https://doi.org/10.1109/ hpca.2013.6522328
9. Kuleshov, S.V., Tsvetkov, O.V.: Active data in digital software-defined systems. Informatsionno-izmeritelnye i upravlyayushchie sistemy 6, 12–19 (2014). (in Russia)
10. Alexandrov, V.V., Kuleshov, S.V., Zaytseva, A.A.: Active data in digital software defined systems based on SEMS structures. In: Gorodetskiy, A. (ed.) Smart Electromechanical Systems. Studies in Systems, Decision and Control, vol. 49, pp. 61–69. Springer, Cham (2016)
11. Samad, T., Bay, J.S., Godbole, D.: Network-centric systems for military operations in urban terra. The role of UAVs. J. Proc. IEEE. 95(1), 4118473, 92–107 (2007). https://doi.org/10. 1109/jproc.2006.887327
12. Lysenko, O.I., Valuiskyi, S.V., Tachinina, O.M., Danylyuk, S.L.: A method of control by telecommunication airsystems for wireless AD HOC networks optimization. In: IEEE 3rd International Conference Actual Problems of Unmanned Aerial Vehicles Developments (APUAVD) Proceedings, Kyiv, Ukraine, pp. 182–185, October 2015

13. Ono, F., Ochiai, H., Miura, R.: A wireless relay network based on unmanned aircraft system with rate optimization. IEEE Trans. Wirel. Commun. **PP(99)**, 7562472 (2016). https://doi.org/10.1109/twc.2016.2606388

14. Kuleshov, S.V., Zaytseva, A., Aksenov, A.Y.: The conceptual view of unmanned aerial vehicle implementation as a mobile communication node of active data transmission network. Int. J. Intell. Unmanned Syst. **6**(4), 174–183 (2018). https://doi.org/10.1108/IJIUS-04-2018-0010

15. Turing, A.M.: Correction to: on computable numbers, with an application to the entscheidungs problem. Proc. London Math. Soc. Ser. **2**(43), 544–546 (1938)

16. Afanasyev, A.L., Garmonov, A.V., Kashchenko, G.A.: Analysis of security threats and secure routing protocols in MANET networks. In: Radiolocation, Radio Navigation, Communication: Materials of the XX International Scientific and Technical Conference (RLNC-2014), vol. 2. Voronezh: Publishing house SPC «CAKBOEE» OOO, pp. 846–857 (2014). (in Russia)

17. Anjum, F., Mouchtaris, P.: Security for Wireless Ad-hoc Networks. Wiley, Hoboken (2007)

18. Irshad, S., Halabi, B.H., Jamalul-Lail, A.M., Iftikhar, A., Daniyal, A.: Classification of attacks in vehicular ad hoc network (VANET). INFORMATION Int. Interdisc. J. **16** (5), 2995–3004 (2013)

19. Gohale, V., Gosh, S.K., Gupta, A.: Classification of Attacks on Wireless Mobile Ad Hoc Networks and Vehicular Ad Hoc Networks, 196–217. CRC Press (2011)

20. Zegzhda, D., Ivanov, P.V., Moskvin, D.A., Kubrin, D.S.: Actual security threats for vehicular and mobile ad hoc networks. Autom. Control Comput. Sci. **52**, 993–999 (2018). https://doi.org/10.3103/S0146411618080308

# The Platform of the Industrial Internet of Things for Small-Scale Production in Mechanical Engineering

Igor G. Chernorutsky, Pavel D. Drobintsev[✉],
Vsevolod P. Kotlyarov, Alexey A. Tolstoles, and Alexey P. Maslakov

Peter the Great St. Petersburg Polytechnic University, St. Petersburg, Russia
pavel.drobintsev@gmail.com, vpk@spbstu.ru

**Abstract.** The paper deals with the problem of creating an industrial Internet of Things (IoT) platform for a small-scale machine-building site, which is important because this type of production is characterized by an imbalance between the time of technological preparation of production and the production process itself. The developed platform is focused on wide applicability by adapting to technological routes for the creation of hardware parts of varying complexity for different structures of production equipment and varying resources. The platform concept allows a manifold reduction of the complexity of the design work for and creation of technological processes for small-scale or individual production based on promising directions of the modular technologies development.

**Keywords:** Internet of things · Small-scale production · Mechanical engineering · Software development

## 1 Introduction

Industry 4.0 [1] today is the paradigm of future production. Its essence lies in the organization of the material production of goods and services based on networks that integrate the exchange of information between sensors, data ports, control devices and other terminal objects using intelligent, operational and strategic technological process management, i.e., the so-called Internet of Things [2].

The paper considers the issue of creating an IoT platform for a small-scale machine-building site.

The platform concept is based on the automation of the development of a technological route (TR) for creating a specific part by means of its automated assembly from standard parameterized modules (SPMs) recorded in a database. Routes are optimized for the cost and price criteria. Then, using digital modeling, the calculation of the optimal schedule for the execution of the set of TRs based on the resources of the production site containing CNC machines, robots, warehouses and support personnel is provided.

D. G. Arseniev et al. (Eds.): CPS&C 2019, LNNS 95, pp. 500–512, 2020.
https://doi.org/10.1007/978-3-030-34983-7_49

An important feature of the developed platform is its orientation towards the implementation of modular technologies of engineering production [3], which are adapted to the production technology of specific parts and production resources of a specific small-scale production.

## 2  State of the Art

When creating an intelligent platform, it is not enough only to provide modern equipment for a factory; it is necessary to ensure the efficiency of its work [4]. In the area of small enterprises with small-scale production, it is necessary to quickly create production plans that can change depending on the state of the process equipment and manufactured products, and the implementation of plans should be effectively automated.

Usually, the tasks of operational planning and automated production management are carried out by the manufacturing execution systems (MES) [5, 6]. They occupy an intermediate place in the hierarchy of enterprise management systems, between the level of information collection from equipment in workshops provided by supervisory control and data acquisition (SCADA) systems [7] and the level of operations over a large amount of administrative, financial and accounting information provided by the enterprise resource planning (ERP) [5] systems. Nowadays, there are three most popular largest solutions on the Russian market: PHOBOS system, YSB.Enterprise. Mes system, and PolyPlan system. PHOBOS is traditionally used in large and medium-size mechanical engineering enterprises. YSB.Enterprise.Mes originated from the woodworking industry and focuses on the sector of medium and small enterprises. The PolyPlan system is positioned as an operational scheduling system for automated and flexible manufacturing in engineering [8].

However, with all the attractiveness of such systems, due to the extensive set of functions provided and deep integration into the production processes at the enterprise at all stages, their practical implementation is a whole multipart and expensive project that not all enterprises, especially small-scale and individual productions, can afford. In addition to this, in order to work effectively with MES, high qualification of its operator is required.

The developed solution given in this work is designed to solve a narrower class of problems - to simplify the technological preparation of production for a small-scale mechanical engineering site, which can be based on the introduction of operative digital modeling and analysis of the technological process of the production site.

## 3  IoT Automation Object Architecture

A promising direction in the organization of automation of small-scale mechanical engineering production sites is their focus on network-centric control and management. The advantages of this approach are indisputable provided that complex network-centric systems will have high reliability of operation and flexible control of technological processes and routes [9]. Figure 1 shows an example of a network-centric architecture of a production site.

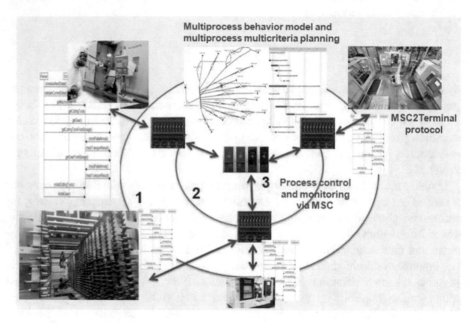

**Fig. 1.** Three levels of network-centric control of a mechanical engineering working site

The levels depicted are as follows:

- The level of technological operations for CNC machines, robots and other objects that provide control actions and collect data on the state of network objects.
- The level of technological processes (management of technological routes, which contain sequences of technological operations).
- The level of multi-criteria hierarchical optimization and production planning.

# 4    Technological Route Formalization

Formalization of the manufactured part is carried out by the technologist on the basis of its drawing. They should highlight the processing modules in the drawing - the basic surfaces (BSes) and compound modules (CMs), and provide the description of each with a variety of design and technological parameters, such as their geometry, dimensional accuracy, surface hardness, processing method, necessary equipment, cutting tools, cutting modes, etc. [10].

For these purposes, the automated workplace of the technologist (AWT) is used (Fig. 2).

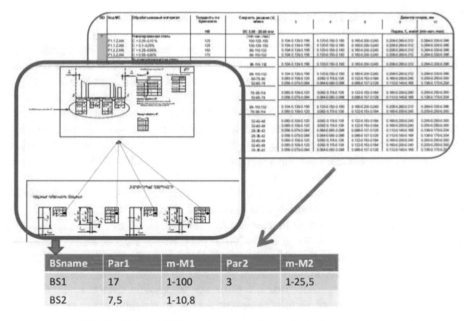

**Fig. 2.** Using AWT to select a cutting tool and its parameters

In addition to the surface modules, modules for the technological route of manu-facturing the part, the modules for equipment and gear, the modules for instrumental adjustment, and the modules for measuring instruments are described in a similar way. The technologist obtains the necessary parameters from the drawing, reference catalogs or other documentation. For a number of parameters, ranges of possible values are specified.

AWT allows the technologist to automate the grouping of surfaces with similar accuracy and processing tolerance into phases, their aggregation by usage of similar tools in the processing, and as a result, to get the order of application of the processing operations, i.e., the technological route.

In the approach presented here we use the MSC language [11, 12] for the encoding of the technological route. MSC is a standardized language for describing behaviors using message exchange diagrams between parallel-functioning objects (CNC, robots). The main unit of the diagram in its textual form is a line starting with the name of a processing stage of elementary surface followed by its parameters. The resulting parameterized line takes the following form, for example:

Turning (stageNumber, surfaceType1, surfaceType2, surfaceNumber, [machine1, ma-chine2, machine3, machine4, machine5], [cuttingTool1, cuttingTool2, cutting Tool3], workpieceParams.code, blockParams.number, numberInBlock, group Number, num-berInGroup);

## 5  Technological Route Characteristics Estimation

Using a formalized TR in the MSC form allows the technologist to obtain preliminary estimates of its time and cost. Formulas stored in the database are used to calculate them for each processing stage, the fragment of the set of such formulas is shown in Table 1. The relative estimate of the route (Fig. 3) can be obtained as the sum of the estimates of each operation on each individual surface module that make the route, which is sufficient for ranking alternative solutions on the choice of the parameters of the route. To obtain absolute values, it suffices to use the multiplicative and additive correction factors obtained on the basis of statistical estimates of the technological processes of a particular production.

The route can be optimized. By changing the parameters of the route within the allowable ranges and re-calculating the indicators of processing time and cost, the technologist can get a solution that meets the limitations of the management on a particular job or get the Pareto-optimal solution [13]. However, it should be noted that the mentioned optimization is valid provided that the production by the route is carried out without taking into account the current state and restrictions on the resources of the production site. Obtaining more realistic estimates is possible with the help of simulation modeling of the distribution of resources for the routes simultaneously performed at the production site.

**Table 1.** Formulas for turning time calculations

| Formulas | Parameters description |
|---|---|
| $T_m = \frac{L}{n \cdot s} \cdot i$ | $T_m$ – machining time |
| | $L$ – estimated length of processing in mm |
| | $n$ – workpiece rounds per minute |
| | $s$ – cutter feed per round in mm |
| | $i$ – the number of passes of the cutter |
| $L = l + l_1 + l_2$ | $l$ – the length of the workpiece in the feed direction, mm |
| | $l_1$ – cutting–in length of the tool |
| | $l_2$ – the length of the tool exit, mm |
| $n = \frac{1000 \cdot v}{\pi \cdot d}$ | $v$ – the speed of the cutting, mm per minute |
| | $d$ – the diameter of the processed workpiece, mm |
| $i = h/t$ | $h$ – the amount of overmeasure in mm |
| | $t$ – cutting depth in mm |

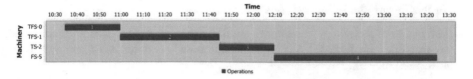

**Fig. 3.** Illustration of a technological route consisting of 4 operations performed on 4 machines

# 6 Digital Modeling of Technological Processes at the Production Site

The digital model of the production site simulates the implementation of the production of different batches of parts by different specified technological routes. The site model is built on the basis of information on the resources of the production site (CNC machines, transport robots, warehouses, staff, etc.) which include amounts of time for their usages, stated by estimations of the operations on the basic surfaces of the detail. The size of the batch of parts is also associated with the route.

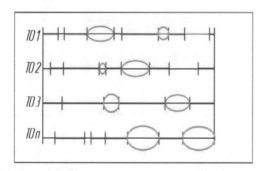

**Fig. 4.** Example of the schedule of a production site

The model uses the method of dynamic priorities to simulate the workload of resources of the production site and determine the duration of the implementation of the technological process for orders.

The result of simulation modeling is a schedule for the implementation of the technological process, which provides estimates of the time to manufacture a batch of parts in accordance with each route (Fig. 5). These estimates may differ from the original by amounts of downtime and amounts of time for changeover operations required when using multiple machines to implement the route. An example of a fragment of the schedule with such delays taken into account is shown in Fig. 4. The corresponding delays in the use of technological equipment are marked by ovals.

A set of estimates of the time of execution of the route can be analyzed for the fulfillment of certain criteria and restrictions characterizing the conditions of the order.

In this regard, the following tasks can be solved:

- Estimation of the minimal amount of additional resources that need to be allocated, so that the total time for the implementation of the route is no more than the specified value $T_0$.

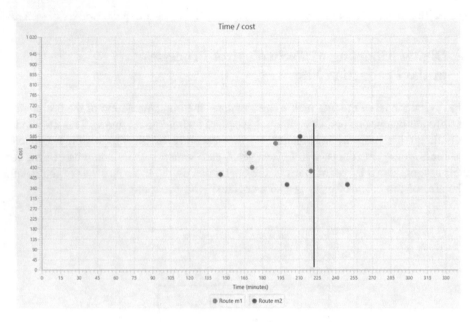

**Fig. 5.** Time and cost estimations example of two technological routes

- Redistribution of processing tools between individual operations in order to minimize the total implementation time of the route (optimal transfer of resources from non-critical operations to critical ones).
- The use of time reserves $T_0 - T$ arising when the calculated time $T$ of the implementation of the route is less than the specified value $T_0$ - in order to further improve the process.

In the process of implementing a specific work schedule (in a certain sense, optimal), various unforeseen failures are possible: machine breakage, shortage of components, unforeseen delays in performing individual operations, etc. Therefore, the management system should continuously monitor the entire process and should have a mode for operative changing of the schedule for the implementation of the remaining work in the new environment in order to optimize it. Thus, it turns out that it is necessary to correct the process of implementing the set of necessary operations in real time taking into account the requirements set for optimization and the formulated criteria of optimality.

For this reason, it is advisable to refer to the principles of network-centric management and methods of coordination in hierarchical systems. It is also advisable to use the methods of hierarchical construction of Pareto sets at various technological levels [13, 14].

# 7  Reliability and Variability of the Technological Route

The reliability of network-centric production systems is one of the defining characteristics of their performance. The platform implements three complementary approaches to its provision:

- Behavior analysis based on formal symbolic verification
- Operations history monitoring
- Control and monitoring of states of the system.

The first one is based on the proof of correctness of the technological route in the symbolic verification process, which checks the acceptable ranges of the parameters, as well as the correctness of order of the whole route operation sequence [15]. The MSC representation of the original route is automatically converted to the equivalent UCM diagram [16] (Fig. 6). The figure shows the following types of routes:

- A – route with alternatives
- B – route without alternatives
- C – route for the terminal equipment controller
- D – a part of the route A from its start to the first alternative
- E – a part of the route A showing the second alternative.

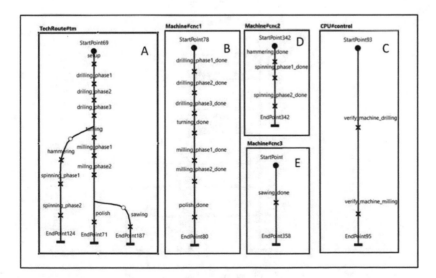

**Fig. 6.** Models of technological routes

The use of alternative branches can be quite flexible: they can be used to define alternative ways of implementing the same route using different tools or process equipment, the conditions for termination of the main route by a time-out signal or by a signal of an identified error. Essentially, TRs with alternatives are trees representing the sets of possible non-alternative TRs.

With the help of such trees of routes, the variability of technological solutions that can be used in the process of manufacturing a product can be described. Managing the transition to an alternative branch is set by a specially formulated condition and is carried out in the dynamics of the production process by the controller, which identifies the trigger conditions for the selection of the corresponding transition. Regardless of

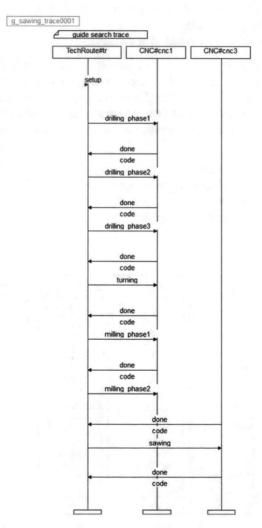

**Fig. 7.** Route example in the form of an MSC diagram describing the interaction of the controller and two machines

the choice of alternatives, the actual process follows the specific MSC diagram, generated from the route belonging to the TR tree (Fig. 7). It follows that the tree is a static description of all possible TRs.

Reliability is ensured by systematic applying of the following procedure in the process of creating technological software:

- Proof of the correctness of the formal model of the TR tree and the fixation of admissible areas of values of the parameters and attributes of the TR corresponding to their correct behavior in the process of symbolic verification.
- Proof of the completeness of behavioral technological scenarios in the process of symbolic verification, as a result of which the conditions of a possible violation of permissible areas or conditions of transition to the next TR operation are identified. Such situations should be corrected or closed using an alternative branch for error handling and a corresponding change in the TR.
- Analysis of the behavior of all operational modes defined by the TR tree, and generation of protective shells that control and prevent all outputs of behavioral scenarios going beyond permissible limits due to incorrect input information, failures or defects.

History monitoring approach provides security at the transport level of the network-centric system. Readers-Writers Flow Model [17, 18] (RWFM) is used to dynamically label the transactions of the platform and derive the constraints to be satisfied by various components and their interactions to preserve the desired security and privacy requirements. Based on the analysis of the interaction history, a large class of errors common to the transport layer is prevented.

The optimized schedule in the form of MSC diagrams, built on level 3 (Fig. 1), is sent to level 2 controllers, which launch, control and monitor parallel execution of technological operations on terminal objects (machines and robots). Service personnel also receive control information in human-readable form to their smartphones.

Monitoring is carried out on the basis of information received from level 1 sensors and dynamically identifies events violating the execution of the plan.

After analysis, the level 2 controller decides whether to continue the TR, or to use the adjustment provided by the technological program, or to transfer information to the third level about the need for re-planning.

## 8   Operational Card Generation

Operational card (Fig. 8) is the main document for the technologist and site workers, which presents the sequence of actions for the manufacturing of parts and the necessary technical equipment (cutting tools, fixtures and clamping devices, etc.). An automatic generation of the operational card is based on optimized TR, which seriously reduces the complexity and time of technological preparation of small-scale production [19]. The main document describing the format of the card is GOST 3.1404-86 [20].

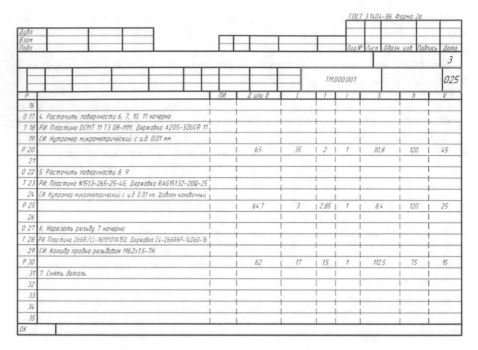

**Fig. 8.** An example of a fragment of the generated operational card

# 9   Conclusions

The proposed solution to the problem of technological preparation of the production for small-scale manufacturing is implemented in the form of a platform with the following features:

1. The platform adapts to specific production conditions: equipment, resources, orders, and personnel.
2. The platform provides optimal characteristics of specific production TRs in accordance with a selected set of criteria for the success of a route.
3. The work provides a reliable organization of production TR by a complete consideration of various situations and environmental conditions.
4. The platform provides an increase in labor productivity of small-scale production due to:

   - A significant reduction in the period of technological preparation for the manufacturing of new products;
   - Automatic generation of technological documentation and control programs; monitoring of production according to its optimized plans;
   - Operative updating of plans and tracking of their implementation.

At the moment, the platform is a working prototype, ready to demonstrate the listed properties and being in the process of accumulating typical solutions for metalworking tasks of small-scale ship-repair enterprises. Implementations on production sites are planned for 2019–2020 years.

**Acknowledgements.** The work was financially supported by the Ministry of Education and Science of the Russian Federation in the framework of the Federal Targeted Program for Research and Development in Priority Areas of Advancement of the Russian Scientific and Technological Complex for 2014-2020 (14.584.21.0022, ID RFMEFI58417X0022).

# References

1. Abdulbarieva, E., et al.: High-tech computer engineering: a review of markets and technologies. St.Petersburg, Publishing House of the Polytechnic University Press (2014). 110 p
2. Yuan, J., et al.: Uncertainty measurement and prediction of IoT data based on Gaussian process modeling. Nongye Jixie Xuebao, **46**(5) (2015)
3. Bazrov, B.: Modulnaya tekhnologiya v mashinostroenii [Modular technology in me-chanical engineering]. Moscow, "Mashinostroenie" ["Mechanical engineering"] (2001). 366 p. (in Russian)
4. Solkin, A.: Sposoby avtomatizatsii sozdaniya upravlyayushchikh programm dlya metal-lorezhushchego oborudovaniya s ChPU [Ways to automate the creation of control programs for metal-cutting equipment with CNC]. Volzhsky University after V.N. Tatischev Gazette, **19**(2), 165–168 (2012). (in Russian). https://cyberleninka.ru/article/n/sposoby-avtomatizatsii-sozdaniya-upravlyayuschih-programm-dlya-metallorezhuschego-oborudovaniya-s-chpu. Accessed 29 Apr 2019
5. Frolov, E., Zagidullin, R.: MES-sistemy, kak oni est ili evolyutsiya sistem planirovaniya proizvodstva (chast II) [MES as they are or the evolution of the production planning systems (part II)] (2007). (in Russian). http://www.fobos-mes.ru/stati/mes-sistemyi-kak-oni-est-ili-evolyutsiya-sistem-planirovaniya-proizvodstva.-chast-ii.html. Accessed 29 Apr 2019
6. Andreev, E., Kutsevich, I., Kutsevich, N.: MES-sistemy: vzglyad iznutri [MES: a look form the inside]. Moscow, RTSoft (2015). 240 p. (in Russian)
7. Davidyuk, Y.: SCADA-sistemy na verkhnem urovne ASUTP [SCADA systems at the top level of advanced process control systems]. Intelligent Enterprise, **30**(13), (2001). (in Russian). https://www.iemag.ru/platforms/detail.php?ID=16479. Accessed 29 Apr 2019
8. Leondes, C.T.: Computer-Aided Design, Engineering, and Manufacturing Systems Techniques and Applications. FL CRC Press, Boca Raton (2001)
9. Voinov, N., Chernorutsky, I., Drobintsev, P., Kotlyarov, V.: An approach to net-centric control automation of technological processes within industrial IoT systems. Adv. Manuf. **5** (4), 388–393 (2017)
10. Garaeva, Y., Zagidullin, R., Tsin, S.: Rossiiskie MES-sistemy, ili Kak vernut proizvodstvu optimizm [Russian MES or how to return optimism to production]. SAPR i grafika [CAD an graphics]. (11) (2005). (in Russian). https://sapr.ru/article/14614 Cited 29 Apr 2019
11. Recommendation ITUT Z. 120. Message Sequence Chart (MSC), November (2000). https://www.itu.int/rec/T-REC-Z.120. Accessed 29 Apr 2019
12. Rudolph, E., Graubmann, P., Gabowski, J.: Tutorial on message sequence charts. Comput. Networks ISDN Syst.-SDL MSC **28**, 1629–1641 (1996)

13. Chernorutsky, I.: Decision Making Tools. "BHV", St. Petersburg (2005). 418 p. (in Russian)
14. Drobintsev, P., Chernorutsky, I., Kotlyarov, V., Kotlyarova, L., Tolstoles, A., Khrustaleva, I.: Net-centric Internet of Things for industrial machinery workshop. In: Proceedings of the 4th Ural Workshop on Parallel, Distributed, and Cloud Computing for Young Scientists, Yekaterinburg, Russia, pp. 112–122 (2018). http://ceur-ws.org/Vol-2281/paper-12.pdf. Accessed 29 Apr 2019
15. Baranov, S., Kotlyarov, V., Letichevsky, A., Drobintsev, P.: The technology of automation verification and testing in industrial projects. In: Proc. of St.Petersburg IEEE Chapter, International Conference, 18–21 May, St.Petersburg, Russia, pp. 81–86 (2018)
16. Z.151: User requirements notation (URN) – Language definition. https://www.itu.int/rec/T-REC-Z.151/en. Accessed 29 Apr 2019
17. Narendra Kumar, N.V., Shyamasundar, R.K.: A complete generative label model for lattice-based access control models. In: Proceedings of the 15th International Conference on Software Engineering and Formal Methods (SEFM 2017), Trento, Italy, September 4–8, LNCS, vol. 10469, pp. 35–53 (2017)
18. Shyamasundar, R.K., Narendra Kumar, N.V., Muttukrishnan, R.: Information-flow control for building security and privacy preserving hybrid clouds. In: Proceedings of the IEEE Data Science and Systems Conference, Sydney, Australia, 12–14 December, pp. 1410–1417 (2016)
19. Eizenakh, D., Cherepovskii, D., Kotlyarov, V.: Sistema generatsii operatsionnoi karty tekhnologicheskogo protsessa dlya melkoseriinogo mashinostroitel'nogo proizvodstva [The system for generation of the operating card for the technological process for a small-scaled mechanical engineering production], *Sovremennye tekhnologii v teorii i praktike program-mirovaniya* [Modern technologies in the theory and practice of programming] conference pro-ceedings, St.Petersburg, Russia (2019). 46 p. (in Russian)
20. GOST 3.1404-86 Edinaya sistema tekhnologicheskoi dokumentatsii (ESTD). Formy i prav-ila oformleniya dokumentov na tekhnologicheskie protsessy i operatsii obrabotki rezaniem [Unified system for technological documentation (USTD). Forms and rules for paperwork on technological processes and machining operations]. (in Russian) http://docs.cntd.ru/document/1200012135. Accessed 29 Apr 2019

# 3D Hand Movement Measurement Framework for Studying Human-Computer Interaction

Toni Kuronen[1]($\boxtimes$), Tuomas Eerola[1], Lasse Lensu[1], Jukka Häkkinen[2], and Heikki Kälviäinen[1]

[1] Computer Vision and Pattern Recognition Laboratory,
LUT University, Lappeenranta, Finland
{Toni.Kuronen, Tuomas.Eerola, Lasse.Lensu,
Heikki.Kalviainen}@lut.fi
[2] Institute of Behavioural Sciences, University of Helsinki, Helsinki, Finland
jukka.hakkinen@helsinki.fi

**Abstract.** In order to develop better touch and gesture user interfaces, it is important to be able to measure how humans move their hands while interacting with technical devices. The recent advances in high-speed imaging technology and in image-based object tracking techniques have made it possible to accurately measure the hand movement from videos without the need for data gloves or other sensors that would limit the natural hand movements. In this paper, we propose a complete framework to measure hand movements in 3D in human-computer interaction situations. The framework includes the composition of the measurement setup, selecting the object tracking methods, post-processing of the motion trajectories, 3D trajectory reconstruction, and characterizing and visualizing the movement data. We demonstrate the framework in a context where 3D touch screen usability is studied with 3D stimuli.

**Keywords:** High-speed video · Hand tracking · Trajectory processing · 3D reconstruction · Video synchronization · Human-computer interaction

## 1 Introduction

In the human-computer interaction (HCI) research, it is necessary to accurately record hand and finger movements of test subjects in tasks related to user interfaces. Advances in gesture interfaces, touch screens and augmented and virtual reality have brought new usability concerns that need to be studied in a natural environment and in an unobtrusive way [21]. Data gloves with electromechanical, infrared or magnetic sensors can measure the hand and finger location with high accuracy [5]. However, such devices affect the natural hand motion and cannot be considered feasible solutions when pursuing natural HCI. Consequently, image-based solutions which provide an unobtrusive way to study and to track human movement and tenable natural interaction with the technology have become a pronounced subject of research interest.

Commercially available off-the-shelf measurement solutions such as Leap Motion and Microsoft Kinect do not allow frame rates high enough to capture all the nuances of rapid hand movements. Moreover, Leap Motion limits the hand movement to a relatively small area. The field of view of the sensor is an inverted pyramid, with an

© Springer Nature Switzerland AG 2020
D. G. Arseniev et al. (Eds.): CPS&C 2019, LNNS 95, pp. 513–524, 2020.
https://doi.org/10.1007/978-3-030-34983-7_50

angle of 150° in the left-right direction and an angle of 120° in the front-back direction, and the measurement distance ranges from 25 mm to 60 cm above the device [6]. Kinect is sufficient for detecting arm and full-body gestures, but it is imprecise for accurate finger movement measurements [13]. Furthermore, such commercial solutions lack the inspecting capability of a camera-based system and do not allow further analysis of hand pose beyond the limitations of the sensors.

An alternative approach for accurate recording of fast phenomena including rapid and subtle hand movements is high-speed imaging. High-speed videos provide the basis for building a system that is more versatile than the existing black-box solutions. From the implementation viewpoint, high-speed imaging requires more light than the conventional imaging to allow short exposure times, which imposes additional demands on the measurement setup. Moreover, the bright illumination can disturb the user performing the HCI experiment since it reduces the perceived contrast and, making it difficult to see the stimulus. Thus, careful planning of the measurement setup is important to ensure that the conditions for the interaction are as natural as possible.

To record hand movements in 3D with a camera-based measurement, at least two cameras with different viewing angles are required. However, a setup consisting of multiple high-speed cameras is both expensive and difficult to build. This motivates to use of a normal-speed camera in addition to the high-speed camera to provide the depth information for reconstructing the 3D trajectories.

Recent progress in object-tracking techniques has made it possible to automatically determine motion trajectories from videos. Gray-scale high-speed imaging is commonly used to keep illumination requirements at a reasonable level, and, consequently, the use of hand-tracking methods relying specifically on color information becomes impractical. These matters motivate the utilization of general object trackers. For example, in [11], several object-tracking methods were compared using high-speed videos, and the best methods were found to be suitable for the problem of measuring hand movements in the context of HCI.

The main problem with using existing object tracking methods for accurate measurement of hand and finger movements is that they were developed for applications where high spatial accuracy is not crucial, as the research focus was on developing more computationally efficient and robust methods. For these methods, losing the target is considered a much more severe problem than a small spatial shift of the tracking window. This is not the case in hand trajectory measurement based on high-speed videos where small hand movements between the frames and a controlled environment help to achieve robustness. Thus, high spatial accuracy is the main concern. Even small errors in the spatial locations can lead to large fluctuations in the speed and acceleration determined from the location data. Therefore, existing tracking algorithms alone are insufficient for the accurate measurements of hand movements and further processing of the hand position data is needed.

Raw trajectory data contain small spatial location fluctuations that can make calculation of accurate velocities and accelerations impossible. Smoothing raw trajectory data with an appropriate filtering method provides a solution for small irregularities in the trajectory data without compromising the tracking results [16]. After smoothing, it is possible to compute the velocities and accelerations, i.e., the first and second derivatives of the position, with greater accuracy.

To process large amounts of video data, it is advantageous to automatically detect tracking failures, i.e., cases where the tracking is lost or the tracking window drifts away from the target. When a failure is detected, either the tracking can be repeated with a different tracking method, or the incorrect trajectory can be excluded from further analysis. A common approach to detect failures is to use backtracking and compare the tracked target to an earlier sample of the object (e.g., [7, 12, 23]). Such methods perform well when the tracking is lost causing large displacements between the tracking result and the actual object location. However, when a tracker slowly drifts away from the target, failure detection with backtracking methods becomes more challenging.

To address the above issues and requirements, we present a multi-camera framework for measuring hand movements in HCI studies, focusing on touch and gesture user interfaces. The framework is developed for a measurement setup consisting of a high-speed camera and a normal-speed camera with different viewing angles. The high-speed camera makes it possible to detect fast and subtle changes in the trajectories, while the normal-speed camera provides the necessary additional information to construct the 3D trajectories. The framework includes the construction of the measurements setup, selection of object-tracking methods, detection of tracking failures, postprocessing of the trajectories, and characterization and visualization of the movement data. The framework is generic in nature, and in this work, it is demonstrated with an application in which 3D touch screen usability is studied with 3D stimuli.

## 2  Overview of the Framework

An overview of the proposed hand movement measurement framework for the HCI studies is shown in Fig. 1. The dashed line in the figure represents the use of the camera calibration results in the computation of the real-world features. The first step is to design and to build the measurement setup which comprises cameras, illumination, a display, and other interacting devices, and the required hardware for triggering and storing recordings. The main considerations when designing an HCI measurement setup are that it should not interfere with the usability of the user interface and that it should offer a natural setting for test subjects performing the selected HCI task. This aspect is particularly important in the design of the illumination because a high-speed imaging setup requires a lot of light and bright illumination can disturb test subjects. Moreover, the illumination should not result in flickering on the recorded videos, so the available flicker-free light sources should be used, including LED light panels with reliable and constant power sources and Hydrargyrum medium-arc iodide (HMI) lamps, where flicker can be avoided by using electronic ballast that operate at high frequencies.

The second step in the framework is to geometrically calibrate the cameras by determining the intrinsic and extrinsic camera parameters to obtain the mapping from the image point locations to the real-world coordinates. The pinhole camera model parameters can be determined by using known and imaged interest point coordinates of an imaging target designed for calibration. These parameters can then be used to transform the image point locations to real-world coordinates via a perspective

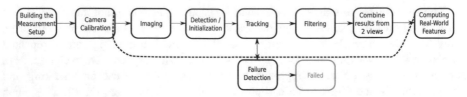

**Fig. 1.** Overview of the measurement framework

projection. The pinhole camera model can be further enhanced by taking into account the lens and sensor distortions.

After imaging the object of interest, the object needs to be detected before its movement can be tracked. In a typical controlled HCI study, the hand or finger movement starts from a static trigger box or another predefined location. However, if the initial position is unknown, a detection component is needed before tracking. The detection can be performed using state-of-the-art detection methods, such as Faster R-CNN [19] or YOLO9000 [18]. If the background is static, a simple method, such as frame differencing or background subtraction, can be used for the detection. The initialization of the object position has an important role in the tracking process since a typical tracking method utilizes the initial position to generate the object model used for tracking.

Tracking is applied in order to follow the position of the detected or otherwise initialized target object while it is moving. In general, the idea is to repeatedly estimate the transformation of an object from time step $t$ to $t + 1$, i.e., from one image frame to the next one. In most cases, the transformation is simply the translation of an object. However, there are situations where a more advanced motion model that takes into account, e.g., rotation, skew, and scale changes, is required. An extensive comparison of object-tracking algorithms for measuring hand movements in the HCI study is presented in [16] with the high-speed videos and in [17] with the normal-speed videos.

Extracting higher-level features from the tracking results can be challenging [16]. Although a list of center locations of an object over time produced by tracking is usable for tasks such as checking the position of an object at a certain time, sub-pixel accuracy is preferred when derived quantities, such as velocity or acceleration, are required. Typical object trackers operate at the pixel level and the resulting trajectory often contains noise. The desired level of spatial accuracy and the noise cause challenges for the determination of derived quantities, such as velocity and acceleration. High-speed videos can be challenging where movements between the frames are very small (often less than a pixel). Consequently, filtering of the trajectories is required. Finally, to reconstruct hand trajectories in 3D, the tracking results from two views obtained from the normal and high-speed videos are combined and various features of the trajectories are computed.

## 3  Stereoscopic 3D Touch Display Experiment

The framework is demonstrated with an HCI experiment using a stereoscopic 3D touch screen setup. In the experiment, test subjects were advised to perform intentional single-finger pointing actions from a trigger-box toward a target that was on a different parallax than others on the touch screen. Hand movements were recorded with a high-speed camera and a normal-speed camera. The trigger-box and the touch screen were placed on a table as shown in Fig. 2. The flow of the experiments was controlled by a middleware program specially coded for the experiments. A detailed description of the setup can be found in [15].

Similar to earlier pointing action research, e.g., [3], the experiment focused on studying intentional pointing actions. The stimuli were generated by a 3D display with the touch screen to evaluate the effect of different parallaxes, i.e., perceived depth. This arrangement enables study of (potential) conflict between visually perceived and touch-based sensations of the depth.

The 3D stereoscopic touch screen experiment was executed as follows. 20 test subjects conducted 4 different sessions of pointing actions with different parameters. These were divided into nine blocks. The test image contained a fixation cross in the middle of the screen and 10 rectangular blocks around it in a circle formation. The aim of the experiment was to locate and touch the target that appears on a different parallax to the others in the test image.

The high-speed videos were recorded at 500 fps and 800×600 resolution. The normal-speed videos were recorded using interlaced encoding with 50-field rate and 1440×1080 (4:3) resolution. For deinterlacing the normal-speed videos, yet another deinterlacing filter (yadif) [1] was utilized with field-to-frame conversion producing double frame rate (50 fps) videos. In total, 2597 pointing actions were recorded with both cameras.

### 3.1  Camera Calibration

To calibrate the cameras, a standard calibration board with 26.5 mm checker-board patterns was used. A set of calibration images was captured and used to compute the

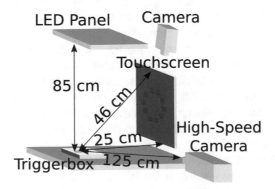

**Fig. 2.** 3D touch display experiment

intrinsic camera parameters. The Camera Calibration Toolbox for Matlab [2] was used to perform the calibration, as it is a robust and well-established calibration tool, based on [24] and [9].

## 3.2  Hand Tracking

Since trackers specifically designed for hand tracking rely on color information whereas gray-scale imaging is used in this work the selected state-of-the-art general object trackers were utilized. Based on the comprehensive evaluation on the same video dataset reported in [16], Kernelized Correlation Filters (KCF) based tracker [10] was selected for tracking in high-speed videos. The tracking window was initialized by a manually placed initial bounding box on the trigger box button image.

The normal-speed videos were processed with motion detection near the monitor area. The motion detection was performed using background subtraction (frame differencing). The detected motions were used to obtain the location of the finger tip which was further used to initialize the tracking window for the normal-speed videos. Comprehensive evaluation of state-of-the-art object trackers for finger tracking from normal-speed videos with the presented experimental setup has been provided in [17]. Based on the results, the KCF tracker extended by a scale estimation and color-names features (KCF2) [22] was selected as a normal-speed video tracker for the final measurement framework.

## 3.3  Trajectory Post-processing

*Failure Detection.* In situations where a highly robust tracking system is required or massive datasets are processed, there is a need for a failure detection system to identify failed trajectories as it was identified in [14]. One of the methods to detect tracking failures is to use backtracking to estimate the trajectory from the current point to the beginning of the tracking, or another earlier point of time, and to check if the backtracked trajectory matches the original "forward-tracked" trajectory [7]. Other methods include gathering samples of the earlier appearances of the object and comparing them to the currently tracked window using similarity measures [12, 23].

Typically, failures are easier to detect when the drift is large. However, when the tracker slowly drifts on the target, it is more difficult to detect the failure, and the above-mentioned methods become unreliable, especially if high spatial accuracy is desired. In HCI studies, the end point of the trajectory is often known. Moreover, in some studies, the start point of the trajectory is also known, for example, a trigger-box button. In the touch screen experiment, for example, the point on the screen that the test subject touches is known, and this information can be used to implement a reliable method to detect failures in tracking. When failures are detected, either the tracking can be repeated with another tracking method, or the incorrect trajectory can be excluded from the further analysis.

If the end position of the trajectory is unknown, a reliable backtracking or drifting detection method should be applied. In the HCI studies, such methods as good features to track [20] and metrics for the performance evaluation of video object segmentation and tracking without the ground-truth [4], based on earlier templates of an object, work relatively well since the target object is usually a hand or a finger. These objects contain well identifiable features that can be used to detect if the tracker loses the target. Moreover, the object detection methods used for the tracker initialization can be applied for the last frames to test if the end point of the tracked trajectory contains the correct object.

It should be noted that many of the tracking failure cases could be avoided by giving the test subjects precise instructions and by ensuring sufficient practice before the actual data collection, so that the test subjects are comfortable with the task. Erroneous behavior by test subjects can include, for example, test subjects withdrawing their hand from the touching position before the recording ends, incorrect positioning of the hand in the beginning of an individual test, and obstruction of the pointing finger with other fingers.

In the 3D touch display experiment, the tracking failure detection method needed to be able to reliably process a large amount of trajectories. The implemented failure detection system was based on the fact that the trajectory had to end within a specific area of the projected touch screen point. If the correct end point was not reached with the default gray-level features used by the KCF tracker, the tracking was repeated with more computationally demanding HOG features. If the tracking failed again, the trajectory was considered as incorrect and was excluded from the further analysis.

*Trajectory Filtering.* Based on the results of the high-speed trajectory filtering in [16], the LOESS filter was also selected for the normal-speed trajectories. The filtering window size was selected to be the same 80 ms as in high-speed case. This translates into 4 window size samples with 50 fps. Comparison between different window sizes can be found in [17].

*Video Synchronization.* In order to automatically align the normal-speed videos with the high-speed videos, the ratio of the framerates and the delay (difference between the camera-produced time information) were determined in [15]. The synchronization process uses timestamps from the high-speed videos and known starting time of the normal-speed videos to coarsely align the videos and to identify blocks of corresponding actions from both videos. The known location from both views are used to set up an event which can be then used to align the video sequences accurately. A more detailed explanation of the video synchronization method is given in [15].

*Reconstruction of 3D Trajectories.* To obtain a 3D trajectory, the 2D trajectories estimated using the calibrated cameras with a different viewpoint need to be combined. The task of computing a 3D trajectory from multiple 2D trajectories is essentially equivalent to the process of 3D scene reconstruction. For this purpose, we utilized the 3D reconstruction method presented in [8]. Detailed explanation of the 3D reconstruction and results are available in [17] and [15].

# 4  Data Analysis

## 4.1  3D Trajectory Reconstruction

The success rate of the finger tracking was measured as the proportion of trajectories which reached the predefined end points. For the high-speed videos, the end points were the touch target areas reprojected onto the image plane, and for the normal-speed videos, the defined end point was the trigger-box button. 77% of point actions were tracked correctly from the high-speed videos and 69% from the normal-speed videos. In total, 1237 (62%) of the pointing actions were correctly tracked from both videos and were aligned correctly. Since there was no ground truth for the 3D trajectories, the 3D reconstruction accuracy was assessed by using the re-projection error measure [8]. The mean re-projection error over all the trajectory points used from 1172 videos in the 3D reconstruction experiment was 31.2 pixels. This corresponds to approximately 10 mm in the real world.

## 4.2  Trajectory Features

When hand movements are considered in HCI studies, the most important measurements are the velocities and accelerations of the hand [3]. Velocity and acceleration can be computed as the first and second derivatives of the position with respect to time of using the tracked hand trajectories and Euclidean distances. Trajectory filtering makes it possible to compute the velocity using differences in consecutive trajectory points, and the acceleration can be computed using consecutive filtered velocity points.

For visualization purposes, velocity and acceleration curves can be plotted with respect to either time or position. For example, in the 3D touch screen experiment, the distance from the fingertip to the monitor surface is a useful measurement. It should be noted that this measurement fails to capture a movement that occurs in a plane parallel to the screen which is not crucial when simple intentional pointing actions are studied.

In a typical experiment, individual movement trajectories vary considerably. To detect small recurring events and phenomena, as well as to identify inter-subject behavioral differences in slightly different tasks, it is important to be able to analyze and visualize a large number of trajectories. One option is to determine the average position, velocity, and acceleration curves. In order to do this, the different trajectories need to be normalized in such a way that the mean values could be computed for a certain moment of time or a certain position.

In the 3D touch screen experiment, the trajectories were normalized, so that all of them were at a distance of 250 mm from the touch screen measured from the initial finger position at the trigger-box. The visualization of the grouped acceleration and speed plots was used to detect submovement intervals of the trajectories similarly to [3]. The primary submovement starts with the initial acceleration, and ends when there is a sign change from negative to positive. This is the starting point of the secondary submovement of intentional pointing actions where minor adjustments to the trajectory are made and the movement is fixed to the final target position. Similar submovement events can be seen in the visualization of the 3D touch screen experiment in Fig. 3. The results indicate that differences in acceleration and speed of the pointing actions when using different parallaxes are small but, nevertheless, observable. Figure 4 shows the velocity and acceleration curves for the last 25 mm before the touch display. There are small differences in velocities and accelerations in movements towards different disparities, so parallax information seems to slightly affect the hand trajectories.

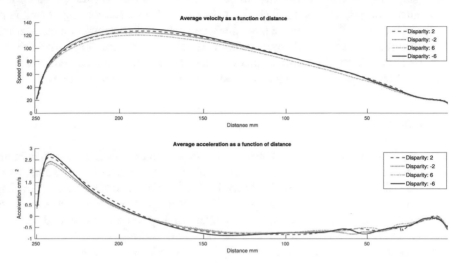

**Fig. 3.** Averaged results of all test subjects performing the 3D touch screen experiment with disparities −6, −2, 2, and 6.

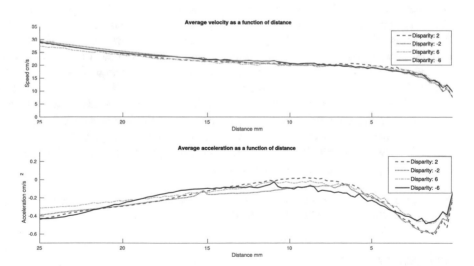

**Fig. 4.** Last 25 mm of the averaged velocity and acceleration curves with disparities −6, −2, 2, and 6

In [15], eleven features were computed from the obtained trajectories: the mean velocity, median velocity, maximum velocity, maximum velocity during the 2nd submovement, maximum acceleration during the 2nd submovement, mean velocity during the 2nd submovement, and mean acceleration during the 2nd submovement. Moreover, a two-sample T-test with 5% significance level was used to analyze the trajectory features. As expected, it was concluded in [15] that the small disparity changes, 2 and −2, had only minor impact on the hand movements according to the computed features, whereas the disparity values (6 and −6) had more significant impact to the movements. Moreover, the large positive disparity 6 (the target object in front of the screen) seemed to have a more prominent effect on the pointing actions than the others. Furthermore, the velocity features seem to be better than the acceleration features to distinguish the pointing actions towards different disparity values.

## 5   Conclusion

A framework for measuring hand movements, in particular, pointing actions, in human-computer interaction situations using a multiple-camera system containing a high-speed and normal-speed cameras was introduced. Suitable object trackers to perform finger tracking in both high and normal-speed videos were proposed based on earlier comprehensive studies. The selected KCF tracker for high-speed videos and KCF2 for normal-speed videos perform well in the given hand-tracking task, achieving low error rates and operating at high processing speeds. In order to process large amounts of videos, we proposed a tracking failure detection method that excludes incorrect trajectories from further analysis, including cases where the test subject failed to follow the given instructions. By using trajectory filtering, the tracked trajectories could be smoothed to obtain reliable acceleration and velocity curves for visualization purposes. Finally, a method to construct a 3D trajectory from two 2D trajectories was proposed. The framework provides real-time processing speeds. Video loading, tracking, filtering and visualization included, the processing speed of the framework was on average around 100 fps for high-speed videos. The framework was demonstrated in a context where 3D touch screen usability was studied with 3D stimuli. Some feature correlation with different parallaxes were already detected, but deeper analysis of the effects of different parallaxes on the trajectories is planned for the future research. The work provides valuable information about the suitability of general object tracking methods for high-speed hand-tracking while producing appropriate velocity and acceleration features computed from filtered tracking data.

**Acknowledgements.** The authors would like thank Dr. Jari Takatalo for his efforts in implementing the experiments and producing the data for the research.

## References

1. FFmpeg (2018). https://mpeg.org/. Accessed 01 May 2018
2. Camera calibration toolbox for Matlab (2016). http://www.vision.caltech.edu/bouguetj/calib_doc. Accessed 03 Feb 2016

3. Elliott, D., Hansen, S., Grierson, L.E.M., Lyons, J., Bennett, S.J., Hayes, S.J.: Goal-directed aiming: two components but multiple processes. Psychol. Bull. **136**(6), 1023–1044 (2010)
4. Erdem, C.E., Tekalp, A.M., Sankur, B.: Metrics for performance evaluation of video object segmentation and tracking without ground-truth. In: Proceedings of 2001 International Conference on Image Processing, pp. 69–72 (2001)
5. Erol, A., Bebis, G., Nicolescu, M., Boyle, R.D., Twombly, X.: Vision-based hand pose estimation: a review. Comput. Vis. Image Underst. **108**(1–2), 52–73 (2007)
6. Guna, J., Jakus, G., Pogacnik, M., Tomazic, S., Sodnik, J.: An analysis of the precision and reliability of the leap motion sensor and its suitability for static and dynamic tracking. Sensors **14**(2), 3702–3720 (2014)
7. Hariyono, J., Hoang, V.D., Jo, K.H.: Tracking failure detection using time reverse distance error for human tracking. In: International Conference on Industrial, Engineering and Other Applications of Applied Intelligent Systems, pp. 611–620 (2015)
8. Hartley, R.I., Zisserman, A.: Multiple View Geometry in Computer Vision, 2nd edn. Cambridge University Press, Cambridge (2004)
9. Heikkilä, J., Silven, O.: A four-step camera calibration procedure with implicit image correction. In: Conference on Computer Vision and Pattern Recognition, Washington, DC, USA, p. 1106 (1997)
10. Henriques, J.F., Caseiro, R., Martins, P., Batista, J.: High-speed tracking with kernelized correlation filters. IEEE Trans. Pattern Anal. Mach. Intell. **37**(3), 583–596 (2015)
11. Hiltunen, V., Eerola, T., Lensu, L., Kälviäinen, H.: Comparison of general object trackers for hand tracking in high-speed videos. In: International Conference on Pattern Recognition (ICPR), pp. 2215–2220 (2014)
12. Kalal, Z., Mikolajczyk, K., Matas, J.: Forward-backward error: automatic detection of tracking failures. In: 20th International Conference on Pattern Recognition, pp. 2756–2759 (2010)
13. Khoshelham, K., Elberink, S.O.: Accuracy and resolution of kinect depth data for indoor mapping applications. Sensors **12**(2), 1437–1454 (2012)
14. Kuronen, T.: Moving object analysis and trajectory processing with applications in human-computer interaction and chemical processes. Ph.D. thesis, Lappeenranta University of Technology (2018)
15. Kuronen, T., Eerola, T., Lensu, L., Kälviäinen, H.: Two-camera synchronization and trajectory reconstruction for a touch screen usability experiment. In: Advanced Concepts for Intelligent Vision Systems, pp. 125–136 (2018)
16. Kuronen, T., Eerola, T., Lensu, L., Takatalo, J., Häkkinen, J., Kälviäinen, H.: High-speed hand tracking for studying human-computer interaction. In: Image analysis. Proceedings of 19th Scandinavian Conference, SCIA 2015, Copenhagen, Denmark, 15–17 June 2015, pp. 130–141. Springer, Cham (2015)
17. Lyubanenko, V., Kuronen, T., Eerola, T., Lensu, L., Kälviäinen, H., Häkkinen, J. Multi-camera finger tracking and 3D trajectory reconstruction for HCI studies. In: Advanced Concepts for Intelligent Vision Systems, pp. 63–74 (2017)
18. Redmon, J., Farhadi, A. Yolo9000: better, faster, stronger. arXiv preprint, arXiv:1612.08242 (2016)
19. Ren, S., He, K., Girshick, R., Sun, J.: Faster R-CNN: towards real-time object detection with region proposal networks. IEEE Trans. Pattern Anal. Mach. Intell. **39**(6), 1137–1149 (2017)
20. Shi, J., Tomasi, C.: Good features to track. In: Conference on Computer Vision and Pattern Recognition, pp. 593–600 (1994)
21. Valkov, D., Giesler, A., Hinrichs, K.: Evaluation of depth perception for touch interaction with stereoscopic rendered objects. In: ACM International Conference on Interactive Tabletops and Surfaces, pp. 21–30 (2012)

22. Vojir, T.: Tracking with Kernelized Correlation Filters (2017). https://github.com/vojirt/kcf/. Accessed 01 May 2018
23. Wu, H., Chellappa, R., Sankaranarayanan, A.C., Zhou, S.K.: Robust visual tracking using the time-reversibility constraint. In: IEEE 11th International Conference on Computer Vision (ICCV), pp. 1–8 (2007)
24. Zhang, Z.: Flexible camera calibration by viewing a plane from unknown orientations. In: International Conference on Computer Vision, pp. 666–673 (1999)

# Functional Modeling of an Integration Information System for Building Design

Alexander Bukunov[(✉)]

Peter the Great St. Petersburg Polytechnic University, Saint Petersburg, Russia
sasbukunov@yandex.ru

**Abstract.** The processes at the construction stages are characterized by considerable complexity due to the large amount of information contained in the design and construction documentation, and its interaction. Information management requires engineering data management systems and an object information model. Technology of building information modeling is based on the development and use of a virtual model of a construction object in the form of a three-dimensional information model and related documents. The article analyzes the Polterovich-Tonis evolutionary model as applied to information modeling processes in Russia. The functional approach to the modeling of complex project management systems in construction and visual modeling of construction objects is considered. Links between various design and calculation systems, estimated systems, scheduling and resource management systems are revealed. A formal description of the conceptual model of an integration information system for design options in construction and a strategy of data exchange between the system for automation of accounting and control and the BIM-system of visual presentation was developed.

**Keywords:** Functional modeling · Building information modeling (BIM) · Virtual model · Information system

## 1 Introduction

Building Information Modeling (BIM) is actively applied in the construction industry for visualizing, modeling and imitating the design and characteristics of buildings [1–3]. While modeling, exchange of information between all interested parties, design coordination, task coherence, collision detection, and monitoring of the management process are carried out. The technology is based on the development and use of a virtual building model in the form of an information parametric model and related documents. Such a model is created at the early stages of the project, gradually being updated with information. The implementation of BIM contributes to the access of used information, controlled coordination and monitoring of processes [4–6].

© Springer Nature Switzerland AG 2020
D. G. Arseniev et al. (Eds.): CPS&C 2019, LNNS 95, pp. 525–535, 2020.
https://doi.org/10.1007/978-3-030-34983-7_51

## 2 Rationales of the Research

Parametric modeling is based on an object-oriented approach that allows to reuse elements of an object with attributes and constraints while building models. Building visualization reduces rework. Due to BIM, it is possible to analyze and model building characteristics, for example, cost estimates, energy consumption, lighting analysis, etc. [7]. Analysis of the performance of the building provides a functional assessment of the building models before the construction [1]. This allows to choose the most cost-effective and sustainable solution when comparing alternative project options. It is advisable to create information processing systems in the design of construction in the form of integrated databases (DB). This is related to the expansion and increase of structural complexity of the processed data, growth of the range of data users and the increase in the functional requirements for information processing systems [8]. Most of the proposed approaches, based on traditional data models (hierarchical, network and relational), complicate the process of creating information processing systems due to the complexity of describing information about the construction objects. Currently, multidimensional data models are intensively developed and focused on storing and processing large volumes of time-varying information.

Information modeling in industry and construction is investigated in the works of Succar, Steel [9, 10], Kennan [11], Codd [12, 13], Chen [14], and in the works of Russian scientists Ginzburg, Talapov [15, 16], Venderov [17], Ignatova [18], Antonov et al. [19, 20]. The methods proposed in these works do not allow to fully solve the problems of organizing an enterprise information flow system.

The backlog of construction in terms of the intensity of integration of information systems can be explained by its specificity related to the diversity of financial and economic activities of its enterprises and features of the production cycle, where a significant number of projects are at various stages of implementation [20]. The choice of the research topic is stipulated by the scientific and practical significance of the problems noted above.

## 3 Aims of the Study

The aim of the study is to increase the efficiency of the information management process at the design stage of construction through the method of functional visual modeling and the formation of the knowledge base for the design, constructing and management tasks. To achieve this goal, the study addressed the following tasks: evaluation of the concept of horizontal BIM transfer in terms of the theory of Polterovich-Tonis [21] as applied to construction; studying of the functional approach to modeling of complex control systems for the creation of design solutions in construction; analysis of visual modeling of construction objects. In the course of the work, connections between various design and settlement systems, cost estimates, scheduling systems and resource management systems were identified. The features of the knowledge base for design tasks were defined; a formal description of a conceptual model of an integrated information environment for design decisions in construction and a data exchange strategy between the system for automating accounting and management and the BIM-system of visual presentation was developed.

## 4   Evolutionary Model Suggested by Polterovich and Tonis

At the turn of the twentieth and twenty first centuries, the use of technology transfer strategies has begun in a number of industries: automotive, electronics, communications and connections, production of building materials, etc. The study focused on the BIM process of technological transfer. The diffusion model of distribution is one of the main models of innovation. Let's consider a differential equation to describe such a model.

$$dX(t)/dt = bX(t) \cdot [N - X(t)]/N \tag{1}$$

where $X(t)$ is the number of entities that have implemented innovation by time $t$; $N$ is the number of potential consumers of innovation; $b$ is the parameter of the speed of innovation.

The solution of Eq. (1) is shown in Fig. 1. The graphical solution looks like an S–curve or logistician. This pattern of the development of the innovation market is considered to be the main one.

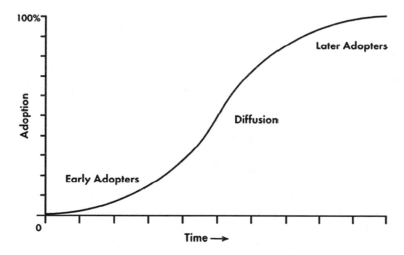

**Fig. 1.** The diffusion of innovations

Due to the specifics of the construction industry, transfer of innovative technologies at industry enterprises is described using a two-vector model, according to which the diffusion of innovations occurs in two directions: horizontal and vertical. In organizations with monotypic products, the spread of innovation occurs along the vertical vector y. In construction associations, innovations are spread along the horizontal vector x in accordance with the engineering process [4].

The processes of borrowing and birth of innovations are related according to the evolutionary model of Polterovich–Tonis [21]. The main conclusions based on the model are as follows: availability of advanced technologies and their development are becoming more expensive over time; to reach a new technological level, it is necessary to study already developed methods [22].

Companies adopting developments of the technology leader increase their innovativeness. Rapid implementation of innovations by the industry leader automatically activates imitation strategies of other market players. For the majority of competitors, it is unprofitable to invest in development at the stage of their backlog [22]. Figure 2 shows the dependence of the specific capital costs from the achieved level of technological development $i$.

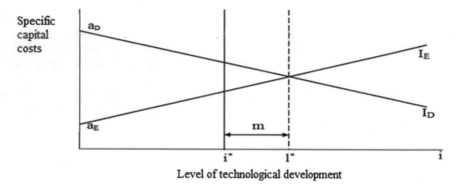

**Fig. 2.**  Investment (technological) dichotomy

To describe these dependencies, the authors of [22] proposed the following formulas:

$$I_{iE} = 0.413 + 0.624i, \qquad (2)$$

$$I_{iD} = 0.997 - 0.325i, \qquad (3)$$

where $I_{iE} = K_E/x_0$ and $I_{iD} = K_D/x_0$ are unit costs for borrowing and the creation of new technologies, respectively; $K_E$ are the capital costs of borrowing new technologies; $K_D$ are capital costs of creating new technologies; $x_0$ is the volume of production in physical terms at the initial moment of time; $i$ is the achieved level of technological development.

The Polterovich–Tonis model is built on the dichotomy of investment regimes based on two multidirectional effects: adaptation and learning. When the level of technological development ($i$) approaches the technological boundary ($i^*$), a transition from an adaptation development strategy to an innovation strategy takes place. As applied to BIM, the analysis of the model showed that the effects of borrowing and creation produce a technological frontier, in the area of which the borrowing regime will change to the mode of creating technologies through a technological leap.

According to this theory, prior to switching from one development path to another, the transfer of BIM technologies to Russia is required. An essential condition for transition from one level to another through a technological leap are objective preconditions developed in the course of effective training based on modern foreign training programs implementing BIM and using the experience of BIM implementation, as this creates a border where borrowing will be inevitably replaced by creation of technologies.

## 5  Methods and Modeling Methodology

Simulation methods, one of which is BIM, make it possible to solve problems of analyzing large systems, including assessment tasks: variants of the system structure, the effectiveness of various system control algorithms, the effects of changes in various system parameters. The solution obtained with the help of BIM corresponds to certain parameters of the system, which have to be repeatedly changed to reach the optimal solution while the modeling process [23, 24]. Consequently, the imitation model does not allow to calculate the optimal state of the system, but only makes it possible when conducting numerical experiments on a computer, to study the system, predict its development, bring its structure or characteristics closer to optimal ones.

The functional Structured Analysis and Design Technique (SADT) methodology for structural analysis and design of business processes is one of the reviewed methodologies and analyzed approaches, used in data and object modeling [25]. The development of SADT is related to the creation of the ICAM (Integrated Computer Aided Manufacturing) Definition (IDEF) methods for analyzing and studying the structure of the organization of processes and the connections between information flows. The essential requirement in the development of the methodology is the possibility of effective exchange of information among all the participants of the modeling process [26].

The use of graphical notations by structural methods is used to bring the construction process, as a complex system, to the desired state. For structural analysis, it is important to display: (1) the functional structure of the system, sequence of processes, transfer of information between the elements of the functional structure; (2) the connections between the data; (3) dynamic system behavior. An object-oriented approach to modeling processes in construction is developing as a set of classes and objects of the domain based on the use of the Unified Modeling Language (UML).

The proposed integration functional model of building design management is based on the information model of the project and the joint integrated work of the project team on it. It is not quite correct to speak of a pure functional model, as we do not use functional relations in it. But since the functional model is represented graphically (block diagrams of the system operation process, work flow diagrams), it is more expedient to use the term of "structural-functional model" in the field of modeling and system analysis [26].

# 6  Methods of Information Platform Development

During the study, the interaction of elements in the operation of a complex system, one of which was the design management system in construction, was considered as the result of mutual influences of system elements that received input signals and produced an output. When modeling our complex system of designing a construction object, we, at first, performed decomposition and then, each part considered separately, we combined the information together.

To improve management of the design in construction on the basis of the structural-functional approach, a model of an integration information environment for the design preparation of buildings was developed. The structural-functional model of a single information space for an automated system for designing building production is presented in Fig. 3; it contains a description of the system of objects using interrelated computer tables, flowcharts, diagrams, and visual objects that display the structure and the connections between the elements of the object. It is suggested to single out two basic informational structures in the design of a construction object: a system for automating accounting and control and a BIM system connected by the most important information flows. Graphical presentation of information has a number of features that make its use preferable. Such features are an informative capacity of graphic objects, high speed of perception of images, visibility of connections between graphic objects (for example, relative positioning).

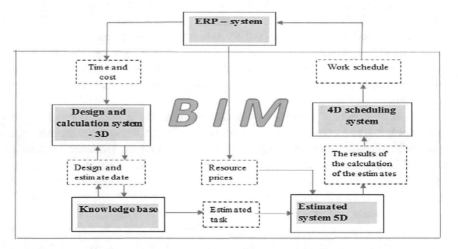

**Fig. 3.** BIM-system in the structure of a single information space of the construction industry

The use of Enterprise Resource Planning system (ERP) in the construction is difficult due to the length of the production cycle. The feature of ERP solutions for construction companies is that a large number of documents, orders, deliveries, contracts can be involved in a single project; different stages of the project are carried out in parallel. ERP is implemented to improve the quality of planning, reduce inventory, increase flow rate and turnover, produce in time, and increase the profitability of projects [27, 28].

Implementation of modern projects is impossible without adequate planning, effective control and management. For these purposes, informational project management tools based on well-known mathematical methods for calculating critical works, estimating the used volume and analyzing software risks are typically used. In the unified environment of the ERP construction project, the system is connected through information flows to the components of the BIM system, i.e., design and calculation systems for the architecture, structures and engineering networks, cost estimate system, scheduling and knowledge base (Fig. 1). The knowledge base links project engineering models and acts as a transmitter of the design decisions to estimated and financial decisions and a data accumulator for the decision-making analytical center [29].

Quantities and components of the model are extracted for all structures from the 3D model of the building. Then the interrelationships of resources, duration of work, norms of development and definition of logical relations between them are set. Calculation of the data for the formation of the construction schedule and automatic connection of the components of the 3D model with the parameters of work for creating the schedule complete the creation of a 4D model. A link with a database of similar previous projects is established [30].

4D models expand the capabilities of 3D models and create additional benefits, primarily due to the fact that they contain a work plan in the form of a calendar-network diagram. The result is a visual work plan, which helps to improve the understanding of team members. One of the main advantages of such models is the ability to test and improve existing versions of the project work plan. Synthesis of the calendar schedule and model of the building allows to check with the help of special tools how accurate the building construction process has been [30].

When conducting research, the method of regression determination of the duration of the stages of works [31] in the form of a quadratic regression dependence was used.

$$T = X + Y \cdot F + Z \cdot N^2 \tag{4}$$

$N$ is the number of floors, $F$ is the total area, $T$ is the cycle duration.

According to the results of processing statistical arrays of brick, monolithic and panel buildings, coefficients $X$, $Y$ and $Z$ included in the regression dependence (4), are presented in Table 1.

It is possible to link each structural element and, equipment with a temporary stage and create a schedule of work using the classifier. Then it is possible to watch all the processes in dynamics, identify inconsistencies or positions to optimize the construction. We have an opportunity to make a fairly wide range of data that cannot directly affect the building model but significantly affects the construction. This is the location of the crane, the number of cars that can drive through the construction site per day, and much more. Everything put together allows us to identify possible shortcomings in logistics and correct them at a stage when the construction process has not yet begun.

With the help of plug-ins in BIM, it is possible to carry out calculation of estimates, using the integration into a single model of design programs and audits using APIs from various applications. The special software environment of the knowledge base

**Table 1.** Coefficients of regression equations for calculating the duration of work cycles

| Type of building | Cycle of works | X | Y | Z |
|---|---|---|---|---|
| Brick | Zero cycle | 0.885 | 0.000067 | 0.00000 |
| | The above-ground part | 2.622 | 0.000417 | 0.00284 |
| | Underground part | 1.505 | 0.000047 | 0.00000 |
| Monolithic | Zero cycle | 0.570 | 0.000106 | 0.00000 |
| | The above-ground part | 3.041 | 0.000364 | 0.00354 |
| | Underground part | 0.837 | 0.000118 | 0.00000 |
| Panel | Zero cycle | 0.856 | 0.000049 | 0.00000 |
| | The above-ground part | 2.284 | 0.000168 | 0.00265 |
| | Underground part | 0.872 | 0.000049 | 0.00000 |

allows to combine project data from different systems into a single estimated project. Through the knowledge base software environment, a connection is made with the 3D model for transmitting results of the cost calculations. The generated estimate task is transferred to the 4D model (estimated software complex).

The package of documents on the object is formed at the output of the estimated system. Having an opportunity to programmatically create a model is important for an estimated model audit. The 3D-model allows attaching a passport with attributed data to an object, forming a reference search system. This creates an archive of documentation with a convenient document search interface.

The scheduling system of the construction project determines the work tasks, their duration, the sequence of their execution; it calculates the construction technology and evaluates the necessary resources. When typical projects are created, previously created documents can be used as templates. This makes it possible to determine the types of project work, calculate their duration, and carry out project scheduling more effectively.

The technology of creating a unified environment requires access of various groups of specialists (engineers, estimators, architects, designers, builders, analysts and experts) to the same model representations. Moreover, it is necessary to separate access to project information and exclude unauthorized access and change of project information. The architecture of the integrated environment should provide for information exchange of data among different categories of users.

The data exchange strategy, in accordance with the software, allows integrating calculations into a 3D model. The transfer of parametric information of the geometry of objects is possible with the use of application-programming interfaces (API) managed with programming languages (VBA, C #, etc.). The model on a united Bentley and Autodesk platform has the advantages. The closed file format allows data exchange between applications, which ensures effective coordination of information.

The initial information in the construction industry after the design stage is the operational input information, regulatory reference information, the output information of engineering and economic training in the construction organization. Tools for extracting information from an information system today are data warehouses, knowledge bases, online analytical processing of OLAP applications and Data Mining tools.

The impossibility of simultaneous automating of all the functional procedures of the construction stages due to their algorithmic complexity requires the architecture of the integrated environment to allow gradual addition of new functional and management modules as they are developed and has a modular structure with the option to add, update and delete software modules when implementing corresponding procedures.

## 7  Results

As a result of the research:

- A new scheme of communication between designers, builders and users of structures was elaborated. Communication was carried out at the model level through its various parts, using the information already included in the model and stored in files containing data of models, unique for these systems, though having their own, often incompatible formats. Three ways to solve the issue are proposed: a single IFC description language, direct APIs, development of a single platform.
- The approach to virtual information modeling is considered. It puts virtual computer models into an intermediate position between mathematical models and simulated physical objects.
- A model integrating BIM and a resource management system is proposed. It provides an integrated platform using the information modeling and supporting data mining through creation of a storage for building data.
- It was concluded that BIM borrowing would be inevitably replaced by creation of own technologies in Russia based on the analysis of the evolutionary model suggested by Polterovich and Tonis with regard to BIM processes.

## 8  Conclusion

Information modeling brings design and construction to a new technological level. It makes it possible to increase the efficiency of the decisions made but requires transfer of the new information to digital form, as well as its repeated use at various stages of the object's life cycle.

BIM allows you to collect information on building requirements, planning, design, construction and operation and use it to make decisions regarding the management of objects. In this case, all departments can embed relevant information about the project into a single model. For example, the estimates, the schedule of the project, information about the management of the object are included in a single building model. Information affects modeling, visualization, and simulation, which helps identify problems associated with the design, construction, and operation before they actually occur. Parametric BIM with the accumulation of information about the life cycle of the building expands to 4D, 5D, 6D, etc. Preliminary processing of information through the entire life cycle of buildings is important for effective management of objects. This will increase the efficiency while reducing project costs and duplication of efforts. Although it takes more time to create a unified model, its benefits outweigh the costs.

# References

1. Eastman, C., Teicholz, P., Sacks, R., Liston, K.: BIM Handbook, 2nd edn. Wiley, Hoboken (2011)
2. Azhar, S., et al.: Building information modeling for sustainable design and leed (R) rating analysis. Autom. Constr. **20**(2), 217–224 (2011)
3. Azhar, S., et al.: Building information modeling (BIM): now and beyond. Australas. J. Constr. Econ. Build. **12**(4), 15–28 (2012)
4. Tardif, M., Smith, K.D.: Building Information Modeling: A Strategic Implementation Guide. Wiley, Hoboken (2009)
5. Gudgel, J.: Building Information Modeling: Transforming Design and Construction to Achieve Greater Industry. McGraw-Hill SmartMarket Report, p. 42 (2014)
6. Abbasnejad, B., Moud, H.: BIM and basic challenges associated with its definitions, interpretations and expectations. Int. J. Eng. Res. Appl. **3**(2), 287–294 (2013)
7. Manning, R., Messner, J.: Case studies in BIM implementation for programming of healthcare facilities 13 (Special Issue – Case Studies of BIM Use). ITcon – IT in Construction. ITcon (2008)
8. Creswell, J.W.: Qualitative Inquiry and Research Design: Choosing among Five Approaches, 3rd edn. Sage Publications, London (2013)
9. Succar, B.: Building information modelling framework: a research and delivery foundation for industry. Autom. Constr. **18**(3), 357–375 (2009)
10. Steel, J., Drogemuller, R., Toth, B.: Model interoperability in building information modelling. Softw. Syst. Model. **11**, 99–109 (2012)
11. A Forum for BIM Collaboration. https://buildingsmart.fi/en/common-bim-requirements-2012. Accessed 21 Feb 2019
12. Codd, E.A.: Relational model of data for large shared databanks. Commun. ACM **6**, 377–387 (1970)
13. Codd, E.: Recent investigations in relational data base systems. In: Proceedings of IFIP Congress 1974, pp. 1017–1021. North-Holland Pub. Co., Amsterdam (1974)
14. Pin-Shan, C.P.: The entity-relationship model-toward a unified view of data. ACM Trans. Database Syst. **1**(1), 9–36 (1976)
15. Ginzburg, A.V.: BIM-technologies throughout the life cycle of a construction object. Inf. Resour. Russ. **153**(5), 28–31 (2016)
16. Talapov, V.: Fundamentals of BIM. Introduction to building information modeling, 392 p. DMK Press Publ., Moscow (2011). (in Russian)
17. Vendorov, A.M.: CASE-technology. Modern methods and means of designing information systems, p. 98. Finance and Statistics, Moscow (2009)
18. Ignatov, E. B., Elsheikh A.M.: Compilation of a 4D construction schedule based on BIM. Nat. Tech. Sci. **9–10**(77), 268–272 (2014)
19. Antonov, A., Emelyanov, A., Khrapkin, P.: Using CAD systems of various configurations. CAD Graph. **6**, 25–38 (2015)
20. Ryndin, A., Tuchkov, A.: Project data management systems in the field of industrial and civil construction: our experience and understanding. CAD Graph. **2**, 35–51 (2013)
21. Polterovich, V., Tonis, A.: Innovation and Imitation at Various Stages of Development. New Economic School, Moscow (2003)
22. Polterovich, V., Tonis, A.: Innovation at various stages of development: a model with Capital. Working Paper No 048. New Economic School, Moscow (2005)
23. Green BIM. How Building Information Modeling is Contributing to Green Design and Construction. McGraw-Hill Construction. (2010)

24. BIM for the Terrified: A guide for manufacturers. NBS,19 (2013)
25. Rogozov, Y., Sviridov, A.S.: The concept of building an enterprise information model. Moscow. Technocentre **5**, 25–31 (2004)
26. Matveev, A.A., Novikov, D.A., Tsvetkov, A.V.: Models and methods of project portfolio management, p. 206. PMSOFT, Moscow (2005)
27. Starostova, E.N.: ERP-systems for general contractors. Head Constr. Organ. **4**, 34–39 (2010)
28. Jiao, Y., et al.: A cloud approach to unified lifecycle data management in architecture, construction & facilities management: integrating BIMs and SNS. Adv. Eng. Inform. **27**(2), 173–188 (2013)
29. Bukunov, A.S.: Lifecycle management of a construction object based on information modeling technology. In: System Analysis in Design and Management (SAEC): sb. scientific tr. XXII International Scientific-Practical Conference Part 1, St. Petersburg, 22–24 May 2018, pp. 324–330. Polytechnic Publishing House. University (2018)
30. Bukunov, A.S., Bukunova, O.V.: Integration of blockchain technology and information modeling of real estate objects. In: BIM — Modeling in Construction and Architecture Problems: Materials of the All-Russian Scientific and Practical Conference, pp. 45–51. SPSUKU, St. Petersburg (2018)
31. Bolotin, S.A., Dadar, A., Ptukhina, I.S.: Simulation of scheduling in the programs of information modeling of buildings and regression detailing the norms of the duration of construction. Eng. Constr. J. **7**, 82–86 (2011)

# EEG-Based Brain-Computer Interface
# for Control of Assistive Devices

Nikolay V. Kapralov$^{(\boxtimes)}$, Jaroslav V. Ekimovskii,
and Vyacheslav V. Potekhin

Peter the Great St. Petersburg Polytechnic University, Saint Petersburg, Russia
nikolay.kapralov@gmail.com

**Abstract.** The study describes an approach for supervisory control of a limb prosthesis and a mobile robot based on a non-invasive brain-computer interface. Key applications of the system are the maintenance of immobilized patients and rehabilitation procedures. An interface performs imaginary hand movement decoding using electroencephalographic signals. The decoding process consists of several steps: (1) signal acquisition; (2) signal preprocessing (filtering, artefact removal); (3) feature extraction; (4) classification. The study is focused on obtaining the best accuracy of decoding by comparing different feature extraction and classification methods. Several methods (Riemannian geometry-based) were tested offline. Furthermore, online testing of control capabilities using in-house data was performed.

**Keywords:** EEG · Brain-computer interface · Assistive devices

## 1 Introduction

Nowadays, elder and disabled people face serious social problems, but their life can be improved by means of assistive technologies, such as limb prostheses or mobile robots [1]. TurtleBot [2] is a mobile robot that possesses a number of superior qualities, compared to other solutions – it is cheap, multifunctional and efficient. This robot can act as a prototype of a robotic cart, which can deliver water or medicine to patients in a hospital, or a wheelchair. These applications are similar in terms of control commands, therefore a physically realistic model of TurtleBot is suitable for designing and testing control approaches, such as direct and scenario control.

Due to the physical limitations, patients should not control devices using joysticks or keyboards, and a promising approach is to use mind to control a device. Brain-computer interfaces (BCI) are a kind of devices that analyze brain state (e.g., concentration or meditation) or detect patterns of several imaginary actions (hand grasp, finger flexion, etc.), yielding a command which can be used for control [3, 4]. In this study we investigate the possibilities of controlling a limb prosthesis and a mobile robot in the simulated environment by imagination of left and right hand movement.

Several methods to study brain function exist, including functional magnetic resonance imaging (FMRI), magnetoencephalography (MEG), electroencephalography (EEG), electrocorticography (ECoG), near-infrared spectroscopy (NIRS), and so on. EEG is non-invasive, relatively inexpensive and has very high temporal resolution, so

© Springer Nature Switzerland AG 2020
D. G. Arseniev et al. (Eds.): CPS&C 2019, LNNS 95, pp. 536–543, 2020.
https://doi.org/10.1007/978-3-030-34983-7_52

this method is the most suitable for real-time signal analysis. On the other hand, the spatial resolution and signal-to-noise ratio of EEG are low, thus signal processing and classification becomes more difficult.

Machine learning methods are capable of solving the problem of classification and are widely used in the field of BCI. The accuracy of the classification highly depends on the complexity of the classification and the feature extraction methods, as it is very important to extract useful information from EEG signals. Basic methods, such as logistic regression or Support Vector Machine, generally do not perform well in the field of EEG classification, but recently a classifier based on the Riemannian geometry was introduced [5, 6], demonstrating high accuracies on publicly available datasets. This study is focused on applying this classifier to online control of assistive technologies.

## 2   Control Approaches

Brain computer interface yields a command corresponding to the type of decoded imaginary movement – left or right hand. External devices, such as wheelchair of limb prosthesis, should be controlled by these two commands. In this paper two approaches for control are considered – direct control, where commands are affecting the behavior of motors of robot, and scenario control, where commands are used to launch one of pre-defined scenarios for robot.

Direct control is the simplest approach in terms of implementation, because control commands are directly sent to the motors. On the other hand, it requires high accuracy of classifier to obtain adequate control possibilities. In case of limb prosthesis, the fingers are bent when the control signal comes from the BCI. The mobile robot can be guided to the destination point with two commands corresponding to the left and right hand imaginary movement respectively: (1) turn clockwise by 15°; (2) move one meter forward.

Scenario control is another approach to control the moving platform. In this approach, the user chooses a pre-defined scenario, which includes a destination point and intermediate stops if needed. For example, after the activation of the scenario 'bring me water' mobile robot should move to the intermediate point where it grab a glass of water and the destination point should be the patient's bed. Platform can move automatically according to the scenario using the knowledge of environment and sensor values. This approach is simpler for the user, but requires implementation of Simultaneous Localization and Mapping (SLAM) algorithms to perform navigation and obstacle avoidance.

Simultaneous Localization and Mapping (SLAM) is definitely characterized as quest of localization and mapping at the same time that provides by sensors of mobile robot. In practice, these two issues cannot be fathomed freely of one another.

*Localization.* The mobile robot uses a particle filter (Monte Carlo Localization, MCL) to localize itself in the environment. This filter is a version of Markov Localization, which works on the principle of localization based on the probability distribution of every possible position the robot can occupy on the map [7]. Some algorithms make the mapping more productive. The Tree-Map algorithm [8] performs productive redesigns with a tree-based subdivision of the map into feebly related parts. The Smoothing and Mapping (SAM) algorithm [9] approaches have been investigated as a viable alternative to extended Kalman filter (EKF)-based solutions to the problem. In particular, approaches that have been looked at factorize either the associated information matrix or the measurement Jacobian into square root form and algorithm to permit hierarchical disintegration into sub-maps [10].

*Path Planning.* As a solution to this problem, it was decided to use movebase concept [11]. To use the movebase concept, we need to have a global planner and a local planner. First one checks if the given goal is an obstacle, and if so, it picks an alternative goal close to the original one by moving back along the vector between the robot and the goal point. For navigation it uses Dijkstra's algorithm to find a global path with minimum cost between the start point and end point. Local planner function makes the decision to execute the command of movement or change it. The algorithm is responsible for the sudden obstacles that may arise on the way.

## 3    Assistive Devices

We used a 3D-printed anthropomorphic arm as a prototype of a limb prosthesis. An individual servomotor controlled each finger. In turn, Arduino Uno controlled the position of servomotors and, therefore, the extent of finger bend.

The Kinect sensor was created by Microsoft Company. In the last 5 years, more and more solutions for using it in robotic sensory have appeared [12]. The main reason for this popularity is its low cost and good efficiency in solving problems where computer vision, lidar and sonar technologies are required. Kinect combines the technologies of RGB-D cameras [13] that can be remotely controlled from a Personal Computer, thanks to a number of libraries and Software Development Kits (SDKs) implemented by both Microsoft and third parties. Kinect device had good results through the investigation of outdoor navigation, with the aim to create an integrated low-cost GNSS and photogrammetric navigation solution [14].

The principle of building the platform software and hardware

• ROS

Used in building software for the platform, it is no longer an operating system in its usual sense, but a framework over the Linux platform. The construction of the program architecture in the POC consists of exchanges of messages between nodes, which are executable program code written in one of the programming languages (Python, C, etc.).

• Visualization in rviz and gazebo

It has the necessary functionality for building robotic systems. The rviz software package allows you to visualize all the data coming from the sensors and display the location map and obstacles. The gazebo package is designed to simulate robots and their environments. The created model and environment for the experiment are presented in Fig. 1.

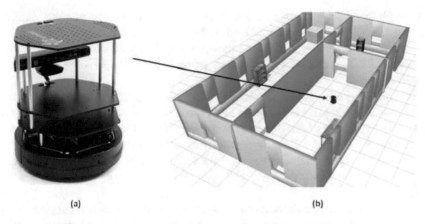

(a)         (b)

**Fig. 1.** Turtlebot (a) in a playground simulated in Gazebo (b)

## 4 BCI Structure

The EEG signals were recorded using a wireless Smart BCI EEG headset of the "Mitsar" Company. We used 21 electrodes, which were located according to the 10–20 system with referent electrodes on ears. Subjects performed two types of motor imagery tasks: grasp movements of their left and right hands. Tasks were split into blocks as shown in Fig. 3. Each block contained two series of trials: synchronizing trials, when the subjects had to perform grasp movements simultaneously with the presented sound and to learn the rhythm, and the trials with imagination of the same movements according to the rhythm but without sound presentation.

As a result, each subject executed at least 70 real and 120 imaginary movements within each task. Each trial lasted for 1200 ms with fixed 100 ms intervals between trials. EEG signals were recorded with the sampling rate of 250 Hz and later processed as a set of trials.

We adhered to the conception that execution of the real and the respective imaginary movements took approximately the same time [15, 16]. The instruction for motor imagery was aimed at initiating kinesthetic feelings in a subject's hand [17] (Fig. 2).

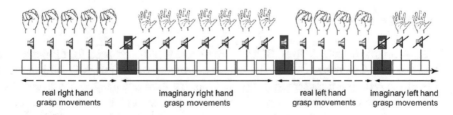

| real right hand grasp movements | imaginary right hand grasp movements | real left hand grasp movements | imaginary left hand grasp movements |

**Fig. 2.** Experimental paradigm

A notch band-stop filters at 50 and 100 Hz to prevent power line inference. Recorded signals were band-pass filtered in 8–30 Hz by fifth-order Butterworth filter. Trials containing artefacts, such as eye movements, slow down and fasten waves, and fragments of EEG signals with an amplitude of more than 100 μV were excluded from the analysis.

Classification problem can be formally described as follows: each trial $X \in \Re^{E \times T}$, where $E$ is the number of electrodes and $T$ is the number of time samples, corresponds to left or right imaginary hand movements $- y \in [- 1, 1]$. Given a set of $N$ trials, which were recorded during the training session for each subject, we need to develop a classifier which would determine the class for trials in the testing/control session. This would allow controlling devices with person's imaginary left and right hand movement.

Recently, a new approach for motor imagery EEG classification taking into account the Riemannian geometry was introduced [18]. For each trial, a sample covariance matrix (SCM) can be estimated using (1):

$$C = \frac{1}{T - 1} XX^T.$$    (1)

SCMs are symmetric and positive definite (SPD), and the space of these SPD $E \times E$ matrices forms a differentiable manifold $M$. At each point (covariance matrix), $C$ there exists a tangent space $TcM$, which is locally homomorphic to the manifold; therefore, distance computations in the manifold can be approximated by similar computations in the tangent space. Classic machine learning methods, such as Support Vector Machine (SVM), can be used after projection, as the tangent space is Euclidean. The projection is equivalent to using a kernel according to (2) [19]:

$$k\left(C_i, C_j\right) = \mathrm{tr}\left[\mathrm{logm}\left(C_{ref}^{-0.5} C_i C_{ref}^{0.5}\right) \mathrm{logm}\left(C_{ref}^{-0.5} C_j C_{ref}^{0.5}\right)\right],$$    (2)

where logm denotes the logarithm of the matrix and $C_{ref}$ is a reference point where tangent space is computed. A common choice for $C_{ref}$ is a geometric mean of covariance matrices of all trials. The combination of SVM and Riemannian-based kernel was used for EEG signal classification.

## 5   Experiments

Publicly available dataset 2a from BCI Competition IV [20] was used for offline testing of the classifier. The dataset contains EEG recordings from nine healthy subjects. The results of the testing are presented in Fig. 3. Classification accuracy varies from 60 up to 95% for different subjects; therefore human factor should be taken into account during the evaluation – some people are less capable of controlling the BCI or need more training sessions. Also, the importance of cutting low frequencies is shown, as it leads to significantly higher accuracy values.

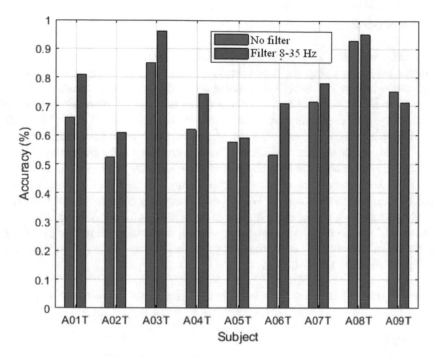

**Fig. 3.** Classification accuracy for BCI Competition IV dataset 2a

During the experiment, 2 successful strategies were identified, one for each principle – for direct and scenario control. Since the accuracy of classification is not 100%, strategies that can significantly reduce the magnitude of the error have been considered. For direct control, a choice of commands with double confirmation was used; the management team was formed only when the same action was classified 3 times in a row. This approach, although it reduces the speed of platform management, creates conditions for adequate management. It is also worth noting that in this approach, the mental reaction of the operator to the correctness or incorrectness of recognition by the classifier of control commands has a strong influence. This circumstance contributes greatly to classification errors. For scenario control, the precision element must be even higher. Increased accuracy can be achieved by focusing on one of the classes for a certain time. For the experiment, 2 scenarios were prepared, each class was assigned its own class. If when classifying, one of the scenarios stood out, then it was required to confirm the selected scenario within 10 s. The confirmation was formed by the accumulation of 3 signals assigned to the class scenario. After 10 s, another class stood out and so on, until the choice of one of them was confirmed.

# 6  Conclusion

This article outlined the essence and principle of controlling by EEG-based brain-computer interface. Using the new classifier based on the Riemannian geometry, a classification accuracy of 80% was achieved on publicly available dataset 2a from BCI Competition IV. With direct testing to control the arm and the mobile platform model the accuracy was expectedly decreased and amounted to about 65%. With direct control, such accuracy does not allow to perform control tasks; therefore, an approach to the control with the accumulation of target commands for their execution was presented. In this mode, the accuracy of the command execution was as follows: the next 20% commands were executed, 75% command were not executed, 5% the opposite command were executed. The degree of accuracy obtained can be improved; a very strong influence on the deterioration of the classification makes the difference in the backward visual communication during training and testing. In further experiments, it is worthwhile to change the approach at the training stage, so that the subject learns visually on what the test will take.

**Acknowledgements.** The work was financially supported by RFBR grant 16-29-08296.

# References

1. Achic, F., Montero, J., Penaloza, C., Cuellar, F.: Hybrid BCI system to operate an electric wheelchair and a robotic arm for navigation and manipulation tasks. In: 2016 IEEE Workshop on Advanced Robotics and Its Social Impacts (ARSO), pp. 249–254 (2016). https://doi.org/10.1109/arso.2016.7736290
2. Boulcher, S.: Obstacle detection and avoidance using TurtleBot. Research Assistantship Report, Department of Computer Science, Rochester Institute of Technology, Monroe County, NY, USA, p. 6 (2012)
3. Gundelakh, F., Stankevich, L., Sonkin, K.: Mobile robot control based on noninvasive brain-computer interface using hierarchical classifier of imagined motor commands. MATEC Web Conf. **161** 03003 (2018). https://doi.org/10.1051/matecconf/201816103003
4. Wolpaw, J.R., Wolpaw, E.W.: Brain-Computer Interfaces: Principles and Practice. Oxford University Press, New York (2012)
5. Congedo, M., Barachant, A., Bhatia, R.: Riemannian geometry for EEG-based brain-computer interfaces; a primer and a review. Brain Comput. Interfaces **4**(3), 155–174 (2017). https://doi.org/10.1080/2326263X.2017.1297192
6. Barachant, A., Bonnet, S., Congedo, M., Jutten, C.: Riemannian geometry applied to BCI classification. In: 9th International Conference Latent Variable Analysis and Signal Separation (LVA/ICA 2010), Saint-Malo, France, pp. 629–636 (2010). https://doi.org/10.1007/978-3-642-15995-4_78
7. Fox, D., Burgardy, W., Dellaert, F., Thrun, S.: Monte Carlo Localization: efficient position estimation for mobile robots. School of Computer Science, Carnegie Mellon University, Pittsburgh, PA, USA and Computer Science Department III University of Bonn, Bonn, Germany (1999)
8. Frese, U.: Treemap: an O (log n) algorithm for indoor simultaneous localization and mapping. Auton. Robots **21**, 103–122 (2006)

9. Dellaert, F., Kaess, M.: Square Root SAM: simultaneous localization and mapping via square root information smoothing. Int. J. Robot. Res. **25**, 1181–1203 (2006)
10. Ni, K., Steedly, D., Dellaert, F.: Tectonic SAM: exact, out-of-core, submap-based SLAM. In: 2007 IEEE International Conference on Robotics and Automation (2007)
11. ROS. http://wiki.ros.org/move_base
12. McDonald, S., Salimi, P.: Motion planning of intelligent robots. BSc Qualifying Project Report, Worcester Polytechnic Institute, Worcester, MA, USA, p. 20–21 (2014)
13. Cong, R., Winters, R.: How does the Xbox Kinect work. https://www.jameco.com/jameco/workshop/howitworks/xboxkinect.html
14. Pagliaria, D., Pinto, L., Reguzzoni, M., Rossi, L.: Integration of kinect and low-cost gnss for outdoor navigation. Int. Arch. Photogramm. Remote Sens. Spat. Inf. Sci. **XLI-B5**, 565–572 (2016). https://doi.org/10.5194/isprs-archives-XLI-B5-565-2016
15. Decety, J., Michel, F.: Comparative analysis of actual and mental movement times in two graphic tasks. Brain Cogn. **11**, 87 (1989)
16. Sirigu, A., Duhamel, J.R., Cohen, L., et al.: The mental representation of hand movements after parietal cortex damage. Science **273**(5281), 1564 (1996)
17. Neuper, C., Scherer, R., Reiner, M., Pfurtscheller, G.: Imagery of motor actions: differential effects of kinesthetic and visual motor mode of imagery in single trial EEG. Cogn. Brain. Res. **25**(3), 668 (2005)
18. Barachant, A., Bonnet, S., Congedo, M., Jutten, C.: BCI signal classification using a Riemannian-based kernel. In: 20th European Symposium on Artificial Neural Networks, Computational Intelligence and Machine Learning (ESANN 2012), Bruges, Belgium, pp. 97–102 (2012)
19. Wang, E.-Y., Guo, W., Dai, L.-R., et al.: Factor analysis based spatial correlation modeling for speaker verification. In: ISCSLP 2010, Tainan, China, pp. 166–170 (2010)
20. Brunner, A.C., Leeb, R., Mueller-Putz, G.R., Schloegl, A., Pfurtscheller, G.: BCI Competition 2008 – Graz data set (2008)

# Proactivity and Subsidiarity as the Basic Principles of Digital Transformation of State Interaction with Citizens and Businesses

Galina S. Tibilova[1(✉)], Andrey V. Ovcharenko[1],
and Anastasiya V. Potapova[2]

[1] "St. Petersburg Information and Analytical Centre", Saint Petersburg, Russia
secretar@iac.spb.ru
[2] St. Petersburg State Forestry University named after S.M. Kirov,
Saint Petersburg, Russia
public@spbftu.ru

**Abstract.** This paper describes two stages of digital transformation of state interaction with citizens and business. The first stage is the construction of e-government, conversion of interaction to electronic form; the second stage is human-centered digital transformation based on proactivity and subsidiarity.

**Keywords:** Proactivity · Subsidiarity · Public services · E-government · Digital economy

## 1 Introduction

Nowadays, the national program of Digital Economy is being implemented in Russia; this implies the integration of digital technologies into all spheres of life. The most important component of the digital economy is digitization of public administration, including interaction between the state and citizens and businesses.

The digital transformation of public administration is no news: it has been successfully implemented over the past twenty years. According to the UN study "E-government 2018", the Russian Federation (hereinafter referred to as the RF) is among the countries with very high EGDI (e-government development index, i.e., the indicator that includes the development of online transactions, trends in open government and mobile services, as well as public involvement in the provision of innovative public services), including:

- a very high OSI (online services index, is a composite indicator of using information and communication technologies by states in the provision of public services);
- a very high EPI (e-participation index based on electronic information or online availability of information; electronic and/or online public hearings and electronic decision-making or direct involvement of citizens in decision-making processes).

© Springer Nature Switzerland AG 2020
D. G. Arseniev et al. (Eds.): CPS&C 2019, LNNS 95, pp. 544–553, 2020.
https://doi.org/10.1007/978-3-030-34983-7_53

Since 2016, the Russian Federation has risen by three positions in the countries' ranking for the development of e-government, moving from a group with high EGDI to a group with a very high EGDI.

St. Petersburg, in its turn, is traditionally one of the leaders in the country in the development of e-government by various federal rankings, and St. Petersburg's software and hardware solutions for e-government have repeatedly received awards in all-Russian and international competitions.

Today, e-government in St. Petersburg includes:

1. A well-developed infrastructure, including the Distributed Regional Data Processing Center (DRDPC), the Unified Multiservice Telecommunications Network of Executive Authorities of St. Petersburg (UMTN), and the Certification Center of Executive Authorities of St. Petersburg;
2. A complex of integrated state information systems, which can be divided into the following main groups:

   - infrastructure systems, e.g., the Inter-Agency System of Electronic Interaction of St. Petersburg (IASEI);
   - sectoral systems: healthcare, education, transport, tourism, etc.;
   - intersectoral systems, including the Integrated System of Information and Analytical Support for the activities of Executive Authorities of St. Petersburg (IS IAS) and the Inter-Agency Automated Information System for Providing in St. Petersburg state and municipal services in electronic form (IAIS ESS);
   - supporting systems, for example, the Classification System of St. Petersburg (CS St. Petersburg) and the Unified System of Electronic Document Management and Office Management of the Executive Authorities in St. Petersburg (USEDO).

The most important component of e-government is the interaction between the Executive Authorities (EA), citizens and businesses in the provision of public services, automated within the framework of the IAIS ESS, including the Portal of State and Municipal Services of St. Petersburg.

Public service is a process of decision-making by the Executive Authorities, aimed at providing certain material goods to citizens in a declarative manner, that is, when citizens personally apply to EA and submit an application. Examples of public services may include registration of property rights, the appointment and payment of social benefits, civil registration, etc.

In terms of digitalization, public service is a structured management task that requires information, collection, updating and storage of which within the framework of decision maker (DM) system is not appropriate, and in some cases impossible. To ensure the solution of such problems, information from various sources external to the DM system, geographically distributed and independent from a legal and technical point of view, is required. This information, belonging to various organizations, can be stored in information systems that are not integrated with each other, or it can be non-existant in electronic form. Thus, in order to ensure the decision-making process, the decision maker's information system must refer to external sources of information for the necessary information in accordance with a specific algorithm, which depends on the specific decision-making process. After making a decision, the sequence of obtaining information arrays from external sources [7], formed in the implementation of this algorithm, ceases to exist as a sequence, transforming into a distributed set of information arrays that are not interconnected.

The digital transformation of the interaction of the state with citizens and the business can be divided into two stages:

- the first stage is the construction of e-government (2001–2018);
- the second stage is a person-centered digital transformation based on the principles of proactivity and subsidiarity (2019–2024).

## 2   The Construction of E-Government (2001–2018)

The purposes of building e-government were:

1. Ensuring the provision of state services to applicants in electronic form, improving the quality of service for applicants;
2. Ensuring the implementation of the one-window principle in the territory of St. Petersburg, namely, the principle that:

   - interdepartmental information interaction is hidden from the applicant, all necessary measures are carried out by the authorities independently;
   - there is one entry point for the applicant where they apply for the provision of the services and where get the results;

3. Reduction of paper workflow;
4. Minimizing the budgetary costs of St. Petersburg;
5. Increasing the transparency of activities of the Executive Authorities of St. Petersburg and local governments and organizations.

The main problem of designing the IAIS ESS in the first stage was structuring of a set of public services and their prioritization from the point of view of conversion into electronic form.

Prioritization was to be carried out taking into account:

- Public and EA needs, since the creation of e-government is aimed, inter alia, at obtaining social effect;
- The possibility of obtaining certain information in electronic form, since the activities of many agencies that provide public services and individual documents were not automated;
- Resource constraints on the elaboration of an information system.

For the structuring and prioritization of public services, a system-target approach was applied [1]. The following features were used:

- the result of a service;
- the type of legislation regulating the provision of the service;
- the organization that makes the decision on the provision of the service;
- the role of the organization that makes the decision on the service provision.

Expert assessments were applied according to the criteria of "significance of a group of services for citizens" and "significance of a group of services for organizations providing these services", as well as objective characteristics of services according to the following criteria [2, 3]:

- number of normative documents regulating the realization of the decision-making process (characterizes the significance of this process for higher instances);
- number of changes of normative documents regulating this process for a certain period (characterizes the modifiability of processes);
- number of organizations taking part in the decision-making process (characterizes the need to automate collection of information for consumers of the decision-making result);
- relative decision time (the ratio of the minimum time for making a decision, namely, the time required for making a decision, to the maximum time for making a decision, that is, the time allotted for making a decision; it characterizes the need to automate the collection of information for decision makers).

These criteria have been formalized. A mathematical model was developed that allows ranking public services in order to further study their information processes. This model represented a multicriteria problem of integer linear programming with Boolean variables and it was solved by convolution criteria.

The next task was to study the information processes that do the selected public services. The study of information processes suggested:

- identification of information files required for decision-making, as well as their sources;
- selection of information process options that are of priority and available for automation, since within each service, depending on the life situation and the applicant's characteristics, several information processes are defined.

To solve this problem, a process-oriented approach was applied. The need for some information arrays in the provision of public services is often determined by the results of the analysis of other arrays. This allows us to represent the information process in the form of a directed graph, the vertices of which are information arrays - as an n-step task, with each step being completed and independent from the others.

Financial constraints and restrictions on the availability of various information files in electronic form were taken into account, as well as the ultimate purpose of automation is achieving social effect, improving the ease of providing public services for the applicant and decision makers.

Two main classes of characteristics of the information array were identified:

- characteristics of the need for an information array (probability of claiming an information array);
- accessibility characteristics of the information array.

The probability of claiming information was evaluated at each step of the decision-making process. The probability of claiming an information array can be determined statistically (for example, on the basis of statistics on the provision of a government service, it can be concluded which categories of applicants eligible for this service most often apply for it) and expertly.

The availability of the information array was assessed qualitatively and quantitatively. For a qualitative assessment of the availability of information arrays, a scale was used that included only two values: 1 and 0, that is, the source is basically available in the existing conditions, and the source in the existing conditions is completely unavailable.

For a quantitative assessment of availability, the estimated cost of work on the automation of a particular source of information was used, namely:

- elaboration of an automated workplace providing a web-based interface that allows authorized persons of an organization to respond to requests from the information support system by providing relevant information arrays;
- elaboration of a web service to provide the necessary information array on the IS side of the organization providing the information array, and a web service on receiving this information array on the IS side of information support in distributed information arrays;
- development of a web service for receiving an information array from the organization's IS providing this information.

The task of dynamic programming was set, the target criterion of which took into account both the probability of demand and the availability of information arrays.

In the formulation of the dynamic programming problem, the following definitions were adopted:

- the System $(S)$ – decision-making process (public service);
- step $j$ $(j = 1, 2, \ldots n)$ – the stage of collecting information that is necessary for making a decision. For example, for public services, the following steps can be distinguished (the same for all services):
- identification of the applicant;
- identification of the object of the application;
- confirmation of the applicant category;
- additional conditions for the provision of public services.

The number of conditions may vary for different services; however, if this step is in the service, it means that it is there for all its variants.

- state of system $(X_j^i)$ – information array (document) i, that is required for one or another variant of the service at the i-th step.

Each information array (system state) within the model is characterized by two parameters [8]:

- theoretical availability of the i-th information array (document) in electronic form (), estimated by experts. Wherein, $q(i_j) = \{1; 0\}$. Each expert should evaluate only whether the information array is available in electronic form in the existing conditions (that is, whether it can be obtained in electronic form) or not;
- probability of claiming information array (document) or claiming information array (document) $i$ at step $j - p_j(i_j)$;
- control $(u)$ – the choice of an information array $i$ at a particular step $j$.

The information potential of the chosen information process variant calculated by A. A. Denisov's criterion was used as the objective function in determining the optimal variant of the information process. Information potential is defined as [2, 4, 5]:

$$H^B(i) = -\sum_{j=1}^{n} q(i_j) \log_2(1 - p_j(i_j)), \; i_j \in I(j) \tag{1}$$

where $n$ – total number of steps in the information process, $j$ – serial number of a specific step, $i_j$ – serial number of a specific information array, which is selected on a specific step $j$; however, this number belongs to the set of numbers of information arrays $I(j)$ which may be required at this step; $q(i_j)$ – theoretical availability of the i-th array at the j-th step estimated by experts, while $q(i_j) = \{1; 0\}$, and $p_j(i_j)$ are the probability of claiming information array $i$ at the specific step $j$ (normalized to 1 at each step).

Thus, a set of models and approaches was created that allowed for optimal control of the transfer of public services to the electronic form. Similar approaches were used to automate state functions of the Executive Authorities and other activities within their competency.

## 3  Human-Centered Digital Transformation

During the implementation of the first stage, infrastructure, industry, inter-sectoral and state information systems were created. However, at present, approaches to digitalization require a revision due to:

- new needs: national purposes and strategic development objectives until 2024;
- new opportunities: the development of Big Data, the Internet of things and artificial intelligence [6, 9].

One of the problems of the existing digitalization is the sectoral isolation of information resources. Each department independently forms the requirements for the creation and development of information systems, focusing only on internal needs and their own authorities.

Sectoral isolation:

- increases corruption risks, reduces the transparency of EA;
- makes it difficult to provide informational support for citywide administrative tasks that are intersectoral (cross-cutting);
- reduces the speed of response of city authorities to emerging problem situations;
- increases budgetary costs for the integration of disparate information systems and the automation of intersectoral processes;
- reduces the controllability of digitalization by shifting the center of competence to sectoral EA.

Current challenges require a cross-sectoral approach and smart solutions. In this regard, the digitization of state interaction with citizens and businesses, namely the provision of public services, can be called the "smart specialization" of St. Petersburg, since the city has all necessary prerequisites for introducing "smart" solutions into this sphere, which are in the priority of the city leadership and are presently created.

The main principle of modern digital transformation of the interaction of the state with citizens and businesses can be identified as human centrism; that means focusing on the interests and needs of citizens and businesses.

Digital transformation is carried out in three main directions:

1. Proactivity

At present, the process of providing public services begins at the moment when a citizen applies to the EA with and submits a package of documents. The package of documents is reviewed by the decision maker, after which a decision is made about the provision or non-provision of the relevant service, as well as, in some cases, the amount of the service (for example, the amount of payment). If a citizen does not know that he/she is entitled to a service, he/she will not receive it.

However, the Message of the President to the Federal Assembly in 2019 said that public services should be provided proactively. Proactive provision of public services means that the initiator of the process is the state, which, on the basis of the analysis of the citizens' data known to it, offers itself to receive the requests for services it is entitled to provide.

The concept of "proactivity" is closely related to the concept of super service. Super service is a combination of public services that a citizen is supposed to have when a life situation happens (birth of a child, moving to another region, etc.). In the future, all government services included in the super service should be provided on the basis of one application, which is formed proactively.

2. Expanding the boundaries of interaction from public services to public services and other commercial and non-commercial electronic services.

Interaction of state with citizens and businesses needs to be understood more broadly than regulatory public services. It is necessary to create and develop social electronic services that citizens could use in their daily activities, as well as enable businesses to promote their services in a trusted, state-guaranteed digital environment.

It is obvious that other services besides public services should also be provided to citizens proactively.

3. Direct involvement of citizens and businesses into solving citywide issues

Tools should be created for direct involvement of citizens and businesses in solving citywide issues, creating new projects, developing the smart city and solving any beautification matters, as well as the integration with social networks.

In terms of technical implementation, the principle of proactivity is the most difficult one.

Proactivity is carried out on the basis of:

- data that the state knows about a citizen (or business member, hereinafter referred to as the user).

During the creation of e-government, these data were converted to electronic form and are currently stored in various industry information systems. It is important to note that some of these data do not change (for example, year of birth), and some need to be regularly updated (for example, marriage data). Thus, it is required not only to identify these data and their sources, but also to determine the frequency of their updating;

- data on the user's activity, on what services they received earlier, on what services they used. This data may come from industry-specific systems and accumulate in the electronic public services system.
- data on the user's activity in social networks.

For the implementation of proactivity, a constantly updated and filled up digital user profile is created, integrated with state information systems, commercial companies' systems and social networks.

At the same time, the restrictions under which the public services were transferred to the electronic form do not lose their relevance for the implementation of the principle of proactivity. First of all, these are restrictions on the availability and veracity of information. So, currently, information from state information systems of the Federal Executive Authorities, for example, the Ministry of Internal Affairs of Russia, is difficult to access, since the Ministry of Internal Affairs of Russia is extremely reluctant to conclude agreements on information interaction with regions. In its turn, information from social networks has low veracity, and that also needs to be considered while using it in the proactive providing services.

As a result, and also taking into account the need for major changes in the regulatory framework at the federal and regional levels (including the processing of all administrative rules), proactivity in St. Petersburg should be implemented in three stages:

Stage 1. "Smart" assistant. At this stage, due to the incomplete availability and veracity of information, proactivity will be implemented in the form of recommendations, suggestions to the user to consider the possibility of receiving one or another service. At this stage, the system will not generate an application or make any actions that would entail legal consequences.

Stage 2. "Smart" assistant with the generation of the application. At this stage, the availability and veracity of the information is higher, so the system, along with the recommendation, can automatically generate a draft application and make preliminary approval for the citizen to receive the service.

Stage 3. Proactive Super Services. At this stage, information is available and reliable, so the system automatically generates one application for the entire pool of services the citizen is entitled to when a life situation happens. For certain services, e.g., those that a citizen must and not just can receive in accordance with the law, the declarative form at this stage can be completely abolished and the request and decision will be generated automatically.

This way, proactivity supposes the implementation of subsidiary management of the provision of public services at the system level. At present, an official makes a decision on the service, which significantly affects the convenience and speed of decision-making, contributes to strengthening the bureaucracy. With the introduction of proactivity, it is supposed to transfer decision-making on public services to the system level, leaving to the official only the functions of a controller.

Currently, a methodology is being developed to transfer the providing of public services to the proactive mode, creating a "smart" assistant. In the framework of the methodology, it is assumed that at the first stage of introducing proactivity, in the conditions of low availability and veracity of some data, the service can be recommended to the user only with a certain probability.

The composition of the characteristics of information arrays is expanding:

- the need for an information array (the probability of claiming an information array, was used at the first stage);
- availability of the information array (used at the first stage);
- veracity of the information array (entered at the second stage);
- the degree of influence of the information array on the decision for the recommendation of a service (introduced at the second stage).

The latter characteristics is associated with the requirements for the applicant. These requirements are well-formalized and described in administrative rules, and the applicant must meet a specific set of criteria in order to receive a public service. According to the rules, the match should amount to 100%; however, in the conditions of incomplete and unreliable arrays, when it is impossible to know all necessary data with 100% accuracy, these criteria have a different influence on the decision to recommend a service to the user.

## 4   Conclusion

Within the framework of the methodology, mathematical models for optimizing the management of the transfer of public services to electronic forms will be expanded to reflect the new characteristics required for the introduction of proactivity and subsidiarity. It is also important to note that the regulatory framework for the provision of public services often changes, including changes in the criteria to be met by applicants. In this regard, in the future, it is planned to use a self-learning system and elements of artificial intelligence to calculate the probability with which services should be recommended to users. The final implementation of proactivity and subsidiarity (Phase 3 "Proactive Super Services") is scheduled to be completed by 2024.

## References

1. Volkova, V.N., Denisov, A.A., Temnikov, F.E.: Methods of Formalized Representation of Systems: Tutorial. Publishing House of Polytechnic University, St. Petersburg (1993). (In Russian)
2. Volkova, V.N., Denisov, A.A.: Basics of the Theory of Systems and Systems Analysis: Textbook. Publishing House of Polytechnic University, St. Petersburg (2005). (In Russian)
3. Denisov, A.A.: Introduction to Information Analysis of Systems: The Text of Lectures. Leningrad Polytechnic Institute, Leningrad (1988). (In Russian)

4. Denisov, A.A.: Information Bases of Management. Energoatomizdat, Leningrad (1983). (In Russian)
5. Denisov, A.A.: Information in Control Systems: Study Guide. Leningrad Polytechnic Institute, Leningrad (1980). (In Russian)
6. Shvab, K. The Fourth Industrial Revolution [Chetvertaya promyshlennaya revolyutsiya], 208 p. "E" Publ., Moscow (2017). (In Russian)
7. Volkova, V.N., Denisov, A.A.: Theory of Systems and Systems Analysis: Tutorial. [Teoriya sistem i sistemnyi analiz: ucheyunik], 462 p. Yurait Publ., Moscow (2017). (In Russian)
8. Denisov, A.A.: Modern Problems of System Analysis: Textbook [Sovremennye problemy sistemnogo analiza: uchebnik], 3rd edn. St. Petersburg, Publishing House of Polytechnic University, 304 p. (2008). (In Russian)
9. Volkova, V.N., Kudryavtseva, A.S.: Models for managing the innovation activities of industrial enterprises. [Modeli dlya upravleniya innovatsionnoy deyatel'nost'yu promyshlennogo predpriyatiya]. Open Educ. J. [Otkrytoye obrazovaniye] 22 (4), 64–73 (2018). https://doi.org/10.21686/1818-4243-2018-4-64-73. (In Russian)

# Neural Network Compensation of Dynamic Errors in a Robot Manipulator Programmed Control System

Yan Zhengjie[1], Ekaterina N. Rostova[2], and Nikolay V. Rostov[1(✉)]

[1] Peter the Great St. Petersburg Polytechnic University, Saint Petersburg, Russia
yanzhengjie1019@gmail.com, rostovnv@mail.ru
[2] St. Petersburg Institute for Informatics and Automation of the Russian
Academy of Science, Saint Petersburg, Russia
rostovae@mail.ru

**Abstract.** The subject of consideration in this paper is a programmed control system of a robot manipulator. Mathematical description of the control system was presented taking into account the nonlinear dynamics of the robot mechanism. Synthesis of multivariable compensators of dynamic errors for a prototype control system was carried out. Computer models of the control system with synthesized compensators were developed using MATLAB package. The results of teaching of neural network compensators are given for a programmed trajectory of the robot gripper. Comparative analysis of dynamic errors in the prototype system and the system with neural network compensators was conducted.

**Keywords:** Robot manipulator · Programmed control system ·
Neural network · Nonlinear multivariable compensators · Simulation ·
Dynamic analysis · Dynamic errors

## 1 Introduction

Since the 1980s, the problems of robot manipulator control have been drawing a lot of attention. One challenge of achieving acceptable performance (such as small trajectory errors and disturbance rejection) might come mainly from the fact of high nonlinearity and dynamic coupling of robot joints [8, 23].

At the early stage of model-based robot control design, the system's dynamics and physical parameters are usually required to be accurately known, and some control design methods, such as computed torque control and inverse dynamics control, could work well based on an accurate model [13, 14]. But it is difficult to have an exact mathematical model of a robot because of the presence of uncertainties. Also, due to the effect of various payloads, the adequate model-based method might be hard to obtain.

Recently, in order to improve robot control system characteristics, neural network calculators have been used in the design of control systems for robot manipulators [6, 11, 17, 18]. Also, in robot drives, they can be used as nonlinear, quasi time-optimal

© Springer Nature Switzerland AG 2020
D. G. Arseniev et al. (Eds.): CPS&C 2019, LNNS 95, pp. 554–563, 2020.
https://doi.org/10.1007/978-3-030-34983-7_54

regulators [9]. In computer numerical control (CNC) systems, neural network interpolators of robot links can be used instead of traditional spline interpolators [20, 21].

This work is devoted to the investigation of CNC systems for robot manipulators
with neural network compensators of dynamic errors in the drives of robot links that
occur when the drives reproduce program trajectories on specified trajectories of the
robot gripper due to torque loads caused by the complex nonlinear dynamic coupling of
the drives. A four-link robot manipulator operating in an angular coordinate system
was chosen as a specific control object.

The aim of the work is the synthesis of structures and the training of nonlinear
multidimensional neural network compensators of dynamic errors occurring on typical
program trajectories of the robot gripper.

The main issues considered in the work are as follows:

- mathematical modeling of CNC systems with prototype compensators of dynamic
  errors
- training of neural network compensators of errors caused by torque loads of various
  types
- study of CNC systems containing neural network compensators
- comparative analysis of dynamic errors that occur in the systems with prototype and
  neural network compensators.

## 2   Mathematical Models of Nonlinear Dynamic Compensators

Generally speaking, multidimensional compensators of dynamic errors are described
by nonlinear expressions corresponding to the dynamic model of the robot presented in
the form of the Lagrange equations:

$$A(q)\ddot{q} + B(q,\dot{q},)\dot{q} + C(q) = Q_d - Q_L \tag{1}$$

where $(q, \dot{q}, \ddot{q})$ are $N \times 1$ vectors of generalized coordinates of positions, speeds and
accelerations of the robot links; $N$ is the number of robot links.

The main loads of the robot drives are elements of the vectors in the left part of
Eq. (1), where:

$Q_{iner} = A(q)\ddot{q} - N \times 1$ vector of inertia torques or forces caused by accelerated
motion of links; $A(q) - N \times N$ kinetic energy matrix of the robot mechanism;

$Q_{cor} = B(q,\dot{q},)\dot{q} - N \times 1$ vector of coriolis and centrifugal torques or forces;
$B(q,\dot{q}) - N \times N$ matrix;

$Q_{grav} = C(q) - N \times 1$ vector of gravitational and other potential torques or forces.

In the right part of Eq. (1): $Q_d - N \times 1$ vector of torques or forces generated by the
robot drives; $Q_d - N \times 1$ vector of additional loads arising in the drives due to friction
in the joints and the action of external forces on the gripper.

The trajectories of the robot links in the CNC systems are calculated by solving the
inverse kinematics problems in the base points of the robot gripper trajectory; then they
are interpolated using spline polynomials. Therefore, using the program values of the
position, speed, and acceleration vectors of the links $(q_p, \dot{q}_p, \ddot{q}_p)$, it is possible to

calculate the program values of the torque or force loads that the link actuators must overcome:

$$Q_{iner} = A(q_p)\ddot{q}_p, Q_{cor} = B(q_p, \dot{q}_p,)\dot{q}_p, Q_{grav} = C(q_p), \qquad (2)$$

$$Q_{ff} = Q_{iner} + Q_{cor} + Q_{grav}. \qquad (3)$$

Expressions (2) and (3) can be directly used for compensation of dynamic errors in the robot CNC systems with torque drives of links and PID-regulators of the robot link positions [8]. Figure 1 shows the corresponding functional diagram of the system with dynamic compensators included in the feedforward (FF) circuit of the control system.

However, if there are additional loads $Q_L$, the system with FF torque control might have big dynamic errors, which can be reduced by using an additional linear compensator:

$$U_{pid} = K_p e + K_i \int edt - K_d \dot{q} + K_{com}\ddot{q}_p, \qquad (4)$$

where $K_{com} = \text{diag}\{K_{com,i}\}$ – the matrix of the linear compensator coefficients.

Nonlinear compensators can be included into the feedback (FB) of the control system together with a PID controller or a more complex nonlinear regulator. In this case, FB compensators use feedback signals of real positions and speeds of the robot links measured by corresponding sensors:

$$Q_{iner} = A(q)U_{pid}, Q_{cor} = B(q, \dot{q},)\dot{q}, Q_{grav} = C(q), \qquad (5)$$

$$Q_{fb} = Q_{iner} + Q_{cor} + Q_{grav}, \qquad (6)$$

where vector $U_{pid}$ is considered as the vector of real accelerations.

Figure 2 shows a functional diagram with nonlinear FB compensators included in the closed loop of the system.

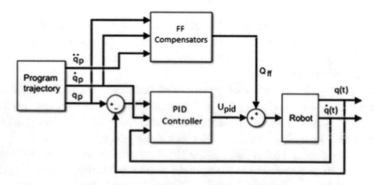

**Fig. 1.** Control system with feedforward compensators

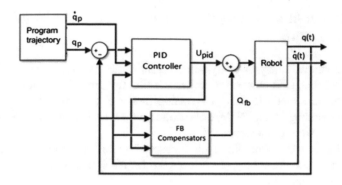

**Fig. 2.** Control system with feedback compensators

Multivariable compensators (5) and (6) perform the linearization of the nonlinear dynamics of the robot described by Eq. (1), and, thereby, ensure more stable work of the robot drives.

## 3  Typical Program Trajectories of the Robot Gripper

Figure 3 shows the robot animations with typical trajectories of the gripper – a straightline one with trapezoidal projections of the velocity $V_x(t)$, $V_v(t)$, $V_z(t)$ and with a more complex, helical (spiral-shaped) one.

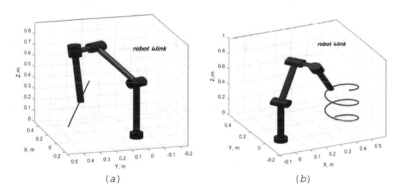

**Fig. 3.** Program trajectories of the robot gripper: $a$ – rectilinear, $b$ – helical

Further on, the analysis of dynamic errors is carried out on the helical gripper trajectory according to the following expression:

$$E = \sqrt{(X_p - X_r)^2 + (Y_p - Y_r)^2 + (Z_p - Z_r)^2}. \tag{7}$$

where the coordinates of program and real gripper trajectories are used.

## 4   Simulation of a Prototype Control System with Nonlinear Compensators

For the comparative analysis of dynamic errors Simulink-models of prototype CNC systems with FF and FB compensators were built. The processes in the systems under consideration were modeled using a dynamic model of a 4-link robot with parameter values corresponding to the PUMA-560 robot, in which frictions in the joints of the links were not taken into account [3].

Figure 4 shows dynamic errors for 2 cases: 1 – for the system only with a PID-regulator, without compensators (dash line); 2 – for the system with an additional linear FF compensator (solid line).

Figure 5 shows dynamic errors in the system with a linear compensator (blue solid line) and additional nonlinear FF compensators of different types.

Figure 6 shows dynamic errors in the system with a linear compensator (blue solid line) and additional nonlinear FB compensators of different types.

As it can be seen from Figs. 5 and 6, the smallest dynamic errors are enabled when using a linear compensator together with all nonlinear ones.

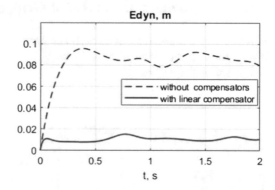

**Fig. 4.** Errors in the system without nonlinear compensators

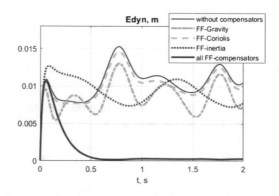

**Fig. 5.** Errors in the system with nonlinear feedforward compensators

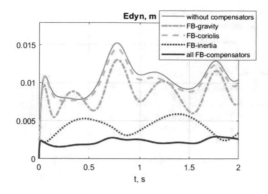

**Fig. 6.** Errors in the system with nonlinear feedback compensators

## 5   Training of Nonlinear Neural Network Compensators

The training of neural network compensators was carried out according to program values $(q_p, \dot{q}_p, \ddot{q}_p)$ and calculated load torques $(Q_{iner}, Q_{cor}, Q_{grav})$. Radial-basis networks were used as neural network compensators. The training was carried out with the use of functions from Neural Network Toolbox in MATLAB. Figure 7 represents the macroblock with NN compensators.

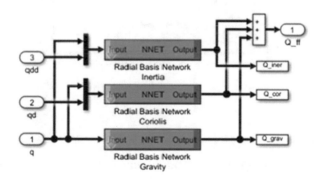

**Fig. 7.** Macroblock of neural network compensators

## 6   Simulation of a Control System with Neural Network Compensators

Figures 8 and 9 represent Simulink models of control systems with neural network FF and FB compensators.

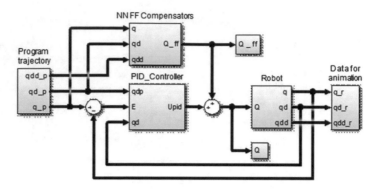

**Fig. 8.** The model of control system with neural network feedforward compensators

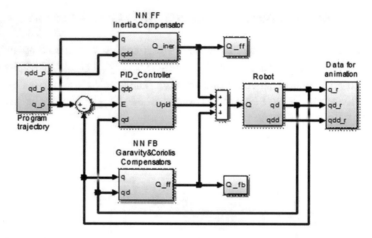

**Fig. 9.** The model of control system with neural network feedback and feedforward compensators

In Fig. 10 the estimation of dynamic errors is shown for three cases:

1 – for the system with neural network FF compensators, corresponding to Fig. 8 (green dash line);

2 – for the system with neural network FB Gravity and FB Coriolis compensators, corresponding to Fig. 9 (blue dash line);

3 – for the system with neural network FB Gravity, FB Coriolis compensators and additional FF Inertia compensator (red solid line).

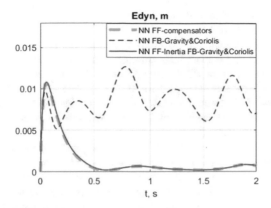

**Fig. 10.** Errors in the systems with neural network compensators

Thus, based on the estimation of dynamic errors in Fig. 10, we can conclude that the combination of neural network feedback and feedforward compensators allows to get just as small errors as the errors in prototype control systems.

## 7  Conclusion

Neural network compensators of dynamic errors are an alternative to prototype non-linear compensators, the structures and parameters of which are quite specific and described by complex nonlinear equations.

Training of neural network compensators should be carried out offline, on the basis of the results of computer simulation of prototype compensators for specific program trajectories of the robot gripper. The shortest learning time is achieved by using neural networks with radial basis neurons.

The investigation of control systems with multivariable prototype and neural network compensators confirms their high accuracy achieved by compensating for the nonlinear dynamics of the robot mechanism.

For program realization of nonlinear neural network compensators, fast DSP-type microprocessors are required, but the technical implementation of such compensators requires additional, more detailed consideration.

## References

1. Bhattacharjee, T., Bhattacharjee, A.: A study of neural network based inverse kinematics solution for a planar three joint robot with obstacle avoidance. Assam Univ. J. Sci. Technol.: Phys. Sci. Technol. **5**(2), 1–7 (2010)
2. Chiddarwar, S.S., Babu, N.R.: Comparison of RBF and MLP neural networks to solve inverse kinematic problem for 6R serial robot by a fusion approach. Eng. Appl. Artif. Intell. **23**, 1083–1092 (2010)

3. Corke, P.I.: Robotics, Vision and Control. Fundamental Algorithms in MATLAB, 2nd edn. Springer, Heidelberg (2017)
4. Corke, P.I.: Robotics Toolbox for MATLAB. Release 10 (2017)
5. Dixon, W.E.; Moses, D.; Walker, L.D.; Dawson, D.M.: A Simulink-based robotic toolkit for simulation and control of the PUMA 560 robot manipulator. In: Proceedings of IEEE/RSJ International Conference on Intelligent Robots and Systems, A, pp. 2202–2207 (2001)
6. Duc, M.N, Trong, T.N.: Neural network structures for identification of nonlinear dynamic robotic manipulator. In: Proceedings of IEEE International Conference on Mechatronics and Automation, pp. 1575–1580 (2014)
7. Farzam, T., Nafise, F.R.: Robust control of a 3-DOF parallel cable robot using an adaptive neuro-fuzzy inference system. In: Artificial Intelligence and Robotics (IRANOPEN) (2017)
8. Ignatova, E.I., Lopota, A.V., Rostov, N.V.: Robot Motion Control Systems. Computer-Aided Design. Polytechnic Publishing Center, St. Petersburg (2014)
9. Ishmuratov, V.N., Rostov, N.V.: Computer training of neural network quasi time-optimal digital regulators. In: Proceedings of International Scientific and Practical Conference, pp. 134–136. Polytechnic Publishing Center, St. Petersburg (2007)
10. Islam, S., Liu, X.P.: Robust sliding mode control for robot manipulators. Proc. IEEE Trans Ind Electron. **58**, 2444–2453 (2011)
11. Kim, Y.H., Lewis, F.L.: Neural network output feedback control of robot manipulators. Proc. IEEE Trans Robot Autom. **15**, 301–309 (1999)
12. Kumar, N., Panwar, V., Borm, J.H., Chai, J.: Enhancing precision performance of trajectory tracking controller for robot manipulators using RBFNN and adaptive bound. Appl. Math. Comput. **231**, 320–328 (2014)
13. Lee, G.W., Cheng, F.T.: Robust control of manipulators using the computed torque plus H∞ compensation method. IEEE Proc. Control Theory Appl. **143**(1), 64–72 (1996)
14. Lewis, F.L., Abdallah, C.T., Dawson, D.M.: Control of Robot Manipulators. Macmillan, New York (1993)
15. Lewis, F.L., Jaganathan, S., Yesildirek, A.: Neural Network Control of Robot Manipulators and Nonlinear Systems (1999)
16. Santibañez, V., Camarillo, K., Moreno-Valenzuela, J., Campa, R.: A practical PID regulator with bounded torques for robot manipulator. Int. J. Control Autom. Syst. **8**(3), 544–555 (2010)
17. Seshagiri, S., Khalil, H.K.: Output feedback control of nonlinear systems using RBF neural networks. Proc. IEEE Trans Neural Netw. **11**, 69–79 (2000)
18. Singh, H.P, Sukavanam, N, Panwar, V.: Neural network based compensator for robustness to the robot manipulators with uncertainties. In: Proceedings of International Conference on Mechanical and Electrical Technology, pp. 444–448 (2010)
19. Tai, N.T., Ahn, K.K.: A RBF neural network sliding mode controller for SMA actuator. Proc. Int J Control Autom. Syst. **8**, 1296–1305 (2010)
20. Terpukhov, S.Yu., Rostov, N.V.: Neural network interpolation of program trajectories of links of a robot manipulator. In: Proceedings of International Scientific and Practical Conference, pp. 70–72. Polytechnic Publishing Center, St. Petersburg (2010)
21. Yan, Z., Rostov, N.V.: Error analysis of neural network interpolators of program trajectories of links of a robot manipulator. In: Proceedings of ComCon 2018, pp. 114–119. Polytechnic Publishing Center, St. Petersburg. (2018)
22. Yurevich, E.I.: Fundamentals of Robotics. BHV, St. Petersburg (2005)
23. Zenkevich, S.L., Yuschenko, A.S.: Fundamentals of Robot Manipulator Control. MSTU by name N. E. Bauman, Moscow (2005)

24. Zhang, Y., Zhu, H., Lv, X., Li, K.: Joint angle drift problem of Puma560 robot arm solved by a simplified LVI-based primal-dual neural network. In: Proceedings of IEEE International Conference on Industrial Technology (2008)
25. Zhang, Y., Wang, J.: Obstacle avoidance for kinematically redundant manipulators using a dual neural network. Proc. IEEE Trans Syst. Man Cybern. Part B Cybern. **34**(1), 752–759 (2004)

# A Framework for the Analysis of Resource-Flows in the Extended Manufacturing Network Based on Cyber-Physical Infrastructure

Aleksandr E. Karlik, Vladimir V. Platonov,
and Elena A. Yakovleva[✉]

St. Petersburg State University of Economics, St. Petersburg, Russia
karlikl@mail.ru, vladimir.platonov@gmail.com,
helen7199@gmail.com

**Abstract.** The paper deals with the cyber-physical system as an infrastructure that enables extended inter-organizational networks connecting manufacturing facilities of many independent firms (of network nodes) in the production process up to a global scale. It presents a basic analytical framework that allows to study such network as a system of the highest complexity that enables the heterogeneous resource-flows between nodes and emerges from the interaction of cyber-physical and anthropogenic systems. In the focus of attention is the following: equipment and inventory flows of digitally enabled physical objects that can be fully automated in the cyber-physical system; information and knowledge flows which involves the challenge of extracting information from big data; resource allocation between manufacturing facilities of different firms connected via the network. Besides, the article considers the anthropogenic system as another key regulator of the manufacturing network necessary not only for dealing with cyber-physical effects but also for regulating the idiosyncratic resource flows: of human capital and trust.

**Keywords:** Cyber-physical system · Inter-organizational network ·
Anthropogenic system · Manufacturing · Resource-flows · Industrial
equipment · Know-how · Big data · Resource allocation

## 1 Introduction

A cyber-physical system (CPS) represents an information technology concept, which implies the integration of computing resources and physical processes [11, 22]. The CPS is a multifunctional structure of sensors and actuators to register and manipulate the surrounding physical environment by communicating with distributed data management systems via the Internet of Things (IoT), and, in the manufacturing, with the virtual twin of the manufacturing equipment [9, 14]. The IoT is the major disruptive innovation that makes it feasible to implement CPSs for global the automation of the production processes. In industrial organization perspective, it is important that a CPS may not be a part of any external network being an intra-organizational system. Example is a fully computerized system of a nuclear power plant. However, it is the application of cyber-physics as an infrastructure of the

© Springer Nature Switzerland AG 2020
D. G. Arseniev et al. (Eds.): CPS&C 2019, LNNS 95, pp. 564–572, 2020.
https://doi.org/10.1007/978-3-030-34983-7_55

extended inter-organizational networks in manufacturing (EIONM) that creates the powerful trend often referred to as the fourth industrial revolution [18]. It goes far beyond cloud and big data analytics, 3D printing, robotics and so forth [1], allowing decentralized control of manufacturing equipment and communications between real and virtual objects and processes in a global system [9]. The EIONM based on the cyber-physical infrastructure are in the focus of this publication. In its application to the EIONM, the CPSs virtually eliminate the firm boundaries in the production process and provide for an unprecedented opportunity to induce the rise in the productivity due to the combination of resources of many businesses whose equipment will be integrated within the production system and further on, across the material flow cycle (resource supply – manufacturing – consumption – recycling) to the end users. The aim of this publication is to present an analytical framework to account for the peculiarities of the EIONM in manufacturing as of an organizational device that channels resource-flows. The paper is organized as following. In the first part, it provides a simplified framework to consider the EIONM based on the cyber-physical infrastructure. Then it introduces a more detailed view of the major resource-flows. In the final part, it presents the approach for the tangible resource allocation within the EIONM.

## 2    A Simplified Framework for the Analysis of Manufacturing Network Supported by the Cyber-Physical Infrastructure

The CPS brings more convenience and flexibility for controlling the system, as well as extra complexity with emergent uncertainty and dynamics that challenge the coordination [20]. According to the resource-based view of the firm the elimination of the boundaries between companies poses an organizational challenge to maintain resource flows in the production process and the problem of manageability of extended networks deteriorates in the hightechnology environment [10]. Before the advent of the Industry 4.0, the connectivity of industrial companies had already significantly increased [5], as well as the methodology to analyze the inter-organizational network as loci of resources explaining the networks with the causal logic of the resource flows had been formed [3, 7]. Cyber-physical infrastructure helps to build an extended manufacturing network spreading far beyond the organizational boundaries. Introduction of the CPS contributes to the growth of such networks in width, up to the global level with the help of the EIONM, as well as in depth, with increasing volumes and frequencies of transactions. Presented framework considers the cyber-physical and anthropogenic systems as subsystems of the EIONM (Fig. 1).

Anthropogenic system is a necessary part of the manufacturing network due to several reasons. The first reason is that the human capital and trust flows, at least in the near future, will be controlled mainly by the anthropogenic system. The second reason are the cyber-physical effects that require further development of the business process management system outside the CBS [19]. Finally, in particular cases, the digitalization is more effective when it supports anthropogenic system rather than fully automates the production process. For instance, human operators could manually control 3D machines or, in the end point of manufacturing cycle, customers could order 3D printing services from a cloud [12].

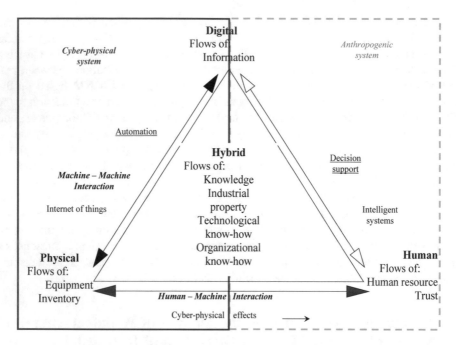

**Fig. 1.** Framework for the analysis of manufacturing network supported by the cyber-physical infrastructure

## 3  Resource Flows in the Network's Anthropogenic System

The formation of the EIONM based on cyber-physical infrastructure logically starts with the creation of an anthropogenic system, then comes the CPS. The first flows in the emerging inter-organizational network are those of trust [16], and the digital and physical systems can facilitate this kind of flows lesser than any other, i.e., they regulated by the anthropogenic system rather than the CPS. The formation of *trust flows* is a prerequisite for the movement of confidential information needed to instigate the process of network formation [6]. Opinions and judgments about the reliability of a partner and characteristics of its behavior are transmitted among participants of the emerging network. Trust flows between industrial partners is a prerequisite for the emergence of the manufacturing network's other flows, as well as for the introduction of the CPS that comes next. Trust exists when it is embedded in the network [8]; once the trust is embedded, the formal or informal network is set-up. Then human capital flows, information flows, the CPS and all other resource flows come.

In most cases, the *human capital flows* in a manufacturing network are not associated with changes of permanent employment. They are carried out in the course of the inter-company division of labor. Traditionally, that was primarily associated with business trips, but now it is increasingly carried out remotely: based on the application of digital technologies [17], the CPS can take part in the control over these flows but it will not control them completely [1], hopefully.

# 4 Resource Flows in the Network's Cyber-Physical System

Following the implementation of the CPS, the information and data turn into critically important manufacturing assets. The proposed framework differentiates the structured and unstructured datasets, as well as *the information flows* and *knowledge flows*. The data are symbols that reflect the properties of real objects and processes. Information is extracted from data, passing through a cognitive filter, which is knowledge, i.e., the previous information accumulated by the individual [2]. In the perspective of the cyber-physics and machine learning, this understanding should be augmented to include the knowledge accumulated by physical and digital systems. According to this understanding, the CPS infrastructure enables the information flows by preserving, retrieving, filtering and processing the data, as well as facilitate the knowledge flows, i.e., the information accumulated earlier. Finally, it allows to transform information into formal instructions [17] for the manufacturing equipment. Such instructions are obtained either purely on the machine basis within the CPS, or during human – machine interaction. Distinctive characteristics of the big data are high volume, velocity and variety that set the highest requirements for technology and analytical methods to retrieve information [4]. The knowledge tacitness and cognitive barriers, such as the cognitive diversity, further impedes the information flows in the extensive inter-organizational networks [15].

The generation of information and knowledge flows is a prerequisite to generate *equipment and inventory flows* through getting the link between physical manufacturing equipment and its virtual twin [9]. The cyber-physical infrastructure fundamentally changes the control of the tangible flows of intelligent objects equipped with sensors and microcontrollers. The connection to the IoT makes control over the flows of fixed assets, raw materials, work in process and other inventories in the EIONM fully automated. The IoT changes the paradigm of the shared use of the manufacturing facilities in industry through organizational and managerial innovations that implement coordination technologies. Direct interaction of the equipment and inventories connected via the IoT will decrease the need for the investments and the physical transfer of the equipment due to sharing the manufacturing facilities. At the same time, it enhances the volume and speed of the inventory circulation. Hence in the analytical framework of how the tangible flows should be divided into equipment flows and inventory flows. The resulted growth of the value added will be conditioned by the cost cuts in industrial logistics due to the decrease in the expenses on product handling and the scale effect from the possibility to form larger batches, where each item of cargo is equipped with remote sensors and globally searchable via the IoT. Another effect on the value added by the CPS will take place in the final part of the value chain conditioned by the additional value created for the customer using the finished good connected via the IoT.

The proposed framework differentiates the *industrial property* from the *know-how flows* because the CPS has different impact on these flows. The industrial property flows are referred to as the transfer of the results of the intellectual activity legally protected with patents and transferable on the basis of license agreements in the form of patent licenses, given the results of the intellectual activity are registered as inventions, utility models or industrial designs. The potential impact of the cyber-physical infrastructure of a particular EIONM on the industrial property flows is limited because its protection is secured outside the network system. At the same time, since the function of the industrial property protection within the network is highly formalized and relatively unsophisticated due to its auxiliary nature, the remnant functions within the network could be automated within the implementation of the CPS.

Know-how, unlike the industrial property, is not protected by patents: it is a trade secret transferred mainly in the formal and informal agreements built on trust. Many functions of the know-how transfer would hardly be automated at all. Nevertheless, the introduction of the CPS can dramatically facilitate the know-how flows in two ways. First, the cyber-physical infrastructure assists with keeping the trade secrets within the EIONM with the network security solutions [21]. Second, the CPS prevents misuse of know-how by other network participants within the network with a block-chain technology.

The proposed model further divides know-how flows into the flows of production and organizational know-how. Technological know-how flows are related to the sharing knowledge, skills and competencies, which allow to add value in the production process: maximize the outcome with minimum resource inflows. These flows should be synchronized with tangible resource flows but pose a problem for automation. The flow of organizational know-how is the transfer of organizational abilities purposely developed by a company, explicitly embodied in regulations, manuals and other documentation. The major role in the facilitation of organizational know-how flow will be played by cognitive systems for decision support, such as participative cognitive mapping, but it more refers to the anthropogenic rather than to the cyber-physical system via human-human and human-machine interactions.

## 5  Allocation of Tangible Resources in the Manufacturing Network

In the final part of the paper, we consider the significant organizational task in the EIONM based on the cyber-physical infrastructure. It is the task of resource allocation among manufacturing facilities of independent companies interacting within the CPS of the EIONM (elementary control nodes (EA)). It is describes with the production program, links and nodes. Equipped with sensors, actuators and connection technologies, EA communicate with the network virtual decision-making center (VRC) to apply for the resources needed for their production programs (Table 1).

**Table 1.** Parameters and variables of the network resource allocation model

| Name | |
|---|---|
| Number of nodes | $N$ |
| Resource module, $i = 1 \ldots N$ | $S_i$ |
| Application of the i-th EA for the resource | $s_i$ |
| Resource directed to i-th node | $x_i$ |
| Need $i$-th network node in the resource | $r_i$ |
| Node priority (execution of operation on i node), $i = 1, \ldots N$ | $A_i$ |
| Production capacity of the program | $R$ |

The network's production capacity is equal to R. The CPS allocates resources within the EIONM's manufacturing facilities in order to meet requirements of the production program taking into account the resource constraints (Fig. 2).

**Network nodes $r_i$, i=1...N, and orders:**

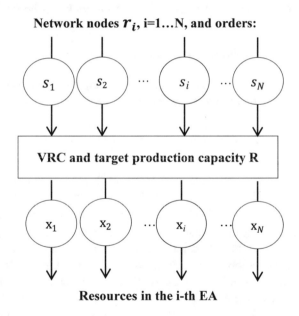

**Resources in the i-th EA**

**Fig. 2.** Two-tier organizational system for the EIONM

Table 2 presents the quantitative outcomes of the proposed procedure for the resource allocation in accordance with four methods. In the absence of a resource shortage: $\sum_{i=1}^{n} S_i \leq R$, the solution of the VRC is to satisfy of all received applications, i.e. each node receives the required amount of resources: $x_i = s_i$, $i = 1, \ldots N$.

In the event of a resource shortage: $\sum_{i=1}^{n} S_i > R$, or the sum of all applications exceeds the network production capacity.

**Table 2.** Quantitative outcomes of the application of the allocation methods

| Node | Applications in units of time | Methods of resource allocation when it is limited, equal to 30,000 operations in units of time | | | | | | | |
|------|------|------|------|------|------|------|------|------|------|
| | | Direct priority | | Reverse priority | | Competitive mechanism | | Open control mechanism | |
| 1 | 5000 | 4286 | 85,7% | 4383 | 87,7% | 5000 | 100,0% | 5000 | 100,0% |
| 2 | 3000 | 2571 | 85,7% | 3395 | 113,2% | 3000 | 100,0% | 3000 | 100,0% |
| 3 | 7000 | 6000 | 85,7% | 5186 | 74,1% | 7000 | 100,0% | 5333 | 76,2% |
| 4 | 4000 | 3429 | 85,7% | 3920 | 98,0% | 4000 | 100,0% | 4000 | 100,0% |
| 5 | 6000 | 5143 | 85,7% | 4801 | 80,0% | 6000 | 100,0% | 5333 | 88,9% |
| 6 | 8000 | 6857 | 85,7% | 5544 | 69,3% | – | – | 5333 | 66,7% |
| 7 | 2000 | 1714 | 85,7% | 2772 | 138,6% | 2000 | 100,0% | 2000 | 100,0% |
| Total | 35000 | 30000 | | 30000 | | 27000 | | 30000 | |

Under the mechanism of direct priority decision-making, the decision rule is as follows (1):

$$x_i = \min\{s_i, \ \gamma \times A_i \times s_i\}, \ \text{where } i = 1, \dots N. \tag{1}$$

Reverse priority mechanism is based on the principle that the less the need for an $i$-node resource is, the more efficient is its usage (2).

$$X_i = \min\left\{S_i, \ \gamma \frac{a_i}{s_i}\right\}, \ \text{where } i = 1, \dots N. \tag{2}$$

The distribution coefficient $\gamma$, as in condition (*), presents a graph (Fig. 3) of the function $x_i = x_i(s_i)$ that demonstrates that the maximum is reached at $s_i^*$, defined as $s_i^* = \gamma \frac{A_i}{s_i}$ or $s_i^* = \sqrt{\gamma A_i}$. The application of the reverse priority mechanism for the VRC allows to eliminate unjustified requests from the node (that prevents the resource shortage deterioration) and the situation $s_i^* > r_i$ does not occur. The CPS, according to the algorithm of the equilibrium organizational strategy $s_i^*$, fully meets the needs of all manufacturing facilities distributed in the EIONM. On the negative side, the volume of the allocated resources might be less than the real needs $r_i$, due to the cyber-physical effect of the following kind: the VRC does not receive reliable information about the real shortage of a particular resource $\left(\sum_{i=1}^{n} r_i\right) - R$.

Competitive Mechanism. The principle of action is to ban the rejections of applications from the EA because resources are needed to support the critical production requirements of the EIONM system as a whole. With this approach, those applications that are completely provided with a resource win the competition, but other applications remain in the system.

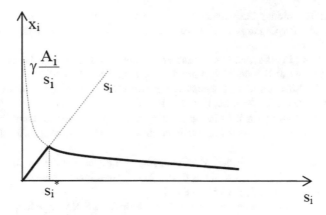

**Fig. 3.** Graph of the function $x_i = x_i(s_i)$

## 6   Conclusions

Implementation of the CPS provides a strong impetus to the expansion of cooperative networks in manufacturing. This trend will bring to life more and more EIONMs representing the systems of the highest complexity, which secure the interaction of heterogeneous resources, both physical and digital. The methodology for the analysis of these systems that is to be developed will be necessarily interdisciplinary. This publication lays a basic analytical framework where the loci for analysis are heterogeneous resource flows controlled by both the cyber-physical and anthropogenic systems. The future research avenue will be to develop an approach to analyze the EIONM as a system [13, 22] emerging from the interaction between cyber-physical and anthropogenic systems.

**Acknowledgements.** With the support of the Russian Foundation for Basic Research, grant 19-010-00257 A. Methodology for the analysis of the industrial enterprises and intangible industries in the conditions of information society and digitalization.

## References

1. Agolla, J.E.: Human capital in the smart manufacturing and Industry 4.0 revolution. In: Petrillo, A., Cioffi, R., De Felice, F. (eds.) Digital Transformation in Smart Manufacturing, pp. 41–58. IntechOpen, London (2018)
2. Boisot, M.: Knowledge Assets: Securing Competitive Advantage in the Information Economy. Oxford University Press, New York (1998)
3. Castells, M.: The Rise of the Network Society, 2nd edn. Wiley-Blackwell, West Sussex (2010)
4. De Mauro, A., Greco, M., Grimaldi, M.: A formal definition of big data based on its essential features. Libr. Rev. 1(3), 122–135 (2016)

5. Ebers, M.: Explaining inter-organizational network formation. In: Ebers, M. (ed.) The Formation of Inter-Organizational Networks, pp. 3–40. Oxford University Press, Oxford (1999)
6. Friedman, R.A., Podolny, J.: Differentiation of boundary spanning roles: labor negotiations and implications for role conflict. Adm. Sci. Q. **37**(1), 28–47 (1992)
7. Gnyawali, D., Madhavan, R.: Cooperative networks and competitive dynamics: a structural perspective. Acad. Man Rev. **26**(3), 431–445 (2001)
8. Henttonen, K., Blomqvist, K.: Managing distance in a global virtual team: the evolution of trust through technology-mediated relational communication. Strateg. Change **14**(2), 107–119 (2005)
9. Illmer, B., Kaspar, J., Vielhaber, M.: Cyber-physical effects on the virtual commissioning architecture. In: DS 87-5 Proceedings of the 21st International Conference on Engineering Design (ICED 17), vol. 5, pp. 169–178 (2017)
10. Karlik, A.E., Platonov, V.V., Iakovleva, E.A., Shirokov, S.N.: Experience of cooperation between St. Petersburg universities and industrial enterprises. In: Proceeding of IEEE 5th Forum Strategic Partnership of Universities and Enterprises of Hi-Tech Branches, Science. Education. Innovations, St. Petersburg, pp. 9–11. IEEE (2017)
11. Lee, E.A., Seshia, S.A.: Introduction to Embedded Systems: A Cyber-Physical Systems Approach, 2nd edn. MIT Press, Boston (2017)
12. Liu, X., Shahriar, M.-D., Nahian, S.M., Sunny, A., Leu, M.C., Hu, L.: Cyber-physical manufacturing cloud: architecture, virtualization, communication, and testbed. J. Manuf. Syst. **43**(2), 352–364 (2017)
13. Mittal, S., Risco-Martin, J.L.: Simulation-based complex adaptive systems. In: Mittal, S., Durak, U., Ören, T. (eds.) Guide to Simulation-Based Disciplines: Advancing Our Computational Future, pp. 127–150. Springer, Berlin (2017)
14. Monostori, L.: Cyber-physical production systems: roots, expectations and R&D challenges. Procedia CIRP **17**, 9–13 (2014). https://doi.org/10.1016/j.procir.2014.03.115
15. Platonov, V., Bergman, J.-P.: Cross-border cooperative network in the perspective of innovation dynamics. In: Wang, J. (ed.) Intelligence Methods and Systems Advancements for Knowledge-Based Business, pp. 150–169. IGI-Global, Hershey (2013)
16. Ring, P.S.: Processes facilitating reliance on trust in inter-organizational networks. In: Ebers, M. (ed.) The Formation of Inter-Organisational Networks. Oxford University Press, Oxford (1999)
17. Rowley, J.: The wisdom hierarchy: representations of the DIKW hierarchy. J. Inf. Sci. **33**(2), 163–180 (2007)
18. Schwab, K.: The Fourth Industrial Revolution. Crown Business, New York (2017)
19. Seiger, R., Huber, S., Heisig, P., Assmann, U.: A framework for self-adaptive workflows in cyber-physical systems. In: Becker, S., Bogicevic, I., Herzwurm, G., Wagner, S. (eds.) Software Engineering and Software Management, pp. 125–126. Gesellschaft für Informatik, Bonn (2019)
20. Talcott, C.: Cyber-physical systems and events. In: Wirsing, M., Banâtre, J.-P., Hölzl, M., Rauschmayer, A. (eds.) Software-Intensive Systems and New Computing Paradigms, vol. 5380, pp. 101–115. Springer, Berlin (2008)
21. Tuptuk, N., Hailes, S.: Security of smart manufacturing systems. J. Manuf. Syst. **47**, 93–106 (2018)
22. Volkova, V., Chernyy, Y., Leonova, A.: Development of information technologies and their emergence in concepts of cyberphysical system. J. Appl. Inform. **14**(1), 68–81 (2019)

# New Approach to Feature Generation by Complex-Valued Econometrics and Sentiment Analysis for Stock-Market Prediction

Dmitry Baryev[✉], Igor Konovalov, and Nikita Voinov

Peter the Great St. Petersburg Polytechnic University, St. Petersburg, Russia
baryev@yahoo.com, rogee.nok@gmail.com,
voinov@ics2.ecd.spbstu.ru

**Abstract.** The theory of complex-valued econometrics makes it possible to generate qualitatively new features that can be used in machine learning algorithms. Our study reveals the task of determining the long-term dependence of future companies' stock prices from a time-generated feature, i.e., a calculated tonality coefficient gained by methods of semantic analysis of texts from social networks. Data was gathered from the Twitter platform with the use of Big Data ETL-scenarios. The resulting data sets were used to train machine learning algorithms designed to work with Big Data technologies. A semantic coefficient was calculated on the basis of aggregated estimates for each day, with the further application of the methods of complex-valued econometrics. To demonstrate the new approach of feature generation, a complex-valued linear regression model based on the semantic coefficients and stock markets data was constructed. The outcome obtained by the new approach was compared with existing solutions in terms of accuracy. Finally, we demonstrate a possible route for impacting improvements of the existing algorithms for trading strategies using the complex-valued regression.

**Keywords:** Machine learning · Sentiment analysis · Complex-valued modeling · NLP · Big Data · ETL · Spark · Python · PySpark · Mongo DB · Twitter · Stock markets

## 1 Introduction

Machine learning is actively used to predict stock prices [1]. Millions of players attempt to maximize profits by selling and buying assets on stock exchanges every day [2]. According to the reports of the U.S. Securities and Exchange Commission [3], around 4,000 brokers are registered in the USA alone. Most of their clients prefer to use automated systems that allow them to buy and sell securities in seconds based on their own predictions. The share of such systems in the market in 2004 was already ¾ of all participants [4]. As a result, the fulfilment of predictions based only on time series becomes an increasingly difficult task. Millions of bidders using various algorithms (trend following strategies [5], weighted average pricing [6], determining local

© Springer Nature Switzerland AG 2020
D. G. Arseniev et al. (Eds.): CPS&C 2019, LNNS 95, pp. 573–582, 2020.
https://doi.org/10.1007/978-3-030-34983-7_56

shortages [7], etc.) gradually reduce the role and, accordingly, the possible profit from the dependencies they found. Thus, it becomes especially important to analyse the external information, not directly related to the process of trading on stock exchanges. One of the sources of such information can be the Internet. A text on the Internet has its own tonality – an emotional indicator expressing the author's attitude towards an object of the utterance. The main purpose of this work is to describe a new approach of feature generation that can be used in order to predict future values based on the regression model. This model uses the complex-valued coefficient, characterising the possible connection between statements in social networks in a natural language.

## 2   Data Acquisition and Conversion

The three possible sources of data were identified in this research: dedicated forums of brokers trading on financial exchanges, news aggregators, and social networks. The social network Twitter, which has more than 300 million active users [8] and an extensively documented API [9], were identified as a preferable source of information. For the subsequent analysis, a dataset of 1.8 million messages was collected from February 24 to April 20, 2016. It contains the references to "Brexit" – the process of the UK leaving the EU that resulted in a series of economic shocks [10].

For the pre-processing data sources, the open-source tool Hydrator was used to download full texts [11] by identifiers. For establishing an ETL module, the Bonobo framework [12] was used for receiving raw JSON-files. The processed data is loaded into the MongoDB database. The final format of the data is presented in Table 1.

In order to support continuous data processing, the Digital Ocean cloud platform [13] was used. The ETL scenarios, which were performing the data processing continuously, were run after the preliminary settings.

**Table 1.** Data format

| A | Record data | B | Composite field «user» |
|---|---|---|---|
| id | Record identifier | name | Account name |
| user | User | location | User's location |
| date | Publication date | verified | Verified account |
| text | Message text | desc | Account description |
| hashtags | Hashtags | followers | Number of followers |
| retweets | Number of retweets of message | name | Account name |

# 3  Semantic Analysis Model

Big Data is a term which describes data sets with high rates of volume, diversity and speed of appearance. Gigabytes of Twitter messages are an example of Big Data. Such large amounts of information are difficult to be handled with traditional methods of data processing.

The Apache Spark [14] open-source framework was chosen as the main data processing tool; it is included in the Apache Hadoop ecosystem [15]. Spark provides multiple access to the data stored in memory for processing information in RAM [16]. Programming language Python was chosen for Spark.

## 3.1  Data Preprocessing

The whole process of handling data by machine learning methods must be divided into two parts in order to obtain the resulting set of values for the tonality of messages from Twitter:

- Construction of the model and its training on the ready marked up data set.
- Applying the trained model to the target message set.

Nowadays, several data sets exist on the Internet–Twitter messages classified into groups: positive, negative and neutral. The most extensive collection is one of 1.6 million classified messages made by Stanford University [17].

For further work, data in the form of a special format - DataFrame, was loaded into Spark. Spark DataFrame is a distributed two-dimensional collection organized as a set of named columns.

In the subsequent processing of this set by machine learning methods using regular expressions, links to users and other sites were removed from all messages and hashtags were turned into plain text.

The pre-processing of the received data was enclosed in a pipeline – a set of the following text processing algorithms.

- Tokenizer performs a process of analyzing the input sequence of characters into recognized groups—lexemes, in order to get identified sequences—tokens;
- StopWordsRemover word filter clears the input from frequently used meaningless words.
- Hashing TF-IDF hashing vectorizer of the TF-IDF statistical measure is used to assess the importance of a word in the context of a document.
- String Indexer is applied to the resulting word frequency vectors.

The entire data set was divided into two subsamples: training and test sets with the ratio of 95% to 5%. The procedure described above was applied to both samples.

## 3.2  Training of the Machine Learning Model

Logistic regression (LR) [18] was chosen as the main algorithm for the implementation of the machine learning model. The LR-model was first applied in a training set containing both the pre-processing pipeline results and classification of the corresponding messages. The model trained on that sample was then applied to the test set in order to measure its accuracy.

The resulting classification, based on the obtained posterior probabilities of the belonging of objects to two classes of tonality, was compared with test indicators. The F-measure (3) – an aggregate criterion that represents the harmonic mean for the precision (1) and recall (2) metrics [19] – was used as the main metric for the model.

$$precision = \frac{TP}{TP + FP} = 0.7774, \tag{1}$$

$$recall = \frac{TP}{TP + FN} = 0.7925, \tag{2}$$

$$F - measure = 2 \cdot \frac{precsion \cdot recall}{precsion + recall} = 0.7849. \tag{3}$$

### 3.3  Applying the Model

Brexit affects a huge number of people and is closely connected with the sphere of finance – the results of the referendum have already influenced the European market structure.

In order to extend the functionality of Spark, SQLContext was created. It is an entry point for working with Spark [20]. Spark SQL allows performance of data processing. A dedicated data structure DataFrame is used.

Filtered data from the MongoDB database is fed to the input of the trained model. In order to improve the accuracy by using resulting probability, it is possible to define the boundaries of belonging to classes (negative, neutral and positive): 0.2 and 0.8, which is shown in Fig. 1. Negative (0–0.2), neutral (0.2–0.8) and positive (0.8–1).

The results of the partitioning are then aggregated using the sum of values of each group for each day. PySpark DataFrame is converted there to Pandas DataFrame for further work within the traditional data size.

## 4  Complex-Valued Feature Generation

The results of tonality analysis and stock trading were visualized. The dependence of the amount of each key on time is presented in Fig. 2 in the form of a stacked bar graph. It is shown that the amount of data is unstable and has constant outliers. This can be explained by a sharp social agenda of the event. For instance, there was an act of terrorism in Brussels [21] on March 22, 2016, which resulted in a flurry of messages.

**Fig. 1.** Boundaries of selection of classes of Twitter messages tonality

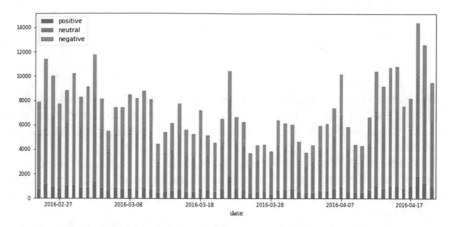

**Fig. 2.** Stacked bar graph of positive, neutral and negative semantics.

Three tonality indicators were aggregated into one coefficient (4), which is proposed by authors:

$$k = \frac{pos - neg}{pos + neut + neg}.$$ (4)

As stated above, there is a huge variety of stock trading algorithms. The main purpose of this study is to demonstrate a new approach of feature generation that can be used in order to predict more accurate future values in trading strategies. The main distinguishing feature of this proposed approach is using the complex-valued econometric theory for feature generation.

The complex-valued econometrics theory is a vast field for researches. The main provisions of this theory were described several years ago [22], but this area is still under development.

The received semantic coefficients and relevant indicators of the shares (adjusted closing price and volume per day) were compared for the purposes of the feature generation. The FTSE (Financial Times Stock Exchange 100 Index) shared stock index was taken. This index displays the British stock markets status.

In order to carry out the complex-valued analysis, the volume and the closing price were brought to dimensionless form and centered. The target variable in our analysis is the closing price. The remaining two quantities, the volume and the semantics, were converted into a complex variable, with the former being its real part, and the latter being the imaginary one.

To check the correctness of the choice of values, it is necessary to perform a correlation analysis of the obtained values. However, as shown in the theory of complex-valued econometrics [22], the generally accepted methods for calculating pair correlation are not applicable if there are interrelations between the real and imaginary parts. In this regard, the complex correlation coefficient was calculated (5):

$$r_{cXY} = \frac{\sum_t (y_{rt} + iy_{it})(x_{rt} + ix_{it})}{\sqrt{\sum_t (y_{rt} + iy_{it})^2 \sum_t (x_{rt} + ix_{it})^2}}. \quad (5)$$

In order to obtain a complete picture of the pair dependence, iteratively for the offset from one quantity to another, both coefficients were found: the complex and standard Pearson coefficients, implemented in the NumPy package [24]. The results are presented in Fig. 3.

As can be seen in Fig. 3, the difference between the two coefficients is small, with a slight shift of the complex one. This fact might be explained by the appearance of hidden relationships between stock prices and the volume and semantics manifesting over the time shifts.

In this case, the highest negative correlation is achieved with a slight forward and backward shifts. Thus, according to the data obtained, we can speak about some possible influence of the complex-valued components on the trading process several days before the fact of trading and vice versa – several days after.

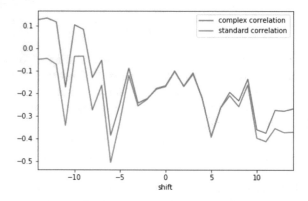

**Fig. 3.** Dependence of complex correlation and Pearson coefficient on the time shift

Therefore, some relationship between the generated complex value and the target index – the closing price – was proved. In that regard, the generated feature can be estimated in terms of possible practical impact in improving the existing algorithms for trading strategies. To this end, two types of construction were carried out by one of the most common regression analysis models – linear regression.

As a standard LR-model, the object LinearRegression without fit interception from the Scikit-Learn tools library [25] was used. The input parameters for this function are the same two columns, which are the basis of the generated complex-valued feature – the *volume* and *semantic coefficient*.

According to the complex-valued econometrics theory [22], in case of a relation between the real and imaginary parts of a complex value, the standard linear regression equation, as well as the standard correlation coefficient, does not work appropriately.

In that regard, the following formula (6) was used to build a complex linear regression model. Since all values are centered, the formula for the parameter of the model is reduced to:

$$(b_0 + ib_1) = \frac{\sum_t (y_{rt} + iy_{it})(x_{rt} + ix_{it})}{\sum_t (x_{rt} + ix_{it})^2}. \tag{6}$$

Generally, in stock market price predicting analysis, forecast for the next day is the most significant, because longer-term forecasts, in one way or another, will have to be adjusted by incoming values in the future. Therefore, in order to evaluate the generated features, both models will be trained on identical iterative forward moving samples with different lengths.

RMSE (Root-mean-square error) was chosen as the comparison metric for two linear regression models. In this case, RMSE is implemented as the root results of the *mean_squared_error* [26] function from Scikit-Learn tools library.

Figure 4 shows an example of the comparison of RMSE between the complex and the standard linear regression models. As can be seen from the figure, the complex model error line usually lies below the standard regression model, i.e., the complex model RMSE is smaller. The average RMSE of the complex model is 0.007869, and that of the standard model is 0.009752. Thus, in this case, the complex linear regression model, most likely, will give a better result than the standard regression.

In order to verify the stability of the obtained results, an iterative resizing of the training set should be carried out and the mean RMSE for a current training sample size calculated. A dependency graph of the resulting mean RMSE on the size of the training sample is presented in Fig. 5.

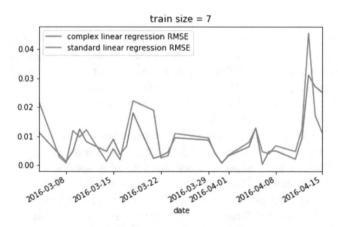

**Fig. 4.** RMSE of complex and standard regressions

Figure 5 shows how RMSE of the complex and standard linear regression models changes depending on the sample size. It is easy to see that a complex regression model

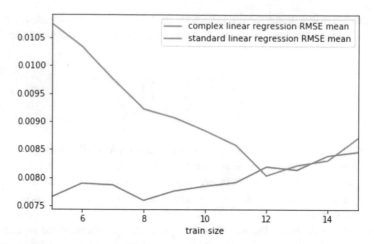

**Fig. 5.** Dependence of mean RMSE of complex and standard regressions on a train sample size

based on the generated complex feature appears to be better than the standard linear regression. The difference between RMSE reaches more than 40% with a small size of the training sample (5 elements). With a gradual increase of the training sample size, RMSE of the standard model decreases, while the complex one slowly increases. After that, their RMSEs becomes approximately equally ascending.

In addition to the RMSE-case, all described earlier actions were performed with MSE (mean square error) and MAE (mean absolute error) metrics. All obtained results were similar to the RMSE-case.

## 5    Conclusion

The paper presents new approaches of generation features that can be used in regression tasks to improve existing or create new trading strategies. The implementation of the new approaches of feature generation was carried out using the example of determining the long-term dependence of future companies' stocks prices on a time-generated feature – a calculated tonality coefficient gained by methods of semantic analysis of texts from social networks.

For semantic analysis, Big Data ETL scenarios were used to aggregate and pre-process Twitter messages associated with Brexit, from a micro-blogging platform.

The methods of the complex-valued econometrics theory were applied to the generated complex feature. The complex correlation coefficient between the generated feature and the closing price was calculated. It appears that negative correlations between the feature and the closing price are the strongest several days before and after the trading. We have compared the complex correlation coefficient with the real-valued one and found that the discrepancy in the results of the complex and standard correlation increases with the shift of the feature over time.

The accuracy of the complex and standard linear regression models with the proposed feature was compared on the test data set. We found that the complex-valued regression model based on the generated complex feature performs better in terms of RMSE than the standard linear regression.

**Acknowledgements.** The study was supported by the Russian Foundation for Basic Research, Grant No. 19-010-00610\19 "Theory, Methods and Techniques for Forecasting Economic Development by Autoregressive Models of Complex Variables."

# References

1. Chan, E.: Algorithmic Trading: Winning Strategies and Their Rationale, 656 p. Wiley, Hoboken (2013). ISBN 978-1118460146
2. Harris, L.: Trading and Exchanges: Market Microstructure for Practitioners, 1st edn., 304 p. Oxford University Press, Oxford (2002). ISBN 978-0195144703
3. Company Information about Active Broker-Dealers. U.S. Securities and Exchange Commission. https://www.sec.gov/help/foiadocsbdfoiahtm.html
4. MacGregor, A.: As automated trading takes over markets, rational human investors matter even more. http://www.abmac.com/industry-insight/as-automated-trading-takes-over-markets-rational-human-investors-matter-even-more
5. Merello, S., Ratto, A.P., Oneto, L., Cambria, E.: Predicting future market trends: which is the optimal window? In: Oneto, L., Navarin, N., Sperduti, A., Anguita, D. (eds.) Recent Advances in Big Data and Deep Learning. INNSBDDL 2019. Proceedings of the International Neural Networks Society, vol. 1. Springer, Cham (2020)
6. Yang, R., He, J., Xu, M., Ni, H., Jones, P., Samatova, N.: An intelligent and hybrid weighted fuzzy time series model based on empirical mode decomposition for financial markets forecasting. In: Perner, P. (ed.) Advances in Data Mining. Applications and Theoretical Aspects, ICDM 2018. Lecture Notes in Computer Science, vol. 10933. Springer, Cham (2018)
7. Galimberti, J.K., Suhadolnik, N., Da Silva, S.: Comput. Econ. **50**, 393 (2017). https://doi.org/10.1007/s10614-016-9591-2
8. Twitter's Q3 earnings by the numbers. Fast Company. https://www.fastcompany.com/90256723/twitters-q3-earnings-by-the-numbers
9. Makice, K.: Twitter API: Up and Running. Learn How to Build Applications with the Twitter API, 416 p. O'Reilly, Sebastopol (2009). ISBN 978-0596154615
10. Brexit and the UK's Public Finances. Institute for Fiscal Studies (IFS Report 116), May 2016. https://www.ifs.org.uk/uploads/publications/comms/r116.pdf
11. Hydrator by samdark. https://github.com/samdark/hydrator
12. Bonobo. Data-processing for humans. https://www.bonobo-project.org
13. Tagliaferri, L.: DigitalOcean eBook: How to Code in Python. DigitalOcean, New York City. ISBN 978-0-9997730-1-7
14. Karau, H., Konwinski, A., Wendell, P., Zaharia, M.: Learning Spark: Lightning-Fast Big Data Analytics, 304 p. O'Reilly, Sebastopol (2015). ISBN 978-5-97060-323-9
15. Frampton, M.: Mastering Apache Spark, 318 p. Packt Publishing Ltd., Birmingham (2015). ISBN 978-1783987146
16. Karau, H.: High-Performance Spark: Best Practices for Scaling and Optimizing Apache Spark, 358 p. O'Reilly, Sebastopol (2017). ISBN 978-1491943205

17. Go, A., Bhayani, R., Huang, L.: Twitter Sentiment Classification using Distant Supervision. https://cs.stanford.edu/people/alecmgo/papers/TwitterDistantSupervision09.pdf

18. Lyman Ott, R., Longnecker, M.T.: An Introduction to Statistical Methods and Data Analysis, 1296 p. Cengage Learning (2015). ISBN 978-1305269477

19. Hackeling, G.: Mastering Machine Learning with Scikit-Learn Paperback, 238 p. Packt Publishing Ltd., Birmingham (2014). ISBN 978-1783988365

20. Gulati, S., Kumar, S.: Apache Spark 2.x for Java Developers: Explore Big Data at Scale Using Java APIs, 350 p. Packt Publishing Ltd., Birmingham (2017). ISBN 978-1787126497

21. Brussels explosions: What we know about airport and metro attacks. BBC News. https://www.bbc.com/news/world-europe-35869985

22. Sergey, S.: Complex-Valued Modeling in Economics and Finance, 318 p. Springer, New York (2012)

23. FTSE UK Index Series. https://www.ftse.com/products/indices/uk

24. numpy.corrcoef – NumPy v1.16 Manual. https://docs.scipy.org/doc/numpy/reference/generated/numpy.corrcoef.html

25. sklearn.linear_model.LinearRegression – Scikit-Learn 0.20.3 Documentation. https://scikit-learn.org/stable/modules/generated/sklearn.linear_model.LinearRegression.html

26. sklearn.metrics.mean_squared_error – Scikit-Learn 0.20.3 Documentation. https://scikit-learn.org/stable/modules/generated/sklearn.metrics.mean_squared_error.html

# Flexographic Printing of Optical Multimodal Y-Splitters for Optical Sensor Networks

Keno Pflieger[1,2] and Ludger Overmeyer[1,2(✉)]

[1] Institute of Transport and Automation Technology, Hannover, Germany
{Keno.Pflieger,Ludger.Overmeyer}@ita.uni-hannover.de
[2] Cluster of Excellence PhoenixD (Photonics, Optics, and Engineering – Innovation Across Disciplines), Hannover, Germany

**Abstract.** Flexographic printing on Polymethylmethacrylat (PMMA) substrates is a promising technology for the cost-effective production of large-scale optical networks. It allows for numerous applications in the fields of data transmission, sensing and point-of-care systems. This article introduces the flexographic printing of optical multimode Y-splitter as a fundamental element for these networks. It is printed with an optical grade acrylic polymer on flexible PMMA substrates. We investigated geometric and optical properties of splitters with angles up to 20° and achieved attenuations due to the Y-splitter below 1.3 dB at 638 nm.

**Keywords:** Waveguide · Y-splitter · Printing · Flexographic

## 1 Introduction

Modern sensor technology developed quite far. Often, process parameters must be controlled spatially resolved in order to have a stable process window. An example is the temperature controlling on the surface of an engine. For this purpose, discrete sensors are mounted close to the component surface. Often these cannot be attached directly to the component, but only in close proximity. This can lead to measurement errors. Furthermore, electrical communication systems are prohibited in some environments because of flammable or explosive substances.

The concept of planar optronic systems (POS) offers a solution to these technological limitations. They consist of a two-dimensional optical network with an array of sensors [1]. These purely optical sensors are capable of measuring temperature [2], strain [3] or chemical concentrations [4]. Additionally, printed optical sensors on flexible substrates can be easily mounted on most surfaces, creating dense sensor matrices with a high integrability due to their minimal height and flexible substrate. Furthermore, the heat dissipation and the electromagnetic compatibility of optical waveguide is preferable compared to classical electronic sensors.

Nevertheless, this vision implies the manufacturability of printed interconnecting devices like Y-splitters. Otherwise, a high number of optical devices would be necessary to replace the bulky electrical devices. This would not reduce the footprint of the measurement setup.

© Springer Nature Switzerland AG 2020
D. G. Arseniev et al. (Eds.): CPS&C 2019, LNNS 95, pp. 583–591, 2020.
https://doi.org/10.1007/978-3-030-34983-7_57

## 2  Theoretical Considerations for Printed Optical Waveguides

Optical waveguides are capable of carrying light waves along a defined path with very low signal losses. They consist of a concentric two-layer system. The inner phase is called core and the name of the outer phase is cladding. The propagation of light in waveguides is based on total reflection due to the higher refractive index of the core material in contrast to the outer cladding. Today, optical fibres are used for the digital long-range communication between two network nodes. They build the backbone of the modern information society. Additionally, optical waveguides can be used for large-scale optical networks.

Wolfer et al. (2014) showed that such structures can be produced by flexographic printing. Those waveguides form dome-like shapes because of surface tension. The cross sections of these waveguides are characterised by parabolas [5].

For the creation of large optical networks, the ability to create printed optical splitters is necessary. A splitter is an optical element that takes a single optical input and creates multiple optical outputs. This work presents a way to produce 1:2-splitter.

## 3  Methods for Waveguide Creation

For the production of Y-splitters a flexographic printing machine was used. Flexographic printing is a relief printing method. It uses three cylinders that are transferring the polymer from one onto another and finally onto the substrate.

The first cylinder is called anilox roller. Its surface is engraved with a honeycomb structure. The function is to collect liquid polymer from a chambered doctor blade. On the perimeter of the second cylinder, an invariable printing plate with elevated structures on the surface is mounted. If both cylinders get in contact, polymer will be transferred from the first roller to the second one on the elevated structures. From there, the structure is mirror-inverted transferred onto the substrate. Finally, the polymer is cured by a UV-radiation source.

Since the transferred volume is limited, the possible structure height is also limited. Subsequently, the structure needs to be overprinted multiple times in order to create a waveguide structure. This work uses 10 printing runs (Fig. 1).

### 3.1  End Facet Preparation

In order to measure the Y-splitter attenuation precisely the coupling losses have to be low. Otherwise, it is hard to distinguish between losses directly linked to the splitter and losses at the waveguide facets. In this paper, the method of breaking is used [6]. For repeatable breaks, the substrate is punctuated in straight lines perpendicular to the waveguide. The locally weakened substrate marks the axis on which the break occurs. A laser cutter was used for this task. Finally, the substrate was broken manually along the punctuated line. With this method, one gets very clean facets along a fixed axis.

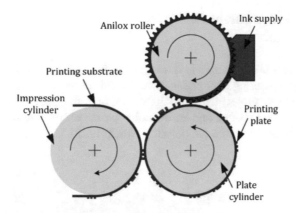

**Fig. 1.** Schematic representation of flexographic printing process [6]

## 3.2    Construction of Optical Power Splitter

For the flexographic printing process, a printing plate is required. On the printing plate, an inverted version of the desired image is engraved. These plates are made of a photosensitive polymer, which is processed with a laser in order to create the mask.

Figure 2 shows the structure of the splitters on the printing plate. The splitter is made of an input line and two output lines with a line width d of 200 μm. At the intersection point of both lines, the line width grows to 400 μm. The output lines are at an angle α. α varies from 2° to 20°.

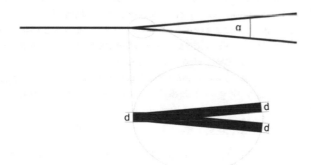

**Fig. 2.** Structure used for the creation of the printing plate. The black parts are elevated on the plate. The line width d is 200 μm. The angle α varies from 2° to 20°

# 4   Optical Measurement Setup for Waveguide Characterization

The quality of the printed splitters is characterized by their optical losses. These losses are quantified by measuring their attenuation. For this purpose, an optical measurement setup which allows to project a circular light spot onto the waveguides end facets has been installed. A beam profiler and an Ulbricht sphere can be placed at the waveguide end. These detectors determine the output power (Fig. 3).

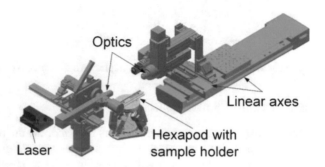

**Fig. 3.** The used setup is capable of measuring the attenuation of optical fibers. As a light source can a laser or a LED is used. For the power measurement an Ulbricht sphere and a beam profiler are available

In a first step, the sample is positioned with the help of a hexapod and linear axes, such that the laser source and the optical detector are aligned. There are two optical detectors available. The first one is an Ulbricht sphere for measuring the total power, which is emitted from the waveguide and the substrate. The second detector, a beam profiler, captures the spatial power distribution in a relative scale. These two detectors combined are capable of measuring the total power coming from the sample and the portion emitted from the waveguide compared to the scattered light.

## 4.1   Calculation of the Attenuation

Power losses in a waveguide occur due to absorption and scattering. These losses in a Y-splitter have multiple reasons:

- coupling losses at the facet surface due to reflections;
- absorption losses by the waveguide material itself;
- impurities in the bulk material;
- geometrical defects, for example, scratches or bubbles on the surface.

Optical losses of a sample are always a combination of those four effects. The attenuation $A$ is in this context defined as the quotient of the input power $P_i$ to the output power $P_0$:

$$A = 10 \log_{10} \frac{P_1}{P_0}. \tag{1}$$

The attenuations given in this paper are compensated by the coupling losses, losses due to the material absorption and impurities. Only losses caused by the geometrical deviation of Y-splitters from the ideal tube-like geometry are considered. This is achieved by measuring first the attenuation a1 of a straight waveguide with the same length. It was created in the same printing charge. Secondly, by measuring the attenuation a2 of the y-splitter and finally subtracting both values:

$$A_s = A_2 - a_1. \tag{2}$$

### 4.2 Calculation of Separation Circles

Since flexographic printing uses hydrostatic effects to create the desired structure, the results are directly linked to the surface tension. This means there are some constrains to the freedom of design. One of those constrains is the inability to print sharp edges. Sharp-edged surfaces are not a stable condition. They will turn in a sphere like surface (see Fig. 4). This has negative consequences for the light guiding properties of Y-splitters. The round surface offers a larger scattering cross-section than a sharp edge. In the consequence, the sphere radius or its 2D representation should be small. In order to compare different approaches, those radiuses should be determined. This paper defines it as the circle, which fits in the gap between both arms in a specific height. The used height for fitting is half-way between the highest and the lowest points of the confocal microscope measurement.

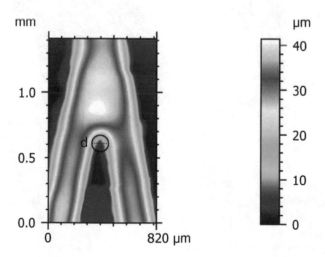

**Fig. 4.** Measurement of Y-splitter with a confocal microscope. The height is color-coded. The separation circle and its diameter $d$ is marked

## 5   Structural Investigation of Printed Y-Splitters

In Fig. 5 is the confocal measurement of different manufactured splitters drawn. It shows the height depending on the position. The height is drawn on the z-axis and it is additionally color-coded. The used linewidth of the printing plate is 200 μm, but the linewidth of the manufactured structures is between 350 μm and 400 μm. This is due to the printing process itself. The printing plate is pressed against the substrate during the printing process. The liquid polymer is squeezed out from the gap between the printing plate and the substrate. In the consequence, the area to the left and right from the line gets wetted. This leads to a higher real line width compared to the printing plate. This phenomenon can be seen all four samples.

The 2°-sample has a very low angle α. Therefore, the gap between both output waveguide arms is narrow over a long distance. This leads to multiple connections between both arms. These form when the surfaces of both arms touch. Both volumes open up and form a combined surface. The newly formed channels widen over time.

Additionally, one arm has a greater height. This means that it has more volume accumulated. This is expected since both arms are connected through the flow channels and the system tries to minimize its surface. This should lead in a liquid, non-cured phase to the consumption of one arm by the other. Nevertheless, the sample was cured, thus the consumption process has been frozen.

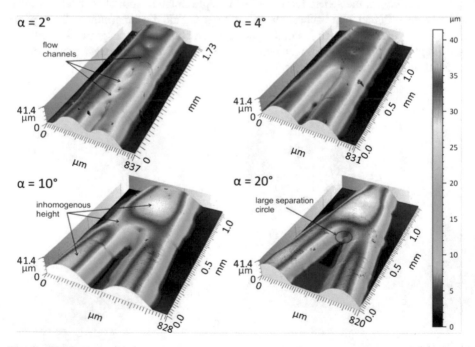

**Fig. 5.** Three-dimensional confocal measurements of Y-splitters with a specified line width of 200 μm. The opening angle α is 2°, 4°, 10° and 20°

In addition, the point where the arms separate is shifted by 10 mm along the waveguide axis from the location on the template. Since the printed structures have a higher line width than on the template and the separation angle is small, it takes a long distance before the waveguides separate. With a diameter of 48 μm, the separation circle is small.

The 4°-sample has a homogenous height and both arms are sharply separated. Due to the bigger angle, the shift of the separation point is only 2 mm. The height of the two printed arms is in the same range as the input arm. This is an important factor for a good photoconductivity. The separation circle diameter is 65 μm.

The 10° and 20°-samples look similar. Both have sharply separated arms, and there is no shift of the separation point. The separation circle is large for both samples. The diameter is 107 μm for the 10°-sample and 141 μm for the 20°-sample. This is significantly more than for the first two samples. Furthermore, the waveguide height is inhomogeneous. Both have the highest point before the arms are separating from each other. This point is where the waveguide has its greatest line width. Right next to the separation point is the lowest height of the investigated waveguide.

## 6 Measurement of Attenuation

Figure 6 shows the splitter losses depending on the splitter angle. These values are compensated for conduction losses. These losses are 1.28 dB/cm. The measured attenuations of the manufactured Y-splitter have a high degree of variation. The best

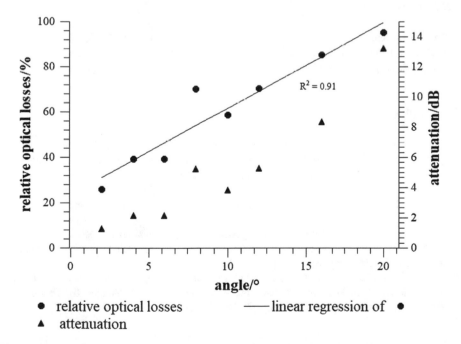

**Fig. 6.** Measurement of the relative losses (black curve) and attenuation (grey curve) depending on the splitter angle compensated for the conduction losses

results show losses of 26.9% at an angle of 2° and the most attenuation occurs at a high angle of 20°. 95.3% of incoming power was lost in this splitter. The loss function shows a linear behaviour.

# 7   Discussion

The results show that a small angle between both output arms is highly desirable. Even though the arms of the 2°-sample where over a wide range not completely separated, it had the lowest attenuation of all samples. This gives the motivation to make some modifications in the design of the printing plate in order to ensure a clean and predictable point of separation. One idea is to create such a modification that could be a curvature of the output arms. This would increase the distance between the two arms rapidly. It would reduce these scattering centers and increase the photoconductivity.

This inhomogeneous height distribution in combination with the big separation circle makes high separation angles for the production of Y-splitters quite unattractive. These findings confirm the thesis, that small separation angles are desirable.

Furthermore, it must be investigated how to reduce the separation circle. This is one of the main scattering centers. It needs to be investigated how to reduce the separation angle below 2°. This should create even smaller separation circles and consequently reduce the scattering.

In addition, it would be conceivable to work on the tip of the splitter with a second subtractive process to reduce the rounding even further. This could be, for example, an ablative laser process.

**Acknowledgments.** Funded by the Deutsche Forschungsgemeinschaft (DFG, German Research Foundation) under Germany's Excellence Strategy within the Cluster of Excellence PhoenixD (EXC 2122, Project ID 390833453).

# References

1. Overmeyer, L., Wolfer, T., Wang, Y., Schwenke, A., Sajti, L., Roth, B., et al.: Polymer based planar optronic systems. In: Proceedings of LAMP2013 – the 6th International Congress on Laser Advanced Materials Processing, Niigata, Japan (2013)
2. Xiao, Y., Pichler, E., Hofmann, M., Bethmann, K., Köhring, M., Willer, U., et al.: Towards integrated resonant and interferometric sensors in polymer films. Proc. Technol. **15**, 692–702 (2014)
3. Sherman, S., Zappe, H.: Printable Bragg gratings for polymer-based temperature sensors. In: 2nd International Conference on System-Integrated Intelligence. Challenges for Product and Production Engineering, vol. 15, pp. 703–710 (2014)
4. Kelb, C., Reithmeier, E., Roth, B.: Foil-integrated 2D optical strain sensors. In: 2nd International Conference on System-Integrated Intelligence. Challenges for Product and Production Engineering, vol. 15, pp. 711–716 (2014)

5. Wolfer, T., Bollgruen, P., Mager, D., Overmeyer, L., Korvink, J.G.: Flexographic and inkjet printing of polymer optical waveguides for fully integrated sensor systems. Proc. Technol. **15**, 522–530 (2014)
6. Wolfer, T., Bollgruen, P., Mager, D., Overmeyer, L., Korvink, J.G.: Printing and preparation of integrated optical waveguides for optronic sensor networks. Mechatronics **34**, 119–127 (2016)

# Complex Expert Assessment as a Part of Fault Management Strategy for Data Storage Systems

Mikhail B. Uspenskij, Svetlana V. Shirokova<sup>(✉)</sup>,
Olga V. Mamoutova, and Vladimir A. Zhvarikov

Peter the Great St. Petersburg Polytechnic University, St. Petersburg, Russia
{mikhail.uspenskiy, olga.mamoutova}@spbpu.com,
swchirokov@mail.ru

**Abstract.** Fault management strategy for a data storage system usually takes into account methods of error and anomaly detection, fault diagnosis, root cause of faults analysis and failure prediction. There is a variety of methods in the field of reliability and fault tolerance that are applicable for an enterprise data storage fault management system. However, contemporary implementations of fault management strategies lack automated support for incorporation of particular expert knowledge. Informational approach as an element of complex expert assessment is one of the ways to engage expert knowledge in the fault management strategy. The methodology presented in the paper presents a new comprehensive approach to the problem of fault management with a data storage system as an application example.

**Keywords:** Enterprise data storage · Fault diagnosis · Expert systems · Decision support systems · Management systems

## 1 Introduction

The ideal state of a storage system is autonomous operation without the need for human intervention. In the presence of faults, the desired autonomy can be achieved only partially: integrated diagnostic tools can detect and predict problems, but only embedded redundancy schemes can be employed to allow automatic repair. The rest of occurred or predicted critical situations still need reactive technical support with proper knowledge of what changes need to be made [1].

A decision support system intents to support this last step of a health lifecycle: an appropriate fault containment procedure should be recommended after the list of potential faults and their causes is derived from diagnostic information. The recommended action can be to run certain control and configuration operations, to enable a self-healing procedure or to engage an operator or a technician. Due to a big amount of deciding factors, a decision support system for enhanced decision making should heavily use expert knowledge [8].

Complex expert assessment methods significantly increase the objectivity of direct expert assessments and are usually applied during technical, technological or

D. G. Arseniev et al. (Eds.): CPS&C 2019, LNNS 95, pp. 592–600, 2020.
https://doi.org/10.1007/978-3-030-34983-7_58

organizational innovations in presence of increased requirements for a thorough analysis of problem situations, design exploration and project components [4]. However, when failures occur in a data storage system and it is necessary to make immediate decisions in response to the situation, application of complex expert assessment methods for implementation of a decision support system is also desirable [12, 20, 21].

More objective assessments allow us to obtain methods that are based on dividing large uncertainties into smaller ones, which makes them better explorable. Hence it is advisable to apply the methods of structuring and methods of system analysis of goals. Such methods include methods that were developed to evaluate the components of hierarchical structures: a method of complex expert procedure by detailing the evaluation criteria and taking into account their weights, proposed in the PATTERN method, and the pairwise comparison method developed by T. Saati. If the system is multi-level, or when making a decision requires taking into account several levels of problem, it becomes convenient to use the G.S. Pospelov solving matrix method: the idea behind this method is to perform the stratified (level-by-level) dismemberment of a large uncertainty into smaller ones [5–7].

In this paper complex expert assessment methods based on the goals structuring and Denisov's informational assessments are studied and applied. This allows making a more objective analysis of problems with uncertainty by including information models for analyzing the situation and taking into account the mutual influence of the evaluated components. In order to implement this complex methodology as a part of a diagnostic software suit the authors propose to utilize an ontological approach both to provide experts with data for assessment, and to store the obtained results to be used in a decision support system [22].

## 2  Decision Support Model

### 2.1  Input and Output Data

In general, a storage system consists of several subsystems, which can provide data path functions or/and control path functions. Key functional components of a storage system that can be analyzed during diagnostics are storage processors, disks and various network devices. Additionally, for a highly available storage system, each subsystem employs some type of redundancy scheme [5].

From a diagnostics point of view we consider three general types of fault for any component: a failure, when the component can no longer perform its functions and requires a repair; an error, when the component still operates, but lacks a full performance; and a pre-failure (predicted failure), when the component operates without explicit failure symptoms, but shows some signs of a future component failure. We expect that failures and errors can be diagnosed based on symptoms in actual monitoring data, while pre-failures can be detected using data science methods. Table 1 gives some typical examples of components' properties and possible errors [10, 12].

Root cause analysis, performed during system diagnosing, can potentially point to a reason of a fault or an error: it could be a software error, hardware defects, and ageing,

**Table 1.** Components of a storage system and its characteristics

| Component | Examples of properties (data type) | Error mode examples |
|---|---|---|
| Storage processor | memory and processor capacity available for user processes (real); storage volume count (int); storage pool count (int); | capacity depletion; performance degradation; software hangup; |
| Disk | capacity (real); operating lifecycle (real); | sector failure; controller error; |
| Networking device | bandwidth (float); connectivity (int); connection type (enum); | throughput reduction; configuration error in software; firmware errors; traffic congestion; |

configuration errors or environmental factors. In some cases, a definite reason for a fault cannot be determined [8].

Depending on how bad the problems are and how close to the actual failure the storage system is each status can be characterized with a severity level: ok, warning, vulnerable and critical. If no problems are detected, then the overall system health is "ok". Depending on a set of found problems an aggregate status of the storage system can be further classified by the worst possible outcome as one of the following: user access denial, data loss and quality of service loss. For example, if a storage pool loses one disk, the severity level is "warning", if it loses $n$ out of $k$ disks, then the severity level is "vulnerable", and if more than $n$ disks are lost, then the severity level is "critical". In all those cases the aggregate status can be classified as "data loss".

When diagnosing shows a severity level of an aggregate system status other than "ok", an appropriate recovery action has to be performed in order to contain the fault. In some cases, depending on implemented redundancy schemes, the decision to activate said redundancy scheme is obvious and the corresponding recovery action can even be performed in a self-healing mode. In other cases, the decision demands an intervention into the operating system and engaging with technical support. It implies some kind of trade-off and the decision becomes not so obvious. Examples of these external actions are component repair or replacement, system reconfiguration, remote or on-site extensive diagnostics and external data back-up.

As a result, a comprehensive decision process should take into account all available information, which can be grouped into four categories:

- Features of the action that is being decided upon: a monetary cost, an expected efficiency in terms of system recovery or data availability, and risks in terms of the attainability of the expected result [9, 11, 22].
- Results of diagnostics: a severity level of estimated status of the system and its components, identified the root cause.

- Data, that a diagnosis is based on: aggregated health and performance monitoring data, detected symptom, recognized fault, obtained prediction with a certain time frame, and confidence level.
- Features of components that are considered problematic: its system functions and properties, current and average utilization, employment in redundancy scheme, etc.

It should be noted that a particular set of available data can vary depending on a particular implementation of diagnostics for a particular storage system. However, types and groups of heterogeneous data, as well as types of external actions can be expected to be consistent for a wide range of systems.

## 2.2 Justification of the Need for an Information Approach

Direct expert assessments not only carry inherent narrowly subjective features from individual experts, but also possess collective subjective features, which do not disappear when processing the results of a survey and can even be further strengthened by the methods for improving the coherence of expert opinions. Therefore, to obtain more reliable estimates it is necessary to apply an approach to improve the methods of processing the survey results, highlighting and taking into account the rare and contradictory opinions of experts and increasing the objectivity of the survey [24–27].

Methods to organize complex expert assessment use information assessments, facilitate the calculation of a generalized assessment and provide the opportunity to take into account several criteria, reducing the assessment of many criteria to a single information assessment. In addition, when using the information approach, it is possible to evaluate the project in time and taking into account the dynamics and progress of the project and allow refining the estimates by taking into account the mutual influence of the evaluated components [4, 13–15, 23].

## 2.3 Evaluation Criteria

When deciding on possible actions (for example, the need to contact the technical service), it is necessary to take into account heterogeneous characteristics.

In accordance with the informational approach, for each evaluation criterion, an expert evaluates the degree of satisfaction (i.e. the probability of achieving the goal) and the probability of using the criterion [6]. With these two probabilities, the potential (i.e. significance) of the criterion can be calculated [24–27].

$$H_i = -q_i \log(1 - p_i), \tag{1}$$

where $H_i$ is information potential of the $i$-th criterion, bit, $p_i$ is the probability of achieving the goal when using the $i$-th criterion, $q_i$ is the probability of using the $i$-th evaluation criterion to achieve the appropriate goal.

For a storage system, this analysis is possible within a certain period of time (system parameters) by comparing changes in information assessments over time. There are two ways to measure $H_i$:

- in terms of the probability of $p_i$, that is, in our case, this is a probabilistic assessment of the forecast – "the problem is predicted";
- by means of deterministic characteristics of the perceived information, that is, on the basis of specific indicators – "the problem is detected".

Using two methods for determining $H_i$ (the potential, the possibilities of a given outcome; for example, contacting technical support) makes it possible to calculate $n_i$:

$$n_i = J_i/H_i, \tag{2}$$

where $n_i$ means the scope of the concept of the $i$-th object under study; information potential $H_i$ is calculated through probability $p_i$; and $J_i$ is measured ($J_i$ is information about the number of the $i$-th object under study, taking into account the significant step of changes in its value $\Delta A_i$: $J_i = A_i/\Delta A_i$).

Then, by estimating the predictive $p_i$ we can get an estimate (in terms of percentage) both on the basis of the forecast (prediction of the problem situation) and on the basis of the found problems.

## 2.4   Development of a Complex Examination Model

The first branch is characterized by the deterministic characteristics of the perceived information, that is, on the basis of specific indicators – "the problem is detected", using current data monitoring of a storage system [16–19].

The second branch evaluates $H_i$ through the likelihood of $p_i$, that is, in our case, a probabilistic assessment of the forecast – "the problem is predicted".

As a result of the assessment, we see that on the basis of current indicators, 90% are due to the fact that a disk replacement is necessary. The same version when making a decision can be taken when using the predictive approach – 64%.

This approach forms the basis of a prototype decision support system. Figure 1 shows the model (implemented as a part of research on developing a model of complex expert assessment methods and decision making algorithm for preventing fault occurrence in storage systems, see acknowledgments) for conducting a complex examination.

Since the general approach identifies problem situations using both the system state deducted from current monitoring data and forecasted system state based on historical data, hence the resulting decision support system produces more informed decisions using both the predictive decision-making option and the current monitoring data.

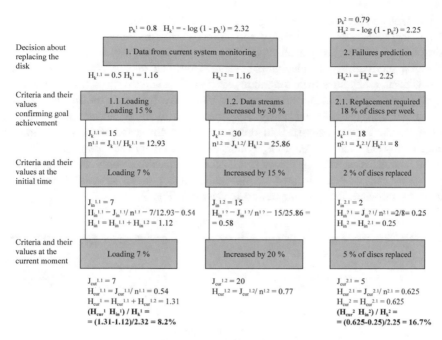

**Fig. 1.** The model for conducting a complex examination based on the information approach of A.A. Denisov

## 3    Implementation as a Part of Storage Array Fault Diagnosing Software

The developed approach to decision making is implemented as a decision support service in the monitoring and diagnosis software suite. This software has a microservice architecture and provides a set of services required for storage array fault diagnosis and prediction:

- Storage array parameter value collection service;
- Data exchange service;
- Fault diagnosis service, based on machine learning (including classification and anomaly detection algorithms) and ontological storage array model. Such implementation of fault diagnosis service allows applying deterministic model-based algorithm to detect system states unambiguously described by a certain set of parameter values and stochastic machine learning-based algorithms for others [3].
- Fault prediction service based on machine learning (mostly classification algorithms);
- Notification service;
- Data visualization service (provides graph plotting functionality);
- Storage service, providing collected data storage, ontological model storage, configuration storage, etc. [20].

Parameter collection service aggregates data from different data sources, such as system logs, temperature, humidity and vibration sensors outputs, operation system utility dumps, storage array health monitoring, topology and layout services. Parameter collection service has an independent instance running on each cluster node, which collects parameter values from that node using local and cluster level data sources. Data exchange service and data storage services also have instances on all cluster nodes. They synchronize data, collected by the parameter collection service instances, on all cluster nodes. This is necessary in order to build a full parametric description of a storage array on each cluster node, which can be used by fault diagnosis, fault prediction, and decision making services. All other services have only one instance at the same time, and they are located on the same cluster node, selected by the cluster managing software. Service interaction is shown in Fig. 2.

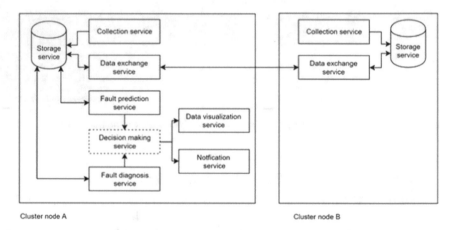

**Fig. 2.** Service interaction scheme

This picture also contains decision-making service that converts the output of fault detection services to the input of user interaction services using decision-making approach introduced in this article.

The decision-making service uses an ontological model of a storage array stored in a graph database, namely Dgraph [2], just like the fault diagnosis service, extending it with a set of nodes and predicates needed to support the decision-making algorithm.

The ontology stores classes, which represent events in a fault-prone system and appropriate managing actions [3]. After performing diagnosis and prediction procedures implemented algorithm determines the best suitable management actions by iterating in a top-down direction through the nodes that represent faulty storage array components.

# 4   Conclusion

This paper describes a new complex methodology to solve a comprehensive problem of fault management in a data storage system. The suggested approach uses elements of complex expert assessment methods and allows making correct decisions regarding management actions relying both on current monitoring data, and predictions based on retrospective data. A prototype of an automated decision support system is implemented as a part of a storage array fault diagnosing software.

**Acknowledgments.** The research is carried out with the financial support of the Ministry of Science and Higher Education of the Russian Federation within the framework of the Federal Program "Research and Development in Priority Areas for the Development of the Russian Science and Technology Complex for 2014–2020". The unique identifier is RFMEFI58117 X0023.

# References

1. Tate, J., Beck, P., Ibarra, H.H., Kumaravel, S.: IBM: Redbooks 2018. Introduction to Storage Area Networks. IBM Redbooks (2018)
2. Dgraph documentation. Dgraph official site. https://docs.dgraph.io/. Accessed 6 May 2019
3. Mamoutova, O.V., Shirokova, S.V., Uspenskij, M.B., Loginova, A.V.: The ontology-based approach to data storage systems technical diagnostics. In: E3S Web Conferences, vol. 91, p. 08018 (2019)
4. Bolsunovskaya, M.V., Shirokova, S.V., Loginova, A.V., Gintciak, A.M.: IT project team management based on a network-centric model. In: XVII Russian Scientific and Practical Conference on Planning and Teaching Engineering Staff for the Industrial and Economic Complex of the Region (PTES), pp. 165–168 (2018). https://doi.org/10.1109/PTES.2018.8604232
5. Volkova, V.N., Emelyanov, A.A. (eds.): Systems theory and systems analysis in the management of organizations [Teoriya system i sistemnyi analiz v upravlenii organizatsiyami], Moscow, 847 p. (2013). (in Russian)
6. Denisov, A.A.: Introduction to information systems analysis [Vvedenie v informatsionnyi analiz sistem], Leningrad, 67 p. LPI (1980). (in Russian)
7. Denisov, A.A.: Modern problems of system analysis [Sovremennye problemy systemnogo analiza], 293 p. (2008). (in Russian)
8. Koren, I., Mani Krishna, C.: Fault-Tolerant Systems, 1st edn. Morgan Kaufmann Publishers Inc., San Francisco (2007)
9. Desyatirikova, E.N., Belousov, V.E., Fedosova, S.P., Ievleva, A.A.: DSS design for risk management of projects. In: Proceedings of the 2017 International Conference "Quality Management, Transport and Information Security, Information Technologies", IT and QM and IS 2017, St. Petersburg, pp. 492–495 (2017). https://doi.org/10.1109/itmqis.2017.8085869
10. Warwick, K., Tham, M.T.: Failsafe Control Systems. Chapman & Hall, London (1991)
11. Wood, R.: J. Magn. Magn. Mater. **321**(6), 555–561 (2009)
12. Zhu, B., et al.: Proceedings of the IEEE 29th Symposium on Mass Storage Systems and Technologies (MSST), pp. 1–5 (2013)

13. Wildani, A., et al.: Proceedings of the IEEE International Symposium on Modeling, Analysis & Simulation of Computer and Telecommunication Systems, MASCOTS 2009, pp. 1–11 (2009)
14. Yang, T., Jiang, H., Feng, D., Niu, Z., Zhou, K., Wan, Y.: Proceedings of the IEEE International Symposium on Parallel & Distributed Processing (IPDPS), pp. 1–12 (2010)
15. Xin, Q., Schwarz, T.J.E.: Proceedings of the 13th IEEE International Symposium on Modeling, Analysis, and Simulation of Computer and Telecommunication Systems, pp. 125–134 (2005)
16. Wylie, J.J., Bakkaloglu, M., Pandurangan, V., Bigrigg, M.W., Oguz, S., Tew, K., Williams, C., Ganger, G.R., Khosla, P.K.: Selecting the right data distribution scheme for a survivable storage system. School of Computer Science, Carnegie Mellon University (2001)
17. Venkatesan, V.: Reliability Analysis of Data Storage Systems: These pour l'obtention du grade de docteur es sciences [Defence site: Йcole polytechnique fйdйrale de Lausanne], Lausanne, 165 p. (2012)
18. Venkatesan, V., Iliadis, I.: Proceedings of the Ninth International Conference on Quantitative Evaluation of Systems, pp. 209–219 (2012)
19. Al-Mamun, A., Narayanan, I., Wang, D., Sivasubramaniam, A., Fathy, H.K.: Proceedings of the American Control Conference (ACC), pp. 3206–3211. IEEE (2016)
20. Ganin, D.V., Klimov, R.V.: Features of simulation of reliability of distributed data storage systems [Osobennosti modelirovaniya nadezhnosti raspredelennykh sistem khraneniya dannykh]. Herald NGIEI (Herald of Nizhny Novgorod State Institute of Engineering and Economics) (Nizhny Novgorod region, Knyaginino, Nizhny Novgorod State Institute of Engineering and Economics Publ.) 74(7), 18–25 (2017). (in Russian)
21. Klimov, V.R.: Modern problems of design, production and operation of radio systems [Sovremennye problemy proektirovaniya, proizvodstva i ekspluatacii radiotekhnicheskikh system] 9(1–2), 148–150 (2015). (in Russian)
22. Igumnov, A.V., Sarajishvili, S.E.: St. Petersburg Polytechnic University Journal of Engineering Science and Technology Series "Informatics. Telecommunications. Management" 193(2), 99–109 (2014)
23. Kantor, O.G., Spivak, S.I.: Informatics and its application [Informatika i ee primenenie] 8(2), 111–121 (2014)
24. Volkova, V.N., Gorelova, G.V., Efremov, A.A., et al.: Systems and processes modelong [Modelirovanie sistem i protsessov. Praktikum], 295 p. Yurait Publ., Moscow (2016). (in Russian)
25. Volkova, V.N.: Theory of information systems. Textbook [Teoriya informacionnykh sistem. Uchebnoe posobie], 300 p. St. Petersburg Polytechnic University Publ., St. Petersburg (2014). (in Russian)
26. Volkova, V.N., Cherny, Yu.Yu.: The laws of information processes in open systems. In: Rethinking L. Von Bertalanffy Proceedings of the International Conference on SARC 2016, pp. 95–108 (2016)
27. Volkova, V.N., Kozlov, V.N., Mager, V.E., Chernenkaya, L.V.: Classification of methods and models in system analysis. In: Proceedings of the 20th IEEE International Conference on Soft Computing and Measurements, SCM 2017, St. Petersburg, ETU "LETI", pp. 183–186 (2017)

# Cyber-Physical System as the Development of Automation Processes at All Stages of the Life Cycle of the Enterprise Through the Introduction of Digital Technologies

Arina Kudriavtceva[✉]

St. Petersburg, Russia

**Abstract.** Nowadays the complexity of automation processes is increasing. As a result, there is a need for distributed automation systems that are necessary for work in conditions of limited control in real time and communication in production processes. Cyber-physical systems imply a fully synergistic integration of computing and control with physical devices and processes. Furthermore, introduction of the cyber-physical system into the enterprise's automation systems will help to combine automation process control and automation production control and the enterprise as a whole, will help create a controlled system, from order to implementation. In this paper, a cyberphysical approach to the design of a distributed automation system is considered. This approach allows integrating control, communication, and calculations at all stages of the product life cycle. The basis of the cyber-physical system is the introduction of digital technologies, as all innovations are provided and improved by computing power and data analytics. To assess the usefulness of introducing technologies for creating a cyber-physical system, it is proposed to use informational assessments by A.A. Denisov.

**Keywords:** Cyber-physical system · Distributed automation system · Automation process the enterprise · Informational assessments of A.A. Denisov

## Introduction

We are on the cusp of the fourth industrial revolution, which is becoming an increasingly "cyber-physical system". The fourth industrial revolution is a synthesis of the latest technologies (from decoding information recorded in human genes to nanotechnology, from renewable energy resources to quantum computing) and their interaction in physical, digital and biological systems [6].

The relevance of the problem is confirmed by the president of the World Economic Forum K. Schwab. In his book *The Fourth Industrial Revolution* (2016, 2017) [6] he describes in detail the changes in technology that are happening in our time. In his research, K. Schwab focuses on the concept of Industry 4.0. The fourth industrial revolution is also called digital. A distinctive feature of the revolution - convergence of technology and erasing the boundaries between the digital, biological and physical areas.

© Springer Nature Switzerland AG 2020
D. G. Arseniev et al. (Eds.): CPS&C 2019, LNNS 95, pp. 601–607, 2020.
https://doi.org/10.1007/978-3-030-34983-7_59

Industry 4.0 is a combination of technologies: PLM, Big Data, Smart Factory, Cyber-physical systems [5], the Internet of Things, Interoperability, which allow you to create an effective business model of an enterprise. To build and improve the basic strategy is necessary to increase the competitiveness of the enterprise [5], and for enterprises of the federal level - to improve the country's economy. High efficiency is achieved mainly due to the rational management of automation systems for physical operations of production and related processes integrated into a single information space.

The priority of the development of industrial enterprises based on the use of digital technologies is currently the key. Our country has developed and implements relevant documents regulating the introduction of digital technologies and cyber-physical systems: "Strategy for the development of the information society in the Russian Federation for 2017–2030" (approved by Presidential Decree of May 9, 2017); Program "Digital Economy of the Russian Federation" »(Approved by order of the Chairman of the Government of the Russian Federation dated July 28, 2017). The listed decrees are necessary to execute including at the industrial enterprises.

With the introduction of digital technologies, production relations change significantly; there is a synergistic integration of computing and control with physical devices and processes [10]. The result is a creation of cyber-physical system.

There are a number of problems with the use of CPS, which are investigated by different authors. Firstly, the modeling methodology for distributed automation software with integrated physical processes [7]. From a design point of view, software models should cover various hardware platforms, execution semantics and fieldbus communication [10]. Secondly, the problem of evaluating innovative technologies that will form the basis of the cyber-physical system.

There are various interpretations of the concept of "cyber-physical system". This paper discusses the use of the cyberphysical system for the development of automation at all stages of the enterprise's life cycle based on the introduction of digital technologies, and offers estimates for choosing innovative technologies for its creation.

# 1 Cyber-Physical System for the Development of Automation at All Stages of the Enterprise Life Cycle Through the Introduction of Digital Technologies

The Fig. 1 shows the main technologies that can become the basis for creating a cyber security system for the development of automation at all stages of the enterprise's life cycle based on the introduction of digital technologies.

One of the most important works, allowing to understand the essence and structure of the cyber-physical system, is the book by Lee E.A. and Seshia S.A. "Introduction to Embedded Systems. Cyber-physical systems approach" [9]. Simplified cyber-physical system of the enterprise, according to the authors, can be represented schematically, as shown in Fig. 2.

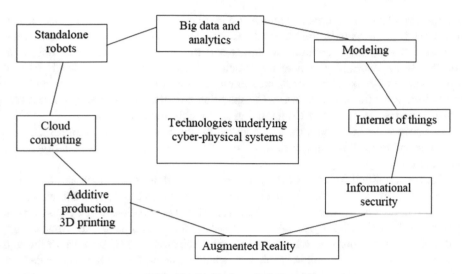

**Fig. 1.** Technology Industry 4.0

The CPS structure consists of three parts. The first part - the physical layer - the physical part of the cyber-physical system, which in its implementation does not use computer or digital networks. At an industrial enterprise, mechanical parts or human operators may be responsible for this level. The second part of the cyber-physical system includes: one or more computing platforms, which consist of sensors, actuators, one or more computers, and possibly one or more operating systems. The third part is the network structure that provides mechanisms for the interaction of computers. Together, the platforms and the network structure are the "cyber" part of the cyber-physical system [9].

Figure 2 shows two networked platforms each with its own sensors and/or actuators. The action taken by the actuators affects the data provided by the sensors through

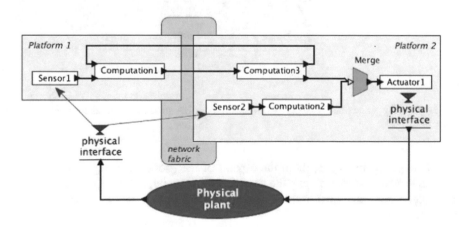

**Fig. 2.** Example structure of a cyber-physical system.

the physical plant. In the figure, Platform 2 controls the physical plant via Actuator 1. It measures the processes in the physical plant using Sensor 2. The box labeled Computation 2 implements a control law, which determines based on the sensor data what commands to issue to the actuator. Such a loop is called a feedback control loop. Platform 1 makes additional measurements using Sensor 1, and sends messages to Platform 2 via the network fabric. Computation 3 realizes an additional control law, which is merged with that of Computation 2, possibly preempting it.

Constantly tracking real-time feedback from physical processes, control systems must respond to actions within real-time constraints in order to connect intelligence at the device level [5].

So, the simplified structure of the cyber-physical system is presented and described in detail. However, for an industrial enterprise, it is of interest to design and implement such a system. Lee E.A. and Seshia S.A. represent the creation of a cyber-physical system, as shown in Fig. 3. The process consists of three main parts: modeling, design and analysis. Modeling is a process of a deeper understanding of the system through the construction of its simplified structure. Models imitate the system and reflect its properties. Also, models show how the system behaves. Therefore, design is a structured creation of artifacts [5].

It determines how the system does what it does. Analysis is a process of a deeper understanding of the system through analysis. He indicates why the system does what it does (or cannot do what, according to the structure of the model, it should do). Thus, the design process of a cyber-physical system iteratively moves between the three parts listed.

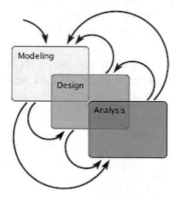

**Fig. 3.** Creating embedded systems requires an iterative process of modeling, design, and analysis.

As suggested in Fig. 3, these three parts of the process overlap, and the design process iteratively moves among the three parts. Normally, the process will begin with modeling, where the goal is to understand the problem and to develop solution strategies.

The process may progress quickly to the design phase, where we begin selecting components and putting them together (motors, batteries, sensors, microprocessors, memory systems, operating systems, wireless networks, etc.). An initial prototype may reveal flaws in the models, causing a return to the modeling phase and revision of the models [5].

If earlier information technologies were used to automate the management of an enterprise and technological processes, then using digital technologies it is possible to combine automated management of technological processes and automated management of production and the enterprise as a whole. By introducing digital technologies in all areas of the enterprise, it is possible to create a single managed system, from an order to implementation.

The introduction of individual elements of modern IT systems can also help to achieve improved staff productivity. Modern systems allow to reduce the number of errors made by employees due to the optimal supply of raw materials, to ensure a more complete load of production equipment, to reduce production waste, etc. [3].

## 2 Application in the Selection of Innovative Technologies for CPS Informational Assessments of A.A. Denisov

To create a cyber-physical system, it is necessary to select and implement innovative technologies. Often, innovations are introduced not one by one, but jointly with other technologies, as a result of which the technologies are layered. The combination of technologies leads to an emergent effect, and the technologies themselves are called emergent [8]. The term emergent technologies (emerging technologies, from emerge – to appear), means the emergence of new properties as a result of the combination of technologies. The use of such technologies will bring new results in production.

To obtain and process expert assessments of the proposed innovative technologies, it is proposed to use informational assessments by A.A. Denisov. In the works of A.A. Denisov proposed a framework for analyzing heterogeneous systems of arbitrary structure, based on a consistent dialectical disclosure of the material-informational dualism of all things. The author considers the approach and method for further independent system knowledge of the world [3].

The main advantage of the method is the ability to get estimates from individual experts for each evaluated innovation, who offer it and know its capabilities better. It is impossible to apply traditional expert assessments based on the organization of collective expertise and averaging the opinions of experts. Such an application can lead to one of the drawbacks of collective assessments, which level out the highly subjective features of assessments, but reinforce collectively subjective ones. Moreover, it is also impossible to form a group of specialists who know all the evaluated technologies equally well and are able to compare them.

The use of information models of the 1st type is based on an assessment of the degree of technology influence on the implementation of the enterprise's goals in the analyzed period of development. In accordance with the information approach for assessing each technology, assessments of the degree of satisfaction (i.e., the probability of achieving the goal) and the probability of using are introduced, and the potential (significance) of the technology is calculated:

$$H = - \sum_{i=1}^{n} q_i \log(1 - p').$$  (1)

Information models of the 2nd type are based on a comparative analysis of complex systems during a certain initial period of their design (implementation, development) by comparing changes in information assessments over time. In this case, estimates are obtained from individual experts who are competent in the relevant field of activity of the enterprise.

When applying information models of the 2nd type, based on a comparative analysis of complex systems during a certain initial period of their design by comparing changes in information estimates over time, two methods of measuring $H_i$ are used:

1) through the probability $p_i$;
2) through deterministic characteristics of the perceived information.

The latter method involves two approaches.

In statics at some point in the introduction of technology (taking the average averaging)

$$H_i = J_i/n_i.$$  (2)

$J$ – information on the number of innovations, measured in relative units, i.e., $J_i = A_i/\Delta A_i$, where $\Delta A_i$ – minimum number of technologies of the $i$-th type, which determines the unit of measurement [1].

Models of the 3rd type describe the assessment of situations described by information equations, taking into account the mutual influence of technology:

$$\begin{aligned} H_1 &= f(H_{11}, H_{12}, H_{13}), \\ H_2 &= f(H_{21}, H_{22}, H_{23}), \\ H_3 &= f(H_{31}, H_{32}, H_{33}). \end{aligned}$$  (3)

$H_1, H_2, H_3, \ldots$ – significance (essence) of the 1st, 2nd, 3rd, etc. technology;
$H_{11}, H_{22}, H_{33}, \ldots H_{ii}, \ldots$ – the intrinsic importance of technology in the absence of other technologies that affect its value;
$H_{12}, H_{13}, H_{21}, \ldots H_{ij}, \ldots$ – change in the value of the $i$-th technology when the $j$-th innovation is on the market [2].

Information estimates A.A. Denisov will allow technology to be assessed at the subsystem level and the system itself, whose subgoals are initiated by its own (internal) needs, motives, and programs that constantly arise in the developing system [4].

The advantage of informational estimates is that informational estimates provide convenient processing of estimates, the ability to combine probabilistic estimates with quantitative deterministic characteristics. This contributes to increasing the objectivity and reliability of estimates, and, in addition, allows to obtain changes in the degree of influence of subgoals based on changes in measured deterministic parameters, factors, funds for the implementation of the objectives of the enterprise (organization) [4].

Estimates of $p_i'$ in models by A.A. Denisov is obtained from individual experts for each evaluated innovation, who offer it and know its capabilities better.

The results of the estimates obtained are recommended to be presented in the form of histograms for further processing.

## Conclusion

Cyber-physical systems integrate networked embedded systems with physical processes. A comprehensive abstract modeling language to cover existing models of computation, communication networks as well as physical plant models is one of the key feature in cyber-physical system research. Co-simulation of control software, plant model and networking is achieved by concurrent model of computations in cyber-physical systems. This paper discusses the cyberphysical approach to the design of a distributed automated system that allows you to integrate control, communication, calculations at all stages of the product life cycle. The simplified scheme of the cyber-physical system, as well as the main stages of its design, are considered. Since the basis of the cyberphysical system is the introduction of digital technologies, it is proposed to use the informational assessments of A.A. Denisov.

## References

1. Volkova, V.N., Loginova, A.V., Yakovleva, E.A.: St. Petersburg, Polytechnic University Publishing House St. Petersburg, 246 p. (2014)
2. Volkova, V.N., Leonova, A.E.: When developing projects for inclusion in the plan of a research and production organization. Probl. Manag. Soc. Syst. **8**(12), 220–224 (2015)
3. Denisov, A.A.: Modern Problems of System Analysis: A Textbook, 3rd edn., 291–293. Publishing House of the Polytechnic University, St. Petersburg (2008)
4. Denisov, A.A.: Management Information Basics / A.A. Denisov. - L.: Ener-goatomizdat, 1983. Kudryavtseva A.S. Models of comparative analysis of innova-tions for the Admiralty Shipyards JSC System analysis of projects and management of the collection of scientific papers of the XXI International Scientific and Practical Conference June 29–30 (2017)
5. Masyutin, S.A.: The basic strategy of the company in the transition to the concept of "Industry 4.0" Report from the meeting of the Committee for Foundry and Forging and Press Production (2017)
6. Schwab, K.: The Fourth Industrial Revolution: Translation from English, 208 p. "E" Publishing House, Moscow (2017)
7. Eidson, J., Lee, E.A., Matic, S., Seshia, S.A., Zou, J.: Distributed real-time software for cyber-physical systems. Proc. IEEE **100**(1), 45–59 (2012)
8. Emerging Technologies. https://www.huffpost.com/entry/five-emerging-technologie_b_10396122
9. Lee, E.A., Seshia, S.A.: Introduction to Embedded Systems, Cyber-Physical Systems Approach (2011). http://LeeSeshia.org. ISBN 978-0-557-70857-4
10. Vyatkin, V.: IEC 61499 as enabler of distributed and intelligent automation: state-of-the-art review. IEEE Trans. Ind. Inform. **7**(4), 768–781 (2011)

# Method of Classification of Fixed Ground Objects by Radar Images with the Use of Artificial Neural Networks

Anton V. Kvasnov[✉]

Peter the Great St. Petersburg Polytechnic University, Saint Petersburg, Russia
kvasnov_av@spbstu.ru

**Abstract.** The article considers the method of classification for ground stationary objects. As the source of data are used radar pictures of the land, which were received using the air of radio-electronic monitoring systems in the synthetic-aperture radar (SAR) mode. For the discovered ground objects, estimates of their characteristics are evaluated with the mutual orientation, geometric features of the location taken into account. The resulting data set allows creating a training sample, which is used to build an Artificial Neural Network. As a result, the Artificial Neural Network is able to classify the detected groups of objects with a given probability. In the process of modeling, the method used software module Image Processing Toolbox Matlab 2016, which allows evaluating the raster images of radar portraits. Software Neural Network Toolbox Matlab was used to form an artificial neural network. The obtained simulation results showed the possibility of application and using the technique in air radio electronic systems of monitoring the ground situation.

**Keywords:** Radar object classification · Radar image · Artificial neural networks · Synthetic-aperture radar

## 1 Introduction

At the moment, one of the significant tasks of creating air radar systems for monitoring of the earth's surface is the problem of classification of detected objects [1, 2]. The use of such systems has obvious advantages over optical systems, i.e., the infrared radar and aerial photography. First, the performance of such complexes does not depend on the level of natural light. Second, the operation of the radar equipment is not affected by meteorological conditions.

At the same time, there are a number of problems associated with the reception and processing of radar images. These include the potential accuracy in the evaluation of detected objects (artifacts), as well as the existence of noise in the antenna-receiving path as a result of interfering and diffraction effects on the electromagnetic wave propagation [3, 4].

In order to obtain high resolution radar images, the use of an antenna with a large aperture size is required. Placement of these antennas on the aircraft is often not possible; therefore, to provide for adequate resolution. the synthetic-aperture radar

© Springer Nature Switzerland AG 2020
D. G. Arseniev et al. (Eds.): CPS&C 2019, LNNS 95, pp. 608–616, 2020.
https://doi.org/10.1007/978-3-030-34983-7_60

(SAR) can be used [5]. Consequently, there is ability to detect small objects and increase the potential accuracy of the radar.

The high-resolution radar image allows identifying a number of essential features that can be used to classify a particular group of objects. By classification we mean the possibility to associate with one of a number of sets (classes) which are distinguished by one or more criteria, irrespectively of whether there is any prior knowledge of the class membership or class boundaries [6]. The value of the assessed features is an important task in the classification process. According to these data, a learning sample (database) formed. This database allows generating and training the neural network [7]. There is a number of articles, which are dedicated classification problems in SAR-systems by using neural network [8–10].

The general approach to the classification problem of fixed ground objects by radar images can carry out according to the following steps [11, 12]:

- Object (artifact) detection: select a group of objects to be classified
- Filtering detected objects in the background noise
- Estimation of valuable features for classification of detected objects
- The formation of the learning sample (database)
- Development a neural network to classify objects.

The article purpose is to simulate the method of radar images selection on the Earth's surface background and its implementation for object classification using neural networks.

## 2  Problem Statement and Assumptions

The following assumptions were made during the research:

1 – Availability of detected object groups depends on the selected area, which must be classified
2 – Radar resolution exceeds the size of the detected objects
3 – Type of class configuration on the ground is based on expert evaluation.

Thus, the mathematical formulation of the classification problem can be represented as follows:

Let $X$ be a set of object descriptions, $Y$ be a set of object classes. There is an unknown target dependency on the objects of the finite learning sample $X^m = \{(x_1, y_1), \ldots, (x_m, y_m)\}$. An algorithm $a : X \rightarrow Y$ that can classify a random object classes $y \in Y$ has to be formed.

## 3  Implementation Method

Let's consider the method of implementation at the example of airport object classification. The initial data were obtained by the onboard radar, from several angles in the SAR-mode [13] (see Table 1).

**Table 1.** Characteristics of the onboard radar

| Feature | Value |
|---|---|
| Radar designation | AN/APG-81 |
| Type | Airborne dual mode radar |
| Frequency range | 8–12.5 GHz |
| Antenna type | Active Electronically Scanned Array (AESA) |

The possibility of the SAR-mode allows processing high-resolution images for a group of classified object (see Fig. 1).

The above picture highlights a number of detected objects that can be associated with the following classes:

1 – A group of small aircraft (3 classes);
2 – Single big aircraft (2 classes);
3 – Infrastructure facility (building) (1 class);
4 – Unidentified object (1 class).

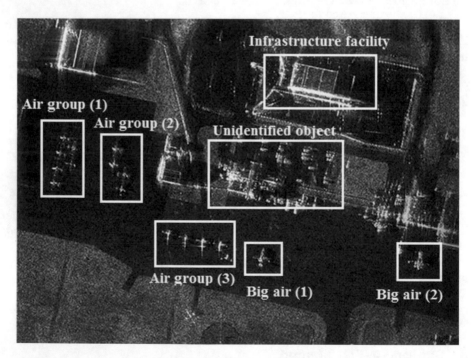

**Fig. 1.** Radar image of the airport

The vector of classified objects will be

$$Y_4 = \begin{bmatrix} y_{air\_group} & y_{big\_air} & y_{building} & y_{unidentified} \end{bmatrix}^T.$$  (1)

The presented classes may have specific features that allow identifying patterns in their subsequent occurrence. For instance, it can be a quantity of objects on radar images which is being classified as a "group". Second, a specific cruciform shape on the radar images can belong to the "aircraft class". Third, the same orientation of the group and the distance between objects of this group is also indicative. Finally, it may be the size of the radar image in pixels.

The infrastructure facility (buildings) is characterized by a rectangular form. If it is not possible to allocate any valuable features, the object remains not initialized. Thus, the characteristic features of the classes will be:

$$X_5 = \begin{bmatrix} x_{number} & x_{form} & x_{orientation} & x_{size} & x_{interval} & x_{distance} \end{bmatrix}^T.$$  (2)

Radar picture processing showed that the considered image can be studied by the cluster analysis method and represented by individual pixels. The Matlab Image Processing Toolbox was used to select specific marks on the map (see Fig. 2).

Based on the processed data in the Image Processing Toolbox, a learning sample was formed (see Table 2).

**Table 2.** A learning sample

| | Number | Form | Orientation | Size | Interval | Distance | | | | | | |
|---|---|---|---|---|---|---|---|---|---|---|---|---|
| | | | | | | Air group (1) | Air group (2) | Air group (3) | Big aircraft (1) | Big aircraft (2) | Building | Unidentified |
| Air group (1) | 6 | Cross | −18 | 46 | 18 | 0 | 72 | 184 | 274 | 459 | 334 | 255 |
| | | | −13 | 46 | 21 | | | | | | | |
| | | | −90 | 45 | 20 | | | | | | | |
| | | | 0 | 43 | 18 | | | | | | | |
| | | | −24 | 36 | 20 | | | | | | | |
| | | | – | 55 | 20 | | | | | | | |
| Air group (2) | 4 | Cross | 0 | 42 | 21 | 72 | 0 | 127 | 210 | 392 | 272 | 202 |
| | | | 84 | 48 | 20 | | | | | | | |
| | | | 51 | 50 | 19 | | | | | | | |
| | | | – | 65 | 19 | | | | | | | |
| Air group (3) | 4 | Cross | 92 | 76 | 23 | 184 | 127 | 0 | 89 | 258 | 246 | 682 |
| | | | – | 68 | 24 | | | | | | | |
| | | | 88 | 110 | 22 | | | | | | | |
| | | | 77 | 89 | 22 | | | | | | | |
| Big aircraft (1) | 1 | Cross | 45 | 126 | – | 274 | 210 | 89 | 0 | 198 | 224 | 115 |
| Big aircraft (2) | 1 | Cross | 88 | 155 | – | 459 | 392 | 258 | 198 | 0 | 260 | 205 |
| Building | 1 | Rectangle | 71 | 2740 | – | 334 | 272 | 246 | 224 | 260 | 0 | 114 |
| Unidentified | 1 | – | 0 | 4484 | – | 255 | 202 | 682 | 115 | 205 | 114 | 0 |

This database was used for training a Neural Network. Subsequently, the simulation of Artificial Neural Network in the Neural Network Toolbox Matlab software package was based on the Levenberg-Marquardt algorithm [14–16]. Levenberg-Marquardt algorithm is an alternative to the Gauss-Newton method for finding the minimum of a function $F(x)$ that is a sum of squares of nonlinear functions

$$F(x) = \frac{1}{2} \sum_{i=1}^{m} f_i^2(x). \tag{3}$$

Let the Jacobian of $f_i(x)$ be denoted as $J_i(x)$, then the Levenberg-Marquardt method searches in the direction given by the solution $P_k$ to the equations

$$\left(J_k^T J_k + \lambda_k I\right) P_k = -J_k^T f_k, \tag{4}$$

where $\lambda_k$ are non-negative scalars and $I$ is the identity matrix.

The simulation result is the generation of a neural network with 60 hidden layers in the Neural Network Toolbox Matlab software package, shown in Fig. 3.

The results of the learning sample for 4 epochs were used to calculate the performance factor. The training data was taken for model convergence of the objective function: training, validation, testing (see Fig. 4).

**Fig. 2.** Cluster analysis of radar image

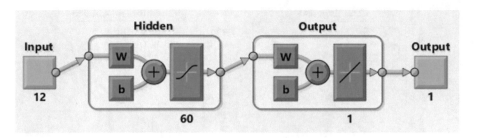

**Fig. 3.** Neural network for radar images classification

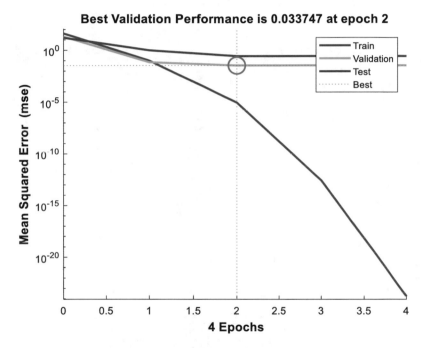

**Fig. 4.** MSE for 4 Epochs

The classification assessment of the selected classes was carried out by the regression method, because it allows ranking the answers and choosing their maximum values. Figure 5 presents linear regression as a method of reconstructing the dependency between the output data $Y^n$ and the input data $X^m$.

As a linear regression result, the assessments at each object class were obtained. Feature values $X_5 = [x_1, \ldots, x_5]$ were entered into the model of the neural network. The output value of the object class was estimated as a deviation from the input value $Y_5 = [y_1 = 1, \ldots, y_7 = 7]$ (see Table 3).

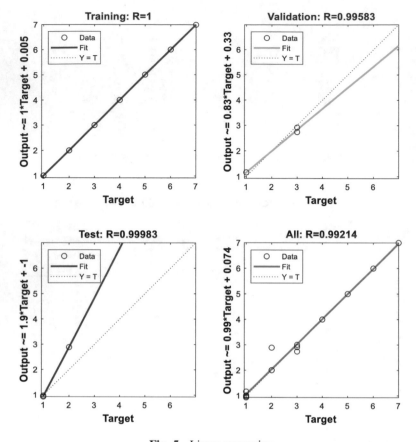

**Fig. 5.** Linear regression

**Table 3.** Simulation result of the neural network

|  | Input value | Output value | Accuracy |
|---|---|---|---|
| Air group (1) | 1 | 0.9293 | 0.07 |
| Air group (2) | 2 | 1.4475 | 0.44 |
| Air group (3) | 3 | 3.4528 | 0.45 |
| Big aircraft (1) | 4 | 4.0041 | 0.001 |
| Big aircraft (2) | 5 | 5.8169 | 0.81 |
| Building | 6 | 5.8902 | 0.11 |
| Unidentified | 7 | 7.0000 | 0.00 |

The resulting data showed that the neural network is capable of false classifying of individual objects. This applies to classes:

Air group (2) – $P_2 = 0.45$;
Air group (3) – $P_3 = 0.45$;
Big aircraft (2) – $P_5 = 0.81$.

This is due to the small area of the objects being estimated for Air group (2) and Air group (3) (see Fig. 2); The Big aircraft (2) is the only representative image and has no characteristic cross-shaped feature of the aircraft.

# 4 Conclusions

The article deals with the classification method of fixed ground objects by radar images using neural networks. As a modeling result of this process, the following conclusions are obtained:

- Radar image objects whose dimensions allow carrying out cluster analysis can be classified. The neural network is also potential to analyze group objects for which additional characteristic features are introduced.
- It is necessary to take into account the radar features in order to form a learning sample. For instance, it is the area of the radar image, its orientation, geometric characteristics (the number of objects, the distance between them).
- Simulation model of airport object classification showed that most accurately can be identified group objects ($P < 0.45$), the least accurately single objects ($P < 0.81$).

# References

1. Kouemou, G. (ed.): Radar Technology. InTech Publishers, Rijeka (2010)
2. Jordanov, I., Petrov, N.: Intelligent radar signal recognition and classification. In: Abielmona, R., Falcon, R., Zincir-Heywood, N., Abbass, H. (eds.) Recent Advances in Computational Intelligence in Defense and Security. Studies in Computational Intelligence, vol. 621. Springer, Cham (2016)
3. Skolnik, M.I.: Introduction to Radar Systems. McGraw Hill, Boston (2001)
4. Rob, M.: Fundamentals of radar signal processing (Richards, M.A. 2005) [Book review]. IEEE Sig. Process. Mag. **26**, 100–101 (2009). https://doi.org/10.1109/MSP.2009.932123
5. Cumming, I.G., Wong, F.H.: Digital Processing of Synthetic Aperture Radar Data: Algorithms and Implementation. Artech House, Norwood (2005)
6. Anderson, S.J.: Target classification, recognition and identification with HF radar. In: Proceedings of the NATO Research and Technology Agency Sensors and Electronics Technology, Panel Symposium SET–080/RSY17/RFT: "Target Identification and Recognition Using RF Systems", Oslo, Norway, 11–13 October 2004, pp. 1–20. RTO–MP–SET–080, Oslo (2004)
7. Krizhevsky, A., Sutskever, I., Hinton, G.E.: ImageNet classification with deep convolutional neural networks. Neural Inf. Process. Syst. **25** (2012). https://doi.org/10.1145/3065386
8. Ding, J., Chen, B., Liu, H., Huang, M.: Convolutional neural network with data augmentation for SAR target recognition. IEEE Geosci. Remote Sens. Lett. **13**(3), 364–368 (2016)
9. Chen, S., Wang, H., Xu, F., Jin, Y.: Target classification using the deep convolutional networks for SAR images. IEEE Trans. Geosci. Remote Sens. **54**(8), 4806–4817 (2016)
10. Chen, S.-Q., Zhan, R.-H., Hu, J.-M., Zhang, J.: Feature fusion based on convolutional neural network for SAR ATR. ITM Web Conf. **12**, 05001 (2017). https://doi.org/10.1051/itmconf/20171205001

11. Kvasnov, A., Shkodyrev, V., Arsenyev, D.: Method of recognition the radar emitting sources based on the naive Bayesian classifier. WSEAS Trans. Syst. Control **14**, 112–120 (2019)
12. Ali, S.A., Venkata Ramaiah, K.: SAR image classification by multilayer back propagation neural network. Int. J. Mod. Trends Sci. Technol. **1**(2), 36–39 (2015)
13. Stimson, G.W.: Introduction to Airborne Radar, 2nd edn. SciTech Publishing Inc, Mendham (1998)
14. Nocedal, J., Wright, S.J.: Numerical Optimization, 2nd edn. Springer, New York (2006). ISBN 978-0-387-30303-1
15. Levenberg, K.: A method for the solution of certain problems in least squares. Q. Appl. Math. **2**, 164–168 (1944)
16. Marquardt, D.: An algorithm for least-squares estimation of nonlinear parameters. SIAM J. Appl. Math. **11**, 431–441 (1963)

# Intellectual Cognitive Technologies
# for Cyber-Physical Systems

Galina V. Gorelova[(⊠)]

Engineering - Technological Academy of the Southern Federal University,
Rostov-on-Don, Russia
gorelova-37@mail.ru

**Abstract.** The paper presents information on the cognitive modeling of complex systems and provides considerations on its place within the framework of
research and development of cyber-physical systems. Theoretical and practical
results of the application of cognitive modeling of complex systems can be
attributed to the intellectual decision support systems in the socio-economic and
industrial areas. Mathematical and software apparatus of cognitive modeling of
complex systems is designed to describe the structure and behavior of a complex
system in the face of uncertainty, implements an interdisciplinary approach to
decision-making problems. For the future, it can be recommended for use in the
direction of "Cognitive city". A number of results of cognitive modeling of a
regional socio-economic system are presented. The cognitive map of the system,
the results of the study of its structural and dynamic properties, the results of
scenario modeling and recommended management decisions are presented.

**Keywords:** Cyber-physical system · Cognitive modelling · Complex system,
properties · Behavior · Simulation modeling

## 1 Introduction

In recent decades in the world, the transition from one technological mode to another
entails the unification of once scattered advances in science and technology into a kind
of a whole, defined by the concept of "cyber-physical systems". Without arguing over
the concept of "cyber-physical systems", let us take as a basis the feature of integrating
computing resources with physical entities of any kind, both natural and man-made,
highlighted in the definition of fusion of the Internet of people, things and services.
Figure 1 shows a diagram explaining the content of the concept of "cyber-physical
systems". The emergence and development of cyber-physical systems (CPS) is associated with the advantages of integration – combining individual components into large
systems, such as the Internet of Things, the Worldwide Sensor Net, the Smart Habitats
(Smart Building Environment), the future defense systems, etc. The process by systems
of "capturing" by CPS of different activities and industries in many countries is very
fast, and a lag in this direction of scientific and technological progress can adversely
affect the economy of any country. CPS in general affect the interests of the whole

© Springer Nature Switzerland AG 2020
D. G. Arseniev et al. (Eds.): CPS&C 2019, LNNS 95, pp. 617–631, 2020.
https://doi.org/10.1007/978-3-030-34983-7_61

society and should be considered in at least two aspects - technical and socio-cultural. This means that their development will entail inevitable changes in society and this must be taken into account.

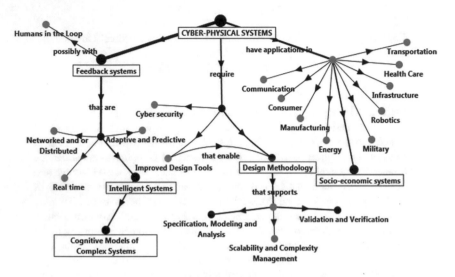

**Fig. 1.** Conceptual map of the basic elements of cyber-physical systems

The emergence of all new areas of application of the ideas of cyber-physical systems leads to the prediction of such states of development in the future as "cyber-physical society" [1–3]. Existing and future smart cities seem unthinkable without cyber-physical systems.

The time has come when a person's cognitive abilities are not enough; people are no longer able to cope with the amount of information required to make decisions in complex systems, and it becomes necessary to transfer some CPS actions by removing the human out of the control loop (the human out of the loop). But still, humans should retain some control loops. CPS can enhance a person's analytic abilities, so there is a need to create new-level interactive systems that keep humans in the control loop (the human in the loop). And in these conditions, it is desirable to have tools that enhance humans' cognitive abilities, helping them to develop and make informed, effective management decisions. Means are in the field of cognitive sciences, including such areas as artificial intelligence and intelligent systems. Such a "contact" of cognitive sciences with cyber-physical systems has already been noted in the disclosure of the concept of CPS; intelligent systems are included in the scope of cyber-physical systems. The diagram in Fig. 1 marks this fact.

The paper proposes to include in the scheme of ideas about CPS as means for enhancing humans' cognitive abilities cognitive modeling of complex systems and relevant cognitive information technologies. Following the works of [4–9] and [10–12] in SFU, methodology and software products for cognitive modeling of complex systems were developed [13–24]. In Fig. 1, the framework highlights the concepts that determine, in the opinion of the author, the place of cognitive modeling of complex systems in the general scheme of CPS.

## 2  Brief Outline of the Cognitive Modeling of Complex Systems

Complex systems (economic, social, sociotechnical, political, etc.), operating under conditions of various kinds of uncertainty, have properties and obey laws that make it difficult to understand, describe, predict their behavior, manage them, adapt to them. Weakly structured problems that are inherent in complex systems require the use of specific methods to solve them. Cognitive modeling of complex systems that imitates the structure and behavior of such systems makes it possible to structure knowledge of an expert in the subject area; describe and explain the system; analyze its properties using the developed cognitive model; develop possible scenarios for the development of the system under the influence of control and disturbing factors.

Figure 2 shows a diagram of the stages of cognitive modeling of complex systems.

Cognitive modeling of complex systems is done in stages. Perhaps one of the most serious stages is the stage which can be called the stage of pre-project research and which is least subject to formalization and automation. Determining the purpose and content of a cognitive study, designed, for example, for a projected cyber-physical system, the choice of experts and necessary information and its sources, devices that are needed in the future, are poorly formalized decision-making procedures. The most formalized stages are the subsequent stages of cognitive modeling. Note that the development stage and the image of a cognitive model in the form of a cognitive map (a sign-oriented graph) is traditional among many researchers. The subsequent stages of cognitive modeling, combined in one software system [23], are original.

It is better to start the development of a cognitive model of a system with a cognitive map, as the most mathematically simple. As is known [4–12], a cognitive map is a sign-oriented graph

$$G = \,<V, E>, \tag{1}$$

in which $V = \{v_i\}$ is a set of vertices (concepts, objects, entities), $i = 1, 2, \ldots k$, $E = \{e_{ij}\}$ is a set of arcs reflecting the relationship between the vertices. The influence of vertex vi on vj in the situation under the study may be considered positive, when an increase (decrease) in one factor leads to an increase (decrease) in another, and negative, when an increase (decrease) in one factor leads to a decrease (increase) in another, or is non-existant in a particular situation.

Depending on the object, the purpose of the research and available information, a cognitive model of different complexity can be developed. The initial model of the system in the form of a cognitive map can be successively transformed into a more complex mathematically (and by content) cognitive model [12], for example, of the type of a parametric vector functional graph (2), in which $G$ denotes a cognitive map, $X$ denotesa plurality of parameters of vertices, $\theta$ denotes the space of parameters of vertices; $F = (X, E) = f(x_i, x_j, e_{ij})$ is the functional of the transformation of arcs, can take the form of the function $f_{ij}$, the weighting factor $w_{ij}$, the membership function

$$\Phi_n = \langle G, X, F, M \rangle. \tag{2}$$

Complex systems can be represented by hierarchical cognitive models [19]:

$$IG = \langle G_k, G_{k+1}, E_k \rangle, \quad k \geq 2, \tag{3}$$

in which $G_k$ is $k$-level cognitive map, $G_{k+1}$ is $(k + 1)$-level cognitive map, $E_k$ denotesrelations between vertices of different levels.

Figure 3 schematically depicts the idea of a hierarchical cognitive model.

## 3   Example of Cognitive Modeling

### 3.1   Creating a Cognitive Model

The analysis of a number of aspects of the regional socio-economic system given below was made on the cognitive model of the top level of the cognitive maps hierarchy (Fig. 4). At this level, names of most of the vertices are generalizations of the components of these concepts, which are revealed to specific statistical and expert indicators at the lower levels. Monitoring these indicators and entering data about them into the model is further required for real-time operation. At the top level under consideration, there is a "design" of the general structure of the system and an analysis of its general structural properties and behavior. Such cognitive models determine the structure of the knowledge base of intelligent decision-support systems in the relevant subject area [14, 19]. The cognitive model in Fig. 4 was obtained using the software system CMSS (Cognitive Modeling Software System) [23].

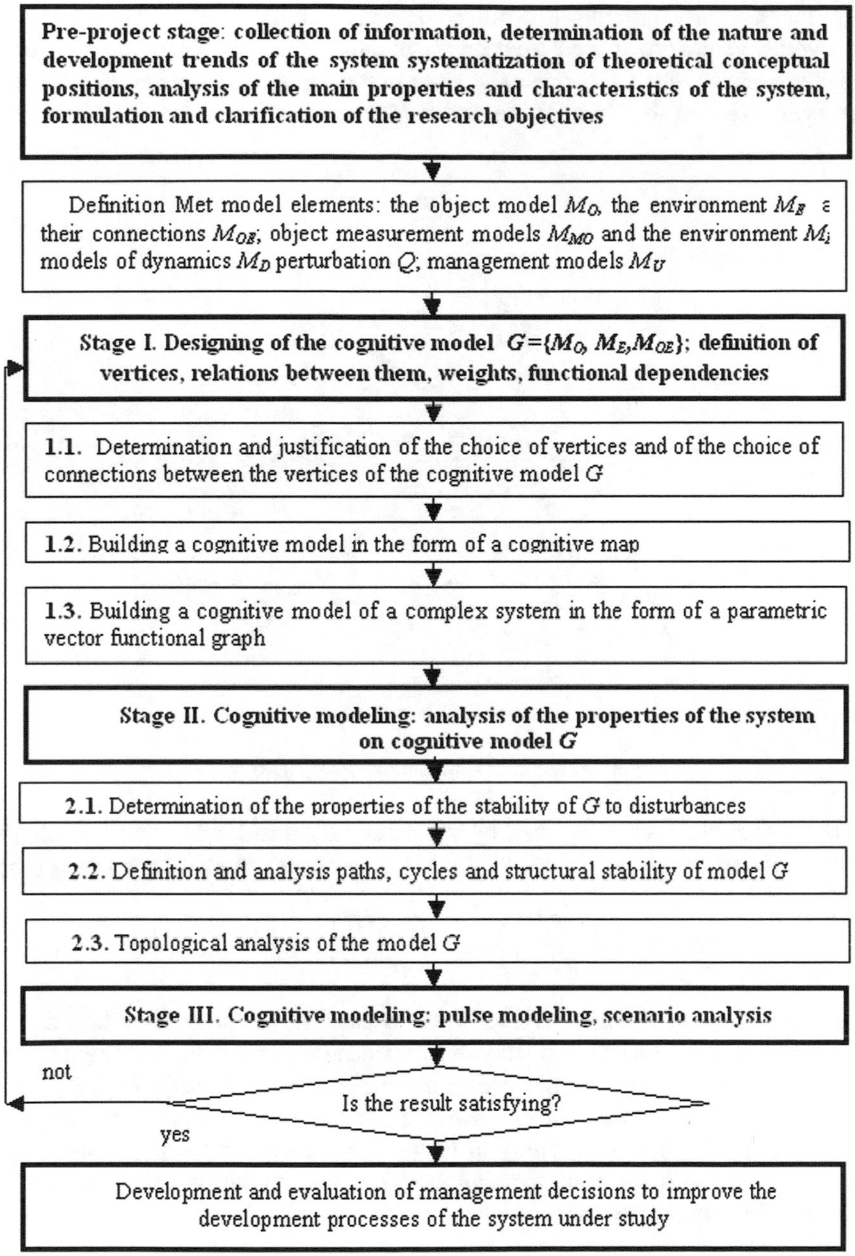

Pre-project stage: collection of information, determination of the nature and development trends of the system systematization of theoretical conceptual positions, analysis of the main properties and characteristics of the system, formulation and clarification of the research objectives

▼

Definition Met model elements: the object model $M_O$, the environment $M_B$ ε their connections $M_{OB}$; object measurement models $M_{MO}$ and the environment $M_i$ models of dynamics $M_D$ perturbation $Q$; management models $M_U$

▼

Stage I. Designing of the cognitive model $G=\{M_O, M_E, M_{OE}\}$; definition of vertices, relations between them, weights, functional dependencies

▼

1.1. Determination and justification of the choice of vertices and of the choice of connections between the vertices of the cognitive model $G$

▼

1.2. Building a cognitive model in the form of a cognitive map

▼

1.3. Building a cognitive model of a complex system in the form of a parametric vector functional graph

▼

Stage II. Cognitive modeling: analysis of the properties of the system on cognitive model $G$

▼

2.1. Determination of the properties of the stability of $G$ to disturbances

▼

2.2. Definition and analysis paths, cycles and structural stability of model $G$

▼

2.3. Topological analysis of the model $G$

▼

Stage III. Cognitive modeling: pulse modeling, scenario analysis

▼

not ◄——  Is the result satisfying?

yes

▼

Development and evaluation of management decisions to improve the development processes of the system under study

**Fig. 2.** Scheme of stages of cognitive modeling of complex systems

In Fig. 4, solid lines indicate positive arcs (when an increase/decrease in the signal at the vertex $V_i$ leads to an increase/decrease in the signal at the vertex $V_j$), a dotted line indicates negative arcs (an increase/decrease in the signal at the vertex $V_i$ leads to the decrease/increase in the signal at the vertex $V_j$).

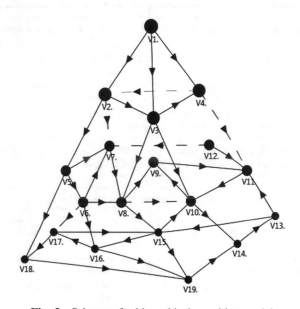

**Fig. 3.** Scheme of a hierarchical cognitive model

The cognitive model of Fig. 4 is, by mathematical content, a functional graph, since its block of vertices $V_0, V_1, V_2, V_3, V_4, V_6$ the latter are functionally connected [15]:

$$Q_i = \text{sign}_s\left(\frac{2F_i(t)}{q_F X(t)}\right) \text{sign}_s\left(\frac{2S_i(t)}{q_S X(t)}\right) \text{sign}_s\left(\frac{2}{q_R R_i(t)}\right), \tag{4}$$

where $q_F$ is the coefficient of quality of life depending on the level of security with fixed assets, $q_S$ is coefficient of quality of life depending on the level of wage security, $q_R$ is coefficient of quality of life depending on population density, $F_i(t)/X(t)$ is the number of fixed assets in accordance with the current price level per one inhabitant of the region, $S_i(t)/X(t)$ is the quantity of goods that a person in the region can buy for wages. The development of a cognitive model completes the first stage of cognitive modeling of complex systems.

## 3.2   Cognitive Model Path Analysis

At the second stage of cognitive modeling, analysis of the model's paths and cycles, its structural properties and stability properties is carried out. The analysis of paths (cause-effect chains) allows, first, to verify whether the model contradicts to the real system and, second, to assess the possibility and effectiveness of various ways of achieving the goal or subgoals. Figure 5 shows one of the possible paths from a vertex to another vertex; there are 30 paths of the kind in total.

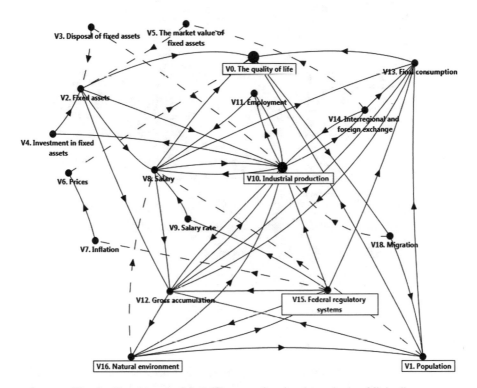

**Fig. 4.** Cognitive model $G$ "Factors of regional standards of living"

Let's analyze the selected path $V_{13} \rightarrow V_{12} \rightarrow V_{16} \rightarrow V_{15} \rightarrow V_{10} \rightarrow V_{11} \rightarrow V_8 \rightarrow V_0$. It can be interpreted as follows: if final consumption starts to grow, this will trigger an increase in gross savings, which can positively affect the natural environment, and this will positively affect federal regulatory systems which will contribute to the development of production without causing damage to the natural environment; the growth of production contributes to the growth of employment, which, taking into account the entire previous chain, can lead to an increase in wages and an improvement in the quality of life of the population in this aspect. That is, within the framework of this "collective" name of the vertices, there is no violation of theoretical ideas about a possible causal relationship between them. But for a reliable conclusion about the entire cognitive map, it is desirable to analyze as many model paths as possible (CMSS allows to isolate any paths) for logical consistency and relevance to reality.

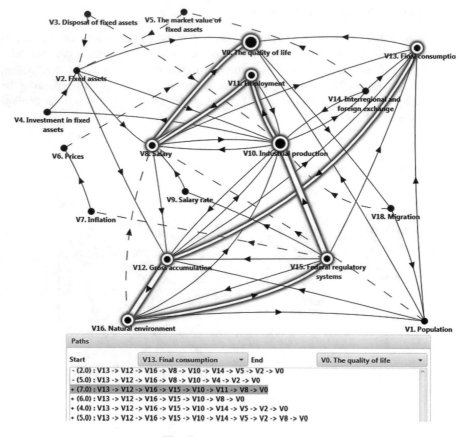

**Fig. 5.** Paths from $V_{13}$ to $V_0$

## 3.3 Analysis of Cognitive Model Cycles

To judge the noncontradiction of the reality model, it is also necessary to analyze both the cycles of the cognitive model as a whole and separately the cycles of accelerators and stabilizers of the processes in the system. Figure 6 shows the cycles of a cognitive model $G$.

Figure 6 shows the result of a computational experiment to determine the cycles of a cognitive model and highlights one of the positive cycles and one of the negative cycles. Note, the cycle of positive feedback is called a cycle in which there are no negative arcs or the number of negative arcs is even; a negative feedback cycle is a cycle that has an odd number of negative arcs.

Analysis of the cycles of model $G$ showed that in this system there are 252 cycles, of which 117 are negative and stabilizing. This indicates that the analyzed system is structurally stable [6, 7].

## 3.4  Disturbance Resistance Analysis

In addition to the analysis of structural stability, it is necessary to analyze the resistance of the model to disturbances. For this, it is necessary to calculate the roots of the characteristic equation of the adjacency matrix RG of the graph $G$. Table 1 shows a fragment of such calculations.

**Table 1.** Fragment of calculation of the roots of the characteristic equation of the matrix $R_G$

| # | Real part | Imagiary part | Module (2.6881) |
|---|-----------|---------------|-----------------|
| 0 | 2.6881    | 0.0           | 2.6881          |
| 1 | -1.8742   | 0.0           | 1.8742          |
| 2 | 1.1456    | 0.0           | 1.1456          |
| 3 | 0.9341    | 0.4003        | 0.9341          |
| 4 | 0.9341    | -0.4002       | 0.9341          |
| 5 | -0.9052   | 0.8401        | 0.9052          |
| 6 | -0.9052   | -0.8401       | 0.9052          |

According to the stability criterion, $|M| < 1$ [6, 7, 19], where |M| is the eigenvalue of the RG matrix (the root of the characteristic equation) maximal in absolute value. Since in this case $|M| = 2.6881$, we can assume that $G$ is not impulsively resistant to perturbations.

## 3.5  Scenario Analysis, Prediction of Possible Ways of Development

Scenario analysis is performed by pulsed simulation of the distribution of disturbances along the cognitive map [6, 12]. Before conducting a pulse simulation, it is necessary to plan the computational experiment, that is, decide which vertices or set of perturbations-impulses $Q = \{q_i\}$ of a certain size and sign and in what sequence. Pulse simulation gives an answer to the question: "What will happen if …?" Each such option is a model scenario for the development of situations. The situation at the moment of time $t_n$ (at the $n^{th}$ simulation step) is characterized by a set of impulse values at the vertices of the cognitive model and the corresponding vector $Q_n$.

We give examples of pulse simulation in several scenarios.

Scenario No.1. Let production begin to develop, set the momentum $q_{10} = +1$, the vector of perturbations $Q = \{q_1 = 0; \ q_2 = 0; \dots q_{10} = +1; \dots q_{18} = 0\}$.

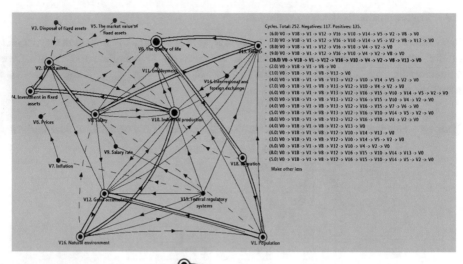

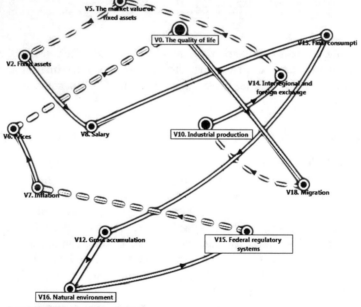

**Fig. 6.**  Cognitive card cycles

**Table 2.** Calculation of the impulse process for scenario No.1

| Step<br>Vertex | 0 | 1 | 2 | 3 | 4 | 5 | 6 | 7 | 8 | 9 | 10 |
|---|---|---|---|---|---|---|---|---|---|---|---|
| V0. The quality of life | 1 | 1 | 1 | 2 | 7 | 15 | 47 | 118 | 333 | 868 | 2387 |
| V15. Federal regulatory systems | 0 | 0 | 0 | 0 | 1 | 2 | 12 | 26 | 92 | 224 | 664 |
| V10. Industrial production | 0 | 1 | 1 | 5 | 9 | 32 | 70 | 216 | 526 | 1513 | 3876 |
| V14. Interregional and foreign exchange | 0 | 0 | 1 | 1 | 5 | 9 | 32 | 70 | 216 | 526 | 1513 |
| V11. Employment | 0 | 0 | 1 | 1 | 5 | 9 | 32 | 70 | 216 | 526 | 1513 |
| V8. Salary | 0 | 0 | 1 | 2 | 6 | 13 | 33 | 74 | 207 | 508 | 1412 |
| V12. Gross accumulation | 0 | 0 | 1 | 2 | 11 | 24 | 80 | 198 | 572 | 1473 | 4097 |
| V13. Final consumption | 0 | 0 | 0 | 3 | 5 | 24 | 51 | 175 | 408 | 1225 | 3072 |
| V1. Population | 0 | 0 | 0 | 0 | 1 | 3 | 18 | 40 | 138 | 341 | 996 |
| V16. Natural environment | 0 | 0 | 0 | 1 | 2 | 12 | 26 | 92 | 224 | 664 | 1697 |
| V2. Fixed assets | 0 | 0 | 0 | 1 | 2 | 6 | 14 | 41 | 102 | 286 | 742 |
| V6. Prices | 0 | 0 | 0 | 0 | 0 | 0 | −1 | −2 | −12 | −26 | −92 |
| V7. Inflation | 0 | 0 | 0 | 0 | 0 | −1 | −2 | −12 | −26 | −92 | −224 |
| V4. Investment in fixed assets | 0 | 0 | 1 | 1 | 5 | 9 | 32 | 70 | 216 | 526 | 1513 |
| V5. The market value of fixed assets | 0 | 0 | 0 | −1 | −1 | −5 | −9 | −32 | −70 | −216 | −526 |
| V3. Disposal of fixed assets | 0 | 0 | 0 | 0 | 0 | 0 | 0 | 0 | 0 | 0 | 0 |
| V9. Salary rate | 0 | 0 | 0 | 0 | 0 | 1 | 2 | 12 | 26 | 92 | 224 |
| V18. Migration | 0 | 0 | 0 | 0 | 1 | 6 | 14 | 46 | 117 | 332 | 867 |

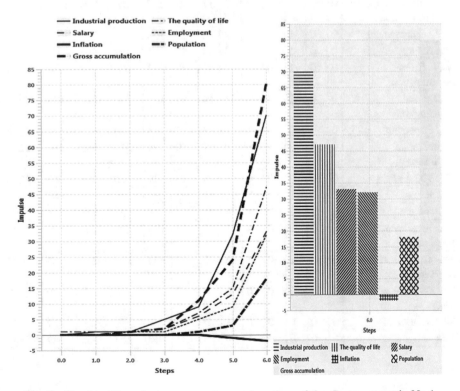

**Fig. 7.** Graphs of impulse processes at several vertices of the $G$ map, scenario No.1

The results of 10 cycles of pulse simulation are presented in Table 2. Note that the modeling is carried out until the development trend of situations becomes obvious. In this case, 10 steps were enough. Figure 7 shows the graphs of the pulse processes in several vertices and histograms of the corresponding values of the pulses at the last simulation step. The graphs are constructed according to the calculations in Table 2.

Scenario No.2. Let's suppose that production is developing $q_{10} = +1$, inflation is falling $q_7 = -1$, but inter-regional and foreign economic exchange is also falling, $q_{14} = -1$; disturbance vector $Q = \{Q = \{q_1 = 0; \ldots q_7 = -1; \ldots q_{10} = +1; \ldots q_{14} = -1; \ldots q_{18} = 0\}$.

The results of 10 cycles of pulse simulation are presented in Table 3.

The graphs are constructed according to the calculations in Table 3.

**Table 3.** Calculation of the impulse process for scenario No.2

| Step / Vertex | 0 | 1 | 2 | 3 | 4 | 5 | 6 | 7 | 8 | 9 | 10 | 11 | 12 |
|---|---|---|---|---|---|---|---|---|---|---|---|---|---|
| V0. The quality of life | 0 | 0 | 0 | 1 | 4 | 5 | 25 | 48 | 153 | 365 | 1062 | 2686 | 7537 |
| V15. Federal regulatory systems | 0 | 0 | 0 | 0 | 1 | 0 | 8 | 7 | 50 | 84 | 317 | 691 | 2160 |
| V10. Industrial production | 0 | 1 | 0 | 4 | 2 | 18 | 23 | 108 | 201 | 705 | 1584 | 4841 | 11918 |
| V14. Interregional and foreign exchange | 0 | -1 | 0 | -1 | 3 | 1 | 17 | 22 | 107 | 200 | 704 | 1583 | 4840 |
| V11. Employment | 0 | 0 | 1 | 0 | 4 | 2 | 18 | 23 | 108 | 201 | 705 | 1584 | 4841 |
| V8. Salary | 0 | 0 | 1 | 1 | 3 | 5 | 15 | 27 | 96 | 197 | 637 | 1520 | 4431 |
| V12. Gross accumulation | 0 | 0 | 1 | 0 | 7 | 7 | 42 | 77 | 267 | 607 | 1843 | 4536 | 12984 |
| V13. Final consumption | 0 | 0 | -1 | 2 | 0 | 15 | 14 | 94 | 145 | 595 | 1220 | 3971 | 9386 |
| V1. Population | 0 | 0 | 0 | 0 | 1 | 1 | 12 | 12 | 75 | 132 | 470 | 1056 | 3222 |
| V16. Natural environment | 0 | 0 | 0 | 1 | 0 | 8 | 7 | 50 | 84 | 317 | 691 | 2160 | 5227 |
| V2. Fixed assets | 0 | 0 | 0 | 0 | 0 | 3 | 5 | 19 | 40 | 130 | 308 | 905 | 2288 |
| V6. Prices | 0 | 0 | -1 | -1 | -1 | -1 | -2 | -1 | -9 | -8 | -51 | -85 | -318 |
| V7. Inflation | 0 | -1 | -1 | -1 | -1 | -2 | -1 | -9 | -8 | -51 | -85 | -318 | -692 |
| V4. Investment in fixed assets | 0 | 0 | 1 | 0 | 4 | 2 | 18 | 23 | 108 | 201 | 705 | 1584 | 4841 |
| V5. The market value of fixed assets | 0 | 0 | 1 | 0 | 1 | -3 | -1 | -17 | -22 | -107 | -200 | -704 | -1583 |
| V3. Disposal of fixed assets | 0 | 0 | 0 | 0 | 0 | 0 | 0 | 0 | 0 | 0 | 0 | 0 | 0 |
| V9. Salary rate | 0 | 0 | 0 | 0 | 0 | 1 | 0 | 8 | 7 | 50 | 84 | 317 | 691 |
| V18. Migration | 0 | 0 | 0 | 0 | 1 | 4 | 5 | 25 | 48 | 153 | 365 | 1062 | 2686 |

Comparing the simulation results of two scenarios (Tables 2 and 3, Figs. 7 and 8), we see:

- the development of industrial production in the region in any case generates positive trends in the system for all indicators (there are tendencies for the growth of desirable indicators at all peaks and tendencies for falling in unwanted $V_5$, $V_6$, $V_7$ in both scenarios);
- if industrial production develops but begins to be hindered by a decrease in inter-regional and foreign economic exchange and inflation (scenario No.2), then the indicators become lower than under scenario No. 1.

A full-scale study of the potential of a region requires modeling and analyzing more scenarios; the example illustrated the possibility of doing this. The next stage of the research is the stage of determining the worst, best, realistic and optimal scenarios for the development of the system. Choosing the best scenario can be done by applying appropriate decision criteria [24].

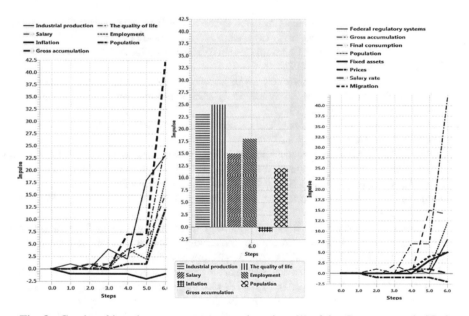

**Fig. 8.** Graphs of impulse processes at several vertices (1) of the $G$ map, scenario No.2

Figure 8 shows the graphs of the pulse processes in several vertices and histograms of the corresponding values of the pulses on the last simulation step shown.

# 4   Conclusion

The experience of many years of applying the developed methodology of cognitive modeling of various complex systems has shown the possibilities and advantages of a cognitive approach to their research and decision making. We hope that further use of the theory and practice of cognitive modeling of complex systems within the framework of the Intelligent Systems direction will find its development in the field of cyberphysical systems in such applications as an intelligent cognitive city keeping people in the control loop.

# References

1. Zhuge, H.: Interactive semantics. Artif. Intell. **174**, 190–204 (2010)
2. Zhuge, H.: Cyber physical society. In: The 1st Workshop on Cyber Physical Society, in Conjunction with the 6th International Conference on Semantics, Knowledge and Grids, Ningbo, China (2010)
3. Zhuge, H., Xing, Y.: Probabilistic resource space model for managing resources in cyberphysical society. IEEE Trans. Serv. Comput. **5**, 404–421 (2011)
4. Axelrod, R.: The Structure of Decision: Cognitive Maps of Political Elites, p. 395. Princeton University Press, Princeton (1976)
5. Atkin, R.H.: Combinatorial Connectivies in Social Systems. An Application of Simplicial Complex Structures to the Study of Large Organisations. Interdisciplinary Systems Research, Birkhäuser, Basel (1997)
6. Casti, J.: Connectivity, complexity and catastrophe in large-scale systems. Chichester-New York-Brisbane-Toronto, 216 p. (1979)
7. Roberts, F.: Discrete Mathematical Models with Applications to Social, Biological, and Environmental Problems. Englewood Cliffs, Prentice-Hall, New Jersey 559 p. (1976)
8. Eden, C.: Cognitive mapping. Eur. J. Oper. Res. **36**, 1–13 (1998)
9. Langley, P.: Cognitive architectures: research issues and challenges. Cogn. Syst. Res. **10**(2), 141–160 (2009)
10. Abramova, N., Portsev, R.: Reflexive approach to multi-subject situations in cognitive mapping. In: 18th IFAC Conference on Technology, Culture and International Stability (TECIS2018) IFAC Papers OnLine, vol. 51–30, pp. 516–521 (2018)
11. Avdeeva, Z.K., Kovriga, S.V.: On governance decision support in the area of political stability using cognitive maps. In: 18th IFAC Conference on Technology, Culture and International Stability (TECIS2018) IFAC Papers OnLine, vol. 51–30, pp. 498–503 (2018)
12. Kulba, V.V., Kononov, D.A., Kovalevsky, S.S., Kosyachenko, S.A., Nizhegorodtsev, R.M., Chernov, I.V.: Scenario analysis of the behavior dynamics of socio-economic systems, Moscow, IPU RAN, 122 p. (2002)
13. Gorelova, G.V., Zakharova, E.N., Radchenko, S.A.: Study of semi-structured problems of socio-economic systems: a cognitive approach, 332 p. Publishing House of RSU, Rostov-on-Don (2006)
14. Gorelova, G.V., Melnik, E.V.: Cognitive modules of intelligent decision support in information management systems. In: Systems analysis in economics. Materials of the scientific-practical conference, pp. 45–49 (2010)

15. Gorelova, G.V., Maslennikova, A.V.: Simulation based on cognitive methodology and system dynamics, analysis of the system "South of Russia". In: Tr. scientific-practical conference "System Analysis in the Economy 2012", pp. 50–65 (2012)
16. Gorelova, G.V., Zhertovskaya, E.V., Yakimenko, M.V.: Innovative modernization of the national and regional economies: prerequisites, principles and priorities, 143 p. Ed.-SFU, Rostov-on-Don (2014)
17. Gorelova, G.V., Ryabtsev, V.N.: Architecture modeling and dynamics of the geopolitical regions of the modern world: a cognitive approach (zone "Black Sea-Caucasus-Caspian"), 374 p. SFU publishing house, Rostov-on-Don (2014)
18. Gorelova, G.V., Kolodenkova, A.E.: Safety assessment of information and control systems of nuclear power plants using cognitive modeling. Technosphere Saf. Technol. **62**(4), 339–348 (2015)
19. Gorelova, G.V., Pankratova, N.D. (eds.) Innovative development of socio-economic systems based on foresight and cognitive modeling methodologies. Publishing House Naukova Dumka, Kiev 464 p. (2015)
20. Gorelova, G.V.: Cognitive approach to the development and justification of management decisions in organizational systems. Systems analysis in design and management: Collection of scientific papers of the XX International. scientific and practical conference, Part 1, pp. 67–72 (2016)
21. Gorelova, G., Pankratova, N.D.: Strategy of complex systems development based on the synthesis of foresight and cognitive modelling methodologies. Syst. Anal. Intell. Comput. IEEE (SAIC) **08**(12), 1–6 (2018)
22. Gorelova, G., Kalinichenko, A.: Toolkit of cognitive research of large systems and the risk of a human factor. In: Fourth International Forum on Cognitive Modeling, Tel Aviv, Israel (2018)
23. Program for cognitive modeling and analysis of socio-economic systems at the regional level. Certificate of state registration of computer programs No. 2018661506 dated 07.09 (2018)
24. Gorelova, G.V.: Models of decision making when designing and managing the objects under conditions probability uncertainty. News SFU, Technical Sciences (1), 177–188 (2019)

# Blending Traditional and Modern Approaches to Teaching Control Theory

Inna A. Seledtsova[1,2] and Leonid Chechurin[1,2(✉)]

[1] Peter the Great Saint Petersburg Polytechnic University,
Saint Petersburg, Russia
[2] LUT University, Lappeenranta, Finland
{Inna.Seledtcova,Leonid.Chechurin}@lut.fi

**Abstract.** Control Theory serves as a fundamental background for a number of popular paradigms, including the cyber-physical one. The contents have been standardized over the decades of teaching, but new digital technologies and market practical skills demand to raise the questions on how the course is to be taught. The goal of the report is to share the results of 3 years (2015–2018) of experimenting on the transition from classical to project-based blended design of the Modern Control and Automation course. The course is part of the curricula of the SPbPU MSc degree programme "Management of Innovative Processes (Innovatika)"; 75 students were taught. Now, the course is represented by online and offline studying of the basic theoretical aspects and hands-on development of a hardware device in a project group. Each element of the course, its three-year evolution, analysis of the collected data, results of surveys and recommendations for adaptation of the used tools for teaching control theory in the frames of another educational program or university are described in the article.

**Keywords:** Control theory · Blended learning · Project-based learning

## 1 Introduction

The world is about to enter a new technological paradigm, namely, Industry 4.0. Cyber-physical systems with Control Theory in their foundation are in the basis of this paradigm. The contents of Control Theory have been standardized over the years of teaching. Teachers from all over the world know a number of answers to the question, "What theoretical aspects should be included in a course curricula?" However, new digital technologies and market practical skills demand to raise questions on how the course is to be taught. Modern higher education is looking for active learning course designs, flipped classrooms, raising digital software skills, project-based learning and focuses on competencies. The goal of this report is to share the 3-year experience on the transformation of the Modern Control and Automation course from the traditional (in-class) format to a blended and project-based one.

© Springer Nature Switzerland AG 2020
D. G. Arseniev et al. (Eds.): CPS&C 2019, LNNS 95, pp. 632–642, 2020.
https://doi.org/10.1007/978-3-030-34983-7_62

## 2 Blended and Project-Based Approach

The analysis of the approaches used in teaching Control Theory and similar disciplines allows to determine the following current trends and their combinations:

– Use of remote labs [1–4]
– Use of online lectures [5, 6]
– Use of blended learning [7–9]
– Use of project-based and group work approach [10–13].

The Modern Control and Automation course is part of the curricula of the MSc 27.04.05 degree programme "Management of Innovative Processes (Innovatika)". This course is being taught during one academic term for the first-year MSc students. This course was presented as a classical theoretical course on control theory with a number of in-class lectures, homework, midterm tests and a final exam till 2016/2017 academic year. The first attempt of a blending and project-based approach implementation to teaching this course was made in 2016. The 2018/2019 academic year is the year of the third version of the course transformation. Current course components are shown in Table 1. Two key features of the course are: the blended content and project-based learning.

Blended learning is a combination of face-to-face and online teaching and learning [14]. The use of blended of approach allows:

To switch passive learning to active learning because it requires more attention and time from students side [15].

To provide the facility to learn and access materials in different modes [14].

To provide the opportunity to use various tools for education and use the online tools in a traditional classroom-based format [16].

Project-based learning is a method providing for the students to learn a range of skills and subject matters in the process of creating their own projects. Students work in groups and bring their own experiences, abilities, learning styles, and perspectives to the project [17]. This approach allows not only to achieve higher involvement of students in the educational process but also provides a number of opportunities [18]:

• Experiential learning
• Using more than just disciplinary knowledge to succeed: teamwork, project management, leadership, communication, etc.
• Building educational infrastructure around the interdisciplinary projects
• Community involvement.

It is important to mention that students of the reviewed MSc program have intensive theoretical base on how to establish the process of engineering projects development. Therefore, in spite of the fact that control theory is not the core discipline for students of this programme, we assume that the existing knowledge base in fundamental disciplines and project management allows them to study control theory by doing their own projects.

Project work within the course on modern control and automation is organized in the following way:

(1) Teams forming (3–6 students per team)
(2) Choosing and researching on an idea of a device which then teams would develop. This preliminary study comprises technical feasibility study, market research, and analysis of the necessary components for realization.
(3) Preliminary presentation of the project idea
(4) Financing (100 EUR per team)
(5) Development of the device
(6) Project and product presentation. Final presentation assumes offline presentation of the results and final project reports that consist of the *feasibility study*, *mathematical model* of the system, *simulation* in Simulink, description of the *assembling*, *development* and *programming*, description of the *project management tools* (such as project planning, risk management, etc.), project *financing* and *purchasing* processes, analysis of the *achieved results and marketing research* [19, 20].

As the results of the 3 year-experience, 16 projects have been developed. Examples of the developed devices are shown on Fig. 1. Comparative characteristics of the 3 year-experience (2016/2017, 2017/2018, 2018/2019 academic years) is shown in Table 2.

An important component of the course is the platform on which communication between students and the teacher is carried out. We recommend using two types of communication channels: one of them is required for communication among the students of the course and the teacher [21], the second type of communication channels is used for interactions among a certain team members [22]. Anyway, every student is able to join any channel of any team to be aware of the progress or share comments.

**Table 1.** Content of the Modern Control and Automation course

| No | Components | Goals | Details |
|---|---|---|---|
| 1 | Theory | To provide basic overview of control theory concepts | Lectures consist of two parts: online and offline |
| 1.1 | Online | To provide an opportunity to be prepared for an offline lecture in advance<br>To provide continuous remote inflow of micro lectures<br>To support a certain project development by providing additional theoretical materials (*just-in-time teaching*) | Online part of the course consists of the online book, online video lectures from other courses, on-demand webinars (via Skype), and daily support in the course through the communication channel.<br>Online components are shared among students using the communication channel |
| 1.2 | Offline | To provide for face-to-face discussions of the most difficult lectures<br>To provide for offline discussions of theoretical fundamentals of each project | Offline part of the course consists of monthly intensive lectures with discussions of the most difficult theoretical aspects and working on projects |

*(continued)*

**Table 1.** (*continued*)

| No | Components | Goals | Details |
|---|---|---|---|
| 2 | MatLab practice | To provide basic understanding on building system models and simulation of their work in Simulink | Every product that is developed by students in the frames of the project work must be modelled and stabilized in Simulink |
| 3 | Project work | To increase involvement of students in learning<br>To establish learning as an on-going process<br>To demonstrate application of control theory in real-life projects | Project work assumes development of a hardware device in a project group based on the theoretical part of the course. The developed device is presented at the end of the course |
| 4 | Communication | To provide the channel for daily interaction between students and a teacher | The communication channel is based on social media tools such as: vk.com, Telegram, etc. It is assumed that there are two types of communication channels. One of them is for all students. All of the course updates and online theoretical content are shared in this channel. Another channel is meant as a certain team's channel. It is used for the team members interactions, sharing specific theoretical content applied to a certain project. But everybody can connect to the team channel to check what is happening and provide feedback |
| 5 | Exam | To check the progress in the theoretical part of the course | The exam is a theoretical check on the course content. It assumes answering questions in groups or individually |

The social network VK.com (or any other social network with a similar organization of participants' involvement and information storage may be considered) was used during the first and the second years of this course implementation. Telegram (or any other messenger may be considered) was used during the third year. Social networks and messengers were chosen as a communication platform for two reasons:

- Habitual interface for students
- Continuous and stable habit of daily use.

These reasons allow to reduce the time for onboarding students in the communication channels and start to use them as soon as they are created. The comparative analysis of two platforms for communication *for educational process* is given in Table 3.

**Fig. 1.** Examples of developed products: a – robot, b – quadcopter, c – laser toy for a cat

We can see from the comparative analysis that messengers provide a *higher level of involvement* of students. The level of involvement was defined based on the *average number of messages* per 1 team and per 1 student, *average time spent* on the platform per 1 student (Table 2). Time of reaction to the message (*average time between messages*) is lower in messengers than in social networks. It also is remarkable that the messenger provides for a *higher level of viewings* of the messages placed by a teacher or a student. It means that placing any educational content in messengers means its *higher coverage among students*.

**Table 2.** Comparative analysis of the course implementations

| No | Criteria | 2016/2017 | 2017/2018 | 2018/2019 |
|----|----------|-----------|-----------|-----------|
| *Overall numbers* | | | | |
| 1 | Number of students | 25 | 26 | 29 |
| 2 | Number of teams | 5 | 5 | 6 |
| 3 | Main features | Blended type of learning and project work were implemented for this course | Some parts of theory were represented by the online courses from other universities | Marketing analysis and searching for channels of distribution were implemented as a mandatory requirement |

*(continued)*

**Table 2.** (*continued*)

| No | Criteria | 2016/2017 | 2017/2018 | 2018/2019 |
|----|----------|-----------|-----------|-----------|
| *Theory* | | | | |
| 4 | Course type | Blended | Blended | Blended, intensive involvement of the course teacher in daily online discussions of all projects |
| 5 | Language | English | English | English |
| *Projects* | | | | |
| 6 | Number of accepted projects | 4 | 5 | 6 |
| 7 | Average money spent, euro | 95 | 89 | 83 |
| 8 | Level of application of project management tools (based on a qualitative analysis of the final project reports and communications in the teams' channels) | High | Medium | Low |
| *Communication* | | | | |
| 9 | Platform | vk.com (1 community for common discussions/announcements and 1 community for each team) | | Telegram (1 channel for common discussions/announcements and 1 channel for each team) |
| 10 | Average number of messages per 1 team in teams' communities/channels | 15 | 6 | 43 |
| 11 | Average number of messages per 1 student | 9 | 4 | 17 |
| 12 | Average time spent on the platform per 1 student, minutes | 76 | 32 | 113 |

But there is an important weakness of messengers that doesn't allow to organize transparent storage and information search in different channels. Strengths and weaknesses of each communication channel will be considered for the further process of implementation of communication platforms.

**Table 3.** Comparative analysis of the communication platforms

| No | Criteria | vk.com | Telegram |
|---|---|---|---|
| 1 | Statistics analysis | Limited, mostly manual analysis | Automated |
| 2 | Level of involvement in discussions* | Low | High |
| 3 | Average time between messages, hours | 9,7 | 3,3 |
| 4 | Average level of views per 1 post, % (number of unique views/total number of students) | 41,2% | 81,7% |
| 5 | Ease of navigation | Transparent navigation; it is possible to create separate subchannels for discussing a certain issue, task, link, news | Nontransparent navigation, all discussions are in the same channel; it is difficult to delineate discussions on different topics and to find everything related to a particular topic |

## 3  Results and Prospects

The Modern Control and Automation course was completed by 80 students; 16 projects were developed. According to the feedback from students, 82.5% of students said that the project-based approach was much more productive for studying new content than the standard approach. The main difficulties of the course identified by students were:

- Lack of time for project development
- Insufficient level of interest of some team members
- Insufficient level of previous knowledge of control theory
- Imbalanced distribution of the team members' contribution: it was either in the technical part or in management.

One of the main prospects of the further development of the course is its improvement and placement on the educational CEPHEI platform (Cooperative E-learning Platform for Industrial Innovation, co-funded by Erasmus+ Programme, *cephei.eu*). Placement of this course on the CEPHEI platform will allow involving industries and their cases in the process of idea generation for new projects.

# 4    Adaptation of the Course

On the basis of the 3-year experience of implementation of the course, we suggest a step-by-step adaptation guide for the stages of preparation and analysis of the course and its implementation for similar educational programs curricula. Supporting tools are suggested for some parts of this guide.

Stage "Preparation"

### Platform
Where will the course content be placed? Where will students communicate?

*Supporting tools: university LMS (learning management system), messengers (WhatsApp, Telegram, etc.), social networks (vk.com, Facebook).*

### Course Design
It is recommended to use blended classroom approach. It means that theoretical lectures are taught both in online and offline modes, project work is performed in offline and online modes (the ratio is managed by students).

It is recommended to develop the course curriculum and to determine the following set of parameters for each part:

A. *Theory. Online or offline lectures?*
What materials will be used for online lectures?
– Existing teacher's materials
– Existing materials from other universities/courses
– New materials that will be created by the teacher (new video lectures, webinars, screen recordings, etc.)
   B. Type of quizzes and other tasks, midterm assignments, examination process
*Supporting tools: Google forms, Quizizz, Typeforms, Quizzes mechanism included to your LMS.*
   C. Type of students' projects: technical scopes, theoretical scopes, time scopes, financial scopes
   D. Supporting materials and hands-on for sharing among the students (books, papers, links for sharing, etc.)

### Project Schedule
Recommended project plan is shown in Table 4. This plan is shared among the students in the beginning of the course and all project events take place according it.

### Budgeting
What scheme of financing will be used? There are several options:

– Pre-financing from the University
– Pre-financing from partner companies or companies that provide their cases for solving during the course

The scheme of post-financing after presentations of the project results might also be considered; however, it has never been applied by the authors.

It is also important to determine the criteria and timing of financing. It is recommended to provide financing after the midterm presentation of the preliminary project work (after the weeks 4–7, Table 4).

### Requirements Development

A. Determine the requirements for a project, mid-term and final project report [19, 20]

B. Determine the requirements for exam passing

C. Establish the points of the preliminary progress check and conditions for their passing

Stage "Analysis"

### Collect the Feedback from the Students

Ask questions according to examples [23] that allow to analyze the feedback from the audience and to make improvements for future courses.

*Supporting tools: Google Forms, Typeforms, Questionnaires included to your LMS. Provide the feedback for students*

**Table 4.**  Project activities, outputs and deadlines

| Week | Activity | Controlled output |
|---|---|---|
| 1 | *Kick-off meeting* | |
| 2 | Concept design | Patent. A sketch of the design system, its description, |
| 3 | Concept design | references (to papers, patents, videos, other documents), basic ideas |
| 4 | Detailed design | Project plan, devices in math model |
| 5 | Detailed design | Required fields: |
| 6 | Detailed design | 1. Detail description of the project, drawings (preferably in |
| 7 | Detailed design | professional software), 2. Technical feasibility study (why do you think it will work in theory, math model for the object, description of the controller, why the closed loop system will be stable? How the designed feedback will be implemented in the hardware, choice of sensors/measurements and controller) 3. Simulation models of the plant, controller and closed loop system (Matlab+Simulink) 4. List of required components, prices and suppliers 5. Total budget of the project (upper limit – 100,- euro) 6. Marketing and economic study |
| 8 | *Approval of the project plan/Money allocation* | |
| 9 | Purchasing | Hardware devices. The hardware assembly, all elements in |
| 10 | Purchasing/assembling | their places. A controller is programmed |
| 11 | Assembling/programming | |
| 12 | *Testing* | |

(*continued*)

**Table 4.** (*continued*)

| Week | Activity | Controlled output |
|------|----------|-------------------|
| 13   | Tuning/reporting | Final report. Detail description of activities, documentation, |
| 14   | Reporting | time sheets of the team members, receipts, user manual for |
|      |          | the device, video report, photo-gallery, etc. |
| 15   | *Commissioning/show* | |
| 16   | Bonus week | |

## 5  Conclusions

The use of new tools in teaching control theory allows not only to achieve a more flexible course curriculum but also provide for a huge opportunity for the analysis of the collected data and adaptation and improvement of the course for a specific audience and its requirements on the basis of this analysis. Application of blended learning helps to make teaching more interactive and saturated. Application of the project-based approach allows to make teaching more targeted and to form practical understanding of the control theory concepts application.

**Acknowledgments.**  This report was partially supported by CEPHEI project of the ERASMUS+ EU framework.

## References

1. Dobriborsci, D., Bazylev, D., Margun, A.: Teaching students the basics of control theory using NI ELVIS II. In: Uskov, V., Howlett, R., Jain, L. (eds.) SEEL 2017. Smart Innovation, Systems and Technologies, vol. 75. Springer, Cham (2018)
2. Rojko, A., Hercog, D.: Teaching of robot control with remote experiments. In: Tzafestas, S. (ed.) Web-Based Control and Robotics Education. Intelligent Systems, Control and Automation: Science and Engineering, vol. 38. Springer, Dordrecht (2009)
3. Casini, M., Prattichizzo, D., Vicino, A.: A Matlab-based remote lab for control and robotics education. In: Tzafestas, S. (ed.) Web-Based Control and Robotics Education. Intelligent Systems, Control and Automation: Science and Engineering, vol. 38. Springer, Dordrecht (2009)
4. Doğan, B., Erdal, H.: System control through the internet and a remote access laboratory implementation. In: Leung, H., Li, F., Lau, R., Li, Q. (eds.) ICWL 2007. LNCS, vol. 4823, pp. 532–541. Springer, Heidelberg (2008)
5. Tzafestas, S.G.: Teaching control and robotics using the web. In: Tzafestas, S. (ed.) Web-Based Control and Robotics Education. Intelligent Systems, Control and Automation: Science and Engineering, vol. 38. Springer, Dordrecht (2009)
6. Almusawi, A.R.J., Dulger, L.C., Kapucu, S.: J. Braz. Soc. Mech. Sci. Eng. **40**, 437 (2018)
7. Seiler, S., Sell, R.: Comprehensive blended learning concept for teaching micro controller technology utilising homelab kits and remote labs in a virtual web environment. In: Pan, Z., Cheok, A.D., Müller, W., Iurgel, I., Petta, P., Urban, B. (eds.) Transactions on Edutainment X. Lecture Notes in Computer Science, vol. 7775. Springer, Heidelberg (2013)

8. Köttgen, L., Winter, S., Schröder, S., Richert, A., Isenhardt, I.: Integrating blended learning – on the way to an excellent didactical method-mix for engineering education. In: Jeschke, S., Isenhardt, I., Hees, F., Henning, K. (eds.) Automation, Communication and Cybernetics in Science and Engineering 2015/2016. Springer, Cham (2016)

9. Rogado, A.B.G., Conde, M.J.R., Miguelánez, S.O., Riaza, B.G., Peñalvo, F.J.G.: Efficiency assessment of a blended-learning educational methodology in engineering. In: Lytras, M.D., et al. (eds.) TECH-EDUCATION 2010. Communications in Computer and Information Science, vol. 73. Springer, Heidelberg (2010)

10. Moorthi, M.N., Vaideeswaran, J.: Overview of effective and efficient learning model project-based learning (PBL). In: Natarajan, R. (eds.) Proceedings of the International Conference on Transformations in Engineering Education. Springer, New Delhi (2015)

11. Jou, M., Wu, M.J., Wu, D.W.: Development of online inquiry environments to support project-based learning of robotics. In: Lytras, M.D., Carroll, J.M., Damiani, E., Tennyson, R. D. (eds.) WSKS 2008. Lecture Notes in Computer Science, vol. 5288. Springer, Heidelberg (2008)

12. Tavares, R.: A self-reflection on the importance of project activities in engineering education. In: Auer, M., Guralnick, D., Uhomoibhi, J. (eds.) ICL 2016. Advances in Intelligent Systems and Computing, vol. 544. Springer, Cham (2017)

13. Béres, I., Kis, M.: Flipped classroom method combined with project based group work. In: Auer, M., Guralnick, D., Simonics, I. (eds.) ICL 2017. Advances in Intelligent Systems and Computing, vol. 715. Springer, Cham (2018)

14. Kaur, M.: Blended learning - its challenges and future. Procedia – Soc. Behav. Sci. **93**, 612–617 (2013)

15. Lopez, V., Pérez-López, M., Lázaro, C.: Blended learning in higher education: students' perceptions and their relation to outcomes. Comput. Educ. **56**, 818–826 (2011)

16. Thorne, K.: Blended Learning: How To Integrate Online and Traditional. Kogan Page, London (2003)

17. Jadhav, H.S., Patil, S.N.: Mini projects: a new concept of transformation of teaching-learning process. In: Natarajan, R. (eds.) Proceedings of the International Conference on Transformations in Engineering Education. Springer, New Delhi (2015)

18. Patil, S.R., Thombare, D.G., Kulkarni, S.S.: Positive influence of association of technical students organizations with professional bodies on academics: a case study of SAE RIT. Indian TechnEduc. **34**(2), 67–75 (2011). H.S. Jadhav and S.N. Patil

19. Example of the final report, 2016/2017 academic year. https://vk.com/doc53407714_439764616?hash=a736d264122ff1988a&dl=03c73a074ef256fda3. Accessed 02 May 2019

20. Example of the final report, 2017/2018 academic year. https://vk.com/doc50300661_456249274?hash=4bf675981aeecc1d4b&dl=13479ce744c31ee52c. Accessed 02 May 2019

21. Example of the course group, 2017/2018 academic year. https://vk.com/club153880806. Accessed 30 Apr 2019

22. Example of the team communication channel, 2016/2017 academic year. https://vk.com/in_progress_team. Accessed 27 Apr 2019

23. Example of the final questionnaire. https://docs.google.com/forms/d/e/1FAIpQLScEtkE cCZctwORfGqKmeAsyieLdPqcsLRHxM9WSUF1Wbx_a8w/closedform. Accessed 28 Apr 2019

# Systematic Approach to Education of Specialists for a New Technological Paradigm

Sergey G. Redko[1(✉)], Nadezhda A. Tsvetkova[1],
Inna A. Seledtsova[1,2], and Sergey A. Golubev[1]

[1] Peter the Great St. Petersburg Polytechnic University, Saint Petersburg, Russia
redko_sg@spbstu.ru
[2] LUT University, Lappeenranta, Finland
Inna.Seledtcova@lut.fi

**Abstract.** The world is about to enter a new technological evolution with cyber-physical systems in its basis. One of the vital questions is exploring of new approaches to training specialists capable to develop, contribute and maintain corporate and industrial infrastructure in the new technological framework. The main goal of this research is to show approach to complex training of qualified specialists in the upcoming economic and technological paradigm in terms of three aspects: whom to teach, what to teach, and how to teach. Interrelation and the impact of these three aspects on training highly demanded professionals in the field of cyber-physical systems and control are shown in the article. The main emphasis is made on the question "How to teach?"

**Keywords:** Digital education · Industry-based education · New forms of teaching · Educational platform · Knowledge management · In-demand skills · Cyber-physics systems

## 1  Introduction

The world is on the threshold of the fourth industrial revolution, Industry 4.0. Today, most people believe that digital transformation is inevitable in almost all spheres of life and economics. Fully digital industry based on modern information technology is appearing, including: digital manufacturing, virtualization of production functions that is accompanied by the formation of a shared economy, change in the functionality of devices without making changes in them as physical objects, by changing the technologies of their control, etc. [1].

Cyberphysical systems provide a technological basis for the transition to a new economic paradigm. The essence of cyberphysical systems consists in that they connect the physical processes of production or other processes that require practical implementation of continuous control in real time with the use of software and electronic systems [2].

© Springer Nature Switzerland AG 2020
D. G. Arseniev et al. (Eds.): CPS&C 2019, LNNS 95, pp. 643–650, 2020.
https://doi.org/10.1007/978-3-030-34983-7_63

The search for new approaches to training of specialists able to develop and maintain corporate and industrial infrastructure in a new technological environment has already become a topical issue.

The aim of this work is to demonstrate an approach to comprehensive training of qualified specialists in the new economic paradigm in terms of three aspects: who to teach, what to teach, and how to teach. The primary focus in the research is on the question of "How to teach?"

## 2 Whom to Teach?

It is necessary to improve the education system, as it has to provide the digital economy with competent personnel. At the same time, the transformation of the labor market, which should be based on the requirements of the digital economy, is inevitable. The key factor in the successful transition to this stage is education at all levels, from schools to universities, with the transition to continuous adult education.

In order to make effective use of human potential for the digital economy, it is necessary to provide training *for the widest possible range of citizens* who could be involved in productive activities in accordance with their skills and mobility. This is possible by providing flexible forms of employment, including full-scale distance employment. This range of full participants will include various categories of citizens, from schoolchildren and university students to pensioners.

## 3 What to Teach?

An important factor in the successful development of the digital economy is the contents of training. Now most educational programs are focused on training in the narrow sense of the word, which is the transfer of information, knowledge and some skills. Rather than skills, *it is necessary to develop habits for digital technology and related views*. They can be referred to as competence chains: they consist of super-professional competencies (beliefs, habits, lifestyle, general cultural competencies), professional competencies and basic competencies [7].

As it was noted at the international technology forum in April 2016, the word profession has completely lost its meaning. Profession as a concept ceased to exist long ago, but there are clusters of competence sets. Modern students need to have basic technological skills, to be able to manage a project and make their own decisions, to work in complex interdisciplinary teams.

In our opinion, the ability to perform project-oriented activities is one of these key core competencies. The trend towards the convergence of engineering and managerial education leads to a steady decline in demand for managers with traditional management education and engineers with classic engineering education [4]. The interdisciplinarity of educational courses and programmes is beginning to play a key role. Interdisciplinarity refers not only to joint consideration of problems by specialists from different branches of science and technology, who see them from different sides, but also, and above all, the synthesis of knowledge from different areas aimed at obtaining a novel solution to a complex problem and supporting the implementation of solutions from their conceptualization to implementation and commercialization [5].

## 4 How to Teach?

Training for digital technologies must definitely be implemented with the use of digital technologies. Suitable digital technologies already exist: these are distance learning environments.

Continuous education should become the norm and one of the pillars of the entire education system. It can be built around continuing education networks which unite suppliers and consumers of individual courses and groups of courses (modules), certifiers (specialists and organizations conducting competence assessment), navigators (specialists and organizations providing information and career support to consumers). Traditional educational organizations should interact with these networks both as course providers and as consumers [3]. The involvement of industries in the development of educational programs content and the definition of a set of skills that professionals must have to find employment is becoming an increasingly clear requirement from the industries [6].

SPbPU is one of the participants of CEPHEI international project, the Cooperative E-learning Platform for Industrial Innovation (co-funded by Erasmus+Programme). One of the goals of this project is creation of new training courses, the practical relevance of which is confirmed by the industrial sector, using modern educational formats (such as digital education, blended learning, etc.) [9].

As an example of such a course, the Fundamentals of Project-Based Activities course might be considered. One of its modifications will be presented on the CEPHEI project educational platform. The Fundamentals of Project-Based Activities course is aimed at the formation of an interdisciplinary modern skill: project activity skill [8]. Currently, the Open Education national platform [11] presents 12 courses, and Coursera [12] offers 17 courses from Russian universities related to projects implementation and project management. The main feature of the Fundamentals of Project-Based Activities course is that the emphasis is not on specific management processes but on the step-by-step solving of complex problems from any area with the help of the project approach: what tools from project management can be used to solve the problem at all of its stages from the search for ideas to the implementation of solutions and in which ways.

Currently, the Fundamentals of Project-Based Activities course is already provided in four formats, and the work is underway on the creation of the fifth version of the course for the CEPHEI project. Table 1 shows the comparative characteristics for each version of the course, indicating the technical facilities for its implementation and the specific features of the content.

**Table 1.** Versions of the course on the Fundamentals of Project-Based Activities

| Course components | For students from SPbPU | For students and teachers from other universities | For everybody | For specific companies | For CEPHEI project |
|---|---|---|---|---|---|
| Start year | 2017 | 2018 | 2019 | 2018 | 2020 |
| Number of students | >6000 annually | 500 | 5500 | 500 | To be determined |
| Platform | Project.Spbstu (Moodle) | Project.Spbstu (Moodle) | OpenEdu (edx) | Project.Spbstu (Moodle) | CEPHEI (edx) |
| Course type | Blended with instructors | Blended with instructors | Online self-passed | Online/blended Self-passed/with instructors | Online/blended Self-passed/with instructors |
| Supervising | Yes | On-demand | Yes | On-demand | On-demand |

## 4.1    Educational Course in SPbPU

The course was introduced for all areas of bachelor's degree and specialist degree training at St. Petersburg Polytechnic University and is mandatory for students of the second year (project.spbstu.ru). The course is taken annually by more than 6,000 students in all areas of training: technical, humanitarian, economic. It provides for the formation of necessary universal competencies in the development and implementation of projects of various types, teamwork and communication, systems thinking, self-organization and self-development.

The main principle of the course is learning through practice. Accordingly, the students are given the task to study the theoretical material and to complete the project: implement an idea and get the result during the term. It is important to note that students are not limited in the choice of subjects: the project can be of any nature: research, engineering, business, social or creative. As a result, more than 500 projects are carried out simultaneously by teams consisting of students studying in any areas of specialization.

The distance education platform makes it possible to organize interaction with Customers to collect and agree on the proposed subjects of projects, form teams for specific projects and appoint teachers/coaches for the teams, track the process of tasks completion in the practical part of the course (project), etc. Thus, the control of the knowledge of the theoretical material through the passage of automated tests and monitoring the implementation of the practical part, that is, the implementation of the project, is performed.

As the crucial aspect of the practical part is teamwork, each student is awarded two grades:

- Evaluation of the work of the whole team by the teacher/coach on the portal for the performance of group tasks (templates and presentations).
- Individual assessment of the team members by their project manager based on the personal contribution to the work on the template/presentation.

Individual assessment of each student is a product of the team points and the personal contribution. As a result, a graphical representation of the learning result of each student is automatically generated on the portal at the end of the course and can later be used as a digital representation of the student's competencies for the development of educational and professional recommendations. The presentation includes an assessment by 12 indicators (Fig. 1).

## 4.2 Network Program for Partner Universities

The next stage of the course development was a network-based program for partner universities [10]. Thanks to this initiative, not only SPbPU students but also students from other universities can take the course. The theoretical and practical part, as well as the evaluation for this version of the course, was based on the experience of its implementation at SPbPU.

## 4.3 Open Education National Platform

The third form of course implementation is a national platform called Open Education (openedu.ru/course/spbstu/OPD/). The main distinguishing feature of this version of the course from the previous two versions is that a student is more independent in mastering materials. Thus, students are completely independent in determining their pace of dealing with their lectures and assignments. Assessment is implemented in the form of automated testing: two intermediate tests during the course and one final test at the end. The role of the course team in this implementation is to communicate with the audience, advising on theoretical issues at the forum.

As the survey shows, the main audience of this version of the course is students *aged 30 and over*, and the vast majority of the students have jobs (66.7%), including 22.6% working in the field of science/education, 15.1% in IT, 12.1% in construction, 10.4% in industry, and 9% in marketing.

The main expectations in respect of the course are distributed as follows:

- gaining knowledge about the approaches used in project activities - 74.2%
- applying knowledge in practice - 48.9%
- systematization of existing knowledge - 41.6%
- checking their existing knowledge - 21.8%

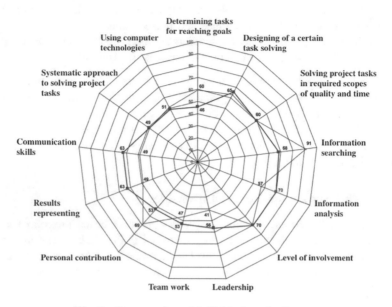

**Fig. 1.** Team grade and individual grade diagram

## 4.4    A Course for Enterprises

The course for enterprises is deployed on the Moodle platform. The course is developed both for a specific company and for a group of companies (for example, those united by industry), which makes it possible:

(1) To adapt the materials: the course may include company's materials related to project activity. Such materials can include standards, procedures, templates, and cases.
(2) To update the control of the development of the course not only by adding new questions on advanced materials, but also by the ability to check the open-type tasks.
(3) For the company's management to monitor the progress on the course, collect analytics on the competencies of their employees.

## 4.5    CEPHEI Platform

It is planned to create a new format of the course within the CEPHEI project, and the course will be based on existing materials modified in accordance with the needs of the enterprises. The main participants of the course will be representatives of industries interested in the development of project management skills of their employees. At the same time, all the existing content of the course will be adapted to the actual needs of enterprises (based on the surveys and interviews with the company' representatives), and the individual and group project work of the students will be based on real cases from companies, with feedback from the curators of the course from SPbPU.

## 5 Conclusions

The Fundamentals of Project-Based Activities course in different forms of its implementation provides an opportunity to develop an interdisciplinary skill which is in great demand for all possible categories of students: starting from the open version of the course on the national platform and ending with the version which has highly specialized content for the needs of enterprises and is presented on the CEPHEI platform. All this is becoming possible due to the use of modern means of distance education.

An important aspect of the presented course implementations is the possibility not only to identify leaders among all the students, i.e., people who are able to lead the team and to ensure the implementation of the idea with a specific result, but also to obtain in a digital form the indicators for evaluating the work of each student in terms of different indicators, which can later be used to form a portfolio for potential employers, as well as to identify individual characteristics and needs for the development of a more personalized educational and professional trajectory.

An important advantage of the use of the online learning tools under consideration is the broad scale of student coverage. Actually, almost 20,000 students have taken the course over the time of its existence. The analysis of the statistics of the course makes it possible to flexibly implement changes in the script and the contents of the course, to optimize, for example, such parameters as the duration of the video, the format of quizzes for different target audiences, and quickly receive feedback on the changes introduced. That, in its turn, makes it possible to significantly improve the educational process efficiency.

The presented approach is promising for comprehensive training of the widest possible range of specialists in accordance with the requirements of the new technological paradigm.

**Acknowledgments.** This report was partially supported by CEPHEI project of ERASMUS+EU framework.

## References

1. Voskov, L.S., Rolich, A.Y.: Internet of things and cyber-physical systems. Concept of master student program. HSE, Moscow (2017)
2. Wolf, W.: Cyber-physical systems. Computer **42**, 88–89 (2009)
3. Shmelkova, L.V.: Specialists for digital economics: Look to the future. Additional professional education in the country and in the world, vol. 8, pp. 1–4 (2016)
4. Gitelman, L.D., Kozhevnikov, M.V.: Education for managers paradigm to technological breakthrough in economics. Econ. Reg. **14**(2), 433–449 (2018)
5. Gitelman, L.D., Sandler, D.G., Gavrilova, T.B., Kozhevnikov, M.V.: Complex systems management competency for technology modernization. Int. J. Des. Nat. Ecodyn. **12**(4), 525–537 (2017)

6. Gielen, G.: Final overview of the expected competencies of future nano-electronics engineers (2011). http://cordis.europa.eu/docs/projects/cnect/1/257051/080/deliverables/001-Eurodots D12.pdf. Accessed 15 Apr 2019
7. Breslav, E.P.: How to enhance digital economics in your company today. Qual. Innov. Educ. **4**(143), 44–54 (2011)
8. Redko, S.G., Golubev, S.A., Tsvetkova, N.A., Its, T.A., Surina, A.V.: Project activities foundation. SPbPU, p. 84 (2018)
9. Cooperative E-learning Platform for Industrial Innovation. https://www.cephei.eu. Accessed 15 Apr 2019
10. Politech and UrFU will start cooperative educational program, 21 January 2019. https://www.spbstu.ru/media/news/partnership/polytech-urfu-launch-joint-educational-program/. Accessed 15 Apr 2019
11. List of courses on OpenEDU: https://openedu.ru/course. Accessed 15 Apr 2019
12. List of courses on Coursera: https://www.coursera.org/courses. Accessed 15 Apr 2019

# Some Results of the Analysis of 3 Years of Teaching of a Massive Open Online Course

Sergey A. Nesterov$^{(\boxtimes)}$ and Elena M. Smolina

Peter the Great St. Petersburg Polytechnic University, St. Petersburg, Russia
nesterov@saiu.ftk.spbstu.ru, smolensk9595@mail.ru

**Abstract.** The paper describes the results of a massive open online course (MOOC) "Data management" on the Russian platform of Open Education openedu.ru. Some approaches to the analysis of the results of distance learning, including data mining, are discussed. The produced analysis of the results of studying MOOC helps to understand students and their reasons for leaving the course. This could be taken into account during the renewal of the course. If some of the tasks are too difficult for a certain group of students, these tasks could be changed or an additional training material could be given before those. The offered method gives an opportunity to suggest new interesting topics and tasks that could be added to the course in those weeks when students are dropping out en masse. For some groups of students, additional courses for preliminary training could be recommended. The next task which we'll try to solve is classification: we'll try to predict if the course will be completed by the student or not based on their results during the first weeks. The results of such prediction may help to keep students in the course.

**Keywords:** MOOC · Higher education · E-learning · Data mining

## 1 Introduction

Nowadays, e-learning becomes more and more popular. One of the forms of e-learning are massive open online courses (MOOC) that are broadly used all other the world. MOOC platforms accumulate a large amount of data about activities of course participants and the results of their training. This data needs further analysis in order to develop recommendations for improving the educational process. For that purpose, data mining methods and algorithms could be used. This led to the appearance of a special area in data mining, i.e., data mining of the educational process (educational data mining), which develops data research methods for the education sector. For example, it can help to identify the most difficult topics of the course, assess the correctness of test tasks, predict the future results of the students, and so on [1–3].

One of the most essential problems of MOOCs is that only a small percent of students who enrolled in the course pass it through. Data mining methods could help in understanding the causes of this problem [4]. In this paper, we also will discuss this topic.

D. G. Arseniev et al. (Eds.): CPS&C 2019, LNNS 95, pp. 651–657, 2020.
https://doi.org/10.1007/978-3-030-34983-7_64

## 2   Course Description

Peter the Great St. Petersburg Polytechnic University (SPbPU) is one of several leading Russian universities that are actively involved in the development of MOOCs. 12 courses which were developed at SPbPU are published on the Coursera and 48 courses published on the Russian "Open Education" portal (openedu.ru).

This paper describes results of the analysis of grade reports of the MOOC "Data Management" from the "Open Education" [5]. It was developed in the summer of 2016 and was first launched in September 2016. Sessions of the course started once a semester, in September and February, and by now, it has been given for five times.

The duration of the "Data Management" course is 16 weeks. Each week is devoted to a new topic:

1. Introduction. Database system architecture.
2. Steps of database design. Overview of the basic data models.
3. The relational data model: basic structures and constraints.
4. Relational algebra.
5. Normalization: first, second and third normal forms.
6. Boyce-Codd normal form and senior normal forms.
7. Entity-Relationship model, ER diagrams, IDEF1x notation.
8. IDEF1x and IE notations, the transformation from logical to the physical model.
9. SQL: history, data types, some functions, basic DDL statements.
10. SQL: DML statements.
11. SQL. SELECT statement: simple queries, selecting data from several tables.
12. SQL. SELECT statement: subqueries.
13. Views.
14. Transactions.
15. Programming in database: variables, operators, temporary tables.
16. Programming in database: stored procedures, functions, cursors, triggers.

Each topic of the course contains video lectures, practical tasks, and a short test marked in the grade report as homework. After the 8th week, there is a midterm exam, and at the end of the course, there is the final exam. The final grade of the course is calculated from the average result of the week tests (homework) and midterm and final exam results. For the first session of the course, coefficients were 30% for homework, 35% for the midterm exam and 35% for the final exam. After the analysis of the first results, another scale was used for other sessions: 20% for homework, 20% for the midterm exam and 60% for the final exam.

## 3   Data Visualization and Brief Analysis

The R language was chosen as a tool for data analysis. It has many different packages for statistical data processing, visualization and machine learning [6, 7]. After downloading from the "Open Education" portal, each report was imported into the R as a data frame. Missed data was relabeled to be in the readable by the R environment format.

For each session of the course, the percent of students who had enrolled in the course but would not complete any tasks was calculated. The results are summarized in Table 1. In the five sessions, only 31%, 32%, 23%, 19%, 23% of students completed at least one assignment.

**Table 1.** Percent of students who did not complete any tasks

| Course session | Number of enrolled students | Number of students who have completed at least one task | Percent of students who did not perform any tasks |
|---|---|---|---|
| Fall 2016 | 2547 | 798 | 69 |
| Spring 2017 | 1572 | 499 | 68 |
| Fall 2017 | 1823 | 427 | 77 |
| Spring 2018 | 1504 | 279 | 81 |
| Fall 2018 | 2346 | 529 | 77 |

As it is shown in Table 1, the largest number of students registered for the course at the first session (fall 2016). It could be explained by the interest of students in a new course. Also, a large number of students registered at the 2018 fall session could be related with the fact that due to organizational difficulties the end date for enrollment was changed and, as a result, the period for enrollment was much longer than usual.

Figure 1 shows the number of students who performed tasks during the first session of the course. The task number is shown along the x-axis and the number of students along the y-axis. 798 students performed the first task, but only 435 students continued to perform the tasks of the second week. The number of active students continued to decrease from week to week, with a slight jump upwards at the intermediate exam.

The picture was the same for other sessions of the MOOC. A rather large number of students enroll for the course and only between 1/3 and 1/5 of them perform the first assignment. Then the number of active students decreases weekly, and after the mid-term exam, the number of such students becomes practically stable.

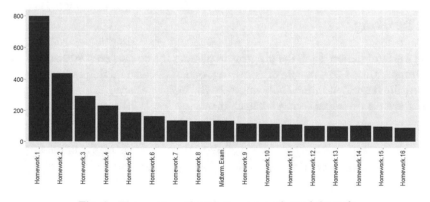

**Fig. 1.** The number of students who performed the task

Figure 2 presents a dependency between the number of students who passed the weekly test and the number of students who completed the test of the first week. For all sessions, there is a sharp decline after the first week. This could be due to many reasons, for example:

- the course material is too difficult to study;
- lack of motivation to learn;
- the student did not plan to study, but signed up to look through the material;
- the course was too long.

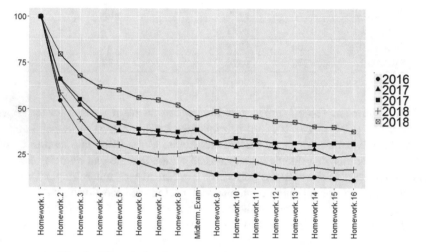

**Fig. 2.** The relative decrease in the number of active students

It also could be noticed that after the middle of the course, the decrease in the number of active students is insignificant. So we can say that most of the students who completed the intermediate exam continued to study until the end of the course.

## 4   Clustering

Using the information about the progress in passing the course, students can be divided into groups. This is an example of clustering, one of the data mining tasks, which could be formally described in the following way [8, 9].

If $I$ is a set of students of a MOOC, then

$$I = \{i_1, i_2, \ldots, i_j, \ldots, i_n\} \tag{1}$$

Each of the students is described by a set of attributes

$$i_j = \{x_1, x_2, \ldots, x_k, \ldots, x_m\} \tag{2}$$

$$x_k \in \{v_k^1, v_k^2, \ldots\}$$

So the task is to group this set of objects in such a way that objects in the same group (cluster) are more similar to each other than to those in other clusters.

$$c = \{c_1, c_2, \ldots, c_k, \ldots, c_g\} \tag{3}$$

$$c_k = \{i_j, i_p | i_j, i_p \in I, d(i_j, i_p) < \sigma\}$$

Here $d(i_j, i_p)$ is a metric or distance function that defines the distance between each pair of elements of a set.

We chose the number of clusters based on the analysis of the sum of squared errors (distance between each point and the mean of its cluster) [7, 9]. For all course sessions, according to this indicator, 4 clusters were chosen. Cluster analysis was done using the k-means algorithm. We were interested in grouping students who started their learning, so, before clustering, all students who had not completed at least one task were removed from the set.

Figure 3 shows the average results of weekly tests and the midterm exam for each of the 4 clusters that were formed for the 2018 spring session. For the rest of the course sessions, the clustering results were approximately the same.

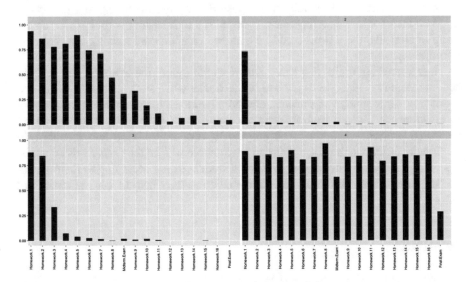

**Fig. 3.** Average results of weekly tests and the midterm exam for each cluster (spring 2018)

The final exam at the MOOC "Data management" is free only for the students of SPbPU (those students passed the exam in the class). Other students of the MOOC, if they want to pass the exam and get a certificate from openedu.ru, must pay for proctoring, and only a few of them did that. So, in our analysis, we did not take into account the results of the final exam.

As it's showed in Fig. 3, the following groups of students can be distinguished:

(1) Students who actively studied until the middle of the course, and then looked at the materials from time to time and passed the tests;
(2) Students who were active only during the first week and dropped out after it;
(3) Students who were active during the first 2-3 weeks;
(4) Students who regularly studied in the course.

So if we want to get more students who would finish the course, we, first of all, need to work more closely with those who are in the first cluster. This work could be, for example, in a form of additional emails after the midterm exam.

Also, we tried to use hierarchical agglomerative clustering with different parameters. But for our data, we got different results for each session of the course and we could not find a suitable interpretation for this difference. We assume that this was due to the fact that hierarchical clustering algorithms are more sensitive for outliers in the dataset [7]. So in our case, the k-means algorithm is a better choice.

## 5   Conclusion and Perspectives

The analysis of the results of the MOOC studying helps to understand students and their reasons for leaving the course. This could be taken into account during the renewal of the course. The results of the clustering analysis will help to group students based on their academic performance. If some of the tasks are too difficult for a certain group of students, these tasks could be changed or additional training materials given before those.

New interesting topics and tasks, which would attract the attention and interest of students, could be added to the course in the weeks when students are dropping out en masse. For some groups of students, additional courses for preliminary training could be recommended. If we have information about the results of the students on other connected courses, which they took previously, it could be taken into account [10].

The next task which we are going to try to solve will be the classification: we'll try to predict if a course will be completed by a student or not according to their results of the few first weeks. The results of such prediction may help to keep students in the course.

## References

1. Sweeney, M., Lester, J., Rangwala, H., Johri, A.: Next-term student performance prediction: a recommender systems approach. JEDM **8**(1), 22–51 (2016)
2. Algarni, A.: Data mining in education. Int. J. Adv. Comput. Sci. Appl. (IJACSA) **7**(6) (2016). https://doi.org/10.14569/IJACSA.2016.070659

3. Manjarres, A.V., Sandoval, L.G.M., Suárez, M.J.S.: Data mining techniques applied in educational environments: literature review. Digit. Educ. Rev. (33), 235–266 (2018). https://www.scopus.com/inward/record.uri?eid=2-s2.0-85049298943&partnerID=40&md5=fd38f ad9463b8644491b22f64f5cd377

4. Yang, D., Kraut, R., Rose, C.: Exploring the effect of student confusion in massive open online courses. JEDM **8**(1), 52–83 (2016)

5. Nesterov, S.A., Andreeva, N.V.: Data management (Massive open online course). https://openedu.ru/course/spbstu/DATAM/. (in Russian)

6. Lantz, B.: Machine Learning with R, 2nd edn. Packt Publishing, Birmingham (2015). ISBN 9781784393908

7. Bruce, A., Bruce, P.: Practical Statistics for Data Scientists. O'Reilly Media, Sebastopol (2017). ISBN 9781491952955

8. Barsegyan, A.A., Kupriyanov, M.S., Stepanenko, V.V., Kholod, I.I.: Methods and Models for Data Analysis: OLAP and Data Mining. BHV-Petersburg, St. Petersburg (2004). (in Russian)

9. Grus, J.: Data Science from Scratch: First Principles with Python. O'Reilly Media, Sebastopol (2015). ISBN 9781491901427

10. Kogan, M.S., Gavrilova, A.V., Nesterov, S.A.: Training engineering students for understanding special subjects in English: the role of the online component in the experimental ESP course. In: 2018 IV International Conference on Information Technologies in Engineering Education (Inforino), pp. 23–26. IEEE Xplore, Moscow, October 2018 (2018). https://doi.org/10.1109/inforino.2018.8581837. https://ieeexplore.ieee.org/document/8581837/

# Flipped Classroom Design: Systems Engineering Approach

Iuliia Shnai[1] and Leonid Chechurin[1,2(✉)]

[1] LUT University, Lappeenranta, Finland
Leonid.Chechurin@lut.fi
[2] Peter the Great St. Petersburg Polytechnic University, St. Petersburg, Russia

**Abstract.** Flipped or inverted classroom is one of the upside down pedagogies, which combines different learning theories. Its learning design constitutes already well-known components: video lectures and activities, both structured in a novel way. The proposed research targets are to increase the understanding and scalability of flipped classroom implementation. It is an attempt to plan the course transition in a more effective way. To achieve this goal, 9 window design approaches and the Flipped Classroom Design Approach (FCDA) are used. The preliminary results describe the initial a flipped classroom model with its connections, impacts and a list of parameters as one element. The model is adapted to the transition of the course of System Modelling from the Traditional to a flipped form. Further on, it can be used by teachers as an algorithm at the stage of flipped classroom planning, design, development and implementation. Practically, the current results appear in a guided FCDA approach with the template and supporting online learning materials.

**Keywords:** Flipped classroom · Education · Guidelines · Systems engineering

## 1 Introduction

A decision to move from classical PPT-based lecturing, exercises and seminars to the flipped form of learning heavily supported by electronic technologies and communication is a fundamental shift of the paradigm. The chance is it could be used to rethink many if not all aspects of teaching. Otherwise, the electronic tools would just retrofit the conventional approach, the same way as we could use a plane to drive along a highway.

However, there still is a limited amount of clear guidelines or supporting systems, which an implementer can adopt to improve learning outcomes. For instance, a teacher decides to flip a course after reading articles, instructions in the Internet, discussions with colleagues; first, he/she decides to develop videos. Having recorded 90-min videos, the teacher realizes that students come to the class unprepared. Therefore, the teacher explores how to improve the course or just finds this approach fruitless and does not use it anymore. Besides that, there is a lot of misconceptions behind the course redesign. The question is, how to support teachers in building flipped classroom scenarios and implementing them more efficiently? Is there any systematic approach to the flipped classroom design?

D. G. Arseniev et al. (Eds.): CPS&C 2019, LNNS 95, pp. 658–669, 2020.
https://doi.org/10.1007/978-3-030-34983-7_65

In this paper, the systems engineering is used to the flipped classroom design in order to consider it as a system and extend the understanding of the flipped classroom design.

The goals of the study are the following:

1. To introduce the Flipped Classroom Design Approach (FCDA) which the authors have formed via literature review and experiments on the basis of systems engineering design.
2. To provide more information about course transition to digital and, specifically, flipped form in a more efficient and automated way.
3. To develop a plan for a course transition of System modelling course.

## 2  Background

Flipped classroom is not a completely new approach to education but rather an innovatively structured combination of already existing educational methods. Digital lectures before the class frees time for accumulating knowledge in the class. Functionally, the preparation relates to the first steps of Bloom's taxonomy (McLaughlin et al. 2014; Jensen et al. 2014; Gilboy et al. 2015), and refers to the "engage, explore and explain" in 5E learning cycle (Jensen et al. 2014). For students it offers flexibility, mobility and familiarity with self-study. The second part belongs to higher stages of Bloom's taxonomy and "elaboration and evaluation" targets of knowledge in 5E learning cycle (Jensen et al. 2014). The second part supports active teaching with cooperative, collaborative, peer-assisted and problem-based learning (Bishop and Verleger 2013). The flipped classroom model turns teachers into guides and mentors. Digital lectures in conjunction with activities provide for a more individualized approach to students' preferences and needs rather than solely teacher's interest. Overall, the flipped classroom constitutes a variety of student-centered learning theories and methods (Bishop and Verleger 2013), which support different learning styles.

Majority of flipped classroom studies reveals the theoretical or experiment-based justifications of the of flipped classroom impact on learning outcomes. Authors provide guidelines, suggestions, revealing common practices of successful design supported by the improved learning outcomes. Despite the high complexity and multidisciplinarity of the approach, the flipped classroom is mainly considered from solely the pedagogical side and has never been approached to as a system.

Systems engineering is "an interdisciplinary approach and means to enable the realization of successful systems" (INCOSE 2018). The field stems from 1940 Bell Telephone Laboratories. First time it was taught in 1950 at MIT by Mr. Gilman, Director of Systems Engineering at Bell (INCOSE, 2018). According to the international council on systems engineering (INCOSE, 2018) that was this mixed approach "when a system is considered as a combination of system elements, systems thinking acknowledges the primacy of the whole (system) and the primacy of the relation of the interrelationships of the system elements to the whole". Systems engineering process is a top-down problem-solving approach. Defense Acquisition University Press (2001) it starts from the development phase, which involves concept description, system level, subsystem and component level consideration with performance descriptions. Following that, the systems engineering process is applied. According to the standards ISO/IEC/IEEE 15288:2015, there is "a set of processes and associated terminology from an engineering viewpoint. These processes involve requirements analysis, functional analysis and design synthesis and can be applied at any level in the hierarchy of a system's structure".

There are several tools to help in this disruptive design challenge and all of them are rooted in system analysis, sometimes packaged as a wide range of doctrines from dialectics to systems engineering or, in a more vulgar version, the theory of inventive problem solving. The main taking out of all of them is the framework of analysis of a system in the context of time and environment, understanding the hierarchy of its elements and the available resources, and assigning the goals of the design. Among them TRIZ (Theory of Inventive Problem Solving) analysis can be used for the course description in a systematic way (Altshuller and Shapiro 1956). One of its tools is a 9-screen scheme.

## 3 Method

Designing the flipped classroom with the systems engineering approach means system development and application of the systems engineering processes. This primarily refers to evaluation of the course, considering it as not a separate element but a part of the whole system dismantling it also into subsystems and elements. One of the TRIZ tools, the "9-screen scheme" from TRIZ by G. Altshuller is applied together with the FCDA (Flipped classroom design approach) which is developed on the basis ADDIE model (Branson et al. 1975). It constitutes 3 primary stages:

1. Analysis of context and resources.
2. Design and Development.
3. Implementation and Evaluation.

## 4 Results

As the result, the flipped classroom is described using systematic design tool screen scheme (Table 1) and the Flipped Classroom Design Approach (FCDA). In this paper, a part of context analysis of the System Modelling course (Table 2), a course transition plan by components from the traditional to flipped one (Table 3), and the final plan for the System modelling course transition (Table 4) are presented.

"System modelling" is taken as an exemplary course. It is taught in LUT University and is an optional master-level course which is a part of the Global Management of Innovation Technologies (GMIT) program. The course is equal to 6 ECTS and approximately 156 h. The course is aimed to introduce analytical models and model complex systems and develop skills in modelling and modelling results analysis (analytical and numerical), primarily by using Matlab. The course has a low amount of enrollments and a significant percent of students that do not finish the course. One of the reasons for that is a lack of interest and motivation of students at the master level to refresh the ground mathematical knowledge and their inability easily to accomplish the Matlab tasks without previous experience with this software. Therefore, it was planned to redesign and rebrand the course in an innovative form.

## 4.1  9-Screen Scheme

The 9-screen scheme lets us estimate the system in 9 different windows, where the first raw refers to the supersystem, the second to the main system, and the third to the subsystems and components. Three columns attribute different time periods (Table 1).

Currently, the System Modelling course is taught primarily in the traditional form with integration of different digital elements from MOOCs or other courses. It is in transition to the new flipped form. Therefore, it is of special interest to assume how it could be transformed already now to the flipped form, and what will happen with the course in the future. How the knowledge and skills can be delivered via new technologies and approaches? On the supersystem level, the course is not widely integrated into the existing digital environments. However, it is delivered through the basic learning management system (Moodle) which is a closed system of LUT University. It is planned to integrate the course in a more open learning platform to deliver it to students outside of the university and make it a part of the online learning platform for Industrial Innovation (CEPHEI) or other digital ecosystems. As an assumption, the future supersystem, such as a traditional university, can change significantly to a more monopolised and decentralised learning distribution system.. The fast- growing emergence of new private providers of education is a trend which supports this assumption already now. Regarding the level of the subsystems and components, new elements appear to and should substitute the current ones. Systems modelling primarily consists of the traditional elements evenly delivered in the learning management system and it is planned to involve new components, such as videos, games, boards, etc.

## 4.2  Flipped Classroom System Context

Following that, we will focus on the evaluation of the Flipped Classroom System and basically assume which context should be taken into account within the course transition. The teacher, student, course content and classroom are parts of the flipped classroom design context. The scheme of the described parameters presented in Table 2.

**Table 1.** 9-screen scheme for the System Modelling course

| Time period | Past | (Almost) Present | Future |
|---|---|---|---|
| Supersystem, Environment | University of 20th century, School, MSc Program,… | University of 21st century, Globalization, Open eLearning Platforms, Communication environment (social media, messengers, apps, hackathons, crowdsourcing,…) | Uber-University (?), Demonopolization of professional education, Learning block chains,… |
| System in focus | System Modelling course | Flipped System modelling Course | Skill/Knowledge/Employment Deliver service of the System Modelling course (?) |
| System components, subsystems | Topics, Lectures, Seminars, Exercises, Project, Tests, Eexams | Modules, Videos, Quizzes, Discussion boards, L.learning data analytics | Navigation tools, Teaching tools, Teaching environment, …(?) |

### 4.3    Course Context

Class size is used for the amount of students in the class per one instructor. It is not possible to define a standard number of students, due to the culture, field, students, teachers and other specifics on the country and university levels. The flipped online preparation part does not depend on the amount of students; quite the opposite, it gives a potential for personalised learning on a massive scale. The videos can be watched by as many people as desired without any limits. Moreover, an increased time for the face-to-face activities supports the class-size reduction, thus the increased amount of students divided into groups can study without any loss of efficiency (Baepler et al. 2014).

Prior to the video production, instructors should decide if it is reasonable to develop their own new materials or to use the already existing. It is worthwhile to check availability and access to the course-related video materials among the colleagues and in open sources. Basically, if the field of study is generic e.g., an introduction to mathematics, there is a bunch of free and high-quality videos hosted on the Khan or Ted platforms, YouTube channels and other repositories. The use of existing materials can significantly reduce the invested time of and financial burden on teachers and universities (Galway et al. 2014). Taking into account duration of lectures, exploitation and cost-effectiveness of the videos, lectures can be assessed in advance (Shnai and Kozlova 2016). Authors confirm that development of video materials is crucially resource-consuming in the first year with the decrease in the second year (Ferreri and O'Connor 2013; McLaughlin et al. 2014), especially for the preparation of lectures.

In the case of the System Modelling course, the class size is rather small anyway and, therefore, there is no need to decrease the class size. The course refers to the general field and various already developed materials can be integrated from other

MOOCs. The cost effectiveness is taken into account and video materials will be primarily gathered rather than developed from the scratch (Table 2).

## 4.4 Students' Parameters

*Familiarity* of students with the flipped classroom and other related blended- learning concepts is one of the factors which the instructor should take into account before the flip. Although the effect is not proven in the follow-up studies, students indicated their being unfamiliar with the flipped classroom concept as one of the barriers for learning (Shnai 2017). Teachers state that providing instructions for students about the flipped classroom can lead to a better understanding of the process and easier navigating (Gilboy et al. 2015; Kim et al. 2014; Mason et al. 2013). Students' background and level of understanding vary. Therefore, the basic knowledge should be checked in advance and the level of materials adapted. Learning styles affect the flipped classroom preparation part in a way that requires adaptation to the diversity of ways for students to learn in a better way (Bishop and Verleger 2013).

## 4.5 Classroom Parameters

Researchers mention that transition from the traditional teaching approach to active pedagogy is supported by physical rearrangement of the class from an amphitheater-type room (Baepler et al. 2014). Standard black board or a projector are utilised for the introduction of activities, discussions and exercises. Activities, group work, presentations, simulations, quizzes and surveys require laptops or personal mobile devices. Being surrounded by even the modest "touchable", "triable", "deployable" hands-on equipment can lead to more active participation and knowledge scaffolding. The System Modelling course is a party simulation based on and taught in computer lab arrangements (Table 2).

## 4.6 Teacher's Parameters

In case if the teacher explores the flipped classroom design without any assistance, the 4 core parameters should be analysed in advance in order to succeed. The first one is to identify the teacher's experience with creating a blended learning design and specifically the flipped classroom. Lack of skills and confidence are also mentioned as barriers, on the top of which is the lack of time (Shnai 2017). Primarily, video development is affected by that; however, in case the teacher feels experienced and comfortable with recording, it encourages him/her to develop the video (Bergmann and Sams 2012). In other cases, assistance or existing materials can be helpful.

**Table 2.**  Context evaluation for "Systems Modelling Course"

| Teacher | | Students | |
|---|---|---|---|
| Time available | Limited | Composition of Students: | International |
| Skills and experience | Experience and technology | Grade Level/Age | Master's degree level |
| Technology competency | | Background | Different background lack of refreshed mathematical knowledge required |
| Role/Vision | Students | Familiarity with FC, Online, Digital learning | Not familiar |
| Support & Standards | | Technology & Room Arrangement | |
| Available support | Limited available support | Technical equipment and software | PC equipment with Matlab software |
| Available policy & standards | Limited policy & standards | Room arrangement | Computer lab room set up |

## 4.7  Flipped Classroom Components and Subsystems

Flipped classroom disassembling is a step toward the detailed understanding of the design process. Flipped classroom elements with similar function can be replaced. For instance, to transfer knowledge, videos or texts, or audioscans can be deployed. Therefore, all the options should be considered and the best ones selected, in respect to the desired learning outcomes.

In the literature, researchers describe inverted classroom design variations, elements and their connections, impacting on the results. It is worth to mention that not all of them are defined by the controlled studies or rigorous focused analysis. Primarily, they are distinguished on the basis of personal observations and results of students' surveys.

According to the systemic way, the flipped classroom environment is a system, which consists of two subsystems. The preparation part can be implemented through visual, audio, textual materials, simulations (O'Flaherty and Phillips 2015); the learning one should be controlled and supported by quizzes, assignments, exercises. In-class part should start from the discussion to support the connection between the online and offline elements. The variety of activities can involve discussions, quizzes, assignments, group works, labs, etc.

Planning a flipped classroom for the System Modelling course, the main change will comprise the involvement of extended preparation which can be optional for students who have the required knowledge and are available to accomplish the test in the beginning of the class. Therefore, the self-study guided part will take more time. Online modules include videos, texts and other elements from other online courses. Substituting the lecture leaves free time for the activities and discussion (Table 3).

**Table 3.** System Modelling course transition from the traditional to a flipped one

| Moment of educational process | Before class | In class | After class |
|---|---|---|---|
| Traditional form | – | Lectures (36 h) In class assignments (mostly with Matlab) Projects (20 h) | Homeworks, Assignments (mostly with Matlab) (100 h) |
| Flipped form | Online modules (30 h) | Tests (14 h) Summary & Discussions (16 h) In class assignments (mostly with Matlab) projects and games (26 h) | Homeworks, Assignments (mostly with Matlab) (70 h) |

### 4.8    Overall Plan of System Modelling Course Transition Based on the FCDA

The overall plan of the redesigned course is presented in Table 4. Among the primary activities which should be done for course, the overall transition activities to the Flipped form are as follows:

1. To extend the gallery of materials to satisfy students with different learning styles and backgrounds
2. To deliver the materials before the class and check the understanding in the beginning of the class
3. To integrate the course on a learning platform and involve more students from outside
4. To develop mini tutorials specific for the course
5. To rename the course and provide a corresponding description

## 5    Discussion

Let us speculate on the problem of classical course (re)design as we are about to move from the "Past" to "Present". To approach this design task systematically, we need to understand the current educational landscape (environment), as well as available (present) technologies or tools or components on which our new design can be built of. There are critical questions that immediately arise:

1. Whom we are (re)making this course for? For only "hostages:" my university students are bounded to take the course because they need credits and degrees, or for the whole world? If only for the "hostages", is there any sense to care much about the quality and contents, because it will be "for the internal use only"?
2. If the contents of the course is worth digitizing it, investing lots of efforts into shooting and editing the videos, for example, the development of automated quizzes and tests? If the contents is so unique that it cannot be found anywhere in the world... Does the world need it? And if the subject is quite ordinary, for example,

Automation Basics or Basics of Coding, then why is it not grounded on the excellent open materials available from other universities that are the leaders in the field, or from professors who are the legends in the subject or even Nobel Prize winners? Then we can focus on active elements of learning, such as consulting on projects, exercises, case studies, bringing the industry in to the class, etc.

3. Do we want to cultivate the knowledge and skills that are necessary in practice? If so, why the assignments and project tasks for the course are not retrieved from the "outside of the class", from businesses, industry and society? Internet-based learning makes it so easy to do: just a couple of clicks can summon an experienced engineer or businessman to sit in your virtual classroom commenting on what is taught and how students are progressing. If there is no much request from the outside, but instead, there are difficulties with finding real problems, could it mean that what we are teaching is not in the high demand?

4. Should the designed course be ready for scaling? Are we prepared to see one day not 20, but 200 or even 2000 students enrolled in the course? If not, why would we create a course that is a priori unlikely to be popular?

The experience and intensive discussions on how to address these questions resulted in the following speculations that laid the ground for our design principles:

(a) We should think in terms of global relevance, see the designed course as not only a part of the curriculum of a specific educational program of a university but also as an independent product, an application;

(b) To rely massively on open availability of a large number of materials that may be course elements (video materials, texts, quizzes, experimental beds, data, etc.,), build a course as a gallery, providing rather support, navigation, counseling, trainee assessment s than "teaching" per se.

(c) To use the network-based course for maximum communication of its materials and students with the "outside world", i.e., with experts, employers, potential clients, customers, task providers, as well as with similar courses (students and teachers) from other countries;

(d) To create software skills, that are relevant for the 21$^{st}$ century that basically means the ability of a student to work efficiently in mixed face-to-face and virtual realities, actively use the modern tools of communications and be emotionally wise in both domains; to have the competences to organize the work of multinational remote teams in comfortable digital environment with full documentation, reporting and other tools of project management.

This study is an attempt to approach the learning design from the systematic point of view. After providing the 9-window analysis, context description and decomposition of the flipped classroom settings, the functioning of a flipped classroom becomes clearer. In course planning, transition for Systems Modelling had a number of steps and related issues. The first issue is to make the flipped classroom fit the environment. Initially, attention should be paid to 4 defined categories, like the student, teacher, course, and classroom parameters (Table 2). After that, the design can be assumed in a more detailed way. First, it is necessary to observe the content delivery forms before the class, like videos textual materials and controlling them quizzes and exercises

(Table 3). Next, the general plan should be presented (Table 4). This paper attempts to uncover gaps in the flipped classroom design and emphasize the elements and parameters which require special attention in the course re-design.

**Table 4.** Overall System Modelling course plan description

| Moment of educational process | Before class | In class | | | | After class |
|---|---|---|---|---|---|---|
| Elements | Online modules | Intro tests | Summary and discussions | In class assignments (mostly with Mathlab) projects and games | | Final project |
| Duration | (30 h) | (14 h) | (16 h) | (26 h) | | (70 h) |
| Aims & learning outcomes | Theory elements which refresh basic mathematical knowledge required to course accomplishment | Check Students understanding of the preparation materials | Learn basic mathematical language for System Modeling discussed in a wide range of phenomena with the Homecourt in economy and demography. Input-Output models for static and dynamic multivariable systems of various orders, Linear and nonlinear, Stable and unstable | | | |
| Design & Technology | Integrated from other recourses and platforms and gathered on the Thinkific platform. Components include Video, Texts, Tutorials which suit different levels and learning styles | Test is delivered via quizzes and every student answers from their own devices | Students accomplish them on laptops alone or in small groups | Students compete in the tasks related to course face-to-face or via laptops | | Accomplishing the projects in the group |
| Evaluation | 0% | 10% | 50% | | | 40% |

## 6  Conclusion

The connections between outcomes of a flipped classroom design and its subsystems should be traced down and proven, identifying more design guidelines. Thus, the teacher can build the design in the most effective way based on the expected results. The initial step toward extended understanding provided in this paper for one of the blended learning designs, a flipped classroom, will be the ground for a future work. The experimental planning for a system modelling course is made based on the already existing templates.

# References

Altshuller, G., Shapiro, R.: Psychology of Inventive Creativity. Vopr. Psikhologii (Issues Psychoilogy), no. 6 (1956)

Baepler, P., Walker, J.D., Driessen, M.: It's not about seat time: blending, flipping, and efficiency in active learning classrooms. Comput. Educ. **78**, 227–236 (2014). https://doi.org/10.1016/j.compedu.2014.06.006

Bergmann, J., Sams, A.: Flip Your Classroom: Reach Every Student in Every Class Every Day. International Society for Technology in Education, Washington, DC (2012)

Bishop, J.L., Verleger, M.A.: The flipped classroom: a survey of the research. In: 120th American Society for Engineering Education Annual Conference and Exposition, Atlanta, GA (2013)

Branson, R.K., Rayner, G.T., Cox, J.L., Furman, J.P., King, F.J., Hannum, W.H.: Interservice procedures for instructional systems development. (5 vols.) (TRADOC Pam 350-30 NAVEDTRA 106A). U.S. Army Training and Doctrine Command, Ft. Monroe, August 1975 (1975). (NTIS No. ADA 019 486 through ADA 019 490)

Ferreri, S.P., O'Connor, S.K.: Instructional design and assessment: redesign of a large lecture course into a small-group learning course. Am. J. Pharm. Educ. **77**(1), 13 (2013). Article 1

Galway, L.P., Corbett, K.K., Takaro, T.K., Tairyan, K., Frank, E.: A novel integration of online and flipped classroom instructional models in public health higher education. BMC Med. Educ. **14**, 181 (2014). https://doi.org/10.1186/1472-6920-14-181

Gilboy, M.B., Heinerichs, S., Pazzaglia, G.: Enhancing student engagement using the flipped classroom. J. Nutr. Educ. Behav. **47**(1), 109–114 (2015). https://doi.org/10.1016/j.jneb.2014.08.00

Jensen, J.L., Kummer, T.A., Godoy, P.D.D.M.: Improvements from a flipped classroom may simply be the fruits of active learning **14**, 1–12 (2014). http://doi.org/10.1187/10.1187/cbe.14-08-0129

Kim, S., Khera, O., Getman, J.: The experience of three flipped classrooms in an urban university: an exploration of design principles. Internet High. Educ. **22**, 37–50 (2014)

The International Council on Systems Engineering (INCOSE). https://www.incose.org

ISO standards. https://www.iso.org/standard/63711.html

Defense Acquisition University Press: Systems engineering fundamentals (2001). https://ocw.mit.edu/courses/aeronautics-and-astronautics/16-885j-aircraft-systems-engineering-fall-2005/readings/sefguide_01_01.pdf

Mason, G., Shuman, T., Cook, K.: Comparing the effectiveness of an inverted classroom to a traditional classroom in an upper-division engineering course. IEEE Trans. Educ. **56**(4), 430–435 (2013)

McLaughlin, J.E., Roth, M.T., Glatt, D.M., Gharkholonarehe, N., Davidson, C.A., Griffin, L.M., Mumper, R.J.: The flipped classroom: a course redesign to foster learning and engagement in a health professions school. Acad. Med. **89**(2), 236–243 (2014). https://doi.org/10.1097/acm.0000000000000086

O'Flaherty, J., Phillips, C.: The use of flipped classrooms in higher education: a scoping review. Internet High. Educ. **2015**(25), 85–95 (2015). https://doi.org/10.1016/j.iheduc.2015.02.002

Shnai, I., Kozlova, M.: Resource and profitability assessment of transition to flipped video-based lecturing. In: IAFOR Conference, Lappeenranta University of Technology, School of Business and Management (2016)

Shnai, I.: Systematic review of challenges and gaps in flipped classroom implementation: towards future model enhancement. In: Proceeding of the 16th European Conference on eLearning, Lappeenranta University of Technology, School of Business and Management (2017)

# Analysis of Students' Performance in an Online Discussion Forum: A Social Network Approach

Arnob Islam Khan, Vasilii Kaliteevskii, Iuliia Shnai, and Leonid Chechurin[✉]

LUT University, Lappeenranta, Finland
{arnob.khan, vasilii.kaliteevskii, iuliia.shnai, leonid.chechurin}@lut.fi

**Abstract.** In the new era of digitalization, the education sector is experiencing changes in terms of the learning design, teaching methods, engagement of the learners and integration of technology. Flexibility of online courses is attracting more and more students to the learning platforms every day. The learning activity of students on the online platforms generates enormous amount of data. Practically every click can be traced and described. Learners are viewing the video lectures and digital lessons as a passive form of learning and most of the active learning is taking place in the form of online discussions. Therefore, in order to measure the students' active engagement in an online course, it is essential to evaluate the communication channels. However, the assessment methods for online discussion remain limited. The objective of the work is to provide a ranking of students, based on their participation in the online discussion forum. It provides an opportunity for the teachers to automatically assess students' performance quantitatively based on systematic approach. In this paper, network centrality measures are utilized to rank the students based on their interactivity. Text analytics is applied in association with sentiment analysis to assess meaningfulness of each student's communication. The method was tested on the online course data of "Systematic Creativity and TRIZ basics" at LUT University, Finland. This work was partially supported by CEPHEI project of the ERASMUS+ EU framework.

**Keywords:** Social-network analysis · Online-discussion forums · Online learning · Sentiment analysis · Neuro-linguistic programming · NLP

## 1 Introduction

Online learning opens up great opportunities for students and teachers from around the world to increase learning efficiency [3].

By mastering the skill of translating courses to online or flipped format, teachers get considerable advantages in scaling their courses without actually limiting the number of students to the classroom and their own participation during the course. At the same time, students also enjoy flexibility in learning with respect to time and place during the course, and receive the material in a more familiar form which is easier to interact with [1,14].

© Springer Nature Switzerland AG 2020
D. G. Arseniev et al. (Eds.): CPS&C 2019, LNNS 95, pp. 670–680, 2020.
https://doi.org/10.1007/978-3-030-34983-7_66

With such advantages, increasing popularity of online learning [4] opens up new research problems, such as how to evaluate the student performance in a flipped classroom or online learning [11].

Since the course teacher is not able to question the audience during the lecture or conduct personal interaction while teaching online, he/she does not have many options for measuring students' activity/participation during the course. Traditional assessment methods, such as attendance, homework, and quiz, may not be very suitable, especially with a large number of enrolled students. Furthermore, if the online course is asynchronous, and not periodical with fixed deadlines, it can be extremely difficult to manually check homework. However, data generated from students' learning activity on an online platform can be analyzed to evaluate student performance on the course. Thus, in this article it is proposed to use a number of metrics adopted from the social network theory to evaluate students' engagement in the online discussion forum. Metrics such as Centrality measures, Degree Centrality, Closeness Centrality and Betweenness Centrality provide a highly accurate assessment of how meaningful a student is in dialogue, with how many people he/she has been interacting, and what is his/her social influence in such dialogues.

Thus, such metrics can serve as a very efficient addition to the traditional assessment method, e.g., test assessment of students, or be an independent form of assessment.

The objective of the analysis was to provide a score (a set of scores) and ranking for each of the students based on their participation in the online discussion forum. This will help the teachers to assess students' performance in the discussion more quantitatively, based on certain criteria.

## 2 Related Works

Social Network Analysis method provides explicit mathematical definition to describe the characteristics of the members of a community and the underlying network [5]. These characteristics are evaluated based on the relation among actors (members). Scott [18] and Hanneman [13] discussed the introductory analysis of Social Network Analysis.

Wasserman and Faust [10] elaborate based on the various values of the network structures. Researchers from different domains utilize network analysis to represent the community interests, e.g., citation network [17], World Wide Web [12], food webs, biochemical networks.

SNA (Social Network Analysis) has also been used in analyzing relations within families [22], analysis of the Military C4ISR (Command, Control, Communications, Computers and Intelligence, Surveillance, and Reconnaissance) network [8] analyzing terrorist network. Literature suggests that students' active engagement is significant for learning and one of the principal component of effective teaching [7, 9]. Discussion forum is the main instrument to facilitate active engagement of learners in online teaching. Several studies indicate the application of SNA in learning environment. Researchers from University of Alberta, Canada discuss the application of Social Network Analysis to measure learners' interaction [15].

Application of SNA has also been studied in another research in the clinical online discussion forum by the NICHE Research Group, Dalhousie University, Canada [19]. In this research, they analyzed the communication pattern using SNA to understand how a community member shares experiential knowledge. In another research, students' contribution and response have been analyzed using SNA for asynchronous online discussion forum [23]. The authors of [23] proposed an intervention from the analytics as an integral part of the course, and initial results illustrate changes in students' interaction in the forum.

## 3   Method Description

The hypothesis of the approach proposed in the paper is as follows: "Degree of interaction of each student in an online discussion forum correlates with the number of meaningful words."

The whole system was coded in the software environment R using its extensive library for Social Network Analysis and text mining.

**Centrality Measures**

Social Network Analysis is employed on the students' discussion to depict the social relationships consisting of nodes, which represents the individual actors in the network and ties or edges illustrating the relationship between the individuals. The network is then illustrated as a social network diagram.

The different Centrality measures of the Network are calculated to explain the students' activity pattern [16]. Usually, centrality measures refer to indicators, which identify the most important vertices within a graph. In this paper, *Degree Centrality, Betweenness Centrality, Closeness Centrality* and *Eigenvector Centrality measures* have been used for assessment of the students' engagement. Table 1 below illustrates the implication of the Centrality measures in this scope of the work.

**Table 1.**  Interpretation of each Centrality measures in this scope of the work

| Name of the Centrality measure | Description |
| --- | --- |
| Degree Centrality | The fundamental intuition is that nodes with more links in a network are more influential and significant. In our case, the higher is the degree of node, the more interactive the student is [20] |
| Closeness Centrality | It is a measure for calculating information-spreading time from one node to another in a sequential manner. In our case, it indicates the responsiveness of students [19] |
| Betweenness Centrality | In this work, High Betweenness represents that a learner has posted in more frequently, thus creating more nodes, meaning creating more opportunity for discussion [2] |
| Eigenvector Centrality | In this case, students having higher value in eigenvector centrality represent higher influence in the course discussion forum [5, 20]. |

## Natural Language Processing

The text of each user's discussion was analyzed using natural language processing with a view to identify the meaningful words. The first step was data cleaning by changing everything to lowercase, removing punctuations, numbers, white spaces, the brackets, replacing numbers with textual forms, and replacing abbreviations, contractions, and symbols (see Fig. 1).

The number of meaningful words for each user is counted. In the next step, lexicon-based sentiment analysis (see [21]) was conducted and each user was assigned a sentiment score in terms of positivity based on their meaningful words.

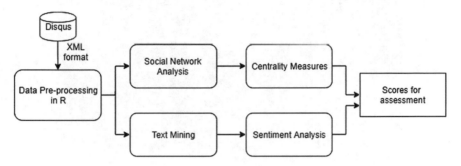

**Fig. 1.** Architecture diagram

## Experiment and Results

**Dataset:** The dataset consists of students' communication from the online course of "Systematic Creativity and TRIZ" which was part of the Winter School-2019 program. The students participated in the course via an online platform, and online discussion was facilitated through a disqus blog. The discussion of the students were exported from the disqus platform in xml format, after that the xml file was preprocessed using R, and the dataset was transformed into the edge format for further Social Network Analysis. The final dataset contains the following columns:

- **Source:** The person who initiated a discussion
- **Target:** The person whom the communication was intended for
- **Author messages list:** All the messages from each individual users

## Network Visualization

The network data of the discussion forum is processed using *igraph* package of R. The edge list is obtained after preprocessing of the data and converted to the *igraph* format using *graph.edgelist*. The original names of the participants is recoded to maintain the data integrity and privacy (see Fig. 2).

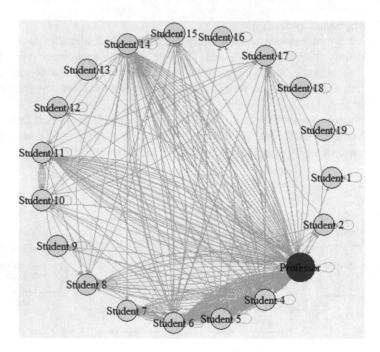

**Fig. 2.** Social Network of the forum

## Centrality Measures' Scores

The scores of each participant for their engagement in the discussion forum were based on *Degree Centrality*, *Closeness Centrality*, *Betweenness Centrality* and *Eigenvector Centrality*, as it is illustrated in Fig. 3.

In this work, *Degree Centrality* illustrates how interactive (central) a student is in the forum. *Betweenness centrality* indicates frequency (number) of posts by each student and *Eigenvector centrality* depicts the degree of influence for each student.

Since the objective of this work is to provide specific quantitative measures to the instructors for assessment of the students' activity in an online discussion forum, these measures are further analyzed by employing Principal Component Analysis (PCA) in order to identify the measure that explains the highest variance. Figure 4 illustrates the results of PCA, and it can be seen that it may be reasonable to use only *Degree centrality* or *Eigenvector centrality* to assess the students' performance.

| | degC | sw_closeness | star_betweenness | star_eigen |
|---|---|---|---|---|
| Professor | 1.00000 | 1.0000000 | 1.0000000000 | 1.000000000 |
| Student 6 | 0.58750 | 0.9391964 | 0.5377442580 | 0.939401788 |
| Student 4 | 0.51250 | 0.9391964 | 0.2488763705 | 0.776956620 |
| Student 14 | 0.43750 | 0.8842770 | 0.0845156539 | 0.677584813 |
| Student 11 | 0.26875 | 0.8505245 | 0.0502762800 | 0.401931330 |
| Student 5 | 0.19375 | 0.8344271 | 0.0221146992 | 0.265065442 |
| Student 10 | 0.17500 | 0.8344271 | 0.0106620074 | 0.208930810 |
| Student 7 | 0.15625 | 0.8344271 | 0.0104004717 | 0.189535256 |
| Student 8 | 0.15000 | 0.8188175 | 0.0041247495 | 0.159537932 |
| Student 15 | 0.15000 | 0.8188175 | 0.0029909095 | 0.134835042 |
| Student 17 | 0.11250 | 0.8036739 | 0.0022328644 | 0.123044377 |
| Student 2 | 0.05625 | 0.7747036 | 0.0009918042 | 0.045121413 |
| Student 9 | 0.01875 | 0.7091287 | 0.0000000000 | 0.027040493 |
| Student 12 | 0.01875 | 0.0000000 | 0.0000000000 | 0.022229552 |
| Student 16 | 0.01250 | 0.0000000 | 0.0000000000 | 0.018431582 |
| Student 1 | 0.00625 | 0.0000000 | 0.0000000000 | 0.008843787 |
| Student 13 | 0.00625 | 0.0000000 | 0.0000000000 | 0.008843787 |
| Student 18 | 0.00000 | 0.0000000 | 0.0000000000 | 0.000000000 |
| Student 19 | 0.00000 | 0.0000000 | 0.0000000000 | 0.000000000 |

**Fig. 3.** Scoring of the students based on Centrality measures

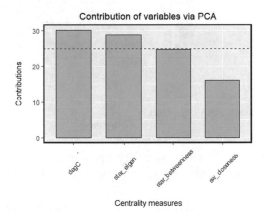

**Fig. 4.** Contribution of different Centrality measure based on PCA

In the next step, the post of each student is analyzed and after data cleaning, the number of meaningful words has been counted (see Fig. 5).

Then the meaningful words have been scored based on positive sentiment using lexicon-based sentiment analysis method [21] (see Fig. 6).

Based on these Centrality measures, word count and sentiment score, each student was ranked, and score for him/her was given (see Figs. 7 and 8). These score and ranking can be used as a single assessment tool for the performance measurement of students in an online discussion forum.

| Student 6 | Student 14 | Professor | Student 4 | Student 10 | Student 11 | Student 5 | Student 8 | Student 17 | Student 7 |
|---|---|---|---|---|---|---|---|---|---|
| 2178 | 2051 | 2029 | 1453 | 696 | 694 | 577 | 529 | 296 | 284 |
| Student 12 | Student 2 | Student 15 | Student 19 | Student 9 | Student 18 | Student 16 | Student 13 | Student 1 | |
| 247 | 128 | 112 | 82 | 77 | 32 | 23 | 9 | 3 | |

**Fig. 5.** Number of meaningful words posted by each student

| author name | Postive Sentiment Ratio |
|---|---|
| Professor | 59.53 |
| Student 6 | 54.19 |
| Student 4 | 50 |
| Student 14 | 55 |
| Student 11 | 61.36 |
| Student 5 | 68.09 |
| Student 10 | 67.92 |
| Student 7 | 66.67 |
| Student 8 | 77.55 |
| Student 15 | 90.91 |
| Student 17 | 61.54 |
| Student 2 | 83.33 |
| Student 9 | 71.43 |
| Student 12 | 42.86 |
| Student 16 | 100 |
| Student 1 | 100 |
| Student 13 | 0 |
| Student 18 | 0 |
| Student 19 | 50 |

| | Degree Centrality | Closeness Centrality | Betweenness | Eigenvector centrality | Number of Words |
|---|---|---|---|---|---|
| 1 | Professor | Professor | Professor | Student 6 | Student 6 |
| 2 | Student 6 | Student 4 | Student 4 | Professor | Student 14 |
| 3 | Student 4 | Student 6 | Student 6 | Student 14 | Professor |
| 4 | Student 14 | Student 14 | Student 14 | Student 4 | Student 4 |
| 5 | Student 11 | Student 5 | Student 11 | Student 11 | Student 10 |
| 6 | Student 5 | Student 2 | Student 5 | Student 5 | Student 11 |
| 7 | Student 10 | Student 7 | Student 10 | Student 7 | Student 5 |
| 8 | Student 7 | Student 8 | Student 8 | Student 8 | Student 8 |
| 9 | Student 8 | Student 11 | Student 17 | Student 15 | Student 17 |
| 10 | Student 15 | Student 17 | Student 2 | Student 10 | Student 7 |
| 11 | Student 17 | Student 10 | Student 15 | Student 17 | Student 12 |
| 12 | Student 2 | Student 15 | Student 7 | Student 2 | Student 2 |
| 13 | Student 9 | Student 9 | Student 1 | Student 12 | Student 15 |
| 14 | Student 12 | Student 1 | Student 9 | Student 9 | Student 19 |
| 15 | Student 16 | Student 12 | Student 12 | Student 16 | Student 9 |
| 16 | Student 1 | Student 13 | Student 13 | Student 1 | Student 18 |
| 17 | Student 13 | Student 16 | Student 16 | Student 13 | Student 16 |
| 18 | Student 18 | Student 18 | Student 18 | Student 18 | Student 13 |
| 19 | Student 19 | Student 19 | Student 19 | Student 19 | Student 1 |

**Fig. 6.** Positive sentiment ratio by each student

**Fig. 7.** Positive sentiment ratio by each student

## Pearson Correlation of Different Centrality Measures

In order to test our hypothesis, the Pearson correlation is calculated for different measures (see Fig. 9).

The results depict strong positive correlation between *Degree centrality* and *Eigenvector Centrality* measures, which stands for determining influential nodes/students of the blog. Both measures also have high positive correlation with the number of meaning words.

The significance level of both measures was tested. The correlation between *Degree centrality* and *Number of words* returns a P-value of $5.90e-8$ (less than .05), and the P-value for correlation between *Eigenvector centrality* and *Number of words* is

| | degC | sw_closeness | star_betweenness | star_eigen | Words_count | Sentiment_positive |
|---|---|---|---|---|---|---|
| Professor | 1.00000 | 1.0000000 | 1.0000000000 | 1.000000000 | 2029 | 59.53 |
| Student 6 | 0.58750 | 0.9391964 | 0.5377442580 | 0.939401788 | 2178 | 54.19 |
| Student 4 | 0.51250 | 0.9391964 | 0.2488763705 | 0.776956620 | 1453 | 50.00 |
| Student 14 | 0.43750 | 0.8842770 | 0.0845156539 | 0.677584813 | 2051 | 55.00 |
| Student 11 | 0.26875 | 0.8505245 | 0.0502762800 | 0.401931330 | 694 | 61.36 |
| Student 5 | 0.19375 | 0.8344271 | 0.0221146992 | 0.265065442 | 577 | 68.09 |
| Student 10 | 0.17500 | 0.8344271 | 0.0106620074 | 0.208930810 | 696 | 67.92 |
| Student 7 | 0.15625 | 0.8344271 | 0.0104004717 | 0.189535256 | 284 | 66.67 |
| Student 8 | 0.15000 | 0.8188175 | 0.0041247495 | 0.159537932 | 529 | 77.55 |
| Student 15 | 0.15000 | 0.8188175 | 0.0029909095 | 0.134835042 | 112 | 90.91 |
| Student 17 | 0.11250 | 0.8036739 | 0.0022328644 | 0.123044377 | 296 | 61.54 |
| Student 2 | 0.05625 | 0.7747036 | 0.0009918042 | 0.045121413 | 128 | 83.33 |
| Student 9 | 0.01875 | 0.7091287 | 0.0000000000 | 0.027040493 | 77 | 71.43 |
| Student 12 | 0.01875 | 0.0000000 | 0.0000000000 | 0.022229552 | 247 | 42.86 |
| Student 16 | 0.01250 | 0.0000000 | 0.0000000000 | 0.018431582 | 23 | 100.00 |
| Student 1 | 0.00625 | 0.0000000 | 0.0000000000 | 0.008843787 | 3 | 100.00 |
| Student 13 | 0.00625 | 0.0000000 | 0.0000000000 | 0.008843787 | 9 | 0.00 |
| Student 18 | 0.00000 | 0.0000000 | 0.0000000000 | 0.000000000 | 32 | 0.00 |
| Student 19 | 0.00000 | 0.0000000 | 0.0000000000 | 0.000000000 | 82 | 50.00 |

**Fig. 8.** Scoring of students based on different measures

| | Degree Centrality | Closeness | Betweenness | Eigenvector | Words_count |
|---|---|---|---|---|---|
| Degree Centrality | 1 | 0.6368003 | 0.9198131 | 0.9625202 | 0.9111444 |
| Closeness | 0.6368003 | 1 | 0.4022015 | 0.646138 | 0.611013 |
| Betweenness | 0.9198131 | 0.4022015 | 1 | 0.8332505 | 0.7623324 |
| Figenvector | 0.9625202 | 0.646138 | 0.8332505 | 1 | 0.9680857 |
| Words_count | 0.9111444 | 0.611013 | 0.7623324 | 0.9680857 | 1 |

**Fig. 9.** Pearson correlation of different measures

$1.191638e-11$ (less than .05). Thus, both of the correlations are statistically significant for a 5% confidence interval, which supports our hypothesis that degree of meaningful words lead to active participation in the discussion forum.

However, the instructor can use either degree centrality or eigenvector centrality for simplicity, as both of the parameters are highly correlated with each other (see Fig. 10).

## Scatterplot Matrix

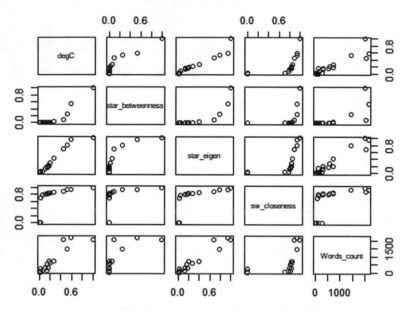

**Fig. 10.** Scatter plot of correlation

## Conclusion and Future Work

This work illustrates the initial results of developing a method to evaluate students engagement during an online course and, thereby, an alternative method for evaluating students' performance. The approach has its theoretical background based on the methods of Social Network Analysis, and its metrics, such as Centrality Measures, Degree Centrality, Closeness Centrality and Betweenness Centrality. In this work, the blog data from disqus is analyzed using network analysis and text analytics by open source tool R. This work can be replicated for any course using disqus for online discussion. This work can also be replicated for any online discussion if the input data is formatted as edge list of a network. The method should be tested for the different fields, since in the provided research the open-ended questions were widely used.

The limitation and future scope of this work includes:

- In the sentiment analysis, few students/users have zero score, as their used words did not hit any similarity measures with any words of the used lexicon. It is possible to develop specific lexicon for different subjects/courses based on subject area related topics. Moreover, students' behavior/communication pattern may change if they know the approach of the assessment. Many students may try to post irrelevant texts frequently to obtain higher scores. This limitation also supports the idea of developing subject-specific lexicon to assess the quality and relevance of students' posts in the blog.

- These metrics along with the data from LMS may be aggregated to predict students' performance and retention rate. Due to the training purpose of the classifier, the large number of observation may be required.

The method should be tested for more courses comprising different pedagogic environments and subject matters to identify wide usability and other possible limitations of the method. In general, these metrics with proper visualization provide better means to the online educators/teachers/instructors to speculate and assess the engagement of students in the discussion forum.

**Acknowledgments.** This work was partially supported by CEPHEI project of ERASMUS+ EU framework which is an ongoing project focusing on digitalization of industrial innovation-related contents. The authors hope to conduct further research and extend this work in future with possible support from this project.

# References

1. Antonova, N., Shnai, I., Kozlova, M.: Flipped classroom in the higher education system: a pilot study in Finland and Russia. New Educ. Rev. **48**(2), 17–27 (2017). https://doi.org/10.15804/tner.2017.48.2.01
2. Barthélemy, M.: Betweenness centrality in large complex networks. Eur. Phys. J. B **38**(2), 163–168 (2004). https://doi.org/10.1140/epjb/e2004-00111-4
3. Beichner, R.J., et al.: The student-centered activities for large enrollment undergraduate programs (SCALE-UP) project. Physics **1**(1), 1–42 (2007). https://doi.org/10.1093/schbul/sbp059
4. Bergmann, J., Sams, A.: Remixing chemistry class: two Colorado teachers make vodcasts of their lectures to free up class time for hands-on activities. Learn. Lead. Technol. **36**, 22–27 (2009)
5. Bonacich, P.: Power and centrality: a family of measures. Am. J. Sociol. (2002). https://doi.org/10.1086/228631
6. Bonacich, P.: Some unique properties of eigenvector centrality. Soc. Netw. **29**(4), 555–564 (2007). https://doi.org/10.1016/j.socnet.2007.04.002
7. Bryson, C., Hand, L.: The role of engagement in inspiring teaching and learning. Innov. Educ. Teach. Int. **44**(4), 349–362 (2007). https://doi.org/10.1080/14703290701602748
8. Dekker, A.: Applying social network analysis concepts to military C4ISR architectures. Connections **24**, 93–103 (2002)
9. Early, S.L.: Book review student engagement techniques: a handbook for college faculty. J. Sch. Teach. Learn. **11**(1), 155–157 (2011)
10. Faust, K., Wasserman, S.: Social Network Analysis: Methods and Applications, Structural Analysis in the Social Science. Cambridge University Press, New York (1995)
11. Ferguson, R.: Learning analytics: drivers, developments and challenges. Int. J. Technol. Enhanced Learn. **4**(5/6), 304 (2013). https://doi.org/10.1504/ijtel.2012.051816
12. Ferrara, E.: Measurement and analysis of online social networks systems. In: Encyclopedia of Social Network Analysis and Mining (2018). https://doi.org/10.1007/978-1-4939-7131-2_242
13. Hanneman, R.A., Riddle, M.: Introduction to Social Network Methods. Network (1998). https://doi.org/10.1109/78.700969

14. Prober, C.G., Khan, S.: Medical education reimagined: a call to action. Acad. Med. **88**(10), 1407–1410 (2013). https://doi.org/10.1097/ACM.0b013e3182a368bd
15. Rabbany, R., et al.: Collaborative learning of students in online discussion forums: a social network analysis perspective. In: Studies in Computational Intelligence (2014). https://doi.org/10.1007/978-3-319-02738-8_16
16. Rajaraman, A., Ullman, J.D.: Mining of Massive Datasets (2011). https://doi.org/10.1017/cbo9781139058452
17. Rosvall, M., Bergstrom, C.T.: Maps of information flow reveal community structure in complex networks. Proc. Natl. Acad. Sci. U.S.A. (2008). https://doi.org/10.1073/pnas.0706851105
18. Scott, J., et al.: Social network analysis: an introduction. In: The SAGE Handbook of Social Network Analysis (2015). https://doi.org/10.4135/9781446294413.n2
19. Stewart, S.A., Abidi, S.S.R.: Applying social network analysis to understand the knowledge sharing behavior of practitioners in a clinical online discussion forum. J. Med. Internet Res. **14**(6), 170 (2012). https://doi.org/10.2196/jmir.1982
20. Suraj, P., Roshni, V.S.K.: Social network analysis in student online discussion forums. In: 2015 IEEE Recent Advances in Intelligent Computational Systems, RAICS 2015, pp. 134–138, December 2016. https://doi.org/10.1109/raics.2015.7488402
21. Taboada, M., et al.: Lexicon-based methods for sentiment analysis. Comput. Linguist. **37**(2), 267–307 (2011). https://doi.org/10.1162/COLI_a_00049
22. Widmer, E.D., Lafarg, L.-A.: Boundedness and connectivity of contemporary families: a case study. Connections **22**, 30–36 (1999)
23. Wise, A., Zhao, Y., Hausknecht, S.: W11-Learning analytics for online discussions: a pedagogical model for intervention with embedded and extracted analytics. In: Proceedings of the Conference on Learning Analytics (2013). https://doi.org/10.1145/2460296.2460308

# Reference Model of Service-Oriented IT Architecture of a Healthcare Organization

Igor V. Ilin, Anastasia I. Levina$^{(\boxtimes)}$, and Aleksandr A. Lepekhin

Peter the Great St. Petersburg Polytechnic University, St. Petersburg, Russia
alyovina@gmail.com

**Abstract.** The healthcare system nowadays rapidly moves towards digital transformation: digital technologies influence not only technical support of medical processes, but change business models of healthcare organizations. Introduction of new technologies into the architecture of healthcare system requires clear understanding of all the elements of the existing IT architecture and its relations in order to combine effectively the requirements of medical processes and possibilities of digital technologies. The IT architecture is a set of application components that realize a set of IT-services, which in turn supports the system of business processes of a healthcare organization. The paper describes the transparent and coherent reference model of the IT architecture and reference list of IT-services based on the functional structure of the healthcare organization. The reference model can be adopted to further detailed design of the applications of a particular healthcare organization.

**Keywords:** Digital technologies · IT architecture · IT services · Healthcare organization

## 1 Introduction

Nowadays trends in the healthcare industry development and modern digital technologies determine the emergence of a new type of medical organization – the "Smart Hospital". This term refers to a medical organization that implements modern principles of healthcare, while using modern digital technologies to implement these principles and improve medical and economic efficiency.

The leading concepts, under the influence of which modern healthcare system is formed, are: value medicine, personalized medicine, the concept of Health 4.0. Value medicine is a result-oriented approach to the organization of the system of medical care, involving the choice of patient management method, which allows to achieve better results at lower costs, including those from the patient point of view. Value medicine focuses on the value of the procedure for the patient and the cost-effectiveness of medical interventions, which are not taken into account in classical evidence-based medicine. Personalized medicine involves the selection and organization of an individual patient's trajectory of treatment, based on their individual characteristics [3]. The use of such tactics in routine clinical work is expectedly

© Springer Nature Switzerland AG 2020
D. G. Arseniev et al. (Eds.): CPS&C 2019, LNNS 95, pp. 681–691, 2020.
https://doi.org/10.1007/978-3-030-34983-7_67

accompanied by a number of issues and limitations: it requires personalized collecting and processing data about each patient, a certain level of IT safety and data protection, and ensuring data availability.

The concept of Health 4.0 implies the use of modern digital technologies (Big Data, Internet of things, blockchain, telemedicine, predictive analytics, machine learning, etc.) to increase the economic and medical efficiency and availability of medical care. [12, 15]. The implementation of the principles of personalized medicine involves the effective collection, processing, analysis of a large amount of primary data about each patient in real time. Traditional medical organizations are not able to cope with the increasing flow of information, because the existing management architecture, including the architecture of information systems and applications, does not provide interfaces for interaction with modern digital technologies [13].

Modern management technologies, including digital ones, have significant potential in solving a number of problems on the way to providing more affordable, cost-effective and high-quality medical care. In view of existing trends in the development of health care, the urgent task is to develop such a model of medical organization that implements the principles of value and personalized medicine, uses the capabilities of the Health 4.0 technologies, and in the mean time enables prompt and flexible responses to dynamically changing conditions of the external environment [5].

Requirements for compliance with the principles of value-based and personalized medicine should be reflected in the business process model of a medical organization [3]. Health 4.0 and related technologies set the requirements for the IT architecture and the technological infrastructure of medical organizations. Thus, there is a need of an appropriate methodology which allows to form a management system for medical organizations as a coherent whole of interconnected domains, such as business process system, IT systems and applications structure, IT services and hardware. Modern management science pays serious attention to the development of particular elements of the management system (business processes, information systems, services), and also underlines the importance of their integration (the concept of enterprise architecture, business engineering). However, the absence of industry-specific methodologies for integrated management system design, in particular, in the field of healthcare, should be mentioned. This sets a problem of development a methodology for the formation of a corporate architecture of medical organizations, based on the functional structure of activity and including a model of a process system, IT architecture, service architecture, and technological architecture.

The paper is devoted to the formation of a reference model of the service-oriented IT architecture of medical organizations based on the functional model of activity. Such a model can be adapted to medical organizations of different specializations and scale of activity and is aimed to reduce time of business and IT consultants to formulate requirements for the IT architecture of medical organizations. The approaches and models described in the paper are based on the authors' experience in research projects in the field of improving the healthcare management system.

## 2  Methodology

The methodological basis of the study is an architectural approach to the design of management systems [4, 9–11], which promotes an integrated approach to the formation of business management systems, in which such elements as business processes, organizational structure, functional structure, information systems and applications, IT services, IT infrastructure are considered in their interrelation and interaction. There is a number of enterprise architecture researches, devoted to the development of the theory and practice of specific elements of the enterprise architecture and the development of industry-specific methods and models of architecture solutions, including the specifics of the enterprise architecture models in view of digital transformation. The most related to the current study are [1, 7, 8, 16].

The formation of industry specific methods and models of architecture solutions and a reference model of a medical organization that implements the principles of value and personalized medicine and the following trends in the development of modern digital technologies is possible using the approaches and methods of the enterprise architecture discipline. The enterprise architecture is a coherent whole of such elements of the management system as business processes, functional and organizational structures, material and cash flows, information systems and applications, data, workflow, technological infrastructure objects [9]. Reforming the operations of medical organizations, due to the need to introduce modern management technologies in accordance with the concept of enterprise architecture, should be carried out systematically, taking into account the interrelations and interdependencies of all elements of the organization's management system.

One of the key ideas of the enterprise architecture discipline is its service orientation. The main idea of the service-oriented approach to enterprise management is to present the final result of the enterprise's operations in the form of services for customers [9, 10]. The set of services provided to external customers determines the list of key business processes of the enterprise. To ensure the quality of the business services provided, the processes implementing these services must be defined, and formalized; they must be transparent and measurable. Similarly, the relationship between business processes and their IT support, as well as IT processes and IT infrastructure objects, is built on the service principle. Such a model of the formation of an enterprise management system is called service-oriented architecture (SOA) [9]. This term, which originally appeared in the IT area, is currently used in relation to the principles of the formation of the entire enterprise architecture.

Service-oriented architecture is a set of design principles that allow us to consider units of functionality in the form of provided and consumed services. The idea of providing system services to other systems and their users, born in the area of software development, began to be widely used in business disciplines.

Service orientation logically leads to the representation of the enterprise's architecture as a model consisting of various layers, where the concept of service is the main link between various layers. Alignment of business strategy and IT strategy and the corresponding alignment of the business layer and the application layer of the enterprise architecture is one of the main management issues nowadays. The main strategic task of information technologies is to support the achievement of long-term goals and solving operational problems of business through the implementation of projects to introduce information systems and technologies that ensure the achievement of competitive advantages and economic benefits, information support for business processes, carried out on the necessary quality level, improvement of the organization of IT services in terms of efficiency, productivity and cost optimization.

To create requirements for IT services of information systems and applications, one of the widely known approaches and models can be used, for example, the IBM Component Business model [6] or SWEBoK [2].

To identify the landscape of IT systems, the functional-oriented approach was used. The functional model, as a key element of the business architecture, is the starting point for the analysis and reengineering of the company's management system, as it:

1. Provides insight into company operations;
2. Provides a visual model for analyzing, benchmarking and identifying optimization potential;
3. Determines the organizational structure, information, material and cash flows;
4. Serves as a basis for identifying IT support requirements, forming requirements for IT services and then shaping the IT architecture landscape.

The example of the functional model for a medical organization can be found in the reference [3].

## 3   Results

The possibilities of applying the technologies of Health 4.0 concept set the requirements for the IT support and technological infrastructure. Thus, a methodology is needed to form a management system for medical organizations, which allows complex development of such elements of a medical organization management system as a system of business processes, architecture of information systems and applications, and their hardware. The formation of such a methodology and reference model of a medical organization that implements the principles of value and personalized medicine is possible with a service-oriented approach. The information systems architecture is a set of application components that implement a set of information technology services, which, in their turn, support the system of business processes of a medical organization.

The baseline services of a medical organization, which allow further development of IT infrastructure, are those that cover the basic functions of organizations and its key processes. In this research, a case-study analysis was done in order to create the baseline functionality of a medical organization.

An analysis of IT support for specific medical organizations has shown that the structure of the application landscape can be represented using a set of application components that implement key IT services. These applications are:

1. Electronic medical record (including dental medical card)
2. The system of providing outpatient care
3. The system of providing inpatient care

4. Clinical monitoring system
5. Anesthetic monitoring system
6. The accounting system
7. Personnel management system
8. POS-system pharmacy
9. Laboratory Medical Information Systems

The prerequisites of formation of the described set of applications is the need to create an IT infrastructure, which would allow further technological development of a medical organization by covering all basic functions and processes. In order to analyze whether this coverage is comprehensive, the analysis on functionality of the described applications was provided.

The functionality of each of the application components, mentioned above, is described below.

### 3.1 Electronic Medical Record

An electronic medical record is a component of an application containing electronic medical records equivalent to paper medical records maintained by the attending physician. They store patient data, including the history of the disease, diagnoses, prescribed medications, a list of receptions and consultations, hospitalization records, services provided, vaccinations made, allergies to medications, and examinations results. An electronic medical record implements such key services as obtaining information about each patient, using data for setting up diagnostics, prescribing drugs, comparing current and past tests, etc. The IT service model of electronic medical record is shown in Fig. 1.

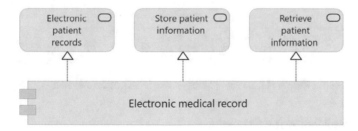

**Fig. 1.** IT service model of an electronic medical record

### 3.2 Outpatient Care System

Outpatient care involves the provision of medical care that does not include round-the-clock medical supervision and treatment. The functionality of the outpatient care system includes registration of both the patients and their visits, ordering, planning, assigning medical services and recording their implementation. In addition, the system registers referrals for hospitalization and fee-based services, it also contains the necessary medical documents and regulatory reference information for outpatient care. The IT service model of the system of the Outpatient care system is shown in Fig. 2.

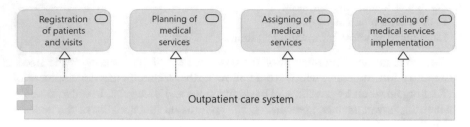

**Fig. 2.** IT service model of the outpatient care system

### 3.3  Inpatient Care System

Inpatient medical care is provided to patients who need emergency medical care or constant monitoring. The functionality of the inpatient care system includes the registration of patients, creation of medical documents, maintenance of data on patients staying in the hospital departments, transfer of patients, services rendered to them, registration of the departure of patients from the hospital, and also maintenance of regulatory and reference information. The IT service model of the Inpatient care system is shown in Fig. 3.

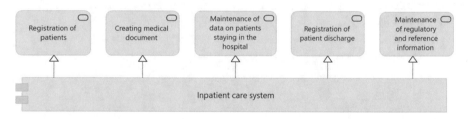

**Fig. 3.** IT service model of the inpatient care system

### 3.4  Clinical Monitoring System

Clinical monitoring involves continuous observation and monitoring of the patient's condition by recording biological signals in real time and evaluating the body's diagnostic indicators and serves to identify complications, deviations of indicators from the norm, and prevent hazards during treatment. Methods of monitoring include primarily the monitoring of indicators of the central nervous system, cardiovascular system, and respiratory function. Figure 4 demonstrates the IT service model of the clinical monitoring system.

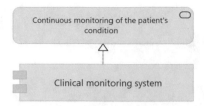

**Fig. 4.** IT service model of the clinical monitoring system

## 3.5    Anesthetic Monitoring System

The anesthetic monitoring system is intended for continuous monitoring, ongoing control and assessment of the patient's physiological indicators during an operation under anesthesia. The methods of anesthetic monitoring include pulse oximetry, recording and processing of ECG, capnometry, non-invasive measurement of arterial pressure, invasive measurement of hemodynamic parameters, analysis of inhalation oxygen concentration and others. As a result of the monitoring, an anesthesia map is generated, containing all the data necessary for analyzing anesthesia. The IT service model of the anesthetic monitoring system is shown in Fig. 5.

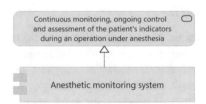

**Fig. 5.** IT service model of the anesthetic monitoring system

## 3.6    Accounting System

In the accounting system all business transactions and incoming funds are recorded, the income and expenditures are monitored and planned, internal and external reports are generated, accounting for budget execution and analytical accounting of fixed assets, financial and settlement transactions, and ratios are conducted, etc. Figure 6 shows the IT service model of the accounting system.

## 3.7    Personnel Management System

The personnel management system involves, first of all, personnel records, including the keeping of personal files of employees, the accounting of working time, payrolls, staffing management, preparation of orders, selection and transfer of employees, and preparation of the necessary statistical reporting. The IT service model of the Personnel Management System is shown in the Fig. 7.

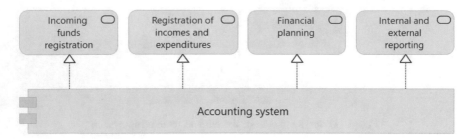

**Fig. 6.** IT service model of the accounting system

**Fig. 7.** IT service model of the personnel management system

## 3.8  POS Pharmacy System

This system enables maintaining operational records of the range of pharmacy products in various product groups. Automation is implemented through the use of barcodes and POS programs installed at the checkout nodes. Thanks to the POS-system, the inventory of goods, the registration of goods receipt, inventory control, inventory taking, planning the purchase of goods assortment are carried out. The system also determines the place of storage of products, selects possible analogues, tracks the shelf life. The IT service model of the POS pharmacy system is shown in the Fig. 8.

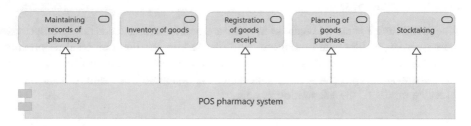

**Fig. 8.** IT service model of the inpatient care system POS pharmacy system

## 3.9  Laboratory Medical Information Systems

This application component implements such IT services as the introduction of laboratory tests upon the request, marking the materials taken, fixing the tube number, printing the necessary reports. This application has the potential of direct data downloading. This component of the application also implements the services of displaying the main indicators of the laboratory's activity on the studies performed, patients, orders, financial, etc. Figure 9 demonstrates the IT service model of the Laboratory Medical Information Systems.

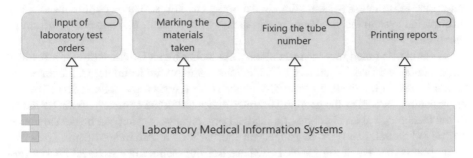

**Fig. 9.** IT service model of the laboratory medical information systems

The described above list of application components and their services is not exhaustive and may differ depending on a particular medical organization. However, these components of applications and their services are basic for IT support of a medical organization and allow register and analyze important data, which can further on provide for the following results, that are strategic for a medical organization:

- improving the quality of patient care to meet the needs of current and future consumers of medical services;
- optimization of treatment results for each individual patient;
- the empowerment of health workers to achieve the best results in the delivery of health services;
- improving the operational efficiency of the medical organization in order to release some resources for the implementation of innovative solutions and increase the quality of patient care;
- applying data-based innovations to assist caregivers and scientists in conducting experimental studies in medicine and bioengineering.

# 4   Discussion and Conclusion

The information systems architecture is a multitude of application components that implement a set of IT services that, in their turn, support the medical organization's business process system. The paper describes the required functionality of information systems of a medical organization, containing a list of possible IT services. The proposed IT service models with the necessary adaptation can be used as a basis for analyzing and reengineering IT services and for designing the IT architecture of specific medical organizations. Such a model is especially relevant for medical organizations in view of ongoing trend for active development of the concept of Health 4.0, declaring the transformation of the health sector based on the use of modern digital technologies [14]. The reference functional model of activity and its elements (which are the models of IT architecture and IT services) is the basis for the introduction of any changes, including the introduction of digital technologies.

Concerning implementation of the described model, it is important to mention that enterprise information systems define technological infrastructure requirements. Each information system has hardware requirements with which it can correctly perform its functions.

It should be noted that when modeling the technological architecture of a medical organization, we faced a number of limitations, namely: the technological architecture of the medical organization essentially depends on the security policy adopted in the organization, as well as the general IT strategy adopted in the company. A single model of the technological architecture of a medical organization, which can be considered as an "as is" model, can only be formed on the basis of a certain set of principles and assumptions that will form its basis. Otherwise, the technology architecture is extremely variable depending on the organization chosen. The technological architecture of the "as is" model of the medical organization architecture can be based on the following principles:

- Client-server application architecture (for unified access to the information base from any computer of the organization)
- Increased attention to data protection (to protect patients' personal data)

It is also important to note that the equipment is not part of the technological architecture of a medical organization in an "as is" model. The integration of medical equipment is one of the stages in the development of a medical organization within the paradigm of Industries 4.0. And this is a topic for a potential future research.

**Acknowledgments.** The reported study was funded by RFBR according to the research project No. 19-010-00579.

# References

1. Anisiforov, A., Dubgorn, A.: Organization of enterprise architecture information monitoring. In: Proceedings of the 29th IBIMA Conference on Education Excellence and Innovation Management through Vision, pp. 2920–2930 (2017)
2. Bourque, P., Fairley, R.E.: SWEBOK V 3.0. Guide to the Software Engineering Body of Knowledge. IEEE Computer Society (2014). http://www.computer.org/ieeecs-swebokdelivery-portlet/swebok/SWEBOKv3.pdf?token=2w9BvZs11g4JIyEOnIVmlSSuV-qnolruB. Accessed 01 Feb 2016
3. Dubgorn, A., Ilin, I., Levina, A., Borremans, A.: Reference model of healthcare company functional structure. In: Proceeding of the 33rd IBIMA Conference, pp. 5129–5137 (2019)
4. Greefhorst, D., Proper, E.: Architecture Principles. The Cornerstones of Enterprise Architecture. Springer, Heidelberg (2011)
5. Gunasekaran, A., Ngai, E.W.T.: Managing digital enterprise. Int. J. Bus. Inf. Syst. **2**, 266 (2007). https://doi.org/10.1504/IJBIS.2007.011979

6. IBM Business Consulting Services: Component Business Model (2006). https://www.ibm. com/industries/financialservices/ru/banking/pdf/fss_component_business_modeling_Rus_ N_2.pdf. Accessed 23 Jan 2016

7. Ilin, I., Levina, A., Abran, A., Iliashenko, O.: Measurement of enterprise architecture (EA) from an IT perspective: research gaps and measurement avenues. In: ACM International Conference Proceeding Series Part F131936, pp. 232–243 (2017)

8. Ilin, I.V., Anisiforov, A.B.: Improving the efficiency of projects of industrial cluster innovative development based on enterprise architecture model. WSEAS Trans. Bus. Econ. **11**, 757–764 (2014)

9. Lankhorst, M.: Enterprise Architecture at Work. Modeling Communication and Analysis. Springer, Heidelberg (2017)

10. Op't Land, M., Proper, E., Waage, M., Cloo, J., Steghuis, C.: Enterprise Architecture. Creating Value by Informed Governance. Springer, Heidelberg (2009)

11. Opengroup: TOGAF (2019). https://www.opengroup.org/. Accessed 18 Jan 2019

12. Oswald, G., Kleinemeier, M. (eds.): Shaping the Digital Enterprise. Springer, Cham (2017). https://doi.org/10.1007/978-3-319-40967-2

13. Ilin I.V., Frolov K.V., Lepekhin A.A.: From Business processes model of the company to software development: MDA business extension. In: Proceedings of the 29th IBIMA 2017, pp. 1157–1164 (2017)

14. Uhl, A., Gollenia, L.A.: Digital Enterprise Transformation: A Business-Driven Approach to Leveraging Innovative IT. Taylor & Francis, London (2016)

15. Ustundag, A., Cevikcan, E.: Industry 4.0: Managing the Digital Transformation. Advanced Manufacturing. Springer, Cham (2018)

16. Zimmermann, A., Schmidt, R., Sandkuhl, K., Wissotzki, M., Jugel, D., Mohring, M.: Digital enterprise architecture – transformation for the Internet of Things. In: IEEE 19th International Enterprise Distributed Object Computing Workshop. Presented at the 2015 IEEE 19th International Enterprise Distributed Object Computing Workshop (EDOCW), Adelaide, Australia, pp. 130–138. IEEE (2015). https://doi.org/10.1109/EDOCW.2015

# IT-Architecture Development Approach in Implementing BI-Systems in Medicine

Oksana Yu. Iliashenko, Victoria M. Iliashenko, and Alisa Dubgorn$^{(\boxtimes)}$

Peter the Great St. Petersburg Polytechnic University, Saint Petersburg, Russia
ioyl20878@gmail.com, alissa.dubgorn@gmail.com

**Abstract.** One of the main key aspect of management in medical organizations is information support by different analytical systems. For this reason, the number of medical organizations that decide on the implementation of Business Intelligence class systems increases. BI-system allows for the data operational monitoring on the medical organizations management, simplifies the entire process of forming analytical reports in the company. BI system is the most complete source of information on current activities at medical organizations. The BI-system implementation at an enterprise involves the modernization of the business processes system, services architecture, the development of IT architecture and information exchange model. One of the important tasks in the information and analytical system implementation is its integration into the existing architectural solution at the enterprise. This paper provides the development of a medical organization's IT architecture and a corresponding information exchange model when solving problems of integrating BI system into the general IT architecture of an enterprise.

**Keywords:** Business intelligence · BI · BI-system · IT architecture · Medicine

## 1 Introduction

In the modern world, the leading ideological concepts under the influence of which the modern health care system is being formed are value medicine, personalized medicine, the concept of Health 4.0. To implement the existing trends in the development of healthcare, a change in the management system is required. An important task is the need for a quick and efficient response to the dynamically changing conditions of the external and internal environment. The question is not just, "What are we supposed to be doing right now?" but, "Where are the current and potential problem areas that will prevent our medical business from succeeding?" One of the ways to do this is to use business intelligence (BI) capabilities. Business intelligence allows get information or knowledge, based on the data obtained from external sources and the results of the clinic's activities. This allows you to obtain the necessary information for the decision-making leaders of medical organizations and, finally, to strengthen the competitive advantages of enterprises. Using business intelligence systems is possible when fulfilling a number of requirements for automating business processes and data that will be used for analysis.

D. G. Arseniev et al. (Eds.): CPS&C 2019, LNNS 95, pp. 692–700, 2020.
https://doi.org/10.1007/978-3-030-34983-7_68

A lot of attention was paid to the formation of requirements for the IT services of medical organizations. For example, the electronic health record (EHR) [9], a service that allows to provide continuous patient care. A large number of studies are devoted to the possibilities of applying solutions of well-known vendors in the medical field: IBM, Microsoft, SAP, Russian Solutions of the Netrika Company [17].

The article pays a lot of attention to the application of business analysis tools to the tasks of the healthcare system. The problems of using Big Data processing technologies in the analysis [8] are considered. The tasks of using BI systems in the implementation of medical logistics management models in a medical organization [20] are investigated. These studies confirm the importance and relevance of using BI systems in solving problems of medical organization managing. However, there are questions about changing the existing IT architecture for the application of BI systems.

## 2  Literature Review

Business Intelligence (BI) transmits various data that are necessary for enterprise managers to analyze the company's activities from enterprises to information systems. At an early stage of BI systems development, scientists were focused on the concept of "data warehouse", as an intermediate component between Operational systems (On-Line Transactions Processing Systems, OLTP systems) and BI systems [3]. Then the concept of "data marts" appeared as a key component for data visualization.

BI systems architecture is closely related to data warehouse architecture. Historically, we have two main approaches to data warehouse architecture.

- An approach proposed by Ralf Kimball. He offered an Architecture named Data Warehouse Bus [10]. There is a standardized dimensional model for all users. It includes both data and multidimensional cubes. Queries descend to progressively lower levels of detailing without being reprogrammed by the user or application designer. This architecture bases on dimensional models of business processes (not corresponding to a business measurement or event), not business departments [19];
- An approach proposed by Bill Inmon. The IT architecture of Bill Inmon is named Corporate Information Factory (CIF) [6, 7]. According to the proposed concept, the marts are tailored by business department/function with dimensionally structured summary data [1, 2].

In practice, specialists usually use a hybrid approach. It combines advantages of two previous approaches.

Liya [15] presented the SOA-ITPA architecture. It allows to provide services to other components of the application via a network communication protocol.

In the available literature, examples of the successful applications of BI systems in various sectors are given. For example, finance, healthcare activities, retail networks, telecommunications, transport [4, 5], etc.

We have many examples of the successful application of the BI systems in healthcare sector. You can find individual solutions, but currently, there is no reference model that would allow you to understand the interaction of the main dataflow, business and IT services, and options for interaction of information systems.

# 3   Research Methodology

We carried out the analysis of the existing IT architecture and the developing of the target IT architecture based on the architectural approach [13, 16]. We analyzed the existing IT architecture and information exchange models, formed the requirements for the target IT-architecture, i.e., the requirements for business, IT and infrastructure services.

# 4   Results

To achieve the goal, related with BI system integration into the existing architectural solution in the medical organization, an approach was proposed. It consists of the following stages:

- explore the IT architecture that exists at the enterprise today;
- form the requirements to business and IT and infrastructure services to implement a BI system in the existing IT architecture;
- analyze the existing BI-systems and make a choice of the platform;
- explore the BI-system architecture on the selected platform;
- suggest possible ways to integrate the BI system and the existing architectural solution;
- formulate the criteria on which the choice of integration technology is based;
- evaluate each of the alternatives according to the established criteria, taking into account their significance;
- after analyzing the estimates obtained, determine the most rational way of introducing the BI system to the enterprise;
- develop the target IT architecture.

**Business Services Requirements**
The key business service requiring the use of the BI system is the possibility of obtaining analytical reports on key aspects of the organization's activities in various ways:

- reports on financial results of activity: data on the amount of income and expenses of the medical organization in the context of analytical codes of income (credits), expenses as of January 1 of the year following the reporting year;
- financial statements: balance sheets;
- statistical medical reporting: a report on the structure and co-relation of diseases, a report on the results of patient treatment, internal reporting on the assignment of hospital beds, etc.;
- reporting on the development of the personnel of the medical organization.

These business services allow to get a reliable assessment of the results of the medical organization's activities, to identify ways of more rational use of funds and their most efficient distribution.

The implementation of business services is carried out through the functioning of relevant IT services [14]. IT services should be implemented taking into account the requirements for BI systems and the specifics of the medical organizations' activities as well.

**IT Services Requirements**

IT services requirements include:

- integration requirements: the availability of integration components for the implementation of relationships among individual modules of the system;
- access requirements: providing access to data and access to system data processing functions;
- interface requirements: availability of a unified interface environment, the use of the principle of "data navigation";
- methodological requirements: the use of modern standards in the medical field in the development of semantically interoperable information systems in order to obtain a result that corresponds to the modern level of IT technologies development in medicine.

**Options for Selecting BI Systems**

Based on the formulated requirements for business and IT services, we can offer options for choosing BI systems for solving business analytical problems at medical organizations.

In order to make choice, a survey was conducted among responsible persons who make decisions on the implementation of the analytical system at the organization. Based on the information collected, the requirements for the future architecture were formulated, and ten parameters were identified, with the help of which the system integration technologies were evaluated [11, 12, 18].

*Quantitative*

- Total Cost of Ownership (TCO) of BI system implemented in the IT architecture of the company;
- time spent on implementing the solution;
- the speed of downloading data from ERP to BI system;
- the speed of downloading data from MIS to BI system;
- potential threats of remake.

*Qualitative*

- ability to integrate various information systems;
- ensuring the required data quality, their integrity;
- low complexity of technical implementation and maintenance;
- resilience to the information system failures.

**Possible Integration Options for ERP, MIS and BI Systems**

In our study, we consider three main ways of possible integration of ERP and BI systems, MIS and BI systems: the main advantages and disadvantages of each of the integration technologies are indicated:

1. *Using the Excel Upload Reporting System*

Advantages: low cost; easy use; doesn't require special skills in tools for analysis easy implementation; minimal intervention by external experts.

Disadvantages: low degree of the data loading process automation; low data security; the need to create new files, which leads to a high probability of system failures, and, as a result, insufficient data integrity and completeness; the need to create new requests to the ERP database depending on users' requirements.

2. *Making Connection of BI System Directly to DBMS Using Standard ODBC/OLE DB Interface*

Advantages: supports external data sources (OLE DB); works with 32-bit and 64-bit drivers (ODBC); relative cheapness of decisions; do not require additional modules; no violation of data integrity and completeness; constant access to ERP database.

Disadvantages: encrypting the names of tables and fields during the formation of the ERP database leads to difficulties in creating queries to the DBMS and the ratio of the names of directories and registers ERP with tables in ERP database.

3. *Uploading Data from ERP and MIS Systems by Xtract BI Software*

Advantages: the opportunity to extract data from ERP system tables and view it; use Xtract BI business application programming interface (BAPI) component to access data from BAPIs and protocol function modules and directly use the output in BI system; support for using dynamic SQL statements with variables; data extraction can be processed in packets to handle Big Data.

Disadvantages: requires additional cost of using and connecting additional modules.

The connector's functionality includes:

- generation of data models for various BI-systems;
- creating a direct sample from epy DBMS for loading data model tables;
- automating the process of assigning data fields that are correct and understandable to users and IT professionals for the names that are ready to be used in the headers of the visual interface of a BI application, including that from the point of view of organizing associative links;
- the connector acts as a reference service for data structures and relationships of ERP tables.

Summing up, it can be argued that the ERP, MIS and BI integration system using special connectors can accurately process and convert data company in the implementation of operational monitoring of sales of the enterprise.

For IT specialist which is engaged in the formation of data models for BI applications, there are 3 levels of work in the browser:

- Level 1: provides basic functionality for selecting data from ERP and MIS in the BI system.
- Level 2: creates complex SQL-views for tabular objects of the ERP-configuration, which can be used both for the automatic creation of data models and for the automatic creation of data models at the tracking level.
- Level 3: level of automation, the ability to create a variety of data models for various BI-applications, without taking into account access to the database, based on SQL representations.

Figure 1 shows one of the options for the model of uploading data from the ERP system to the BI system depending on the type of data source. Each of the above mentioned methods allows to upload data for further aggregation, depending on the type and source of data.

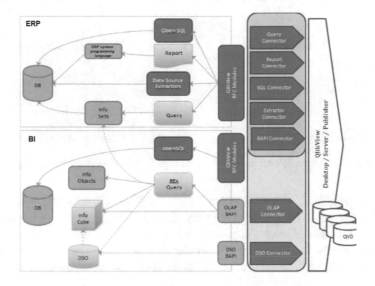

**Fig. 1.** The process of extracting data from ERP modules through use connector.

The target IT-architecture with BI and ERP and MIS systems integration. We propose to develop the target IT-architecture based on the integration of the ERP system, MIS system and BI system using a special connector which will eliminate the previous shortcomings in the organization of information exchange and automate the processes of extracting, converting and loading data as much as possible. In our opinion, it should allow to provide reliable data for data analysis services of collaborative BI system and to improve accuracy and efficiency of data analysis services.

This will ensure the implementation of operational monitoring of the medical organization's main indicators. The target IT architecture model with the Archimate language is presented in Fig. 2.

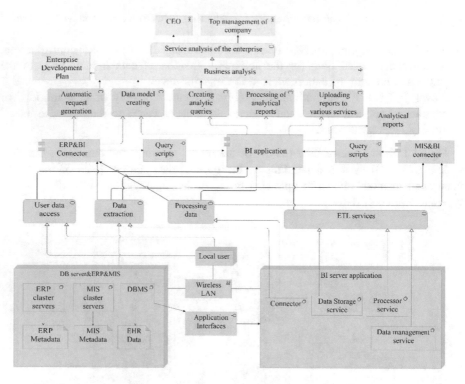

**Fig. 2.** The target IT-architecture with BI and ERP and MIS systems integration

Now the user or developer does not interact with the ERP or MIS systems at any stage of the process. User actions are only related to working in the BI application. Users can view reports, make required selections, create additional visualizations or stories, and upload the necessary data in the right formats. The stages of extracting, processing and uploading data into a BI application have changed significantly.

## 5   Conclusion

The BI system implementing involves the formation of requirements for business and IT services, a change in the medical organization's IT architecture, analysis of ways to integrate the BI system with the MIS and ERP systems. We analyzed the possibilities of using BI systems in the management of a medical organization, formulated the requirements for IT architecture when implementing BI systems. A referent IT architecture model of medical organization with ERP, MIS and BI systems integration has been proposed.

# 6  Discussion

In the future, it is planned to consider the issues of system improving by using different technologies of data processing, concerning specialty data formats and new analytical modules implementation. Also, we are planning to consider the issue concerning the possibility of implementing individual microservices in the integration of various systems using cloud platforms. It can contribute to the improvement of various aspects of medical organizations' activities and the consistent medical organizations development as a whole.

**Acknowledgments.** The reported study was funded by RFBR according to the research project № 19-010-00579.

# References

1. Ariyachandra, T., Watson, H.J.: Key factor in selecting a data warehouse architecture. Bus. Intell. J. **10**(2), 19–26 (2005)
2. Ariyachandra, T., Watson, H.J.: Which data warehouse architecture is most successful. Bus. Intell. J. **11**(1), 4–6 (2006)
3. Devlin, B.A., Murphy, P.T.: An architecture for a business and information system. IBM Syst. J. **27**(1), 60–80 (1988)
4. Iliashenko, O.Yu., Levina, A.I., Dubgorn, A.: Using business intelligence technologies for evaluation of road network organization. In: Proceedings of the 30th IBIMA 2017, pp. 2144–2155 (2017)
5. Ilin, I., Iliashenko, O.Yu., Levina, A.: Reengineering of high-tech and specialized medical care delivery process for telemedicine system implementation. In: Proceedings of the 29th International Business Information Management Association Conference IBIMA, pp. 1822–1831 (2017)
6. Inmon, W.H.: DW 2.0 Architecture for the Next Generation of Data Warehousing. Morgan-Kaufman, Burlington (2008)
7. Inmon, B.: Data Lake Architecture: Designing the Data Lake and Avoiding the Garbage Dump. Technics Publications, Basking Ridge (2016)
8. Yuen-Reed, G., Mojsilović, A.: The role of big data and analytics in health payer transformation to consumer-centricity. In: Weaver, C., Ball, M., Kim, G., Kiel, J. (eds.) Healthcare Information Management Systems Cases, Strategies, and Solutions, 4th edn, pp. 399–420. Springer, Cham (2016)
9. Cresswell, K.M., Sheikh, A.: Key Advances in Clinical InformaticsTransforming Health Care Through Health Information Technology. Academic Press, Cambridge (2017). https://doi.org/10.1016/b978-0-12-809523-2.00002-9. Accessed 1 May 2019
10. Kimball, R., Ross, M.: The Data Warehouse Toolkit: The Definitive Guide to Dimensional Modeling, 3rd edn. Wiley, Indianapolis (2013)
11. Kroenke, D.M., Auer, D.J., Yoder, R.C., Vandenberg, S.L.: Database Processing: Fundamentals, Design, and Implementation, 15th edn. Pearson, New York (2018)
12. Kroenke, D.M., Auer, D.J., Vandenberg, S.L., Yoder, R.C.: Database Concepts. Pearson, New York (2017)
13. Lankhorst, M.: Enterprise Architecture at Work: Modelling, Communication, Analysis, 4th edn. Springer, Heidelberg (2017)

14. Lepekhin, A.A., Borremans, A.D., Iliashenko, O.Yu.: Design and implementation of IT services as part of the "Smart City" concept. In: MATEC Web of Conferences, International Science Conference on Business Technologies for Sustainable Urban Development, SPbWOSCE 2017, 13 June 2018, vol. 170 (2018). Paper number 01029
15. Liya, W., Barash, G., Barash, G., Bartolii, C.: A service-oriented architecture for business intelligence. In: IEEE International Conference on Service-Oriented Computing and Applications (SOCA 2007), pp. 279–285 (2007)
16. Official website of the company OPENGROUP. TOGAF Standart. https://www.opengroup.org/. Accessed 18 Jan 2019
17. Official website of the company Netrika. https://netrika.ru/. Accessed 1 May 2019
18. Troyansky, O., Gibson, T., Leichtweis, C.: QlikView Your Business. An Expert Guide to Business Discovery with QlikView and Qlik Sens. Wiley, Hoboken (2015)
19. Schaffner, J., Jacobs, D., Eckart, B., Brunnert, J., Zeier, A.: Towards enterprise software as a service in the cloud. In: Proceedings of the 2nd IEEE Workshop on Information and Software as Services (WISS), pp. 52–59 (2010)
20. Liu, T., Shen, A., Hu, X., Tong, G., Gu, W.: The application of collaborative business intelligence technology in the hospital SPD logistics management model. Iran. J. Public Health **46**(6), 744–754 (2017)

# BI Capabilities in a Digital Enterprise Business Process Management System

Dmitry Yu. Mogilko[1], Igor V. Ilin[2], Victoria M. Iliashenko[2(✉)], and Sergei G. Svetunkov[2]

[1] AO "NII TM", St. Petersburg, Russia
[2] Peter the Great St. Petersburg Polytechnic University, St. Petersburg, Russia
vmil206@yandex.ru

**Abstract.** The paper is devoted to the study of business intelligence capabilities in the business process management system during the digital transformation of an enterprise. Information flows digitization of relevant processes is presented from the standpoint of a systematic approach based on functional and temporal decomposition. This approach to the digitized data management contributes to increasing the efficiency of designing a digital business and IT enterprise architecture.

**Keywords:** Digital economy · Architecture of system · Architecture description

## 1 Introduction

The "breakeven" of investments in the Business Process Management System (BPMS) arises with the beginning of the enterprise business processes system measurement with a view to their strategic improvement. This solution is especially important during the period of digital business transformation. Therefore, the task of developing analytical indicators of a business processes model is one of the most important and complex in the development of a BPMS.

The development of the Business Process Management System can be divided into three main stages [9]:

- "Organization" – the goal is to increase the coherence of actions of participants in business processes on the basis of rules for interaction and standardization of material and information flows;
- "Automation" – the goal is to reduce the complexity of performing routine operations and to increase the pace of work execution on the basis of the program prescription for unified actions;
- "Controlling and motivation" – the goal is to improve the quality of performance of tasks based on monitoring the parameters of the process, timely analysis of the causes and prevention of risks of their deviation from the target standards, as well as quality improvement based on the formation of motivational feedback and increasing the satisfaction of participants and consumers of the business-process.

© Springer Nature Switzerland AG 2020
D. G. Arseniev et al. (Eds.): CPS&C 2019, LNNS 95, pp. 701–708, 2020.
https://doi.org/10.1007/978-3-030-34983-7_69

In the general case, the "analytics" can be considered as a process and the result of a logical analysis of the subject domain by means of its sequential decomposition in terms of solving managerial tasks.

If the activity of the enterprise (hierarchical system of business processes) acts as the object of analysis, then in this case the analytics can be presented with a set of data for solving managerial problems at the strategic, tactical and operational levels (decomposition).

## 2  Business Process System

### 2.1  Levels of Control (Decomposition)

A competence center for working with analytics in the organizational structure can be the following posts [1]:

- CDO (Chief Data Officer) – Director for the development and implementation of data management strategies, standardization of management technologies and data management;
- CAO (Chief Analytics Officer) – director for strategic analysis and use of (large) data for analytical forecasting and commercial activities.

Information service working with analytics is provided through BI-systems (Business Intelligence) – information systems of business analysis, combining analytical, monitoring and reporting tools to support the adoption of managerial decisions. BI tasks fall into three categories:

- exploring and analyzing data;
- monitoring ongoing data flows through dashboards;
- communicating insights to others, both inside and outside a company [6].

Technical means for visualization of analytical information are performance dashboards, which can be related by content and purpose to the following types [5]:

- strategic – "control" the achievement of strategic goals;
- tactical – "analysis" of the implementation of processes and projects at the unit level;
- operational – "monitoring" of the basic operations at the level of employees.

In accordance with the methodology of functional modeling [11], the "Activity" of an enterprise is sequentially decomposed into "Processes" (business functions) and then to "Operations".

Levels of control (decomposition) by analogy with the parameters of the optimization task have the following functions:

- the strategic level: "Activity" is a kind of "abstract" model for planning the vector of the goal and determining the priorities for the development of the business system;
- the tactical level: "Processes" (business functions) are designed to ensure the balance of the system (based on functional budgets) in the development process, as

well as for timely response and adaptation to the critical changes in the external and internal environment of the enterprise;

- the level of "Operations" is a "real" management object (based on strategic objective function and tactical budget constraints), where parameters of actions are measured and improved.

## 2.2    The Level of Decomposition "Activity"

The level of decomposition "Activity" is characterized by the following objectives:

- business strategy is ensuring the sustainable operation of the enterprise based on the preservation and development of competitive advantages in the long-term perspective;
- center of investments (by financial structure of the enterprise) is an achievement of targeted profitability of investments on the basis of an investment portfolio formation taking into account profitability, risks and the volume of the enclosed capital;
- profit center (by financial structure of the enterprise) is an achievement of the target profit on the basis of positioning "price-quality".

The strategic vector of the business goal can be represented in the basic strategic coordinates of Michael Porter and the "price-quality" coordinates of the corresponding positioning strategy (as it was proposed in [8]):

- cost leadership (quantitative growth due to business scaling);
- differentiation (qualitative development based on the best value proposition);
- focusing (concentration on a relatively narrow segment of consumers on the basis of service leadership in relations with the client).

It is assumed that in a "stable" state of the market the "best quality" has the maximum price (and vice versa). The choice of the direction of the strategic target vector (in these coordinates) is a managerial decision that can be adopted based on the results of the SWOT analysis (Strengths, Weakness, Opportunities, Threats) of internal environment of the enterprise, while:

- factors of the external environment are characterized by the parameters of the market situation, such as: supply-demand ratio; stability and cyclical state; development trends; level of business activity; level of commercial risk; intensity of competition;
- factors of the internal environment are characterized by functional components of the business strategy of the enterprise, including: commodity-market (width and depth of product supply, product quality level, demand level and coverage ratio of sales markets); technological (level of progressiveness and efficiency of technologies, capacity of production equipment, duration of the production cycle); resource (level of reserves and quality of raw materials, flexibility of relationships with suppliers); social (efficiency of organizational and role structure, level of specialization and qualification of personnel, involvement in decision-making, division of values and corporate culture of behavior, payment and economic stimulation of

labor); financial and investment (level of current assets, investment directions and level of investments, sources of financing, repayment period and cost of borrowed capital); management (level of maturity of project management and business processes, status and effectiveness of strategic, tactical and operational management).

Based on the internal environment factors, the Key Success Factors (KFU) are determined, with the help of which the Vision of the enterprise is digitized and the objectives of the strategic map are formulated.

### 2.3  Model of the Theory of Constraints

Financial analytics (the data and the result of a decision) of strategic level management tasks can be represented by a Theory of Constraints (TOC) model, which includes the following parameters [4]:

- T – pass (marginal profit = income – variable costs);
- I – investment costs in stocks (raw materials and infrastructure capacities);
- OE – operating expenses for maintenance (regular);
- NP = (T – OE) – net profit (activities for the reporting period);
- ROI = (NP/I) – return on invested capital.

The business strategy can be formalized with help of the Strategic Goals and Portfolio of Development Projects, while the "Finance" prospect is a consequence of the "Clients" perspective, the reasons for which are the "Processes" and "Personnel" perspectives.

The causal logic of the strategic goal map model has the following TOC justification [10]:

- "profitable" strategic goals (financial perspective) are achieved by the level of the formed demand of the consumer segment and the capacity of a certain type of products (client perspective), while:
  - the formed demand and capacity of the trade channel ensure the target sales volume and duration of the commercial transaction period, which in turn determines the requirements for the capacity and duration of the operational cycle of business processes (process perspective);
  - the volume of sales depends on the target level of margins (margins), which determines the requirements for the quality of business processes (process perspective);
- "expenditure" strategic goals (financial perspective) are due to the volume of investments in the reserves of the capacity of the business process infrastructure and the regular costs of ensuring the readiness and quality of the business process, while:
  - requirements for ensuring the availability of infrastructure and the quality of business processes determine the requirements for the qualification of personnel (client perspective).

The second component of the business strategy is the Portfolio of development projects, when forming the composition and time horizon of its realization (sequence and priorities for achieving strategic goals), it is expedient to take into account 2 criteria [7]:

- "breakeven": long-term revenues from achieving goals - "Clients" should cover the costs of achieving the goals in the perspectives of "Processes" and "Personnel";
- "profitability" (in TOS-parameters): return on investment of the enterprise's activities "after" the implementation of the Project Portfolio should be no less than "before" its implementation.

Thus, the factors of the external and internal environment of the enterprise's activities can be assigned to and monitored by the strategic analysts (indicators) at the strategic level, and a qualitative (ordinal) scale of measurement of their aggregated values can be used.

### 2.4    The "Process" (Business Function) Level of Decomposition

Practically, the most difficult is the transition from the strategic to the operational level of management; it involves not only functional but also temporary decomposition. To implement this transition, the tactical level of budget management is used, which is a kind of "control amount" of the accuracy of translating the strategic objectives of "Activity" into the indicators of "Operations" through functional budgets [9].

In the general case, functional budgets do not fully correspond to the standard functional processes (business functions) of the enterprise, so a direct estimate of the cost of processes can be carried out based on the principles of the ABC-costing method (by operating the calculation of indirect costs) by means of quantitative and temporary drivers coefficients) of transferring the cost of labor and infrastructural resources to operations of business processes and even to enterprise products [2].

The level of decomposition "Process (business function)" is characterized by the following targets:

- functional strategy is creating opportunities to achieve goals of the business strategy based on the effective performance of business functions and the resulting inter-action in the system of business processes of the enterprise;
- the revenue center (the financial structure of an enterprise) is the achievement of target sales values based on stimulating the demand of consumer segments, ensuring the availability of products at a certain price and quantity;
- the center of costs (by the financial structure of an enterprise) is the achievement of target expenditures based on the effective use of resources and the reduction of losses.

Thus, the articles of income and expenditure of functional budgets and financial responsibility centers of an enterprise can be referred to the management analytics (indicators) of the tactical level.

## 2.5   The "Operation" Level of Decomposition

The following targets are typical for the "Operation" level of decomposition:

- the operational strategy is to create opportunities to achieve the goals of the business strategy based on the effective use of enterprise resources in the development, production and sales chain of competitive goods and services;
- effectiveness is to achieve target values by the quantity and quality of development, production and sales of goods and services;
- operativeness is to achieve target values for the duration of the operational cycle of development, production and sales of goods and services;
- resource intensity is to achieve target values for the cost of performing all operations necessary for the development, production and sales of goods and services.

The operational target vector can be represented in the "efficiency-productivity" coordinates, where:

1. Efficiency = Effectiveness/Resource intensity.
2. Productivity = Effectiveness/Operativeness.

The operational strategy is formulated with respect to the value stream, which is a set of completed docked actions that together create some products that have customer value for the client [3]. As part of the implementation of the operational strategy, it is advisable to present the value stream with a model at the level of the corresponding functional processes operations.

Operational indicators (quality) can be grouped by types:

- effectiveness (productiveness) is the compliance with the requirements and expectations of consumers;
- operativeness is the duration and timeliness of completion of operations;
- resource intensity is the cost of maintenance of readiness and performance of operations;
- controllability is the level of perfection of current management and improvement opportunities of an operational chain of value.

Indicators of the efficiency, resource intensity and manageability of business process operations can be assigned to and monitored by the operational analytics (indicators) of the operational level, and qualitative and quantitative scales of measurement of their values can be used.

# 3   Results and Discussion

An example of analytical indicators for different levels of management is shown in Table 1, while at the operational level the accomplishment of tasks for processes and project activities is assessed separately. Therefore, it is possible to identify the main indicators for levels to assess the effectiveness of the enterprise with the applied approach.

Table 1. Analytical indicators for different levels

| Measurement (group) | Strategy (KSF) | Tactics (CFR) | Processes | Projects |
|---|---|---|---|---|
| Effectiveness | Market share | Marginal profit | Defect level | Degree of customer requirements fulfillment |
| Operativeness | Duration of the operating cycle | Period of accounts receivable | Order lead time | Share of timely completed projects |
| Resource intensity | Investments in customers' training | Cost of production | The cost of automated workplaces | Degree of compliance with budget constraints |
| Efficiency | Return on investments | Profitability by customers | Sales conversion | Profitability of the project |
| Productivity | Workforce productivity | Debt turnover | The proportion of time losses in the working cycle | Payback period |
| Dynamics | Growth rates in sales | Moving average price | Dynamics of inconsistencies | Dynamics of innovations |

Business process management is one of the key tasks in a company development. Contrasting approaches to the description of business processes of companies is one of the main problems of modern business intelligence. The management system allows to use different approaches for different processes within one organization, not opposing them to each other. To identify the factors of systemic efficiency in further research, it is necessary to identify key success factors in each subsystem and determine indicators for their condition.

## 4 Conclusion

The business intelligence opportunities in the digital enterprise business process management system was reviewed. From the position of the system approach, digitized data management was considered on the basis of functional and time decomposition. The key performance indicators of the digital company were described and correlated with the main levels of the digital enterprise management.

**Acknowledgments.** The reported study was funded by RSCF according to the research project № 19-18-00452.

# References

1. Anderson, C.: Creating a Data-Driven Organization: Practical Advice From the Trenches, 1st edn. O'Reilly Media Inc., Sebastopol (2015)
2. Cheng, G.: Application research on the activity-based costing of value chain accounting. In: Proceedings of the Second International Conference on Future Information Technology and Management Engineering (FITME), Sanya, China, pp. 412–414. IEEE (2009). https://doi.org/10.1109/FITME.2009.109
3. Chopra, S.K., Kummamuru, S.: A value proposition framework adopting concepts from the field of cybernetics. In: Proceedings of the 2014 IEEE Conference on Norbert Wiener in the 21st Century (21CW), Boston, MA, USA, pp. 1–7. IEEE (2014). https://doi.org/10.1109/NORBERT.2014.6893944
4. Cox, J.F., Schleier, J.G. (eds.): Theory of Constraints Handbook. McGraw- Hill, New York (2010)
5. Eckerson, W.W.: Performance Dashboards: Measuring, Monitoring, and Managing your Business, 2nd edn. Wiley, New York (2011)
6. Fisher, D., Drucker, S., Czerwinski, M.: Business intelligence analytics [guest editors' introduction]. IEEE Comput. Graph. Appl. **34**, 22–24 (2014). https://doi.org/10.1109/MCG.2014.86
7. Hannach, D.E., Marghoubi, R., Dahchour, M.: Project portfolio management towards a new project prioritization process. In: Proceedings of the 2016 International Conference on Information Technology for Organizations Development (IT4OD), Fez, Morocco, pp. 1–8. IEEE (2016). https://doi.org/10.1109/IT4OD.2016.7479281
8. Ilin, I., Levina, A., Iliashenko, O.: Enterprise architecture approach to mining companies engineering. In: MATEC Web of Conferences, vol. 106 (2017). https://doi.org/10.1051/matecconf/201710608066
9. Ilin, I.V., Frolov, K.V., Lepekhin, A.A.: From business processes model of the company to software development: MDA business extension. In: Proceedings of the 29th International Business Information Management Association Conference - Education Excellence and Innovation Management through Vision 2020: From Regional Development Sustainability to Global Economic Growth, pp. 1157–1164 (2017)
10. Kumaran, S.R., Othman, M.S., Yusuf, L.M.: Applying theory of constraints (TOC) in business intelligence of higher education: a case study of postgraduates by research program. In: Proceedings of 2015 International Conference on Science in Information Technology (ICSI Tech), Yogyakarta, pp. 147–151. IEEE (2015). https://doi.org/10.1109/IC-SITech.2015.7407794
11. Li, Q., Chen, Y.-L.: Modeling and Analysis of Enterprise and Information Systems: From Requirements to Realization. Higher Education Press/Springer, Beijing/Berlin (2009)

# The Role of Cyber-Physical Systems in Human Societies: Systemic Principles of Organizational Management

Svetlana Ye. Shchepetova[1](✉) and Olga A. Burukina[2]

[1] Department of Systems Analysis in Economy at the Financial University under the Government of the Russian Federation, Moscow, Russia
seshch.fa.ru@gmail.com
[2] Department of Organisational Communication, University of Vaasa, Vaasa, Finland
burukinaolga@yahoo.com

**Abstract.** The world of today has been facing an incredible variety of opportunities along with unseen before challenges and potential dangers, the latter rooted both in the unpredictable capabilities of IT and artificial intelligence and untamed human rivalry. The research methodology is qualitative, including discourse and content analysis, literature review and conceptual modelling, along with hypothesis testing and theoretical generalization. The paper unfolds the authors' professional perception of the CPS integration in business organizations (primarily smart organizations) and highlights the challenges facing the mankind in the near future, including the danger of potential rivalry between humans and artificial intelligence within or beyond CPS. The paper offers the authors' conceptual model of organizational management in the 21st century, based on systems approach and the notion of the CPS integration in business organizations in the near future and a ground for further discussion within the contemporary systemic paradigm and socio-economic cybernetics as its integral part.

**Keywords:** Systemic paradigm · Industry 4.0 · Complex socio-economic and cyber-physical systems · Socio-economic cybernetics · Systemic principles of management and control

## 1 Introduction

At the present stage of technology development within the framework of Industry 4.0, the development of CPS is critical as part of national interests, and the rapid adoption of CPS in developed countries, which received government support, is explained very simply: CPS are critical for ensuring national security and further technological development in post-industrial economies [1].

The information age, which replaced the industrial age, marked a breakthrough in science and technology, but no matter how great the importance of IT is, new technologies, even the best breakthrough ones, do not produce a variety of goods and

© Springer Nature Switzerland AG 2020
D. G. Arseniev et al. (Eds.): CPS&C 2019, LNNS 95, pp. 709–715, 2020.
https://doi.org/10.1007/978-3-030-34983-7_70

services – all that forms the habitat of modern humans of a post-industrial era, while this routine is still people's responsibility, from which no innovations can free them.

However, scientific and technological progress is a complex phenomenon, fraught with ambiguous consequences, the negative impact of which the mankind has experienced in the 20th century [2]. The current stage of scientific and technological progress is characterized with unprecedented speed of technological development, at which it is difficult enough to predict goals that are essentially static [3]. In the process of developing a system, the original goal set during the creation of the system may change (or be completely replaced by another one), which leads to the phenomenon of shifted goals [4].

As a result of shifted goals, scientific and technological progress can create challenges, and not only technological, but also cognitive and psychological ones, which may result in both oppositions of technologies (for example, artificial intelligence) and humanity and oppositions of various parts of humanity to one another.

## 2  Methodology

Basing on systems approach to analysing the present development level and future prospects of organizational management, this pilot research has encompassed the following scope of methods: discourse analysis and literature review [5], content analysis, hypothesis testing, conceptual modelling [6] and theoretical generalization.

For the purposes of the research, discourse analysis and content analysis were applied to the thematic research discourse focused on CPS and the prospects of their future integration in the life of human societies [7].

Basing on the understanding of conceptualization by Sterman who believed that "conceptualization is at once the most important and least understood of all modelling activities," [8] the authors have used conceptual modelling, which is seen as formal description of core aspects of business organizations in their close interrelation with CPS in the near future.

So, this pilot research aimes at identifying the management domains where cyber-physical systems are very likely to be integrated in the organizational management under the condition of systems approach applied as the conceptual fundamental.

The goal of the research and research questions have prompted two hypotheses:

*Hypothesis 1* – CPS can play an increasingly important role in the human societies in the near future and a considerable part of organizational management (primarily the scope of production management) is most likely to become their full responsibility.

*Hypothesis 2* – systems approach is the only relevant methodology able to contribute most to sustainability attainment and further management.

# 3 Are STP a Panacea?

The information and knowledge industry did not aim at reducing production – it was alternatively transferred to third world countries, which turned into 'factory countries,' on which the population of post-industrial countries has grown increasingly dependent. The return of independence by the leading post-industrial powers, as Bloomer correctly indicates, can be possible at a qualitatively new level due to scientific and technical progress [9].

Nevertheless, the scope of CPS is much broader: cyber-physical systems allow people creating a qualitatively new health care, transport, energy and other key sectors of the economy. At the same time, it is necessary to recognize that innovation is not the only fundamental difference between CPS and embedded systems that preceded them, in spite of their external similarity [10].

Cyber-physical systems integrate the cybernetic principle, computer hardware and software technologies, and qualitatively new executive mechanisms built into their environment and able to perceive its changes, respond to them, learn and adapt [11]. The key element of CPS is the model used in the control system, and how it relates to reality depends on the efficiency of the cyber-physical system.

Cyber-physical systems ensure the joint operation of elements of cybernetic and physical spaces, integrating computing resources. Often, cyber-physical systems maintain real-world processes and provide operational control of objects on the Internet of things, allowing physical devices to perceive the environment and change it [12].

Smart Machines make it possible to optimize all production primarily by creating a unified system in which machines can exchange data between themselves in real time: the exchange between equipment located directly in the production area and in the logistics chain, including business systems, suppliers and consumers; transfer of information about their condition to service personnel. In this case, the production equipment, receiving information about the changed requirements, will itself make adjustments to the process [13].

The ability to "make people's lives better and simpler" with the help of these systems can be perfectly illustrated just by the example of 'smart' cities. Singapore has repeatedly been recognized by various researchers as the smartest of the 'smartest' cities on the planet, and its government goes even further working on a project of 'smart nation' (Smart Nation is the name of the urban development programme of Singapore) [14]. A number of startups jointly create solutions for Singapore that cover almost all areas of citizens' life – from law enforcement and automatic fixation of violations to the management of the transport system and energy resources, water supply and health care. And it yields results, for example, the traffic management system alone can save Singapore's drivers tens of thousands of hours every year [15].

Another version of the 'smart' approach to urban development is the city of Masdar in the UAE, which is being built near Abu Dhabi in the Rub el Khali desert. Masdar should become an 'eco-city,' which fully meets its needs with the help of renewable energy sources, completely recycles all waste and abandons traditional modes of transport in favour of public and personal autonomous transport. Naturally, the latest technologies, including cyber-physical systems, will be used to effectively manage resources and traffic in Masdar [16].

# 4  Phenomenon of Shifted Goals

Many systems are dynamic in their essence, and they can develop at such a rate that correct goal setting becomes very problematic. For correct goal-setting and avoidance of negative consequences of the phenomenon of shifting targets, the motivation of the animated system components is of great importance. Thus, a person, being a part of the system, acts to achieve the goals of the system, but in addition, s/he may have other personal goals that do not coincide or contradict the goals of the system (for example, earn money, climb the career ladder) [17].

The phenomenon of shifting goals is more characteristic of those elements that have a certain power. A person, as an element of the system, being at the head of management of one of the subsystems of the source system can use the capabilities of this system and his power to achieve his personal goals [18].

Thus, goals can overlap, polarize, form a dilemma, and then the choice of the direction of the system's development depends on the human factor. As regards to the challenges arising from the scientific and technical progress with cyber-physical systems, one of them may be dismissal of the so-called 'redundant professionals.'

# 5  Systemic Principles of Organizational Management

As operational processes, along with supply chains, procurement and finance will be soon maintained by computer-centered systems including artificial intelligence systems; the system of management engaging active participation of humans will mainly comprise the following kinds of management:

- Knowledge management (including management of innovations),
- Change management (including risk management),
- Quality management (which can be computerized to a large extent).

As companies are viewed as open interconnected purposive systems consisting of numerous business sections/departments and/or as complex systems of decision-making process [19], the most appropriate concept used as a basis for their analysis aimed at further effective management is systems management [20].

The authors consider the systems approach to management to be a scenario that plays a very important role in creating coordinative relations between all related business systems, as systems management is believed to be an approach closest to reality. The systems approach to management is the key to coordinate all the processes in a large company, and define the importance of individual procedures in the firm [21].

Therefore, systems approach provides really enormous possibilities for unifying management theory. According to Barnard – the pioneer of applying systems approach to management, systems approach allows keeping a balance between conflicting forces and events, with responsible leadership increasing the executives' effectiveness [22].

Basing on the systems approach to organizational management, the following model was built to illustrate the authors' vision of man-engaged organizational management of the near future (Fig. 1 below).

**Fig. 1.** Conceptual model of systemic principles of management in the 21st century

The shared domains provide optimal conditions for the development of the following kinds of management:

1 – Knowledge management

2 – Innovation management

3 – Project management

4 – Strategic management

5 – Sustainability management, as sustainability can only be achieved as a result of joint parallel efforts invested in every of the highlighted domains of organizational management.

The proposed model of organizational management built upon systemic principles highlights interrelations between the management domains that will face a development boost in the 21st century and will still require human engagement and their close interrelation with CPS.

## 6   Conclusions and Discussion

Summing up the above considerations, the authors have come to the following conclusions:

1. CPS will keep their development and get more deeply integrated in the life of human societies. However, cyber-physical systems should be created not only and not so much for the purpose of external control, management and coercion, but for

mutual assistance in the development and solving socio-economic problems [23]. Human life of contemporary people, particularly in post-industrial economies, is full of many soft factors that cannot be measured and unambiguously interpreted; that is why the highly complicated diverse human life cannot and should not be fully digitized.

2. The hope that technological progress alone will solve social and economic problems of the mankind is a mere utopia. The NTP unfolding before us has reached an unprecedented speed and level; however, the current essential problems remain unsolved. At the same time, more and more restrictions and external control measures are introduced into people's life [24]. The two tendencies combined may in the long run lead to any sort of cyber-slavery, if not enough attention is paid to socio-economic and spiritual aspects of daily life. The main principle to guide the development of technologies in the 21st century and beyond can be formulated in this way: 'Technology was made for man, not man for technology.'

3. The authors' concept of organizational management applied to business organizations rests on the perception of systems approach as the most adequate concept, closest to reality and the understanding of business organizations as highly complicated open interconnected purposive systems and the future smart communities [25, 26].

4. The authors' conceptual model of systemic principles of organizational management in the 21st century has highlighted the prominent domains of organizational management, which presume human engagement along with integration of CPS.

5. Any technology can be used to harm or benefit, as it depends on ideas and intentions of humans, their interests, goals and aspirations. However, no matter how sophisticated technologies can be, opponents or competitors will always seek for tools and opportunities to "hack" them for the benefit of free riders. Besides, when people act under duress, they tend to achieve external compliance only laying ground for shifted goals and functions.

Theoretically, basing on literature review and discourse analysis both hypotheses have been proved, as business organizations will be increasingly turning into smart organisations, which tend to apply as much technology as possible, including artificial intelligence and cyber-physical systems based on it. The proposed conceptual model of a CPS-driven management system of organizational management needs further theoretical development and practical testing, as well as the formulated hypothesis need practical testing and substantiation.

**Acknowledgement.** The article has partially summed up the results of the research entitled "System attributes of the digital economy as an environment for the development of innovative processes in Russia" funded on the state order of the Financial University under the Government of the Russian Federation, Moscow, Russia.

# References

1. Hellinger, A., Seeger, H., acatech (eds.) Cyber-Physical Systems: Driving Force for Innovation in Mobility, Health, Energy and Production. acatech Position Paper, National Academy of Science and Engineering (2011)
2. Tegmark, M.: Life 3.0: Being Human in the Age of Artificial Intelligence, p. 384. Vintage, New York (2018). Reprint edition

3. Galimberti, U.: Man in the age of technology. J. Anal. Psychol. **54**(1), 3–17 (2009)
4. Ayala, F.J.: Evolution vs. creationism. Hist. Philos. Life Sci. **28**(1), 71–82 (2006)
5. Piccarozzi, M., Aquilani, B., Gatti, C.: Industry 4.0 in management studies: a systematic literature review in management studies: a systematic literature review. Sustainability **10** (10), 3821 (2018)
6. Quadri, I., Bagnato, A., Brosse, E., Sadovykh, A.: Modeling methodologies for cyber-physical systems: research field study on inherent and future challenges. Ada User J. **36**(4), 246–252 (2015)
7. Leonhard, G.: Technology vs. Humanity: The Coming Clash Between Man and Machine. Series: FutureScapes (Book 2), p. 208. Fast Future Publishing (2016)
8. Sterman, J.D.: Testing behavioural simulation models by direct experiment. Discussion Memorandum, MIT System Dynamics Group Literature Collection, Cambridge (1986)
9. Bloomer, P.H.: Technology and Human Rights. Business and Human Rights Resource Centre. https://www.business-humanrights.org/en/technology-and-human-rights
10. Sanfelice, R.G.: Analysis and design of cyber-physical systems. a hybrid control systems approach. In: Rawat, D., Rodrigues, J., Stojmenovic, I. (eds.) Cyber-Physical Systems: From Theory to Practice. CRC Press (2016)
11. Sanislav, T., Miclea, L.: Cyber-physical systems – concept, challenges and research areas. Control Eng. Appl. Inform. **14**(2), 28–33 (2012)
12. Cyber-Physical Systems Summit Report (2008). http://varma.ece.cmu.edu/summit/
13. Belfiore, M.: Embedded technologies: power from the people. In: Smithsonian Magazine, August 2010. https://www.smithsonianmag.com/science-nature/embedded-technologies-power-from-the-people-1090564/
14. Xia, F., Ma, J.: Building smart communities with cyber-physical systems (2011). https://arxiv.org/ftp/arxiv/papers/1201/1201.0216.pdf
15. Gonzalez, M.C., Hidalgo, C.A., Barabasi, A.-L.: Understanding individual human mobility patterns. Nature **453**(5), 779–782 (2008)
16. Cyber-physical systems in the contemporary world. Toshiba blog. Habr.com, 1 January 2019. https://habr.com/ru/company/toshibarus/blog/438262/
17. Jain, R., Singh, V., Gao, M.: Social life networks. In: Proceedings of the Workshop on Social Media Engagement (2011)
18. National Science Foundation: Cyber-Physical Systems (2011). http://www.nsf.gov/funding/pgm_summ.jsp?pims_id=503286
19. Simon, H.A.: Administrative Behavior (4th expanded edition; first edition 1947). The Free Press, New York (1997)
20. Veyrat, P.: Technology, People and Processes in knowledge management. HEFLO.com (2016). https://www.heflo.com/blog/bpm/technology-people-and-processes/
21. Zhonga, R.Y., Xua, X., Klotzb, E., Newmanc, S.T.: Intelligent manufacturing in the context of industry 4.0: a review. Engineering **3**(5), 616–630 (2017)
22. Gabor, A., Mahoney, J.T.: Chester Barnard and the Systems Approach to Nurturing Organizations, January 2010. http://www.business.illinois.edu/Working_Papers/papers/10-0102.pdf
23. Acemoglu, D., Restrepo, P.: The race between man and machine: implications of technology for growth, factor shares, and employment. Am. Econ. Rev. **108**(6), 1488–1542 (2018)
24. Tech Companies Are Addicting People! But Should They Stop? Nir & Far. https://www.nirandfar.com/tech-companies-addicting-people-stop/
25. Schuh, G., Kramer, L.: Cybernetic approach for controlling technology management activities. In: 48th CIRP Conference on Manufacturing Systems – CIRP CMS 2015 (2015). Procedia CIRP **41**, 437–442 (2016)
26. Saucedo, J., Lara, M., Marmolejo, J., Salais, T., Vasant, P.: Industry 4.0 framework for management and operations: a review. J. Ambient Intell. Humaniz. Comput. **9**(3), 789–801 (2017)

# Assessment of Success Measures and Challenges of the Agile Project Management

Tatiana Kovaleva[✉]

Peter the Great St. Petersburg Polytechnic University, St. Petersburg, Russia
kovaleva_tg@spbstu.ru

**Abstract.** Since every project is unique, there is no ideal project management system suitable for each type of projects. However, during the existence of the project management, many effective approaches, methods and standards have been created. One of these methodologies is the agile one, which was originally created for software development. That is why this methodology is most often used in the IT industry. However, this does not prevent the usage of the agile method (hereinafter Afile) by many non-technical teams. Non-IT companies quickly discovered the benefits of using flexible thinking and some Agile practices that can help businesses achieve better performance. The most important advantage of Agile is its flexibility and adaptability. It can be adapted to almost any conditions and processes of the organization. That is what determines its current popularity and explains why so many systems for different areas have been created on its basis. At the same time, the results of transformation do not always coincide with the expectations. This paper sets out the success measures with the Agile projects. The challenges that companies most often encounter are evaluated and the dependence of the trend in the popularity of Agile techniques on the success of the methodology implementation is studied based on the large-scale surveys of Agile at companies in Russia and worldwide.

**Keywords:** Agile method · Project management · Software engineering

## 1 Introduction

A project is a sequence of unique, complex, and connected activities that have one goal and purpose and that must be completed by a specific time, within certain budget, and according to a specification [1, 2]. A project comprises a number of activities that must be completed in some specified order, or sequence. An activity is a defined chunk of work [3].

There is no perfect project management system suitable for each type of projects. Also, there is no system that would suit every manager and would be convenient for all team members. However, during the existence of project management, many effective approaches, methods and standards that can be adopted have been created.

It can be quite difficult to set up a team. Even if it is established, there is a risk that all efforts would be in vain, because the requirements for the necessary result often change. However, it is possible to significantly simplify the work on the project and

© Springer Nature Switzerland AG 2020
D. G. Arseniev et al. (Eds.): CPS&C 2019, LNNS 95, pp. 716–725, 2020.
https://doi.org/10.1007/978-3-030-34983-7_71

learn how to manage it, thus increasing the efficiency of the team, using the flexible project management system called Agile.

The most important advantage of Agile is its flexibility and adaptability. It can adapt to almost any conditions and processes of the organization. This is what determines its current popularity and the high number of systems in different areas that have been created on its basis.

One of the principles of Agile: "Responding to Change Over Following a Plan" [4]. Rapid and relatively painless response to changes is the reason why many large companies are trying to make their processes more flexible.

At the same time, the results of transformation do not always coincide with the expectations. The aim of this study is to reveal challenges that companies most often encounter with, set out the success measures with the Agile projects, and carry out a comparison of Agile expectations and achievements in Russia and worldwide.

## 2  Research Method

The paper is based on several surveys on Agile that took place in both Russian and global companies.

The 13th annual State of Agile survey was carried out by CollabNet VersionOne Company which specialises in software development and provides delivery solutions [5]. The most recent survey was conducted between August and December 2018. The survey collected responses from 1319 individuals who were working in a broad range of industries in the global software development community. The location of organizations where the respondents were from is presented in Fig. 1.

The annual large-scale Agile study in Russian organizations was conducted from March 18 to October 13, 2018 [6]. The survey involved 1228 people from more than 80 cities including: 53% from Moscow; 17% from Saint Petersburg; and 3.5% from Novosibirsk and Ekaterinburg * (each).

Both questionnaires were given to software practitioners, such as business analyst, product manager, trainer, development team member (architect, developer, QA, tester, UI or UX designer), project manager, development leader (vice president, director, manager), ScrumMaster, DevOps, C-Level executive.

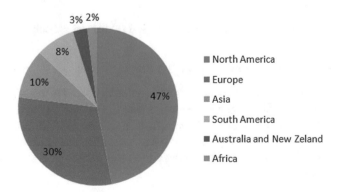

**Fig. 1.** The residence of global Agile report respondents [5]

# 3   Results and Discussion

## 3.1   Agile Methodologies Used in Companies

Among the Agile approaches, Scrum and Scrum/XP Hybrid (64%) continue to be the most common Agile methodologies used by respondents' organizations both in Russia (see Fig. 2a) and globally (see Fig. 2b).

In Russia, the share of application of hybrid Agile approaches (combination of existing approaches) amounts to 28%, which is 2 times higher than the world average.

According to the global survey, notable changes in Agile techniques and practices that respondents said their organization used were the Release planning (57% this year compared to 67% last year) and Dedicated customer/product owner (57% this year compared to 63% last year).

The top 5 Agile techniques used in the world are:

- Daily standup – 86%
- Sprint/iteration planning – 80%
- Retrospectives – 80%
- Sprint/iteration review – 80%
- Short iterations – 67%.

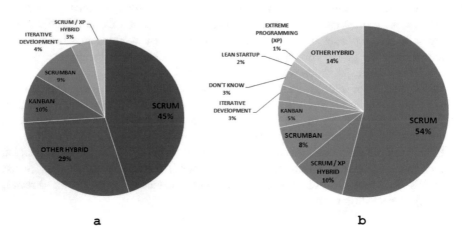

**Fig. 2.** Agile methodologies used in Russia (a) [6] and globally (b) [5]

## 3.2   Goals and Achievements of Agile Transformations

The results of the transformation do not always coincide with the expectations. IT companies are more confident than others to achieve their goals. In telecommunication; the difference between goals and achievements is the biggest so far, which is not surprising, because this industry has moved to Agile just recently. Goals and achievements of agile transformations in three different sectors of Russia are presented in Tables 1, 2 and 3 [6].

According to the survey replies, it has become much easier to manage distributed teams in telecommunication companies. In finance, business/IT alignment is a key goal that is successfully achieved.

Increasing employee motivation after switching to Agile exceeds expectations regardless the industry or the company's experience in Agile (by 14–26%). Product quality improvement is approximately the same in all industries: 41–46% of respondents said that the quality had increased significantly. However, representatives of such industries as IT and finance, where Agile has been used for a long time, impose high quality requirements (about 55%), so that their expectations for this indicator were not met.

With the growth of experience in Agile, companies begin to set other goals. In companies that have already implemented Agile, the popularity of the goal "to improve the quality of products" is higher by 48%, to increase the predictability of supply – by 56%, to improve the engineering culture – by 111%.

Regardless of their experience in Agile, companies improve the transparency of project management (exceeding the expectations from 16% to 28%).

The global data on Agile goals and achievements are about the same:

- In the list of actually achieved benefits, the first 2 places are occupied by the same indicators as in Russia: the Ability to manage changing priorities and Project visibility.
- Software quality is also a significant goal (the fifth most popular in the world), but this goal, as well as in Russia, is quite rarely implemented close to the expectations. Only 47% of responds in the world and 45% in Russia have managed to do that.
- The increase in speed on average around the world is observed more often (62% against 55% in Russia), but – unlike Russia – it is much lower than the expectations.
- The goal of accelerating software delivery in the world is set by 74% of respondents, while in Russia by 56%.

**Table 1.** Goals and achievements of Agile transformations in Software development companies [6]

| Top 5 goals | Top 5 achievements |
|---|---|
| Enhance ability to manage changing priorities | Enhance ability to manage changing priorities |
| Enhance software quality | Improve project visibility |
| Accelerate product delivery | Enhance delivery predictability |
| Improve project visibility | Accelerate product delivery |
| Enhance delivery predictability | Improve team morale |

**Table 2.** Goals and achievements of Agile transformations in Finance companies [6]

| Top 5 goals | Top 5 achievements |
|---|---|
| Accelerate product delivery | Improve business/IT alignment |
| Improve business/IT alignment | Improve project visibility |
| Enhance ability to manage changing priorities | Accelerate product delivery |
| Enhance product quality | Enhance ability to manage changing priorities |
| Increase productivity | Improve team morale |

**Table 3.** Goals and achievements of Agile transformations in Telecommunication companies [6]

| Top 5 goals | Top 5 achievements |
| --- | --- |
| Accelerate product delivery | Improve project visibility |
| Increase productivity | Enhance ability to manage changing priorities |
| Enhance ability to manage changing priorities | Better manage distributed teams |
| Improve project visibility | Accelerate product delivery |
| Improve business/IT alignment | Increase productivity |

The increase in accelerating product delivery, which in Russian Agile seems relatively low, is due to the fact that the average experience of Agile organizations in Russia currently is 2.6 years, while in the world it is about 4 years.

Next, we should consider only those respondents whose companies have already implemented Agile. The main business result of Agile – the speed growth - is realized on average by 2/3 of companies. However, the result depends, among other things, on the approach chosen (see Fig. 3).

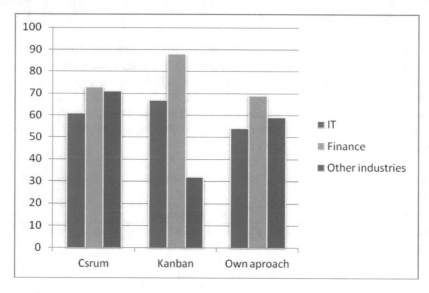

**Fig. 3.** Acceleration product delivery by industry and Agile approach [6]

Among others, Agile approaches include scrum and kanban. Scrum approach divides the work process into equal sprints – usually, these are periods from a week to a month, depending on the project and team. Before the start, tasks for this sprint are formulated, and at the end the results are discussed, and then the team starts a new sprint. It is very convenient to compare sprints with each other, which allows to manage work efficiency [7].

The main goal of kanban approach is to balance different specialists within the team. There are no roles of product owner and scrum master in kanban: the whole team is equal. The business process is not divided into universal sprints, but by the stages of implementation of specific tasks: "Planned", "Developed", "Tested", "Completed", etc. [8].

The success of Agile and, especially, Kanban in finance can be related to the fact that large banks before the introduction of these approaches had very bad "Time to Market" indicators, which were relatively easy to improve with the help of Agile and Kanban.

More rarely, growth in speed is observed by team members who have begun to work on Kanban in non-IT industries. This is due to a combination of several factors: these industries have recently begun to use flexible approaches, including Kanban, and the effect of Kanban usually grows with the increasing number of practices, and that takes time.

IT companies, on the other hand, are on average the most mature in terms of Agile. For such companies, it is characteristic and justified to develop their own Agile-approach (own processes based on a combination of Scrum, Kanban, etc.). A proprietary approach in IT companies correlates noticeably with their maturity, so it is not surprising that most often (69%) growth rates are noted by teams practicing their approaches.

Scrum more often than other approaches leads to an increase in speed (69% observe a noticeable increase), as well as an improvement in all 7 of the most popular performance indicators.

# 4    Estimation of Agile Achievements for the Next Year

The method of least squares is often used to generate estimators and other statistics in regression analysis. In order to predict the Agile achievements for the next year, least squares regression was used. The goal of the least squares method is to minimize the common quadratic error between the values of y and $\hat{y}$. To determine for each point, the error $\hat{y}$, the least squares method minimizes:

$$\sum_{i=1}^{n} (y_i - \hat{y}_i)^2,$$

where $n$ is the number of ordered pairs around the line that most closely matches the data.

In order to make estimations, additional data was taken from the last 5 years of global Agile surveys [5, 9–12]. "Years" will be an independent variable, and "Percentage of responses" – dependent variable that should be predicted.

Using the method of least squares, determine the equation that best matches the data by calculating the value of a, the segment on the axis y, and the slope of the line b:

$$b = \frac{n \sum\limits_{i=1}^{n} x_i y_i - \left(\sum\limits_{i=1}^{n} x_i\right)\left(\sum\limits_{i=1}^{n} y_i\right)}{n \sum\limits_{i=1}^{n} (x_i)^2 - \left(\sum\limits_{i=1}^{n} x_i\right)^2},$$

$$a = y_{av} - b x_{av},$$

where $x_{av}$ is the average value of $x$, an independent variable; $y_{av}$ is the average value of $y$, an independent variable.

The result of the estimations is shown in Table 4. The forecast for 2019 shows that the tendency on getting less percentage for the achievements will keep on.

**Table 4.** Agile achievements worldwide, % [5, 9–12]

| Agile achievements | 2014 | 2015 | 2016 | 2017 | 2018 | 2019 (expected) |
|---|---|---|---|---|---|---|
| Ability to manage changing priorities | 87 | 87 | 88 | 71 | 69 | **64,8** |
| Project visibility | 82 | 84 | 83 | 66 | 65 | **60,4** |
| Business/IT alignment | 75 | 77 | 76 | 65 | 64 | **64,8** |
| Team morale | 79 | 81 | 81 | 62 | 64 | **58,7** |
| Delivery speed/time to market | 77 | 80 | 81 | 61 | 63 | **58,3** |
| Increased team productivity | 84 | 85 | 83 | 61 | 61 | **53,8** |
| Project predictability | 79 | 81 | 75 | 49 | 52 | **41,4** |
| Project risk reduction | 76 | 78 | 74 | 47 | 50 | **40,1** |
| Engineering discipline | 72 | 73 | 68 | 43 | 42 | **32,6** |
| Managing distributed teams | 59 | 62 | 61 | 40 | 39 | **33,6** |
| Software maintainability | 68 | 70 | 64 | 33 | 34 | **22,3** |

**Table 5** Agile achievements in Russia, % [6, 13]

| Agile achievements | 2017 | 2018 |
|---|---|---|
| Ability to manage changing priorities | 52 | 72 |
| Project visibility | 54 | 73 |
| Business/IT alignment | 43 | 54 |
| Team morale | 44 | 57 |
| Delivery speed/time to market | 42 | 55 |
| Increased team productivity | 39 | 52 |
| Project predictability | 37 | 50 |
| Project risk reduction | 24 | 37 |
| Engineering discipline | 36 | 41 |
| Managing distributed teams | 26 | 35 |
| Software maintainability | 31 | 37 |

Unlike world statistics, companies in Russia are getting higher results on Agile achievements (see Table 5) [6, 13].

Due to the insufficient amount of data, it is impossible to predict next year achievements in Russia. Though, already now, Russia is ahead of the world on 3 parameters and, according to the estimated forecast, this number will be at least doubled.

# 5   Challenges Experienced While Adopting Agile

Answers of respondents to the question of what difficulties they or their companies faced during the implementation of Agile showed significant differences between companies of different sizes (see Fig. 4) [5, 6]. There were no major differences between industries.

Regardless of the industry and size of a company, the first place among the difficulties is the "lack of experience", which is inevitable at this stage of Agile development in Russia.

"Organizational culture at odds with agile values" acts as the most noticeable obstacle for large companies whose employees noted this in 48% of cases, especially in banks (58%). Also, it is in large companies that the factors "Pervasiveness of traditional development methods" (37%), "Regulatory compliance or government issue" (35%) and "Weak support from the management of the company" (28%) often interfere.

At medium and small-size organizations, employees refer to "Inconsistent processes and practices across teams" as a key barrier (39%). In small businesses, the implementation of Agile is seriously hampered mostly by this problem, as well as insufficient training (42%) and lack of experience that is natural for all the companies.

Thus, the key success factors for small companies in most cases are:

1. Implementation Agile approaches in full and not partially.
2. Investment of resources in Agile employee trainings.

Meanwhile, large companies on the way to Agile have to solve difficult tasks to change corporate culture and simplify various processes.

The median percentage of answers to all questions shows that small companies (median 16%) have fewer difficulties than medium companies (median 26%) and medium companies have fewer difficulties than large companies (median 29%).

Most of the difficulties in implementing Agile are noted by employees of organizations ranging in size from 500 to 2000 people (for them, the median is 32%, compared to 28% for the largest companies with a size of over 20000 employees).

According to the 13th Global Agile Report [5], the top three responses cited as challenges to adopting and scaling agile practices indicate that internal culture remains an obstacle for the success in many organizations.

In order to overcome challenges, the companies should use the following tips:

- Internal Agile coaches
- Executive sponsorship
- Company-provided trainings
- Consistence practices and processes across teams
- Implementation of a common tool across the team.

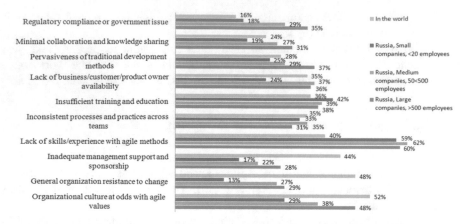

**Fig. 4.** Challenges experienced adopting Agile [5, 6]

## 6  Conclusion

95% of respondents reported on at least some of their agile projects having been successful with 48% reporting that most or all of their agile projects were successful. Organizations indicated the three measures of success for Agile transformations that have remained the same over the last few years (Customer/user satisfaction, Business value and On-time delivery).

Agile method is usually expected to accelerate the supply and delivery of products to the market. In Russia, this goal was achieved in 66% of study participants who reported that their organization had already implemented Agile.

Two of the most noticeable improvements that employees have seen since the transition to Agile are the transparency of work and the management of frequently changing priorities. The reason for this lies in the process itself based on Scrum, Kanban or other Agile approaches: all these approaches involve both visualization of work and iteration, the results of which make it easier to change priorities than in a non-flexible process.

Companies in Russia are getting higher results on Agile achievements distinct from the world statistics. According to the estimated forecast, next year, Russia should be ahead of the rest global companies on at least 6 parameters: ability to manage changing priorities, project visibility, project predictability, engineering discipline, managing distributed teams, software maintainability.

Internal culture remains an obstacle for success in many organizations. In order to overcome that, five tips were suggested based on the best practices in companies.

## References

1. Wysocki, R.K.: Effective Project Management, Traditional, Agile, Extreme, 6th edn. Wiley, Hoboken (2011)

2. Stare, A.: Agile project management. A future approach to the management of projects? Dyn. Relat. Manag. J. **2**(1), 21 (2013)
3. Guleria, P.G., Sharma, V., Arora, M.: Development and usage of software as a service for a cloud and non-cloud based environment - an empirical study. Int. J. Cloud Comput. Serv. Sci. (IJ-CLOSER) **2**(1), 50–58 (2013)
4. Fowler, M., Highsmith, J.: The Agile manifesto. Softw. Dev. **9**(8), 28–35 (2001)
5. 13th annual state of Agile report. CollabNet VersionOne (2019)
6. Agile v Rossii 2018: Otchet o ezhegodnov issledovanii. [Adgile in Russia 2018. Annual survey report] ScrumTrek (2018). (Russian)
7. Schwaber, K.: Agile Project Management with Scrum, pp. 8–20. Microsoft Press, Redmond (2004)
8. Kniberg, H., Skarin, M.: Kanban and Scrum - Making the Most of Both, pp. 10–30. Lulu.com, Morrisville (2010)
9. 12th annual state of Agile report. CollabNet VersionOne (2018)
10. 11th annual state of Agile report. CollabNet VersionOne (2017)
11. 10th annual state of Agile report. CollabNet VersionOne (2016)
12. 9th annual state of Agile report. CollabNet VersionOne (2015)
13. Agile v Rossii 2017: Otchet o ezhegodnov issledovanii. [Adgile in Russia 2017. Annual survey report]. Отчет о ежегодном исследовании. ScrumTrek (2017). (Russian)

# The "Digital Economy of the Russian Federation" Programme: An Analysis Based on Architecture Description

Alexander Danchul[✉]

Moscow City University of Management of the Moscow Government,
Sretenka Street, 28, 107045 Moscow, Russia
adan5l@yandex.ru

**Abstract.** A conceptual model of the system architecture description extended in comparison with the standard ISO/IEC/IEEE 42010:2011 is considered. Based on the analysis of the text of the "Digital economy of the Russian Federation" Programme, elements of its architectural description associated with the relations of different concepts are constructed. An architectural configurator for the description of the programme is proposed. The main relations, tables of which should be compiled and analyzed when using this configurator, are specified.

**Keywords:** Digital economy · Architecture of system · Architecture description

## 1 Architecture Description

A conceptual model of a software-intensive system architecture description was introduced in ISO/IEC/IEEE 42010:2011 [1]. The basic idea of this conceptual model is selection of a list of stakeholders with specific concerns in relation with a system-of-interest, drawn up in the form of viewpoints, established agreements necessary for the further creation, interpretation and use of architectural representations (view) of the system. In terms of the systems theory, this basic idea means that active complex systems with several subjects of activity with a shared object of activity are considered.

Sets of certain types of models (model kind) correspond to different stakeholders' points of views on the system. Sets of models of these types of components form a representation of the system (view) corresponding to this point of view, reflecting a certain set of system properties and relationships.

Possibilities and limitations on the use of this conceptual model of architecture description for the description of various architecture frameworks were considered in [2]. To remove these limitations, in particular, to support the possibility of describing multidimensional architectures in [3, 4], it was proposed to introduce a new group of architectural description elements into the conceptual model, the names of which and the names of their relations, are shown in italics in Fig. 1.

© Springer Nature Switzerland AG 2020
D. G. Arseniev et al. (Eds.): CPS&C 2019, LNNS 95, pp. 726–734, 2020.
https://doi.org/10.1007/978-3-030-34983-7_72

The main one is the concept of an "architectural frame" as part of the architecture description, reflecting the architecture of the system in terms of its specific subsystems described at different levels and their coordination, as well as structuring concerns of stakeholders in accordance with their programme.

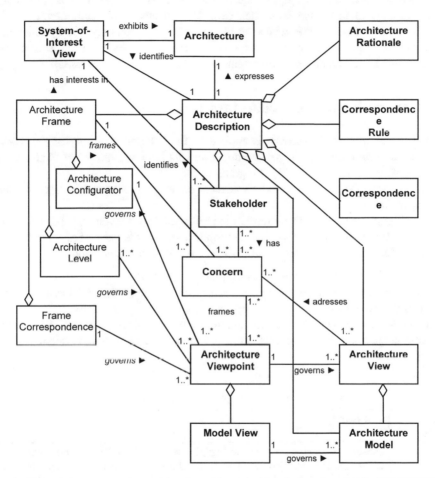

**Fig. 1.** Extended conceptual model of architecture description based on ISO/IEC/IEEE 42010:2011

The architectural frame includes three elements of the architectural description that governs the architecture viewpoint:

- the "architecture configurator" class which defines sets of aspects used by different stakeholders;
- the "architecture level" class which defines a set of levels of descriptions used for various aspects;
- the "frame correspondence" class is derived from the "correspondence" class and it reflects the relationship between the elements of the architectural framework.

The first dimension is based on the concept configurator used in the systems theory, which means a list of aspects (sets of properties) of the system describing it. In particular, if an aspect is understood as a set of properties of the system corresponding a viewpoint on it of certain specialists or groups (stakeholders), then the set of models reflecting the system in this aspect can be identified with the view. It is important to note that all participants can have their own descriptions of the configurator.

The second dimension is related to the regularity of hierarchy: the system can be described in more detail at a lower level of hierarchy. As a result, sets of two-dimensional hierarchical aspect representations (subsystems) are formed.

## 2    Analysis Based on Architecture Description

The above architectural approach is applied to the "Digital economy of the Russian Federation" Programme [5] (Programme 1) and its later version – the National Programme with the same name [6] (Programme 2).

The aims of Programme 1, and, therefore, concerns of its executors (stakeholders), are shown in Table 1.

The digital economy in Programme 1 is represented by 3 levels (L1, L2, L3), which from the standpoint of architectural description define the part of the Architecture Configurator – a set of top-level aspects used by different participants:

- L1 – markets and economic sectors (fields of activity);
- L2 – platforms and technologies, where competencies for the development of markets and sectors of the economy (spheres of activity) are formed;
- L3 – the environment that creates conditions for the development of platforms and technologies and effective interaction of market players and economic sectors (spheres of activity).

**Table 1.** Aims of the digital economy in Programme 1

| Aim code | Aim description |
|---|---|
| A1 | Creation of an ecosystem of the digital economy of the Russian Federation in which data in digital form is a key factor of production in all spheres of social and economic activity and in which effective interaction is provided, including cross-border, business, scientific and educational communities, the state and citizens |
| A2 | Creation of the necessary and sufficient conditions of institutional and infrastructural character, removal of existing obstacles and restrictions for creation and (or) development of hi-tech businesses and prevention of the emergence of new obstacles and restrictions, both in traditional branches of economy and in new branches and hi-tech markets |
| A3 | Improving competitiveness in the global market of both individual sectors of the Russian economy and the economy as a whole |

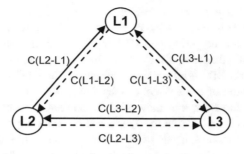

**Fig. 2.** The relationship between top-level aspects of the Programme 1

Correspondence between these elements of the architectural framework, partially defined in the Programme 1, is shown in Fig. 2 solid arrows.

C(L2-L1) – L2 forms competencies for the development of the Level 1.

C(L3-L1) – L3 creates conditions for effective interaction of the Level 1 subjects.

C(L3-L2) – L3 creates conditions for the development of the Level 2

Inverse relations C(L1-L2), C(L1-L3), C(L2-L3) in Programme 1 are not explicitly explained. Since Programme 1 was focused on the two lower levels of the digital economy (L2 and L3), the first two inverse relations are not mandatory for its presentation.

But a lack of the relation C(L2-L3) that is a generalized description of the reverse impact of platforms and technologies on the environment that creates conditions for their development is a significant drawback of Programme 1.

As can be seen from Table 2, there is no one-to-one correspondence of the aims of Programme 1 and its aspects (the levels of the market economy at which they are achieved). Given that Programme 1 has focused on the near future only at the levels L2 and L3, the aim A3 (improving competitiveness in the global market) can only be seen as prospective or indirect.

**Table 2.** The correspondence the levels of the digital economy (aspects) and the aims of Programme 1

| Aims, aspects | A1 | A2 | A3 |
|---|---|---|---|
| L1 | – | – | + |
| L2 | + | – | – |
| L3 | + | + | – |

The above description refers to the top (first) architectural level of aspect descriptions. Let's move on to the second architectural level.

At the first level of the digital economy L1, the Programme 1 was to be supplemented with new sections related to the implementation of certain applied areas in the sectors of the economy (spheres of activity), primarily in the following areas (applied directions, AD): AD1 – healthcare, AD2 – smart cities, AD3 – public administration.

At the levels L2 and L3 of the digital economy, the Programme 1 defines and specifies the goals and objectives in the five basic directions (BD) of the development of the digital economy for the period up to 2024: BD1 – normative and legal regulation, BD2 – personnel and education, BD3 – formation of research competences and technological groundwork, BD4 – information infrastructure, BD5 – information security.

An explicit division of these directions by levels L2 and L3 in the Programme 1 could not be found. However, the analysis of the content of their goals and objectives, as well as indicators of the Programme 1, made it possible to present it in the Table 3.

The results of a similar analysis of the correspondence the aims of Programme 1 and the goals of the basic directions are presented in Table 4. The aim A3 in Table 4 is not shown, since according to Table 2 it is achieved at the level L1, not related to the basic directions.

**Table 3.** The correspondence the levels of the digital economy and the basic directions of the Programme 1

| Directions, aspects | BD1 | BD2 | BD3 | BD4 | BD5 |
|---|---|---|---|---|---|
| L2 | – | – | + | + | + |
| L3 | + | + | + | – | + |

**Table 4.** The correspondence the aims of the Programme 1 and the goals of the basic directions of the Programme 1

| Goals of the basic directions aims | GBD1 | GBD2 | GBD3 | GBD4 | GBD5 |
|---|---|---|---|---|---|
| A1 | – | – | + | – | – |
| A2 | + | + | + | + | + |

Thus, the formed system of goals of the Programme cannot be represented as a tree of goals (see Fig. 3).

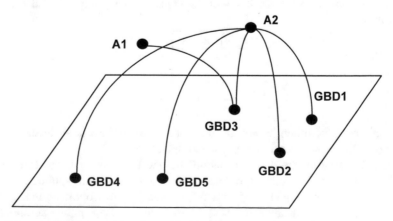

**Fig. 3.** The correspondence of the aims of the Programme 1 and the goals of the basic directions of the Programme 1

The next step in the description should be to obtain a table of relations between the basic directions, taking into account the lower levels, i.e., their tasks and milestones presented in the road map of the Programme 1. This task is not covered in this paper.

The Programme 1 identified 9 end-to-end digital technologies:
T1 – big data;
T2 – neurotechnology and artificial intelligence;
T3 – blockchains;
T4 – quantum technologies;
T5 – new production technologies;
T6 – industrial Internet;
T7 – robotics and sensor components;
T8 – wireless technologies;
T9 – virtual and augmented reality technologies.

Programme 1 is intended to develop as an "industrial" digital platform, as a platform for delivery of public services and technological platforms for research and development in each area of "cross-cutting" technologies, such as digital platforms work with the data. Note that the concepts of "end-to-end technology" and "digital platform" were not defined either in Programme 1 itself or in other official documents. There is no common understanding of those in the scientific and technical literature.

In connection with the above, the Architectural Configurator of the program description should include four aspects:

(1) industrial (spheres of activity) with applied directions AD1–AD3 и others when they appear;
(2) technological, describing the end-to-end technologies T1–T9;
(3) the platform, describing the digital platforms created within the framework of the Program;
(4) basic directions of BD1–BD5 development.

The architectural description should also include tables of relations between the

– end-to-end digital technologies T1–T9 and applied directions AD1–AD3 (and others when they appear);
– end-to-end digital technologies T1–T9 and basic directions BD1–BD5;
– technological platforms (to be developed) and applied directions AD1–AD3 (and other);
– technological platforms (to be developed) and the basic directions BD1–BD5;
– applied directions AD1-AD3 (and others) and basic directions BD1–BD5.

The emergence of a new version of Programme 1, i.e., National "Digital economy of the Russian Federation" Programme [6, 9, 11, 15] with the validity term until the end of 2024 (Programme 2) is connected, first of all, with signing the Decree of the RF President "On the national aims and strategic goals development of the Russian Federation for the period up to 2024" of May 7, 2018 [18]. The decree defines 3 aims and 9 goals of the Programme 2. Thus, changes affected the top level of the architectural description, as is shown in Table 5. Analysis of Table 5 shows that the aims of the Programme 2 are much narrower and more specific than those of the Programme 1.

**Table 5.** The aims of the digital economy in the Programme 1 and in the Programme 2

| Aims | Programme 1 | Aims | Programme 2 |
|------|-------------|------|-------------|
| A1 | The creation of an ecosystem of the digital economy of the Russian Federation | A3 | The use of mainly domestic software in state institutions |
| A2 | Creation of institutional and infrastructure conditions for the creation and development of high-tech businesses | A2 | Building a sustainable and secure information and telecommunications infrastructure |
| A3 | Improving the competitiveness of the Russian economy in the global market | A1 | The increase of internal costs for the development of the digital economy |

The levels of representation of the digital economy in the Programme 2 are not distinguished, which does not allow for an explicit definition of the configurator as a set of top-level aspects used by different participants, as well as the relationship between these aspects and the objectives of the Programme 2.

Six Federal projects implemented within the framework of the Programme 2 correspond to the first 6 tasks formulated in the above-mentioned Decree of the RF President. In terms of the Programme 1, they correspond to the five basic directions of the digital economy development (BD1–BD5) and the applied direction AD3, up to the correction of names. The applied directions AD1 (healthcare) and AD2 (smart cities) are built on individual actions under these Federal projects [14, 16, 17].

Table 6 illustrates the results of analysis of the correspondence of the aims of the Programme 2 and the goals of Federal projects (GFP1–GFP6).

**Table 6.** The correspondence of the aims of the Programme 2 and the goals of Federal projects

| Goals of the Federal projects, aims | GFP1 | GFP 2 | GFP3 | GFP4 | GFP5 | GFP6 |
|-------------------------------------|------|-------|------|------|------|------|
| A1 | − | + | + | + | + | − |
| A2 | − | + | − | + | + | − |
| A3 | − | + | − | + | + | + |

## 3  Conclusion and Future Work

The paper is devoted to the analysis of the text of the "Digital economy of the Russian Federation" Programme; an architectural configurator for the description of the Programme is proposed, and a conceptual model of the system architecture description extended in comparison with the standard ISO/IEC/IEEE 42010:2011 is analyzed.

The next step in the description should be to obtain a table of relations between the Federal projects, taking into account the lower levels, i.e., tasks and activities presented in their passports.

The above conclusions on the Configurator and the tables of relations for the Programme 1 are also valid for the Programme 2. The inclusion of these relationships in the architectural description of the Programme 2 is intended to improve the formulation and solution of the coordination issues.

# References

1. ISO/IEC/IEEE 42010:2011: Systems and software engineering – Architecture description. ISO/IEC, 2011. IEEE (2011). https://www.iso.org/standard/50508.html
2. Danchul, A.N.: Meta-description models of architecture [Modeli metaopisanii arhitektury]. In: Proceedings of the 18th Russian scientific-practical conference Enterprise Engineering and Knowledge Management, vol. 1, pp. 100–108. State University of Economics Statistics and Informatics, Moscow, Russian Federation (2015). (In Russian)
3. Danchul, A.N.: The meta-model of the description of the multidimensional architecture of a system [Metamodel' opisaniya mnogomernoy arhitektury sistemy]. In: Proceedings of the 19th Russian Scientific-Practical Conference Enterprise Engineering and Knowledge Management, pp. 16–22. Plekhanov Russian University of Economics, Moscow, Russian Federation (2016). (In Russian). http://conf-eekm.ru/wp-content/uploads/2016/01/eekm16.pdf. Accessed 21 May 2019
4. Danchul, A.N.: Models and meta models describing the architecture of a complex active system [Modeli i metamodeli opisaniya architectury slozhnoy aktivnoy sistemy]. In: Proceedings of the 20th International Scientific and Practical Conference Systems Analysis in Engineering and Control, SAEC 2019, vol. 1, pp. 72–82. Peter the Great St. Petersburg Polytechnic University Publishing House (2016). (In Russian)
5. Programme "Digital Economy of the Russian Federation": Approved by Order No. 1632 –R of the Government of the Russian Federation, 28 July 2017, which became invalid in accordance with Order No. 195 –R of the Government of the Russian Federation, 12 February 2019. (In Russian). http://static.government.ru/media/files/9gFM4FHj4PsB79I5v7yLVuPgu4bv R7M0.pdf. Accessed 21 May 2019
6. Passport of the National program "Digital economy of the Russian Federation": Approved by the Presidium of the Russian Presidential Council for Strategic Development and National Projects (Minutes No. 16 of 24 December 2018). (In Russian). http://static. government.ru/media/files/urKHm0gTPPnzJlaKw3M5cNLo6gczMkPF.pdf. Accessed 21 May 2019
7. Volkova, V.N., Denisov, A.A.: Systems Theory and System Analysis: Textbook [Teoriya Sistem i Sistemnyi Analiz.] URAIT Publishing House, Moscow (2010). (In Russian)
8. Volkova, V.N., Loginova, A.V., Desyatirikova, E.N., et al.: Simulation modeling of a technological breakthrough in the economy. In: Proceedings of the 2018 IEEE Conference of Russian Young Researchers in Electrical and Electronic Engineering, ElConRus 2018, Saint Petersburg Electrotechnical University "LETI", St. Petersburg-Moscow, Russian Federation, 29 January–1 February 2018, Vol. 2018-January, pp. 1293–1297 (2018). https://doi.org/10.1109/eiconrus.2018.8317332
9. Shafigullina, A.V., Akhmetshin, R.M., Martynova, O.V., et al.: Analysis of entrepreneurial activity and digital technologies in business. In: Advances in Intelligent Systems and Computing, Springer Nature, ISSN: 2194–5357, 908: Proceedings of the Conference on Digital Transformation of the Economy: Challenges, Trends and New Opportunities, Samara, Russian Federation, 29–31 May 2018, pp. 183–188 (2020). https://doi.org/10.1007/978-3-030-11367-4_17

10. Vishnyakova, A.B., Golovanova, I.S., Maruashvili, A.A., et al.: Current problems of enterprises' digitalization. In: Advances in Intelligent Systems and Computing, Springer Nature, ISSN: 2194–5357, 908: Proceedings of the Conference on Digital Transformation of the Economy: Challenges, Trends and New Opportunities, Samara, Russian Federation, 29–31 May 2018, pp. 646–654 (2020). https://doi.org/10.1007/978-3-030-11367-4_62

11. Papaskiri, T., Kasyanov, A., Ananicheva, E.: On creating digital land management in the framework of the program on digital economy of the Russian Federation. IOP Conference Series: Earth and Environmental Science, 274 (1): International Scientific and Practical Conference on Agrarian Economy in the Era of Globalization and Integration 2018, AGEGI 2018, Moscow, Russian Federation, 24–25 October 2018 (2019). https://doi.org/10.1088/1755-1315/274/1/0120927

12. Ismagilova, L.A., Gileva, T.A., Galimova, M.P., Glukhov, V.V.: Digital business model and smart economy sectoral development trajectories substantiation Lecture Notes in Computer Science (including subseries Lecture Notes in Artificial Intelligence and Lecture Notes in Bioinformatics), 10531 LNCS: 17th International Conference on Next Generation Teletraffic and Wired/Wireless Advanced Networks and Systems, NEW2AN 2017, 10th Conference on Internet of Things and Smart Spaces, ruSMART 2017 and 3rd International Workshop on Nano-Scale Computing and Communications, NsCC 2017, St. Petersburg, Russian Federation, 28–30 August 2017, pp. 13–28 (2017). https://doi.org/10.1007/978-3-319-67380-6_2

13. Mingaleva, Z., Mirskikh, I.: The problems of digital economy development in Russia. In: Advances in Intelligent Systems and Computing, 850: International Conference on Digital Science, DSIC 2018, Budva, Montenegro, 19–21 October 2018, pp. 48–55 (2019). https://doi.org/10.1007/978-3-030-02351-5_7

14. Rudskoy, A.I., Borovkov, A.I., Romanov, P.I., Kolosova, O.V.: Ways to reduce risks when building the digital economy in Russia. Educational Aspect. Vysshee Obrazovanie v Rossii 28(2), 9–22 (2019). https://doi.org/10.31992/0869-3617-2019-28-2-9-22

15. Alpackaya, I., Alpackiy, D.: Perspectives and consequences of implementation and development digital economy. In: MATEC Web of Conferences, 193: International Scientific Conference Environmental Science for Construction Industry, ESCI 2018, Ho Chi Minh City, Viet Nam, 2–5 March 2018 (2018). https://doi.org/10.1051/matecconf/201819305087

16. Borremans, A.D., Iliashenko, O.Yu., Zaychenko, I.M.: Digital economy. IT strategy of the company development. In: MATEC Web of Conferences, 170: International Science Conference on Business Technologies for Sustainable Urban Development, SPbWOSCE 2017, Peter the Great St. Petersburg Polytechnic University, St. Petersburg, Russian Federation, 20–22 December 2017 (2018). https://doi.org/10.1051/matecconf/201817001034

17. Bataév, A.V., Gorovoy, A.A., Mottaeva, A.: Evaluation of the future development of the digital economy in Russia. In: Proceedings of the 32nd International Business Information Management Association Conference, IBIMA 2018 – Vision 2020: Sustainable Economic Development and Application of Innovation Management from Regional expansion to Global Growth 2018, Seville, Spain, 15–16 November 2018, pp. 88–101 (2018). ISBN 978-099985511-9

18. The Strategy of the Information Society Development in the Russian Federation up to 2017–2030. (Approved by the Decree of the President of the Russian Federation No. 203 of May 9, 2018). (In Russian)

# Uniformity of Adaptive Control of Socio-Economic and Cyber-Physical Changes Using the Lingua-Combinatorial Model of Complex Systems

Tatyana S. Katermina[1], Elena A. Yakovleva[2(✉)],
Vladimir V. Platonov[2], and Aleksandr E. Karlik[2]

[1] Nizhnevartovsk State University, Nizhnevartovsk, Russia
nggu-lib@mail.ru
[2] St. Petersburg State University of Economics, St. Petersburg, Russia
helen7199@gmail.com, vladimir.platonov@gmail.com,
karlikl@mail.ru

**Abstract.** This paper presents the conceptual framework for revealing the subject-object interactions based on the advances in the theory of complex systems, methodology of the system engineering, artificial intelligence, decentralized platforms, applied in the expert systems for supporting management decisions. This research considers the impact of the global socio-cultural cycle on the development of society, the sustaining a stable state of infrastructure, the dynamism of homeokinetic equilibrium when managing innovations during the development of the master strategic plan, maintaining the system in a zone of the adaptation maximum in a flow of change. Additionally, this paper introduces the risks and threats framework for the anticipated problematic situations, which are identified by applying the cognitive approach and propositions of the adaptive control theory. The guidelines for the integrated integrity of systems and the cyber-physical interactivity in technological ecosystems are developed with application of lingua-combinatorial model and distributed registry technology in the context of maintaining the social and economic security.

**Keywords:** Indeterminacy · Lingua combinatorial model · Technological ecosystems · Risks · Adaptive control

## 1 Introduction

According to the development markers for technological progress and economic relations, all management approaches can be conditionally divided into two phases. The first of these is the "pre-semiotic" management approach which is used in the second and partially in the third technological paradigms of economics of the cycle theory by N.D. Kondratiev and S.Y. Glaziev. The second is the semiotic management approach in which cognitive management technology, the logic-linguistic model (LLM), and the lingua-combinatorial model (LCM) have successfully been applied. This includes not only technology for processing information from a distributed

D. G. Arseniev et al. (Eds.): CPS&C 2019, LNNS 95, pp. 735–745, 2020.
https://doi.org/10.1007/978-3-030-34983-7_73

registry, artificial intelligence, and NBICS-technology; it also includes the modern theory of adaptive control and the current guidelines for an adaptation maximum in managing complex systems in various realms of economic digitalization and the proclaimed scientific breakthrough developing a financial and industrial policy, and improving the coordination of information network interactions. This paper describes the logic-linguistic model of decision-making processes for strategic problems and situations, which were defined in presidential decrees, a government report, and a report of the Accounts Chamber. The research relies on the scientific works of D.A. Pospelov, Y.I. Klykov, B.L. Kukor, G.V. Klimenkov, L.S. Bolotova, V.N. Volkova, A. E. Karlik, V.V. Rokhchin, M.B. Ignatiev, A.A. Denisov, and others.

With the system development theory already as a guide, it should be said that the main conceptual approach currently is the application of the theory of patterns for semistructured complex systems. This allows the common features for various types of systems to be determined [1]. One of these features is the presence of elements of complex problematic situations (PS), which call for the exact explanation of the subject domain (ontology) and the need to "measure the efficacy" of propositional formulas (as linguistic variables), Boolean expressions and their relationships, and management thinking (identifying meaning) [2–4]. Subsequently, in order for the socio-economic system (business) to work effectively, the elements mentioned need to be identified and eliminated in due course, or their impact on economic growth, and the threat of developmental delays of scientific and technological progress and of breakthrough innovations needs to be prevented.

## 2 The "Pushing" Force of Innovation Based on Developing the Global Socio-Cultural Cycle

The management thinking of the leaders of an organization (management structure) is aimed at perceiving the threats of arising problems, thereby forming an advanced management strategy. Here we are guided by the works of D.A. Pospelov, A.E. Karlik, B.L. Kukor, and L. Bolotova, and we introduce corresponding concepts into the ontology of strategic management, e.g., anticipation—foreseeing threats of problematic situations arising, or preventing or resolving them. The occurring risks and threats of anticipating problematic situations should be measured using methods of fuzzy logic and soft computing based on a detailed model of the subject domain and model of the management structure, which should be described through cognitive linguistics. In this way, the occurrence of problematic situations manifests as regular imbalances, bottlenecks, conflicts, etc., the depth and scale of which depend on the structure of the management elements and their correlations.

This method of logic-linguistic modeling has existed for nearly 20 years, and a number of software tools and models have been created to recognize threats of problematic situations, all of which were costly projects at the federal level. It should be noted that these kinds of problematic situations can also occur in a digital economy, innovation management, and the financial sector. The "pushing" force of innovations based on the development of the socio-cultural cycle can be represented as shown in Fig. 1.

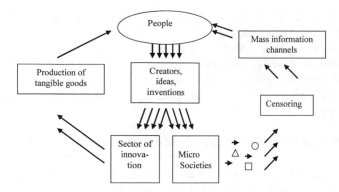

**Fig. 1.** Pushing force of innovations

By examining the development of the socio-cultural cycle, we can identify the following problems which form the basis for identifying semantic models and the vector of linguistic variables with corresponding features for describing the subject domain of the research. The cognitive approach can be applied when determining the risk/threat of a problematic situation occurring.

The conceptual approach to creating a risk management system, taking into account the possibilities of developing the SES, should ensure a level of development of its security potential that can facilitate the synthesis of existing and accumulated opportunities [18–22]. Strategic management implies the presence of strategic control and three interdependent decision-making processes: planning, coordinating, and organizing. The important feature of planned solutions is the conscious support of economic equilibria; for coordinated decisions it is the alignment of goals and interests, while organized decisions are defined by their support of stable equilibrium.

## 3 The Approach for Modeling Socio-Economic and Cyber-Physical Systems Considering the "Greenification" of the Economy

Socio-economic, ecological, and cyber-physical systems demand constant control, diagnostics, and adjustment, as they are vulnerable to a large number of external influences. More often than not, these systems belong to the semistructured kind and are poorly formalized, indicating that a special modeling and managing device needs to be used in order for them to be effectively controlled.

A large number of approaches exists for solving problems of analyzing and modeling semistructured and poorly formalized systems. They can conditionally be divided into "hard" and "soft" approaches. The first group includes methods developed within the possibility theory, cybernetics, system engineering, and conceptual modeling; the second includes cyclic planning, situational leadership [3], sociotechnical systems, interactive planning, and others. The widely used methods of the cognitive approach can be included in the "hard" category (homeostasis), as well as in the "soft" (social governance, sociotechnical systems) [5–7].

Experimental research methods and statistical methods for processing their results will be used when analyzing the accuracy and computational efficacy of the algorithms developed for various purposes. The cognitive modeling of semistructured socio-economic and sociotechnical cyber-physical systems was also applied during the research [8, 9].

## 4  Indeterminacy and Lingua-Combinatorial Modeling

Mathematical models only exist for a small number of real systems. The systems are first and foremost described using natural language. Ignat'yev [1, 2, 8, 10] developed a method of lingua-combinatorial modeling which provides a transition from description through natural language to mathematical equations. For example, given the phrase (1)

$$Word_1 + Word_2 + Word_3 \tag{1}$$

In this phrase we mark the words and only imply their meanings. The meaning is not marked in the current natural language structure. The following form is proposed for the concept of meaning (sense):

$$Word_1 * Sense_1 + Word_2 * Sense_2 + Word_3 * Sense_3 = 0 \tag{2}$$

We will denote the words as $A_i$ for "appearance", and the meanings as $E_i$ for "essence"; the asterisk * denotes the process of multiplication. Thus, Eq. (2) can be presented as

$$A_1 * E_1 + A_2 * E_2 + A_3 * E_3 = 0 \tag{3}$$

Equations (2) and (3) are models of the phrase (1). Forming these equations and equating them to zero is known as the process of polarization.

The lingua-combinatorial model is a ring of algebra, which uses three operations - addition, subtraction, and multiplication - in accordance to the axioms of algebra. Equation (3) can be solved either relative to $A_i$ or relative to $E_i$ by introducing a third group of variable–arbitrary coefficients $U_S$:

$$\begin{aligned} A_1 &= U_1 * E_2 + U_2 * E_3 \\ A_2 &= - U_1 * E_1 + U_3 * E_3 \\ A_3 &= - U_2 * E_1 - U_3 * E_2 \end{aligned} \tag{4}$$

or

$$\begin{aligned} E_1 &= U_1 * A_2 + U_2 * A_3 \\ E_2 &= - U_1 * A_1 + U_3 * A_3 \\ E_3 &= - U_2 * A_1 - U_3 * A_2 \end{aligned} \tag{5}$$

where $U_1$, $U_2$, $U_3$ are the arbitrary coefficients, which can be used to solve various problems of manifolds (3).

Suppose there are three variables and two control conditions:

$$A_1^1 E_1 + A_2^1 E_2 + A_3^1 E_3 = 0$$
$$A_1^2 E_1 + A_2^2 E_2 + A_3^2 E_3 = 0 \tag{6}$$

In these equations $A_i^j$ are the characteristics, $i$ is the number of characteristic, $j$ is the number of the control condition. $E_i$ - change of i-th characteristic.

The corresponding matrix is:

$$A_1^1 A_2^1 A_3^1$$
$$A_1^2 A_2^2 A_3^2 \tag{7}$$

Then the change $E_i$ in characteristics $A_i^j$ is defined as:

$$\begin{cases} E_1 = u_1 D_{23}^1 \\ E_2 = -u_1 D_{13}^2 \\ E_3 = u_1 D_{12}^3 \end{cases} \tag{8}$$

$D_{23}^1$, the determinant of the minor of the matrix (7), is calculated by excluding columns 2 and 3.

The upper index 1 of $D_{23}^1$ denotes the equation number in system (8).

$$D_{23}^1 = \begin{vmatrix} A_2^1 & A_3^1 \\ A_2^2 & A_3^2 \end{vmatrix} = A_2^1 * A_3^2 - A_3^1 * A_2^2$$

$$D_{13}^2 = \begin{vmatrix} A_1^1 & A_3^1 \\ A_1^2 & A_3^2 \end{vmatrix} = A_1^1 * A_3^2 - A_3^1 * A_1^2 \tag{9}$$

$$D_{12}^3 = \begin{vmatrix} A_1^1 & A_2^1 \\ A_1^2 & A_2^2 \end{vmatrix} = A_1^1 * A_2^2 - A_2^1 * A_1^2$$

In general, if we have $n$ variables and $m$ manifolds and restrictions, then the number of arbitrary coefficients $S$ will be equal to the number of combinations of $n$ by $m + 1$:

$$S = C_n^{m+1}, n > m \tag{10}$$

The number of arbitrary coefficients is a measure of indeterminacy and adaptability. Lingua-combinatorial modeling can rely on analyzing the whole corpus of natural language texts. This is a time-consuming task for extracting meanings for supercomputers, but it can also be used by relying on key words in specific fields, allowing new models for specific areas of knowledge to be obtained. The number of arbitrary coefficients is a measure of indeterminacy and adaptability.

Lingua-combinatorial modeling can also be used by relying on key words in specific fields, allowing new models for specific areas of knowledge to be obtained.

The procedure of lingua-combinatorial modeling starts with identification of the key words in a particular subject area, then combining the key words into phrases of type (1), and finally solving the equivalent systems of equations with arbitrary coefficients.

Accounting for past experience Past experience can be integrated into the model by using additional restrictions. For example, when modeling a city, the additional restrictions correspond to the statistical data for 2018, 2017, 2016, and other years.

$$
\begin{matrix}
A_1^1 & A_2^1 & \cdots & A_7^1 \\
A_1^1 & A_2^2 & \cdots & A_7^2 \\
\cdots & \cdots & \cdots & \cdots \\
A_1^1 & A_2^5 & \cdots & A_7^5
\end{matrix}
\tag{11}
$$

An example of forming this type of system is shown in Fig. 2. This system has blocks of discrete input time, which may correspond to the indicators of process dynamics in models of cities, organisms, movement of lithospheric plates, and other complex dynamic systems. Figure 3 shows a graph on the dependency of the variables of a dynamic system on time.

In this case, the structure of the equivalent equations will contain 6 arbitrary coefficients $U_1, U_2, \ldots U_6$

$$
dA_1/dt = E_1 = U_1 D_{2345}^1 + U_2 D_{2346}^1 + \ldots + U_5 D_{3456}^1
$$
$$
\vdots \tag{12}
$$
$$
dA_6/dt = E_6 = U_2 D_{1324}^6 + U_3 D_{1235}^6 + \ldots + U_6 D_{2345}^6
$$

where $D_{2345}^1$ is a determinant composed of columns 2, 3, 4, 5, 6 of the matrix (11).

Management and planning it is clear from the analysis of the matrix of arbitrary coefficients that in this case we can easily manipulate any two coefficients and thereby control any two variables. For example, we can use this method to adjust the coefficients so that variable A1 will increase (Fig. 3), which is important for managing and, accordingly, for planning.

The system model assumes the blocks of discrete input signals that can be assigned to the indicators of the dynamics of processes to build the model of a city, organism, movement of lithospheric plates, and other complex dynamic systems.

For example, we can put zero coefficients U2, U3, U4, U5, U6, and the coefficient $U_1 = D_{2345}$, and the variable $A_1$ will increase. If you put the coefficient $U_6 = -D_{2345}$, the variable $A_6$ will decrease, etc., which is important for management and planning respectively.

Thus, a way of taking past experience into account in forecasting was demonstrated. In relation to modeling, this can be data on the subsystems of the body in different modes, and in relation to the modeling of lithospheric plates, this can be the coordinates of the reference points of the plates at different times. Application software packages are developed to manipulate matrices of arbitrary coefficients.

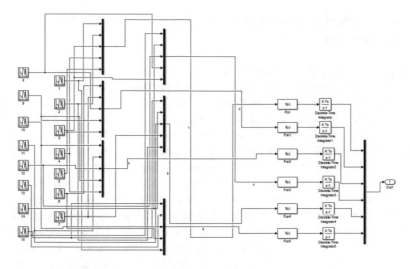

**Fig. 2.** Schematic of system model with six indeterminate coefficients and additional restrictions for integrating past experience

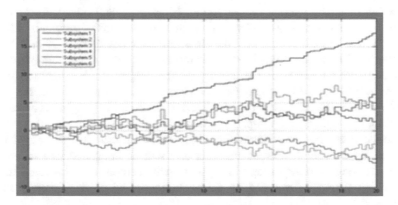

**Fig. 3.** Graph of dependency of the variables of a dynamic system on time

Technology of distributed registry the lingua-combinatorial models can be technologically based on the use of information models of distributed data, in particular, a distributed registry [11, 12]. The large amount of data which has been accumulated as of now does not necessarily mean that it should all be stored in one place. This database can be split between several network nodes or computing devices. Each node receives information from the other nodes and saves a complete copy of the registry. The nodes are updated independently of each other.

# 5 Cyber-Physical Ecosystem and Risks of Introducing Technologies

An integral part of modern socio-economic and socio-technical systems is the many kinds of computer technologies. The term "technological ecosystem" has become generally accepted recently to describe this component [13–16]. Biologists use this term to describe the community of living organisms interacting with non-living components of their environment as a system. Considering this definition, the term "ecosystem" can be used to describe the new era of technological development. When applied to technology, "ecosystem" describes a system of products or platforms supplemented with applications. Ecosystems relate to the dynamic interaction between people, equipment, systems, and services. This is an optimized flow of ideas and how they are combined, divided, transformed, and increased together within one system. By combining independent technologies into a wider system of use, companies reform whole markets and change the way we depend on digital devices. For this reason, companies should plan to develop their own equipment, software, and services based on how it will fit into the existing technological ecosystem of the client, while the client should be concerned with how the integrated innovation will flow into this ecosystem.

In order to develop the concept of the embeddability of modern technology [17], such as big data, the internet of everything, blockchain and others, into the socio-economic, ecological, and cyber-physical systems, and the business models of financial intermediation, the main positions of the algorithm for recognizing risks and threats of occurring problematic situations need to be presented. Likewise, a system of qualitative and quantitative indicators of the activity of the subject (as a complex system of decision-makers) and of the object (as a complex socio-economic and cyber-physical system) needs to be organized for management when forming a digital economy using the concept of sustainable development and network-centric management. The system suggested will be used for determining the assessment criteria for recognizing the importance of network-centric management and will be integrated into the model of scenario planning. In order to form the system, new methods and effective algorithms for searching for causality based on the semantic model of (big) data need to be implemented.

When introducing innovations into the technological ecosystem, the possible risks and threats need to be taken into consideration. This can be done by using a logic-linguistic approach. The possible risks and threats are examined for every problematic situation. A matrix of risks and threats is formed, where the impact of each risk on the other is assessed on a pre-determined scale. Out of the constructed matrix, the prevalent risks and threats that have gotten the maximum number of points are chosen. These are critical risks which companies should pay attention to first.

# 6   Conclusions

The modern cognitive methods of adaptive management and modeling and, in particular, linguistic-combinatorial modeling addresses to several important challenges of dealing with poorly structured systems, such as complex socio-technological systems. Unlike previous approaches, with introduction of the proposed methodology, the task for the experts of a specific field would be to define the key concepts of the model and to determine the statistical data for representing these concepts.

This paper further develops the conceptual framework for revealing the subject-object relations by application of the advance theory of the complex systems and methods of system engineering, artificial intelligence, decentralized platforms, and other expert methods for supporting management decisions. It provides the guidelines for maintaining the system integrity and the cyber physical interactivity in technological ecosystems by applying the Lingua-combinatorial model and distributed registry technology. This framework helps to consider in more details the dynamics of the homeokinetic equilibrium when managing innovations during the strategic planning, keeping the system in a zone of adaptation maximum in a flow of change and, finally, to present an approach for dealing with risks and threats by anticipating problematic situations, which are identified by applying the cognitive approach and adaptive control theory. The approach to implementation of the risk management system takes into account the perspective of the industrial development for ensuring a level of development to manage current, as well as the future, risks and exposures. The proposed system is designed to develop the assessment criteria within the methodology of the network-centric management and will be integrated into the model of scenario planning. In order to implement the system, the new methods and effective algorithms for searching for causality with application of the semantic model to process the big data need to be implemented. The perspective of the development of information technology depends on the correct combination of human capabilities, information and computing technology. In other words, it requires the development of hybrid systems. Another research avenue is the research on the development of the human capital to provide the prerequisites for the implementation of the new system including adaptive capabilities, genetic memory, intuition, benevolence.

**Acknowledgements.** With the support of the Russian Foundation for Basic Research, grant 18-010-00971 A. "Research of new forms of inter-firm interaction and organization in the real sector in the conditions of the information and networked economy".

# References

1. Volkova, V.N., Denisov, A.A.: Teoriya sistem i sistemnyy analiz (System Theory and Systems Analysis). Yurayt, Moscow (2010)
2. Ignat'yev, M.B., Katermina, T.S.: Kontrol' i korrektsiya vychislitel'nykh protsessov v real'nom vremeni na osnove metoda izbytochnykh peremennykh (Control and Correction of Computational Processes in Real Time Based on the Method of Redundant Variables). NVSU, Nizhnevartovsk (2014)

3. Pospelov, D.A.: Situatsionnoye upravleniye: Teoriya i praktika (Situational Management: Theory and Practice). Nauka, Moscow (1986)
4. Uyemov, A.I.: Sistemnyy podkhod i obshchaya teoriya system (The System Approach and the General Theory of Systems). Mysl', Moscow (1978)
5. Minzoni, A., Mounoud, E., Niskanen, V.A.: A case study on time-interval fuzzy cognitive maps in a complex organization. In: Proceedings of IEEE International Conference on Cognitive Infocommunications (CogInfoCom 2017), St. Peters-burg (2017)
6. Francisco, J.: Ruiz de Mendoza Ibáñez: Conceptual complexes in cognitive modeling. Rev. Esp. Lingüística Apl.: Span. J. Appl. Linguist. **30**(1), 299–324 (2017)
7. Heathcote, A., Brown, S., Wagenmakers, E.J.: An introduction to good practices in cognitive modeling. In: An Introduction to Model-Based Cognitive Neuro-Science, pp. 25–48. Springer, New York (2015)
8. Ignat'yev, M.B., Katermina, T.S.: Sistemnyy analiz kiberfizicheskikh struktur (System analysis of cyberphysical structures). In: Sistemnyy analiz v proyektirovanii i upravlenii: sbornik nauchnykh trudov XXI Mezhdunarodnoy nauchno-prakticheskoy konferentsii, pp. 15–24. FGAOU VO SPbPU, St. Peters-burg (2017)
9. Bott, O., Augurzky, P., Sternefeld, W., Ulrich, R.: Incremental generation of answers during the comprehension of questions with quantifiers. Cognition **166**, 328–343 (2017)
10. Ignat'yev, M.B., Katermina, T.S.: Sistemnyy analiz problemy upravleniya khaosom (System analysis of the problem of chaos control). In: Sistemnyy analiz v proyektirovanii i upravlenii: sbornik nauchnykh trudov XX Mezhdunarodnoy nauchno-prakticheskoy konferentsii. FGAOU VO SPbPU, St. Petersburg (2016)
11. Third, A., Domingue, J.: Linked data indexing of distributed ledgers. In: Proceedings of the 26th International Conference on World Wide Web Companion (WWW 2017 Companion), pp. 1431–1436 (2017)
12. Ølnes, S., Ubacht, J., Janssen, M.: Blockchain in government: benefits and implications of distributed ledger technology for information sharing. Gov. Inf. Q. **34**(3), 355–364 (2017)
13. Briscoe, G., Sadedin, S., De Wilde, P.: Digital ecosystems: ecosystem-oriented architectures. Nat. Comput. **10**, 1143–1194 (2011)
14. Mayer, S., Hodges, J., Yu, D., Kritzler, M., Michahelles, F.: An open semantic framework for the industrial internet of things. IEEE Intell. Syst. **32**(1), 96–101 (2017)
15. Li, W., Badr, Y., Biennier, F.: Digital ecosystems: challenges and prospects. In: Proceedings of the International Conference on Management of Emergent Digital EcoSystems (MEDES 2012), pp. 117–122. ACM, Addis Ababa (2012)
16. Ion, M., Danzi, A., Koshutanski, H., Telesca, L.: A peer-to-peer multidimensional trust model for digital ecosystems. In: 2008 2nd IEEE International Conference on Digital Ecosystems and Technologies, pp. 461–469 (2008)
17. Shkodyrev, V.P.: Technical systems control: from mechatronics to cyber-physical systems. In: Gorodeskiy, A.E. (ed.) Smart Elecromechanical Systems. Studies in Systems, Decision and Control, vol. 49, pp. 3–6. Springer, Cham (2016)
18. Kukor, B.L. Yakovleva, E.A.: Ob informatsionno-kommunikatsionnoy sisteme strategich-eskogo upravleniya ekonomikoy (On the information and communication system of strategic management of the economy). In: Sistemnyy analiz v proyektirovanii i upravlenii: sbornik nauchnykh trudov XXI Mezhdunarodnoy nauchno-prakticheskoy konferentsii, vol. 2, pp. 19–25. FGAOU VO SPbPU, St. Petersburg (2017)

19. Karlik, A.E., Kukor, B.L., Dymkovets, I.A., Yakovleva, E.A.: Aktualizatsiya osobennostey razrabotki sistemy strategicheskogo upravleniya ekonomikoy Rossii (Updating the features of the development of a system of strategic management of the Russian economy). Mezhdunarodnaya konferentsiya po myagkim vychisleniyam i izmereniyam, vol. 2, pp. 303–306 (2017)
20. Toomey, J.: MRP II: Planning for Manufacturing Excellence. Springer, Cham (2013)
21. Lunn, T., Neff, S.: MRP: Integrating Material Requirements Planning and Modern Business. Business One Irwin, Homewood (1992)
22. Ray, R.: Enterprise Resource Planning. McGraw-Hill Education, New York (2011)

# An Expert System as a Tool for Managing Technology Portfolios

Alla Surina, Danila Kultin, and Nikita Kultin[✉]

Peter the Great St. Petersburg Polytechnic University, Saint Petersburg, Russia
surina_av@spbstu.ru, kultin_nb@spbstu.ru,
dankultin@yandex.com

**Abstract.** The technology portfolio of a company is the basis for its competitiveness. When formulating a development strategy, the correct choice of priorities, which are embodied in projects for the development and implementation of innovative technologies, is of great importance. Decisions on the development or introduction of technology relate to the field of strategic management. Strategic decisions are characterized by a high degree of uncertainty, which often would not allow the use of analytical methods. The analysis shows that under the conditions of uncertainty and dynamic external environment, the portfolio approach to the development of a strategy for the technological development of an enterprise is the most promising. Technology portfolio management is an essential component of the enterprise management strategy. In practice, the simplest models and methods of technology portfolio management are used. Involvement of experts reduces management efficiency. As part of the management system of the technological portfolio of the enterprise, it is proposed to use an expert system. With the help of an expert system, it is possible, for example, to assess the commercial potential of a technology as part of the portfolio. Knowledge in the expert system should be organized in the form of inference rules, which makes the system flexible, making it easy to adjust the accumulation of knowledge in the process of using the system.

**Keywords:** Technology portfolio · Portfolio management · Enterprise development strategy · Decision support system · Expert system · Artificial intelligence

## 1 Introduction

Innovations are increasingly viewed as an essential component of development strategy and ensuring the longterm competitiveness of an enterprise. For the formation of such a strategy, the selection of priorities for the technological development of enterprises, which are embodied in specific projects for the development and implementation of modern innovative technologies, is of great importance.

It should be borne in mind that innovation is a non-linear set of processes, and a company developing a new technology often creates it in collaboration with suppliers, potential consumers and even competitors. To resolve contradictions, the selection and adaptation of the technology should be attributed to the area of strategic management decisions. These solutions are characterized by such features as high uncertainty,

D. G. Arseniev et al. (Eds.): CPS&C 2019, LNNS 95, pp. 746–753, 2020.
https://doi.org/10.1007/978-3-030-34983-7_74

including the parameters of technology implementation, and the irreversibility of costs associated with the creation or acquisition of a technology.

These features impose special requirements on decision-making procedures; therefore, a portfolio approach to the development of an enterprise's technological strategy seems to be the most promising tool, especially in a dynamic external environment.

Technology portfolio management is an essential component of any technology management strategy. The technology portfolio and project portfolio are inherently close concepts. A project portfolio is a set of projects, not necessarily technologically dependent, implemented by an organization in the context of resource constraints and ensuring the achievement of the strategic goals of the enterprise [1]. In its turn, a technology portfolio is a combination of technologies that a company possesses.

Technology in a broad sense is application of scientific knowledge to solve practical problems. It includes ways of working, its mode and sequence of actions [2], which allows considering a particular technology being implemented or planned for use at an enterprise as an independent project.

Therefore, it is legitimate to assume that technology portfolio management tools are largely consistent with the approaches used in project portfolio management.

In practice, as a rule, in the management of a portfolio of projects, the simplest approaches are used [3], which do not take into account the possible dependence between projects, which is a significant drawback when making relevant decisions. Therefore, when choosing an approach to the management of a technology portfolio of a company, it is necessary to take into account the possible dependence of the technologies implemented and expected to be implemented. Such a dependence can manifest itself not only in the sequence (order) of the technologies being implemented, but also in the form of intellectual property for innovation (technology), the method of commercialization, the scope and level of economic benefit.

## 2 Stages of Project Portfolio Organization

The most important task of managing a technology portfolio is to increase the competitiveness of technologies in the portfolio; therefore, when developing an appropriate tool, it is necessary to take into account not only managerial, but also technical and technological factors.

Both groups of factors, which can be conventionally defined as "technological" and "managerial", can be used as criteria for selecting specific tools when managing an organization's technology portfolio.

The implementation of this process will allow the organization to achieve the ultimate goal of building a fully managed technology portfolio by incorporating all new projects, initiatives and investments. The process of project portfolio management includes two stages (or steps):

1. Stage of project portfolio formation, which includes identification and selection of projects, as well as portfolio balancing, which allows to create a project portfolio that is optimal in terms of achieving the company's strategic goals implemented within the allocated resources, and optimal in terms of the level of risks and types of projects.

2. Stage of project portfolio implementation, including monitoring and control processes, necessary, first of all, to ensure that the portfolio as a whole achieves its goals.

The key task of portfolio formation is its optimization, ensuring the achievement of strategic and current business goals, including coordination of the composition, project time and volume of available resources of an enterprise with the goals of its business within a portfolio; maximization of project revenues and portfolios based on them; identifying dependencies between projects and conflicts that arise between them, and assessing their impact on the effectiveness of project portfolios; modeling of alternative project portfolios and selection of the most appropriate ones for implementation.

Monitoring project portfolios and effective dynamic management of projects and resources involves solving such tasks as adapting to changing conditions and replanning project portfolios; management of the priorities of projects in the portfolio; periodic review of plans and budgets of project portfolios; dynamic inclusion in the portfolios of new projects or elimination of completed or discontinued, and redistribution of released resources to other projects, etc. In other words, the main goal of this stage is to ensure effective control of the effects of random events (risks and uncertainties) during the implementation of the portfolio.

As we see, the problem of effective management of project portfolios is very diverse, requiring a good knowledge of the methods of analysis and an integrated approach when solving it.

In the process of portfolio management, most decisions are made under conditions of uncertainty, based on insufficient, incomplete or inaccurate information about the current state of the technology and its development prospects, which makes it difficult to use analytical methods of analysis. The quality of decisions is determined by the intuition and experience of the decision-makers.

The management decision is preceded by an examination. The examination process comprises the stages of preparing, conducting and processing the results. In the process of preparing the examination, evaluation criteria are formulated and a pool of experts is formed. At the implementation stage, experts evaluate the proposed solutions. At the final stage, the results are processed and a consolidated solution is formulated. Thus, the expertise, as an element of the management decision-making process, is a long and expensive exercise [6]. The quality of expertise is largely determined by the qualifications of specialists acting as experts.

Time and material costs of expertise can be reduced, the influence of the human factor can be reduced by automating and using the expert system as a decision-making tool [7]. An expert system is a computer program that, on the basis of the knowledge embedded in it, can give reasonable advice and offer a reasonable solution to the problem [8].

With regard to the task of managing a technology portfolio, the expert system can be used both at the stage of portfolio formation and at the management stage.

At the stage of forming a portfolio of technologies, the following tasks can be solved:

- assess the commercial potential of the technology;
- evaluate competing technologies;

- evaluate technology relatedness;
- estimate the cost of the project to create (implement) the technology;
- estimate the duration of the project to create (implement) the technology;
- evaluate technology prospects;
- assess technology risks;
- advise on the inclusion of technology in the portfolio.

At the stage of technology portfolio management, the following tasks can be solved:

- assess the current state of the technology (degree of compliance with planned indicators);
- evaluate competing technologies;
- give a recommendation on the development of technology;
- make a recommendation to stop using the technology.

## 3  The Structure of the Expert System for the Project Portfolio Management

The core of the expert system is the knowledge base in the subject area. The knowledge base contains information about the objects of the subject area. In expert systems, semantic networks, frames, and inference rules are used to represent knowledge.

Semantic networks and frames are used to solve research problems of artificial intelligence.

The rules in general form are expressions of the form *IF condition TO conclusion*. The rules reflect the course of the expert's reasoning and provide a natural way of describing the decision-making process.

In the rule-based expert system, knowledge is separated from the program code that implements the conclusion derivation process. Under this way of organizing, an expert system makes it easy for changes to the knowledge base.

In addition to the knowledge base (KB) containing the inference rules and facts, the elements of the expert system are the inference engine (IE), the user interface (UI), the developer interface (DI), and the explanatory system (EpS). The user interface provides for the user interaction with the expert system during the consultation process. The knowledge engineer interface has access to the knowledge base, which allows you to adjust the behavior of the expert system. The explanatory system provides a display of a chain of logical inference. In a minimal configuration, an expert system can consist of a knowledge base, an output mechanism, and a developer interface. The structure of the expert system is shown in Fig. 1.

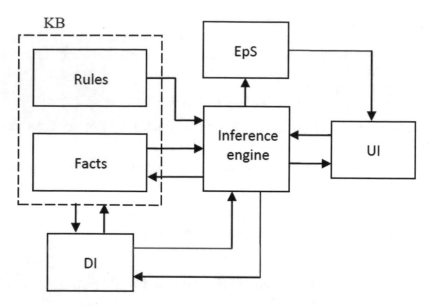

**Fig. 1.** The structure of the expert system on the rules

There are quite a few freely distributed solutions that allow you to create an expert system based on these rules. To solve the problem of creating an expert system of technology portfolio management, the platform Expert 2.0 was used [9].

The knowledge base in Expert 2.0 is represented by a set of rules, which are generally written as:

rule (N)

*Object1* = *Value1*, cf1 = *k1*

*Object2* = *Value2*, cf2 = *k2* . . .

*Object(k)* = *Valuek*, cf = *k(k)*

then *Object(i)* = *Valuei*, cf = *k(i)*;

where:

Rule, then and cf are keywords of the rules presentation language used by the rules interpreter (output mechanism) Expert 2.0;

*Object(i)* and *Value(i)* are respectively the object of the subject area and its value; cfi is the degree (coefficient) of confidence, an integer in the range from 0 to 100 corresponding to the confidence that the state of the object is characterized by the specified value.

Table 1 describes the objects of the expert system for solving the problem of forming a portfolio of technologies.

**Table 1.** Objects of the expert system of technology portfolio formation

| Object | Description | Possible values |
|---|---|---|
| Commercial potential | Probability of making a profit from the application of technology | Low, medium, high enough, high, very high |
| Security | Degree of protection of technology against copying by competitors | Patent, know-how, industrial design, is absent |
| Complexity | Level of technology required to reproduce a product by competitors | High, medium, low |
| Production process | Impact on the production process of the enterprise | Does not require changes, requires minor changes, requires significant changes |
| Personnel | Personnel costs associated with attracting new and advanced staff skills | Significant, acceptable, insignificant |
| Innovation potential | Influence of technology on the change in the innovation potential of an enterprise | Not affect, not significant, significant |
| Market | Technology market | Creates a new, expands current |
| Investment | Cost of acquiring assets (tangible and intangible), research and development, modernization of production | Significant, acceptable, insignificant |
| Financing | Source of financing a project for creating (implementing) a technology | Own funds, own funds and a loan, own funds and investors, a loan |
| Risk | Risk of a technology creation project (implementation) | Low, medium, high |
| Impact | Relationship with other candidate technologies for inclusion in portfolio | No, weak, donor, acceptor |
| Portfolio | Decision on the inclusion of technology in the portfolio | Is not, yes, the candidate |

It should be noted that the list of object values, as well as the list of objects themselves, can be expanded in the process of using the expert system.

As an example, Listing 1 shows a fragment of the knowledge base of a research prototype of an expert technology portfolio management system.

Listing 1. Knowledge base fragment
rule (1)
Security = patent
Market = creates_ a_new_one
Difficulty = high
then
Commercial_Potential = very_high;
rule (3)
Security = Industrial_ Sample
Market = expands_current
Difficulty = medium
Then
Commercial _potential = sufficiently_high;
rule (4)
Security  = none
Market = expands _current
Difficulty = medium
then
CommercialPotential = Medium;
rule (6)
Commercial_Potential = Very_High
Investment = significant
InnovativePotential = increase
then
Portfolio = yes,  cf = 75;
rule (8)
Commercial_Potential = Enough_high
Investment = significant
InnovativePotential = increase
then
Portfolio = candidate, cf = 80;
rule (13)
Commercial_Potential = low
then
Portfolio = no;

vl (Portfolio = no, yes, candidate)
vl (Commercial_Potential = low, medium,  high enough,   high,  veryhigh)
vl (Security = patent, know-how, industrial  sample, absent)
qu (complexity = technology_level_to_play)
vl (Difficulty = high, medium, low)
qu (Market = Technology)
vl (Market = creates new, expands current)
qu (Influence = Influence_  on_Other _Technology _portfolio)
vl (Influence = no donor)

# 4 Conclusions

In the paper, the idea of an expert system for managing a technology portfolio was proposed.

The main conclusions on the work are as follows:

- The expert system as an intellectual component could help to reduce the time and material costs of the examination and the influence of the human factor.
- The expert system can be used both at the stage of forming a technology portfolio and at the stage of portfolio management. With the help of an expert system you can: assess the commercial attractiveness of the technology; estimate the duration and cost of the project to create (implement) the technology; evaluate technology prospects; make a recommendation about the inclusion of technology in the portfolio.
- To implement the expert system, you should use the architecture with the representation of knowledge in the form of a set of rules of logical inference, as the most flexible and open to making changes in the knowledge base.

# References

1. Matveyev, A.A., Novikov, D.A., Tsvetkov, A.V.: Modeli i metody upravleniya portfelyami proyektov Moscow, PMSOFT (2005)
2. Nekrasov, S.I., Nekrasova, N.A.: Filosofiya nauki i tekhniki: tematicheskiy slovar. OGU, Orel (2010)
3. Kornilov, S.S.: Metod formirovaniya portfeley tekhnologiy promyshlennykh predpriyatiy: na primere aviatsionno-kosmicheskogo kompleksa Samarskoy oblasti : avtoreferat dis. kandidata ekonomicheskikh nauk: 08.00.05 [Mesto zashchity: Sam. gos. aerokosm. un-t im. S. P. Koroleva] Samara (2007)
4. Tukkel, I.L., et al.: Upravleniye innovatsionnymi proyektami. SPb.: BKhV-Peterburg, 416 p. (2017)
5. Anshin, V.M., et al.: Modeli upravleniya portfelem proyektov v usloviyakh neopredelennosti, 194 p. MATI (2008)
6. Ablyazov, V.I., et al.: Ekspertiza innovatsionnykh proyektov // V kn. Nauchno-tekhnicheskiye vedomosti Sankt-Peterburgskogo gosudarstvennogo politekhni-cheskogo universiteta, № 121, pp. 184–188 (2011)
7. Kultin, N.B.: Ekspertnaya sistema, kak instrument podderzhki uprav-lencheskikh resheniy. Nauchno-tekhnicheskiye vedomosti Sankt-Peterburgskogo gosudarstvennogo politekhnich-eskogo universiteta, № 121, pp. 139–141 (2011)
8. Waterman, D.: A Guide to Expert System. Addison-Wesley Publishing Company (1988)
9. https://www.microsoft.com/store/apps/9PHPDLLRDX4P

# Network Challenges for Cyber-Physical Systems in Training Programmes

Dmitry G. Arseniev[1], Victor I. Malyugin[1],
Vyacheslav V. Potekhin[1(✉)], Hieu Cao Viet[2], Hoang-Sy Nguyen[2],
and Tan Nguyen Ngoc[2]

[1] Peter the Great St. Petersburg Polytechnic University, Saint Petersburg, Russia
{vicerector.int, Slava.Potekhin}@spbstu.ru,
vim@imop.spbstu.ru
[2] Binh Duong University, Thu Dau Mot City, Vietnam
{vhieu, nhsy, nntan}@bdu.edu.vn

**Abstract.** Joint educational programmes offer opportunities to integrate partner universities in the fields of research, development and education in order to solve such issues as improving the quality of teaching, search for new forms of effective interaction between scientific and educational schools of partners, development of innovative educational technologies. The most sought after now is organisation of educational programmes that implement practice-oriented online courses, short- and long-term skills training, implementation of such technologies that are implemented in educational process at the School of Cyber-Physical and Control Systems. A very important step in the development of interaction between Peter the Great St. Petersburg Polytechnic University and Binh Duong University is the creation of a unified educational environment and laboratory of intelligent control systems and further development of the Synergy Project.

**Keywords:** CPS in education · Practice-oriented online courses · Skills training · Joint international educational programmes · Curricula development

## 1 Introduction

International academic cooperation keeps actively develop in the constantly changing world. Interest of Russian students in studying abroad and interest of foreign students to study in Russia is growing. One of the directions of the international academic cooperation is development and implementation of international educational programmes [1–6]. Peter the Great St. Petersburg Polytechnic University and Binh Duong University have a wide experience in developing and implementation of this kind of programmes and development of joint laboratories.

Remote access to laboratory equipment and the development of collaborative projects can overcome the functional fragmentation between people and processes using innovative digital communications. The process approach allows you to control all stages of the development and subsequent operation of cyber-physical systems and other related processes, including the design, electronic systems, firmware, documentation and technological process.

© Springer Nature Switzerland AG 2020
D. G. Arseniev et al. (Eds.): CPS&C 2019, LNNS 95, pp. 754–759, 2020.
https://doi.org/10.1007/978-3-030-34983-7_75

Trends in intellectual automation and the development of cyber-physical systems require a mirror approach not only in the industry but also in the training processes. Therefore, it is important to consider such trends as [7–10]:

- Conversion from the centralised control system structures to distributed intelligent control systems
- Using Ethernet at all levels of automation and cyber-physical systems
- Expanding of use of open IT standards
- Understanding that the world of IT and automation are growing together.

## 2 Synergy Centre

For the multifaceted implementation of various approaches and combinations of curricula design, especially in international programmes, in 2014, a North-West Russian Regional Intercollegiate Education and Research SPbPU-Festo Synergy Centre was founded at Peter the Great St. Petersburg Polytechnic University.

The Centre puts its main task for a more efficient use of the intellectual potential and high-tech equipment, as well as a focus of the educational programmes, improving the efficiency of scientific research in collaboration with leading companies and universities [11, 12].

The activities aimed at developing and implementing educational engineering programmes. Educational programmes of the Centre are based on specific disciplines in the field of control, automation and drive systems, and cyber- physical systems. As part of the educational programmes, modernisation of existing and development of new curricula and programmes are carried out; preparation of the necessary guidelines and manuals, updating of existing and creation of new experimental base and laboratory stands used both in practice-oriented online courses and skills training is going on.

The main objectives of educational programmes are formulated in the approaches to the graduates' competence in the following fields: (1) research activities in the field of advanced intelligent systems and technologies at the national and foreign research centres and in teaching in this field; (2) design and engineering activities on implementing cyber-physical systems and technologies at enterprises; (3) management of the research, design, implementation and commercial operation of modern intelligent systems and technological complexes.

Certainly, acquisition of such competencies is impossible without practical skills. At the present time, the Centre has laboratory equipment allowing to improve students' skills in the design of integrated intelligent control systems and management of complex distributed objects; systems and processes in a large flow of information and lack of predictable control algorithms.

The Centre includes the following laboratories of:

- Intelligent systems for data processing and control systems
- Motion control systems for robotics
- Complex automation and control
- Cyber-physical systems.

## 3 Tasks and Projects for Teams

In addition to the locally available equipment, such as

- A system for the development of the skills in the design of control systems for industrial objects and technological processes (based on SIMATIC SIEMENS PLC)
- A laboratory complex FMS500, provided by the Festo Company
- A stand of a fault-tolerant industrial control system of the upper level
- A system of distributed digital sensors for industrial electronics
- An industrial monitoring system
- A hard- and software complex of embedded vision systems,

a RoboLab network operates in the Centre, allowing students of various universities to remotely use the laboratory equipment and create project teams to work during the programmes.

Students can develop the architecture of the systems which may be different depending on their tasks:

- Monitoring (continuous measurement and control with archiving of the received information)
- Automatic control (in the system with or without feedback)
- Dispatch control (control with the help of a human dispatcher who interacts with the system)
- Through a human machine interface
- Security, etc.

To do this, the concept of a cloud service is applied, in which the integration of ERP/MES/PLM/SCADA systems, etc. can be deployed; see Fig. 1.

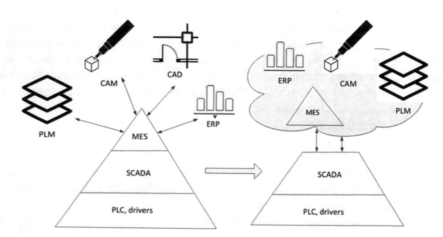

**Fig. 1.** Cyber-physical system configuration using cloud services

The technology of cloud services was implemented at the example of various laboratory complexes shown in Fig. 2.

However, it is not always possible to create a fully automatic system; often, a cyber-physical system requires the intervention of a dispatcher, for which the SCADA systems serve; see Fig. 3. As a rule, these are two-level systems, since at these levels the processes are directly controlled. The specifics of each control system is determined by the software and hardware platform used at each level.

**Fig. 2.** Laboratory complexes with filtration station, reactor and a diagnostic module

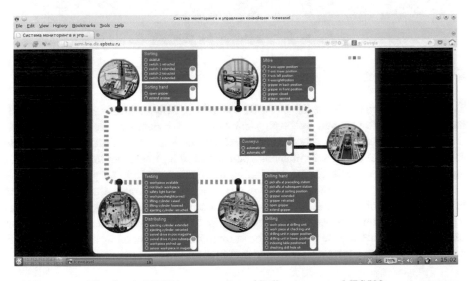

**Fig. 3.** A SCADA system assembly line conveyor MPS500

I Implementation of team projects allows to take into account the work and resources associated with the management and tracking of all processes, such as management of changes in different disciplines, development of cyber-physical systems, starting with the formulation of the problem, coordination of response measures, collection and analysis of the necessary information to make the right and timely decisions.

As an example of joint projects, the Centre hosts competitions in the development of cyber-physical systems; see Fig. 4. Students acquire the following skills:

- Choosing, creating and maintaining complex hardware and software through information systems and networks
- Formulation and solving circuit engineering problems correlated with the choice of elements requirements for specified parameters of cyber-physical systems;
- Installing, testing and using of hardware and software; etc.

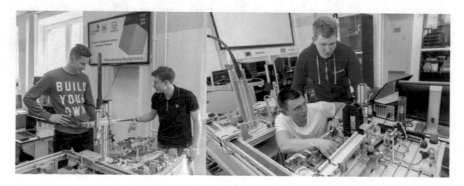

**Fig. 4.** Solving tasks at the competition

## 4   Conclusions

The Synergy Centre at Peter the Great St. Petersburg Polytechnic University works in the areas of intelligent and cyber-physical systems for data processing and control, motion control systems for robotics, complex automation, etc. Main efforts were made for a more efficient use of the intellectual potential and high-tech equipment available at the laboratories, particularly, for developing innovative educational programmes.

The experience of joint programme implementation showed that the developed materials and techniques can be successfully used in educational programmes, such as summer and winter schools, research training programmes for students, international semesters, etc. This can be the basis for innovative programmes in the field of information technology, control systems and automation.

**Acknowledgements.** The article is published with the support of the project Erasmus+ 573545-EPP-1-2016-DE-EPPKA2-CBHE-JP Applied curricula in space exploration and intelligent robotic systems (APPLE) and describes a part of the project conducted by SPbPU.

# References

1. Potekhin, V.V., Shkodyrev, V.P., Potekhina, E.V., Selivanova, E.N.: Double degree programme in engineering education: practice and prospects. In: Proceedings of 41st SEFI Conference (2013)
2. Surygin, A.I., Potekhina, E.V., Potekhin, V.V.: Curriculum design aligned with Russian national and EUR-ACE Standards. In: Proceedings of 40th Annual Conference SEFI Master's Degree Programme in the Field of Information Technology and Computers, Thessaloniki, Greece, pp. 372–373 (2012)
3. Duz, I.A., Emelianova, O.G., et al.: Normative and methodological recommendations on the realisation of joint educational programs at the example of the Russian-Finnish Cross-Border University (CBU) SPBSPU, St. Petersburg (2011)
4. Potekhin, V.V., Panyukhov, D.N., Mikheev, D.V.: Intelligent control algorithms in power industry. EAI Endorsed Trans. Energy Web 3(11), e5 (2017)
5. Potekhina, E.V., Selivanova, E.N., Potekhin, V.V., Shkodyrev, V.P.: Implementation in joint international educational programs. In: SPbPU's Experience in Festo Technologies, Proceedings of 26th DAAAM International Symposium on Intelligent Manufacturing and Automation, pp. 1572–1581 (2015)
6. Arseniev, D.G., Kovalevsky, V.E., Shkodyrev, V.P., Potekhin, V.V.: Multiagent approach to creating an energy consumption and distribution system. In: Proceedings of the VII International Conference International Cooperation in Engineering Education, St. Petersburg, Russia, 2–4 July, pp. 131–140 (2012)
7. Turner, C.J., Hutabarat, W., Oyekan, J., Tiwari, A.: Discrete event simulation and virtual reality use in industry: new opportunities and future trends. IEEE Trans. Human-Mach. Syst. 46(6), 882–894 (2016)
8. Mäkiö, J., Mäkiö-Marusik, E., Yablochnikov, E.: On educating cyber-physical systems in a global environment. In: Proceedings of the International Multiconference Network Cooperation in Science, Industry and Education (NCSIE), St. Petersburg, Russia, 4–6 July, pp. 133–140 (2016)
9. Antila, E., Virrankoski, R.: PAC – a multi-vendor environment based on IEC 61850. In: Proceedings of the International Multiconference Network Cooperation in Science, Industry and Education (NCSIE), St. Petersburg, Russia, 4–6 July, pp. 21–28 (2016)
10. Haskovic, D., Katalinic, B., Zec, I.: Support and learning functions of the intelligent adviser module. In: Proceedings of the International Multiconference Network Cooperation in Science, Industry and Education (NCSIE), St. Petersburg, Russia, 4–6 July, pp. 183–189 (2016)
11. Bobryakov, A., Breido, I., Eliseev, A., Filaretov, V., Kabanov, A., Katalinic, B., Khomchenko, V., Stazhkov, S., Potekhin, V.: Experience of application of network technologies in engineering education. EAI Endorsed Trans. Energy Web Inf. Technol. 5 (16), e5 (2018)
12. Khokhlovskiy, V.N., Potekhin, V.V., Razinkina, E.M.: Role of the SPbPU-FESTO «synergy» centre in the development of engineering competences and skills. EAI Endorsed Trans. Energy Web Inf. Technol. 5(19) (2018)

# Author Index

D. G. Arseniev et al. (Eds.): CPS&C 2019, LNNS 95, pp. 761–763, 2020.
https://doi.org/10.1007/978-3-030-34983-7

Printed in the United States
By Bookmasters